| | | | |
|---|---|---|---|
| $Q_1, Q_2, Q_3$ | quartiles | $t_{\alpha/2}$ | critical value of $t$ |
| $D_1, D_2, \ldots, D_9$ | deciles | $df$ | number of degrees of freedom |
| $P_1, P_2, \ldots, P_{99}$ | percentiles | $F$ | $F$ distribution |
| $x$ | value of a single score | $\chi^2$ | chi-square distribution |
| $f$ | frequency with which a value occurs | $\chi_R^2$ | right-tailed critical value of chi-square |
| $\Sigma$ | (capital sigma) summation | $\chi_L^2$ | left-tailed critical value of chi-square |
| $n$ | number of scores in a sample | $p$ | probability of an event or the population proportion |
| $n!$ | factorial | | |
| $N$ | number of scores in a finite population; also used as the size of all samples combined | $q$ | probability or proportion equal to $1 - p$ |
| | | $\hat{p}$ | sample proportion |
| $\overline{x}$ | mean of the scores in a sample | $\hat{q}$ | sample proportion equal to $1 - \hat{p}$ |
| $\mu$ | (mu) mean of all scores in a population | $\overline{p}$ | proportion obtained by pooling two samples |
| $s$ | standard deviation of a set of sample values | $\overline{q}$ | proportion or probability equal to $1 - \overline{p}$ |
| $\sigma$ | (lower case sigma) standard deviation of all values in a population | $P(A)$ | probability of event $A$ |
| $s^2$ | variance of a set of sample values | $P(A|B)$ | probability of event $A$ assuming event $B$ has occurred |
| $\sigma^2$ | variance of all values in a population | | |
| $z$ | standard score | $_nP_r$ | number of permutations of $n$ items selected $r$ at a time |
| $z_{\alpha/2}$ | critical value of $z$ | $_nC_r$ | number of combinations of $n$ items selected $r$ at a time |
| $t$ | $t$ distribution | | |

FIFTH EDITION

# ELEMENTARY STATISTICS

## FIFTH EDITION

# ELEMENTARY STATISTICS

## Mario F. Triola

Dutchess Community College
Poughkeepsie, New York

**ADDISON-WESLEY PUBLISHING COMPANY**

Reading, Massachusetts • Menlo Park, California • New York
Don Mills, Ontario • Wokingham, England • Amsterdam • Bonn
Sydney • Singapore • Tokyo • Madrid • San Juan
Milan • Paris

Sponsoring Editor:   Lisa Moller
Editorial Assistant:   Diane Honigberg
Senior Production Editor:   John Walker
Production Services:   Proof Positive/Farrowlyne Associates, Inc.
Text Design:   Proof Positive/Farrowlyne Associates, Inc.
Cover Designer and Cover Artist:   Mark Ong
Figure Art:   Mary Burkhardt; Proof Positive/Farrowlyne Associates, Inc.
Margin Essay Artist:   James Staunton
Composition and Film:   Weimer Typesetting Co., Inc.

*Reprinted with corrections June, 1992.*

**Library of Congress Cataloging-in-Publication Data**

Triola, Mario F.
   Elementary statistics / Mario F. Triola.—5th ed.
      p.     cm.
   Includes index.
   ISBN 0-8053-7631-3
   1. Statistics.    I. Title.
QA276.12.T76    1992
519.5—dc20                        91-31105
                                 CIP

2 3 4 5 6 7 8 9 10 – DO – 95 94 93 92

To Marc and Scott

# Preface

## Why Study Statistics?

Modern times create modern problems. Many problems involve the collection and analysis of data in surveys, polls, quality control, market research, medical research, and standardized testing. More employers are seeking job applicants who are better prepared to use statistical methods. As one example, *The New York Times* reported that "when Motorola Inc. introduced a system of quality control at its plant in Arcade, N.Y., it found that many employees lacked the mathematical skills needed to understand the new statistics-based approach." Jay Dean is a Senior Vice President at Young and Rubicam Advertising. In an interview with the author, he said, "If I could go back to school, I would certainly study more math, statistics, and computer science." In another interview, David Hall told the author that, "Right now, American industry is crying out for people with an understanding of statistics and the ability to communicate its use." David Hall is a Division Statistical Manager at the Boeing Commercial Airplane Group. As employees, employers, and as citizens we must learn at least the elementary concepts that constitute the field of statistics. This book is designed to be an interesting and readable introduction to those concepts.

## Audience

This book is an introduction to elementary statistics for students majoring in any field except mathematics. A strong mathematics background is not necessary, but students should have completed a high school algebra course. Although underlying theory is included, this book does not stress the mathematical rigor more suitable for mathematics majors.

In this book, strong emphasis is placed on interesting, clear, and readable writing. Because the many examples and exercises cover a wide variety of different applications, this book is appropriate for many disciplines. The previous editions have been used successfully by hundreds of thousands of majors in psychology, sociology, business, computer science, data processing, biology, education, engineering technology, fine arts, humanities, history, social science, nursing, health, economics, ecology, agriculture, and many others.

# Changes in the Fifth Edition

This fifth edition of *Elementary Statistics* includes all the basic features of previous editions. In response to extensive surveys, almost every section has been modified to some extent. One new feature is the **Writing Projects** near the end of each chapter. They are designed to assist the growing number of instructors who try to help improve critical thinking and writing skills by implementing a "writing across the curriculum" philosophy. Another new feature is the **Videotape** program recommendations at the end of each chapter. We recommend programs from the series *Against All Odds: Inside Statistics*. For information about acquiring these programs, call 1-800-LEARNER.

Our **Feature Interviews** are another new addition to the text. They highlight discussions with professionals who use statistics on the job. These interesting interviews demonstrate the relevance of statistics to students' future careers.

### Beginning-of-Chapter Features

- List of **chapter sections** along with brief descriptions of their contents
- **Chapter problem**
- **Overview** of the chapter, including statement of chapter objectives

### End-of-Chapter Features

- **Vocabulary list** of important terms introduced in the chapter (A glossary of terms is in Appendix D.)
- **Review** of the chapter
- Summary list of important **formulas**
- **Review Exercises**
- **Computer Projects**
- **Applied Projects**

- **Writing Projects**
- **Videotapes:** recommended program from the series *Against All Odds: Inside Statistics*

## Major Content Changes

- New section in Chapter 1: **Statistical Experiments and Sampling**
- New section in Chapter 2: **Exploratory Data Analysis**
- New section in Chapter 11: **Two-Way Analysis of Variance**
- Analysis of variance is now included as a separate chapter.
- Chapters 6 and 7 have been interchanged so that "Estimates and Sample Sizes" now precedes "Testing Hypotheses." Chapters 6 and 7 are designed so that you can cover them in any order.
- Confidence intervals are now included in Chapter 8, "Inferences from Two Samples."
- **Bootstrap methods** and **Statistical Process Control** are now discussed.

## Exercises

- This text now has more than 1600 exercises, many of them involving **real-world data**.
- Exercises are arranged in order of increasing difficulty. Also, exercises are divided into groups A and B, with B types involving more difficult concepts or a stronger mathematical background. In some cases, B exercises introduce a new concept.
- In addition to the regular exercises, there are twenty-two **Computer Projects**, forty-two **Applied Projects**, and thirty **Writing Projects**.

## Other Features

- The **flowcharts** help clarify the more complicated procedures.
- Appendix D contains an expanded **glossary** of important terms.
- Appendix F contains **answers** to all the odd-numbered exercises.
- A **symbol table** is included on the front inside cover for quick and easy reference to key symbols.
- Copies of Tables A-2 and A-3 are included in the rear inside cover for quick and easy reference.
- A detachable **formula/table card** is enclosed for use throughout the course.
- There are now more than 100 of the very popular **margin essays**, including 30 new ones. These short essays illustrate uses of statistics in very real and practical applications. The following is a sample of some of the topics covered.

*Biology:* Were Mendel's experimental data manipulated?

*Business:* How airlines save money by using sampling to determine revenues from split-ticket sales

*Criminology:* How probability is used by forensic experts to prosecute criminals

*Drugs:* How the Salk vaccine was tested

*Ecology:* How a Florida statistical study led to regulations that protect manatees

*Engineering:* How probability is used to make systems more reliable with redundancy of components

*Entertainment:* How it takes seven shuffles before a deck of cards is completely mixed

*Gambling:* Why some lottery number combinations are better choices than others

*Medicine:* How studies often use male subjects only, so that effects on women are often left unknown

*Public Policy:* How a statistical analysis showed that the death penalty doesn't deter murders

*Sports:* What happened to the 0.400 hitters in baseball?

*Surveys:* How one firm used invisible ink to identify respondents in a "confidential" survey

# Computers

This text can be used without any reference to computers. For those who choose to supplement the course with computers, we have included **computer projects** at the end of each chapter.

We also have two different levels of software available.

- **STATDISK** is an easy-to-use statistical software package that does not require any previous computer experience. Developed as a supplement specifically for this textbook, STATDISK is available for the IBM PC, Macintosh, and the Apple IIe systems. This software is provided at no cost to colleges who adopt this text.

- **STATDISK Student Laboratory Workbook** includes instructions on the use of the STATDISK software package. It also includes experiments to be conducted by students.

The STATDISK software and the STATDISK Student Laboratory Workbook have been designed so that instructors can assign computer experiments without using classroom time that may be quite limited. STATDISK includes a wide variety of programs that can be used throughout the course, and the experiments do more than number crunch or duplicate text exercises. They include concepts that can be discovered through computer use. This text includes several sample displays that result from the use of STATDISK.

For those who wish to use **Minitab**, we have included Minitab displays throughout the text. Appendix C summarizes key components of Minitab.

- The Data Sets in Appendix B are now available on disk for use with Minitab and STATDISK.

- **Minitab Student Laboratory Workbook**, designed specifically for this text, includes instructions and examples of Minitab use. It also includes experiments to be conducted by students.
- **Student Edition of Minitab, 2e** is available from Addison-Wesley for the price of a textbook. Based on Version 8.1, it includes program software with data sets developed by Minitab, Inc. and a comprehensive user's manual with tutorials and a reference section written by Robert L. Schaefer of Miami University, Oxford, Ohio, and Elizabeth Farber of Bucks County Community College in Newtown, Pennsylvania.

# Acknowledgments

I extend my sincere thanks for the suggestions made by the following reviewers of the fourth and fifth editions.

James Arnold
University of Wisconsin at Milwaukee

David Bernklau
Long Island University

Neal Brinneman
Huntington College

Arthur J. Daniel
Macomb Community College

Elizabeth Farber
Bucks County Community College

Frank Gunnip
Oakland Community College

Judith Hector
Walters State College

Mike Karelius
American River College

Charles Klein
De Anza College

Mary Ann Lee
Mankato State University

Milton Loyer
Messiah College

David Mathiason
Rochester Institute of Technology

Mary Parker
Austin Community College

David Stout
University of West Florida

Gary Taka
Santa Monica College

Gwen Terwilliger
ComTech, University of Toledo

Peyton Watson
Miami-Dade Community College

I also wish to thank nearly 200 individuals who gave me valuable input for the fifth edition through their responses to our detailed survey. In addition, I appreciate the interview contributions from David Hall of the Boeing Commercial Airplane Group, Barry Cook of Nielsen Media Research, and Jay Dean of Young and Rubicam.

I would like to extend special thanks to Milton Loyer of Messiah College for his valuable assistance in checking and providing solutions for the Instructor's Guide. I also wish to extend my thanks to Lisa Moller, John Walker, Diane Honigberg, Barbara Piercecchi, and the entire Addison-Wesley staff. Finally, I thank Ginny, Marc, and Scott for their support, encouragement, and assistance.

# Supplements

- **Annotated Instructor's Edition** by Mario Triola (contains teaching suggestions and answers in the margins)
- **STATDISK** software for the IBM PC, Version 3.0 (also includes the Appendix B data files)
- **STATDISK** software for the Apple IIe
- **STATDISK** software for the Macintosh (also includes the Appendix B data files)
- **STATDISK Student Laboratory Workbook**
- **Minitab Student Laboratory Workbook**
- **Data Disk** with Appendix B data files for use with Minitab
- **Student Edition of Minitab** (software and manual)—New edition
- **Student Solutions Manual** by Donald K. Mason (provides detailed, worked-out solutions to odd-numbered exercises)
- **Instructor's Guide and Solutions Manual** (includes all answers, detailed solutions to the even-numbered answers, printed test bank, data sets, transparency masters, sample course syllabi)
- **Computer Test Generator** for the IBM PC is available from Addison-Wesley

# To the Student

I strongly recommend the use of a calculator. You should have one that can be used for finding square roots. Such calculators usually have a key labeled $\sqrt{x}$. Also, it should use algebraic logic instead of chain logic. You can identify the type of logic by pressing the buttons

$$2 + 3 \times 4 =$$

If the result is 14, the calculator uses algebraic logic. If the result is 20, the calculator uses chain logic and it is not very suitable for a statistics course. Some inexpensive calculators can directly compute the mean, standard deviation, correlation coefficient, and the slope and intercept values of a regression line. Such keys are usually identified as Mean or $\bar{x}$; S. Dev, SD, or $\sigma_{n-1}$; Corr or r; Slope; and Intcp.

I also recommend that you read the overview carefully when you begin a chapter. Read the next section quickly to get a general idea of the material; reread it carefully. Try the exercises. If you encounter difficulty, return to the section and work some of the examples in the text so that you can compare your solution to the one given.

When working on assignments, first attempt the earlier odd-numbered exercises. Check your answers with Appendix F and verify that you are correct before moving on to the other exercises. Keep in mind that neat and well-organized written assignments tend to produce better results. When you finish a chapter, check the review section to make sure that you didn't miss any major topics. Before taking tests, do the review exercises at the end of the chapters. In addition to helping you review, this will help you cope with a variety of different problems.

You might consider purchasing the *Student Solutions Manual* for this text. Written by Donald K. Mason of Elmhurst College, it gives detailed solutions to many of the odd-numbered text exercises.

M. F. T.
LaGrange, New York
December 1991

# Contents

# Videotape Reference

Listed below are recommended programs from the videotape series **Against All Odds: Inside Statistics** produced by COMAP, available from Annenberg/CPB. The series consists of 26 half-hour programs on 13 videocassettes. For information about obtaining the series, call 1-800-LEARNER.

| Text Chapter | Recommended program from **Against All Odds** |
|---|---|
| 1 | 1: What Is Statistics? |
| | 12: Experimental Design |
| 2 | 2: Picturing Distributions |
| | 3: Describing Distributions |
| 3 | 15: What Is Probability? |
| 4 | 16: Random Variables |
| | 17: Binomial Distributions |
| 5 | 4: Normal Distributions |
| | 5: Normal Calculations |
| | 18: The Sample Mean and Control Charts |
| 6 | 19: Confidence Intervals |
| 7 | 20: Significance Tests |
| | 21: Inference for One Mean |
| 8 | 22: Comparing Two Means |
| | 23: Inference for Proportions |
| 9 | 8: Describing Relationships |
| | 9: Correlation |
| | 11: The Question of Causation |
| | 25: Inference for Relationships |
| 10 | 24: Inference for Two-Way Tables |
| 11 | 13: Blocking and Sampling: Experiments and Samples |
| 12 | 6: Time Series |
| | 7: Models for Growth |
| | 10: Multidimensional Data Analysis |
| | 14: Samples and Surveys |
| | 26: Case Study |

# Essay Contents

FIFTH EDITION

# ELEMENTARY STATISTICS

# Chapter One

## In This Chapter

**1–1** **Overview**

The term **statistics** is defined along with the terms **population, sample, parameter,** and **statistic.** Statistics, its beginning, and its general nature are discussed.

**1–2** **Uses and Abuses of Statistics**

Examples of **beneficial uses** of statistics are presented, along with some of the common ways that statistics are used to **deceive.** Among the examples of abuses cited are the uses of different averages, graphs with modified scales, and the use of objects of volume.

**1–3** **The Nature of Data**

Different ways of **arranging data** are discussed. The four **levels of measurement** (nominal, ordinal, interval, ratio) are defined along with **discrete** and **continuous** data.

**1–4** **Statistical Experiments and Sampling**

The plan of a statistical experiment is discussed along with the importance of sampling. Sampling methods discussed are **random sampling, stratified sampling, systematic sampling, cluster sampling,** and **convenience sampling.** Writing a **project report** is also discussed.

# 1 Introduction to Statistics

## Chapter Problem

Much attention has been given to apparent discrepancies between incomes of men and women. From the Bureau of Labor Statistics we have recent data describing the mean annual earnings of husbands and wives who hold year-round full-time jobs. Figure 1–1 shows three ways to depict the *same* data, with Figures 1–1(b) and (c) drawn to exaggerate the discrepancy. Can you identify the tricks used to create the impression that the differences are more extreme than they really are? We will learn about these and other deceptive techniques in this chapter.

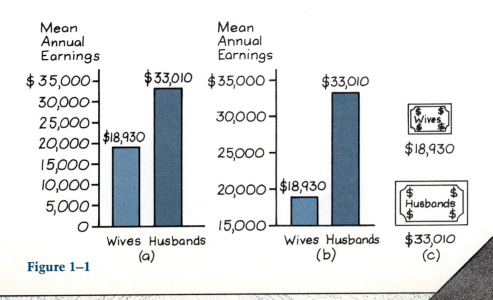

**Figure 1–1**

# 1–1 Overview

The word *statistics* has two basic meanings. We sometimes use this word when we refer to actual numbers derived from data, such as driver fatality rates, consumer price indices, or baseball attendance figures. The second meaning refers to statistics as a subject.

Statistics involves much more than the simple collection, tabulation, and summarizing of data. In this introductory book we will learn how to develop inferences that go beyond the original data and how to form more general and more meaningful conclusions.

> **Definition**
>
> **Statistics** is a collection of methods for planning experiments, obtaining data, and then analyzing, interpreting, and drawing conclusions based on the data.

Chapter 1 describes the general nature of statistics and presents a small sample of beneficial uses as well as some common abuses. Throughout the book we continue to give many examples of ways that the theories and methods of statistics have been used, as well as some of the ways that statistics have been abused. Abuses sometimes stem from ignorance or honest errors, and are sometimes the result of intentional deception. We will provide a few examples of how statistics have been used for deceptive purposes. These examples will help you to become more aware and more critical when you are presented with statistical claims. As you learn more about the acceptable uses of statistics, you will be better prepared to challenge misleading statements.

# Background

In the seventeenth century, a successful store owner named John Graunt (1620–1674) had enough spare time to pursue outside interests. His curiosity led him to study and analyze a weekly church publication, called "Bills of Mortality," that listed births, christenings, and deaths and their causes. Based on these studies, Graunt published his observations and conclusions in a work with the catchy title "Natural and Political Observations Made upon the Bills of Mortality." This 1662 publication was the first real interpretation of social and biological phenomena based on a mass of raw data, and many people feel it marks the birth of statistics.

Graunt made observations about the differences between the birth and mortality rates of men and women. He noted a surprising consistency among events that seem to occur by chance. These and other early observations led to conclusions or interpretations that were invaluable in planning, evaluating, controlling, predicting, changing, or simply understanding some facet of the world we live in.

We say that statistics can be used to predict, but it is very important to understand that we cannot predict with absolute certainty. In fact, statistical conclusions involve an element of uncertainty that can (and often does) lead to incorrect conclusions. It is possible to get 10 consecutive heads when an ordinary coin is tossed 10 times. Yet a statistical analysis of that experiment would lead to the incorrect conclusion that the coin is biased. That conclusion is not, however, certain. It is only a "likely" conclusion, reflecting the very low chance of getting 10 heads in 10 tosses.

In general, mathematics tends to be **deductive** in nature, meaning that specific conclusions are deduced with certainty from general principles or assumptions. Statistics is often inductive in nature, because conclusions are basically generalizations that may or may not correspond to reality. In most branches of mathematics we *prove* results, but in statistics many of our conclusions are associated with different degrees of "likelihood."

In statistics, we commonly use the terms *population* and *sample.*

## Definitions

A **population** is the complete collection of elements (scores, people, measurements, and so on) to be studied.

A **sample** is a subset of a population.

Closely related to the concepts of population and sample are the concepts of *parameter* and *statistic.*

## Definitions

A **parameter** is a numerical measurement describing some characteristic of a *population.*

population
↕
parameter

A **statistic** is a numerical measurement describing some characteristic of a *sample.*

sample
↕
statistic

## WAGE GENDER GAP

■ Many articles note that, on average, full-time female workers earn about 70¢ for each $1 earned by full-time male workers. Researchers at the Institute for Social Research at the University of Michigan analyzed the effects of various key factors and found that about one-third of the discrepancy can be explained by differences in education, seniority, work interruptions, and job choices. The other two-thirds remains unexplained by such labor factors. ■

Of the 315 members of a recent graduating class at a high school in Hyde Park, New York, 205 went on to college. Since 205 is 65% of 315, we can say that 65% went on to college. That 65% is a *parameter* (not a statistic), since it is based on the entire population of graduates at this high school. If we could somehow rationalize that this high school is representative of all high schools in New York State so that we can treat these graduates as a *sample* drawn from a larger population, then the 65% becomes a *statistic*.

Statisticians draw conclusions about an entire population based on the observed data in a sample. Thus they infer a general conclusion from known particular cases in the sample. In most branches of mathematics the procedure is reversed. That is, we first prove the generalized result and then apply it to the particular case. Geometers first prove that, for the population of all triangles, all possess the property that the sum of their respective angles is 180°. They then apply that established result to specific triangles, and they can be certain that the general property will always hold.

As we proceed with our study of statistics, you will learn how to extract pertinent data from samples and how to infer conclusions based on the results of those samples. You will also learn how to assess the reliability of conclusions. Yet you should always realize that, while the tools of statistics enable you to make inferences about a population, you can never accurately predict the behavior of any one individual.

A unique aspect of statistics is its obvious applicability to real and relevant situations. Although many branches of mathematics deal with abstractions that may initially appear to have little or no direct use in the real world, the elementary concepts of statistics have direct and practical applications. A wide variety of these applications will be found throughout this book.

## 1–2 Uses and Abuses of Statistics

Short essays that use real-world examples to illustrate the uses and abuses of statistics appear throughout this book. The uses include many applications in business, economics, psychology, biology, computer science, military intelligence, English, physics, chemistry, medicine, sociology, political science, agriculture, and education. Statistical theory applied to these diverse fields often results in changes that benefit humanity. Social reforms are sometimes initiated as a result of statistical analyses of factors such as crime rates and poverty levels. Large-scale population planning can result from projections devised by statisticians. Manufacturers can provide better products at lower costs through the effective use of statistics in quality control. Epidemics and diseases can be controlled and anticipated through application of standard statistical techniques. Endangered species of fish and other wildlife can be protected through regulations and laws developed in part using statistics.

Educators may discard innovative teaching techniques if statistical analyses show that traditional techniques are more effective. By pointing to lower fatality rates, legislators can better justify laws such as those governing air pollution, auto inspections, seat belt use, and drunk driving. Deposits of oil, natural gas, and coal can be located and evaluated. Retired employees can benefit from financially stable pension plans. Farmers can benefit from the development of better feed and fertilizer mixtures.

Students choose a statistics course for many different reasons. Some students take a required statistics course. Increasing numbers of other students voluntarily elect to take a statistics course because they recognize its value and application to whatever field they plan to pursue.

Apart from job-motivated or discipline-related reasons, the study of statistics can help you become more critical in your analyses of information so that you are less susceptible to misleading or deceptive claims. You use external data to make decisions, form conclusions, and build your own warehouse of knowledge. If you want to build a sound knowledge base, make intelligent decisions, and form worthwhile opinions, you must be careful to filter out the incoming information that is erroneous or deceptive. As an educated and responsible member of society, you should sharpen your ability to recognize distorted statistical data; in addition, you should learn to interpret undistorted data intelligently.

About a century ago, statesman Benjamin Disraeli said, "There are three kinds of lies: lies, damned lies, and statistics." It has also been said that "figures don't lie; liars figure." Historian Andrew Lang referred to using statistics "as a drunken man uses lampposts—for support rather than illumination." Author Darrell Huff has said that "a well wrapped statistic is better than Hitler's big lie; it misleads, yet it cannot be pinned on you." Economist Sir Josiah Stamp said, "The Government is very keen on amassing statistics. They collect them, add them, raise them to the $n$th power, take the cube root and prepare wonderful diagrams. But you must never forget that every one of these figures comes in the first instance from the village watchman, who just puts down what he damn well pleases."

The preceding statements refer to abuses of statistics in which data are presented in ways that may be misleading. Some abusers of statistics are simply ignorant or careless, whereas others have personal objectives and are willing to suppress unfavorable data while emphasizing supportive data. Here are a few examples of the many ways that data can be distorted.

The term *average* refers to several different statistical measures that will be discussed and defined in Chapter 2. To most people, the average is the sum of all values divided by the number of values, but this is only one type of average. For example, given the 10 annual salaries of $25,000, $25,000, $25,000, $26,000, $28,000, $30,000, $32,000, $34,000, $35,000, and $55,000, we can correctly claim that the "average" is either $31,500 (since $31,500 is the sum of the 10 values divided by 10), or $29,000 (since half of the values are above $29,000 and half are below), or $25,000 (since it is the value that occurs most often), or $40,000 (since this value is exactly midway between the lowest and highest salaries listed). This example illustrates that a reported average can be very misleading, since it can be any one of several different numbers. There are no objective criteria that can be used to determine the specific average that is most representative, and the user of statistics is free to select the average that best supports a favored position.

A union contract negotiator can choose the lowest average in an attempt to emphasize the need for a salary increase, while the management negotiator can choose the highest average to emphasize the well-being of the employees. In actual negotiations both sides are usually adept at exposing such ploys, but the typical citizen often accepts the validity of an average without really knowing which specific average is being presented and without knowing how different the picture would look if another average were used. The educated and thinking citizen is not so susceptible to potentially deceptive information; the educated and thinking citizen analyzes and criticizes statistical data so that meaningless or illusory claims are not part of the base on which his or her decisions are made and opinions formulated. If you read that the average annual salary of an American family is $27,735 (the latest available figure from the Bureau of the Census), you should attempt to determine which of the averages that figure represents. If no additional information is available, you should realize that the given figure may be very misleading. Conversely, as you present statistics on data you have accumulated, you should attempt to provide descriptions identifying the true nature of the data.

Many visual devices—such as bar graphs and pie charts—can be used to exaggerate or deemphasize the true nature of data. These are also discussed in Chapter 2. Figure 1–1 uses data from the Bureau of Labor Statistics to show figures that depict the *same data,* but part (b) is designed to exaggerate the difference between the salaries of women and men. By not starting the horizontal axis at zero, part (b) tends to produce a misleading subjective impression. Too many of us look at a graph superficially and develop intuitive impressions based on the pattern we see. Instead, we should scrutinize the graph and search for distortions of the type illustrated in Figure 1–1. We should analyze the *numerical* information given in the graph instead of being misled by its general shape.

Drawings of objects may also be misleading. Some objects commonly used to depict data include moneybags, stacks of coins, army tanks (for military expenditures), cows (for dairy production), and houses (for home construction). When we double the dimensions of a two-dimensional figure, the area is quadrupled. Figure 1–1(c) presents the *same* data, but both the height and width of the smaller dollar bill are 57% of those dimensions of the larger bill. The area of the smaller bill becomes 32% of the area of the larger bill, again creating a misleading impression. The impact created by Figure 1–1(c) will be greater than that of Figure 1–1(a). Figure 1–2 further illustrates how drawings can be used to mislead.

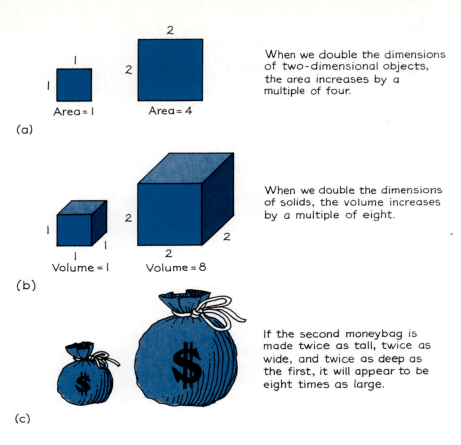

When we double the dimensions of two-dimensional objects, the area increases by a multiple of four.

Area = 1    Area = 4

(a)

When we double the dimensions of solids, the volume increases by a multiple of eight.

Volume = 1    Volume = 8

(b)

If the second moneybag is made twice as tall, twice as wide, and twice as deep as the first, it will appear to be eight times as large.

(c)

**Figure 1–2**

## AIRLINES SAMPLE

■ Airline companies once used an expensive accounting system to split up income from tickets that involved two or more companies. They now use a sampling method whereby a small percentage of these "split" tickets is randomly selected and used as a basis for dividing up all such revenues. The error created by this approach can cause some companies to receive slightly less than their fair share, but these losses are more than offset by the clerical savings accrued by dropping the 100% accounting method. This new system saves companies millions of dollars each year. ■

Let's suppose that school taxes in a small community doubled from one year to the next. By depicting the amounts of taxes as bags of money and by doubling the dimensions of the bag representing the second year, we can easily create the impression that taxes more than doubled (as in Figure 1–2). Another variety of statistical "lying" is often inspired by small sample results. The toothpaste preferences of only 10 dentists should not be used as a basis for a generalized claim such as "Caressed toothpaste is recommended by 7 out of 10 dentists." Even if the sample is large, it must be unbiased and representative of the population from which it comes.

Sometimes the numbers themselves can be deceptive. A mean annual salary of $27,735.29 sounds precise and tends to instill a high degree of confidence in its accuracy. The figure of $27,700 doesn't convey that same sense of precision and accuracy. A statistic that is very precise with many decimal places is not necessarily accurate, though.

Another source of statistical deception involves numbers that are ultimately guesses, such as the crowd count at a political rally, the dog population of Chicago, and the amount of money bet illegally. When the Pope visited Miami, officials estimated the crowd size to be 250,000, but the *Miami Herald* used aerial photos and grids to come up with a better estimate of 150,000. As another example, the Associated Press ran an article in which the roach population of New York City was estimated to be 1 billion. That claim was made by a spokesperson for the Bliss Exterminator Company. The head of New York City's Bureau of Pest Control would not confirm that figure. He said, "I haven't the slightest idea if that's right. We do strictly rats here."

Continental Airlines ran full-page ads boasting better service. In referring to lost baggage, these ads claimed that this is "an area where we've already improved 100% in the last six months." Do you really believe that they no longer lose any baggage at all? That's what the 100% improvement figure actually means.

"Ninety percent of all our cars sold in this country in the last 10 years are still on the road." Millions of consumers heard that commercial message and got the impression that those cars must be well built to last through those long years of driving. What the auto manufacturer failed to mention was that 90% of the cars it sold in this country were sold within the last three years. The claim was technically correct, but it was very misleading.

The preceding examples comprise a small sampling of the ways in which statistics can be used deceptively. Entire books have been devoted to this subject, including Darrell Huff's *How to Lie with Statistics* and Robert Reichard's *The Figure Finaglers*. Understanding these practices will be extremely helpful in evaluating the statistical data found in everyday situations.

## 1–2 Exercises A

1. A graph similar to Figure 1–3 appeared in *Car and Driver* magazine. What is wrong with it?

2. Seventy-two percent of Americans squeeze the toothpaste tube from the top. This and other not so serious findings are presented in *The First Really Important Survey of American Habits*. Those results are based on 7000 respondents from the 25,000 questionnaires that were mailed. What is wrong with this survey?

3. "According to a nationwide survey of 250 hiring professionals, scuffed shoes was the most common reason for a male job seeker's failure to make a good first impression." Newspapers carried this statement based on a poll commissioned by Kiwi Brands, producers of shoe polish. Comment.

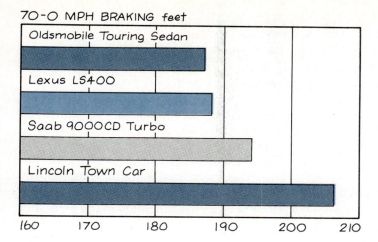

70-0 MPH BRAKING *feet*

**Figure 1–3**

## THE *LITERARY DIGEST* POLL

■In the 1936 presidential race, *Literary Digest* magazine ran a poll and predicted an Alf Landon victory, but Franklin D. Roosevelt won by a landslide. Maurice Bryson notes, "Ten million sample ballots were mailed to prospective voters, but only 2.3 million were returned. As everyone ought to know, such samples are practically always biased." He also states, "Voluntary response to mailed questionnaires is perhaps the most common method of social science data collection encountered by statisticians, and perhaps also the worst." (See Bryson's "The *Literary Digest* Poll: Making of a Statistical Myth," *The American Statistician*, Vol. 30, No. 4.) ■

4. The Australian Minister of Labor stated, "We look forward to the day when everyone will receive more than the average wage." Comment.

5. In a study on college campus crimes committed by students high on alcohol or drugs, a mail survey of 1875 students was conducted. A *USA Today* article noted, "Eight percent of the students responding anonymously say they've committed a campus crime. And 62% of that group say they did so under the influence of alcohol or drugs." Assuming that the number of students responding anonymously is 1875, how many actually committed a campus crime while under the influence of alcohol or drugs?

6. A study conducted by the Insurance Institute for Highway Safety found that the Chevrolet Corvette had the highest fatality rate—"5.2 deaths for every 10,000." The car with the lowest fatality rate was the Volvo, with only 0.6 death per 10,000. Does this mean that the Corvette is not as safe as a Volvo?

7. The Labor Department reported that the median weekly pay of women is about 70% of that of men. One reason for this is discrimination based on sex. Cite a second reason that might help to explain the discrepancy between the salaries of men and women.

8. A study by Dr. Ralph Frerichs (UCLA) showed that family incomes are related to the risk of dying because of heart disease. Higher family income levels corresponded to lower heart disease death rates.
   a. Does this imply that more earned money *causes* the risk of dying of heart disease to be lower?
   b. Identify a factor that could explain the correspondence.

9.  You plan to conduct a telephone survey of 500 people in your region. What would be wrong with using the telephone directory as the population from which your sample is drawn?

10. You plan to conduct a poll of students at your college. What is wrong with polling every 50th student who is leaving the cafeteria?

11. A college conducts a survey of its alumni in an attempt to determine their typical annual salary. Would alumni with very low salaries be likely to respond? How would this affect the result? Identify one other factor that might affect the result.

12. One study actually showed that smokers tend to get lower grades in college than nonsmokers. Does this mean that smoking causes lower grades? What other explanation is possible?

13. A report by the Nuclear Regulatory Commission noted that a particular nuclear reactor was being operated at "below-average standards." What is the approximate percentage of nuclear power plants that operate at below-average standards? Is a below-average standard necessarily equivalent to a dangerous or undesirable level?

14. An employee earning $400 per week was given a 20% cut in pay as part of her company's attempt to reduce labor costs. After a few weeks, this employee's dissatisfaction grew and her threat to resign caused her manager to offer her a 20% raise. The employee accepted this offer since she assumed that a 20% raise would make up for the 20% cut in pay.
    a.  What was the employee's weekly salary after she received the 20% cut in pay?
    b.  Use the salary figure from part *a* to find the amount of the 20% increase and determine the weekly salary after the raise.
    c.  Did the 20% cut followed by the 20% raise get the employee back to the original salary of $400 per week?

15. The first edition of a textbook contained 1000 exercises. For the second edition, the author removed 100 of the original exercises and added 300 new exercises. Which of the following statements about the second edition are correct?
    a.  There are 1200 exercises.
    b.  There are 33% more exercises.
    c.  There are 20% more exercises.
    d.  Twenty-five percent of the exercises are new.

16. What differences are there between the following two statements, and which one do you believe is more accurate?
    a.  Drunk drivers cause about half of all fatal car crashes.
    b.  Of all fatal motor vehicle crashes, about 50% involve alcohol.

17. *Good Housekeeping* magazine reported on studies leading to the discovery that "the average person spends six years eating" (not all at once). Assuming an average life span of 75 years, develop your own estimate and identify how it was obtained. Do you think the stated figure of 6 years is too low, too high, or about right?

18. An article in *Forbes 400* magazine commented on the 400 wealthiest Americans. It noted that the average worth of those who were divorced at least once was $819 million, while those who remained with the same spouse had an average worth of only $617 million. The article stated that if you plan to remain with the same spouse, "make darn sure he or she is worth taking a $200 million hit for." Comment.

19. In an advertising supplement inserted in *Time*, the increases in expenditures for pollution abatement were shown in a graph similar to Figure 1–4. What is wrong with the figure?

20. A *New York Times* article noted that the mean life span for 35 male symphony conductors was 73.4 years, in contrast to the mean of 69.5 years for males in the general population. The longer life span was attributed to such factors as fulfillment and motivation. There is a fundamental flaw in concluding that male symphony conductors live longer. What is it? (*Hint:* How old are males when they are identified as symphony conductors?)

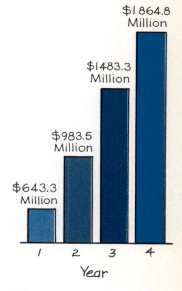

AMERICA'S CHEMICAL INDUSTRY
Capital Expenditures and Operating Costs for Pollution Abatement

$1864.8 Million

$1483.3 Million

$983.5 Million

$643.3 Million

1   2   3   4
Year

**Figure 1–4**

## 1–2  Exercises B

21. A researcher at the Sloan-Kettering Cancer Research Center was once criticized for falsifying data. Among his data were figures obtained from 6 groups of subjects, with 20 individual subjects in each group. These values were given for the percentage of successes in each group: 53%, 58%, 63%, 46%, 48%, 67%. What's wrong?

22. If an employee is given a cut in pay of $x$ percent, find an expression for the percent raise that would return the salary to the original amount.

23. A *New York Times* editorial criticized a chart caption that described a dental rinse as one that "reduces plaque on teeth by over 300%." What does it mean to reduce plaque by over 300%?

24. If you must travel from New York to California, is it safer to go by car or plane? Here are some recent data from the National Safety Council.

|  | Passenger Miles (billions) | Total Deaths |
|---|---|---|
| Car | 2,467 | 22,974 |
| Plane | 278 | 201 |

## 1–3  The Nature of Data

People tend to think of collections of data as lists of numbers, such as the income levels of actors or the batting averages of baseball players. Yet data may be nonnumerical, and even numerical data can belong to different categories with different characteristics. For example, a pollster may compile nonnumeric data such as the sex, race, and religion of voters in a sample.

Numeric data, instead of being in an unordered list, might be data matched in pairs (discussed in Chapters 8, 9, 12), as in the following two tables.

| *Pretraining weights (kg) | 99 | 62 | 74 | 59 | 70 |
|---|---|---|---|---|---|
| Posttraining weights (kg) | 94 | 62 | 66 | 58 | 70 |

| *Pretraining weights (kg) | 99 | 62 | 74 | 59 | 70 |
|---|---|---|---|---|---|
| Pretraining heart rate (beats/min) | 174 | 180 | 171 | 177 | 168 |

Another common arrangement for summarizing sample data is the contingency table (discussed in Section 10–3), such as the one that follows.

|  | Grade A | B | C | D | F |
|---|---|---|---|---|---|
| Math | 494 | 689 | 642 | 415 | 610 |
| Business | 1,895 | 2,048 | 1,675 | 537 | 784 |
| History | 320 | 631 | 585 | 215 | 183 |

Based on data from Dutchess Community College.

In this table the numbers are frequencies (counts) of sample results.

Clearly, the nature of the data can affect the nature of the relevant problem and the method used for analysis. With the paired weight data, a fundamental concern would be whether there is a difference in the pretraining and posttraining weights. Any analysis of these data should attempt to determine whether posttraining weights are significantly less than pretraining weights. With the paired weight/heart rate data, the fundamental concern would be whether some relationship exists between those two factors. This requires a different method of analysis. With the contingency table, the fundamental concern would be whether the grade distribution is independent of the course. This requires yet another method of analysis. As we consider the topics of later chapters, we will see that the structure and nature of the data affects our choice of the method we will use.

Data may be categorized as **attribute** data or **numerical** data. Attribute data consist of qualities, such as political party, religion, or sex. Numerical data consist of numbers representing counts or measurements. We can further categorize numerical data by distinguishing between the discrete and continuous types.

We know what a finite number of values is (1, 2, 3, and so on), but our definition of discrete data also involves the concept of a **countable** number of values. As an example, if we count the number of rolls of a pair of dice before a 7 turns up, we can get any one of the values 1, 2, 3, . . . . We now have an infinite number of possibilities, but they correspond to the counting numbers. Consequently, this type of infinity is called countable. In contrast, the number of points on a continuous scale is not countable and represents a higher degree of infinity. There is no way to count the points on a continuous scale, but we can count the number of times a die is rolled, even if the rolling seems to continue forever.

## Definitions

**Discrete** numerical data result from either a finite number of possible values or a countable number of possible values.

**Continuous** numerical data result from infinitely many possible values that can be associated with points on a continuous scale in such a way that there are no gaps or interruptions.

As an example, you can obtain exact counts of the numbers of Pepsi bottles in different stores, and the results will be 0, or 1, or 2, or 3, and so on. These exact bottle counts are discrete data. But if we measure the amounts (in liters) of Pepsi in different bottles, we could get values such as 1.026 or 0.99 or *any* value in between, and such measurements are continuous data. Similarly, shoe sizes (7, 9½, 8, etc.) are discrete data whereas the actual lengths of feet (10.03 in., 11.738 in., 8.62 in., etc.) are continuous data.

Another common way to classify data is to use four **levels of measurement:** nominal, ordinal, interval, ratio.

## Definition

The **nominal level of measurement** is characterized by data that consist of names, labels, or categories only. The data cannot be arranged in an ordering scheme.

If we associate *nominal* with "name only," the meaning becomes easy to remember. An example of nominal data is the collection of "yes, no, undecided" responses to a survey question. Data at this nominal level of measurement cannot be arranged according to some ordering scheme. That is, there is no criterion by which values can be identified as greater than or less than other values.

## Example

The following are other examples of sample data at the nominal level of measurement.

1. Responses consisting of 12 Democrats, 15 Republicans, and 9 Independents.

2. Responses consisting of 14 students from New York, 17 from California, 8 from Connecticut, and 7 from Florida.

# MEASURING DISOBEDIENCE

■ How are data collected about something that doesn't seem to be measurable, such as people's level of disobedience? Psychologist Stanley Milgram devised an experiment: A researcher instructed a volunteer subject to operate a control board that gave increasingly painful "electrical shocks" to a third person. Actually, no real shocks were given and the third person was an actor. The volunteer began with 15 volts and was instructed to increase the shocks by increments of 15 volts. The disobedience level was the point at which the subject refused to increase the voltage. Surprisingly, two-thirds of the subjects obeyed orders even though the actor screamed and faked a heart attack. ■

## THE CENSUS

■ Every 10 years, the U.S. government undertakes a census intended to obtain information about each American. The last census cost about $10 per person, for a total of $2.5 billion. The results affect over $100 billion in government allocations, seats in Congress, and redistricting of state and local governments. Pollsters use census results for designing samples and weighting survey data. Businesses use census data in selecting plant and office locations. Governments use census data in planning new projects, such as building new schools and roads. ■

It should be obvious that the preceding data cannot be used for calculations because the categories lack any ordering or numerical significance. We cannot, for example, "average" 12 Democrats, 15 Republicans, and 9 Independents. Numbers are sometimes assigned to the different categories, especially when the data are processed by computer. We might find that Democrats are assigned 0, Republicans are assigned 1, and Independents are assigned 2. Even though we now have number labels, those numbers lack any real computational significance. The average of twelve 0s, fifteen 1s, and nine 2s might be 0.9, but that is a meaningless statistic. (0.9 does not represent a liberal Republican!)

### Definition

The **ordinal level of measurement** involves data that may be arranged in some order, but differences between data values either cannot be determined or are meaningless.

### Example

The following are examples of data at the ordinal level of measurement.

1. In a sample of 36 batteries, 12 were rated "good," 16 were rated "better," and 8 were rated "best."

2. In a class of 19 students, 5 required remediation, 10 were average, and 4 were gifted.

3. In a high school graduating class of 463 students, Sally ranked 12th, Allyn ranked 27th, and Mike ranked 28th.

In the first example given, we cannot determine a specific measured difference between "good" and "better." In the third example, we can determine a difference between the rankings of 12 and 27, but the resulting value of 15 doesn't really mean anything. That is, the difference of 15 between ranks of 12 and 27 isn't necessarily the same as the difference of 15 between ranks of 30 and 45. This ordinal level provides information about relative comparisons, but the degrees of differences are not available. We know that a low-income worker earns less than a middle-income worker, but we don't know how much less. Again, data at this level should not be used for calculations.

### Definition

The **interval level of measurement** is like the ordinal level, with the additional property that we can determine meaningful amounts of differences between data. Data at this level may lack an inherent zero starting point. At this level, differences are meaningful, but ratios are not.

Temperature readings of 25° F and 50° F are examples of data at this measurement level. Those values are ordered and we can determine their difference (often called the *distance* between the two values). However, there is no inherent zero point. The value of 0° F might seem like a starting point, but it is arbitrary and not inherent. The value of 0° F does not indicate "no heat," and it is incorrect to say that 50° F is twice as hot as 25° F. For the same reasons, temperature readings on the Celsius scale are also at the interval level of measurement. (Temperature readings on the Kelvin scale are at the ratio level of measurement; that scale has an absolute zero.)

### Example

The following are other examples of data at the interval level of measurement.

1. Years in which IBM stock split
2. Body temperatures (in degrees Celsius) of hospital patients

### Definition

The **ratio level of measurement** is actually the interval level modified to include the inherent zero starting point. For values at this level, differences and ratios are meaningful.

### Example

The following are examples of data at the ratio level of measurement.

1. Heights of pine trees around Lake Tahoe
2. Volumes of helium in balloons
3. Times (in minutes) of runners in a marathon

| Levels of Measurement | | |
|---|---|---|
| Level | Summary | Example |
| Nominal | Categories only. Data cannot be arranged in an ordering scheme. | Voter distribution: 45 Democrats, 80 Republicans, 90 Independents — Categories only. |
| Ordinal | Categories are ordered, but differences cannot be determined or they are meaningless. | Voter distribution: 45 low-income voters, 80 middle-income voters, 90 upper-income voters — An order is determined by "low, middle, upper." |
| Interval | Differences between values can be found, but there may be no inherent starting point. Ratios are meaningless. | Temperatures of steel rods: 45° F, 80° F, 90° F — 90° F is not twice as hot as 45° F. |
| Ratio | Like interval, but with an inherent starting point. Ratios are meaningful. | Lengths of steel rods: 45 cm, 80 cm, 90 cm — 90 cm is twice as long as 45 cm. |

Ratio
↑
Interval
↑
Ordinal
↑
Nominal

Values in each of these data collections can be arranged in order, differences can be computed, and there is an inherent zero starting point. *This level is called the ratio level because the starting point makes ratios meaningful.* Since a tree 50 feet high *is twice* as tall as a tree 25 feet high and 50° F *is not twice* as hot as 25° F, heights are at the ratio level while Fahrenheit temperatures are at the interval level.

Among the four levels of measurement, the nominal is considered the lowest, followed by the ordinal level, the interval level, and the ratio level.

The following is an important guideline. **The statistics based on one level of measurement should not be used for a lower level, but can be used for a higher level.** We can, for example, calculate the average for data at the interval or ratio level, but not at the lower ordinal or nominal levels. An implication of this guideline is that data obtained from using a Likert scale, such as the one that follows, should not be used for such calculations since these data are only at the ordinal level. This guideline is sometimes ignored

| Superior | Good | Average | Poor | Very Poor |
|---|---|---|---|---|
| 1 | 2 | 3 | 4 | 5 |

and Likert scale results are often treated as interval- or ratio-level data, even though they are not, which may result in serious errors. If, for data processing requirements, we assign the numbers 0, 1, and 2 to Democrats, Republicans, and Independents, respectively, and proceed to calculate the average, we are creating a meaningless statistic that can lead to incorrect conclusions.

Chapter 2 will introduce basic methods of dealing with data sets that are primarily at the interval or ratio levels of measurement.

# 1–3  Exercises A

In Exercises 1–10, identify each number as being *discrete* or *continuous*.

1. Among campus vending machines, 14 are found to be defective.
2. Today's records show that 25 employees were absent.
3. The car weighs 1430 kilograms.
4. Among all SAT scores last year, 23 were perfect.
5. Radar indicated that the driver was going 72.4 mi/h.
6. The car stopped in 187.3 ft.
7. Of all students who took the test, 49 failed.
8. Of the respondents, 423 are women.
9. The crew completed refueling in 17.5 minutes.
10. Among the issues traded on the New York Stock Exchange, 327 declined.

In Exercises 11–20, determine which of the four levels of measurement (nominal, ordinal, interval, ratio) is most appropriate.

11. Cars described as subcompact, compact, intermediate, or full-size.
12. Weights of a sample of machine parts
13. Colors of a sample of cars involved in alcohol-related crashes
14. Years in which Republicans won presidential elections
15. Zip codes
16. Social security numbers
17. Total annual incomes for a sample of families
18. Final course averages of A, B, C, D, F
19. Body temperatures (in degrees Fahrenheit) of a sample of bears captured in Wyoming
20. Instructors rated as superior, above average, average, below average, or poor

# 1–3  Exercises B

21. In a final examination for a statistics course, one student received a grade of 50, and another student received a grade of 100.
    a. If we consider these numbers to represent only the points earned on the exam, then the score of 100 is twice that of 50. What is the corresponding level of measurement?

*continued*

## INVISIBLE INK

■ The *National Observer* once hired a firm to conduct a confidential mail survey. Editor Henry Gemmill wrote in a cover letter that "each individual reply will be kept confidential, of course, but when your reply is combined with others from all over this land, we'll have a composite picture of our subscribers." One clever subscriber used an ultraviolet light to detect a code written on the survey in invisible ink. That code could be used to identify the respondent. Gemmill was not aware that this procedure was used and he publicly apologized. Confidentiality was observed as promised, but anonymity was not directly promised and it was not maintained.

■

b.   If we consider these numbers to represent the amount of the subject learned in the course, it is wrong to conclude that the one student knows twice as much as the other. What is the level of measurement in this case?

22.   Many people question what IQ scores actually measure. Assuming that IQ scores measure intelligence, is a person with an IQ score of 150 twice as intelligent as another person with an IQ score of 75?

a.   What does an affirmative answer imply about the level of measurement corresponding to IQ scores?

b.   What does a negative answer imply about the level of that data?

23.   If a recipe requires cooking something at "300° F for three hours," but you decide to cook it at 900° F for one hour, the result will be different. Explain.

24.   The years 1990, 1988, 1972, 1963, and 1984 form a collection of data at the interval level of measurement. Explain.

# 1–4 Statistical Experiments and Sampling

## Identifying Objectives

In designing a statistical experiment, we must begin by determining exactly what question we want answered. Beginning researchers are often overcome by enthusiasm as they set out to collect facts without considering *precisely* what they are investigating and which facts are truly relevant. The original statement of the problem is often too vague or too broad. When this happens, too many different directions are sometimes pursued.

As an example, suppose some group wants to survey public opinion about abortion. What is the population we will sample? Will it be all Americans? Clearly, the opinion of a two-year-old child on this topic is not very relevant. Do we restrict our population to adults? Over 16, over 18, or over 21? Men and women? After determining the exact population of interest, we need to determine the number of people that will be surveyed. (Chapter 6 will discuss methods of determining the sample size.) How do we select the actual subjects? What options do we have? Do we mail a questionnaire, contact them by telephone, or interview them in person? How do we word the questions? In an article on polls, *Money* magazine presented these two questions:

1.   Should there be a constitutional amendment prohibiting abortions?

2.   Should there be an amendment protecting the life of the unborn child?

A majority of those surveyed answered no to the first version, while 20% switched their answers when presented with the second version. Obviously, the wording of the question can dramatically affect the responses. (Can you explain why such different results occurred?)

# Designing the Experiment

In order to obtain meaningful data in an efficient way, we must develop a complete plan for collecting data *before* the collection actually begins. Researchers are often frustrated and discouraged if they learn that the method of collection or the data themselves cannot be used to answer their questions. Will the experiment be conducted on the entire population or will a sample be drawn from the population? The population size usually requires us to make inferences on the basis of sample data. We need to determine the size of the sample and the method of sampling at the very beginning.

In some cases we may be able to conduct a **retrospective** study that involves looking back at past events. For example, a study of car crash fatalities might be based on available records. In other cases, we might use a **prospective** study that tracks groups forward in time. A company might use a prospective study to compare the effects of two new and different advertising campaigns.

# Sampling and Collecting Data

Sampling and data collection usually require more time, effort, and money than the statistical analysis of that data. Careful planning will minimize the expenditure of those precious resources. Take care that the sampling is done according to plan and the data are recorded in a complete and accurate manner.

If you are obtaining measurements of some characteristic from people, realize that you will get better results if you can do the measuring instead of asking the subject for the value. Asking tends to yield a disproportionate number of rounded results. Thus the data are distorted and the sample is flawed.

When conducting a survey, consider the medium to be used. Mail surveys, telephone surveys, and personal interviews are most common, although other methods are used. Mail surveys tend to get lower response rates. Personal interviews are obviously more time consuming and expensive, but they may be necessary if detailed and complex data are required. Telephone interviews are relatively efficient and inexpensive.

Be especially careful when choosing the sampling method. The sample size must be large enough for the required purposes, but even large samples may be totally worthless if the data have been carelessly collected. We now describe some of the more common methods of sampling.

## Random Sampling

In **random sampling,** each member of the population has an equal chance of being selected. Random sampling is also called *representative* or *proportionate* sampling, since all groups of the population should be proportionately represented in the sample.

Random sampling is not the same as haphazard or unsystematic sampling. Much effort and planning must be invested in order to avoid any bias. For example, if a list of all elements from the population is available, the names of those elements can be placed in capsules and put in a bowl; those capsules can then be mixed and samples selected. Computers or tables of random numbers are often used as practical alternatives to bowls of capsules.

A major problem with this approach is the difficulty of finding a complete list of *all* elements in the population. Even when a complete list seems to be available, we must be sure that it is not biased and the resulting sample is not biased. For example, if we decide that voters are to be sampled by selecting numbers from a telephone directory, we automatically eliminate all unlisted numbers and our sample may be biased. People who choose to have unlisted phone numbers may constitute a special interest group when certain issues are raised. This can dramatically affect some results because there are regions with high proportions of unlisted numbers. In Los Angeles, 42.5% of the telephone numbers are unlisted. The percentages for Chicago and San Francisco are 40.9% and 39.5%, respectively. (The data are from Survey Sampling, Inc.) Pollsters commonly circumvent this problem by using computers to generate phone numbers so that all of them become accessible. However, we still run the risk of having a biased sample if we simply ignore those who are unavailable or refuse to comment. Humphrey Taylor, president of the Harris polling company, states that the refusal rate for telephone interviews generally is at least 20%. Ignore those people who initially refuse and you run a real risk of having a biased sample.

A recent example of a nonrandom sample is the book *Women and Love: A Cultural Revolution in Progress* by Shere Hite. She based conclusions on 4500 responses from 100,000 questionnaires distributed to women. Such results are not random, and they may well reflect the opinions of respondents who have strong feelings about the survey topics. That book was widely criticized for its obvious bias and lack of sound statistical methodology. Her sample size of 4500 was certainly large enough, but her method of sampling was seriously flawed.

## Stratified Sampling

After classifying the population into at least two different **strata** (or classes) that share the same characteristics, we draw a sample from each stratum. In surveying views on an Equal Rights Amendment to the Constitution, we might use sex as a basis for creating two strata. After obtaining a list of men and a list of women, we use some suitable method (such as random sampling) to select a certain number of people from each list. If it should happen that some strata are not represented in the proper proportion, then the results can be adjusted or weighted accordingly. Stratified sampling is often the most efficient of the various sampling methods.

## Systematic Sampling

In **systematic sampling** we select some starting point and then select every $k$th element. For example, we use a telephone directory of 10,000 names as our population, and we must choose 200 of those names. We can randomly select one of the first 50 names and then choose every 50th name after that. This method is simple and is used frequently.

## Cluster Sampling

In **cluster sampling** we first divide the population area into sections and then randomly select a few of those sections. We then choose all the members from those sections. For example, in conducting a preelection poll, we could randomly select 30 election precincts and then survey all the people from each of those chosen precincts. This would be much more efficient than selecting 1 person from each of the many precincts in the population area. The results can be adjusted or weighted to correct for any disproportionate representations of groups. Cluster sampling is used extensively by the government and by private research organizations.

## Convenience Sampling

In **convenience sampling** we simply use results that are readily available. In some cases, results may be quite good, while they may be seriously biased in other cases. In investigating the proportion of left-handed people, it would be convenient for a teacher to survey students. Even though the sample is not random, it will tend to be unbiased because left-handedness is not the type of characteristic that would be related to presence in class. But if the same teacher questions the same students about their opinions on federal aid to education, the results will be clearly biased and not representative of the general population. Be very wary of convenience sampling.

The preceding descriptions of different sampling methods are intended to be brief and general. Thoroughly understanding these different methods so that you can successfully use them requires much more extensive study than is practical in a single introductory course of this type.

## Sampling Errors

Even experienced and reputable research organizations sometimes get erroneous results due to biased sampling or poor methodology. In 1948 the Gallup poll was wrong when it predicted Truman would lose to Dewey. That mistake led to a revision of Gallup's polling methods. A quota system had been used to obtain the opinions of a proportionate number of men, women, rich, poor, Catholics, Protestants, Jews, and so on. After the 1948 error, Gallup abandoned the quota system and instituted random sampling based on clusters of interviews in several hundred areas throughout the nation.

# A Professional Speaks About Sampling Error

■ Daniel Yankelovich, in an essay for *Time*, commented on the sampling error often reported along with poll results. He stated that sampling error refers only to the inaccuracy created by using random sample data to make an inference about a population; the sampling error does not address issues of poorly stated, biased, or emotional questions. He said, "Most important of all, warning labels about sampling error say nothing about whether or not the public is conflict-ridden or has given a subject much thought. This is the most serious source of opinion poll misinterpretation." ■

This point cannot be stressed enough: Data collected carelessly can be absolutely worthless, even if the sample is large. It might be convenient to mail thousands of questionnaires and use only those that are returned, but such samples may be seriously biased. It might be even more convenient to set up a 900 phone number that respondents pay to call, but such sample results may be seriously biased. Sample subjects should be selected by the pollster; they should not select themselves as they do with mailed responses or 900 phone surveys.

In analyzing results, it is often helpful to discuss two sources of errors—**sampling errors** and **nonsampling errors.** Sampling errors result from the actual sampling process. They include such factors as the small size of the sample and the fact that no sample can be expected to be a perfect representation of the entire population. Nonsampling errors arise from other external factors not related to sampling, such as a defective measuring instrument.

Nonsampling errors can't be described as objectively as sampling errors, but we should note any factors that might significantly affect the results. These might include a large number of response refusals or missing values, errors in coding or recoding data, or a discovered bias in the sample.

### Project Report

An ideal way to gain insight into statistical methods is to conduct a statistical experiment from beginning to end. It might be advantageous to relate the subject of such an experiment to another course or discipline.

In writing the project report, consider the intended readers. Statisticians will expect specific and detailed results accompanied by fairly complete descriptions of methodology. However, a lay audience would not benefit from this type of report, so a different approach is necessary. Be sure to identify the type of sampling (such as random or systematic). Identify the features of the sampling process that tend to make your sample representative of the population. Also be sure to identify the population from which the sample was drawn. Any survey should be seriously and carefully constructed with consideration for the sensitivities and emotions of those surveyed. If your data source is people, take every necessary precaution and obtain any necessary approval.

## ETHICS IN REPORTING

■ The American Association for Public Opinion Research developed a voluntary code of ethics to be used in news reports of survey results. This code requires that the following be included. (1) Identification of the survey sponsor; (2) Date the survey was conducted; (3) Size of the sample; (4) Nature of the population sampled; (5) Type of survey used; (6) Exact wording of survey questions. Surveys funded by the U.S. government are subject to a prescreening that assesses the risk to those surveyed, the scientific merit of the survey, and the guarantee of the subject's consent to participate. ■

## 1–4  Exercises A

In Exercises 1–12 identify the type of sampling used.

1. A teacher selects every 5th student in the class for a test.
2. A teacher writes the name of each student on a card, shuffles the cards, and then draws 5 names.
3. A teacher surveys all students from each of 12 randomly selected classes.
4. A teacher selects 5 men and 5 women from each of 4 classes.

5. A teacher surveys all students in order to study public opinion.
6. A teacher selects 15 students under 21 years of age and 15 students over 21.
7. A pollster interviews 75 men and 75 women.
8. A pollster interviews all voters in each of 15 randomly selected blocks.
9. A medical researcher interviews all leukemia patients in each of 20 randomly selected counties.
10. A pollster interviews every 50th voter on the listing of all county voters.
11. A pollster uses a computer to generate 150 random numbers, and then interviews the voters corresponding to these numbers.
12. A pollster interviews 15 voters who happen to be waiting at a bus stop.

## 1–4  Exercises B

13. Assume that you are employed by a car manufacturer to collect data on the waist sizes of drivers. Why is it better to obtain direct measurements than to ask people the sizes of their waists?
14. Distinguish between sampling error and nonsampling error.
15. Public opinion is sometimes measured by asking television viewers to call a 900 number. The cost of such a call is usually around 50¢. Identify two different factors that cause the resulting sample to be biased.
16. You plan to estimate the average weight of all passenger cars used in the United States. Is there universal agreement about what a "passenger car" is? Are there any factors that might lead to regional differences among the weights of passenger cars? How can you obtain a sample?
17. Two categories of survey questions are *open* and *closed*. An open question allows a free response, while a closed question allows only a fixed response. Here are examples based on Gallup surveys.

    Open question: What do you think can be done to reduce crime?
    Closed question: Which of the following approaches would be most effective in reducing crime?

    - Hire more police officers.
    - Get parents to discipline children more.
    - Correct social and economic conditions in slums.
    - Improve rehabilitation efforts in jails.
    - Give convicted criminals tougher sentences.
    - Reform courts.

    What are the advantages and disadvantages of open questions? What are the advantages and disadvantages of closed questions? Which type is easier to analyze with formal statistical procedures?
18. Distinguish between a retrospective study and a prospective study.

*Vocabulary List*

Define and give an example of each term.

| | | |
|---|---|---|
| statistics | continuous data | stratified sampling |
| population | nominal | systematic sampling |
| sample | ordinal | cluster sampling |
| parameter | interval | convenience sampling |
| statistic | ratio | sampling errors |
| attribute data | retrospective study | nonsampling errors |
| numerical data | prospective study | |
| discrete data | random sampling | |

 *Review*

This chapter described the general nature of statistics along with some of its uses and abuses, while presenting some very basic concepts dealing with the nature of data and different types of sampling. Section 1–1 discussed statistics as a discipline and defined these very fundamental and important terms: **population, sample, parameter, statistic.** The use of statistics sometimes involves all of the data in a population, and it sometimes involves samples drawn from a population.

Section 1–2 presented uses of statistics as well as several examples of intentional or unintentional abuses. Section 1–3 discussed the effect of different arrangements of data, such as lists or tables. We also noted that some quantitative data are **discrete** while others are **continuous.** Also, data may be categorized according to one of these levels of measurement: **nominal, ordinal, interval, ratio.**

Section 1–4 discussed various aspects of statistical experiments, in particular, the **random, stratified, systematic, cluster,** and **convenience** methods of sampling. It is extremely important to recognize that data collected carelessly may be absolutely worthless. Great care must be taken to ensure that samples are representative of the population from which they are drawn. We concluded with some important points about writing a **report** on a statistics experiment.

*Review Exercises*

1. A consumers' group measures the actual horsepower of a sample of lawn mowers labeled as 12 hp. The sample is obtained by selecting 3 lawn mowers from each manufacturer.

a. Are the values obtained discrete or continuous?

b. Identify the level of measurement (nominal, ordinal, interval, ratio) for the horsepower values.

c. What type of sampling is being used? (random, stratified, systematic, cluster, convenience)

2. A news report states that the police seized forged record albums with a value of $1 million. How do you suppose the police computed the value of the forged albums, and in what other ways can that value be estimated? Why might the police be inclined to exaggerate the value of the albums?

3. In obtaining data on the following, determine which of the four levels of measurement (nominal, ordinal, interval, ratio) is most appropriate.

a. The religions of a sample of voters

b. Movie ratings of 1, 2, 3, or 4 stars

c. The body temperature of runners who just completed a marathon

d. The weights of runners who just completed a marathon

e. Consumer product ratings of "recommended, acceptable, not acceptable"

4. Identify each number as being discrete or continuous.

a. The Minolta Corporation surveyed 703 small business owners.

b. The New York Metropolitan Transit Authority conducted a survey of commuting times, and the first result was 49 minutes.

c. A consumer check of packaging revealed that a container of milk contained 30.4 ounces.

5. A newspaper article reports that a demonstration was attended by "1250 angry protesters." Comment.

6. Identify the type of sampling (random, stratified, systematic, cluster, convenience) used in each of the following.

a. A sample of products is obtained by selecting every 100th item on the assembly line.

b. Random numbers generated by a computer are used to select serial numbers of cars to be chosen for sample testing.

c. An auto parts supplier obtains a sample of all items from each of 12 different randomly selected retail stores.

d. A car maker conducts a marketing study involving test drives performed by a sample of 10 men and 10 women in each of 4 different age brackets.

e. A car maker conducts a marketing study by interviewing potential customers who happen to request test drives at a local dealership.

7. Census takers have found that in obtaining people's ages, they get more people of age 50 than of age 49 or 51. Can you explain how this might occur?

8. In a typical year, about 46,000 deaths result from motor vehicle accidents, according to data from the National Safety Council.
   a. How many deaths would result from motor vehicle accidents in a typical day?
   b. How many deaths would result from motor vehicle accidents in a typical 4-day period?
   c. For the 4 days of the Memorial Day weekend (Friday through Monday), assume that driving increases by 25% and that there are 630 deaths resulting from motor vehicle accidents. Does it appear that driving is more dangerous over the Memorial Day weekend?

 *Computer Project*

Use STATDISK or Minitab or any other available statistics software package to enter the 150 home selling prices in Appendix B. Enter the values in thousands of dollars. For example, the first entry of 179,000 would be entered as 179. Enter the data once, and make corrections only as you enter the data. Answer the following questions before checking your completed list.

1. Obtain a printed copy of the 150 scores and carefully compare them to the original 150 scores in Appendix B. How many errors did you make? (According to *PC Magazine*, "Studies have shown that the process of hand-keying data is associated with an error rate of 11.6 percent.")
2. Obtain a printed copy of the 150 scores after they have been ranked (arranged in increasing order) by the program. Identify the lowest and highest home selling prices. Examination of these values will often reveal errors. For example, if a home selling price of $150,000 is incorrectly entered as 15, it is likely to appear near the top of the ranked list, where we would recognize that a selling price of $15,000 is not realistic.
3. After correcting any errors you made, store the data in a file named SP (for selling prices).

 *Applied Projects*

1. If the above computer project was completed, compare your error rate to the 11.6% rate given by *PC Magazine*. Identify at least two major factors that would tend to make your error rate different than 11.6%.
2. Collect an example from a current newspaper or magazine in which data have been presented in a potentially deceptive manner. Identify

the source from which the example was taken, explain briefly the way in which the data might be deceptive, and suggest how the data might be presented more fairly.
3. Refer to the 150 home selling prices in Appendix B to answer the following.
   a. Identify at least one important factor suggesting that a sample of home selling prices is not representative of the population of all home values.
   b. Is this a sample of discrete data or continuous data? Explain your choice.
   c. Categorize those values according to the appropriate level of measurement (nominal, ordinal, interval, ratio). Explain your choice.
4. Refer to Data Set II in Appendix B to answer the following.
   a. Identify the level of measurement (nominal, ordinal, interval, ratio) for each of the eleven categories.
   b. Does the female/male ratio tend to support or refute the claim that this sample is representative of the population of all adults aged 18–74 years? Explain.

#  *Writing Projects*

1. Assume that your local newspaper included the misleading Figure 1–1(c) in an article comparing salaries of men and women. Write a letter to the editor explaining how that figure misleads, and present an argument for adopting an editorial policy that excludes such misleading graphs.
2. Write a report summarizing one of the videotape programs listed below.

#  *Videotapes*

Programs 1 and 12 from the series *Against All Odds: Inside Statistics* are recommended as supplements to this chapter.

# Chapter Two

## In This Chapter

# 2 Descriptive Statistics

## Chapter Problem

You've just landed a job in beautiful Dutchess County, located in upstate New York. You plan to buy a home, but you don't have a good sense of what homes cost in that area. You know that the cost of comparable homes can vary by large amounts depending on their locations. A home on the beach at Malibu will cost much more than a similar home in a housing development. Because buying a home is such an expensive and important decision, you decide to do some investigation. Table 2–1 lists the actual selling prices of 150 randomly selected homes that were recently sold in Dutchess County (see page 32).

A visual examination of those selling prices may provide some insight, but it is generally difficult to draw meaningful conclusions from a collection of raw data that are simply listed in no particular order. We need to look further into the data and do something with them. As one example, we might add the 150 selling prices and then divide that total by 150. The result will be an average that helps us understand the data. There are several other things we might do. The major objective of this chapter is to develop a variety of methods that will give us more insight into data sets such as the one listed in Table 2–1.

| TABLE 2–1 | Selling Prices (in dollars) of 150 Dutchess County Homes | | | | |
|---|---|---|---|---|---|
| 179,000 | 126,500 | 134,500 | 125,000 | 142,000 | 164,000 |
| 146,000 | 129,000 | 141,900 | 135,000 | 118,500 | 160,000 |
| 89,900 | 169,900 | 127,500 | 162,500 | 152,000 | 122,500 |
| 220,000 | 141,000 | 80,500 | 152,000 | 231,750 | 180,000 |
| 185,000 | 265,000 | 135,000 | 203,000 | 141,000 | 159,000 |
| 182,000 | 208,000 | 96,000 | 156,000 | 185,500 | 275,000 |
| 144,900 | 155,000 | 110,000 | 154,000 | 151,500 | 141,000 |
| 119,000 | 108,500 | 126,500 | 302,000 | 130,000 | 140,000 |
| 123,500 | 153,500 | 194,900 | 165,000 | 179,900 | 194,500 |
| 127,500 | 170,000 | 160,000 | 135,000 | 117,000 | 235,000 |
| 223,000 | 163,500 | 78,000 | 187,000 | 133,000 | 125,000 |
| 116,000 | 135,000 | 194,500 | 99,500 | 152,500 | 141,900 |
| 139,900 | 117,500 | 150,000 | 177,000 | 136,000 | 158,000 |
| 211,900 | 165,000 | 183,000 | 85,000 | 126,500 | 162,000 |
| 169,000 | 175,000 | 267,000 | 150,000 | 115,000 | 126,500 |
| 215,000 | 190,000 | 190,000 | 113,500 | 116,300 | 190,000 |
| 145,000 | 269,900 | 135,500 | 190,000 | 98,000 | 137,900 |
| 108,000 | 120,500 | 128,500 | 142,500 | 72,000 | 124,900 |
| 134,000 | 205,406 | 217,000 | 94,000 | 189,900 | 168,500 |
| 133,000 | 180,000 | 139,500 | 210,000 | 126,500 | 285,000 |
| 195,000 | 97,000 | 117,000 | 150,000 | 180,500 | 160,000 |
| 181,500 | 124,000 | 125,900 | 165,000 | 122,000 | 132,000 |
| 145,900 | 156,000 | 136,000 | 142,000 | 140,000 | 144,900 |
| 133,000 | 196,800 | 121,900 | 126,000 | 164,900 | 172,000 |
| 100,000 | 129,900 | 110,000 | 131,000 | 107,000 | 165,900 |

# 2–1 | Overview

In analyzing a data set, we should first determine whether we know all values for a complete population, or whether we know only the values for some sample drawn from a larger population. That determination will affect both the methods we use and the conclusions we form.

We use methods of **descriptive statistics** to summarize or *describe* the important characteristics of a known set of data. If the 150 values given in Table 2–1 represent the selling prices of *all* homes sold in Dutchess County, then we have known population data. We might then proceed to improve our understanding of this known population data by computing some average or by constructing a graph. In contrast to descriptive statistics, **inferential statistics** goes beyond mere description. We use inferential statistics when we use sample data to make *inferences* about a population.

Suppose we compute an average of the 150 scores in Table 2–1 and obtain a value of $153,775. If exactly 150 homes are sold in Dutchess County and our list is complete, that average of $153,775 is a parameter that describes and summarizes known population data. But those values are the actual selling prices of 150 homes that were randomly selected from a larger population. Treating those 150 values as a sample drawn from a larger population, we might conclude that the average selling price for all Dutchess County homes sold is $153,775. In so doing, we have made an *inference* that goes beyond the known data.

This chapter deals with the basic concepts of descriptive statistics. Chapter 3 includes an introduction to probability theory, and the subsequent chapters deal mostly with inferential statistics. Descriptive statistics and inferential statistics are the two basic divisions of the subject of statistics.

# Important Characteristics of Data

We use the tools of descriptive statistics in order to understand an otherwise unintelligible collection of data. The following three characteristics of data are extremely important, and they can give us considerable insight:

1. Representative score, such as an average
2. Measure of scattering or variation
3. Nature of the distribution, such as bell-shaped

We can learn something about the nature of the distribution by organizing the data and constructing graphs, as in Sections 2–2 and 2–3. In Section 2–4 we will learn how to obtain representative, or average, scores. We will measure the extent of scattering, or variation, among data as we use the tools found in Section 2–5. In Section 2–6 we will learn about measures of position so that we can better analyze or compare various scores. In Section 2–7 we will learn about methods for exploring data sets. As we proceed through this chapter, we will refer to the 150 scores given in Table 2–1, and our insight into that data set will be increased as we reveal its characteristics.

There is one last point that should be made in this overview. When collecting data, we must be extremely careful about the methods we use (common sense is often a critical requirement). If we plan our data collection with care and thoughtfulness, we can often learn much by using simple methods. If our data collection is quick, easy, and without much thought, we may well end up with something that is misleading or worthless. As we consider the methods of this chapter, remember that they will give us misleading results if the sample data are not representative of the population.

## 2–2 Summarizing Data

When beginning an analysis of a large set of scores, such as those listed in Table 2–1, we must often organize and summarize the data by developing tables and graphs. A **frequency table** lists categories of scores along with their corresponding frequencies. Table 2–2 is a frequency table with six classes. The **frequency** for a particular category or class is the number of original scores that fall into that class. The first class in Table 2–2 has a frequency of 7, indicating that there are 7 scores between 1 and 5 inclusive.

### AUTHORS IDENTIFIED

■ In 1787–88 Alexander Hamilton, John Jay, and James Madison anonymously published the famous *Federalist* papers in an attempt to convince New Yorkers that they should ratify the Constitution. The identity of most of the papers' authors became known, but the authorship of 12 of the papers was contested. Through statistical analysis of the frequencies of various words, we can now conclude that James Madison is the *likely* author of these 12 papers. For many of the disputed papers, the evidence in favor of Madison's authorship is overwhelming to the degree that we can be almost certain of being correct. ■

| TABLE 2–2 | |
|---|---|
| Score | Frequency |
| 1–5 | 7 |
| 6–10 | 12 |
| 11–15 | 19 |
| 16–20 | 16 |
| 21–25 | 8 |
| 26–30 | 4 |

The construction of a frequency table is not very difficult, and many statistics software packages can do it automatically. The following are some standard terms used in discussing frequency tables.

### Definitions

**Lower class limits** are the smallest numbers that can actually belong to the different classes. (Table 2–2 has lower class limits of 1, 6, 11, 16, 21, 26.)

**Upper class limits** are the largest numbers that can actually belong to the different classes. (Table 2–2 has upper class limits of 5, 10, 15, 20, 25, 30.)

The **class boundaries** are obtained by increasing the upper class limits and decreasing the lower class limits by the same amount so that there are no gaps between consecutive classes. The amount to be added or subtracted is one-half the difference between the upper limit of one class and the lower limit of the following class. (Table 2–2 has class boundaries of 0.5, 5.5, 10.5, 15.5, 20.5, 25.5, 30.5.)

The **class marks** are the midpoints of the classes. (Table 2–2 has class marks of 3, 8, 13, 18, 23, 28.)

The **class width** is the difference between two consecutive lower class limits (or class boundaries). (Table 2–2 has a class width of 5.)

The process of actually constructing a frequency table involves these key steps:

1.  *Decide on the number of classes your frequency table will contain.* Guideline: The number of classes should be between 5 and 20. The actual number of classes may be affected by convenience or other subjective factors.

2.  *Find the class width* by dividing the number of classes into the range. (The range is the difference between the highest and lowest scores.) Round the result *up* to a convenient number. This rounding up (not off) is not only convenient, but it also guarantees that all of the data will be included in the frequency table. (If the number of classes divides into the range evenly with no remainder, you may need to add another class in order for all of the data to be included.)

$$\text{class width} = \text{round } up \text{ of } \frac{\text{range}}{\text{number of classes}}$$

3.  *Select as a starting point either the lowest score or a convenient value slightly less than the lowest score.* This starting point is the lower class limit of the first class.

4.  *Add the class width to the starting point to get the second lower class limit.* Add the class width to the second lower class limit to get the third, and so on.

5.  *List the lower class limits in a vertical column and enter the upper class limits,* which can be easily identified at this stage.

6.  *Represent each score by a tally in the appropriate class.*

7.  *Replace the tally marks in each class with the total frequency count for that class.*

# NEW DATA COLLECTION TECHNOLOGY

■ Measuring or coding products and manually entering results is one way of collecting data, but technology is now providing us with alternatives that are not so susceptible to human error. Supermarkets use bar code readers to set prices and analyze inventory and buying habits. Manufacturers are increasingly using "direct data-entry" devices, such as electronic measuring gauges connected directly to a computer that records results. A third possibility is voice data entry, such as the Voice Data Logger. It consists of a headset connected to a small box that attaches to the speaker's belt. The box transmits the data to a computer. ■

## Example

Using the preceding steps, we construct the frequency table that summarizes the data in Table 2–1.

*Step 1:* We begin by selecting 10 as the number of desired classes. (Many statisticians recommend that we should generally aim for about 10 classes.)

*Step 2:* With a minimum of $72,000 and a maximum of $302,000 the range is $302,000 − $72,000 = $230,000, so that

$$\text{class width} = \text{round } up \text{ of } \frac{230,000}{10}$$
$$= \text{round } up \text{ of } 23,000$$
$$= \$25,000 \text{ (rounded for its convenience)}$$

*Step 3:* Starting with $72,000 would not be good since the resulting class limits would be inconvenient. Instead, we start with $50,000 since it is a more convenient number.

*Step 4:* Add the class width of $25,000 to the lower limit of $50,000 to get the next lower limit of $75,000. Continuing, the other lower class limits are $100,000, $125,000, and so on.

*Step 5:* These lower class limits suggest these upper class limits.

| $ 50,000 | $ 74,999 |
|----------|----------|
| 75,000 | 99,999 |
| 100,000 | 124,999 |
| • | • |
| • | • |
| • | • |

*Step 6:* The tally marks are shown in the middle of Table 2–3.

*Step 7:* The frequency counts are shown in the extreme right column of Table 2–3. The final table should exclude the tally marks, since they are only a device for determining the frequencies.

Note that the resulting frequency table, as shown in Table 2–3, actually has 11 classes instead of the 10 we began with. This is a result of selecting class limits that are convenient. We could have forced the table to have exactly 10 classes, but we would get messy limits such as $72,000–$96,999.

**TABLE 2–3**

| Selling Price (dollars) | Tally Marks | Frequency |
|---|---|---|
| 50,000–74,999 | | | 1 |
| 75,000–99,999 |卌 \|\|\|\| | 9 |
| 100,000–124,999 | 卌 卌 卌 卌 \|\| | 22 |
| 125,000–149,999 | 卌 卌 卌 卌 卌 卌 卌 卌 卌 \|\| | 47 |
| 150,000–174,999 | 卌 卌 卌 卌 卌 卌 \| | 31 |
| 175,000–199,999 | 卌 卌 卌 卌 卌 \|\|\| | 23 |
| 200,000–224,999 | 卌 \|\|\|\| | 9 |
| 225,000–249,999 | \|\| | 2 |
| 250,000–274,999 | \|\|\| | 3 |
| 275,000–299,999 | \|\| | 2 |
| 300,000–324,999 | | | 1 |

Table 2–3 provides much more useful information by making more intelligible the otherwise unintelligible list of 150 selling prices of homes. Yet this information is not gained without some loss. In constructing frequency tables, we may lose the accuracy of the raw data. To see how this loss occurs, consider the first class of $50,000–$74,999. Table 2–3 shows that there is one score in that class, but there is no way to determine from the table exactly what that score is. We cannot reconstruct the original 150 selling prices from Table 2–3. The exact values have been compromised for the sake of comprehension.

Summarizing data generally involves a compromise between accuracy and simplicity. A frequency table with too few classes is simple but not accurate. A frequency table with too many classes is more accurate but not as easy to understand. The best arrangement is arrived at subjectively, usually in accordance with the common formats used in particular applications. Some of these difficulties will be overcome in Section 2–7, when stem-and-leaf plots are discussed.

When constructing frequency tables, the following guidelines should be followed.

1. *The classes must be mutually exclusive.* That is, each score must belong to exactly one class.
2. *Include all classes,* even if the frequency might be zero.
3. *All classes should have the same width,* although it is sometimes impossible to avoid open-ended intervals such as "65 years or older."
4. *Try to select convenient numbers for class limits.*
5. *The number of classes should be between 5 and 20.*

A variation of the standard frequency table is used when cumulative totals are desired. *The cumulative frequency for a class is the sum of the frequencies for that class and all previous classes.* Table 2–4 is an example of a **cumulative frequency table,** and it corresponds to the same 150 selling prices presented in Table 2–3. A comparison of the frequency column of Table 2–3 and the cumulative frequency column of Table 2–4 reveals that the latter values can be obtained from the former by starting at the top and adding on the successive values. For example, the cumulative frequency of 10 from Table 2–4 represents the sum 1 and 9 from Table 2–3.

| TABLE 2–4 | |
|---|---|
| Selling Price (dollars) | Cumulative Frequency |
| Less than $75,000 | 1 |
| Less than $100,000 | 10 |
| Less than $125,000 | 32 |
| Less than $150,000 | 79 |
| Less than $175,000 | 110 |
| Less than $200,000 | 133 |
| Less than $225,000 | 142 |
| Less than $250,000 | 144 |
| Less than $275,000 | 147 |
| Less than $300,000 | 149 |
| Less than $325,000 | 150 |

Another important variation of the basic frequency table is the **relative frequency table.** The relative frequency for a particular class can be easily found by dividing the class frequency by the total of all frequencies.

$$\text{relative frequency} = \frac{\text{class frequency}}{\text{sum of all frequencies}}$$

Table 2–5 shows the relative frequencies for the 150 home selling prices summarized in Table 2–3. The first class has a relative frequency of 1/150 = 0.007. (Relative frequencies can also be given as percentages.) The second class has a relative frequency of 9/150 = 0.060, and so on. Relative frequency tables make it easier for us to compare distributions of different sets of data by using comparable relative frequencies instead of original frequencies that might be very different. (See Exercise 25.) For large and representative samples, the relative frequency table can also give us approximate probabilities of different values. (Probabilities will be discussed in Chapter 3.)

**TABLE 2–5**

| Selling Price (dollars) | Frequency | Relative Frequency |
|---|---|---|
| 50,000–74,999 | 1 | 0.007 |
| 75,000–99,999 | 9 | 0.060 |
| 100,000–124,999 | 22 | 0.147 |
| 125,000–149,999 | 47 | 0.313 |
| 150,000–174,999 | 31 | 0.207 |
| 175,000–199,999 | 23 | 0.153 |
| 200,000–224,999 | 9 | 0.060 |
| 225,000–249,999 | 2 | 0.013 |
| 250,000–274,999 | 3 | 0.020 |
| 275,000–299,999 | 2 | 0.013 |
| 300,000–324,999 | 1 | 0.007 |

In the next section, we will explore various graphic ways to depict data so that they are easy to understand. Frequency tables are necessary for some of the graphs. The graphs are often necessary for considering the way the scores are distributed, thus frequency tables become important prerequisites for other more useful concepts.

## 2–2  Exercises A

In Exercises 1–4, identify the class width, class marks, and class boundaries for the frequency table identified.

1.  Table 2–6    2.  Table 2–7    3.  Table 2–8    4.  Table 2–9

**TABLE 2–6**

| IQ | Frequency |
|---|---|
| 80–87 | 16 |
| 88–95 | 37 |
| 96–103 | 50 |
| 104–111 | 29 |
| 112–119 | 14 |

**TABLE 2–7**

| Time (hours) | Frequency |
|---|---|
| 0.0–7.5 | 16 |
| 7.6–15.1 | 18 |
| 15.2–22.7 | 17 |
| 22.8–30.3 | 15 |
| 30.4–37.9 | 19 |

**TABLE 2–8**

| Weight (kg) | Frequency |
|---|---|
| 16.2–21.1 | 16 |
| 21.2–26.1 | 15 |
| 26.2–31.1 | 12 |
| 31.2–36.1 | 8 |
| 36.2–41.1 | 3 |

**TABLE 2–9**

| Sales (dollars) | Frequency |
|---|---|
| 0–21 | 2 |
| 22–43 | 5 |
| 44–65 | 8 |
| 66–87 | 12 |
| 88–109 | 14 |
| 110–131 | 20 |

In Exercises 5–8, construct the cumulative frequency table corresponding to the frequency table indicated.

5.   Table 2–6        6.   Table 2–7        7.   Table 2–8        8.   Table 2–9

In Exercises 9–12, construct the relative frequency table corresponding to the frequency table indicated.

9.   Table 2–6      10.   Table 2–7      11.   Table 2–8      12.   Table 2–9

In Exercises 13–16, modify the class limits of the frequency table identified so that the new limits make the table easier to read. Omit the frequencies since they cannot be determined. It may be necessary to change the number of classes.

13.   Table 2–6      14.   Table 2–7      15.   Table 2–8      16.   Table 2–9

In Exercises 17–20, use the given information to find the upper and lower limits of the first class.

17.   Assume that you have a collection of scores representing the selling prices (rounded to the nearest dollar) of used cars ranging from a low of $750 to a high of $19,950. You wish to construct a frequency table with 10 classes.

18.   In a study of client stock transactions, a broker has amounts ranging from $38.50 to $73,568.75. She intends to construct a frequency table with 15 classes.

19.   A scientist is investigating the time (in seconds) required for a certain chemical reaction to occur. The experiment is repeated 200 times and the results vary from 17.3 s to 42.7 s. (You wish to construct a frequency table with 12 classes.)

20.   The Food and Drug Administration wants to replicate an experiment that supposedly shows the effects of a drug. Sample measurements vary between 41.3 mg and 95.6 mg, and you want a frequency table with 14 classes.

21.   The ages of airliners cause some safety and economic concerns. When 40 commercial aircraft are randomly selected in the United States, they are found to have the ages given below (based on data reported by Aviation Data Services). Construct a frequency table with 14 classes.

| | | | | | | | |
|---|---|---|---|---|---|---|---|
| 3.2 | 22.6 | 23.1 | 16.9 | 0.4 | 6.6 | 12.5 | 22.8 |
| 26.3 | 8.1 | 13.6 | 17.0 | 21.3 | 15.2 | 18.7 | 11.5 |
| 4.9 | 5.3 | 5.8 | 20.6 | 23.1 | 24.7 | 3.6 | 12.4 |
| 27.3 | 22.5 | 3.9 | 7.0 | 16.2 | 24.1 | 0.1 | 2.1 |
| 7.7 | 10.5 | 23.4 | 0.7 | 15.8 | 6.3 | 11.9 | 16.8 |

22.   *Time* magazine collected information on all 464 people who died from gunfire in the United States during one week. Here are the ages of 50

men randomly selected from that population. Construct a frequency table with 11 classes.

```
19  18  30  40  41  33  73  25  23  25
21  33  65  17  20  76  47  69  20  31
18  24  35  24  17  36  65  70  22  25
65  16  24  29  42  37  26  46  27  63
21  27  23  25  71  37  75  25  27  23
```

23.  Listed below are the actual energy consumption amounts as reported on the electric bills for one residence. Each amount is in kilowatt-hours and represents a two-month period. Construct a frequency table with 6 classes.

```
728  774  859  882  791  731  838  862  880  831
759  774  832  816  860  856  787  715  752  778
829  792  908  714  839  752  834  818  835  751
837
```

24.  Listed below are the daily sales totals (in dollars) for 44 days at a large retail outlet in Orange County, California. Construct a frequency table with 10 classes.

```
24,145  66,644  52,250  59,708
33,650  29,099  68,945  32,097
37,196  34,423  31,987  36,802
39,897  37,809  35,018  38,366
46,351  41,160  38,275  44,249
65,798  50,857  43,923  61,923
25,597  66,892  55,254  33,310
34,132  30,284  31,995  37,052
37,229  34,804  35,420  38,812
39,921  37,858  38,295  46,285
47,563  42,906  43,952  64,842
```

## 2–2  Exercises B

25.  Table 2–10 is a frequency table of alcohol consumption prior to arrest of male inmates serving a sentence for DWI. Table 2–11 is the corresponding table for females (based on data from the U.S. Department of Justice). (See the following page.) First construct the corresponding relative frequency tables, then use those results to compare the distributions. Note that it is difficult to compare the original frequencies, but it is much easier to compare the relative frequencies.

| TABLE 2–10 | |
|---|---|
| Ethanol Consumed by Men (ounces) | Frequency |
| 0.0–0.9 | 249 |
| 1.0–1.9 | 929 |
| 2.0–2.9 | 1,545 |
| 3.0–3.9 | 2,238 |
| 4.0–4.9 | 1,139 |
| 5.0–9.9 | 3,560 |
| 10.0–14.9 | 1,849 |
| 15.0 or more | 1,546 |

| TABLE 2–11 | |
|---|---|
| Ethanol Consumed by Women (ounces) | Frequency |
| 0.0–0.9 | 7 |
| 1.0–1.9 | 52 |
| 2.0–2.9 | 125 |
| 3.0–3.9 | 191 |
| 4.0–4.9 | 30 |
| 5.0–9.9 | 201 |
| 10.0–14.9 | 43 |
| 15.0 or more | 72 |

26. Listed below are two sets of scores that are supposed to be heights (in inches) of randomly selected adult males. One of the sets consists of real heights actually obtained from randomly selected adult males, but the other set consists of numbers that were fabricated. Construct a frequency table for each set of scores. By examining the two frequency tables, identify the set of data that you believe to be false.

```
70  73  70  72        70  73  70  72
71  73  71  67        71  66  74  76
68  72  67  72        68  75  67  68
71  73  72  70        71  77  66  69
72  68  71  71        72  67  77  75
71  73  69  73        66  76  76  77
71  66  77  67        73  74  69  67
```

27. A table in a recent issue of *USA Today* included the following data that were based on U.S. Bureau of the Census observations. Refer to the five guidelines for constructing frequency tables. Which of them are not followed?

| Age | U.S. Population (millions) |
|---|---|
| Under 5 | 18.1 |
| 5–13 | 30.3 |
| 14–17 | 14.8 |
| 18–24 | 28.0 |
| 25–34 | 43.0 |
| 35–44 | 33.1 |
| 45–54 | 22.8 |
| 55–64 | 22.2 |
| 65–84 | 9.1 |
| 85/older | 2.8 |

28. In constructing a frequency table, Sturges' guideline suggests that the ideal number of classes can be approximated by finding the value of $1 + (\log n)/(\log 2)$, where $n$ is the number of scores. Use this guideline to find the ideal number of classes corresponding to a data collection with the number of scores equal to (a) 50; (b) 100; (c) 150; (d) 500; (e) 1000; (f) 50,000.

# 2–3 Pictures of Data

In the preceding section, we saw that frequency tables transform a disorganized collection of raw scores into an organized and understandable summary. In this section we consider ways of representing data in pictorial form. The obvious objective is to promote understanding of the data. We attempt to show that one graphic illustration can be a suitable replacement for a mass of data.

## Pie Charts

It is a simple and obvious characteristic of human physiology that people can comprehend one picture better than a more abstract collection of words or data. Consider this statement: Among the 19,257 murders in a recent year, 11,381 (or 59.1%) were committed with firearms, 3957 (or 20.5%) were committed with knives, 1310 (or 6.8%) were committed with personal weapons (hands, feet, and so on), 1099 (or 5.7%) were committed with blunt objects, and the remaining 1510 (or 7.8%) were committed with a variety of other weapons (based on FBI data in the *Uniform Crime Reports*). After reading the last sentence, how much did you comprehend? What impressions did you develop? Now examine the **pie chart** in Figure 2–1. Which does a better job of making the data understandable? The abstraction of numbers is overcome by the concrete reality of the slices of the pie.

Pie charts are especially useful in cases involving data categorized according to attributes rather than numbers (data at the nominal level of measurement). The weapons—firearms, knives, blunt objects, and so on—are examples of attributes that are not themselves numbers.

The construction of a pie chart (sometimes called a *circle graph*) is not very difficult. It simply involves the "slicing up" of the pie into the proper proportions. If firearms represent 59.1% of the total, then the wedge representing firearms should be 59.1% of the total. (The central angle should be $0.591 \times 360° = 213°$.)

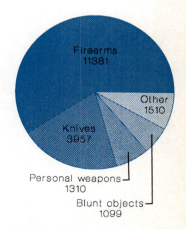

**Figure 2–1**
*Murder Weapons in the United States*

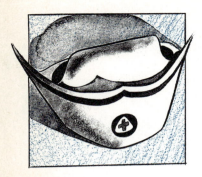

## FLORENCE NIGHTINGALE

■ Florence Nightingale (1820–1910) is known to many as the founder of the nursing profession, but she also saved thousands of lives by using statistics. When she encountered an unsanitary and undersupplied hospital, she improved those conditions and then used statistics to convince others of the need for more widespread medical reform. She developed original graphs to illustrate that during the Crimean War, more soldiers died as a result of unsanitary conditions than were killed in combat. Florence Nightingale pioneered the use of social statistics as well as graphics techniques. ■

# Histograms

A **histogram** is another common graphic way of presenting data. Histograms are especially important because they show the distribution of the data, an extremely important characteristic.

We generally construct a histogram to represent a set of scores after we have completed a frequency table. The standard format for a histogram usually involves a vertical scale that delineates frequencies and a horizontal scale that delineates values of the data being represented. We use bars to represent the individual classes of the frequency table. Each bar extends from its lower class boundary to its upper class boundary so that we can mark the class boundaries on the horizontal scale. (However, improved readability is often achieved by using class limits or class marks instead of class boundaries.) As an example, Figure 2–2 is the histogram that corresponds directly to Table 2–3 in the previous section.

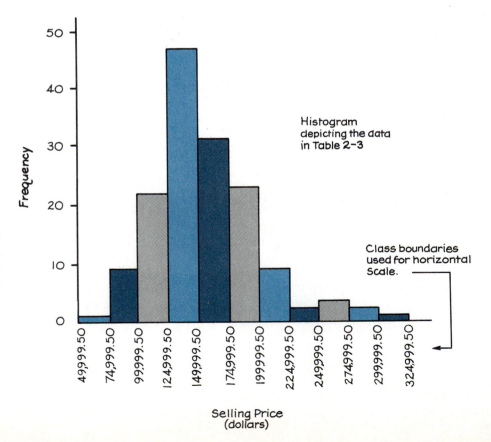

Histogram depicting the data in Table 2–3

Class boundaries used for horizontal scale.

**Figure 2–2**

Before constructing a histogram from a completed frequency table, we must give some consideration to the scales used on the vertical and horizontal axes. The maximum frequency (or the next highest convenient number) should suggest a value for the top of the vertical scale, with 0 at the bottom. In Figure 2–2, we designed the vertical scale to run from 0 to 50. The horizontal scale should be designed to accommodate all the classes of the frequency table. Ideally, we should try to follow the rule of thumb that the vertical height of the histogram should be about 3/4 of the total width. Both axes should be clearly labeled.

A **relative frequency histogram** will have the same shape as a histogram. The horizontal scale will also be the same, but the vertical scale will be marked with *relative frequencies* instead of actual frequencies. Figure 2–2 can be modified to be a relative frequency histogram by labeling the vertical scale as "relative frequency" and changing the values on the vertical scale to range from 0 to about 0.350 or 0.040 (see Figure 2–3). (The highest relative frequency for this data set is 0.313.)

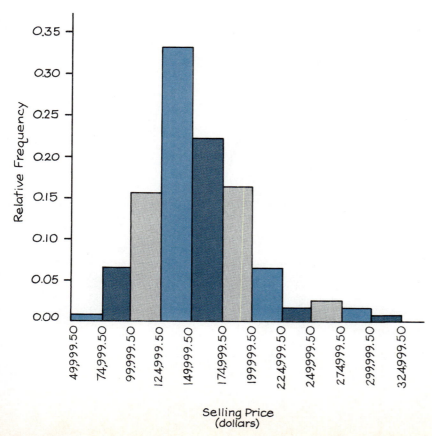

**Figure 2–3**

# Frequency Polygons

A **frequency polygon** is a variation of the histogram with the following changes. (See Figure 2–4.)

1. Bars are replaced by dots that are connected to form a line graph.
2. *Class marks* on the horizontal scale locate the dots directly above them.
3. The line graph is extended at the left and right so that it begins and ends with a frequency of 0.

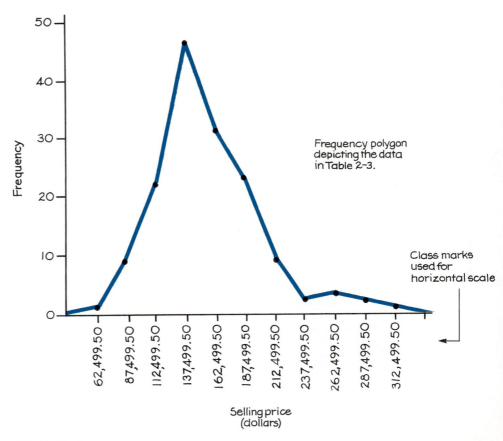

**Figure 2–4**

# Ogives

Another common pictorial display is the **ogive** (pronounced "oh-jive"), or **cumulative frequency polygon.** This differs from an ordinary frequency polygon in that the *frequencies are cumulative.* That is, we add each class frequency to the total of all previous class frequencies, as in the cumulative frequency table described in Section 2–2. The vertical scale is again used to delineate frequencies, but we must now adjust the scale so that the total of all individual frequencies will fit. The horizontal scale should depict the class boundaries. (See Figure 2–5.) Ogives are useful when, instead of seeking the frequency for a given class interval, we want to know how many of our scores are above or below some level. For example, in analyzing selling prices of homes, a buyer might be more interested in the homes that cost less than $185,000 than in the homes that cost between any two particular values. Figure 2–5 shows that approximately 119 (or 79%) of the sampled homes fall below $185,000.

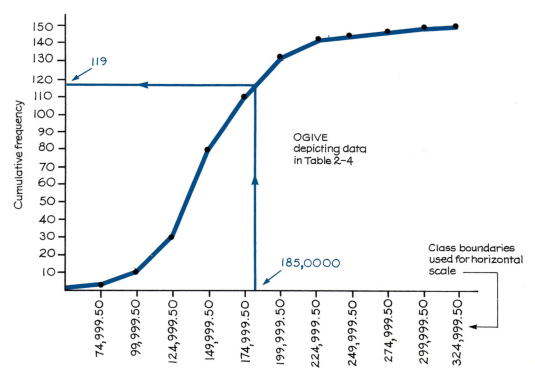

**Figure 2–5**

## Miscellaneous Graphics

Numerous pictorial displays other than the ones just described can be used to represent data dramatically and effectively. Some examples are soldiers, tanks, airplanes, stacks of coins, and moneybags. Although there is almost no limit to the variety of different ways that data can be illustrated, pie charts, histograms, relative frequency histograms, frequency polygons, and ogives are among the most common devices used.

Frequency tables and graphs, such as histograms, make it possible for us to see the distribution of data. In any serious statistical analysis of data, the distribution is a critically important feature. The distribution of responses to an SAT or ACT question might reveal that the question is too easy or too difficult. Any car designer should know something about the distributions of heights and weights of drivers. Many procedures in statistics require a data set with a distribution that is approximately bell-shaped; we must often construct frequency tables and/or graphs before proceeding with many methods in statistics. In the following sections we consider ways of measuring other characteristics of data.

## BEST-SELLER LISTS

■ Listings of the best-selling books are based on sampling. The *New York Times*, for example, solicits book sales figures from about 3000 bookstores in seven regions of the country. *Publishers Weekly* polls about 2000 stores. One book that made the best-seller list is *Confessions of an S.O.B.* by Allen Neuharth; he heads a foundation that spent $40,000 to buy 2000 copies of the book from bookstores around the country. ■

## 2–3  Exercises A

1. Construct a histogram that corresponds to the frequency table given below.

| Airplane Age (years) | Frequency |
|---|---|
| 0.0–5.9 | 10 |
| 6.0–11.9 | 8 |
| 12.0–17.9 | 9 |
| 18.0–23.9 | 9 |
| 24.0–29.9 | 4 |

Data based on information reported by Aviation Data Services.

2. Construct a relative frequency histogram that corresponds to the frequency table in Exercise 1.
3. Construct a frequency polygon that corresponds to the frequency table in Exercise 1.
4. Construct an ogive that corresponds to the frequency table in Exercise 1. Use the ogive to estimate the proportion of airplanes less than 10 years old.

5.  Construct a histogram that corresponds to the frequency table given below.

| Age at Death | Frequency |
|---|---|
| 16–25 | 22 |
| 26–35 | 10 |
| 36–45 | 6 |
| 46–55 | 2 |
| 56–65 | 4 |
| 66–75 | 5 |
| 76–85 | 1 |

Data based on a *Time* magazine study of people who died from gunfire in America during one week.

6.  Construct a relative frequency histogram that corresponds to the frequency table in Exercise 5.

7.  Construct a frequency polygon that corresponds to the frequency table in Exercise 5.

8.  Construct an ogive that corresponds to the frequency table in Exercise 5. Use the ogive to estimate the proportion of people who are under 30 when they die from gunfire in America.

9.  Construct a histogram that corresponds to the frequency table given below.

| Energy (kWh) | Frequency |
|---|---|
| 700–719 | 2 |
| 720–739 | 2 |
| 740–759 | 4 |
| 760–779 | 4 |
| 780–799 | 2 |
| 800–819 | 2 |
| 820–839 | 8 |
| 840–859 | 2 |
| 860–879 | 2 |
| 880–899 | 2 |
| 900–919 | 1 |

Data based on energy consumption reported on electric bill for 31 two-month periods.

10.  Construct a relative frequency histogram that corresponds to the frequency table in Exercise 9.

11.  Construct a frequency polygon that corresponds to the frequency table in Exercise 9.

12.  Construct an ogive that corresponds to the frequency table in Exercise 9. Use the ogive to estimate the proportion of energy levels that *exceed* 830 kWh.

13.  Construct a histogram that corresponds to the frequency table on the following page.

| Sales (dollars) | Frequency |
|---|---|
| 20,000–24,999 | 1 |
| 25,000–29,999 | 2 |
| 30,000–34,999 | 9 |
| 35,000–39,999 | 14 |
| 40,000–44,999 | 5 |
| 45,000–49,999 | 3 |
| 50,000–54,999 | 2 |
| 55,000–59,999 | 2 |
| 60,000–64,999 | 2 |
| 65,000–69,999 | 4 |

Data based on daily sales for 44 days at a large retail outlet in Orange County, California.

14. Construct a relative frequency histogram that corresponds to the frequency table in Exercise 13.

15. Construct a frequency polygon that corresponds to the frequency table in Exercise 13.

16. Construct an ogive that corresponds to the frequency table in Exercise 13. Use the ogive to estimate the proportion of days that sales *exceed* $57,500.

## 2–3 | Exercises B

17. Use the histogram in Figure 2–6.
    a. Construct the corresponding frequency table. Use the eleven class marks of 15, 16, 17, . . . , 25.

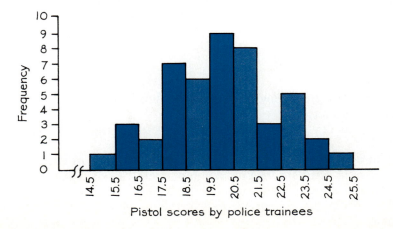

Pistol scores by police trainees

**Figure 2–6**

b.  Construct the corresponding ogive.
c.  Refer to the ogive to estimate how many police trainees scored below 21.
d.  If 17 is a passing score, estimate the percentage of trainees who failed this test.

18.  Use the frequency polygon in Figure 2–7.
a.  Construct the corresponding frequency table.
b.  Construct the corresponding ogive.
c.  Estimate the number of subjects who had pulse rates between 75 and 84.
d.  Estimate the percentage of subjects with pulse rates below 90.

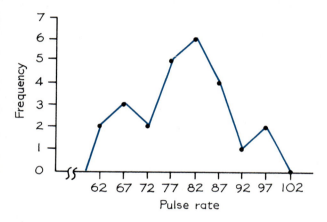

**Figure 2–7**

19.  Given are frequency tables for the first 100 digits in the decimal representation of $\pi$ and the first 100 digits in the decimal representation of 22/7.
a.  Construct histograms representing the frequency tables and note any differences.
b.  The numbers $\pi$ and 22/7 are both real numbers, but how are they fundamentally different?

20.  Using a collection of sample data, a frequency table is constructed with 10 classes, then the corresponding histogram is constructed. How is the histogram affected if the number of classes is doubled, but the same vertical scale is used?

| $\pi$ | | 22/7 | |
|---|---|---|---|
| x | f | x | f |
| 0 | 8 | 0 | 0 |
| 1 | 8 | 1 | 17 |
| 2 | 12 | 2 | 17 |
| 3 | 11 | 3 | 1 |
| 4 | 10 | 4 | 17 |
| 5 | 8 | 5 | 16 |
| 6 | 9 | 6 | 0 |
| 7 | 8 | 7 | 16 |
| 8 | 12 | 8 | 16 |
| 9 | 14 | 9 | 0 |

21.  Given the graphs of the following frequency polygons, construct the graphs of the corresponding ogives.

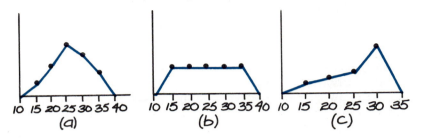

22.  Use the following data to construct a pie chart.

| Job Sources of Survey Respondents | Number |
| --- | --- |
| Help-wanted ads | 56 |
| Executive search firms | 44 |
| Networking | 280 |
| Mass mailing | 20 |

Based on data from the National Center for Career Strategies.

## 2–4  Averages

While frequency tables and graphs are helpful in determining the *distribution* of data, they may also provide some information about another important characteristic of data: an average that consists of a single value that is central or representative of the entire data set. However, we generally need more information than is provided by frequency tables or graphs to determine **averages** or **measures of central tendency.**

An important and basic point to remember is that there are different ways to compute an average. We have already stated that, given a list of scores and instructions to find the average, most people will obligingly proceed first to total the scores and then to divide that total by the number of scores. However, this particular computation is only one of several procedures classified as an average. The most commonly used averages are mean, median, mode, and midrange.

## Mean

The **arithmetic mean** is the most important of all numerical descriptive measurements, and it corresponds to what most people call an *average*. (See Figure 2–8.)

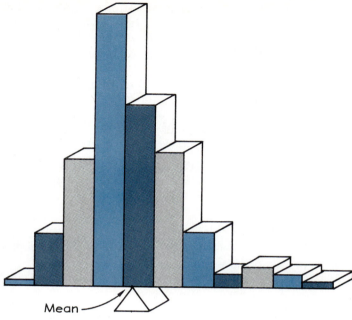

Mean

If a fulcrum is placed at the position of the mean, it will balance the histogram.

**Figure 2–8**

## Definition

The **arithmetic mean** of a list of scores is obtained by adding the scores and dividing the total by the number of scores. This particular average will be employed frequently in the remainder of this text, and it will be referred to simply as the **mean.**

## Example

Find the mean of the quiz scores 2, 3, 6, 7, 7, 8, 9, 9, 9, 10.

## Solution

First add the scores.

$$2 + 3 + 6 + 7 + 7 + 8 + 9 + 9 + 9 + 10 = 70$$

Then divide the total by the number of scores present.

$$\frac{70}{10} = 7$$

The mean score is therefore 7.

Many formulas, developed for different statistics, often involve the Greek letter $\Sigma$ (capital sigma), which indicates summation of values. Letting $x$ denote the value of a score, $\Sigma x$ means the sum of all scores. The symbol $n$ is used to represent the **sample size,** which is the number of scores being considered.

Since the mean is the sum of all scores $(\Sigma x)$ divided by the number of scores $(n)$, we have the following formula.

**Formula 2–1** $$\text{mean} = \frac{\Sigma x}{n}$$

The result can be denoted by $\bar{x}$ if the available scores are samples from a larger population; if all scores of the population are available, then we can denote the computed mean by $\mu$ (mu).

---

### Notation

$\Sigma$    denotes **summation** of a set of values.

$x$    is the **variable** usually used to represent the individual raw scores.

$n$    represents the **number** of scores being considered.

$\bar{x}$    denotes the **mean** of a set of **sample** scores.

$\mu$    denotes the **mean** of all scores in some **population.** (This symbol is the lower case Greek mu.)

---

Applying Formula 2–1 to the preceding quiz scores, we get

$$\text{mean} = \frac{\Sigma x}{n} = \frac{70}{10} = 7$$

According to the definition of mean, 7 is the central value of the ten quiz scores given. Other definitions of averages involve different perceptions of how the center is determined. The median reflects another approach.

# Median

### Definition

The **median** of a set of scores is the middle value when the scores are arranged in order of increasing magnitude. (The median is often denoted by $\tilde{x}$.)

After first arranging the original scores in increasing (or decreasing) order, the median will be either of the following:

1. If the number of scores is *odd*, the median is the number that is exactly in the middle of the list.

2. If the number of scores is *even*, the median is found by computing the mean of the two middle numbers.

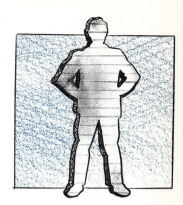

### Example

Find the median of the scores 7, 2, 3, 7, 6, 9, 10, 8, 9, 9, 10.

### Solution

Begin by arranging the scores in increasing order.

2, 3, 6, 7, 7, 8, 9, 9, 9, 10, 10

With these eleven scores, the number 8 is located in the exact middle, so 8 is the median.

### Example

Find the median of the scores 7, 2, 3, 7, 6, 9, 10, 8, 9, 9.

### Solution

Again, begin by arranging the scores in increasing order.

2, 3, 6, 7, 7, 8, 9, 9, 9, 10

With these ten scores, no single score is at the exact middle. Instead, the two scores of 7 and 8 share the middle. We therefore find the mean of those two scores.

$$\frac{7 + 8}{2} = \frac{15}{2} = 7.5$$

The median is 7.5.

## AN AVERAGE GUY

■ The "average" American male is named Robert. He is 31 years old, 5 ft 9½ in. tall, weighs 172 lb, wears a size 40 suit, wears a size 9½ shoe, and has a 34 in. waist. Each year he eats 12.3 lb of pasta, 26 lb of bananas, 4 lb of potato chips, 18 lb of ice cream, and 79 lb of beef. Each year he also watches television for 2567 hours and gets 585 pieces of mail. After eating his share of potato chips, reading some of his mail, and watching some television, he ends the day with 7.7 hours of sleep. The next day begins with a 21-minute commute to a job at which he will work for 6.1 hours. ■

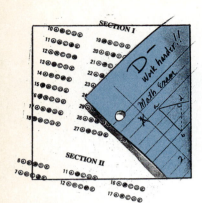

## MISLEADING NORMS

■ Dr. John Cannell, a West Virginia physician, observed that his state's high illiteracy rate was inconsistent with a state report claiming that West Virginia students were performing *above* the national average. Investigations showed that, among other problems, old norms were used as a basis for comparison. By comparing the new test results to those obtained many years ago, all 50 states had results that were above "the national average." Dr. Cannell charged, "The testing industry wants to sell lots of tests, and the school superintendents desperately need help in improving scores." The effect was to create misleading results. ■

# Mode

### Definition

The **mode** is obtained from a collection of scores by selecting the score that occurs most frequently. In those cases where no score is repeated, we stipulate that there is no mode. In those cases where two scores both occur with the same greatest frequency, we say that each one is a mode and we refer to the data set as being **bimodal.** If more than two scores occur with the same greatest frequency, each is a mode and the data set is said to be **multimodal.**

### Example

The scores 1, 2, 2, 2, 3, 4, 5, 6, 7, 9 have a mode of 2.

### Example

The scores 2, 3, 6, 7, 8, 9, 10 have no mode since no score is repeated.

### Example

The scores 1, 2, 2, 2, 3, 4, 5, 6, 6, 6, 7, 9 have modes of 2 and 6, since 2 and 6 both occur with the same highest frequency.

### Example

A town meeting is attended by 40 Republicans, 25 Democrats, and 20 Independents. While we cannot numerically average these party affiliations, we can report that the mode is Republican, since that party had the highest frequency. Among the four averages we are now considering, the mode is the only one that can be used with data at the nominal level of measurement.

# Midrange

## Definition

The **midrange** is that average obtained by adding the highest score to the lowest score and then dividing the result by 2.

$$midrange = \frac{highest\ score\ +\ lowest\ score}{2}$$

## Example

Find the midrange of the scores 2, 3, 6, 7, 7, 8, 9, 9, 9, 10.

$$midrange = \frac{10 + 2}{2} = 6$$

## Example

For the 150 home selling prices listed in Table 2–1, find the values of the (a) mean; (b) median; (c) mode; (d) midrange.

## Solution

a.  The sum of the 150 selling prices is $23,066,256, so that

$$\bar{x} = \frac{\$23,066,256}{150}$$
$$= \$153,775$$

b.  After arranging the scores in increasing order, we find that the 75th and 76th scores are both $144,900, so the median is $144,900. (The scores can easily be arranged in increasing order by constructing a stem-and-leaf plot [see Section 2–7] or by using a computer program such as STATDISK or Minitab.)

c.  The most frequent (five times) selling price is $126,500, so it is the mode. No other value occurs more than four times.

*continued*

**Solution** *continued*

d.  The midrange is found as follows.

$$\text{midrange} = \frac{\text{highest score } + \text{ lowest score}}{2}$$

$$= \frac{302{,}000 + 72{,}000}{2}$$

$$= \$187{,}000$$

We now summarize these results.

| | |
|---|---|
| mean | $153,775 |
| median | $144,900 |
| mode | $126,500 |
| midrange | $187,000 |

## Round-Off Rule

A simple rule for rounding answers is to carry one more decimal place than was present in the original data. We should round only the final answer and not intermediate values. For example, the mean of 2, 3, 5 is expressed as 3.3. Since the original data were whole numbers, we rounded the answer to the nearest tenth. The mean of 2.1, 3.4, 5.7 is rounded to 3.73.

You should know and understand the preceding four averages (mean, median, mode, and midrange). Other averages (geometric mean, harmonic mean, and quadratic mean) are not used as often, and they will be included only in Exercises B at the end of this section (see Exercises 30, 31, and 32).

# Weighted Mean

The **weighted mean** is useful in many situations in which the scores vary in their degree of importance. An obvious example occurs frequently in the determination of a final average for a course that includes four tests plus a final examination. If the respective grades are 70, 80, 75, 85, and 90, the mean of 80 does not reflect the greater importance placed on the final exam. Let's suppose that the instructor counts the respective tests as 15%, 15%, 15%, 15%, and 40%. The weighted mean then becomes

$$\frac{(70 \times 15) + (80 \times 15) + (75 \times 15) + (85 \times 15) + (90 \times 40)}{100}$$

$$= \frac{1050 + 1200 + 1125 + 1275 + 3600}{100}$$

$$= \frac{8250}{100} = 82.5$$

This computation suggests a general procedure for determining a weighted mean. Given a list of scores $x_1, x_2, x_3, \ldots, x_n$ and a corresponding list of weights $w_1, w_2, w_3, \ldots, w_n$, the weighted mean is obtained by computing as follows.

**Formula 2–2**      weighted mean $= \dfrac{\Sigma(w \cdot x)}{\Sigma w}$      where $w$ = weight

That is, first multiply each score by its corresponding weight; then find the total of the resulting products, thereby evaluating $\Sigma wx$. Finally, add the values of the weights to find $\Sigma w$ and divide the latter value into the former.

## Mean from a Frequency Table

Formula 2–2 can be modified so that we can approximate the mean from a frequency table. The home selling price data from Table 2–1 have been entered in Table 2–12, where we use the class marks as representative scores

| TABLE 2–12 | | | |
|---|---|---|---|
| Selling Price (dollars) | Frequency $f$ | Class Mark $x$ | $f \cdot x$ |
| 50,000–74,999 | 1 | 62,499.5 | 62,499.5 |
| 75,000–99,999 | 9 | 87,499.5 | 787,495.5 |
| 100,000–124,999 | 22 | 112,499.5 | 2,474,989.0 |
| 125,000–149,999 | 47 | 137,499.5 | 6,462,476.5 |
| 150,000–174,999 | 31 | 162,499.5 | 5,037,484.5 |
| 175,000–199,999 | 23 | 187,499.5 | 4,312,488.5 |
| 200,000–224,999 | 9 | 212,499.5 | 1,912,495.5 |
| 225,000–249,999 | 2 | 237,499.5 | 474,999.0 |
| 250,000–274,999 | 3 | 262,499.5 | 787,498.5 |
| 275,000–299,999 | 2 | 287,499.5 | 574,999.0 |
| 300,000–324,999 | 1 | 312,499.5 | 312,499.5 |
| Total | 150 | | 23,199,925 |

$\uparrow$ $\Sigma f$      $\uparrow$ $\Sigma(f \cdot x)$

## CLASS SIZE PARADOX

■ There are at least two ways to obtain the mean class size, and they can have very different results. In one college, if we take the numbers of students in 737 classes, we get a mean of 40 students. But if we were to compile a list of the class sizes for each student and use this list, we would get a mean class size of 147. This large discrepancy is due to the fact that there are many students in large classes, while there are few students in small classes. Without changing the number of classes or faculty, the mean class size experienced by students can be reduced by making all classes about the same size. This would also improve attendance, which is better in smaller classes. ■

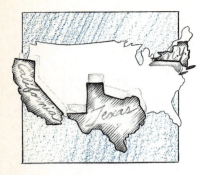

## THE POWER OF YOUR VOTE

and the frequencies as weights. Then the formula for the weighted mean leads directly to Formula 2–3, which can be used to approximate the mean of a set of scores in a frequency table.

**Formula 2–3**     $$\bar{x} = \frac{\Sigma(f \cdot x)}{\Sigma f}$$     where $x$ = class mark
$f$ = frequency

Formula 2–3 is really a variation of Formula 2–1, $\bar{x} = \Sigma x/n$. When data are summarized in a frequency table, $\Sigma f$ is the total number of scores, so that $\Sigma f = n$. Also, $\Sigma(f \cdot x)$ is simply a quick way of adding up all of the scores. Formula 2–3 doesn't really involve a fundamentally different concept; it is simply a variation of Formula 2–1.

We can now compute the weighted mean.

$$\bar{x} = \frac{\Sigma(f \cdot x)}{\Sigma f} = \frac{23,199,925}{150} = \$154,666$$

When we used the original collection of scores to calculate the mean directly, we obtained a mean of $153,775, so the value of the weighted mean obtained from the frequency table is quite accurate. The procedure we use is justified by the fact that a class such as $50,000–$74,999 can be represented by its class mark of $62,499.5 and the frequency number indicates that the representative score of $62,499.5 occurs one time. In essence, we are treating Table 2–3 as if it contained one score of $62,499.5, nine values of $87,499.5, and so on.

## The Best Average

We have stressed that the four basic measures of central tendency (mean, median, mode, midrange) can all be called averages and that the freedom to select a particular average can lead to deception. Consider our 150 home selling prices that yielded these results.

| | | | |
|---|---|---|---|
| mean | $153,775 | mode | $126,500 |
| median | $144,900 | midrange | $187,000 |

Technically, someone could use any of the four preceding figures as the average home selling price. A real estate salesperson might want to encourage a sale by emphasizing that a particular home is a bargain compared to the high average of $187,000. A wise buyer (such as you, after you've taken this course) might try to get a lower price by emphasizing the low average of $126,500. We can see that ignorance of these concepts can be very costly.

Even when we have no self-interest to promote, the selection of the most representative average is not always easy. The different averages have different advantages and disadvantages and there are no objective criteria that determine the most representative average for all data sets. Some of the important advantages and disadvantages are summarized in Table 2–13.

monoxide emissions data (in g/m) are listed for vehicles. The following values are included.

| | | | | | |
|---|---|---|---|---|---|
| 5.01 | 14.67 | 8.60 | 4.42 | 4.95 | 7.24 |
| 7.51 | 12.30 | 14.59 | 7.98 | 11.53 | 4.10 |

11. In "A Simplified Method for Determination of Residual Lung Volumes" (by Wilmore, *Journal of Applied Psychology*, Vol. 27, No. 1), the oxygen dilution was measured (in liters) for subjects and the following sample values were obtained.

| | | | | |
|---|---|---|---|---|
| 1.361 | 1.013 | 1.140 | 1.649 | 1.278 |
| 1.148 | 1.824 | 1.551 | 1.041 | |

12. The following are the ages of motorcyclists when they were fatally injured in traffic accidents (based on data from the U.S. Department of Transportation).

| | | | | | | | | |
|---|---|---|---|---|---|---|---|---|
| 17 | 38 | 27 | 14 | 18 | 34 | 16 | 42 | 28 |
| 24 | 40 | 20 | 23 | 31 | 37 | 21 | 30 | 25 |
| 17 | 28 | 33 | 25 | 23 | 19 | 51 | 18 | 29 |

13. Forty commercial aircraft randomly selected in the United States were found to have the ages given below (based on data reported by Aviation Data Services).

| | | | | | | | |
|---|---|---|---|---|---|---|---|
| 3.2 | 22.6 | 23.1 | 16.9 | 0.4 | 6.6 | 12.5 | 22.8 |
| 26.3 | 8.1 | 13.6 | 17.0 | 21.3 | 15.2 | 18.7 | 11.5 |
| 4.9 | 5.3 | 5.8 | 20.6 | 23.1 | 24.7 | 3.6 | 12.4 |
| 27.3 | 22.5 | 3.9 | 7.0 | 16.2 | 24.1 | 0.1 | 2.1 |
| 7.7 | 10.5 | 23.4 | 0.7 | 15.8 | 6.3 | 11.9 | 16.8 |

14. A student working part time for a moving company in Dutchess County, New York, collected the following load weights (in pounds) for 50 consecutive customers.

| | | | | | | |
|---|---|---|---|---|---|---|
| 8,090 | 3,250 | 12,350 | 4,510 | 8,770 | 5,030 | 12,700 |
| 8,800 | 6,170 | 8,450 | 10,330 | 10,100 | 13,410 | 7,280 |
| 13,490 | 17,810 | 7,470 | 11,450 | 13,260 | 9,310 | 15,970 |
| 7,540 | 7,770 | 6,400 | 14,800 | 14,760 | 6,820 | 11,430 |
| 7,200 | 13,520 | 16,200 | 10,780 | 10,510 | 17,330 | 7,450 |
| 9,110 | 10,630 | 3,670 | 14,310 | 9,140 | 10,220 | 9,900 |
| 11,860 | 12,010 | 12,430 | 8,160 | 26,580 | 4,480 | 6,390 |
| 11,600 | | | | | | |

15. Listed below are the actual energy consumption amounts as reported on the electric bills for one residence. Each amount is in kilowatt-hours and represents a two-month period.

| | | | | | | | | | |
|---|---|---|---|---|---|---|---|---|---|
| 728 | 774 | 859 | 882 | 791 | 731 | 838 | 862 | 880 | 831 |
| 759 | 774 | 832 | 816 | 860 | 856 | 787 | 715 | 752 | 778 |
| 829 | 792 | 908 | 714 | 839 | 752 | 834 | 818 | 835 | 751 |
| 837 | | | | | | | | | |

16. Listed below are the daily sales totals (in dollars) for 44 days at a large retail outlet in Orange County, California.

| | | | | |
|---|---|---|---|---|
| 24,145 | 39,921 | 30,284 | 43,923 | 38,366 |
| 33,650 | 47,563 | 34,804 | 55,254 | 44,249 |
| 37,196 | 66,644 | 37,858 | 31,995 | 61,923 |
| 39,897 | 29,099 | 42,906 | 35,420 | 33,310 |
| 46,351 | 34,423 | 52,250 | 38,295 | 37,052 |
| 65,798 | 37,809 | 68,945 | 43,952 | 38,812 |
| 25,597 | 41,160 | 31,987 | 59,708 | 46,285 |
| 34,132 | 50,857 | 35,018 | 32,097 | 64,842 |
| 37,229 | 66,892 | 38,275 | 36,802 | |

In Exercises 17–24, use the given frequency table. (a) Identify the class mark for each class interval. (b) Find the mean using the class marks and frequencies.

17.

| Distance (in thousands of miles) | Frequency |
|---|---|
| 0–39 | 17 |
| 40–79 | 41 |
| 80–119 | 80 |
| 120–159 | 49 |
| 160–199 | 4 |

Distances traveled by buses before the first major motor failure (based on data from "Large Sample Simultaneous Confidence Intervals for the Multinomial Probabilities Based on Transformations of the Cell Frequencies" by Bailey, *Technometrics*, Vol. 22, No. 4)

18.

| Age (years) | Frequency |
|---|---|
| 0–9 | 44 |
| 10–19 | 42 |
| 20–29 | 33 |
| 30–39 | 30 |
| 40–49 | 27 |
| 50–59 | 22 |
| 60–69 | 18 |
| 70–79 | 9 |

Ages of randomly selected Dutchess County residents (based on data from the *Cornell Community and Resource Development Series*)

19.

| Size of Family (persons) | Frequency |
|---|---|
| 2 | 51 |
| 3 | 31 |
| 4 | 27 |
| 5 | 12 |
| 6 | 4 |
| 7 | 1 |
| 8 | 1 |

Family size for 127 randomly selected families (based on data from the Bureau of the Census)

20.

| Time (years) | Number |
|---|---|
| 4 | 147 |
| 5 | 81 |
| 6 | 27 |
| 7 | 15 |
| 7.5–11.5 | 30 |

Time required to earn bachelor's degree (based on data from the National Center for Education Statistics)

21.

| Age | Number |
|---|---|
| 16–20 | 38 |
| 21–29 | 110 |
| 30–39 | 122 |
| 40–49 | 91 |
| 50–59 | 71 |
| 60–69 | 82 |

Ages of New York drivers involved in accidents (based on data from the N.Y. State Department of Motor Vehicles)

22.

| Number | Frequency |
|---|---|
| 1–5 | 19 |
| 6–10 | 24 |
| 11–15 | 18 |
| 16–20 | 21 |
| 21–25 | 23 |
| 26–30 | 20 |
| 31–35 | 16 |
| 36–40 | 15 |

Frequency table of lottery numbers selected in a 26-week period

23.

| Family Income (dollars) | Number |
|---|---|
| 0–2,499 | 3 |
| 2,500–7,499 | 11 |
| 7,500–12,499 | 16 |
| 12,500–14,999 | 9 |
| 15,000–19,999 | 19 |
| 20,000–24,999 | 19 |
| 25,000–34,999 | 36 |
| 35,000–49,999 | 37 |

Incomes of families (restricted to families with incomes below $50,000 and based on data from the Bureau of the Census)

24.

| Years of Education | Number Who Smoke |
|---|---|
| 0–11 | 195 |
| 12 | 179 |
| 13–15 | 151 |
| 16–20 | 101 |

Number of years of education for a sample of 626 smokers (based on data from the Department of Health and Human Services)

## 2–4 | Exercises B

25. a. Find the mean, median, mode, and midrange for the following lengths (in inches) of wood framing studs that are supposed to be 96 inches long.

    96.05   95.87   96.16   96.06   95.81

   b. Do part *a* after subtracting 95 from each score.
   c. In general, what is the effect of subtracting a constant $k$ from every score? What is the effect of adding a constant $k$ to every score?
   d. Find the mean, median, mode, and midrange of the following estimates of wealth (in dollars) of people from the list of the *Forbes 400* richest Americans.

    1,700,000,000   1,400,000,000   505,000,000   2,150,000,000
    876,000,000    475,000,000   390,000,000

   e. Do part *d* after dividing each score by 1,000,000.
   f. In general, if every score in a data set is divided or multiplied by some constant $k$, what is the effect on the mean, median, mode, and midrange?

26. a. A student receives quiz grades of 70, 65, and 90. The same student earns an 85 on the final examination. If each quiz constitutes 20% of the final grade while the final makes up 40%, find the weighted mean.
   b. A student earns the grades in the accompanying table. If grade points are assigned as A = 4, B = 3, C = 2, D = 1, F = 0 and the grade points are weighted according to the number of credit hours, find the weighted mean (grade-point average) rounded to three decimal places.

| Course | Grade | Credit Hours |
|---|---|---|
| Math | A | 4 |
| English | C | 3 |
| Art | B | 1 |
| Physical education | A | 2 |
| Biology | C | 4 |

27. Compare the averages that comprise the solutions to Exercises 3 and 4. Do these averages discriminate or differentiate between the two lists of scores? Is there any apparent difference between the two data sets that is not reflected in the averages?

28. Using the data from Exercise 3, change the first score (74) to 900 and find the mean, median, mode, and midrange of this modified data set. How does the extreme value affect these averages?

29. Data are sometimes transformed by the process of replacing each score $x$ with $\log x$. Does

$$\log \bar{x} = \frac{\Sigma \log x}{n}$$

for a set of positive values? Explain.  No

30. To obtain the **harmonic mean** of a set of scores, divide the number of scores $n$ by the sum of the reciprocals of all scores.

$$\text{harmonic mean} = \frac{n}{\Sigma \dfrac{1}{x}}$$

For example, the harmonic mean of 2, 3, 6, 7, 7, 8 is

$$\frac{6}{\dfrac{1}{2} + \dfrac{1}{3} + \dfrac{1}{6} + \dfrac{1}{7} + \dfrac{1}{7} + \dfrac{1}{8}} = \frac{6}{1.4} = 4.3$$

(Note that 0 cannot be included in the scores.)

a. Find the harmonic mean of 2, 3, 6, 7, 7, 8, 9, 9, 9, 10.  5.4
b. A group of students drive from New York to Florida (1200 miles) at a speed of 40 mi/h and returns at a speed of 60 mi/h. What is their average speed for the round trip? (The harmonic mean is used in averaging speeds.)
c. A dispatcher of charter buses calculates the average round trip speed (in miles per hour) for a certain route. The results obtained for 14 different runs are listed below. Based on these values, what is the average speed of a bus assigned to this route?

$$\begin{array}{ccccccc} 42.6 & 41.3 & 38.2 & 42.9 & 43.4 & 43.7 & 40.8 \\ 34.2 & 40.1 & 41.2 & 40.5 & 41.7 & 39.8 & 39.6 \end{array}$$

31. Given a collection of $n$ scores (all of which are positive), the **geometric mean** is the $n$th root of their product. For example, to find the geometric mean of 2, 3, 6, 7, 7, 8, 9, 9, 9, 10, first multiply the scores.

$$2 \cdot 3 \cdot 6 \cdot 7 \cdot 7 \cdot 8 \cdot 9 \cdot 9 \cdot 9 \cdot 10 = 102{,}876{,}480$$

Then take the tenth root of the product, since there are 10 scores.

$$\sqrt[10]{102{,}876{,}480} = 6.3$$

*continued*

The geometric mean is often used in business and economics for finding average rates of change, average rates of growth, or average ratios. The *average growth factor* for money compounded at annual interest rates of 10%, 8%, 9%, 12%, and 7% can be found by computing the geometric mean of 1.10, 1.08, 1.09, 1.12, and 1.07. Find that average growth factor.

32.  The **quadratic mean** (or root mean square, or R.M.S.) of a set of scores is obtained by squaring each score, adding the results, dividing by the number of scores $n$, and then taking the square root of that result.

$$\text{quadratic mean} = \sqrt{\frac{\Sigma x^2}{n}}$$

For example, the quadratic mean of 2, 3, 6, 7, 7, 8, 9, 9, 9, 10 is

$$\sqrt{\frac{4 + 9 + 36 + 49 + 49 + 64 + 81 + 81 + 81 + 100}{10}}$$

$$= \sqrt{\frac{554}{10}}$$

$$= \sqrt{55.4} = 7.4$$

The quadratic mean is usually used in physical applications. In power distribution systems, for example, voltages and currents are usually referred to in terms of their R.M.S. values. Find the R.M.S. of the following power supplies (in volts): 151, 162, 0, 81, −68.

33.  A comparison of the mean and median can reveal information about **skewness,** as illustrated below. Refer to the data given in Exercises 1 through 4 and, in each case, identify the data as being skewed to the left, symmetric, or skewed to the right. Data skewed to the left will have the mean and median to the left of the mode, but the order of the mean and median is not always predictable. Data skewed to the right have the mean and median to the right of the mode, but in unpredictable order.

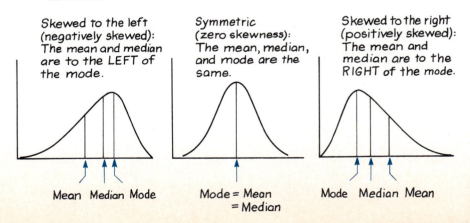

Skewed to the left (negatively skewed): The mean and median are to the LEFT of the mode.

Symmetric (zero skewness): The mean, median, and mode are the same.

Skewed to the right (positively skewed): The mean and median are to the RIGHT of the mode.

Mean  Median  Mode

Mode = Mean = Median

Mode  Median  Mean

34. Frequency tables often have open-ended classes such as the one given below. Formula 2–3 cannot be directly applied since we can't determine a class mark for the class of "more than 20." Calculate the mean by assuming that this class is really (a) 21–25; (b) 21–30; (c) 21–40. What can you conclude?

| Hours Studying per Week | Frequency |
|---|---|
| 0 | 5 |
| 1–5 | 96 |
| 6–10 | 57 |
| 11–15 | 25 |
| 16–20 | 11 |
| More than 20 | 6 |

Time spent studying by college freshmen (based on data from *The American Freshman* as reported in *USA Today*)

35. When data are summarized in a frequency table, the median can be found by first identifying the "median class" (the class that contains the median). Then we assume that the scores in that class are evenly distributed and we can interpolate. This process can be described by

$$(\text{lower limit of median class}) + (\text{class width})\left(\frac{\left(\dfrac{n + 1}{2}\right) - (m + 1)}{\text{frequency of median class}}\right)$$

where $n$ is the sum of all class frequencies and $m$ is the sum of the class frequencies that *precede* the median class. Use this procedure to find the median home selling price by referring to Table 2–3.

36. To find the 10% **trimmed mean** for a data set, first arrange the data in order. Then delete the bottom 10% of the scores and delete the top 10% of the scores and calculate the mean of the remaining scores. Find the 10% trimmed mean for the data in Exercise 14. Also find the 20% trimmed mean for that data set. What advantage does a trimmed mean have over the regular mean?

# 2–5  Dispersion Statistics

The preceding section was concerned with averages, or measures of central tendency, that reflect a certain characteristic of the data from which they came. That is, the averages are supposed to be representative or *central* scores. However, other features of the data may not be reflected at all by the averages. Suppose, for example, that two different groups of 10 students are given identical quizzes, with these results.

| Group A | Group B |
|---------|---------|
| 65 | 42 |
| 66 | 54 |
| 67 | 58 |
| 68 | 62 |
| 71 | 67 |
| 73 | 77 |
| 74 | 77 |
| 77 | 85 |
| 77 | 93 |
| 77 | 100 |

Computing the averages, we get

| | Group A | | Group B |
|---|---------|---|---------|
| mean | = 71.5 | mean | = 71.5 |
| median | = 72.0 | median | = 72.0 |
| mode | = 77 | mode | = 77 |
| midrange | = 71.0 | midrange | = 71.0 |

After comparing the averages, we can see no difference between the two groups. Yet an intuitive perusal of both groups shows an obvious difference: The scores of group B are much more widely scattered than those of group A. This variability among data is one characteristic to which averages are not sensitive. Consequently, statisticians have tried to design statistics that measure this variability, or dispersion. The three basic **measures of dispersion** are range, variance, and standard deviation.

## Range

| Group C | Group D |
|---------|---------|
| 1 | 2 |
| 20 | 3 |
| 20 | 4 |
| 20 | 5 |
| 20 | 6 |
| 20 | 14 |
| 20 | 15 |
| 20 | 16 |
| 20 | 17 |
| 20 | 18 |
| Range = 19 | Range = 16 |

The **range** is simply the difference between the highest value and the lowest value. For group A, the range of 12 is the difference between 77 and 65. The range of group B is $100 - 42 = 58$. This much larger range suggests greater dispersion. Be sure you avoid confusion between the midrange (an average) and the range (a measure of dispersion). The range is extremely easy to compute, but it's often inferior to other measures of dispersion. The rather extreme example of groups C and D should illustrate this point. The larger range for group C suggests more dispersion than in group D. But the scores of group C are very close together while those of group D are much more scattered. The range may be misleading in this case (and in many other circumstances) because it depends only on the maximum and minimum scores.

## Standard Deviation

The next measure of dispersion to be considered is the **standard deviation,** which is generally the most important and useful such measure.

**Formula 2–4** $$s = \sqrt{\frac{\Sigma(x - \bar{x})^2}{n - 1}}$$ sample standard deviation

Formula 2–4 is sometimes defined with a denominator of $n$, but the use of $n - 1$ leads to better estimates of a population standard deviation. You can divide by $n$ if you have all scores for an entire population so that the standard deviation is given by $\sqrt{\Sigma(x - \mu)^2/n}$. However, we will usually deal with sample data, so divide by $n - 1$ as in Formula 2–4. Many calculators do standard deviations, with division by $n - 1$ corresponding to a key labeled $\sigma_{n-1}$ while the key labeled $\sigma_n$ corresponds to division by $n$.

---

### Procedure for Using Formula 2–4

1. Find the mean of the scores $(\bar{x})$.
2. Subtract the mean from each individual score $(x - \bar{x})$.
3. Square each of the differences obtained from step 2. That is, multiply each value by itself. [This produces numbers of the form $(x - \bar{x})^2$.]
4. Add all of the squares obtained from step 3 to get $\Sigma(x - \bar{x})^2$.
5. Divide the preceding total by the number $(n - 1)$; that is, 1 less than the total number of scores present.
6. Find the square root of the result.

---

Why define a measure of dispersion as in Formula 2–4? When measuring dispersion in a collection of data, one reasonable approach is to begin with the individual amounts by which scores deviate from the mean. For a particular score $x$, that amount of **deviation** can be denoted by $x - \bar{x}$. The sum of all such deviations is always zero. To get a measure that isn't always zero, we could take the absolute values as in $\Sigma|x - \bar{x}|$. This leads to the **mean deviation** (or mean absolute deviation).

$$\frac{\Sigma|x - \bar{x}|}{n}$$

However, this approach tends to be unsuitable for the important methods of statistical inference. Instead, we make all of the terms $(x - \bar{x})$ nonnegative by squaring them. Finally, we take the square root to compensate for the fact that the deviations were all squared.

## Variance

If we omit step 6 in the procedure for calculating the standard deviation, we get the **variance,** which is defined as follows.

**Formula 2–5**        $s^2 = \dfrac{\Sigma(x - \bar{x})^2}{n - 1}$        sample variance

By comparing Formulas 2–4 and 2–5, we see that the variance is the square of the standard deviation. The variance will be used later in the book, but we should concentrate on the concept of standard deviation as we try to get a feeling for this statistic. A major difficulty with the variance is that it is not in the same units as the original data. For example, a data set might have a standard deviation of $3.00 and a variance of 9.00 *square* dollars. Since we can't relate well to square dollars, we find it difficult to understand variance.

---

### Notation

$s$    denotes the **standard deviation** of a set of **sample** scores.

$\sigma$    denotes the **standard deviation** of a set of **population** scores. ($\sigma$ is the lowercase Greek sigma.)

$s^2$    denotes the **variance** of a set of **sample** scores.

$\sigma^2$    denotes the **variance** of a set of **population** scores.

Note:    Articles in professional journals and reports often use **SD** for standard deviation and **Var** for variance.

---

### Example

Find the standard deviation for the quiz scores 2, 3, 5, 6, 9, 17.

### Solution

See Table 2–14, where the following steps are executed. It is helpful to use the format of that table.

*Step 1:*  The mean is obtained by adding the scores (42) and then dividing by the number of scores present (6). The mean is 7.0.

*Step 2:*  Subtracting the mean of 7.0 from each score, we get $-5$, $-4$, $-2$, $-1$, 2, and 10. (As a quick check, these numbers must always total 0.)

*Step 3:*  Squaring each value obtained from step 2, we get 25, 16, 4, 1, 4, and 100.

*Step 4:*  The sum of all the preceding squares is 150.

*Step 5:*  There are six scores, so we divide 150 by 1 less than 6. That is, $150 \div 5 = 30$.

*Step 6:*  Find the square root of 30. We get a standard deviation of $\sqrt{30} = 5.5$.

| TABLE 2–14 | | | |
|---|---|---|---|
| | $x$ | $(x - \bar{x})$ | $(x - \bar{x})^2$ |
| | 2 | −5 | 25 |
| | 3 | −4 | 16 |
| | 5 | −2 | 4 |
| | 6 | −1 | 1 |
| | 9 | 2 | 4 |
| | 17 | 10 | 100 |
| Totals: | 42 | | 150 |

$$\bar{x} = \frac{42}{6} = 7.0$$

$$s = \sqrt{\frac{150}{6 - 1}} = \sqrt{\frac{150}{5}} = \sqrt{30.0} = 5.5$$

## Round-Off Rule

As in Section 2–4, we round off answers by carrying one more decimal place than was present in the original data. We should round only the final answer and not intermediate values. (If we must round intermediate results, we should carry at least twice as many decimal places as will be used in the final answer.)

We will now attempt to make some intuitive sense out of the standard deviation. First, we should clearly understand that the standard deviation measures the dispersion or variation among scores. Scores close together will yield a small standard deviation, whereas scores spread farther apart will yield a larger standard deviation. Figure 2–9 shows that as the data spread farther apart, the corresponding values of the standard deviation increase.

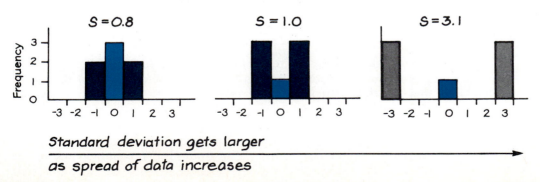

**Figure 2–9**
*All three data sets have the same mean of zero, but the standard deviations are different.*

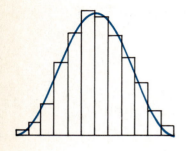

In attempting to develop a sense for the standard deviation, we might consider the following results from **Chebyshev's theorem** (see Exercise 34):

- At least 75% of all scores will fall within the interval from two standard deviations below the mean to two standard deviations above the mean.

- At least 89% of all scores will fall within three standard deviations of the mean.

We might also consider the **empirical rule**, which applies to data having a distribution that is approximately bell-shaped, as in the accompanying figure. Such distributions are important and common. For these bell-shaped distributions, the empirical rule (sometimes referred to as the "68–95–99" rule) states that

- About 68% of all scores fall within *one* standard deviation of the mean.

- About 95% of all scores fall within *two* standard deviations of the mean.

- About 99.7% of all scores fall within *three* standard deviations of the mean.

While Chebyshev's theorem applies to *any* set of scores, the empirical rule applies only to scores with a bell-shaped distribution. For examples of Chebyshev's theorem and the empirical rule, see the following table.

| Given: IQ scores with a mean of 100, a standard deviation of 15, and a distribution that is bell-shaped | |
| --- | --- |
| Chebyshev's Theorem | Empirical Rule |
| | About 68% of all scores are between 85 and 115. |
| At least 75% of all scores are between 70 and 130. | About 95% of all scores are between 70 and 130. |
| At least 89% of all scores are between 55 and 145. | About 99.7% of all scores are between 55 and 145. |

Take the time to study the results in the preceding table. They show us how the standard deviation can be used to get a sense for how data vary. (Note that "within two standard deviations of the mean" translates to "within 2(15) of 100" or "within 30 of 100," which is really from 70 to 130.) According to the empirical rule, an IQ score of 147 would be very rare, since it deviates from the mean by more than three standard deviations.

We will now present two additional formulas for standard deviation, but these formulas do not involve a different concept; they are only different versions of Formula 2–4. First, Formula 2–4 can be expressed in the following equivalent form.

**Formula 2–6**

$$s = \sqrt{\frac{n(\Sigma x^2) - (\Sigma x)^2}{n(n - 1)}}$$

Formulas 2–4 and 2–6 are equivalent in the sense that they will always produce the same results. Algebra can be used to show that they are equal. Formula 2–6 is called the *shortcut* formula because it tends to be convenient with messy numbers or with large sets of data. Also, Formula 2–6 is often used in calculators and computer programs since it requires only three memory registers (for $n$, $\Sigma x$, and $\Sigma x^2$) instead of one memory register for every single score. However, many instructors prefer to use only Formula 2–4 for calculating standard deviations. They argue that Formula 2–4 reinforces the concept that the standard deviation is a type of average deviation while Formula 2–6 obscures that idea. Other instructors have no objections to using Formula 2–6. We have included the shortcut formula so that it is available for those who choose to use it.

For the quiz scores of 2, 3, 5, 6, 9, 17 we could use Formula 2–6 to find the standard deviation as in Table 2–15. Because there are six scores, $n = 6$. From the table we see that $\Sigma x = 42$ and $\Sigma x^2 = 444$, so that

$$s = \sqrt{\frac{n(\Sigma x^2) - (\Sigma x)^2}{n(n - 1)}}$$

$$= \sqrt{\frac{6(444) - (42)^2}{6(6 - 1)}}$$

$$= \sqrt{\frac{2664 - 1764}{30}}$$

$$= \sqrt{\frac{900}{30}} = \sqrt{30} = 5.5$$

| TABLE 2–15 | |
|---|---|
| $x$ | $x^2$ |
| 2 | 4 |
| 3 | 9 |
| 5 | 25 |
| 6 | 36 |
| 9 | 81 |
| 17 | 289 |
| 42 | 444 |
| ↑ $\Sigma x$ | ↑ $\Sigma x^2$ |

We can develop a formula for standard deviation when the data are summarized in a frequency table. The result is as follows.

$$s = \sqrt{\frac{\Sigma f\cdot(x - \bar{x})^2}{n - 1}}$$

We will express this formula in an equivalent expression that usually simplifies the actual calculations.

**Formula 2–7**

$$s = \sqrt{\frac{n[\Sigma(f \cdot x^2)] - [\Sigma(f \cdot x)]^2}{n(n - 1)}}$$

standard deviation for frequency table

where $x$ = class mark
$f$ = frequency
$n$ = sample size

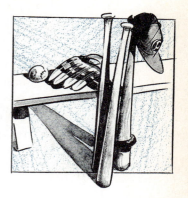

## WHERE ARE THE 0.400 HITTERS?

■ The last baseball player to hit above 0.400 was Ted Williams, who hit 0.406 in 1941. There were averages above 0.400 in 1876, 1879, 1887, 1894, 1895, 1896, 1897, 1899, 1901, 1911, 1920, 1922, 1924, 1925, and 1930, but none since 1941. Are there no longer great hitters? Harvard's Stephen Jay Gould notes that the mean batting average has been steady at 0.260 for about 100 years, but the standard deviation has been decreasing from 0.049 in the 1870s to 0.031, where it is now. He argues that today's stars are as good as those from the past, but consistently better pitchers now keep averages below 0.400. Dr. Gould discusses this in Program 4 of the series *Against All Odds: Inside Statistics.* ■

**TABLE 2–16**

| Score | Frequency $f$ | Class Mark $x$ | $f \cdot x$ | $f \cdot x^2$ or $f \cdot x \cdot x$ |
|---|---|---|---|---|
| 0–8 | 10 | 4 | 40 | 160 |
| 9–17 | 5 | 13 | 65 | 845 |
| 18–26 | 15 | 22 | 330 | 7,260 |
| 27–35 | 3 | 31 | 93 | 2,883 |
| 36–44 | 1 | 40 | 40 | 1,600 |
| Total | 34 | | 568 | 12,748 |

$$s = \sqrt{\frac{n[\Sigma(f \cdot x^2)] - [\Sigma(f \cdot x)]^2}{n(n-1)}}$$

$$= \sqrt{\frac{34(12,748) - (568)^2}{34(34-1)}}$$

$$= \sqrt{\frac{110,808}{1,122}} = \sqrt{98.759} = 9.9$$

Table 2–16 summarizes the work done in applying Formula 2–7 to a frequency table formed by that table's first two columns.

Measures of dispersion are extremely important in many practical circumstances. Manufacturers interested in producing items of consistent quality are very concerned with statistics such as standard deviation. A producer of car batteries might be pleased to learn that a product has a mean life of four years, but that pleasure would become distress if the standard deviation indicated a very large dispersion, which would correspond to many battery failures long before the mean of four years. Quality control requires consistency, and consistency requires a relatively small standard deviation.

Some exercises in the previous section showed that if a constant is added to (or subtracted from) each value in a set of data, the measures of central tendency change by that same amount. In contrast, the measures of dispersion do not change when a constant is added to (or subtracted from) each score. To find the standard deviation for 8001, 8002, . . . , 8009, we could subtract 8000 from each score and work with more manageable numbers without changing the value of the standard deviation, which is a definite advantage. But now suppose a manufacturer learns that the average (mean) readings of a thousand pressure gauges are 10 pounds per square inch (psi) too low. A relatively simple remedy would be to rotate all of the dials to read 10 psi higher. That remedy would correct the mean, but the standard deviation would remain the same. If the gauges are inconsistent and have errors that vary considerably, then the standard deviation of the errors might be too large. If that is the case, each gauge must be individually calibrated and the entire manufacturing process might require an overhaul. Machines might require replacement if they do not meet the necessary tolerance levels. We can see from this example that measures of central tendency and dispersion have properties that lead to very practical and important applications.

## 2–5  Exercises A

In Exercises 1–16, find the range, variance, and standard deviation for the given data.

1. Statistics are sometimes used to compare or identify authors of different works. The lengths of the first 20 words in *The Cat in the Hat* by Dr. Seuss are listed below.

   $$3 \quad 3 \quad 3 \quad 3 \quad 5 \quad 2 \quad 3 \quad 3 \quad 3 \quad 2$$
   $$4 \quad 2 \quad 2 \quad 3 \quad 2 \quad 3 \quad 5 \quad 3 \quad 4 \quad 4$$

2. The lengths of the first 20 words in the foreword written by Tennessee Williams in *Cat on a Hot Tin Roof* are listed below.

   $$2 \quad 6 \quad 2 \quad 2 \quad 1 \quad 4 \quad 4 \quad 2 \quad 4 \quad 2$$
   $$3 \quad 8 \quad 4 \quad 2 \quad 2 \quad 7 \quad 7 \quad 2 \quad 3 \quad 11$$

3. In Section 1 of a statistics class, 10 test scores were randomly selected, and the following results were obtained:

   $$74, 73, 77, 77, 71, 68, 65, 77, 67, 66$$

4. In Section 2 of a statistics class, test scores were randomly selected, and the following results were obtained:

   $$42, 100, 77, 54, 93, 85, 67, 77, 62, 58$$

5. The following values are the distances (in miles) the author's car traveled before it stalled. It took four trips to the Oldsmobile dealer to determine that stalling was caused by a defective crank sensor.

   $$252 \quad 13 \quad 34 \quad 54 \quad 241 \quad 12 \quad 11 \quad 22$$
   $$219 \quad 336 \quad 19 \quad 114 \quad 56 \quad 106 \quad 6$$

6. A police-issue radar gun was used to record speeds for motorists driving through a 40 mi/h speed zone. The results are given below.

   $$41 \quad 40 \quad 43 \quad 53 \quad 36 \quad 47 \quad 40 \quad 38$$

7. The blood alcohol concentrations of 15 drivers involved in fatal accidents and then convicted with jail sentences are given below (based on data from the U.S. Department of Justice).

   $$0.27 \quad 0.17 \quad 0.17 \quad 0.16 \quad 0.13 \quad 0.24 \quad 0.29 \quad 0.24$$
   $$0.14 \quad 0.16 \quad 0.12 \quad 0.16 \quad 0.21 \quad 0.17 \quad 0.18$$

8. The amount of time (in hours) spent on paperwork in one day was obtained from a sample of office managers with the results given below (based on data from Adia Personnel Services).

   3.7   2.9   3.4   0.0   1.5   1.8   2.3   2.4   1.0   2.0
   4.4   2.0   4.5   0.0   1.7   4.4   3.3   2.4   2.1   2.1

9. In "An Analysis of Factors that Contribute to the Efficacy of Hypnotic Analgesia" (by Price and Barber, *Journal of Abnormal Psychology*, Vol. 96, No. 1), the "before" readings (in centimeters) on a mean visual analog scale are given for 16 subjects. They are listed below.

   8.8    6.6    8.4    6.5    8.4    7.0    9.0    10.3
   8.7    11.3   8.1    5.2    6.3    11.6   6.2    10.9

10. In "Determining Statistical Characteristics of a Vehicle Emissions Audit Procedure" (by Lorenzen, *Technometrics*, Vol. 22, No. 4), carbon monoxide emissions data (in g/m) are listed for vehicles. The following values are included.

    5.01    14.67    8.60    4.42    4.95    7.24
    7.51    12.30    14.59   7.98    11.53   4.10

11. In "A Simplified Method for Determination of Residual Lung Volumes" (by Wilmore, *Journal of Applied Psychology*, Vol. 27, No. 1), the oxygen dilution was measured (in liters) for subjects and the following sample values were obtained.

    1.361    1.013    1.140    1.649    1.278
    1.148    1.824    1.551    1.041

12. The following are the ages of motorcyclists when they were fatally injured in traffic accidents (based on data from the U.S. Department of Transportation).

    17    38    27    14    18    34    16    42    28
    24    40    20    23    31    37    21    30    25
    17    28    33    25    23    19    51    18    29

13. Forty commercial aircraft randomly selected in the United States were found to have the ages given below (based on data reported by Aviation Data Services).

    3.2    22.6    23.1    16.9    0.4     6.6    12.5    22.8
    26.3    8.1    13.6    17.0    21.3    15.2    18.7    11.5
    4.9     5.3     5.8    20.6    23.1    24.7     3.6    12.4
    27.3    22.5    3.9     7.0    16.2    24.1     0.1     2.1
    7.7    10.5    23.4    0.7    15.8     6.3    11.9    16.8

14. A student working part time for a moving company in Dutchess County, New York, collected the following load weights (in pounds) for 50 consecutive customers.

| | | | | | | |
|---|---|---|---|---|---|---|
| 8,090 | 3,250 | 12,350 | 4,510 | 8,770 | 5,030 | 12,700 |
| 8,800 | 6,170 | 8,450 | 10,330 | 10,100 | 13,410 | 7,280 |
| 13,490 | 17,810 | 7,470 | 11,450 | 13,260 | 9,310 | 15,970 |
| 7,540 | 7,770 | 6,400 | 14,800 | 14,760 | 6,820 | 11,430 |
| 7,200 | 13,520 | 16,200 | 10,780 | 10,510 | 17,330 | 7,450 |
| 9,110 | 10,630 | 3,670 | 14,310 | 9,140 | 10,220 | 9,900 |
| 11,860 | 12,010 | 12,430 | 8,160 | 26,580 | 4,480 | 6,390 |
| 11,600 | | | | | | |

15. Listed below are the actual energy consumption amounts as reported on the electric bills for one residence. Each amount is in kilowatt-hours and represents a two-month period.

| | | | | | | | | | |
|---|---|---|---|---|---|---|---|---|---|
| 728 | 774 | 859 | 882 | 791 | 731 | 838 | 862 | 880 | 831 |
| 759 | 774 | 832 | 816 | 860 | 856 | 787 | 715 | 752 | 778 |
| 829 | 792 | 908 | 714 | 839 | 752 | 834 | 818 | 835 | 751 |
| 837 | | | | | | | | | |

16. Listed below are the daily sales totals (in dollars) for 44 days at a large retail outlet in Orange County, California.

| | | | | | |
|---|---|---|---|---|---|
| 24,145 | 39,921 | 30,284 | 43,923 | 38,366 | 34,132 |
| 33,650 | 47,563 | 34,804 | 55,254 | 44,249 | 37,229 |
| 37,196 | 66,644 | 37,858 | 31,995 | 61,923 | 50,857 |
| 39,897 | 29,099 | 42,906 | 35,420 | 33,310 | 66,892 |
| 46,351 | 34,423 | 52,250 | 38,295 | 37,052 | 35,018 |
| 65,798 | 37,809 | 68,945 | 43,952 | 38,812 | 38,275 |
| 25,597 | 41,160 | 31,987 | 59,708 | 46,285 | 32,097 |
| 36,802 | 64,842 | | | | |

17. Add 15 to each value given in Exercise 3 and then find the range, variance, and standard deviation. Compare the results to those of Exercise 3 and then form a general conclusion about the effect of adding a constant.

18. Subtract 5 from each value given in Exercise 3 and then find the range, variance, and standard deviation. Compare the results to those of Exercise 3 and then form a general conclusion about the effect of subtracting a constant.

19. Multiply each value given in Exercise 3 by 10 and then find the range, variance, and standard deviation. Compare the results to those of Exercise 3 and then form a general conclusion about the effect of multiplication by a constant.

20. Halve each value in Exercise 3 and then find the range, variance, and standard deviation. Compare the results to those of Exercise 3 and then form a general conclusion about the effect of division by a constant.

In Exercises 21–28, find the variance and standard deviation for the given data.

21.

| Distance (in thousands of miles) | Frequency |
|---|---|
| 0–39 | 17 |
| 40–79 | 41 |
| 80–119 | 80 |
| 120–159 | 49 |
| 160–199 | 4 |

Distances traveled by buses before the first major motor failure (based on data from "Large Sample Simultaneous Confidence Intervals for the Multinomial Probabilities Based on Transformations of the Cell Frequencies" by Bailey, *Technometrics*, Vol. 22, No. 4)

22.

| Age (years) | Frequency |
|---|---|
| 0–9 | 44 |
| 10–19 | 42 |
| 20–29 | 33 |
| 30–39 | 30 |
| 40–49 | 27 |
| 50–59 | 22 |
| 60–69 | 18 |
| 70–79 | 9 |

Ages of randomly selected Dutchess County residents (based on data from the *Cornell Community and Resources Development Series*)

23.

| Size of Family (persons) | Frequency |
|---|---|
| 2 | 51 |
| 3 | 31 |
| 4 | 27 |
| 5 | 12 |
| 6 | 4 |
| 7 | 1 |
| 8 | 1 |

Family size for 127 randomly selected families (based on data from the Bureau of the Census)

24.

| Time (years) | Number |
|---|---|
| 4 | 147 |
| 5 | 81 |
| 6 | 27 |
| 7 | 15 |
| 7.5–11.5 | 30 |

Time required to earn bachelor's degree (based on data from the National Center for Education Statistics)

25.

| Age | Number |
|---|---|
| 16–20 | 38 |
| 21–29 | 110 |
| 30–39 | 122 |
| 40–49 | 91 |
| 50–59 | 71 |
| 60–69 | 82 |

Ages of New York drivers involved in accidents (based on data from the N.Y. State Department of Motor Vehicles)

26.

| Number | Frequency |
|--------|-----------|
| 1–5 | 19 |
| 6–10 | 24 |
| 11–15 | 18 |
| 16–20 | 21 |
| 21–25 | 23 |
| 26–30 | 20 |
| 31–35 | 16 |
| 36–40 | 15 |

Frequency table of lottery numbers selected in a 26-week period

27.

| Family Income (dollars) | Number |
|-------------------------|--------|
| 0–2,499 | 3 |
| 2,500–7,499 | 11 |
| 7,500–12,499 | 16 |
| 12,500–14,999 | 9 |
| 15,000–19,999 | 19 |
| 20,000–24,999 | 19 |
| 25,000–34,999 | 36 |
| 35,000–49,999 | 37 |

Incomes of families (restricted to families with incomes below $50,000 and based on data from the Bureau of the Census)

28.

| Years of Education | Number Who Smoke |
|--------------------|------------------|
| 0–11 | 195 |
| 12 | 179 |
| 13–15 | 151 |
| 16–20 | 101 |

Number of years of education for a sample of 626 smokers (based on data from the Department of Health and Human Services)

29. Which would you expect to have a higher standard deviation: the IQ scores of a class of 25 statistics students or the IQ scores of 25 randomly selected adults? Explain.

30. Which would you expect to have a higher standard deviation: the reaction times of 30 sober adults or the reaction times of 30 adults who had recently consumed three martinis each?

31. Is it possible for a set of scores to have a standard deviation of zero? If so, how? Is it possible for a set of scores to have a negative standard deviation?

32. Test the effect of an *outlier* (an extreme value) by changing the 74 in Exercise 3 to 740. Calculate the range, variance, and standard deviation for the modified set and compare the results to those originally obtained in Exercise 3.

# 2–5 Exercises B

33. Find the range and standard deviation for each of the following two groups. Which group has less dispersion according to the criterion of the range? Which group has less dispersion according to the criterion of the standard deviation? Which measure of dispersion is "better" in this situation: the range or the standard deviation?

    group C:   1, 20, 20, 20, 20, 20, 20, 20, 20, 20
    group D:   2,   3,   4,   5,   6, 14, 15, 16, 17, 18

34. Chebyshev's theorem states that the *proportion* (or fraction) of any set of data lying within $K$ standard deviations of the mean is always *at least* $1 - 1/K^2$, where $K$ is any positive number greater than 1.
    a.  Given a mean IQ of 100 and a standard deviation of 15, what does Chebyshev's theorem say about the number of scores within two standard deviations of the mean (that is, 70–130)?
    b.  Given a mean of 100 and a standard deviation of 15, what does Chebyshev's theorem say about the number of scores between 55 and 145?
    c.  The mean score on the College Entrance Examination Board Scholastic Aptitude Test is 500 and the standard deviation is 100. What does Chebyshev's theorem say about the number of scores between 300 and 700?
    d.  Using the data of part *c*, what does Chebyshev's theorem say about the number of scores between 200 and 800?

35. A large set of sample scores yields a mean and standard deviation of $\bar{x} = 56.0$ and $s = 4.0$, respectively. The distribution of the histogram is roughly bell-shaped. Use the empirical rule to answer the following.
    a.  What percentage of the scores should fall between 52.0 and 60.0?
    b.  What percentage of the scores should fall within 8.0 of the mean?
    c.  About 99.7% of the scores should fall between what two values? (The mean of 56.0 should be midway between those two values.)

36. Find the standard deviation of the 150 home selling prices by using (a) the original set of data given in Table 2–1; (b) the frequency table given in Table 2–3. Then compare the results to determine the amount of distortion caused by the frequency table.

37. For any population of scores having a standard deviation $\sigma$ and range $R$, the relationship $\sigma \le R/2$ must be true. This relationship might be helpful in determining whether a computed value of $\sigma$ is at least a reasonable possibility. Verify this relationship for the data given in Exercise 3. Use

$$\sigma = \sqrt{\frac{\Sigma(x - \mu)^2}{n}}$$

38. If we consider the values 1, 2, 3, . . . , $n$ to be a population, the standard deviation can be calculated by the formula

$$\sigma = \sqrt{\frac{n^2 - 1}{12}}$$

This formula is equivalent to Formula 2–4 modified for division by $n$ instead of $n - 1$, where the values are 1, 2, 3, . . . , $n$.
   a. Find the standard deviation of the population 1, 2, 3, . . . , 100.
   b. Find an expression for calculating the *sample* standard deviation $s$ for the sample values 1, 2, 3, . . . , $n$.

39. Find the mean and standard deviation for the scores 18, 19, 20, . . . , 182. Treat those values as a population (see Exercise 38).

40. Computers commonly use a random number generator, which produces values between 0 and 1. In the long run, all values occur with the same relative frequency. Find the mean and standard deviation for the numbers between 0.00000000 and 0.99999999. (*Hint:* See Exercises 19 and 38.)

41. Given a collection of temperatures in degrees Fahrenheit, the following statistics are calculated:

$$\bar{x} = 40.2 \qquad s = 3.0 \qquad s^2 = 9.0$$

Find the values of $\bar{x}$, $s$, and $s^2$ for the same data set after each temperature has been converted to the Celsius scale.

$$\left[ C = \frac{5}{9}(F - 32) \right]$$

42. a. A set of data yields $\bar{x} = 50.0$ and $s = 8.0$. If we multiply each of the original scores by the constant $b$ and if we then add the constant $c$ to each product, find the values of $b$ and $c$ that will cause the modified data set to yield $\bar{x} = 65.0$ and $s = 12.0$.
   b. If a population has mean $\mu$ and standard deviation $\sigma$, what must be done to each score in order to have a mean of 0 and a standard deviation of 1?

43. a. The **coefficient of variation,** expressed in percent, is used to describe the standard deviation relative to the mean. It is calculated as follows.

$$\frac{s}{\bar{x}} \cdot 100 \qquad or \qquad \frac{\sigma}{\mu} \cdot 100$$

Find the coefficient of variation for the following sample scores: 2, 2, 2, 3, 5, 8, 12, 19, 22, 30.
   b. In the Taguchi method of quality engineering, a key tool is the **signal-to-noise ratio.** The simplest way to calculate this is to divide the mean by the standard deviation. Find this ratio for the sample data given in part *a*.

44. In Exercise 33 from Section 2–4, we introduced the general concept of skewness. Skewness can be measured by **Pearson's index of skewness:**

$$I = \frac{3(\bar{x} - \text{median})}{s}$$

If $I \geq 1.00$ or $I \leq -1.00$, the data can be considered to be *significantly skewed.* Find Pearson's index of skewness for the data given in Exercise 13 and then determine whether or not there is significant skewness.

## 2–6  Measures of Position

There is often a need to compare scores taken from two separate populations with different means and standard deviations. The standard score (or z score) can be used to help make such comparisons.

---

### Definition

The **standard score**, or **z score**, is the number of standard deviations that a given value $x$ is above or below the mean, and it is found by

| *Sample* | | *Population* |
|:---:|:---:|:---:|
| $z = \dfrac{x - \bar{x}}{s}$ | or | $z = \dfrac{x - \mu}{\sigma}$ |

Round $z$ to two decimal places.

---

Suppose two equivalent tests are given to similar groups, but the tests are designed with different scales. Which is better: a score of 65 on test A or a score of 29 on test B? The class statistics for the two tests are as follows:

| Test A | Test B |
|---|---|
| $\bar{x} = 50$ | $\bar{x} = 20$ |
| $s = 10$ | $s = \phantom{0}5$ |

For the score of 65 on test A we get a z score of 1.50, since

$$z = \frac{x - \bar{x}}{s} = \frac{65 - 50}{10} = \frac{15}{10} = 1.50$$

For the score of 29 on test B we get a $z$ score of 1.80, since

$$z = \frac{x - \bar{x}}{s} = \frac{29 - 20}{5} = \frac{9}{5} = 1.80$$

That is, a score of 65 on test A is 1.50 standard deviations above the mean, while a score of 29 on test B is 1.80 standard deviations above the mean. This implies that the 29 on test B is the better score. While 29 is below 65, it has a better *relative* position when considered in the context of the other test results. Later, we will make extensive use of these standard, or $z$, scores.

The $z$ score provides a useful measurement for making comparisons between different sets of data. Quartiles, deciles, and percentiles are measures of position useful for comparing scores within one set of data or between different sets of data.

Just as the median divides the data into two equal parts, the three **quartiles,** denoted by $Q_1$, $Q_2$, and $Q_3$, divide the ranked scores into four equal parts. Roughly speaking, $Q_1$ separates the bottom 25% of the ranked scores from the top 75%, $Q_2$ is the median, and $Q_3$ separates the top 25% from the bottom 75%. To be more precise, at least 25% of the data will be less than or equal to $Q_1$ and at least 75% will be greater than or equal to $Q_1$. At least 75% of the data will be less than or equal to $Q_3$, while at least 25% will be equal to or greater than $Q_3$.

Similarly, there are nine **deciles,** denoted by $D_1$, $D_2$, $D_3$, . . . , $D_9$, which partition the data into 10 groups with about 10% of the data in each group. There are also 99 **percentiles,** which partition the data into 100 groups with about 1% of the scores in each group. (Quartiles, deciles, and percentiles are examples of *fractiles,* which partition data into parts that are approximately equal.) A student taking a competitive college entrance examination might learn that he or she scored in the 92nd percentile. This does not mean that the student received a grade of 92% on the test; it indicates roughly that whatever score he or she did achieve was higher than 92% of those who took a similar test (and also lower than 8% of his or her colleagues). Percentiles are useful for converting meaningless raw scores into meaningful comparative scores. For this reason, percentiles are used extensively in educational testing. A raw score of 750 on a college entrance exam means nothing to most people, but the corresponding percentile of 93% provides useful comparative information.

The process of finding the percentile that corresponds to a particular score $x$ is fairly simple, as indicated in the following definition.

## Definition

$$\text{percentile of score } x = \frac{\text{number of scores less than } x}{\text{total number of scores}} \cdot 100$$

### Example

Table 2–17 (see page 88) lists the 150 home selling prices arranged in order from lowest to highest. Find the percentile corresponding to $100,000.

### Solution

From Table 2–17 we see that there are 10 selling prices less than $100,000, so that

$$\text{percentile of } \$100{,}000 = \frac{10}{150} \cdot 100 = 7 \qquad \text{(rounded off)}$$

The selling price of $100,000 is the 7th percentile.

There are several different methods for finding the score corresponding to a particular percentile, but the process we will use is summarized in Figure 2–10.

### Example

Refer to the 150 home selling prices in Table 2–17 and find the score corresponding to the 35th percentile. That is, find the value of $P_{35}$.

### Solution

We refer to Figure 2–10 and observe that the data are already ranked from lowest to highest. We now compute the locator $L$ as follows.

$$L = \left(\frac{k}{100}\right)n = \left(\frac{35}{100}\right) \cdot 150 = 52.5$$

We answer no when asked if 52.5 is a whole number, so we are directed to round $L$ up (not off) to 53. (In *this* procedure we round $L$ up to the next higher integer, but in most other situations in this book we generally follow the usual process for rounding.) The 35th percentile, denoted by $P_{35}$, is the 53rd score, starting with the lowest. Beginning with the lowest score of $72,000, we count down the list to find the 53rd score of $134,000, so that $P_{35} = \$134{,}000$.

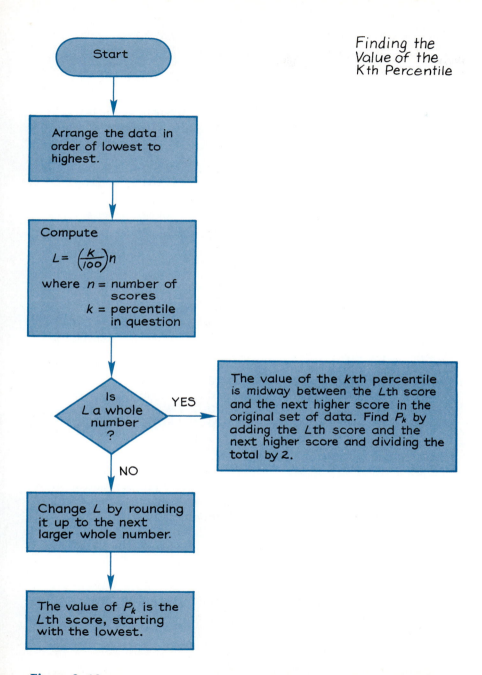

Finding the
Value of the
Kth Percentile

**Figure 2–10**

## INDEX NUMBERS

■ The Consumer Price Index (CPI) is an example of an *index number* that allows us to compare the value of some variable relative to its value at some base period. The net effect is that we measure the change of the variable over some time span. We find the value of an index number by evaluating

$$\frac{current\ value}{base\ value} \times 100$$

The CPI is based on a weighted average of the costs of specific goods and services. Using 1982–84 as a base with index 100, the CPI for a recent year is 124. Goods and services costing $100 in 1982–84 would cost $124 for this year. ■

| **TABLE 2–17** | *Ranked* Selling Prices (in dollars) of 150 Dutchess County Homes | | | | |
|---|---|---|---|---|---|
| 72,000 | 120,500 | 133,000 | 144,900 | 164,900 | 190,000 |
| 78,000 | 121,900 | 133,000 | 145,000 | 165,000 | 190,000 |
| 80,500 | 122,000 | 134,000 | 145,900 | 165,000 | 190,000 |
| 85,000 | 122,500 | 134,500 | 146,000 | 165,000 | 194,500 |
| 89,900 | 123,500 | 135,000 | 150,000 | 165,000 | 194,500 |
| 94,000 | 124,000 | 135,000 | 150,000 | 165,900 | 194,900 |
| 96,000 | 124,900 | 135,000 | 150,000 | 168,500 | 195,000 |
| 97,000 | 125,000 | 135,000 | 151,500 | 169,000 | 196,800 |
| 98,000 | 125,000 | 135,500 | 152,000 | 169,900 | 203,000 |
| 99,500 | 125,900 | 136,000 | 152,000 | 170,000 | 205,406 |
| 100,000 | 126,000 | 136,000 | 152,500 | 172,000 | 208,000 |
| 107,000 | 126,500 | 137,900 | 153,500 | 175,000 | 210,000 |
| 108,000 | 126,500 | 139,500 | 154,000 | 177,000 | 211,900 |
| 108,500 | 126,500 | 139,900 | 155,000 | 179,000 | 215,000 |
| 110,000 | 126,500 | 140,000 | 156,000 | 179,900 | 217,000 |
| 110,000 | 126,500 | 140,000 | 156,000 | 180,000 | 220,000 |
| 113,500 | 127,500 | 141,000 | 158,000 | 180,000 | 223,000 |
| 115,000 | 127,500 | 141,000 | 159,000 | 180,500 | 231,750 |
| 116,000 | 128,500 | 141,000 | 160,000 | 181,500 | 235,000 |
| 116,300 | 129,000 | 141,900 | 160,000 | 182,000 | 265,000 |
| 117,000 | 129,900 | 141,900 | 160,000 | 183,000 | 267,000 |
| 117,000 | 130,000 | 142,000 | 162,000 | 185,000 | 269,900 |
| 117,500 | 131,000 | 142,000 | 162,500 | 185,500 | 275,000 |
| 118,500 | 132,000 | 142,500 | 163,500 | 187,000 | 285,000 |
| 119,000 | 133,000 | 144,900 | 164,000 | 189,900 | 302,000 |

In the preceding example, we can see that there are 52 scores below $134,000, so the definition of percentile indicates that the percentile of $134,000 is $(52/150) \cdot 100 = 34.7$, which is approximately 35. As the amount of data increases, these differences become smaller.

In the preceding example we found the 35th percentile. Because of the sample size, the location $L$ first became 52.5, which was rounded to 53 because $L$ was not originally a whole number. In the next example we illustrate a case in which $L$ does begin as a whole number. This condition will cause us to branch to the right in Figure 2–10.

### Example

Refer to the same home selling prices listed in Table 2–17. Find $P_{70}$, which denotes the 70th percentile.

## Solution

Following the procedure outlined in Figure 2–10 and noting that the data are already ranked from lowest to highest, we compute

$$L = \left(\frac{k}{100}\right)n = \left(\frac{70}{100}\right) \cdot 150 = 105$$

We now answer yes when asked if 105 is a whole number, and we then see that $P_{70}$ is midway between the 105th and 106th scores. Since the 105th and 106th scores are $165,900 and $168,500, we get

$$P_{70} = \frac{165,900 + 168,500}{2} = \$167,200$$

Once these calculations with percentiles are mastered, similar calculations for quartiles and deciles can be performed with the same procedures by noting the following relationships.

| Quartiles | Deciles |
|---|---|
| $Q_1 = P_{25}$ | $D_1 = P_{10}$ |
| $Q_2 = P_{50}$ | $D_2 = P_{20}$ |
| $Q_3 = P_{75}$ | $D_3 = P_{30}$ |
| | $\bullet$ |
| | $\bullet$ |
| | $\bullet$ |
| | $D_9 = P_{90}$ |

Finding $Q_3$, for example, is equivalent to finding $P_{75}$.

Other statistics are sometimes defined using quartiles, deciles, or percentiles. For example, the **interquartile range** is a measure of dispersion obtained by evaluating $Q_3 - Q_1$. The **semi-interquartile range** is $(Q_3 - Q_1)/2$, the **midquartile** is $(Q_1 + Q_3)/2$, and the **10–90 percentile range** is defined to be $P_{90} - P_{10}$.

When dealing with large collections of data, more reliable results are obtained with greater ease when statistics software packages are used. See the following STATDISK and Minitab computer displays for the 150 home selling prices listed in Table 2–1. In the STATDISK display, the results correspond to the values entered as thousands of dollars. For example, $179,000 was entered as 179. This is a common technique used to ease the entry of data, but it does affect some results, such as the variance. STATDISK and Minitab also provide histograms.

# CPI

■ The Consumer Price Index (CPI) is a government measure of inflation. Each month, 300 field workers go to 85 cities and collect price data on thousands of items such as rent, milk, hamburgers, and haircuts. The index is weighted to agree with consumer spending habits. Many labor contracts include a cost of living adjustment (referred to as a "COLA") that is based on the CPI. (See also the essay "Index Numbers.") ■

STATDISK DISPLAY

Sample size = 150

| | | |
|---|---|---|
| Minimum ...... = 72.000 | Sum of scores ... = | 23066.256 |
| Maximum ...... = 302.000 | Sum of squares .. = | 3805003.017 |

MEASURES OF CENTRAL TENDENCY

| | | |
|---|---|---|
| Mean ........ = 153.7750 | Geom. mean .... = | 148.6180 |
| Median ....... = 144.9000 | Harm. mean ..... = | 143.7440 |
| Midrange ...... = 187.0000 | Quad. mean ..... = | 159.2690 |

MEASURES OF DISPERSION

| | | |
|---|---|---|
| Samp. st. dev. .. = 41.6109 | Samp. variance .. = | 1731.4700 |
| Pop. st. dev. .... = 41.4720 | Pop. variance .... = | 1719.9300 |
| Range ........ = 230.0000 | Standard error ... = | 3.3975 |

MEASURES OF POSITION—Quartiles

Q1 = 126.5000          Q2 = 144.9000          Q3 = 179.0000

MINITAB DISPLAY (see also Appendix C)

```
MTB > SET C1
DATA> 179000 126500 134500 125000 142000 164000
DATA> 146000 129000 141900 135000 118500 160000
           •
           •
           •
DATA> 133000 196800 121900 126000 164900 172000
DATA> 100000 129900 110000 131000 107000 165900
DATA> ENDOFDATA
MTB > MEAN C1
    MEAN = 153775
MTB > STDEV C1
    ST.DEV. = 41611
MTB > MEDIAN C1
    MEDIAN = 144900
MTB > SUM C1
    SUM = 23066258
MTB > SSQ C1
    SSQ = 3.805003E+12
MTB > MAXIMUM C1
    MAXIMUM = 302000
MTB > MINIMUM C1
    MINIMUM = 72000
```

The user makes all entries preceded by MTB> or DATA>. The other results are displayed by the Minitab program. For more details on using Minitab, see the *Minitab Student Laboratory Manual and Workbook*, a supplement to this textbook.

# 2–6 Exercises A

In Exercises 1–12, express all $z$ scores with two decimal places.

1. For a set of data, $\bar{x} = 60$ and $s = 10$. Find the $z$ score corresponding to (a) $x = 80$; (b) $x = 40$; (c) $x = 65$.

2. For a set of data, $\bar{x} = 100$ and $s = 15$. Find the $z$ score corresponding to (a) $x = 115$; (b) $x = 145$; (c) $x = 85$.

3. An investigation of the number of hours college freshmen spend studying each week found that the mean is 7.06 h and the standard deviation is 5.32 h (based on data from *The American Freshman*). Find the $z$ score corresponding to a freshman who studies 10.00 hours weekly.

4. A study of the time high school students spend working at a job each week found that the mean is 10.7 h and the standard deviation is 11.2 h (based on data from the National Federation of State High School Associations). Find the $z$ score corresponding to a high school student who works 15.0 h each week.

5. Using the Burnout Measure, sample subjects are found to have measures with a mean of 2.97 and a standard deviation of 0.60 (based on "Moderating Effect of Social Support on the Stress-Burnout Relationship" by Etzion, *Journal of Applied Psychology*, Vol. 69, No. 4). Find the $z$ score corresponding to a subject with a score of 2.00.

6. The heights of six-year-old girls have a mean of 117.8 cm and a standard deviation of 5.52 cm (based on data from the National Health Survey, USDHEW publication 73-1605). Find the $z$ score corresponding to a six-year-old girl who is 106.8 cm tall.

7. For a certain population, scores on the Thematic Apperception Test have a mean of 22.83 and a standard deviation of 8.55 (based on "Relationships Between Achievement-Related Motives, Extrinsic Conditions, and Task Performance" by Schroth, *Journal of Social Psychology*, Vol. 127, No. 1). Find the $z$ score corresponding to a member of this population who has a score of 10.00.

8. For men aged between 18 and 24 years, serum cholesterol levels (in mg/100 ml) have a mean of 178.1 and a standard deviation of 40.7 (based on data from the National Health Survey, USDHEW publication 78-1652). Find the $z$ score corresponding to a male, aged 18–24 years, who has a serum cholesterol level of 249.3 mg/100 ml.

9. Which of the following two scores has the better relative position?
   a. A score of 53 on a test for which $\bar{x} = 50$ and $s = 10$
   b. A score of 53 on a test for which $\bar{x} = 50$ and $s = 5$

10. Two similar groups of students took equivalent language facility tests. Which of the following results indicates the higher relative level of language facility?
   a. A score of 60 on a test for which $\bar{x} = 70$ and $s = 10$
   b. A score of 480 on a test for which $\bar{x} = 500$ and $s = 50$

11. Three prospective employees take equivalent tests of communicative ability. Which of the following scores corresponds to the highest relative position?
   a. A score of 60 on a test for which $\bar{x} = 50$ and $s = 5$
   b. A score of 230 on a test for which $\bar{x} = 200$ and $s = 10$
   c. A score of 540 on a test for which $\bar{x} = 500$ and $s = 15$

12. Three students take equivalent tests of neuroticism with the given results. Which is the highest relative score?
   a. A score of 3.6 on a test for which $\bar{x} = 4.2$ and $s = 1.2$
   b. A score of 72 on a test for which $\bar{x} = 84$ and $s = 10$
   c. A score of 255 on a test for which $\bar{x} = 300$ and $s = 30$

In Exercises 13–16, use the 150 ranked home selling prices listed in Table 2–17. Find the percentile corresponding to the given selling price.

13. $110,000    14. $125,000    15. $175,000    16. $220,000

In Exercises 17–24, use the 150 ranked home selling prices listed in Table 2–17. Find the indicated percentile, quartile, or decile.

17. $P_{15}$      18. $P_5$       19. $P_{80}$      20. $P_{90}$
21. $Q_1$        22. $Q_2$       23. $Q_3$        24. $D_6$

In Exercises 25–28, use the following data to find the percentile corresponding to the given value. These numbers are the actual weights (in pounds) of 50 consecutive loads handled by a moving company in Dutchess County, New York.

| | | | | |
|---|---|---|---|---|
| 8,090 | 17,810 | 3,670 | 10,100 | 17,330 |
| 8,800 | 7,770 | 12,430 | 13,260 | 10,220 |
| 13,490 | 13,520 | 4,510 | 14,760 | 4,480 |
| 7,540 | 10,630 | 10,330 | 10,510 | 12,700 |
| 7,200 | 12,010 | 11,450 | 9,140 | 7,280 |
| 9,110 | 12,350 | 14,800 | 26,580 | 15,970 |
| 11,860 | 8,450 | 10,780 | 5,030 | 11,430 |
| 11,600 | 7,470 | 14,310 | 13,410 | 7,450 |
| 3,250 | 6,400 | 8,160 | 9,310 | 9,900 |
| 6,170 | 16,200 | 8,770 | 6,820 | 6,390 |

25. 5,030       26. 10,220       27. 12,430       28. 14,760

In Exercises 29–36, use the same load weights given above and find the indicated percentile, quartile, or decile.

29. $P_{15}$    30. $P_{20}$    31. $P_{80}$    32. $P_{66}$
33. $Q_1$    34. $Q_3$    35. $D_3$    36. $D_9$

## 2–6 Exercises B

37. Use the ranked home selling prices listed in Table 2–17.
    a. Find the interquartile range.
    b. Find the midquartile.
    c. Find the 10–90 percentile range.
    d. Does $P_{50} = Q_2$? Does $P_{50}$ *always* equal $Q_2$?
    e. Does $Q_2 = (Q_1 + Q_3)/2$?
38. When finding percentiles using Figure 2–10, if the locator $L$ is not a whole number, we round it up to the next larger whole number. An alternative to this procedure is to *interpolate* so that a locator of 23.75 would lead to a value that is 0.75 (or 3/4) of the way between the 23rd and 24th scores. Use this method of interpolation to find $Q_1$, $Q_3$, and $P_{33}$ for these scores:

    16   49   53   58   60   63   63   65   72   80   84   89   92   98

39. Construct a collection of data consisting of 50 scores for which $Q_1 = 20$, $Q_2 = 30$, $Q_3 = 70$.
40. Using the scores 2, 5, 8, 9, and 16, first find $\bar{x}$ and $s$, then replace each score by its corresponding $z$ score. Now find the mean and standard deviation of the five $z$ scores. Will these new values of the mean and standard deviation result from *every* set of $z$ scores?

## 2–7 Exploratory Data Analysis

The techniques discussed in the previous sections can be used to summarize data and find important measures, such as the mean and standard deviation. In summarizing and describing data, we should be careful to avoid overlooking important information that might be lost in our summaries. About 25 years ago, statisticians began to use an approach now referred to as **exploratory data analysis** or **EDA.** Many of the techniques used in this approach are introduced in John Tukey's book *Exploratory Data Analysis* (Addison-Wesley, 1977). EDA is more than simply a collection of new statistical techniques—

it is a fundamentally different approach. With EDA we *explore* data rather than use a statistical analysis to *confirm* some claim or assumption made about the data. For data obtained from a carefully planned experiment with a very specific objective (such as comparing the mpg ratings of two different car models), traditional methods of statistics will probably be sufficient. But if you have a collection of data and a broad goal (such as trying to find out what the data reveal), then you might want to begin with exploratory data analysis. With EDA, the emphasis is on original explorations with the goals of simplifying the way the data are described and gaining deeper insight into the nature of the data. Thus it is easier to identify relevant questions that might be addressed. The table below compares EDA and traditional statistics in three major areas.

| Exploratory Data Analysis | Traditional Statistics |
|---|---|
| Used to *explore* data at a preliminary level. | Used to *confirm* final conclusions about data. |
| Few or no assumptions are made about the data. | Typically requires some very important assumptions about the data. |
| Tends to involve relatively simple calculations and graphs | Calculations are often complex and graphs are often unnecessary. |

In this section we consider two helpful devices used for exploratory data analysis: the stem-and-leaf plot and the boxplot. They are typical of EDA techniques since they involve relatively simple calculations and graphs.

## Stem-and-Leaf Plots

Section 2–3 discussed histograms, frequency polygons, relative frequency polygons, and ogives, which are extremely useful in graphically displaying the distribution of data. By using these graphic devices, we are able to learn something about the data that is not apparent while the data remain in a list of values. This additional insight is clearly an advantage. However, in constructing histograms, frequency polygons, or ogives we also suffer the disadvantage of distorted data. When we transform raw data into a histogram, for example, we lose some information in the process. Generally we cannot reconstruct the original data set from the histogram, which shows that some distortion has occurred. We will now introduce another device that enables us to see the distribution of data without losing information in the process.

In a **stem-and-leaf plot** we sort data according to a pattern that reveals the underlying distribution. The pattern involves separating a number into two parts, usually the first digit and the other digits. The *stem* consists of the leftmost digits and *leaves* consist of the rightmost digits. The method is illustrated in the following example.

### Example

A librarian records the number of daily microfilm uses and compiles the sample data that follow. Construct a stem-and-leaf plot for this data.

| 10 | 11 | 15 | 23 | 27 | 28 | 38 | 38 | 39 | 39 |
| 40 | 41 | 44 | 45 | 46 | 46 | 52 | 57 | 58 | 65 |

### Solution

We note that the numbers have first digits of 1, 2, 3, 4, 5, or 6 and we let those values become the stem. We then construct a vertical line and list the "leaves" as shown. The first row represents the numbers 10, 11, 15. In the second row we have 23, 27, 28, and so on.

| Stem | Leaves |
|------|--------|
| 1 | 015 |
| 2 | 378 |
| 3 | 8899 |
| 4 | 014566 |
| 5 | 278 |
| 6 | 5 |

By turning the page on its side we can see a distribution of these data, which, in this case, roughly approximates a bell shape. We have also retained all the information in the original list. We could reconstruct the original list of values from the stem-and-leaf plot. In the next example we illustrate the construction of a stem-and-leaf plot for data with three significant digits.

### Example

An aeronautical research team investigating the stall speed of an ultra-light aircraft obtained the following sample values (in knots). Construct a stem-and-leaf plot and the graph suggested by that plot.

| 21.7 | 24.0 | 22.4 | 22.4 | 24.3 | 22.3 | 22.6 | 25.2 |
| 24.1 | 21.8 | 23.2 | 23.9 | 23.5 | 23.2 | 23.9 | 23.8 |

*continued*

Note that unlike the previous example, these data are not in order, so two sweeps will be necessary. For the first sweep, we record the data reading across one row at a time. For the second sweep, arrange each row of leaves in order from low to high.

| Stem | Leaves | | Stem | Leaves |
|------|--------|--|------|--------|
| 21. | 78 | | 21. | 78 |
| 22. | 4436 | | 22. | 3446 |
| 23. | 295298 | | 23. | 225899 |
| 24. | 031 | | 24. | 013 |
| 25. | 2 | | 25. | 2 |

Stall speed (knots)

Here we let the stem consist of the two left digits while the leaf is the digit farthest to the right. Again, we can see the shape of the distribution by turning the page on its side and observing the columns of digits above the line.

For some data sets, simplified stem-and-leaf plots can be constructed by first rounding the values to two or three significant digits. If necessary, stem-and-leaf plots can be condensed by combining adjacent rows. They can also be stretched out by subdividing rows into those with the digits 0 through 4 and those with digits 5 through 9. Note that the following two stem-and-leaf plots represent the same set of data, but the plot on the right has been stretched out to include more rows. Also note that such changes may affect the apparent shape of the distribution.

| Stem | Leaves | | | Stem | Leaves | |
|------|--------|--|--|------|--------|--|
| | | | | 51 | | (last digits of 0 through 4) |
| | | | | 51 | 6899 | (last digits of 5 through 9) |
| 51 | 6899 | | | 52 | 034 | (last digits of 0 through 4) |
| 52 | 0347 | expand | | 52 | 7 | (last digits of 5 through 9) |
| 53 | 3788 | → | | 53 | 3 | (last digits of 0 through 4) |
| | | | | 53 | 788 | (last digits of 5 through 9) |

In the preceding example we expanded the number of rows. We can also condense a stem-and-leaf plot by combining adjacent rows as follows.

| Stem | Leaves | | Stem | Leaves |
|------|--------|--|------|--------|
| 50 | 01 | | 50–51 | 01*4 |
| 51 | 4 | | 52–53 | 56*368 |
| 52 | 56 | | 54–55 | 2457*3499 |
| 53 | 368 | | 56–57 | 0127*358 |
| 54 | 2457 | contract | 58–59 | 1269*17 |
| 55 | 3499 | → | | |
| 56 | 0127 | | | |
| 57 | 358 | | (581)  (591) |
| 58 | 1269 | | | |
| 59 | 17 | | | |

In the condensed plot, we separated digits in the "leaves" associated with the numbers in each stem by an asterisk. Every row in the condensed plot must include exactly one asterisk so that the shape of the plot is not distorted.

Another useful feature of stem-and-leaf plots is that their construction provides a fast and easy procedure for ranking data (putting data in order). Data must be ranked for a variety of statistical procedures, such as the Wilcoxon rank-sum test (Chapter 12) and finding the median of a set of data.

# Boxplots

When exploring a collection of numerical data, we want to be sure that we investigate (1) the value of a representative score (such as an average), (2) a measure of scattering or variation (such as the standard deviation), and (3) the nature of the distribution. The distribution of data should definitely be considered, since it may strongly affect the methods we use and the conclusions we draw. In the spirit of exploratory data analysis, we should not simply examine a histogram and think that we understand the nature of the distribution. We should *explore*. As an example, in Figure 2–11(a) we show a computer printout of a histogram in which the triglyceride levels of 20 subjects were entered. The last score was incorrectly entered with an extra zero so that 1600 was used instead of 160. In Figure 2–11(b) we show the computer printout for the corrected data set. From Figure 2–11(b) we conclude that the data are normally distributed, but that was not at all apparent from Figure 2–11(a), even though only one score was incorrect. In this case, the outlier caused a severe distortion of the histogram. In other cases, outliers may be correct values but may continue to disguise the true nature of the distribution through histograms such as the one shown in Figure 2–11(a).

In addition to a histogram, frequency polygon, relative frequency histogram, or ogive, we might also construct a **boxplot** (also referred to as a box-and-whisker diagram). The boxplot reveals more information about how the data are spread out. The construction of a boxplot requires that we obtain the minimum score, the maximum score, the median, and two other values called *hinges*.

## Definitions

The **lower hinge** is the median of the lower half of all scores (from the minimum score up to the original median).

The **upper hinge** is the median of the upper half of all scores (from the original median up to the maximum score).

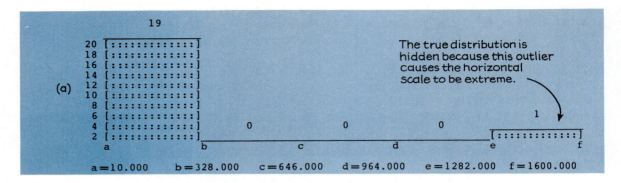

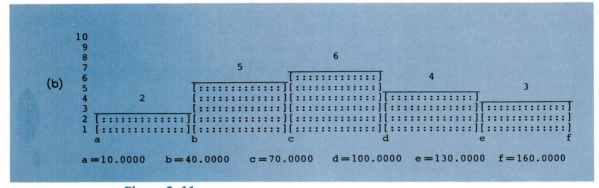

**Figure 2–11**

### Definition

The minimum score, the maximum score, the median, and the two hinges constitute a **5-number summary** of a set of data.

Hinges are very similar to quartiles. In fact, several texts use quartiles instead of hinges for the construction of boxplots. The differences between hinges and quartiles are usually small, especially for larger data sets. Our definition of *hinges* is consistent with John Tukey's definitions. Hinges can be easily found by following these steps:

1. Arrange the data in increasing order.
2. Find the median. (With an odd number of scores, it's the middle score; with an even number of scores, the median is a new score equal to the mean of the two middle scores.)
3. List the lower half of the data from the minimum score up to and including the median found in step 2. The left hinge is the median of these scores.

4.  List the upper half of the data starting with the median and including the scores up to and including the maximum. The right hinge is the median of these scores.

5.  Now list the minimum, the left hinge (from step 3), the median (from step 2), the right hinge (from step 4), and the maximum.

This procedure will vary somewhat with different textbooks. The 20 scores depicted in Figure 2–11(a) are arranged in increasing order and listed below.

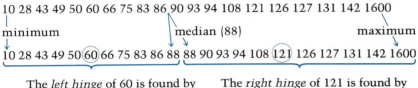

Note that the hinges are different from the quartiles. For this data set, $Q_1 = 55$ and $Q_3 = 123.5$, while the hinges are 60 and 121.

We will now use the 5-number summary (10, 60, 88, 121, 1600) to construct a boxplot. We begin with a horizontal scale as in Figure 2–12(a). We "box" in the hinges as shown and we extend lines to connect the minimum score to a hinge and the maximum score to a hinge. Figure 2–12(b) shows the boxplot for the data set after the incorrect score of 1600 has been corrected to 160.

Note that by the procedures used here, approximately one-fourth of the values should fall between the low score and the left hinge, the two middle quarters are in the boxes, and approximately one-fourth of the values are between the right hinge and the maximum. The diagram therefore shows how the data are spread out. The uneven spread shown in Figure 2–12(a) is in strong contrast to the even spread shown in Figure 2–12(b). Suspecting that triglyceride levels are normally distributed, we would expect to see a boxplot like the one in Figure 2–12(b), whereas the diagram in Figure 2–12(a) would raise suspicion and lead to further investigation. Figure 2–13 shows

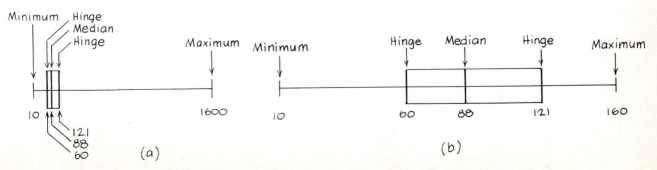

**Figure 2–12**

some common distributions along with the corresponding boxplots. Minitab can be used to create boxplots. (See the sample Minitab display.)

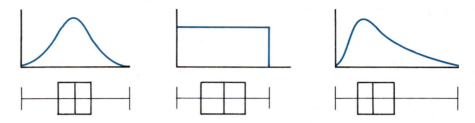

**Figure 2–13**

MINITAB DISPLAY

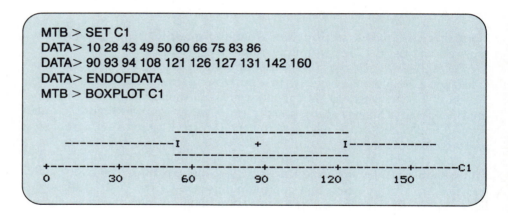

# 2–7   Exercises A

In Exercises 1–4, list the original numbers in the data set represented by the given stem-and-leaf plots.

1.

| Stem | Leaves |
|------|--------|
| 2 | 00358 |

2.

| Stem | Leaves |
|------|--------|
| 1 | 001112278 |
| 2 | 3444569 |
| 3 | 013358 |

3.

| Stem | Leaves |
|------|--------|
| 40 | 6678 |
| 41 | 09999 |
| 42 | 13466 |
| 43 | 088 |

4.

| Stem | Leaves |
|------|--------|
| 68 | 45 45 47 86 |
| 69 | 33 38 89 |
| 70 | 52 59 93 |
| 71 | 27 |

**In Exercises 5–8**, construct the stem-and-leaf plots for the given data sets.

5. High temperatures (in degrees Fahrenheit) for the 31 days in July (recorded at Dutchess Community College):

```
80  68  84  86  85  77  64  81  93  94
97  93  89  82  76  75  83  90  83  84
92  94  90  92  91  84  81  84  79  80
80
```

6. Pulse rates (number of beats per minute) of 20 male statistics students:

```
82  74  77  62  78  58  85  74  66  71
58  80  65  60  54  75  71  74  73  82
```

7. Lot sizes (in acres) of 20 homes sold in Dutchess County:

```
0.75  0.70  0.65  0.75  0.50  0.79  0.50
0.50  0.68  0.65  0.75  0.50  0.50  0.65
0.58  0.80  0.91  0.78  0.68  0.92
```

8. Takeoff distances (in feet) required for a light aircraft:

```
717  716  736  772  740  741  735  735  710  753  757  747
756  715  718  720  726  721  760  769  771  738  721
```

**In Exercises 9–16**, use the given data to construct boxplots. Identify the values of the minimum, maximum, median, and hinges.

9. Ages of selected full-time undergraduate students (in years):

```
17.2,  17.9,  18.6,  18.8,  19.3,  19.3,  20.0,  20.1,  23.4,  26.3
```

10. Monthly rental costs of apartments in one region (in dollars):

```
540,  545,  555,  560,  560,  570,  575,  590,  650,  730
```

11. Blood alcohol contents of drivers given breathalyzer tests:

```
0.02,  0.04,  0.08,  0.08,  0.09,  0.10,  0.10,  0.12,  0.13,  0.19
```

12. Time intervals (in months) between adjacent births for selected families:

```
12.9  13.4  18.3  24.7  31.2  31.3  32.0  32.1  33.4
33.8  34.1  36.2  41.7  41.9  52.5
```

13. Blood pressure levels (in mm of mercury) for patients who have taken 25 mg of the drug Captopril.

```
198  180  142  157  181  183  162  130  170  164
170  173  173  175  195  190  193  157  159  138
```

14. Number of words typed in a 5-min civil service test taken by 25 different applicants.

$$
\begin{array}{cccccccccc}
174 & 181 & 219 & 213 & 213 & 207 & 106 & 111 & 143 & 160 \\
166 & 350 & 183 & 198 & 193 & 190 & 190 & 185 & 220 & 221 \\
229 & 257 & 243 & 281 & 308 & & & & &
\end{array}
$$

15. Time (in hours of operation) between failures for prototypes of computer printers:

$$
\begin{array}{cccccccccc}
34 & 22 & 4 & 9 & 27 & 36 & 12 & 40 & 29 & 32 \\
35 & 25 & 7 & 9 & 26 & 36 & 45 & 43 & 41 & 2 \\
31 & 31 & 30 & 14 & 15 & 18 & 10 & 27 & 38 & 21
\end{array}
$$

16. Construct a boxplot for the data given in the stem-and-leaf diagram.

| Stem | Leaves |
|------|--------|
| 5 | 0 |
| 6 | 6 6 7 8 |
| 7 | 0 2 3 3 4 4 8 9 |
| 8 | 3 3 5 6 6 6 7 7 |
| 9 | 0 2 5 |

# 2–7 Exercises B

17. Make sketches of histograms that correspond to the given boxplots.

(a)

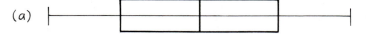

(b)

(c)

18. While the stem-and-leaf plot does provide a graph that retains the exact scores, it does not retain the *order* in which the scores occur. The **digidot** (for digit and dot) plot is a variation that overcomes this disadvantage. Complete the digidot plot in Figure 2–14 by entering the digit

and dot for each given score. (The first dots and digits corresponding to 2.8 and 1.7 have been entered.) Then connect the dots in sequence.

2.8   1.7   2.1   2.5   2.0   1.9   1.6   1.5   1.2   1.2

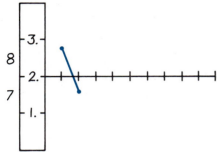

**Figure 2–14**

19. In "Ages of Oscar-Winning Best Actors and Actresses" (*Mathematics Teacher* magazine) by Richard Brown and Gretchen Davis, stem-and-leaf plots and boxplots are used to compare the ages of actors and actresses when they won Oscars. Here are the results for 30 recent and consecutive winners from each category:

```
Actors:     32  51  33  61  35  45  55  39  76  37
            42  40  32  60  38  56  48  48  40  43
            62  43  42  44  41  56  39  46  31  47
Actresses:  80  26  41  21  61  38  49  33  74  30
            33  41  31  35  41  42  37  26  34  34
            35  26  61  60  34  24  30  37  31  27
```

a. Construct a "back-to-back" stem-and-leaf plot for the above data. The first two scores from each group have been entered.

| Actors' Ages | Stem | Actresses' Ages |
|---:|:---:|:---|
|   | 2 | 6 |
| 2 | 3 |   |
|   | 4 |   |
| 1 | 5 |   |
|   | 6 |   |
|   | 7 |   |
|   | 8 | 0 |

b. Using the same scale, construct a boxplot for actors' ages and another boxplot for actresses' ages.

c. Using the results from parts *a* and *b*, compare the two different sets of data and try to explain any difference.

20. The boxplots discussed in this section are often called *skeletal* boxplots. When investigating outliers, a useful variation is to construct boxplots as follows:

   1. Calculate the hinge difference.
      $D = $ (upper hinge) $-$ (lower hinge)
   2. Draw the box with the median and hinges as usual, but when extending the lines that branch out from the box, go only as far as the scores that are within $1.5D$ of the box.
   3. **Mild outliers** are scores above the upper hinge by $1.5D$ to $3D$, or below the lower hinge by $1.5D$ to $3D$. Plot them as solid dots.
   4. **Extreme outliers** are scores above the upper hinge by more than $3D$ or below the lower hinge by more than $3D$. Plot them as small hollow circles.

Figure 2–15 is an example of the boxplot described here. Use this procedure to construct the boxplot for the given scores, and identify any mild outliers or extreme outliers.

3   15   17   18   21   21   22   25   27   30   38   49   68

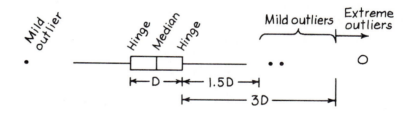

**Figure 2–15**

 *Vocabulary List*

Define and give an example of each term.

| | | |
|---|---|---|
| descriptive statistics | cumulative frequency | ogive |
| inferential statistics |    table | average |
| frequency table | relative frequency table | measure of central |
| lower class limits | pie chart |    tendency |
| upper class limits | histogram | mean |
| class boundaries | relative frequency | median |
| class marks |    histogram | mode |
| class width | frequency polygon | bimodal |

multimodal
midrange
weighted mean
measure of dispersion
range
standard deviation
mean deviation
variance
Chebyshev's theorem

empirical rule
standard score
z score
quartiles
deciles
percentiles
interquartile range
semi-interquartile
  range

midquartile
10–90 percentile range
exploratory data
  analysis (EDA)
stem-and-leaf plot
boxplot
lower hinge
upper hinge
5-number summary

 # Review

Chapter 2 dealt mainly with the methods and techniques of descriptive statistics. This chapter focused on developing the ability to organize, summarize, and illustrate data and to extract from data some meaningful measurements. Section 2–2 considered the **frequency table** as an excellent device for summarizing data, while Section 2–3 dealt with graphic illustrations, including **pie charts, histograms, relative frequency histograms, frequency polygons,** and **ogives.** These visual illustrations help us to determine the position and distribution of a set of scores. Section 2–4 defined the common **averages,** or measures of central tendency. The **mean, median, mode,** and **midrange** represent different ways of characterizing the central value of a collection of data. The **weighted mean** is used to find the average of a set of scores that may vary in importance. Section 2–5 presented the usual **measures of dispersion,** including the **range, standard deviation,** and **variance;** these descriptive statistics are designed to measure the variability among a set of scores. The **standard score,** or **z score,** was introduced in Section 2–6 as a way of measuring the number of standard deviations by which a given score differs from the mean. That section also included the common measures of position: **quartiles, percentiles,** and **deciles.** Finally, Section 2–7 presented **stem-and-leaf plots** and **boxplots** and their use in **exploratory data analysis.**

By now we should be able to organize, present, and describe collections of data composed of single scores. We should be able to compute the key descriptive statistics that will be used in later applications.

Using the concepts developed in this chapter, we now have a better understanding of the 150 home selling prices given in Table 2–1. The histogram allowed us to see the shape of the distribution of those values. We know that the mean is $153,775, the median is $144,900, and the standard deviation is $41,611. The data are summarized in a frequency table and depicted in a histogram, frequency polygon, and ogive. Subsequent chapters will consider other ways of using sample statistics.

---

## Important Formulas

$$\bar{x} = \frac{\Sigma x}{n}$$
Mean

$$\bar{x} = \frac{\Sigma(f \cdot x)}{\Sigma f}$$
Computing the mean when the data are in a frequency table

$$s = \sqrt{\frac{\Sigma(x - \bar{x})^2}{n - 1}}$$
Standard deviation

$$s^2 = \frac{\Sigma(x - \bar{x})^2}{n - 1}$$
Variance

$$s = \sqrt{\frac{n(\Sigma x^2) - (\Sigma x)^2}{n(n - 1)}}$$
Shortcut formula for standard deviation

$$s = \sqrt{\frac{n[\Sigma(f \cdot x^2)] - [\Sigma(f \cdot x)]^2}{n(n - 1)}}$$
Computing the standard deviation when the data are in a frequency table

$$z = \frac{x - \bar{x}}{s} \text{ or } \frac{x - \mu}{\sigma}$$
Standard score or z score

---

## ❓ Review Exercises

1. The values given below are snow depths (in centimeters) measured as part of a study of satellite observations and water resources (based on data in *Space Mathematics* published by NASA). Find the (a) mean; (b) median; (c) mode; (d) midrange; (e) range; (f) variance; (g) standard deviation.

    19,  18,  12,  25,  22,  8,  8,  16

2. A psychologist gave a subject two different tests designed to measure spatial perception. The subject obtained scores of 66 on the first test and 223 on the other test. The first test is known to have a mean of 75 and a standard deviation of 15, while the second test has a mean of 250 and a standard deviation of 25. Which result is better? Explain.

3. The given scores represent the number of cars rejected in one day at an automobile assembly plant. The 50 scores correspond to 50 different randomly selected days. Construct a frequency table with 10 classes.

    | 29 | 58 | 80 | 35 | 30 | 23 | 88 | 49 | 35 | 97 |
    |----|----|----|----|----|----|----|----|----|----|
    | 12 | 73 | 54 | 91 | 45 | 28 | 61 | 61 | 45 | 81 |
    | 83 | 23 | 71 | 63 | 47 | 87 | 36 | 8  | 94 | 26 |
    | 95 | 63 | 86 | 42 | 22 | 44 | 8  | 27 | 20 | 33 |
    | 28 | 91 | 87 | 15 | 67 | 10 | 45 | 67 | 26 | 19 |

4. Construct a relative frequency table (with 10 classes) for the data in Exercise 3.

5. Construct a histogram that corresponds to the frequency table from Exercise 3.

6. For the data in Exercise 3, find (a) $Q_1$, (b) $P_{45}$, and (c) the percentile corresponding to the score of 30.

7. Use the frequency table from Exercise 3 to find the mean and standard deviation for the number of rejects.

8. Use the data from Exercise 3 to construct a stem-and-leaf plot.

9. Use the data from Exercise 3 to construct a boxplot.

10. The values given below are the living areas (in square feet) of 12 homes recently sold in Dutchess County, New York. Find the (a) mean; (b) median; (c) mode; (d) midrange; (e) range; (f) variance; (g) standard deviation.

$$
\begin{array}{cccccc}
3060 & 1600 & 2000 & 1300 & 2000 & 1956 \\
2400 & 1200 & 1632 & 1800 & 1248 & 2025
\end{array}
$$

11. Construct a stem-and-leaf plot for the data in Exercise 10.

12. For the data given in Exercise 10, find the $z$ score corresponding to (a) 1200; (b) 2400.

13. Use the data given in Exercise 10 to construct a boxplot.

14. A supplier constructs a frequency table for the number of car stereo units sold daily. Use that table to find the mean and standard deviation.

| Number Sold | Frequency |
|---|---|
| 0–3 | 5 |
| 4–7 | 9 |
| 8–11 | 8 |
| 12–15 | 6 |
| 16–19 | 3 |

15. Using the frequency table given in Exercise 14, construct the corresponding relative frequency histogram.

16. Using the frequency table given in Exercise 14, construct the corresponding frequency polygon.

17. Using the frequency table given in Exercise 14, construct the corresponding ogive.

18. a. A set of data has a mean of 45.6. What is the mean if 5.0 is added to each score?

    b. A set of data has a standard deviation of 3.0. What is the standard deviation if 5.0 is added to each score?

    c. You just completed a calculation for the variance of a set of scores, and you got an answer of $-21.3$. What do you conclude?

*continued*

d.  True or false: If set A has a range of 50 while set B has a range of 100, then the standard deviation for data set A must be less than the standard deviation for data set B.

e.  True or false: In proceeding from left to right, the graph of an ogive can never follow a downward path.

19.  The following values are the diameters (in nanometers) of virus samples. Find the (a) mean; (b) median; (c) mode; (d) midrange; (e) range; (f) variance; (g) standard deviation.

    175,  183,  168,  191,  181,  183,  170,  174,  184,  181,  182

20.  For the data given in Exercise 19, find the z score corresponding to (a) 175; (b) 182.

21.  Construct a stem-and-leaf plot for the data in Exercise 19.

22.  Use the data given in Exercise 19 to construct a boxplot.

23.  The values given here are the lot sizes (in acres) of 12 homes recently sold in Dutchess County, New York. Find the (a) mean; (b) median; (c) mode; (d) midrange; (e) range; (f) variance; (g) standard deviation.

    0.75  0.26  0.70  0.65  0.75  0.50
    0.40  0.33  3.00  0.50  0.25  1.10

24.  Use the data given in Exercise 23 to construct a boxplot.

 ## *Computer Project*

The NCAA was considering ways to speed up the end of college basketball games. The following values are the lapsed times (in seconds) it took to play the last two minutes of regulation time in the 60 games of the first four rounds of a recent NCAA basketball tournament (based on data reported in *USA Today*). Use STATDISK, Minitab, or any other statistics software package to obtain descriptive statistics (mean, variance, standard deviation, and so on) and a histogram for the given times.

    756   587   929    871   378   503   564   1128   693   748
    448   670   1023   335   540   853   852    495   666   474
    443   325   514    404   820   915   793    778   627   483
    861   337   292   1070   625   457   676    494   420   862
    991   615   609    723   794   447   704    396   235   552
    626   688   506    700   240   363   860    670   396   345

 ## *Applied Projects*

1.  a.  Refer to the data given in the above computer project. Construct a frequency table, histogram, stem-and-leaf plot, and boxplot.

b.  Find the values of the mean, median, mode, midrange, range, standard deviation, variance, minimum, maximum, and the three quartiles.

c.  Using the results from parts *a* and *b* and any other observations, do these times seem to indicate that something should be done to speed up the last two minutes of the game? Support your answer with specific references to the data and any relevant tables, graphs, or statistics.

d.  Identify at least two unfavorable consequences of allowing the last two minutes of regulation time to be delayed too long.

2.  Car designers must consider the "sitting heights" of men and women. Refer to Data Set II in Appendix B and use the methods presented in this chapter to compare the sitting heights of men to those of women. Include references to means, standard deviations, and distributions.

3.  Through observation or experimentation, compile a list of sample data that are at the interval or ratio levels of measurement. Obtain at least 40 values, and try to select data from an interesting or meaningful population.

a.  Describe the nature of the data. What do the values represent?

b.  What method was used to collect the values?

c.  What are some possible reasons why the data might not be representative of the population? That is, what are some possible sources of bias?

d.  Find the value of each of the following: sample size, minimum, maximum, mean, median, midrange, range, standard deviation, variance, and the quartiles $Q_1$ and $Q_3$.

e.  Construct a frequency table, histogram, stem-and-leaf plot, and boxplot.

f.  Write a brief report summarizing any important characteristics of the data.

 # *Writing Projects*

1.  Analyze (with or without a computer) the data in the computer project given above. Write a report summarizing important characteristics (mean, standard deviation, distribution, etc.) and comment on any aspects of the data that are relevant to the issue of speeding up the end of college basketball games. Conclude with a recommendation.

2.  Write a report summarizing one of the videotape programs below.

 # *Videotapes*

Programs 2 and 3 from the series *Against All Odds: Inside Statistics* are recommended as supplements to this chapter.

# Chapter Three

## In This Chapter

**3–1** **Overview**

We identify chapter **objectives** and compare probability and statistics. We see the importance of probability as a discipline in its own right, as well as its importance in statistics.

**3–2** **Fundamentals**

We introduce both the **empirical** and **classical** definitions of probability and apply them to simple events. The **Law of Large Numbers** is also described.

**3–3** **Addition Rule**

We describe the **addition rule** and events that are **mutually exclusive.** In applying the addition rule, we compensate for double-counting in events that are not mutually exclusive.

**3–4** **Multiplication Rule**

We describe the **multiplication rule** and **independent** events.

**3–5** **Complements and Odds**

We describe the **complement** of an event and the process for determining its probability. We also describe the relationship between probability and odds.

**3–6** **Counting**

We describe the **fundamental counting principle, factorial rule, permutations rule,** and **combinations rule** in determining the total number of outcomes for a variety of different circumstances.

# 3 Probability

## Chapter Problem

A quality control manager knows from extensive past experience that her company manufactures VCRs with a 5% rate of defects. An employee team has developed a new technique that supposedly reduces defects. When 20 VCRs are manufactured with the new technique, it is found that none of them are defective. One member of the employee team argues that this is evidence that the new technique is better. "After all, a 5% rate of defects is equivalent to 1 in 20 and there weren't any defects among the 20 VCRs produced." Is it likely that you will get no defects among 20 VCRs if the defect rate is 5%? Assuming that the new manufacturing technique has the same 5% rate of defects, how likely is it that we get no defects among the 20 VCRs? With that result, we can then use the available data to address the employee's claim that the new technique is better.

# 3–1 | Overview

In his book *Innumeracy* (a great little book), John Allen Paulos states that there is better than a 99% chance that if you take a deep breath, you will inhale a molecule that was exhaled in dying Caesar's last breath. In the same morbid spirit, consider this: If Socrates' fatal cup of hemlock was mostly water, then the next glass of water you drink will likely contain one of those exact same molecules.

In a class of 25 students, each is asked to identify the month and day of his or her birth. What are the chances that at least two students will share the same birthday? Our intuition is misleading; it happens that at least two students will have the same birthday in more than half of all classes with 25 students.

In this country of 90 million voters, a pollster needs to survey only 1000 (or 0.001%) in order to get a good estimate of the number of voters who favor a particular candidate.

The preceding conclusions are based on simple principles of probability, which play a critical role in the theory of statistics. All of us now form simple probability conclusions in our daily lives. Sometimes these determinations are based on fact, while others are subjective. In addition to its importance in the study of statistics, probability theory is playing an increasingly important role in a society that must attempt to measure uncertainties. Before firing up a nuclear power plant, we should have some knowledge about the probability of a meltdown. Before arming a nuclear warhead, we should have some knowledge about the probability of an accidental detonation. Before raising the speed limit on our nation's highways, we should have some knowledge of the probability of increased fatalities.

In Chapter 1 we stated that inferential statistics involves the use of sample evidence in formulating inferences or conclusions about a population. These inferential decisions are based on probabilities or likelihoods of events. As an example, suppose that a statistician plans to study the hiring practices of a large company. She finds that of the last 100 employees hired, all are men. Perhaps the company hires men and women at the same rate and the run of 100 men is an extremely rare chance event. Perhaps the hiring practices favor males. What should we infer? The statistician, along with most reasonable people, would conclude that men are favored. This decision is based on the very low probability of getting 100 consecutive men by chance alone.

Subsequent chapters will develop methods of statistical inference that rely on this type of thinking. It is therefore important to acquire a basic understanding of probability theory. We want to cultivate some intuitive feeling for what probabilities are, and we want to develop some very basic skills in calculating the probabilities of certain events (see Figure 3–1).

a. Duncan County voters

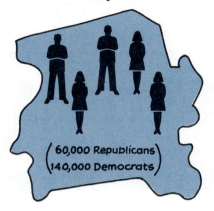

(60,000 Republicans)
(140,000 Democrats)

b. Jackson County voters

(? Republicans)
(? Democrats)

**Figure 3–1**

(a) **Probability.** With a *known population*, we can see that if one voter is randomly selected, there are 60,000 chances out of 200,000 of picking a Republican. This corresponds to a probability of 60,000/200,000 or 0.3. The population of all voters in the county is known, and we are concerned with the likelihood of obtaining a particular sample (a Republican). We are making a conclusion about a sample based on our knowledge of the population.

(b) **Statistics.** With an *unknown population*, we can obtain a fairly good idea of voter preferences by randomly selecting 500 voters and assuming that this sample is representative of the whole county population. After making the random selections, our sample is known and we can use it to make inferences about the population of all voters in the county. We are making a conclusion about the population based on our knowledge of the sample.

## COINCIDENCES?

■ John Adams and Thomas Jefferson (the second and third presidents) both died on July 4, 1826. President Lincoln was assassinated in Ford's Theater; President Kennedy was assassinated in a Lincoln car made by the Ford Motor Company. Lincoln and Kennedy were both succeeded by vice presidents named Johnson. Fourteen years *before* the sinking of the Titanic, a novel described the sinking of the Titan, a ship that hit an iceberg (see Martin Gardner's *The Wreck of the Titanic Foretold?*). Gardner states, "In most cases of startling coincidences, it is impossible to make even a rough estimate of their probability." (See also the margin essay "She Won the Lottery TWICE!") ■

This chapter begins by introducing the fundamental concept of mathematical probability, then proceeds to investigate the basic rules of probability: the addition rule, the multiplication rule, and the rule of complements. It also considers techniques of counting the number of ways an event can occur. The primary objective of this chapter is to develop a sound understanding of probability values that will be used in subsequent chapters. A secondary objective is to develop the ability to solve simple probability problems, which are valuable in their own right as they are used to make decisions and better understand our world.

# 3–2  Fundamentals

In considering probability problems, we deal with experiments and events.

> **Definition**
>
> An **experiment** is any process that allows researchers to obtain observations.

> **Definition**
>
> An **event** is any collection of results or outcomes of an experiment.

For example, the random selection of a letter from a computer data bank is an experiment and the result of a vowel is an event. In this case, the event of a vowel will occur if the letter is *a, e, i, o,* or *u*. We might think of the event of getting a vowel as being a combination of the simpler outcomes corresponding to the individual vowel letters. It is often necessary to consider events that cannot be decomposed into simpler outcomes.

> **Definition**
>
> A **simple event** is an outcome or an event that cannot be broken down any further.

When conducting the experiment of randomly selecting a letter from a data bank, the occurrence of an *e* is a simple event, but the occurrence of a vowel is not a simple event since it can be broken down into the five simple events of *a, e, i, o,* and *u*.

> **Definition**
>
> The **sample space** for an experiment consists of all possible simple events. That is, the sample space consists of all outcomes that cannot be broken down any further.

In the experiment of randomly selecting a letter from a data bank, the sample space consists of the 26 letters of the alphabet.

There is no universal agreement as to the definition of the probability of an event, but among the various theories and schools of thought, two basic approaches emerge most often. The approaches will be embodied in two rules for finding probabilities. The notation employed will relate $P$ to probability, while capital letters such as $A$, $B$, and $C$ will denote specific events. For example, $A$ might represent the event of winning a million-dollar state lottery; $P(A)$ denotes the probability of event $A$ occurring.

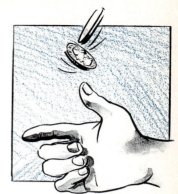

## Rule 1

**Empirical (relative frequency) approximation of probability.** Conduct (or observe) an experiment a large number of times and count the number of times that event $A$ actually occurs. Then $P(A)$ is *estimated* as follows.

$$P(A) = \frac{\text{number of times } A \text{ occurred}}{\text{number of times experiment was repeated}}$$

## Rule 2

**Classical approach to probability.** Assume that a given experiment has $n$ different simple events, each of which has an **equal chance** of occurring. If event $A$ can occur in $s$ of these $n$ ways, then

$$P(A) = \frac{s}{n} \qquad \text{(requires equally likely outcomes)}$$

Note that the classical approach requires equally likely outcomes. If the outcomes are not equally likely, we must use the empirical estimate. Figure 3–2 illustrates this important distinction.

(a) Empirical approach (Rule 1)

When trying to determine $P(2)$ on a "shaved" die, we must repeat the experiment of rolling it many times and then form the ratio of the number of times 2 occurred to the number of rolls. That ratio is our estimate of $P(2)$.

(b) Classical approach (Rule 2)

With a balanced and fair die, each of the six faces has an equal chance of occurring.

$$P(2) = \frac{\text{number of ways 2 can occur}}{\text{total number of simple events}} = \frac{1}{6}$$

**Figure 3–2**

### SENSITIVE SURVEYS

■ Survey respondents are sometimes reluctant to honestly answer questions on a sensitive topic, such as employee theft or sex. Stanley Warner (York University, Ontario) devised a scheme that leads to more accurate results in such cases. As an example, ask employees if they stole within the past year and also ask them to flip a coin. The employees are instructed to answer no if they didn't steal and the coin turns up heads. Otherwise, they should answer yes. The employees are more likely to be honest because the coin flip helps protect their privacy. Probability theory can then be used to analyze responses so that more accurate results can be obtained. ■

When determining probabilities by using Rule 1 (the empirical approach), we obtain an approximation instead of an exact value. As the total number of observations increases, the corresponding approximations tend to get closer to the actual probability. This idea is stated as a theorem commonly referred to as the Law of Large Numbers.

### Law of Large Numbers

As an experiment is repeated again and again, the empirical probability (from Rule 1) of success tends to approach the actual probability.

The Law of Large Numbers tells us that the empirical approximations from Rule 1 tend to get better with more observations. This law reflects a simple notion supported by common sense: In only a few trials, results can vary substantially, but with a very large number of trials, results tend to be more stable and consistent. For example, it would not be unusual to get 4 girls in 4 births, but it would be extremely unusual to get 400 girls in 400 births.

Figure 3–3 illustrates the Law of Large Numbers by showing computer-simulated results. Note that as the number of births increases, the proportion of girls approaches the 0.5 value.

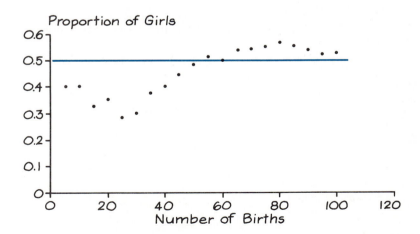

**Figure 3–3**

Many experiments involving equally likely outcomes are so complicated that the classical approach of Rule 2 isn't practical. Instead, we can get estimates of the desired probabilities by using the empirical approach of Rule 1. Computer simulations are often helpful in such cases.

The following examples are intended to illustrate the use of Rules 1 and 2. In some of these examples we use the term *random*.

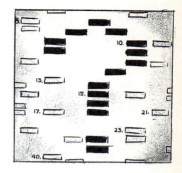

### Definition

In a **random selection** of an element, all elements available for selection have the same chance of being chosen.

This concept of random selection is extremely important in statistics. When making inferences based on samples, we must have a sampling process that is representative, impartial, and unbiased. Also, random selection is different from haphazard selection. Ask people to randomly select one of the 10 digits from 0 through 9, and they tend to favor some digits while ignoring others. Implementation of a random selection process often requires careful and thoughtful planning.

### Example

On an ACT or SAT test, a typical question has 5 possible answers. If an examinee makes a random guess on one such question, what is the probability that the response is wrong?

### Solution

There are 5 possible outcomes or answers, and there are 4 ways to answer incorrectly. Random guessing implies that the outcomes are equally likely, so we apply Rule 2 to get

$$P(\text{wrong answer}) = \frac{4}{5} \quad \text{or} \quad 0.8$$

In the next example, note that it is necessary to first determine the total number of cases before finding the desired probability.

## GUESS ON SATs?

■ Students preparing for multiple-choice test questions are often told not to guess, but that's not necessarily good advice. Standardized tests with multiple-choice questions typically *compensate* for guessing, but they don't *penalize* it. For questions with five answer choices, one-fourth of a point is usually subtracted for each incorrect response. Principles of probability show that in the long run, pure random guessing will neither raise nor lower the exam score. Definitely guess if you can eliminate at least one choice or if you have a sense for the right answer, but avoid tricky questions with attractive wrong answers. Also, don't waste too much time on such questions. ■

### Example

In a study of California drivers, 561 wore seat belts and 289 did not (based on data from the National Highway Traffic Safety Administration). If one of these drivers is randomly selected, what is the probability that he or she wore a seat belt?

### Solution

The total number of subjects in this study is 561 + 289 = 850. With random selection, the 850 subjects are equally likely and Rule 2 applies.

$$P(\text{wearing seat belt}) = \frac{561}{850} = 0.660 \quad \begin{matrix}\text{number wearing seat belts}\\ \text{total number of drivers}\end{matrix}$$

### Example

Find the probability that a couple with 3 children will have exactly 2 boys. (Assume that boys and girls are equally likely and that the sex of any child is independent of any brothers or sisters.)

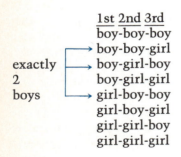

1st 2nd 3rd
boy-boy-boy
boy-boy-girl
boy-girl-boy
boy-girl-girl
girl-boy-boy
girl-boy-girl
girl-girl-boy
girl-girl-girl

exactly 2 boys

### Solution

If the couple has exactly 2 boys in 3 births, there must be 1 girl. The possible outcomes are listed and each is assumed to be equally likely. Of the 8 different possible outcomes, 3 correspond to exactly 2 boys, so that

$$P(2 \text{ boys in 3 births}) = \frac{3}{8} = 0.375$$

### Example

Find the probability of a 20-year-old male living to be 30 years of age.

### Solution

Here the two outcomes of living and dying are not equally likely, so the relative frequency approximation must be used. This requires that we observe a large number of 20-year-old males and then count those who live to be 30. Suppose that we survey 10,000 20-year-old males and find

*continued*

### Solution *continued*

that 9840 of them lived to be 30 (these are realistic figures based on U.S. Department of Health and Human Services data). Then the empirical approximation becomes

$$P(\text{20-year-old male living to 30}) = \frac{9,840}{10,000} = 0.984.$$

This is the basic approach used by insurance companies in the development of mortality tables.

### Example

If a year is selected at random, find the probability that Thanksgiving Day will be on a (a) Wednesday; (b) Thursday.

### Solution

a. Thanksgiving Day always falls on the fourth Thursday in November. It is therefore impossible for Thanksgiving to be on a Wednesday. When an event is impossible, we say that its probability is 0.

b. It is certain that Thanksgiving will be on a Thursday. When an event is certain to occur, we say that its probability is 1.

**The probability of any impossible event is 0.**

**The probability of any event that is certain to occur is 1.**

Since any event imaginable is either impossible, certain, or somewhere in between, it is reasonable to conclude that the mathematical probability of any event is either 0, 1, or a number between 0 and 1 (see Figure 3–4). This property can be expressed as follows:

$$0 \le P(A) \le 1 \qquad \text{for any event } A$$

In Figure 3–4, the scale of 0 through 1 is shown on top, whereas the more familiar and common expressions of likelihood are shown on the bottom.

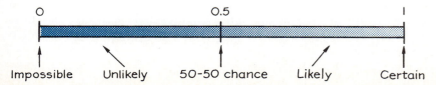

**Figure 3–4**
*Possible Values for Probabilities*

## HOW PROBABLE?

■ How do we interpret terms such as *probable*, *improbable*, or *extremely improbable*? The FAA interprets these terms as follows. Probable: A probability on the order of 0.00001 or greater for each hour of flight. Such events are expected to occur several times during the operational life of each airplane. Improbable: A probability on the order of 0.00001 or less. Such events are not expected to occur during the total operational life of a single airplane of a particular type, but may occur during the total operational life of all airplanes of a particular type. Extremely improbable: A probability on the order of 0.000000001 or less. Such events are so unlikely that they need not be considered to ever occur. ■

# Rounding Off Probabilities

Although it is difficult to develop a universal rule for rounding off probabilities, the following guide will apply to most problems in this text.

> **Either give the exact fraction representing a probability, or round off final decimal results to three significant digits.**

All the digits in a number are significant except for the zeros that are included for proper placement of the decimal point. The probability of 0.00128506 can be rounded to three significant digits as 0.00129. The probability of 1/3 can be left as a fraction or rounded in decimal form to 0.333, but not 0.3.

An important concept of this section is the mathematical expression of probability as a number between 0 and 1. This type of expression is fundamental and common in statistical procedures, and we will use it throughout the remainder of this text. A typical computer output, for example, may involve a $P$-value expression such as "Significance less than 0.001." We will discuss the meaning of $P$-values later, but they are essentially probabilities of the type discussed in this section. We should recognize that a probability of 0.001 (equivalent to 1/1000) corresponds to an event that is very rare in that it occurs an average of only once in a thousand trials.

## $\boxed{3\text{--}2}$ Exercises A

1. Which of the following values *cannot* be probabilities?

$$1.2, \frac{77}{75}, \frac{9}{10}, 0, -\frac{1}{2}, 1, 5, 0.9999, 1.001, \sqrt{2}, \sqrt{\frac{5}{7}}$$

2. a. What is $P(A)$ if event $A$ is certain to occur?
   b. What is $P(A)$ if event $A$ is impossible?
   c. A sample space consists of 14 separate events that are equally likely. What is the probability of each?
   d. On a college entrance exam, each question has 5 possible answers. If an examinee makes a random guess on the first question, what is the probability that the response is correct?

3. When 1084 adults were surveyed by Media General and the Associated Press, 813 indicated support for a ban on household aerosols. Use these survey results to estimate the probability that a randomly selected adult would support such a ban.

4. Among 750 taxpayers with incomes under $100,000, 20 are audited by the IRS (based on IRS data). Use this sample to estimate the probability of a tax return being audited if the income is below $100,000.

5. In a recent year, New York State experienced 68,593 vehicle accidents, with 26,201 of them involving reportable property damage of $600 or more (data from the N.Y. State Department of Motor Vehicles). Use these results to estimate the probability that a random New York State accident results in reportable property damage of at least $600.

6. Among 80 randomly selected blood donors, 36 were classified as group O (based on data from the Greater New York Blood Program). What is the approximate probability that a person will have group O blood?

7. If a person is randomly selected, find the probability that his or her birthday is October 18, which is National Statistics Day in Japan. Ignore leap years.

8. In a recent national election, there were 25,569,000 citizens in the 18–24 age bracket. Of these, 9,230,000 actually voted. Find the empirical probability that a person randomly selected from this group did vote in that national election.

9. In a study of brand recognition, 831 consumers knew Campbell's Soup, while 18 did not (based on data from Total Research Corporation). Use these results to estimate the probability that a randomly selected consumer will recognize Campbell's Soup.

10. A computer is used to generate random telephone numbers. Of the numbers generated and in service, 56 are unlisted, and 144 are listed in the telephone directory. If one of these telephone numbers is randomly selected, what is the probability that it is unlisted?

11. In a survey of U.S. households, 288 had home computers while 962 did not (based on data from Electronic Industries Association). Use this sample to estimate the probability of a household having a home computer.

12. A study of consumer loans found that 37 were defaults while 1383 had all obligations satisfied (based on data from the American Bankers Association). For a consumer loan randomly selected from those studied, find the probability of a default.

13. Data from the National Association of Recording Machines show that among 1500 randomly selected record sales, 135 were in the category of country music. If a record sale is randomly selected, what is the approximate probability that it is a country music record?

14. Data collected by volunteers in the Straphangers Campaign showed that 89 New York City subway cars had broken doors and 286 cars did not. If a car is randomly selected, what is the approximate probability it will have broken doors?

15. A Bureau of the Census survey of 600 people in the 18–25 age bracket found that 237 people smoke. If a person in that age bracket is randomly selected, find the approximate probability that he or she smokes.

16. An Environmental Protection Agency survey of cars originally equipped with catalytic converters found that 280 cars still had their converters and 12 cars had them removed. What is the approximate

# VOLTAIRE BEATS LOTTERY

■ In 1729, the philosopher Voltaire became rich by devising a scheme to beat the Paris lottery. The government ran a lottery to repay municipal bonds that had lost some value. The city added large amounts of money with the net effect that the prize values totaled more than the cost of all tickets. Voltaire formed a group that bought all the tickets in the monthly lottery and won for more than a year. A bettor in the New York State Lottery tried to win a share of an exceptionally large prize that grew from a lack of previous winners. He wanted to write a $6,135,756 check that would cover all combinations, but the state declined and said that the nature of the lottery would have been changed. ■

probability of selecting a car with a removed catalytic converter if the selection is random and is limited to cars originally assembled with converters?

17. Among 400 randomly selected drivers in the 20–24 age bracket, 136 were in a car accident during the last year (based on data from the National Safety Council). If a driver in that age bracket is randomly selected, what is the approximate probability that he or she will be in a car accident during the next year?

18. The U.S. General Accounting Office recently tested the IRS for correctness of answers to taxpayers' questions. For 1733 trials, the IRS was correct 1107 times. Use these results to estimate the probability that a random taxpayer question will be answered correctly.

19. When the allergy drug Seldane was clinically tested, 70 people experienced drowsiness while 711 did not (based on data from Merrell Dow Pharmaceuticals, Inc.). Use this sample to estimate the probability of a Seldane user becoming drowsy.

20. Data provided by the Bureau of Justice Statistics revealed that for a representative sample of convicted burglars, 76,000 were jailed, 25,000 were put on probation, and 2000 received other sentences. Use these results to estimate the probability that a convicted burglar will serve jail time.

21. Blood groups are determined for a sample of people, and the results are given in the accompanying table (based on data from the Greater New York Blood Program). If one person from this sample group is randomly selected, find the probability that the person has group AB blood.

| Blood Group | Frequency |
|:-----------:|:---------:|
| O | 90 |
| A | 80 |
| B | 20 |
| AB | 10 |

22. The following table gives the number of people who receive Social Security benefits in select states (based on data from the Social Security Administration). If one of these recipients is randomly selected, find the probability of getting a Texan.

| State | Number of Social Security Recipients |
|:-----------|:---------:|
| New York | 2,788,649 |
| California | 3,284,313 |
| Florida | 2,196,141 |
| Maine | 198,712 |
| Texas | 1,872,383 |

23. The accompanying table summarizes recent driver convictions for select violations in two counties (the data are from the New York State Department of Motor Vehicles).

|                     | Speeding | DWI |
|---------------------|----------|-----|
| Dutchess County     | 10,589   | 636 |
| Westchester County  | 22,551   | 963 |

If one of the convictions is randomly selected, find the probability that it is for DWI (driving while intoxicated).

24. The stem-and-leaf plot summarizes the time (in hours) managers spend in one day on paperwork (based on data from Adia Personnel Services). Use this sample to estimate the probability that a randomly selected manager spends more than 2.0 hours per day on paperwork.

```
0. | 00
1. | 0578
2. | 00113449
3. | 347
4. | 445
```

25. A couple plans to have 2 children.
    a.  List the different outcomes according to the sex of each child. Assume that these outcomes are equally likely.
    b.  Find the probability of getting 2 girls.
    c.  Find the probability of getting exactly 1 child of each sex.

26. A couple plans to have 4 children.
    a.  List the 16 different possible outcomes according to the sex of each child. Assume that these outcomes are equally likely.
    b.  Find the probability of getting all girls.
    c.  Find the probability of getting *at least* 1 child of each sex.
    d.  Find the probability of getting *exactly* 2 children of each sex.

27. On a quick quiz consisting of 3 true-false questions an unprepared student must guess at each one. The guesses will be random.
    a.  List the different possible solutions.
    b.  What is the probability of answering the 3 questions correctly?
    c.  What is the probability of guessing incorrectly for all questions?
    d.  What is the probability of passing by guessing correctly for *at least* 2 questions?

28. Both parents have the brown-blue pair of eye-color genes, and each parent contributes one gene to a child. Assume that if the child has at least one brown gene, that color will dominate and the eyes will be brown. (Actually, the determination of eye color is somewhat more complex.)
    a.  List the different possible outcomes. Assume that these outcomes are equally likely.
    b.  What is the probability that a child of these parents will have the blue-blue pair of genes?
    c.  What is the probability that a child will have brown eyes?

# 3–2 | Exercises B

29. In Exercise 7 we ignored leap years in finding the probability that a randomly selected person will have a birthday on October 18 (National Statistics Day in Japan).
    a. What is the probability if we assume that a leap year occurs every 4 years? (Express answer as an exact fraction.)
    b. Leap years occur in years evenly divisible by 4, except that they are skipped in 3 of every 4 centesimal years (ending in 00). The years 1700, 1800, and 1900 are not leap years, but 2000 will be a leap year. Find the exact probability for this case and express it as a fraction.
30. Someone has reasoned that since we know nothing about the presence of life on Pluto, either life exists there or it does not, so that we have $P(\text{life on Pluto}) = 0.5$. Is this reasoning correct? Explain.
31. If 2 flies land on an orange, find the probability that they are on points that are within the same hemisphere.
32. Two points along a straight stick are randomly selected. The stick is then broken at those 2 points. Find the probability that the 3 pieces can be arranged to form a triangle. (See also Computer Project at the end of this chapter.)

# 3–3 | Addition Rule

The preceding section introduced the basic concept of probability and considered simple experiments and simple events. Many real situations involve compound events.

---

### Definition

Any event combining 2 or more simple events is called a **compound event.**

---

In this section we want to develop a rule for finding $P(A \text{ or } B)$, the probability that for a single outcome of an experiment, either event $A$ or event $B$ occurs or both events occur. (Throughout this text we use the inclusive *or*, which means either one, or the other, or both. We will *not* consider the exclusive *or*, which means either one or the other, but not both.)

## Notation

$P(A \text{ or } B) = P(\text{event } A \text{ occurs or event } B \text{ occurs or they both occur})$

Let's consider the experiment results summarized in Table 3–1 (based on data from Merrell Dow Pharmaceuticals, Inc.). When exploring data consisting of frequency counts for different categories, it is helpful to arrange the data in a table such as this. (This type of table will be useful when solving several exercises.)

| TABLE 3–1 | | | | |
|---|---|---|---|---|
| | Seldane | Placebo | Control | |
| Drowsiness | 70 | 54 | 113 | 237 |
| No drowsiness | 711 | 611 | 513 | 1835 |
| Totals | 781 | 665 | 626 | 2072 |

The table lists frequencies of subjects in different categories. By adding all of the individual cell frequencies, we see that 2072 subjects are included in this experiment. Also, 237 of them experienced drowsiness, 1835 did not experience drowsiness, 781 took Seldane, 665 took a placebo, and 626 were in a control group. We will use two cases as a basis for developing our rule for determining $P(A \text{ or } B)$.

*Case 1:* If one of the 2072 subjects is randomly selected, the probability of getting someone who took Seldane or a placebo is

$$\frac{781}{2072} + \frac{665}{2072} = \frac{1446}{2072} = 0.698$$

That is, $P(\text{Seldane or placebo}) = P(\text{Seldane}) + P(\text{placebo})$. This might suggest that $P(A \text{ or } B) = P(A) + P(B)$, but consider the next case.

*Case 2:* Suppose one of the 2072 subjects is randomly selected, and we want the probability of getting someone who took Seldane or experienced drowsiness. Following the same pattern of Case 1, we might write

$$\frac{781}{2072} + \frac{237}{2072} = \frac{1018}{2072} = 0.491 \qquad \text{(WRONG!)}$$

## PROBABILITY AND PROSECUTION

■ Probability theory is now being used in forensic science as experts analyze criminal evidence. One particular area of rapid technological advancement is in the analysis of bloodstains. Several years ago, bloodstains could be categorized only according to the four basic blood groups. Current capabilities allow scientists to identify at least a dozen different characteristics. It is estimated that the probability of two people having the same set of these blood characteristics is about $\frac{1}{1000}$, and evidence obtained in this way has already been used to help convict murderers. ■

But this is wrong, because 70 subjects were counted *twice*. One way to compensate for that "double" counting is to subtract the 70 that was included twice. We get

$$P(\text{Seldane or drowsiness}) = \frac{781}{2072} + \frac{237}{2072} - \frac{70}{2072}$$

$$= \frac{948}{2072} = 0.458 \quad \text{(CORRECT)}$$

This last calculation can be expressed as

$$P(\text{Seldane or drowsiness}) = P(\text{Seldane}) + P(\text{drowsiness}) - P(\text{both})$$

This is generalized in the following addition rule.

---

**Addition Rule**

$$P(A \text{ or } B) = P(A) + P(B) - P(A \text{ and } B)$$

where $P(A \text{ and } B)$ denotes the probability that $A$ and $B$ both occur at the same time as an outcome in the experiment.

---

The example in Case 2 addresses this key point: **When combining the number of ways event *A* can occur with the number of ways *B* can occur, we must avoid double counting of those outcomes in which *A* and *B* both happen together.**

Figure 3–5 provides visual illustration of the addition rule. In this figure we can see that the probability of *A* or *B* equals the probability of *A* (left circle) plus the probability of *B* (right circle) minus the probability of *A* and

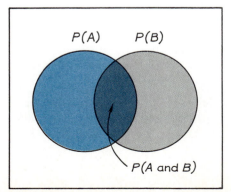

Figure 3–5

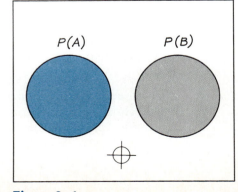

Figure 3–6

*B* (football-shaped middle region). This figure shows that the addition of areas of the two circles will cause double counting of the football-shaped middle region. This is the basic concept that underlies the addition rule. Because of this relationship between the addition rule and the Venn diagram shown in Figure 3–5, the notation $P(A \cup B)$ is often used in place of $P(A$ or $B)$. Similarly, the notation $P(A \cap B)$ is often used in place of $P(A$ and $B)$, so that the addition rule can be expressed as

$$P(A \cup B) = P(A) + P(B) - P(A \cap B)$$

The addition rule is simplified whenever *A* and *B* cannot occur simultaneously, so that $P(A$ and $B)$ becomes zero. Figure 3–6 illustrates that with no overlapping of *A* and *B*, we have $P(A$ or $B) = P(A) + P(B)$. The following definition formalizes the lack of overlapping as shown in Figure 3–6.

> ### Definition
>
> Events *A* and *B* are **mutually exclusive** if they cannot occur simultaneously.

In the experiment of rolling one die, the event of getting a 2 and the event of getting a 5 are mutually exclusive events, since no outcome can be both a 2 and 5 simultaneously. The following pairs of events are other examples of mutually exclusive pairs in a single experiment. That is, within each of the following pairs of events, it is impossible for both events to occur at the same time.

Pairs of mutually exclusive events

{ Manufacturing a defective electronic component
Manufacturing a good electronic component }

{ Selecting a voter who is a registered Democrat
Selecting a voter who is a registered Republican }

{ Testing a subject with an IQ above 100
Testing a subject with an IQ below 95 }

The following pairs of events are examples that are *not* mutually exclusive in a single trial. That is, within each of the following pairs of events, it is possible that both events occur at the same time.

Pairs of events that are not mutually exclusive

{ Selecting a doctor who is a brain surgeon
Selecting a doctor who is a woman }

{ Selecting a voter who is a registered Democrat
Selecting a voter who is under 30 years of age }

{ Testing a subject with an IQ above 100
Testing a subject with an IQ above 110 }

## MONKEY TYPISTS

■ A classical claim is that a monkey randomly hitting a keyboard would eventually produce the complete works of Shakespeare if it could continue to type century after century. Dr. William Bennet used the rules of probability to develop a computer simulation that addressed this problem, and he concluded that it would take a monkey about 1,000, 000, 000, 000, 000, 000, 000, 000, 000, 000, 000 years to reproduce Shakespeare's works. In the same spirit, Sir Arthur Eddington wrote this poem: "There once was a brainy baboon, who always breathed down a bassoon. For he said, 'It appears that in billions of years, I shall certainly hit on a tune.' " ■

We can summarize the key points as follows:

1. To find $P(A$ or $B)$, begin by associating *or* with addition.
2. Consider whether events $A$ and $B$ are mutually exclusive. That is, can they happen at the same time? If they are not mutually exclusive (can happen at the same time), be sure to avoid or compensate for double counting when adding the relevant probabilities.

*Important hint:* If you *understand* the importance of not double counting when you find $P(A$ or $B)$, you don't necessarily have to formally calculate $P(A) + P(B) - P(A$ and $B)$. In finding $P($Seldane or drowsiness$)$ from Table 3–1, you could sum the frequencies for the Seldane column and the drowsiness row by being careful to count each cell exactly once. You would get $70 + 711 + 54 + 113 = 948$, which, when divided by the total number of subjects (2072), will yield the correct probability of 0.458. It's much better to understand what you're doing than to blindly apply a formula.

The following examples further illustrate applications of the addition rule.

---

### Example

Survey subjects are often chosen by using computers to randomly select telephone numbers. Assume that a computer randomly generates the last digit of a telephone number. Find the probability that the outcome is

a. an 8 or 9    b. odd or under 4

### Solution

a. The outcome of 8 and the outcome of 9 are mutually exclusive events. This means that it is impossible for both 8 and 9 to occur together when 1 digit is selected, so $P(8$ and $9) = 0$ and the addition rule is applied as follows:

$$P(8 \text{ or } 9) = P(8) + P(9) - P(8 \text{ and } 9)$$
$$= \frac{1}{10} + \frac{1}{10} - 0$$
$$= \frac{2}{10} \quad \text{or} \quad \frac{1}{5}$$

b. The outcome of an odd number and the outcome of a number under 4 are not mutually exclusive since they both happen if the result is a 1 or a 3. We must compensate for double counting and we get

$$P(\text{odd or under } 4) = P(\text{odd}) + P(\text{under } 4) - P(\text{odd and under } 4)$$
$$= \frac{5}{10} + \frac{4}{10} - \frac{2}{10} = \frac{7}{10}$$

*continued*

> ### Solution continued
>
> In this result, $P(\text{odd}) = 5/10$ because 5 of the 10 digits are odd (1, 3, 5, 7, 9). $P(\text{under 4}) = 4/10$ because 4 of them are under 4 (0, 1, 2, 3). Finally, $P(\text{odd and under 4}) = 2/10$ because 2 of the digits are both odd and under 4 (1 and 3). The correct answer is 7/10.

### Example

Men were once drafted into the U.S. Army according to the random selection of birthdays. If the 366 different possible birthdays are written on separate slips of paper and mixed in a bowl, find the probability of making one selection and getting a birthday in May or November.

### Solution

Let $M$ denote the event of drawing a May date, while $N$ denotes the event of drawing a November date. Clearly, $M$ and $N$ are mutually exclusive because no date is in both May and November. Applying the addition rule, we get

$$P(M \text{ or } N) = P(M) + P(N) = \frac{31}{366} + \frac{30}{366} = \frac{61}{366}$$

The subtraction of $P(M \text{ and } N)$ can be ignored since it is zero.

### Example

Using the same population of 366 different birthdays, find the probability of making one selection that is the first day of a month or a November date.

### Solution

Let $F$ denote the event of selecting a date that is the first of the month. Here $F$ and $N$ are not mutually exclusive because they can occur simultaneously (as on November 1). Applying the addition rule, we get

$$P(F \text{ or } N) = P(F) + P(N) - P(F \text{ and } N)$$
$$= \frac{12}{366} + \frac{30}{366} - \frac{1}{366} \quad \text{— Nov. 1}$$
$$= \frac{41}{366}$$

## NOT TOO LIKELY

■ N. C. Wickramashinghe of University College in Cardiff, Wales, stated that the chance of getting life from a random shuffling of amino acids is about 1 in $10^{40,000}$. He said that this is equivalent to a tornado blowing through a junkyard and creating a jumbo jet in the process. ■

Errors made when applying the addition rule often involve double counting. That is, events that are not mutually exclusive are treated as if they were. One indication of such an error is a total probability that exceeds 1, but errors involving the addition rule do not always cause the total probability to exceed 1.

## 3–3 Exercises A

In Exercises 1 and 2, for each pair of events given, determine whether the two events are mutually exclusive for a single experiment.

1. a. Selecting a person who owns a computer
      Selecting a person who owns a VCR
   b. Selecting a student who gets low SAT or ACT scores
      Selecting a student who succeeds in college
   c. Selecting a person with blond hair (natural or otherwise)
      Selecting a person with brown eyes
   d. Selecting a worker who has reached retirement age of 55
      Selecting a citizen who is too young to vote
   e. Selecting a required course
      Selecting an elective course

2. a. Selecting someone who colors his or her hair
      Selecting someone who has read *The Greening of America*
   b. Selecting an unmarried person
      Selecting a person with an employed spouse
   c. Selecting a high school graduate
      Selecting someone who is unemployed
   d. Selecting a voter who is a registered Democrat
      Selecting a voter who favors the Republican candidate
   e. Selecting a consumer who drivers a car
      Selecting a consumer who subscribes to *Time* magazine

3. If a computer randomly generates the last digit of a telephone number, find the probability that it is odd or greater than 2.

4. A sample consists of 200 business calculators (8 of which are defective) and 150 scientific calculators (9 of which are defective). If 1 calculator is randomly selected from this sample, find the probability that it is a business calculator or is defective.

5. Among 200 seats available on one international airliner, 40 are reserved for smokers (including 16 aisle seats) and 160 are reserved for non-smokers (including 64 aisle seats). If a late passenger is randomly assigned a seat, find the probability of getting an aisle seat or one in the smoking section.

6. A local survey asked 100 subjects for their opinions on a zoning ordinance. Of the 62 favorable responses, there were 40 males. Of the 38 unfavorable responses, there were 15 males. Find the probability of randomly selecting one of these subjects and getting a male or a favorable response.

7. A labor study involves a sample of 12 mining companies, 18 construction companies, 10 manufacturing companies, and 3 wholesale companies. If a company is randomly selected from this sample group, find the probability of getting a mining or construction company.

8. A study of consumer smoking habits includes 200 married people (54 of whom smoke), 100 divorced people (38 of whom smoke), and 50 adults who never married (11 of whom smoke) (based on data from the U.S. Department of Health and Human Services.) If 1 subject is randomly selected from this sample, find the probability of getting someone who is divorced or smokes.

In Exercises 9–12, refer to the data in Table 3–2, which describes the age distribution of Americans who died by accident (based on data from the National Safety Council). In each case, assume that 1 person is randomly selected from this sample group.

9. Find the probability of selecting someone under 5 or over 74.

10. Find the probability of selecting someone between 15 and 64.

11. Find the probability of selecting someone under 45 or between 25 and 74.

12. Find the probability of selecting someone under 25 or between 15 and 44.

| TABLE 3–2 | |
| --- | --- |
| Age | Number |
| 0–4 | 3,843 |
| 5–14 | 4,226 |
| 15–24 | 19,975 |
| 25–44 | 27,201 |
| 45–64 | 14,733 |
| 65–74 | 8,499 |
| 75 and over | 16,800 |

In Exercises 13–16, use the following data: A survey of 400 randomly selected heads of households found 301 people who own cars (116 of whom are women) and 99 other people who don't own cars (59 of whom are women) (based on data from the U.S. Census Bureau). In each case assume that 1 of these 400 survey respondents is randomly selected and find the probability of the given event.

13. A woman is selected.

14. A woman or someone who owns a car is selected.

15. A man or someone who doesn't own a car is selected.

16. A man or someone who owns a car is selected.

**TABLE 3–3**

|  |  | Car accident in last year? | |
|  |  | Yes | No |
| --- | --- | --- | --- |
| Use a cellular phone? | Yes | 23 | 282 |
|  | No | 46 | 407 |

In Exercises 17–20, refer to Table 3–3, which is based on data from AT&T and the Automobile Association of America. In each case assume that 1 of the 758 drivers is randomly selected.

17. Find the probability that the selected driver had a car accident in the last year.
18. Find the probability that the selected driver uses a cellular phone or had a car accident in the last year.
19. Find the probability that the selected driver did not use a cellular phone.
20. Find the probability that the selected driver did not use a cellular phone or did not have an accident in the last year.

In Exercises 21–28, refer to the accompanying figure, which describes the blood groups and Rh types of 100 people (based on data from the Greater New York Blood Program). In each case assume that one of the 100 subjects is randomly selected and find the indicated probability.

21. $P$(group A or group B)
22. $P$(type Rh$^+$)
23. $P$(group A or type Rh$^-$)
24. $P$(group O or type Rh$^+$)
25. $P$(group A or B or AB)
26. $P$(type Rh$^-$)
27. $P$(group O or type Rh$^-$)
28. $P$(group A or group O or type Rh$^-$)

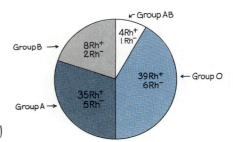

# 3–3 | Exercises B

29. If $P(A$ or $B) = 1/3$, $P(B) = 1/4$, and $P(A$ and $B) = 1/5$, find $P(A)$.
30. Find $P(B)$ if $P(A$ or $B) = 0.6$, $P(A) = 0.6$, and $A$ and $B$ are mutually exclusive events.
31. If $P(A$ or $B) = 79/120$, $P(A) = 11/24$, and $P(B) = 13/40$, find the probability that $A$ and $B$ both occur.

32. If events $A$ and $B$ are mutually exclusive, $P(A) = 0.123$, and $P(B) = 0.456$, find
    a. $P(A$ and $B)$
    b. $P(A$ or $B)$
    c. $P(A$ or not $B)$

33. a. If $P(A) = 0.4$ and $P(B) = 0.5$, what is known about $P(A$ or $B)$ if $A$ and $B$ are mutually exclusive events?
    b. If $P(A) = 0.4$ and $P(B) = 0.5$, what is known about $P(A$ or $B)$ if $A$ and $B$ are not mutually exclusive events?
    c. $P(A$ or $B) = 0.8$ while $P(A) = 0.4$. What is known about events $A$ and $B$?

34. If events $A$ and $B$ are mutually exclusive, and events $B$ and $C$ are mutually exclusive, must events $A$ and $C$ be mutually exclusive? Give an example supporting your answer.

35. How is the addition rule changed if the exclusive *or* is used instead of the inclusive *or*? Recall that the exclusive *or* means either one or the other, but not both.

36. Given that $P(A$ or $B) = P(A) + P(B) - P(A$ and $B)$, develop a rule for $P(A$ or $B$ or $C)$.

# 3–4 | Multiplication Rule

In Section 3–3 we developed a rule for finding the probability that events $A$ or $B$ will occur in a given experiment. We will now develop a rule for finding the probability that events $A$ and $B$ both occur. We begin with a simple example, which will suggest a preliminary multiplication rule. Then we use another example to develop a variation and ultimately obtain a generalized multiplication rule.

Probability theory is used extensively in the analysis and design of tests. Practical considerations often require that standardized tests allow only those answers that can be corrected easily, such as true-false or multiple choice. Let's assume that the first question on a test is a true-false type while the second question is multiple choice with 5 possible answers ($a, b, c, d, e$). We will use the following two questions. Try them!
True of false:

1. The last U.S. census cost about $10 per person.
2. The father of Euclidean geometry is
   a. Gauss
   b. There was no father, only a mother.
   c. Euclid
   d. Triola
   e. None of the above

## PROBABLY GUILTY?

■A witness described a Los Angeles robber as a Caucasian woman with blond hair in a ponytail who escaped in a yellow car driven by a black male with a mustache and beard. Janet and Malcolm Collins fit this description and they were convicted after a college math instructor testified that there is only about 1 chance in 12 million that any couple would have those characteristics. It was estimated the probability of a yellow car is 1/10; the other characteristics were also estimated. The convictions were later reversed when it was noted that no evidence was presented to support the estimated probabilities, and the independence of the characteristics was not established. ■

We want to determine the probability of getting both answers correct by making random guesses. We begin by listing the complete sample space of different possible answers.

| $T,a$ | $T,b$ | $T,c$ | $T,d$ | $T,e$ | 1 case is correct |
|---|---|---|---|---|---|
| $F,a$ | $F,b$ | $F,c$ | $F,d$ | $F,e$ | 10 equally likely cases |

If the answers are random guesses, then the 10 possible outcomes are equally likely. The correct answers are *true* and *c*, so that

$$P(\text{both correct}) = P(\text{true and } c) = \frac{1}{10}$$

Considering the component answers of *true* and *c*, respectively, we see that with random guesses we have $P(\text{true}) = 1/2$ while $P(c) = 1/5$. Recognizing that $1/10$ is the product of $1/2$ and $1/5$, we observe that $P(\text{true and } c) = P(T) \cdot P(c)$ and we use this observation as a basis for formulating the following rule.

## REDUNDANCY

■ Reliability of systems can be greatly improved with redundancy of critical components. Airplanes have two independent electrical systems, and aircraft used for instrument flight typically have two separate radios. The following is from a *Popular Science* article about stealth aircraft: "One plane built largely of carbon fiber was the Lear Fan 2100 which had to carry two radar transponders. That's because if a single transponder failed, the plane was nearly invisible to radar." Such redundancy is an application of the multiplication rule in probability theory. If one component has a 0.001 probability of failure, the probability of two independent components both failing is only 0.000001. ■

**Rule**

Multiplication rule (preliminary)
$$P(A \text{ and } B) = P(A) \cdot P(B)$$

**Example**

Three disk drives are produced and 1 of them is defective. Two are randomly selected for testing, but the first is replaced before the second selection is made. Find the probability that both disk drives are good.

**Solution**

Letting $G$ represent the event of selecting a good disk drive, we want $P(G \text{ and } G)$. With $P(G) = 2/3$, we apply the multiplication rule to get

$$P(G \text{ and } G) = P(G) \cdot P(G) = \frac{2}{3} \cdot \frac{2}{3} = \frac{4}{9}$$

Although the preceding solution is mathematically correct, common sense suggests that product testing should be conducted without replacement of the items already tested. There is always the chance that you could test the same item twice. Also, we cannot be sure of the reliability of any generalization based on a specific case, such as our preliminary multiplication rule. We would be wise to test that rule in a variety of cases to see if any errors arise. We will see that the preliminary rule is sometimes inadequate, as in the improved testing procedure in the next example.

### Example

Let's again assume that we have 3 disk drives, of which 1 is defective. We will again randomly select 2 disk drives, but we will *not* replace the first selection. We will find the probability of getting 2 good disk drives with this improved testing procedure.

### Solution

To understand the sample space better, we will represent the defective disk drive by $D$ and the 2 good disk drives by $G_1$ and $G_2$. Assuming that the first selection is not replaced, we now illustrate the different possible outcomes in Figure 3–7, which is an example of a **tree diagram.** By examining this list of 6 equally likely outcomes, we see that only 2 cases correspond to 2 good disk drives, so that $P(G$ and $G) = 2/6$ or $1/3$. In contrast, our preliminary multiplication rule was used in the preceding example to produce a result of $4/9$. Here, the correct result is $1/3$, so the preliminary multiplication rule does not fit this case.

In this example, the preliminary multiplication rule does not take into account the fact that the first selection is not replaced. $P(G)$ again begins with a value of $2/3$, but after getting a good disk drive on the first selection, there would be 1 good disk drive and 1 defective disk drive remaining, so that $P(G)$ becomes $1/2$ on the second selection.

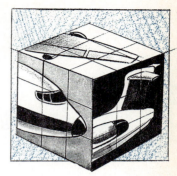

## COMPONENT FAILURES

■ In a report on aircraft component failures, the FAA refers specifically to the multiplication rule as it relates to system independence and redundancy. The FAA states that the most common problem with quantitative analyses presented to it has been the improper treatment of events that are not independent. The FAA states, "The probability of occurrence of two events which are mutually independent may be multiplied to obtain the probability that both events occur using this formula: $P(A$ and $B) = P(A) \cdot P(B)$. This multiplication will produce an incorrect solution if $A$ and $B$ are not mutually independent." ■

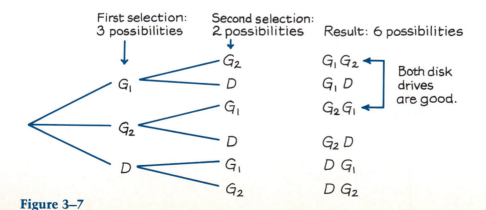

**Figure 3–7**

Here is the key concept of this last example: **Without replacement of the first selection, the second probability is affected by the first result.** There are many other cases in which a probability is affected by another event, or even by additional knowledge you may acquire. The probability of getting to class on time may be affected by the probability of your car starting. An estimated probability of a football team winning the Super Bowl may be affected by news of an injured quarterback. Since this dependence of an event on some other event is so important, we provide a special definition.

### Definition

Two events $A$ and $B$ are **independent** if the occurrence of one does not affect the probability of the occurrence of the other. (Several events are similarly independent if the occurrence of any does not affect the probabilities of the occurrence of the others.) If $A$ and $B$ are not independent, they are said to be **dependent.**

The preliminary multiplication rule holds if the events $A$ and $B$ are independent; in that case $P(B)$ is not affected by the occurrence of $A$. The preceding example suggests the following rule for dependent events.

### Rule

Let $P(B|A)$ represent the probability of $B$ occurring after assuming that $A$ has already occurred. (We can read $B|A$ as "$B$ given $A$.")

**Multiplication rule for dependent events**

$$P(A \text{ and } B) = P(A) \cdot P(B|A)$$

Combining this multiplication rule for dependent events with the preliminary multiplication rule, we summarize the key concept of this section as follows.

### Multiplication Rule

$P(A \text{ and } B) = P(A) \cdot P(B)$ if $A$ and $B$ are **independent.**

$P(A \text{ and } B) = P(A) \cdot P(B|A)$ if $A$ and $B$ are **dependent.**

In the last expression, we can easily solve for $P(B|A)$, and the result suggests the following definition.

> **Definition**
>
> The **conditional probability** of $B$ given $A$ is
>
> $$P(B|A) = \frac{P(A \text{ and } B)}{P(A)}$$

If $P(B|A) = P(B)$, then the occurrence of event $A$ had no effect on the probability of event $B$. This is often used as a test for independence. If $P(B|A) = P(B)$, then $A$ and $B$ are independent events. Another equivalent test for independence involves checking for equality of $P(A \text{ and } B)$ and $P(A) \cdot P(B)$. If they're equal, events $A$ and $B$ are independent. If $P(A \text{ and } B) \neq P(A) \cdot P(B)$, then $A$ and $B$ are dependent events.

In some cases, the distinction between dependent and independent events is quite obvious, while in other cases it is not. Getting heads on a coin and 6 on a die clearly involves independent events. But consider the data in Table 3–4, which summarizes 60 responses to a survey question. Are the events of being a male and answering yes independent? For this group,

$$P(M \text{ and yes}) \neq P(M) \cdot P(\text{yes})$$

24 respondents are males who answered yes

$$\frac{24}{60} \neq \frac{40}{60} \cdot \frac{38}{60}$$

40 males among 60 respondents
38 yes responses out of 60

Since $P(M \text{ and yes}) \neq P(M) \cdot P(\text{yes})$, we conclude that the events of being a male and answering yes are dependent for selections from this group of 60 people. In general, if we can verify or refute the equation $P(A \text{ and } B) = P(A) \cdot P(B)$, then we can establish or disprove the independence of events $A$ and $B$.

| TABLE 3–4 | | | | |
|-----------|---|-----|------|--------|
| | | | Male | Female |
| Do you favor a no-smoking | Yes | | 24 | 14 |
| rule on airliners? | No | | 16 | 6 |

# NUCLEAR POWER

■ MIT scientists once estimated the chance of a serious nuclear energy accident to be about one in a million, but an "event" at Three Mile Island and a major accident at Chernobyl have shown us that the backup and fail-safe systems were not as infallible as we thought. Scientists are carefully reviewing the design of nuclear power plants, including the use of redundant, or backup, components. Compare the two systems above and determine which is likely to be more reliable. Assume that $p$, $q$, $r$, and $s$ are valves and that each system supplies water needed for cooling. ■

## BAYES' THEOREM

■ Thomas Bayes (1702–1761) suggested that probabilities should be revised as we acquire additional knowledge about an event. He formulated what is known as Bayes' theorem. One case of the theorem leads to the following expression. (A more general expression is in many advanced texts.)

$$P(A|B) = \frac{P(A) \cdot P(B|A)}{P(A) \cdot P(B|A) + P(\overline{A}) \cdot P(B|\overline{A})}$$

For example, suppose that 60% of a company's computer chips are manufactured in one factory (denoted by $A$), while 40% are produced in its other factory (denoted by $\overline{A}$). Given a randomly selected chip, the probability it came from the first factory is $P(A) = 0.60$. Suppose we now learn that the randomly selected chip is defective and the defect rates for the two factories

*continued in page 139 margin*

### Example

Using the data in Table 3–4, find the probability of selecting someone from the group who answered yes, if we already know that the person selected is a male.

### Solution

We want to find the probability of getting a yes answer, given that a male ($M$) was selected. That is, we want $P(\text{yes}|M)$.

*First approach:* Use Table 3–4 directly. If we know a male was selected, we know that we have 40 people, and 24 of them answered yes, so that

$$P(\text{yes}|M) = \frac{24}{40} = 0.600$$

*Second approach:*

$$P(\text{yes}|M) = \frac{P(M \text{ and yes})}{P(M)}$$
$$= \frac{24/60}{40/60} = \frac{24}{40} = 0.600$$

Given the condition that a male was selected, the conditional probability $P(\text{yes}|M) = 0.600$.

So far we have discussed two events, but the multiplication rule can be easily extended to several events. In general, **the probability of any sequence of independent events is simply the product of their corresponding probabilities.** The next two examples illustrate this extension of the multiplication rule.

### Example

A couple plans to have 4 children. Find the probability that all 4 children are girls. Assume that boys and girls are equally likely and that the sex of each child is independent of the sex of any other children.

### Solution

The probability of getting a girl is 1/2 and the 4 births are independent of each other, so
$$P(\text{four girls}) = \frac{1}{2} \cdot \frac{1}{2} \cdot \frac{1}{2} \cdot \frac{1}{2} = \frac{1}{16}$$

## Example

At the beginning of this chapter we presented an employee's claim that a new technique for manufacturing VCRs is better because the rate of defects is below 5%, the rate of defects in the past. When 20 VCRs are manufactured with the new technique, there are no defects. Assuming that the new manufacturing technique has the same 5% rate of defects, find the probability of getting no defects among the 20 VCRs.

## Solution

Getting no defects means that all 20 VCRs are good, so we really want the probability of getting 20 good VCRs. Since 5% are defective, it follows that 95% are good so that $P(\text{good VCR}) = 0.95$ and

$$P(20 \text{ good VCRs}) = 0.95 \times 0.95 \times \ldots \times 0.95 \text{ (20 factors)}$$
$$= 0.358$$

If the new manufacturing technique has the same 5% rate of defects as the old method, there is a 0.358 chance of getting no defects among 20 VCRs. The probability of 0.358 is relatively high and it indicates that you could *easily* get no defects when 20 VCRs are manufactured, even if the rate of defects is 5%. We don't yet have enough evidence to support the employee's claim that the new method is better. It might be better, but we would need stronger evidence to conclude that.

The preceding two examples illustrate an extension of the multiplication rule for independent events; the next example illustrates a similar extension of the multiplication rule for dependent events.

## Example

The Federal Deposit Insurance Corporation assigns an accountant to audit 3 different banks to be randomly selected from a list of 5 commercial banks and 7 savings banks. Find the probability that all 3 random selections are commercial banks.

## Solution

Let $A$ be the event of getting a commercial bank on the first selection, while $B$ and $C$ represent the events of getting commercial banks on the

*continued*

*continued from page 138 margin*

are 35% (for $A$) and 25% (for $\overline{A}$) respectively. Given the defective chip, we can now find the probability that it came from factory $A$ as follows. (We denote a defective chip by $B$ for "bad.")

$$P(A|B) = \frac{(0.60)(0.35)}{(0.60)(0.35) + (0.40)(0.25)}$$
$$= 0.677$$

Along with a variety of other applications, Bayes' theorem is used in paternity suits to calculate the probability that a defendant is the father of a child. ∎

**Solution** *continued*

second and third selections. These events are *dependent* because successive probabilities are affected by previous results. We can use the multiplication rule for dependent events to find the probability that the first 2 selections are commercial banks. We get

$$P(A \text{ and } B) = P(A) \cdot P(B|A)$$
$$= \frac{5}{12} \cdot \frac{4}{11} = \frac{20}{132}$$

The probability of the third selection being a commercial bank is 3/10 since, after the first 2 selections, there would be 3 commercial banks among the 10 banks that remain. In summary, we have

$$P(A \text{ and } B \text{ and } C) = P(A) \cdot P(B|A) \cdot P(C|A \text{ and } B)$$
$$= \frac{5}{12} \cdot \frac{4}{11} \cdot \frac{3}{10}$$
$$= \frac{60}{1320} = \frac{1}{22}$$

In the last example we assumed that the events are dependent because the selections are made without replacement. However, it is a common practice to treat events as independent when small samples are drawn from large populations. A common guideline is to assume independence whenever the sample size is at most 5% of the size of the population. When pollsters survey 1200 adults from a population of millions, they typically assume independence, even though they sample without replacement.

When using the multiplication rule for finding probabilities in compound events, tree diagrams are sometimes helpful in determining the number of possible outcomes. (See Figure 3–7.) In a tree diagram we depict schematically the possible outcomes of an experiment as line segments emanating from one starting point. Such diagrams are helpful in counting the number of possible outcomes if the number of possibilities is not too large. In cases involving large numbers of choices, the use of tree diagrams is impractical. However, they are useful as visual aids to provide insight into the multiplication rule. Returning to the situation discussed earlier in this section, if we have the true-false question and the multiple-choice question, we can use a tree diagram to summarize the outcomes. (See Figure 3–8.)

Assuming that both answers are random guesses, all 10 branches are equally likely and the probability of getting the correct pair $(T, c)$ is 1/10. For each response to the first question there are 5 responses for the second. The total number of outcomes is therefore 5 taken 2 times, or 10. The tree diagram in Figure 3–8 illustrates the reason for the use of multiplication.

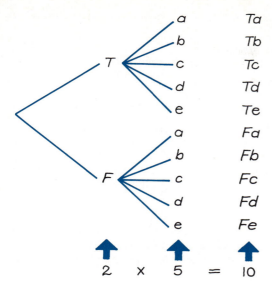

**Figure 3–8**

## 3–4 Exercises A

1. For each of the following pairs of events, classify the two events as independent or dependent.
   a. Randomly selecting a consumer having a credit card
      Randomly selecting a consumer having blue eyes
   b. Making a correct guess on the first question of a multiple choice quiz
      Making a correct guess on the second question of the same multiple choice quiz
   c. Randomly selecting a defective component from a bin of 15 good and 5 defective components
      Randomly selecting a second component that is defective (assume that the same bin is used and the first selection was not replaced)
   d. Events $A$ and $B$, where $P(A) = 0.40$, $P(B) = 0.60$, and $P(A \text{ and } B) = 0.20$
   e. Events $A$ and $B$, where $P(A) = 0.2$, $P(B) = 0.3$, and $P(A \text{ and } B) = 0.06$

2. For each of the following pairs of events, classify the two events as independent or dependent.
   a. Finding your kitchen light inoperable
      Finding your battery-operated flashlight inoperable
   b. Finding your kitchen light inoperable
      Finding your microwave oven inoperable
   c. In a stopover flight from New York to San Francisco, the flight of the first leg (New York to St. Louis) arrives on time. The flight of the second leg (St. Louis to San Francisco) arrives on time.
   d. Events $A$ and $B$, where $P(A) = 0.90$, $P(B) = 0.80$, and $P(A$ and $B) = 0.72$
   e. Events $A$ and $B$, where $P(A) = 0.36$, $P(B) = 0.50$, and $P(A$ and $B) = 0.15$.

3. A store owner has 3 employees who operate independently of each other and who all average a 6% rate of absenteeism. The store cannot open if all 3 employees are absent on the same day. Find the probability that the store cannot open on a particular day. Based on that result, should the owner be concerned about opening?

4. On a TV program (ABC's *Nightline*) on smoking, it was reported that there is a 60% success rate for those trying to stop through hypnosis. Find the probability that for 8 randomly selected smokers who undergo hypnosis, they all successfully stop smoking.

5. A quality control manager sets up a test of equipment to detect defects. A sample of 4 different items is to be randomly selected from a group consisting of 10 defective items and 20 items with no defects. What is the probability that all 4 items are defective?

6. According to a *U.S. News and World Report* article, an analyst for Paine Webber estimated that 35% of the computer chips manufactured by Intel are acceptable. Find the probability of getting all good chips in a batch of 6.

7. The first 2 answers to a true-false quiz are both true. If random guesses are made for those 2 questions, find the probability that both answers are correct.

8. There are 6 defective fuses in a bin of 80 fuses. The entire bin is approved for shipping if no defects show up when 3 randomly selected fuses are tested.
   a. Find the probability of approval if the selected fuses are replaced.
   b. Find the probability of approval if the selected fuses are not replaced.
   c. Comparing the results of parts *a* and *b*, which procedure is more likely to reveal a defective fuse? Which procedure do you think is better?

9. According to the U.S. Bureau of the Census, 62% of Americans over the age of 18 are married. Find the probability of getting 2 married

people (not necessarily married to each other) when 2 different Americans over the age of 18 are randomly selected.

10. According to *Popular Science*, 57% of all chain saw injuries involve arms or hands. If 4 different chain saw injury cases are randomly selected, find the probability that they all involve arms or hands.

11. One couple attracted media attention when their 3 children, who were born in different years, were all born on July 4. Ignoring leap years, find the probability that 3 randomly selected people were all born on July 4.

12. If 2 people are randomly selected, find the probability that the second person has the same birthday as the first. Ignore leap years.

13. Four firms using the same auditor independently and randomly select a month in which to conduct their annual audits. What is the probability that all 4 months are different?

14. A circuit requiring a 500-ohm resistance is designed with five 100-ohm resistors arranged in series. The proper resistance is achieved if all five resistors function correctly. There is a 0.992 probability that any individual resistor will not fail. What is the probability that all five of the resistors will work correctly to provide the 500-ohm resistance?

15. If a death is selected at random, there is a 0.0478 probability that it was caused by an accident, according to data from *Statistical Abstract of the United States*. Find the probability that 5 randomly selected deaths were all accidental.

16. Using U.S. Census Bureau data, we know that 60% of those who are eligible to vote actually do vote. If a pollster surveys 10 people eligible to vote, what is the probability that they all vote?

17. Of 2 million components produced by a manufacturer in one year, 5,000 are defective. If 2 of these components are randomly selected and tested, find the probability that they are both good in each of the following cases.
    a.  The first selected component is replaced.
    b.  The first selected component is not replaced.

18. A sales manager must fly from New York to San Francisco with a change of planes in St. Louis. To arrive on time, both legs of the flight must be on time. TWA Flight 25 from New York to St. Louis has an 80% on-time record, while TWA Flight 17 from St. Louis to San Francisco has a 60% on-time record (based on data from the EAASY SABRE reservation system). What is the probability of arriving on time? (Assume the 2 flights are independent, even though they may be dependent because of common factors such as weather and strikes.)

19. A manager can identify employee theft by checking samples of shipments. Among 36 employees, 2 are stealing. If the manager checks on 4 different randomly selected employees, find the probability that neither of the thieves will be identified.

20. A delivery company has 12 trucks, and when 3 are inspected, it is found that all 3 have faulty brakes. The owner claims that the other trucks all have good brakes and it was just chance that led to selecting the trucks that happened to have faulty brakes. Find the probability of that event, assuming that the manager's claims is correct.

21. In a Riverhead, New York, case, 9 different rape victims listened to voice recordings of 5 different men. All 9 women identified the same voice as that of the rapist. If the voice identifications had been made by random guesses, find the probability that all 9 women would select the same person.

22. A life insurance company issues one-year policies to 12 men who are all 27 years of age. Based on data from the Department of Health and Human Services, each of these men has a 99.82% chance of living through the year. What is the probability that they all survive the year?

23. An approved jury list contains 20 women and 20 men. Find the probability of randomly selecting 12 of these people and getting an all-male jury.

24. A mail-order firm normally experiences a 7% reply rate on a flyer it sends out. There is concern about the status of one batch of 50 flyers that resulted in no returns. What is the probability of this happening by chance if the flyers were delivered and the overall reply rate really is 7%?

In Exercises 25–28, use the data in the accompanying table (based on data from Merrell Dow Pharmaceuticals, Inc.).

|  | Seldane | Placebo | Control |
|---|---|---|---|
| Drowsiness | 70 | 54 | 113 |
| No drowsiness | 711 | 611 | 513 |

25. Find the probability that a subject experienced drowsiness given that he or she took Seldane.

26. Find the probability that a subject experienced drowsiness given that he or she took a placebo.

27. Find the probability that a subject was in the control group given that he or she experienced drowsiness.

28. Find the probability that a subject took Seldane given that he or she experienced drowsiness.

# 3–4 Exercises B

29. Let $P(A) = 3/4$ and $P(B) = 5/6$. Find $P(A \text{ and } B)$ given that
    a.  $A$ and $B$ are independent events
    b.  $P(B|A) = 1/2$
    c.  $P(A|B) = 1/3$
    d.  $P(A|B) = P(A)$

30. Use a calculator or computer to compute the probability (a) that of 25 people, no 2 share the same birthday, and (b) that of 50 people, no 2 share the same birthday.

31. A poll taken on the campus of a large university surveyed student attitudes about a variety of issues. Fifty students (40 males and 10 females) were polled on their involvement in campus activities, and the results showed 20 responses of yes and 30 responses of no. This data is summarized in the table.

    |        | Yes | No |    |
    |--------|-----|-----|----|
    | Male   |     |     | 40 |
    | Female |     |     | 10 |
    |        | 20  | 30  |    |

    a.  If 1 of the 50 students is randomly selected, find the probability of getting a male.
    b.  If 1 of the 50 students is randomly selected, find the probability of getting a female.
    c.  If 1 of the 50 students is randomly selected, find the probability of getting a student who answered yes.
    d.  If 1 of the 50 students is randomly selected, find the probability of getting a student who answered no.
    e.  Assuming that the sex of the respondent has no effect on the response, find the probability of randomly selecting 1 of the 50 polled students and getting a male who answered yes.
    f.  Using the probability from part *e* and the fact that there are 50 respondents, what would you expect to be the number of males who answered yes?
    g.  If the number from part *f* is entered in the appropriate box in the chart, can the numbers in the other boxes then be determined? If so, find them.

32. Do Exercise 31 after changing the data to correspond to the accompanying table.

    |        | Yes | No |    |
    |--------|-----|-----|----|
    | Male   |     |     | 5  |
    | Female |     |     | 45 |
    |        | 35  | 15  |    |

# 3–5 | Complements and Odds

## Complementary Events

In this section we begin by defining complementary events and present one last rule of probabilities that relates to these events. We then examine the relationship between probabilities and odds.

---

### Definition

The **complement** of event $A$, denoted by $\overline{A}$, consists of all outcomes not in $A$.

---

The complement of event $A$ is the event $A$ does *not* occur. If $A$ represents an outcome of answering a test question correctly, then the complementary event $\overline{A}$ represents a wrong answer. Sometimes the determination of the complement of an event is more difficult. For example, if $A$ is the event of getting at least 1 boy in 7 births, the complement of $A$ is the event of getting all girls. (It's *not* the event of getting at least 1 girl.)

The definition of complementary events implies that they must be mutually exclusive, since it is impossible for an event and its opposite to occur at the same time. Also, we can be absolutely certain that either $A$ does or does not occur. That is, either $A$ or $\overline{A}$ must occur. These observations enable us to apply the addition rule for mutually exclusive events as follows:

$$P(A \text{ or } \overline{A}) = P(A) + P(\overline{A}) = 1$$

We justify $P(A \text{ or } \overline{A}) = P(A) + P(\overline{A})$ by noting that $A$ and $\overline{A}$ are mutually exclusive, and we justify the total of 1 by our absolute certainty that $A$ either does or does not occur. This result of the addition rule leads to the following three *equivalent* forms.

---

### Rule of Complementary Events

$$P(A) + P(\overline{A}) = 1$$
$$P(\overline{A}) = 1 - P(A)$$
$$P(A) = 1 - P(\overline{A})$$

---

The first form comes directly from our original result, while the second (see Figure 3–9) and third variations involve very simple equation manipulations. A major advantage of the rule of complementary events is that it can sometimes be used to significantly reduce the workload required to solve certain problems. As an example, let's consider a very ambitious couple planning to have 7 children. We want to determine the probability of getting at least 1 girl among those 7 children. The direct solution to this problem is messy, but a simple indirect approach is made possible by our rule of complementary events. (We assume that $P(\text{boy}) = \frac{1}{2}$ and the births are independent; neither assumption is technically correct, but they will give us extremely good results.) Our solution is summarized in the following example.

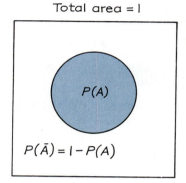

Total area = 1

$P(A)$

$P(\bar{A}) = 1 - P(A)$

**Figure 3–9**

## YOU BET

■ In the typical state lottery, the "house" has a 65% to 70% advantage, since only 35% to 40% of the money bet is returned as prizes. The house advantage at racetracks is usually around 15%. In casinos, the house advantage is 5.26% for roulette, 5.9% for blackjack, 1.4% for craps, and 3% to 22% for slot machines. Some professional gamblers can systematically win at blackjack by using complicated card-counting techniques. They know when a deck has disproportionately more high cards and this is when they place large bets. Many casinos react by ejecting card counters or by shuffling the decks more frequently. ■

---

### Example

Find $P(\text{at least 1 girl in 7 births})$.

### Solution

1. Let $G$ = at least 1 girl in 7 births.
2. $\overline{G}$ = *not* getting at least 1 girl in 7 births
   = getting 7 boys in 7 births
3. $P(\overline{G}) = P(7 \text{ consecutive boys}) = \frac{1}{2} \cdot \frac{1}{2} \cdot \frac{1}{2} \cdot \frac{1}{2} \cdot \frac{1}{2} \cdot \frac{1}{2} \cdot \frac{1}{2} = \frac{1}{128}$
4. $P(G) = 1 - P(\overline{G}) = 1 - \frac{1}{128} = \frac{127}{128}$

   As difficult as this solution might seem, it is trivial in comparison to solutions using a direct approach.

---

A key component of the solution in the preceding example really involves semantics. We must recognize that **the complement of "at least 1 girl" is "no girls," which is equivalent to "all boys."**

One common use of statistics is in quality control of manufactured products. A typical quality control plan involves careful testing of a sample from a large batch. The entire batch is accepted or rejected on the basis of the sample results. The following example illustrates this method.

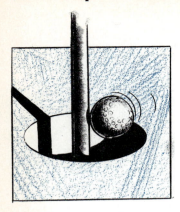

## HOLE-IN-ONE

■ Fundraisers sometimes run golf tournaments with big prizes for a hole-in-one. The National Hole-In-One Association provides insurance, just in case some lucky golfer actually makes one and wins the car, college scholarship, $10,000 cash, or whatever the prize is. The company estimates that when an amateur golfer tees off, the odds against a hole-in-one are 12,600 to 1. The insurance premium begins at $225 and goes up, depending on the number of golfers, the value of the prize, the length of the shot, and the skill involved. ■

### Example

A batch of brake rotors is rejected if at least 1 defect is found when 3 different rotors are randomly selected and tested. Assuming that 1 batch consists of 8 good rotors and 2 defective ones, find the probability that the batch is rejected.

### Solution

Since rejection of the batch occurs when at least 1 defect is selected, we have $P(\text{reject batch}) = P(\text{at least 1 defect})$. We will follow the same steps listed in the preceding example.

1. Let $R$ = getting at least 1 defect among the 3 rotors.
2. $\overline{R}$ = *don't* get at least 1 defect among the 3 rotors
   = all 3 rotors are good
3. $P(\overline{R}) = P(\text{3 consecutive good rotors}) = \dfrac{8}{10} \cdot \dfrac{7}{9} \cdot \dfrac{6}{8} = \dfrac{7}{15}$
4. $P(R) = 1 - P(\overline{R}) = 1 - \dfrac{7}{15} = \dfrac{8}{15}$

The probability of rejecting this particular batch is 8/15.

## Odds

Expressions of likelihood are often given as odds such as 50:1 or "50 to 1." (It is much less common to win a bet given such odds.) The use of odds makes it easier to deal with the money exchanges that result from gambling, but odds are extremely awkward in mathematical and scientific calculations. First, we should know that the likelihood of some event can be expressed in terms of the odds against that event or the odds in favor.

### Definition

The **odds against** event $A$ occurring are the ratio $P(\overline{A})/P(A)$, usually expressed in the form of $a{:}b$ (or "$a$ to $b$") where $a$ and $b$ are integers having no common factors.

The **odds in favor** of event $A$ are the reciprocal of the odds against that event. If the odds *against* event $A$ are $a{:}b$, then the odds in favor are $b{:}a$.

As an example, if $P(A) = 2/5$, then

$$\text{odds against } A = \frac{P(\overline{A})}{P(A)} = \frac{3/5}{2/5} = \frac{3}{2}$$

We express this as 3:2 or "3 to 2." The corresponding odds in favor are 2:3. (See Exercise 20 for conversions from odds into probabilities.)

For bets, the odds against an event represent the ratio of net profit to the amount bet.

**Odds against event $A$ = (net profit) : (amount bet)**

Suppose a bet pays 50:1. If the odds aren't specified as being in favor or against, they are probably the odds against the event occurring. If, by some minor miracle, you should win this 50:1 bet, you would make a profit of $50 for each $1 bet. For example, if you bet $2, your net profit would be $100. You would collect a total of $102, which includes the $100 net profit and the $2 original bet. This really isn't too complicated. But now suppose a circuit has 50:1 odds against failure. What are the odds against 2 such separate and independent circuits both failing? The best way to solve this problem is to first convert the 50:1 odds to the corresponding probability of failure $\left(\frac{1}{51}\right)$. We can then use the multiplication rule for independent events to get

$$\frac{1}{51} \cdot \frac{1}{51} = \frac{1}{2601}$$

This gives the probability of both circuits failing, and is equivalent to odds of 2600:1. There is no easy multiplication rule for odds that allows us to multiply 50:1 by 50:1 to get 2600:1. The point here is that, when performing calculations involving likelihoods, use probability values between 0 and 1, not odds.

## 3–5  Exercises A

In Exercises 1–4, use words to describe the complement of the given event.

1. A manufactured system is defective.
2. Among 5 experiments, none were successful.
3. Among 5 true-false questions, at least 1 answer is false.
4. Among 3 components, there is at least 1 failure.

In Exercises 5–8, determine the probability of the given event and the probability of the complementary event.

5. A defective disk drive is randomly selected from 50 available disk drives, of which 5 are defective. (Only 1 selection is made.)

## LONG-RANGE WEATHER FORECASTING

■A high school teacher in Millbrook, New York, created a controversy by conducting a classroom experiment in which long-range weather forecasts were made by throwing darts. The dart method resulted in 7 correct predictions among 20 winter storms, while a nearby private forecasting service correctly predicted only 6 of the storms. The teacher claimed that while short-range weather forecasts are usually quite accurate, long-range forecasts are not. ■

6. When a multiple choice test question is answered by guessing, the response is correct. (Assume 5 possible answers, of which 1 is correct.)

7. When a computer generates a random number from 1 through 5 (integers only), the result is odd.

8. In a class of 10 men and 15 women, 2 different students are called and they are both women.

9. A blood testing procedure is sometimes made more efficient by combining samples of blood specimens. If 5 samples are combined and the mixture tests negative, we know that all 5 samples are negative. Find the probability of a positive result for 5 samples combined into 1 mixture, assuming that the probability of an individual blood sample testing positive is 0.015.

10. An employee needs to call any 1 of 5 colleagues. Assume that the 5 colleagues are random selections from a population in which 28% have unlisted numbers. Find the probability that at least 1 of the 5 fellow workers will have a listed number.

11. A circuit is designed so that a critical function is properly performed if at least 1 of 4 identical and independent components does not fail. The probability of failure is 0.081. Find the probability that the critical function will be properly performed.

12. Three programmers are given a problem, which they work on independently. The probabilities of each programmer completing a working program by the end of the day are 1/5, 2/5, and 1/4, respectively. Find the probability that a working program will be available by the end of the day.

13. Under certain conditions, there is a 0.015 probability that a randomly selected driver will fail a breathalyzer test for sobriety. If the police test 100 randomly selected drivers, what is the probability of finding at least 1 driver who fails the sobriety test?

14. For a group of medical students given selected case studies, there is a 0.850 probability of a correct diagnosis. For 3 students each given one case, what is the probability of at least 1 incorrect diagnosis?

15. A nuclear weapon will not be misfired if at least 1 of 5 separate and independent fail-safe mechanisms functions properly. The estimated likelihoods of failure for these fail-safe devices are 0.105, 0.200, 0.001, 0.115, and 0.340, respectively. What is the probability that a misfire will not occur?

16. In a recent national election, 52% of the voters were women, according to a *New York Times*/CBC News poll. For a random selection of 4 voters, find the probability of getting at least 1 woman.

# 3–5  Exercises B

17.  a.  Develop a formula for the probability of not getting either $A$ or $B$ on a single trial. That is, find an expression for $P(\overline{A \text{ or } B})$.
   b.  Develop a formula for the probability of not getting $A$ or not getting $B$ on a single trial. That is, find an expression for $P(\overline{A} \text{ or } \overline{B})$.
   c.  Compare the results from parts $a$ and $b$. Are they the same or are they different?

18.  Find the probability that (a) of 25 randomly selected people, at least 2 share the same birthday, and (b) of 50 randomly selected people, at least 2 share the same birthday.

19.  A company has been manufacturing resistors at a cost of 50¢ each, and there is a probability of 0.900 that any such resistor will work. An employee suggests that another process will generate resistors in groups of 4 at a cost of 10¢ for each resistor. There is a 50% failure rate for the resistors produced by this second method. Is it better to produce 4 resistors at 10¢ each, or the one resistor at a cost of 50¢? Explain.

20.  a.  To convert from a probability of an event $P(A)$ to odds against the event, express $P(\overline{A})/P(A)$ as a ratio of two integers having no common factors. If $P(A) = 2/7$, find the odds against event $A$ occurring.
   b.  If the odds against event $A$ are $a{:}b$, then $P(A) = b/(a + b)$. Find the probability of event $A$ if the odds against it are 9:4.
   c.  If the odds against an event are 7:3, what are the odds against this event occurring in all of 3 separate and independent trials?
   d.  In a fair game, all of the money lost by some players is won by others. For one fair game, a \$2 bet results in a net profit of \$18. Find the odds against winning and find the probability of winning.
   e.  The American roulette wheel has 38 different slots numbered 0, 00, and 1 through 36. If you bet on any individual number, the casino gives you odds of 35:1. What would be fair odds if the casino did not have an advantage?
   f.  The actual odds against winning when you bet on "odds" at roulette are 10:9. What is the probability of winning?

# 3–6  Counting

In some probability problems, the biggest obstacle is determining the total number of outcomes. In this section we examine some of the efficient ways this can be done. In addition to their use in probability problems, these counting techniques have also grown in importance because of their applicability to problems relating to computers and programming.

# THE NUMBER CRUNCH

■ Every so often telephone companies split regions with one area code into regions with two or more area codes because the increased number of telephones in the area has nearly exhausted the possible numbers that can be listed under a single code. A seven-digit telephone number cannot begin with a 0 or 1, but if we allow all other possibilities, we get 8 × 10 × 10 × 10 × 10 × 10 × 10 = 8,000,000 different possible numbers! Even so, after surviving for 80 years with the single area code of 212, New York City was recently partitioned into the two area codes of 212 and 718. Los Angeles, Houston, and San Diego have also endured split area codes. ■

Suppose that we use a computer to randomly select 1 of the 2 Rh types (positive, negative) and 1 of the 4 blood groups A, O, B, or AB. Let's assume that we want the probability of getting a positive Rh type and group A blood. We can represent this probability as $P(\text{positive and A})$ and apply the multiplication rule to get

$$P(\text{positive and A}) = \frac{1}{2} \cdot \frac{1}{4} = \frac{1}{8}$$

We could also arrive at the same probability by examining the tree diagram in Figure 3–10. The tree diagram has 8 branches that, by the random computer selection method, are all equally likely. Since only 1 branch corresponds to $P(\text{positive and A})$ we get a probability of 1/8. Apart from the calculation of the probability, this solution does reveal another principle, which is a generalization of the following specific observation: With 2 Rh types and 4 blood groups, there are 8 different possibilities for the compound event of selecting a factor and type. We now state this generalized principle.

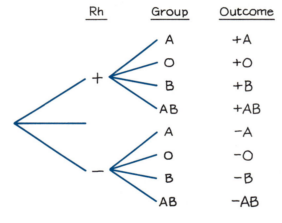

**Figure 3–10**

## Fundamental Counting Rule

For a sequence of 2 events in which the first event can occur $m$ ways and the second event can occur $n$ ways, the events together can occur a total of $m \cdot n$ ways.

The **fundamental counting rule** easily extends to situations involving more than 2 events, as illustrated in the following example.

## Example

In designing a computer, if a byte is defined to be a sequence of 8 bits, and each bit must be a 0 or 1, how many different bytes are possible?

## Solution

Since each bit can occur in 2 ways (0 or 1) and we have a sequence of 8 bits, the total number of different possibilities is given by
$$2 \cdot 2 \cdot 2 \cdot 2 \cdot 2 \cdot 2 \cdot 2 \cdot 2 = 256.$$

The next three rules use the factorial symbol !, which denotes the product of decreasing whole numbers. For example, $5! = 5 \cdot 4 \cdot 3 \cdot 2 \cdot 1 = 120$. By special definition, $0! = 1$. (Many calculators have a factorial key.) Using the factorial symbol, we now present the factorial rule, which is actually a simple case of the permutations rule that will follow.

## Factorial Rule

$n$ different items can be arranged in order $n!$ different ways.

This **factorial rule** reflects the fact that the first item may be selected $n$ different ways, the second item may be selected $n - 1$ ways, and so on.

## Example

Routing problems are extremely important in many applications of this rule. AT&T wants to route telephone calls through the shortest networks. Federal Express wants to find the shortest routes for its deliveries. Suppose a computer salesperson must visit 3 separate cities denoted by $A$, $B$, $C$. How many routes are possible?

## Solution

Using the factorial rule, we see that the 3 different cities ($A$, $B$, $C$) can be arranged in $3! = 6$ different ways. In Figure 3–11 (see the following page), we can see exactly why there are 6 different routes.

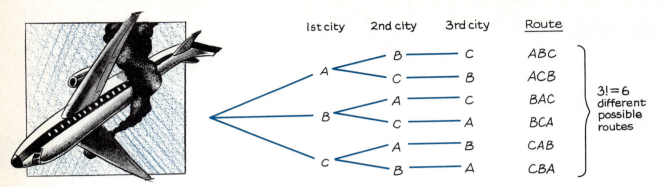

Figure 3–11

### Example

A presidential candidate plans to visit the capital of each of the 50 states. How many different routes are possible?

### Solution

The 50 state capitals can be arranged 50! ways, so that the number of different routes is 50! or 30,414 *followed by 60 zeros*; that is an incredibly large number! Now we can see why the symbol ! is used for factorials!

The preceding example is a variation of a classical problem called the traveling salesman problem. It is especially interesting because the large number of possibilities precludes a direct computation for each route, even if computers are used. The time for the fastest computer to directly calculate the shortest possible route is about

1,000,000,000,000,000,000,000,000,000,000,000,000,000,000 *centuries*!

(A Bell Laboratories mathematician claims that the shortest distance for the 48 contiguous states is 10,628 miles. See *Discover*, July 1985. See also page C1 of the *New York Times* for March 12, 1991.) Clearly, those who use computers to solve problems must be able to recognize when the number of possibilities is so large.

In the factorial counting rule, we determine the number of different possible ways we can arrange a number of items in some type of ordered sequence. Sometimes we don't want to include all of the items available. When we refer to arrangements, we imply that *order* is taken into account. Arrangements are commonly called permutations, which explains the use of the letter *P* in the following rule.

## Permutations Rule

The number of **permutations** (or arrangements) of $r$ items selected from $n$ available items is

$$_nP_r = \frac{n!}{(n-r)!}$$

It must be emphasized that in applying the preceding **permutations rule,** we must have a total of $n$ items available, we must select $r$ of the $n$ items, and we must consider rearrangements of the same items to be different. In the following example we are asked to find the total number of different arrangements that are possible. That clearly suggests use of the permutations rule.

### Example

If an editor must arrange 5 articles in a magazine and there are 8 articles available, how many different arrangements are possible?

### Solution

Here we want the number of arrangements of $r = 5$ items selected from $n = 8$ available articles, so the number of different possible arrangements is

$$_8P_5 = \frac{8!}{(8-5)!} = \frac{8!}{3!} = 6{,}720$$

This permutation rule can be thought of as an extension of the fundamental counting rule. We can solve the preceding problem by using the fundamental counting rule in the following way. With 8 articles available and with space for only 5 articles, we know that there are 8 choices for the first article, 7 choices for the second article, 6 choices for the third article, 5 choices for the fourth article, and 4 choices for the fifth article. The number of different possible arrangements is therefore $8 \cdot 7 \cdot 6 \cdot 5 \cdot 4 = 6720$, but $8 \cdot 7 \cdot 6 \cdot 5 \cdot 4$ is actually $8! \div 3!$. In general, whenever we select $r$ items from $n$ available items, the number of different possible arrangements is $n! \div (n-r)!$, and this is expressed in the permutations rule.

When we intend to select $r$ items from $n$ available items but *do not take order into account*, we are really concerned with possible combinations

# PROMOTION CONTESTS

■ There have been many contests or games designed to promote or sell products, but some have encountered problems. The Beatrice Company ran a contest involving matching numbers on scratch cards with scores from Monday night football games. The game cards had too few permutations and Frank Maggio was able to identify patterns and collect around 4000 winning cards worth $21 million. The contest was canceled and lawsuits were filed. The Pepsi people ran another contest in which people had to spell out their names with letters found on bottle caps, but a surprisingly large number of people with short names like Ng forced the cancellation of this contest also. ■

## SAFETY IN NUMBERS

■ Some hotels have abandoned the traditional room key in favor of an electronic key made of paper and aluminum foil. A central computer changes the access code to a room as soon as a guest checks out. A typical electronic key has 32 different positions that are either punched or left untouched. This configuration allows for $2^{32}$, or 4,294,967,296, different possible codes, so it is impractical to develop a complete set of keys or try to make an illegal entry by trial and error. ■

rather than permutations. That is, when **different orderings of the same items are to be counted separately, we have a permutation problem, but when different orderings are *not* to be counted separately, we have a combination problem** and may apply the following rule.

---

### Combinations Rule

The number of **combinations** of $r$ items selected from $n$ available items is

$$_nC_r = \frac{n!}{(n - r)!\, r!}$$

[Other notations for $_nC_r$ are $\binom{n}{r}$ and $C(n, r)$.]

---

The combinations rule makes sense when we think of it as a modification of the permutations rule. Note that

$$_nC_r = \frac{n!}{(n - r)!\, r!} = \frac{n!}{(n - r)!} \cdot \frac{1}{r!} = {_nP_r} \cdot \frac{1}{r!} = \frac{_nP_r}{r!}$$

Realizing that the combinations rule disregards the order of the $r$ selected items while the permutations rule counts all of the different orderings separately, we see that $_nC_r$ will be a fraction of $_nP_r$. For any particular selection of $r$ items, the factorial rule shows that there are $r!$ different arrangements. We know that any particular selection of $r$ items would be counted as only one combination, but it would yield $r!$ different arrangements. As a result, the total number of combinations will be $1/r!$ of the total number of permutations, and that result is expressed in the combinations rule.

Since choosing between the permutations rule and the combinations rule can be confusing, we provide the following example, which is intended to emphasize the difference between them.

---

### Example

Five students (Al, Bob, Carol, Donna, and Ed) have volunteered for service to the student government.

a. If 3 of the students are to be selected for a special *committee*, how many different committees are possible?

b. If 3 of the students are to be nominated for the offices of president, vice president, and secretary, how many different *slates* are possible?

### Solution

a. When forming the committee, order of selection is irrelevant. The committee of Al, Bob, and Ed is the same as that of Bob, Al, and Ed. Therefore, we want the number of combinations of 5 students when 3 are selected. We get

$$_5C_3 = \frac{5!}{(5-3)!\,3!} = 10$$

Use combinations when order is irrelevant

There are 10 different possible committees.

b. When forming slates of candidates, the order is relevant. The slate of Al for president, Bob for vice president, and Ed for secretary is different from the slate of Bob, Al, and Ed for president, vice president, and secretary, respectively. Here we want the number of permutations of 5 students when 3 are selected. We get

$$_5P_3 = \frac{5!}{(5-3)!} = 60$$

Use permutations when order is relevant.

There are 60 different possible slates.

We stated that the counting techniques presented in this section are sometimes used in probability problems. The following examples illustrate such applications.

### Example

In the New York State lottery, a player wins first prize by selecting the correct 6 number combination when 6 different numbers from 1 through 54 are drawn. If a player selects 1 particular 6 number combination, find the probability of winning.

### Solution

Since 6 different numbers are selected from 54 different possibilities, the total number of combinations is

$$_{54}C_6 = \frac{54!}{(54-6)!\,6!} = \frac{54!}{48!\,6!} = 25{,}827{,}165$$

With only 1 combination selected, the player's probability of winning is only 1/25,827,165.

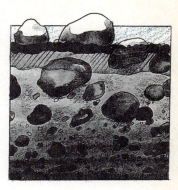

## GROWING ROCKS

■ In a *New York Times* article, James Gleick explained why the larger corn flakes seem to rise to the top of the box and why rocks rise to the surface of land that had been previously cleared of all surface rocks. A computer model was used to simulate mixtures of large and small objects. When such mixtures are shaken, gaps are created, but there are many more smaller gaps than larger ones. There is a greater probability that a smaller object will move downward into those more abundant smaller gaps. Larger objects tend to be forced upward, even if they are heavier. The same principle affects how we mix things like cement and pharmaceuticals. ■

## HOW MANY SHUFFLES?

■After conducting extensive research, Harvard mathematician Persi Diaconis found that it takes seven shuffles of a deck of cards to get a complete mixture. The mixture is complete in the sense that all possible arrangements are equally likely. More than seven shuffles will not have a significant effect, and fewer than seven are not enough. Casino dealers rarely shuffle as often as seven times, so the decks are not completely mixed. Some expert card players have been able to take advantage of the incomplete mixtures that result from fewer than seven shuffles. ■

### Example

A home security device with 10 buttons is disarmed when 3 different buttons are pushed in the proper sequence. (No button can be pushed twice.) If the correct code is forgotten, what is the probability of disarming this device by randomly pushing 3 of the buttons?

### Solution

The number of different possible 3–button sequences is

$$_{10}P_3 = \frac{10!}{(10-3)!} = 720$$

The probability of randomly selecting the correct 3-button sequence is therefore 1/720.

### Example

A dispatcher sends a delivery truck to 8 different locations. If the order in which the deliveries are made is randomly determined, find the probability that the resulting route is the shortest possible route.

### Solution

With 8 locations there are 8!, or 40,320, different possible routes. Among those 40,320 different possibilities, only 2 routes will be shortest (actually the same route in 2 different directions). Therefore, there is a probability of only 2/40,320, or 1/20,160, that the selected route will be the shortest possible route.

In this last example, application of the appropriate counting technique made the solution easily obtainable. If we had to determine the number of routes directly by listing them, we would labor for over 11 hours while working at the rapid rate of one route per second! Clearly, these counting techniques are extremely valuable.

The concepts and rules of probability theory presented in this chapter consist of elementary and fundamental principles. A more complete study of probability is not necessary at this time, since our main objective is to study the elements of statistics, and we have already covered the probability theory that we will need. We hope that this chapter generates some interest in probability for its own sake. The importance of probability is continuing to grow as it is used by more and more political scientists, economists, biologists, actuaries, business executives, and other professionals.

# 3-6 Exercises A

In Exercises 1–16, evaluate the given expressions.

1. $7!$
2. $9!$
3. $\dfrac{70!}{68!}$
4. $\dfrac{92!}{89!}$

5. $(9-3)!$
6. $(20-12)!$
7. $_6P_2$
8. $_6C_2$

9. $_{10}C_3$
10. $_{10}P_3$
11. $_{52}C_2$
12. $_{52}P_2$

13. $_nP_n$
14. $_nC_n$
15. $_nC_0$
16. $_nP_0$

17. Data are grouped according to sex (female, male) and income level (low, middle, high). How many different possible categories are there?

18. How many different ways can 5 cars be arranged on a carrier truck with room for 5 vehicles?

19. A computer operator must select 4 jobs from among 10 available jobs waiting to be completed. How many different arrangements are possible?

20. A computer operator must select 4 jobs from 10 available jobs waiting to be completed. How many different combinations are possible?

21. An IRS agent must audit 12 returns from a collection of 22 flagged returns. How many different combinations are possible?

22. A health inspector has time to visit 7 of the 20 restaurants on a list. How many different routes are possible?

23. Using a word processor, a pollster develops a survey of 10 questions. The pollster decides to rearrange the order of the questions so that any lead-in effect will be minimized. How many different versions of the survey are required if all possible arrangements are included?

24. How many different 7-digit telephone numbers are possible if the first digit cannot be 0 or 1?

25. An airline mail route must include stops at 7 cities.
    a. How many different routes are possible?
    b. If the route is randomly selected, what is the probability that the cities will be arranged in alphabetical order?

26. A 6-member FBI investigative team is to be formed from a list of 30 agents.
    a. How many different possible combinations can be formed?
    b. If the selections are random, what is the probability of getting the 6 agents with the most time in service?

27. How many different Social Security numbers are possible? Each Social Security number is a sequence of 9 digits.

28. A pollster must randomly select 3 of 12 available people. How many different groups of 3 are possible?

29. A union must elect 4 officers from 16 available candidates. How many different slates are possible if 1 candidate is nominated for each office?

30.  a.  How many different zip codes are possible if each code is a sequence of 5 digits?
     b.  If a computer randomly generates 5 digits, what is the probability it will produce your zip code?

31.  A typical combination lock is opened with the correct sequence of 3 numbers between 0 and 49 inclusive. How many different sequences are possible? Are these sequences combinations or are they actually permutations?

32.  One phase of an automobile assembly requires the attachment of 8 different parts, and they can be attached in any order. The manager decides to find the most efficient arrangement by trying all possibilities. How many different arrangements are possible?

33.  A space shuttle crew has available 10 main dishes, 8 vegetable dishes, 13 desserts, and 3 appetizers. If the first meal includes 2 desserts and 1 item from each of the other categories, how many different combinations are possible?

34.  In Denys Parson's *Directory of Tunes and Musical Themes*, melodies for more than 14,000 songs are listed according to the following scheme: The first note of every song is represented by an asterisk (*), and successive notes are represented by $R$ (for repeat the previous note), $U$ (for a note that goes up), or $D$ (for a note that goes down). Beethoven's "Fifth Symphony" begins as *$RRD$. Classical melodies are represented through the first 16 notes. Using this scheme, how many different classical melodies are possible?

35.  A television program director has 14 shows available for Monday night and 5 shows must be chosen.
     a.  How many different possible combinations are there?
     b.  If 650 different combinations are judged to be incompatible, find the probability of randomly selecting 5 shows that are compatible.

36.  A telephone company employee must collect the coins at 40 different locations.
     a.  How many different routes are possible?
     b.  If 2 of the routes are the shortest, find the probability of randomly selecting a route and getting 1 of the 2 shortest routes.

37.  A common lottery win requires that you pick the correct 6-number combination randomly selected from the numbers between 1 and 49 inclusive. Find the probability of such a win and compare it to the probability of being struck by lightning this year, which is approximately $\frac{1}{700,000}$.

38.  A multiple-choice test consists of 10 questions with choices $a$, $b$, $c$, $d$, $e$.
     a.  How many different answer keys are possible?
     b.  If all 10 answers are random guesses, what is the probability of getting a perfect score?

39. The Bureau of Fisheries once asked Bell Laboratories for help in finding the shortest route for getting samples from locations in the Gulf of Mexico. How many different routes are possible if samples must be taken from 24 locations?

40. A manager must choose 5 secretaries from among 12 applicants and assign them to different stations.
    a. How many different arrangements are possible?
    b. If the selections are random, what is the probability of getting the 5 youngest secretaries selected in order of age?

## 3–6  Exercises B

41. The number of permutations of $n$ items when $x$ of them are identical to each other and the remaining $n - x$ are identical to each other is given by

$$\frac{n!}{(n - x)!\, x!}$$

If a sequence of 10 trials results in only successes and failures, how many ways can 3 successes and 7 failures be arranged?

42. Among couples with 12 grandchildren, one is to be randomly selected. What is the probability that the 12 grandchildren will consist of 4 boys and 8 girls? (See Exercise 41.)

43. A common computer programming rule is that names of variables must be between 1 and 8 characters long. The first character can be any of the 26 letters, while successive characters can be any of the 26 letters or any of the ten digits. For example, allowable variable names are A, AAA, and R2D2. How many different variable names are possible?

44. a. Five managers gather for a meeting. If each manager shakes hands with each other manager exactly once, what is the total number of handshakes?
    b. If $n$ managers shake hands with each other exactly once, what is the total number of handshakes?

45. a. How many different ways can 3 people be seated at a round table? (Assume that if everyone moves to the right, the seating arrangement is the same.)
    b. How many different ways can $n$ people be seated at a round table?

46. We say that a sample is *random* if all possible samples of the same size have the same probability of being selected.
    a. If a random sample is to be drawn from a population of size 60, find the probability of selecting any one individual sample consisting of 5 members of the population.

*continued*

b.  Write a general expression for the probability of selecting a particular random sample of size $n$ from a population of size $N$.

47.  Many calculators or computers cannot directly calculate 70! or higher. When $n$ is large, $n!$ can be *approximated* by

$$n! = 10^K \qquad \text{where } K = (n + 0.5)\log n + 0.39908993 - 0.43429448n$$

Evaluate 50! using the factorial key on a calculator and also by using the approximation given here.

48.  The Bureau of Fisheries once asked Bell Laboratories for help in finding the shortest route for getting samples from 300 locations in the Gulf of Mexico. There are 300! different possible routes. If 300! is evaluated, how many digits are used in the result? (See Exercise 47.)

 # Vocabulary List

Define and give an example of each term.

| | |
|---|---|
| experiment | tree diagram |
| event | independent events |
| simple event | dependent events |
| sample space | conditional probability |
| empirical approximation of probability | complement of an event |
| classical approach to probability | odds against |
| random selection | odds in favor |
| compound event | fundamental counting rule |
| addition rule | factorial rule |
| mutually exclusive events | permutations rule |
| multiplication rule | combinations rule |

 # Review

This chapter introduced the basic concept of **probability.** We began with two rules for finding probabilities. Rule 1 represents the *empirical* approach, whereby the probability of an event is approximated by actually conducting or observing the experiment in question.

---

**Rule 1**

$$P(A) = \frac{\text{number of times } A \text{ occurred}}{\text{number of times experiment was repeated}}$$

---

Rule 2 is called the **classical** approach, and it applies only if all of the outcomes are equally likely.

| Rule 2 |
|---|
| $$P(A) = \frac{s}{n} = \frac{\text{number of ways } A \text{ can occur}}{\text{total number of different outcomes}}$$ |

We noted that the probability of any impossible event is 0, while the probability of any certain event is 1. Also, for any event $A$,

$$0 \le P(A) \le 1$$

In Section 3–3, we considered the **addition rule** for finding the probability that $A$ **or** $B$ will occur. In evaluating $P(A \text{ or } B)$, it is important to consider whether the events are *mutually exclusive*—that is, whether they can both occur at the same time (see Figure 3–12).

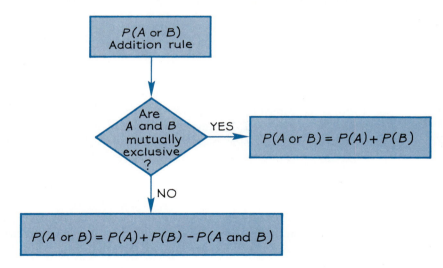

**Figure 3–12**
*Finding the Probability That for a Single Trial,
Either Event* A *or* B *(or Both) Will Occur*

In Section 3–4 we considered the **multiplication rule** for finding the probability that $A$ **and** $B$ will occur. In evaluating $P(A \text{ and } B)$, it is important to consider whether the events are *independent*—that is, whether the occurrence of one event affects the probability of the other event (see Figure 3–13 on the following page).

## THE RANDOM SECRETARY

■ One classic problem of probability goes like this: A secretary addresses 50 different letters and envelopes to 50 different people, but the letters are randomly mixed before he puts them into envelopes. What is the probability that at least one letter gets into the correct envelope? Although the probability might seem like it should be small, it's actually 0.632. Even with a million letters and a million envelopes, the probability is 0.632. The solution is beyond the scope of this text—way beyond. ■

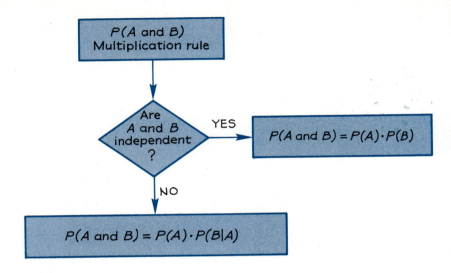

**Figure 3–13**
*Finding the Probability That Event* A *Will Occur in
One Trial and Event* B *Will Occur in Another Trial*

In Section 3–5 we considered **complements and odds.** Using the addition rule, we were able to develop the **rule of complementary events:** $P(A) + P(\overline{A}) = 1$. We saw that this rule can sometimes be used to simplify probability problems. We also defined the **odds against event** $A$ as $P(\overline{A})/P(A)$; the **odds in favor of event** $A$ are the reciprocal of the odds against $A$. Both ratios are normally expressed in the form of $a{:}b$, where $a$ and $b$ are integers having no common factors.

In Section 3–6 we considered techniques for determining the total number of different possibilities for various events. We presented the **fundamental counting rule,** the **factorial rule,** the **permutations** formula, and the **combinations** formula, all summarized below with the other important formulas from this chapter.

Most of the material that follows this chapter deals with statistical inferences based on probabilities. As an example of the basic approach used, consider a test of someone's claim that a quarter is fair. If we flip the quarter 10 times and get 10 consecutive heads, we can make one of two inferences from these sample results:

1. The coin is actually fair and the string of 10 consecutive heads is a fluke.
2. The coin is not fair.

The statistician's decision is based on the **probability** of getting 10 consecutive heads which, in this case, is so small (1/1024) that the inference of unfairness is the better choice. Here we can see the important role played by probability in the standard methods of statistical inference.

---

### Important Formulas

$0 \leq P(A) \leq 1$     for any event $A$

$P(A \text{ or } B) = P(A) + P(B)$     if $A, B$ are mutually exclusive

$P(A \text{ or } B) = P(A) + P(B) - P(A \text{ and } B)$     if $A, B$ are not mutually exclusive

$P(A \text{ and } B) = P(A) \cdot P(B)$     if $A, B$ are independent

$P(A \text{ and } B) = P(A) \cdot P(B|A)$     if $A, B$ are dependent

$P(\overline{A}) = 1 - P(A)$

$P(A) = 1 - P(\overline{A})$

$P(A) + P(\overline{A}) = 1$

odds against event $A = \dfrac{P(\overline{A})}{P(A)} \leftarrow$ usually expressed as $a{:}b$, where $a$, $b$ are integers with no common factors

$m \cdot n$ = the total number of ways two events can occur, if the first can occur $m$ ways while the second can occur $n$ ways

$n!$ = the number of ways $n$ different items can be arranged

$_nP_r = \dfrac{n!}{(n-r)!}$     the number of *permutations* (arrangements) when $r$ items are selected from $n$ available items

$_nC_r = \dfrac{n!}{(n-r)!\, r!}$     the number of *combinations* when $r$ items are selected from $n$ available items

---

# 2  *Review Exercises*

1. A medical study involves the subjects summarized in the accompanying table. If 1 subject is randomly selected, find the probability of getting someone with blood group A or B.

|  | Blood group | | | |
|---|---|---|---|---|
|  | A | O | B | AB |
| Healthy | 7 | 20 | 3 | 9 |
| Ill | 3 | 10 | 2 | 6 |

2. Using the medical data from Exercise 1, if 1 subject is randomly selected, find the probability of getting someone who is healthy or has group O blood.

3. Using the medical data from Exercise 1, if 1 subject is randomly selected, find the probability of getting someone who is ill or has group A blood.

4. Using the medical data from Exercise 1, if 1 subject is randomly selected, find the probability of getting someone with group O blood given that they are healthy.

5. The probability of a hang glider participant dying in a given year is 0.008 (based on the book *Acceptable Risks* by Imperato and Mitchell). If 10 hang glider participants are randomly selected, find the probability that they all survive a year.

6. In an experiment involving smoke detectors, an alarm was triggered after the subject fell asleep. Of the 95 sleeping subjects, 40 were awakened by the alarm, and 55 did not awake. If 1 subject is randomly selected, find the probability that he or she is among those who were not awakened by the alarm.

7. According to Bureau of the Census data, among mothers who begin working, 30.6% have less than a high school education. If 1 working mother is randomly selected, what is the probability that she has at least a high school education?

8. Fifty-five percent of enlisted personnel are married, according to a Department of Defense survey of 89,000 enlisted military personnel. If a crew of 6 enlisted military personnel is formed by random selection, what is the probability that none of them are married?

9. In attempting to gain access to a computer data bank, a computer is programmed to automatically dial every phone number with the prefix 478 followed by 4 digits. How many such telephone numbers are possible?

10. The following table summarizes the ratings for 6,665 films made before the NC 17 rating was introduced in 1990. (The table is based on data from the Motion Picture Association of America.) If one of these movies is randomly selected, find the probability that it has a rating of R.

| Rating | Number |
|--------|--------|
| G | 873 |
| PG | 2,505 |
| R | 2,945 |
| X | 342 |

11. Using the data from Exercise 10, find the probability of randomly selecting 1 of the movies and getting 1 with a rating of G or PG.

12. In a recent national election, 27 million men voted for the Republican candidate, while 19 million men voted for the Democrat. Among women, 27 million voted for the Republican while 27 million voted for the Democrat (based on data from the *New York Times*/CBS News Poll). If a voter is randomly selected from this group, find the probability of getting each outcome:

  a.   a male voter
  b.   a female voter or someone who voted for the Republican

13. In one game of the New York lottery, your probability of winning by selecting the correct 6-number combination from the 54 possible numbers is $\frac{1}{25,827,165}$. What is the probability if the rules are changed so that you must get the correct 6 numbers in the order they are selected?

14. Of 120 auto ignition circuits, 18 are defective. If 2 circuits are randomly selected, find the probability that they are both defective in each case.

  a.   The first selection is replaced before the second selection is made.
  b.   The first selection is not replaced.

15. In the televised NBC White Paper *Divorce*, it was reported that 85% of divorced women are not awarded any alimony. If 4 divorced women are randomly selected, what is the probability that at least 1 of them is not awarded alimony?

16. A pollster claims that 12 voters were randomly selected from a population of 200,000 voters (30% of whom are Republicans) and all 12 were Republicans. The pollster claims that this could easily happen. Find the probability of getting 12 Republicans when 12 voters are randomly selected from this population.

17. If a couple plans to have 5 children, find the probability that they are all girls. Assume that boys and girls have the same chance of being born and that the sex of any child is not affected by the sex of any other children.

18. a.   Find $P(A \text{ and } B)$ if $P(A) = 0.2$, $P(B) = 0.4$, and $A$ and $B$ are independent.

  b.   Find $P(A \text{ or } B)$ if $P(A) = 0.2$, $P(B) = 0.4$, and $A$ and $B$ are mutually exclusive.

  c.   Find $P(\overline{A})$ if $P(A) = 0.2$.

19. A question on a history test requires that 5 events be arranged in the proper chronological order. If a random arrangement is selected, what is the probability that it will be correct?

20. A particular recessive genetic characteristic is present in 235 subjects studied and absent in 755. If 1 of these subjects is randomly selected, find the probability that the genetic characteristic will be present.

21. A group of applicants for a temporary job consists of 8 men and 7 women. If 3 different applicants are randomly selected from this group, find the probability of each event.
    a.   All 3 are women.
    b.   There is at least 1 woman.

22. According to Bureau of the Census data, 52% of women aged 18 to 24 years do not live at home with their parents. If we randomly select 5 different women in that age bracket, what is the probability that none of them live at home with their parents?

23. Evaluate the following.
    a. $8!$    b. $_8P_6$    c. $_{10}C_8$    d. $_{80}C_{78}$

24. On January 28, 1986, the Space Shuttle *Challenger* exploded, killing all seven astronauts aboard. The problem was found to be a failure in a joint of the shuttle's booster rocket. Such a disaster could occur if any 1 of 6 connected joints failed, and the probability of failure for any 1 joint was 0.023. Assuming that the 6 joints are independent, find the probability that at least 1 of them would fail.

25. If a couple plans to have 4 children, find the probability that they are all of the same sex. Assume that boys and girls are equally likely and that the sexes are independent.

26. A pill designed to prevent certain physiological reactions is advertized as 95% effective. Find the probability that the pill will work in all of 15 separate and independent applications.

27. A critical component in a circuit will work properly only if 3 other components all work properly. The probabilities of a failure for the 3 other components are 0.010, 0.005, and 0.012. Find the probability that at least 1 of these three components will fail.

28. If 7 different customers arrive for service in random order, what is the probability that they will be in alphabetical order?

## Computer Projects

1. In many situations, probability problems can be solved by writing a computer program that simulates the relevant circumstances. In determining the probability of winning the game of solitaire, for example, it is easier to program a computer to play solitaire than it is to develop calculations using the rules of probability. The computer can then play solitaire 1000 times and the probability of winning is estimated to be the number of wins divided by 1000. Such simulations often involve the computer's ability to generate random numbers. The subroutine RND is available in many BASIC implementations. One

common use causes an entry of RND(X) to return a value between 0 and 1, such as 0.23560387. That result can be manipulated to produce desired values.

```
10 RANDOMIZE
20 LET R = INT (100 * RND(X))
30 IF R > 37 THEN 20
40 PRINT R
```

In the short program above, we take a value like 0.23560387, multiply by 100 to get 23.560387, and then take only the integer part to get 23. If the result is greater than 37, we go back and try again. Those three lines produce randomly selected values from the list 0, 1, 2, ..., 37. This short program simulates the spinning of a roulette wheel if we stipulate that 37 corresponds to 00.

a.    Enter those lines and run the program 20 times. If you were betting on the number 7, how many times did you win?

b.    In roulette, if you bet $1 on "odd," you win $1 if the result is an odd number. (Remember that in our program, 37 should not win since it represents 00.) Modify the program so that you begin with $10 and you bet $1 on each spin until you either reach $20 or go broke. The final output should indicate whether you reached $20 or went broke first.

c.    A gambler is in Las Vegas with only $10 but must have $20 for the return trip home. Use computer simulations to determine which of the following two strategies is more likely to get him the $20 he needs for the return trip.

    i.    Bet the entire $10 on "odd" for one spin of the roulette wheel.

    ii.   Bet on odd, $1 at a time, until reaching $20 or going broke.

2.    Refer to Exercise 32 in Section 3–2. In the author's humble opinion, that is the most difficult exercise in this book. While a theoretical solution is possible, it is far from easy. Solve that problem by using a computer simulation instead. First, the length of the stick is irrelevant, so assume it's 1 unit long. Use a computer to select 2 random numbers between 0 and 1. Find the length of the longest of the 3 segments; if it is less than 0.5, then a triangle can be formed. Now repeat that experiment many times and record the total number of trials and the number of times a triangle could be formed. The desired probability can be estimated by using the relative frequency definition.

## Applied Projects

1. Referring to Data Set II in Appendix B, estimate the probability of a randomly selected adult being left-handed.

2. In Section 3–1 we stated that at least 2 students will have the same birthday in more than half of the classes with 25 students. Exercise 30 in Section 3–4 requires the theoretical solution to that problem, but many such problems can be solved by developing a **simulation.** Let the days of the year be represented by the numbers from 001 through 365. Now refer to a page randomly selected from a telephone book and record the last 3 digits of 25 telephone numbers, but ignore 000 and any cases above 365. Check to determine whether or not at least 2 of the dates (represented by the 3–digit numbers) are the same. It should take about 30 minutes to repeat this experiment 10 times. Find the estimated probability, which is the number of times matched dates occurred, divided by the number of times the experiment is repeated. (Another approach is to use a computer instead of a telephone book. STATDISK can be used to repeatedly generate 25 numbers between 1 and 365.)

3. In producing an experimental computer chip, a research group experiences a 5% yield, meaning that there is a 0.05 probability of a chip being acceptable. Use a simulation approach (see the preceding applied project) to "generate" chips until getting one that is acceptable. Record the number of chips required to get one that is acceptable. Repeat this experiment 50 times and record the results. Construct a frequency table, histogram, stem-and-leaf plot, and find the mean, median, and standard deviation. Write a brief report describing important characteristics of the data.

4. A psychologist plans to conduct a study of people who are tall and live in relatively small homes. Refer to the data sets in Appendix B to answer the following.
   a. Estimate the probability that a randomly selected adult male is at least 6 ft tall.
   b. Estimate the probability that if a recently sold home is randomly selected, its living area is less than 1500 sq ft.
   c. Assume that a male is randomly selected from a home that was recently bought. Use the results from parts *a* and *b* to estimate the probability of getting someone at least 6 ft tall who lives in a home with a living area less than 1500 sq ft.
   d. In doing the calculations for part *c*, identify any assumptions that were made about the events.

## *Writing Projects*

1.  Assume that you've just received a letter from one of your best friends. He has stated that he is about to spend $39.95 for "The Complete Lottery Tracker and Wheeler," a computer software package that will allow him to track lottery numbers so that he can increase his chances of winning. (Such software packages do exist, and people actually buy them.) Write a letter explaining that such an expenditure would be a waste of money. Try to convince him that lottery number selection involves independent events, so past results will not affect future selections.
2.  Write a report summarizing the videotape program listed below.

 ## *Videotapes*

Program 15 from the series *Against All Odds: Inside Statistics* is recommended as a supplement to this chapter.

Photo courtesy of Nielsen Media Research

## AN INTERVIEW WITH
# BARRY COOK

### SENIOR VICE PRESIDENT AT NIELSEN MEDIA RESEARCH

*Barry Cook is a Senior Vice President and Chief Research Officer at Nielsen Media Research. He has taught at Yale and Hunter College and worked for NBC and the USA Cable Network. He is now in charge of Nielsen's rating system, doing research to better understand how the measurements work, as well as developing new measurement systems.*

**What major trends do you see in the way Americans watch TV?**

In 1985 the average home received 18.8 channels. In 1990 the average home received 33.2 channels. That obviously has an effect on what people choose to view.

**What is your sample size?**

For the national survey we use "people meters" in 4000 homes with about 11,000 people. We increased the sample size because the use of television has changed. Instead of only three major sources of TV (ABC, NBC, CBS), there are now dozens of sources of programming, many of which get only a small piece of the audience. In order to measure those smaller pieces with enough precision, a larger sample is needed. In addition, we also have meter services in 25 markets; the television sets are metered (but not the people) in 250 to 500 homes per market. Nielsen is still very big in the diary business—not for the national audience, but for measuring audiences in the 200 or so separate markets across the country. Those diaries amount to a combined sample size of 100,000, four times a year.

**Have you been experiencing greater resistance to polls and surveys?**

There's no question that we have seen a decline in cooperation in both telephone and in-person contacts with people. It's across the entire survey industry. There are concerns about privacy. The data gathering efforts are being mixed up with sales efforts and that probably is contributing to a decline in the cooperation rate. Also, answering machines make it harder to get through to people.

## Do you weight sample results to better reflect population parameters?

We have a policy against that. We try to represent population parameters by doing the sampling correctly in the first place. We sample in a way that gives an equal probability of selection to all housing units in the 50 states. As a result, there is a known amount of sampling error and there's also an unknown amount of nonsampling error, but we've done validation research to estimate how close the samples are to measures of the population.

> *"We try to represent population parameters by doing the sampling correctly in the first place."*

## What are some of the specific statistical methods you use?

We use a lot of statistics for our own understanding and our clients' understanding of sampling and trends. We get into hypothesis testing when we try to understand why things change. Confidence intervals are very important in interpreting the estimates of the population.

## Could you cite a television programming strategy that is based on survey results?

The most important strategy is called "prime time." The biggest usage of television occurs in the evening hours when most people are home. The most general programming strategy is to put on your best shows when there's the greatest number of people there. With a miniseries, what you get on the first episode serves as almost a cap on what you can get after that. You want to get the maximum possible potential audience for the first installment, so Sunday night does that. ∎

# Chapter Four

## In This Chapter

# 4 Probability Distributions

## Chapter Problem

Statistics is at its best when it is used to benefit humanity in some way. Companies use statistics to become more efficient, increase shareholders' profits, and lower prices. Regulatory agencies use statistics to ensure the safety of workers and clients. In this chapter, one of the many applied examples involves a situation in which cost effectiveness and passenger safety were both critical factors. With new aircraft designs and improved engine reliability, airline companies wanted to fly transatlantic routes with twin-engine jets, but the Federal Aviation Administration required at least three engines for transatlantic flights. Lowering this requirement was of great interest to manufacturers of twin-engine jets (such as the Boeing 767). Also, the two-engine jets use about half the fuel of jets with three or four engines. Obviously, the key issue in approving the lowered requirement is the probability of a twin-engine jet making a safe transatlantic crossing. This probability should be compared to that of three- and four-engine jets. Clearly, such a study involves a thorough understanding of the related probabilities, and the contents of this chapter will enable us to develop such an understanding. We will return to this case study as we develop the relevant principles.

# 4–1  Overview

In Chapter 2 we discussed the histogram as a device for showing the frequency distribution of a set of data. In Chapter 3 we discussed the basic principles of probability theory. In this chapter we combine those concepts to develop probability distributions that are basically theoretical models of the frequency distributions we produce when we collect sample data. We construct frequency tables and histograms using *observed* real scores, but we construct probability distributions by presenting possible outcomes along with their *probable* frequencies.

Suppose a casino manager suspects cheating at a dice table. The manager can compare the frequency distribution of the actual sample outcomes to a theoretical model that describes the frequency distribution likely to occur with fair dice. In this case the probability distribution serves as a model of a theoretically perfect population frequency distribution. In essence, we can determine what the frequency table and histogram would be like for a pair of fair dice rolled an infinite number of times. With this perception of the population of outcomes, we can then determine the values of important parameters such as the mean, variance, and standard deviation.

The concept of a probability distribution is not limited to casino management. In fact, the remainder of this book and the very core of inferential statistics depend on some knowledge of probability distributions. To analyze the effectiveness of a new drug, for example, we must know something about the probability distribution of the symptoms the drug is intended to correct.

This chapter deals with discrete cases, while subsequent chapters involve continuous cases. We begin by examining the concept of a random variable.

# 4–2  Random Variables

In Chapter 3 we defined an experiment to be any process that allows us to obtain observations. Some experiments give us observations that are quantitative (such as weights), while others give us observations that are qualitative (such as colors). Qualitative outcomes can often be expressed with numbers. For example, the qualitative performances of Olympic divers are numerically rated by judges. If we can associate each outcome of an experiment with a single number, then we have a variable whose values are determined by chance. Such variables are called *random variables*.

## Definition

A **random variable** has a single numerical value for each outcome of an experiment.

The values of a random variable are the numbers we associate with the different outcomes that make up the sample space for the experiment. As an example, consider the experiment of randomly selecting 3 consumers. We can let the random variable represent the number of those consumers who recognize the Chrysler logo. That random variable can then assume the possible values of 0, 1, 2, and 3. Even though the actual outcomes are not numbers, we can associate numbers with the results.

## Example

A company gives job applicants a test of mathematical skills. The quiz consists of 10 multiple-choice questions. Let the random variable represent the number of correct answers. This random variable can take on the values of 0, 1, 2, 3, 4, 5, 6, 7, 8, 9, 10.

In Section 1–3 we made a distinction between discrete and continuous data. Random variables may also be discrete or continuous, and the following two definitions are consistent with those given in Section 1–3.

## Definition

A **discrete random variable** has either a finite number of values or a countable number of values.

A **continuous random variable** has infinitely many values, and those values can be associated with points on a continuous scale in such a way that there are no gaps or interruptions.

Random variables that represent counts are usually discrete, while measurements (such as heights, weights, times, temperatures) are usually associated with continuous random variables. In reality, it is not possible to deal with exact values of a continuous random variable, so we round off to a limited number of decimal places.

## PROPHETS FOR PROFITS

■ You can spend a small fortune buying books, computer software, or magazine subscriptions that are supposed to help you select lottery numbers. These "aids" are apparently blind to the fact that lottery numbers are randomly selected and the outcomes are independent of previous results. These aids typically recommend some numbers as "hot" because they have been coming up often. Others are recommended as "due" because they haven't been coming up often. Other approaches involve numerology, astrology, dreams, and numbers that have "appeared or talked to" a seeress, as one book claims. It's all worthless in predicting winning lottery numbers. ■

# Discrete Random Variables

Inferential statistics is often used to make decisions in a wide variety of fields. We begin with sample data and attempt to make inferences about the population from which the sample was drawn. If the sample is very large we may be able to develop a good estimate of the population frequency distribution, but samples are often too small for that purpose. The practical approach is to use information about the sample along with general knowledge about population distributions.

Much of this general information is included in this and the following chapters. Without a knowledge of probability distributions, users of statistics would be severely limited in the inferences they could make. We intend to develop the ability to work with discrete and continuous probability distributions, and we begin with the discrete case because it is simpler.

> ### Definition
>
> A **probability distribution** gives the probability for each value of the random variable.

For example, suppose a drug is administered to 2 patients, the random variable is the number of cures (0, 1, or 2), and the probability of a cure is $\frac{1}{2}$. The probability of 0 cures is $\frac{1}{4}$, the probability of 1 cure is $\frac{1}{2}$, while the probability of 2 cures is $\frac{1}{4}$. Table 4–1 summarizes the probability distribution for this situation. (The given probabilities can be easily verified by listing the 4 cases in the sample space, beginning with "cure-cure.") Ordinarily, the probability distribution of a random variable will be given as a table, such as Table 4–1, or a formula. The formula corresponding to Table 4–1 is $P(x) = 1/2(2 - x)! \, x!$.

| TABLE 4–1 | |
|:---:|:---:|
| $x$ | $P(x)$ |
| 0 | $\frac{1}{4}$ |
| 1 | $\frac{1}{2}$ |
| 2 | $\frac{1}{4}$ |

There are various ways to graph a probability distribution, but we present only the **probability histogram.** Figure 4–1 is a probability histogram that resembles the relative frequency histogram from Chapter 2, but the vertical scale delineates probabilities instead of actual relative frequencies.

In Figure 4–1, note that along the horizontal axis, the values of 0, 1, and 2 are located at the centers of the rectangles. This implies that the rectangles are each 1 unit wide, so the areas of the three rectangles are ¼, ½, and ¼.

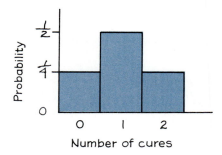

**Figure 4–1**

In general, if we stipulate that each value of the random variable is assigned a width of 1 on the histogram, then the areas of the rectangles will total 1. **We can therefore associate the probability of each numerical outcome with the area of the corresponding rectangle.** This correspondence between probability and area is an important concept that will be used many times in later chapters.

For a discrete random variable, if two events result in different values of the random variable, those events must be mutually exclusive. As an example, if you give a drug to 3 patients, it's impossible to get exactly 2 cures and exactly 3 cures at the same time, so the events of 2 cures and 3 cures are mutually exclusive. Knowing that all the values of the random variable will cover all events of the entire sample space, and knowing that events that lead to different values of the random variable are mutually exclusive, we can conclude that the sum of $P(x)$ for all values of $x$ must be 1. Also, $P(x)$ must be between 0 and 1 for any value of $x$.

---

### Requirements for $P(x)$ to Be a Probability Distribution

**1.** $\Sigma P(x) = 1$      where $x$ assumes all possible values
**2.** $0 \le P(x) \le 1$    for every value of $x$

---

These two requirements for probability distributions are actually direct descendents of the corresponding rules of probabilities (discussed in Chapter 3).

## Is Parachuting Safe?

■ About 30 people die each year as more than 100,000 people make about 2.25 million parachute jumps. In comparison, a typical year includes about 200 scuba diving fatalities, 7000 drownings, 900 bicycle deaths, 800 lightning deaths, and 1150 deaths from bee stings. Of course, these figures don't necessarily mean that parachuting is safer than bike riding or swimming. A fair comparison should involve fatality *rates*, not just the total number of deaths.

   The author, with much trepidation, made two parachute jumps but quit after missing the spacious drop zone both times. He has also flown in a hang glider, hot air balloon, and a Goodyear blimp. ■

### Example

Does $P(x) = x/5$ (where $x$ can take on the values of 0, 1, 2, 3) determine a probability distribution?

### Solution

If a probability distribution is determined, it must conform to the preceding two requirements. But

$$\Sigma P(x) = P(0) + P(1) + P(2) + P(3)$$
$$= \frac{0}{5} + \frac{1}{5} + \frac{2}{5} + \frac{3}{5}$$
$$= \frac{6}{5} \qquad\qquad \Sigma P(x) \neq 1$$

Thus the first requirement is not satisfied and a probability distribution is not determined.

### Example

Does $P(x) = x/10$ (where $x$ can be 0, 1, 2, 3, or 4) determine a probability distribution?

### Solution

For the given function we conclude that

$$P(0) = \frac{0}{10} = 0 \qquad P(3) = \frac{3}{10}$$
$$P(1) = \frac{1}{10} \qquad P(4) = \frac{4}{10}$$
$$P(2) = \frac{2}{10} \qquad \left(\frac{0}{10} + \frac{1}{10} + \frac{2}{10} + \frac{3}{10} + \frac{4}{10} = 1\right.$$
$$\text{so that } \Sigma P(x) = 1\Bigg)$$

The sum of these probabilities is 1, and each $P(x)$ is between 0 and 1, so both requirements are satisfied. Consequently, a probability distribution is determined. The graph of this probability distribution is shown in the probability histogram of Figure 4–2. Note that the sum of the areas of the rectangles is 1, and each rectangle has an area between 0 and 1.

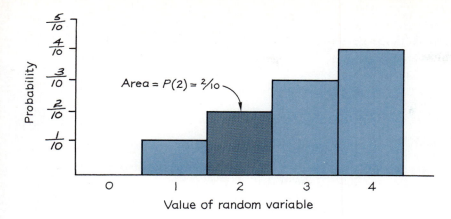

**Figure 4–2**
*Value of Random Variable*

We saw in the overview that probability distributions are extremely important in the study of statistics. We have just considered probability distributions of discrete random variables, and later chapters will consider important probability distributions of continuous random variables.

# 4–2 | Exercises A

In Exercises 1–12, determine whether a probability distribution is given. In those cases where $P(x)$ does not determine a probability distribution, identify the requirement that is not satisfied.

1. In the accompanying table (based on past results), $x$ represents the number of games required to complete a baseball World Series contest.

| $x$ | $P(x)$ |
|---|---|
| 4 | 0.120 |
| 5 | 0.253 |
| 6 | 0.217 |
| 7 | 0.410 |

2. In the accompanying table, $x$ represents the number of children (under 18 years of age) in families. (The table is based on data from the U.S. Census Bureau.)

| $x$ | $P(x)$ |
|---|---|
| 0 | 0.48 |
| 1 | 0.21 |
| 2 | 0.19 |
| 3 | 0.08 |

3.  In the accompanying table, $x$ represents the number of long-distance personal calls an individual makes in one month.

| $x$ | $P(x)$ |
|---|---|
| 0 | 0.32 |
| 1 | 0.08 |
| 2 | 0.12 |
| 3 | 0.09 |
| 4 | 0.07 |
| 5 | 0.06 |
| 6 | 0.04 |

4.  In the accompanying table, $x$ represents the number of employees absent on a given day.

| $x$ | $P(x)$ |
|---|---|
| 0 | 1/5 |
| 1 | 1/5 |
| 2 | 1/5 |
| 3 | 1/5 |
| 4 | 1/5 |
| 5 | 1/5 |

5.  $P(x) = x$ for $x = 0.1, 0.3, 0.6$
6.  $P(x) = x$ for $x = 0, 1/2, 1/4$
7.  $P(x) = 1/3$ for $x = 1, 2, 3$
8.  $P(x) = x^2$ for $x = 0, 1/2, 1/3, 5/6$
9.  $P(x) = x - 0.5$ for $x = 0.5, 0.6, 0.7, 0.8, 0.9$
10. $P(x) = x - 2.5$ for $x = 2, 3$
11. $P(x) = \dfrac{1}{2(2 - x)!x!}$ for $x = 0, 1, 2$
12. $P(x) = \dfrac{3}{4(3 - x)!x!}$ for $x = 0, 1, 2, 3,$

In Exercises 13–20, do each of the following:

   a.  List the values that the random variable $x$ can assume.
   b.  Determine the probability $P(x)$ for each value of $x$.
   c.  Summarize the probability distribution as a table that follows the format of Table 4–1.
   d.  Construct the probability histogram that represents the probability distribution for the random variable $x$. (See Figure 4–2.)
   e.  Indicate the area of each rectangle in the probability histogram of part $d$ so that there is a correspondence between area and probability.

13. A drug is administered to 3 patients and the random variable $x$ represents the number of cures that occur. The probabilities corresponding to 0 cures, 1 cure, 2 cures, and 3 cures are found to be 0.125, 0.375, 0.375, and 0.125, respectively.

14. A manufacturer produces gauges in batches of 3 and the random variable $x$ represents the number of defects in a batch. The probabilities

corresponding to 0 defects, 1 defect, 2 defects, and 3 defects are found to be 0.70, 0.20, 0.09, and 0.01, respectively.

15. A computer simulation involves the repeated generation of a random number. The only possibilities are 0, 1, 2, and they are equally likely.

16. A computer is used to select the last digit of telephone numbers to be dialed for a poll. The possible values are 0, 1, 2, . . . , 9, and they are equally likely.

17. The random variable $x$ represents the number of girls in a family of 3 children. (*Hint:* Assuming that boys and girls are equally likely, we get $P(2) = 3/8$ by examining the sample space of *bbb, bbg, bgb, bgg, gbb, gbg, ggb, ggg.*)

18. The random variable $x$ represents the number of boys in a family of 4 children. (See Exercise 17.)

19. The random variable $x$ represents the number of foul shots a player makes when 2 shots are attempted. Assume that the player hits an average of 85% of his or her free throws, and that the shots are independent. (*Hint:* begin by using the multiplication rule to find $P(0)$ and $P(2)$.)

20. A batch of 20 computer chips contains exactly 6 that are defective. Two different chips are randomly selected without replacement, and the random variable $x$ represents the number of defective chips selected. (*Hint:* Use the multiplication rule to first find $P(0)$ and $P(2)$.)

## 4–2 Exercises B

21. Let $P(x) = 0.4(0.6)^{x-1}$ where $x = 1, 2, 3, \ldots$. Is $P(x)$ a probability distribution?

22. a. Let $P(x) = 1/2^x$ where $x = 1, 2, 3, \ldots$. Is $P(x)$ a probability distribution?
    b. Let $P(x) = 1/2x$ where $x = 1, 2, 3, \ldots$. Is $P(x)$ a probability distribution?

23. According to an analyst for Paine Webber, Intel's yield for computer chips is around 35%. Assume that we have 4 randomly selected chips and that the probability of an acceptable chip is 0.35. Let $x$ represent the number of acceptable chips in groups of 4 and use the multiplication rule and the rule of complements (from Chapter 3) to complete the table so that a probability distribution is determined.

| $x$ | $P(x)$ |
|---|---|
| 0 | |
| 1 | 0.384 |
| 2 | 0.311 |
| 3 | |
| 4 | |

24. Given that $x$ is a continuous random variable with the distribution shown in Figure 4–3, find each of the following.
    a. The total area of the enclosed region
    b. The probability that $x$ is less than 5
    c. The probability that $x$ is less than 7
    d. The probability that $x$ is between 6 and 7
    e. The probability that $x$ is between 3 and 7

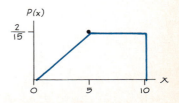

**Figure 4–3**

# $\boxed{4\text{--}3}$ Mean, Variance, and Expectation

In Chapter 2 we saw that there are three extremely important characteristics of data:

1.  Representative score, such as an average
2.  Measure of scattering or dispersion
3.  Nature of distribution, such as bell-shaped

The probability histogram of the previous section can give us insight into the nature of the distribution and, in this section, we will see how to find the mean, standard deviation, and variance for a probability distribution.

Suppose that we want to determine the mean and variance for the numbers of boys that will occur in pairs of independent births. (Assume that boys and girls are equally likely.) We have no specific results to work with, but we do know that the outcomes of 0, 1, and 2 boys have probabilities of 1/4, 1/2, 1/4, respectively. One way to find the mean and variance for the numbers of boys is to pretend that theoretically ideal results actually occurred. We list the different possible outcomes.

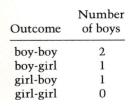

| Outcome | Number of boys |
|---------|:---:|
| boy-boy | 2 |
| boy-girl | 1 |
| girl-boy | 1 |
| girl-girl | 0 |

Since they are equally likely, we can pretend that the 4 results actually did occur. The mean of 2, 1, 1, and 0 is 1, while the variance is 0.5. (The variance is found by applying Formula 2–4. In this use of Formula 2–4 we divide by $n$ instead of $n - 1$ because we assume that we have all scores of the population.) Instead of considering 4 trials that yield theoretically ideal results, we could pretend that a large number of trials (say 4000) yielded these theoretically ideal results:

|  |  | Number of boys | Frequency |
|---|---|:---:|:---:|
| boy-boy | (2 boys 1000 times) ⎤ | | |
| boy-girl | (1 boy 1000 times) ⎥ | 2 | 1000 |
| girl-boy | (1 boy 1000 times) ⎬ | 1 | 2000 |
| girl-girl | (0 boys 1000 times) ⎦ | 0 | 1000 |

Our random variable represents the number of boys in 2 births, so the preceding results suggest a list of 4000 scores that consist of 1000 twos, a total of 2000 ones, and 1000 zeros. The mean of those 4000 scores is again 1, while the variance is again 0.5. In this example, it really makes no difference whether we presume 4 theoretically ideal trials or 4000.

Instead of pretending that theoretically ideal results have occurred, we can find the mean of a discrete random variable by using Formula 4–1.

**Formula 4–1**    $\mu = \Sigma x \cdot P(x)$    Mean for a probability distribution

We should stress that $\Sigma x \cdot P(x)$ is the same as $\Sigma[x \cdot P(x)]$. That is, multiply each value of $x$ by its corresponding probability, and then add the resulting products.

Formula 4–1 is justified by relating $P(x)$ to its role in describing the relative frequency with which $x$ occurs. Recall that the mean of any list of scores is the sum of those scores divided by the total number of scores $n$. We usually compute the mean in that order; that is, first sum the scores and then divide the total by $n$. However, we can obtain the same result by dividing each individual score by $n$ and then summing the quotients. For example, we can find the mean of the preceding 4000 scores by writing

$$\frac{(2 \cdot 1000) + (1 \cdot 2000) + (0 \cdot 1000)}{4000} = 1$$

We can also compute that mean as

$$\mu = \left(2 \cdot \frac{1000}{4000}\right) + \left(1 \cdot \frac{2000}{4000}\right) + \left(0 \cdot \frac{1000}{4000}\right) = 1$$

The right side of this last equation represents $\Sigma\left(x \cdot \frac{f}{n}\right)$, which corresponds to $\Sigma x \cdot P(x)$ in Formula 4–1.

For each specific value of $x$, $P(x)$ can be considered the relative frequency with which $x$ occurs. $P(x)$ takes into account the repetition of specific $x$ scores when we compute the mean. It also incorporates division by $n$ directly into the summation process.

Similar reasoning enables us to take the variance formula from Chapter 2 $(\sigma^2 = \Sigma(x - \mu)^2/N)$ and apply it to a random variable of a probability distribution.

**Formula 4–2**     $\sigma^2 = \Sigma(x - \mu)^2 \cdot P(x)$     Variance for a probability distribution

Again the use of $P(x)$ takes into account the repetition of specific $x$ scores and simultaneously accomplishes the division by $N$. This latter formula for variance is usually manipulated into an equivalent form to facilitate computations, as shown in Formula 4–3.

**Formula 4–3**     $\sigma^2 = [\Sigma x^2 \cdot P(x)] - \mu^2$     (shortcut version of Formula 4–2)

Formula 4–3 is a shortcut version that will always produce the same result as Formula 4–2. Formula 4–3 is usually easier to work with, while Formula 4–2 is easier to understand directly.

To apply Formula 4–3 to a specific case, we square each value of $x$ and multiply that square by the corresponding probability and then add all of those products. We then subtract the square of the mean. The standard deviation $\sigma$ can be easily obtained by simply taking the square root of the variance, as shown in Formula 4–4 on the following page. If the variance is found to be 0.5, the standard deviation is $\sqrt{0.5}$, or about 0.7.

# PICKING LOTTERY NUMBERS

■ In a typical state lottery, you select six different numbers. After a random drawing, any entries with the correct combination share in the prize. Since the winning numbers are randomly selected, any choice of six numbers will have the same chance as any other choice, but some combinations are better than others. The combination of 1, 2, 3, 4, 5, 6 is a poor choice because many people tend to select it. In a Florida lottery with a $105 million prize, 52,000 tickets had 1, 2, 3, 4, 5, 6; if that combination had won, the prize would have been only $1000. It's wise to pick combinations not selected by many others. Avoid combinations that form a pattern on the entry card. ■

**Formula 4–4**          $\sigma = \sqrt{[\Sigma x^2 \cdot P(x)] - \mu^2}$          Standard deviation
for a probability
distribution

   When we calculate the mean from a probability distribution, we get the average (mean) value that we would expect to get if the trials could be repeated indefinitely. We *don't* get the value we expect to occur most often. In fact, we often get a mean value that cannot occur in any one actual trial (such as 1.5 girls in 3 births). The standard deviation gives us a measure of how much the probability distribution is spread out around the mean. A large standard deviation reflects considerable spread, while a lower standard deviation reflects lower variability with values relatively closer to the mean.
   In the following example, we find the mean, variance, and standard deviation of a random variable. We put the calculations in a well organized table.

### Example

United Airlines Flight 470 from Denver to St. Louis has an on-time performance described by the probability distribution given in the first two columns of Table 4–2, where x is the number of on-time flights among 3 independent flights (based on data from the EAASY SABRE reservation system). Find the mean, variance, and standard deviation for the random variable x.

### Solution

In Table 4–2, the two columns on the left side describe the probability distribution, and we create the three columns on the right side for the purposes of the calculations required.

| TABLE 4–2 | | | | |
|---|---|---|---|---|
| $x$ | $P(x)$ | $x \cdot P(x)$ | $x^2$ | $x^2 \cdot P(x)$ |
| 0 | 0.064 | 0 | 0 | 0 |
| 1 | 0.288 | 0.288 | 1 | 0.288 |
| 2 | 0.432 | 0.864 | 4 | 1.728 |
| 3 | 0.216 | 0.648 | 9 | 1.944 |
| Total | | 1.800 | | 3.960 |

*continued*

**Solution** *continued*

Using Formulas 4–1, 4–3, and the table results, we get $\mu = \Sigma x \cdot P(x) = 1.8$ and $\sigma^2 = (\Sigma x^2 \cdot P(x)) - \mu^2 = 3.96 - 1.8^2 = 0.72$, which is rounded to 0.7. The standard deviation is the square root of the variance, so that

$$\sigma = \sqrt{0.72} = 0.8 \qquad \text{(rounded)}$$

We now know that among 3 flights, the mean number of on-time flights is 1.8 while the variance and standard deviation are 0.7 and 0.8, respectively.

An important advantage of these techniques is that a probability distribution is actually a model of a theoretically perfect population frequency distribution. The probability distribution is like a relative frequency distribution based on data that behave perfectly, without the usual imperfections of samples. Since the probability distribution allows us to perceive the population, we are able to determine the values of important parameters such as the mean, variance, and standard deviation. This in turn allows us to make the inferences that are necessary for decision making in a multitude of different professions.

# Expected Value

The mean of a discrete random variable is the theoretical mean outcome for infinitely many trials. We can think of that mean as the expected value in the sense that it is the average (mean) value that we would expect to get if the trials could continue indefinitely. The uses of expected value (also called expectation or mathematical expectation) are extensive and varied, and they play a very important role in an area of application called *decision theory*.

## Definition

The **expected value** of a discrete random variable is $E = \Sigma x \cdot P(x)$

From Formula 4–1 we see that $E = \mu$. That is, the mean of a discrete random variable and its expected value are the same.

As an example, consider the numbers game started many years ago by organized crime groups and now run legally by many organized governments. You can bet that the 3-digit number of your choice will be the winning number selected. Suppose that you bet $1 on the number 327. What is

## DRIVE TO SURVIVE PAST AGE 35

■ Avoid cars—motor vehicle crashes are the leading cause of death among Americans under 35 years of age. If you drive, don't drink—40% of fatally injured drivers are drunk. Drive in large cars—the death rate in the largest cars (1.3 per 10,000 vehicles) is less than half the rate in the smallest cars (3.0 per 10,000 vehicles). Wear safety belts—in a study of 1126 accidents, riders wearing safety belts had 86% fewer life-threatening injuries. ■

your expected value? For this bet there are two simple outcomes: You win or you lose. Since you have the number 327 and there are 1000 possibilities (from 000 to 999), your probability of winning is 1/1000 and your probability of losing is 999/1000. The typical winning payoff is 499 to 1, meaning that for each $1 bet, you would be given $500 and your net return is therefore $499. Table 4–3 summarizes this situation.

| TABLE 4–3 | The Numbers Game | | |
|---|---|---|---|
| Event | $x$ | $P(x)$ | $x \cdot P(x)$ |
| Win | $499 | $\dfrac{1}{1000}$ | $\dfrac{\$499}{1000}$ |
| Lose | $-\$1$ | $\dfrac{999}{1000}$ | $-\dfrac{\$999}{1000}$ |
| Total | | | $\dfrac{-\$500}{1000} = -50¢$ |

From Table 4–3 we can see that when we bet $1 in the numbers game, our expected value is

$$E = \Sigma x \cdot P(x) = -50¢ \qquad \text{(from Table 4–3)}$$

This means that in the long run, for each $1 bet, we can expect to lose an average of 50¢. The house "take" is 50%, which is very high compared to the house take in casino games such as slot machines (13%), roulette (5.25%), or craps (1.4%). In contrast, a **fair game** is one in which the house or other players have no advantage.

In Chapter 10 we will use the concept of expected value to compare actual survey results to expected results; the degree of similarity or disparity will allow us to form some meaningful conclusions.

## $\boxed{4\text{–}3}$ Exercises A

In Exercises 1–8, find the mean, variance, and standard deviation of the random variable $x$.

1. 
| $x$ | $P(x)$ |
|---|---|
| 0 | 0.2 |
| 1 | 0.5 |
| 2 | 0.3 |

2. 
| $x$ | $P(x)$ |
|---|---|
| 0 | 0.2 |
| 1 | 0.7 |
| 2 | 0.1 |

3.

| $x$ | $P(x)$ |
|---|---|
| 1 | 0.15 |
| 2 | 0.45 |
| 3 | 0.35 |
| 4 | 0.05 |

4.

| $x$ | $P(x)$ |
|---|---|
| 5 | 0.20 |
| 6 | 0.10 |
| 7 | 0.45 |
| 8 | 0.25 |

5.

| $x$ | $P(x)$ |
|---|---|
| 5 | 1/4 |
| 10 | 1/4 |
| 20 | 1/4 |
| 50 | 1/4 |

6.

| $x$ | $P(x)$ |
|---|---|
| 5 | 1/20 |
| 10 | 3/20 |
| 20 | 7/20 |
| 50 | 9/20 |

7.

| $x$ | $P(x)$ |
|---|---|
| 2 | 4/30 |
| 4 | 6/30 |
| 6 | 10/30 |
| 8 | 6/30 |
| 10 | 4/30 |

8.

| $x$ | $P(x)$ |
|---|---|
| 2 | 1/5 |
| 4 | 1/5 |
| 6 | 1/5 |
| 8 | 1/5 |
| 10 | 1/5 |

9. If you randomly select a jail inmate convicted of DWI (driving while intoxicated), the probability distribution for the number $x$ of prior DWI sentences is as described in the accompanying table (based on data from the U.S. Department of Justice). Find the mean, variance, and standard deviation.

| $x$ | $P(x)$ |
|---|---|
| 0 | 0.513 |
| 1 | 0.301 |
| 2 | 0.132 |
| 3 | 0.055 |

10. According to an analyst for Paine Webber, Intel's yield for computer chips is around 35%. For a batch of 6 randomly selected computer chips, the accompanying table describes the probability distribution for the number of defects. Find the mean, variance, and standard deviation for the number of defects.

| $x$ | $P(x)$ |
|---|---|
| 0 | 0.002 |
| 1 | 0.020 |
| 2 | 0.095 |
| 3 | 0.235 |
| 4 | 0.328 |
| 5 | 0.244 |
| 6 | 0.075 |

11. Letting $x$ represent the number of credit cards adults have, the accompanying table describes the probability distribution for a certain population (based on data from Maritz Marketing Research, Inc.). Find the mean, variance, and standard deviation for the number of credit cards.

| $x$ | $P(x)$ |
|---|---|
| 0 | 0.26 |
| 1 | 0.16 |
| 2 | 0.12 |
| 3 | 0.09 |
| 4 | 0.07 |
| 5 | 0.09 |
| 6 | 0.07 |
| 7 | 0.14 |

12. According to the Hertz Corporation, 69% of all workers commute in their own cars. The accompanying table describes the probability distribution for the number of workers who commute in their own cars for 8 randomly selected workers. Find the mean, variance, and standard deviation.

| $x$ | $P(x)$ |
|---|---|
| 0 | 0.000 |
| 1 | 0.002 |
| 2 | 0.012 |
| 3 | 0.053 |
| 4 | 0.147 |
| 5 | 0.261 |
| 6 | 0.290 |
| 7 | 0.185 |
| 8 | 0.051 |

13. To settle a paternity suit, 2 different people are given blood tests. If $x$ is the number having Group A blood, then $x$ can be 0, 1, or 2 and the corresponding probabilities are 0.36, 0.48, and 0.16, respectively (based on data from the Greater New York Blood Program). Find the mean, variance, and standard deviation for the number having Group A blood in randomly selected couples.

14. Car headlight manufacturers are concerned about failure rates. One headlight failure is an inconvenience, but if both lights fail, you can't drive at night. Assume that the probabilities of 0, 1, or 2 failures are 0.960, 0.036, and 0.004, respectively. Find the mean and standard deviation for the number of failures.

15. United Airlines Flight 470 from Denver to St. Louis has an on-time performance described as follows: Among 4 independent flights, the probabilities for 0, 1, 2, 3, and 4 on-time flights are 0.026, 0.345, 0.346, 0.154, and 0.129, respectively (based on data from the EAASY SABRE reservation system). Find the mean and standard deviation for the number of on-time flights among the 4.

16. When 5 households are randomly selected, the probabilities of getting 0, 1, 2, 3, 4, or 5 households with VCRs are 0.009, 0.071, 0.221, 0.345, 0.270, and 0.084, respectively (based on data from the U.S. Consumer Electronics Industry). Find the mean and standard deviation for the number of housholds (among 5) with VCRs.

17. On any given weekday, the probabilities of 0, 1, or 2 accidents on a certain highway are 0.650, 0.300, and 0.050, respectively. Find the mean, variance, and standard deviation for the number of accidents in a day.

18. A commuter airline company finds that for a certain flight, the probabilities of 0, 1, 2, or 3 vacant seats are 0.705, 0.115, 0.090, and 0.090, respectively. Find the mean, variance, and standard deviation for the number of vacant seats.

19. Based on past results found in the *Information Please Almanac*, there is a 0.120 probability that a baseball World Series contest will last 4 games, a 0.253 probability that it will last 5 games, a 0.217 probability of 6 games, and a 0.410 probability of 7 games. Find the mean, variance, and standard deviation for the numbers of games that World Series contests last.

20. A computer store finds that the probabilities of selling 0, 1, 2, 3, or 4 microcomputers in one day are 0.245, 0.370, 0.210, 0.095, and 0.080, respectively. Find the mean, variance, and standard deviation for the number of microcomputer sales in one day.

21. If you have a 1/4 probability of gaining $500 and a 3/4 probability of gaining $200, what is your expected value?

22. If you have a 1/10 probability of gaining $1000 and a 9/10 probability of losing $300, what is your expected value?

23. If you have a 1/10 probability of gaining $200, a 3/10 probability of losing $300, and a 6/10 probability of breaking even, what is your expected value?

24. A car wash loses $30 on rainy days and gains $120 on days when it does not rain. If the probability of rain is 0.15, what is the expected value?

25. A contractor bids on a job to construct a building. There is a 0.7 probability of making a $175,000 profit, and there is a probability of 0.3 that the contractor will break even. What is the expected value?

26. A 27-year-old woman decides to pay $156 for a one-year life insurance policy with coverage of $100,000. The probability of her living through the year is 0.9995 (based on data from the U.S. Department of Health and Human Services and AFT Group Life Insurance). What is her expected value?

27. A defendant in a product liability case must choose between settling out of court for a loss of $150,000 or going to trial and either losing nothing (if found not guilty) or losing $500,000 (if found guilty). The attorney estimates that the probability of a "not guilty" verdict is 0.8. Assuming that the defendant decides to go to trial, find the expected value.

28. *Reader's Digest* recently ran a sweepstakes with prizes listed along with the chances of winning: $5,000,000 (1 chance in 201,000,000), $150,000 (1 chance in 201,000,000), $100,000 (1 chance in 201,000,000), $25,000 (1 chance in 100,500,000), $10,000 (1 chance in 50,250,000), $5000 (1 chance in 25,125,000), $200 (1 chance in 8,040,000), $125 (1 chance in 1,005,000), and a watch valued at $89 (1 chance in 3,774).
    a. Compute the expected value.
    b. If the only cost of entering this sweepstakes is a stamp, what is the mean amount that is won or lost?

# 4–3  Exercises B

29. A couple plans to have 5 children. Let the random variable be the number of girls that will occur. Assume that a boy or a girl is equally likely to occur and that the sex of any successive child is unaffected by previous brothers or sisters.

*continued*

| $x$ | $P(x)$ |
|-----|--------|
| 0   |        |
| 1   |        |
| 2   |        |
| 3   |        |
| 4   |        |
| 5   |        |

a. List the 32 different possible simple events.
b. Enter the probabilities in the table at left, where $x$ represents the number of girls in the 5 births.
c. Find the mean number of girls that will occur among the 5 births.
d. Find the variance for the number of girls that will occur.
e. Find the standard deviation for the number of girls that will occur.

30. The variance for the discrete random variable $x$ is 1.25.
    a. Find the variance of the random variable $5x$. (Each value of $x$ is multiplied by 5.)
    b. Find the variance of the random variable $x/5$.
    c. Find the variance of the random variable $x + 5$.
    d. Find the variance of the random variable $x - 5$.
31. A discrete random variable can assume the values 1, 2, . . . , $n$ and those values are equally likely.
    a. Show that $\mu = (n + 1)/2$.
    b. Show that $\sigma^2 = (n^2 - 1)/12$.
       (*Hint:* $1 + 2 + 3 + \cdots + n = n(n + 1)/2$.
       $1^2 + 2^2 + 3^2 + \cdots + n^2 = n(n + 1)(2n + 1)/6$.)
32. Verify that $\sigma^2 = [\Sigma x^2 \cdot P(x)] - \mu^2$ is equivalent to $\sigma^2 = \Sigma(x - \mu)^2 \cdot P(x)$.
    (*Hint:* For constant $c$, $\Sigma cx = c \Sigma x$. Also, $\mu = \Sigma x \cdot P(x)$.)

# 4–4  Binomial Experiments

On January 28, 1986, the Space Shuttle *Challenger* exploded and its seven astronauts were killed. The disaster was apparently caused by the failure of a field-joint O-ring. The Rogers Commission investigated this disaster and concluded that when the *Challenger* was launched in 31° F weather, the chance of a catastrophic failure was at least 13%; if the launch had been delayed until the temperature reached 60° F, the chance of a catastrophic failure would have been around 2%. Methods presented in this section were used in assessing risks in the redesign of space shuttles after the *Challenger* tragedy.

Component failures are examples of a broad category of events that have an element of "twoness." In manufacturing, parts either fail or they don't. In medicine, a new drug either cures someone or it doesn't. In advertising, a consumer either recognizes a product or does not. These situations result in a special type of discrete probability distribution called the *binomial distribution*, which consists of a list of outcomes and probabilities for a binomial experiment.

## Definition

A **binomial experiment** is one that meets all the following requirements:

1. The experiment must have a *fixed number of trials*.
2. The trials must be *independent*.
3. Each trial must have all outcomes classified into *two categories* (even though the sample space may have more than two simple events).
4. The probabilities must remain constant for each trial.

## Example

A manufacturer has a 5% rate of defects when making thermostats, which are produced in batches of 6. Let's assume that the production process involves independent events. That is, the failure of any individual thermostat does not affect the probability of failure for any other thermostats. This situation meets the requirements for a binomial experiment.

1. The number of trials (6) is fixed.
2. The trials are independent (according to the given assumption).
3. Each trial has two categories of outcome: the thermostat was manufactured successfully or it is a failure.
4. The probability of failure (0.05) remains constant for the different thermostats.

## Notation

$S$ and $F$ (success and failure) denote the two possible categories of all outcomes; $p$ and $q$ will denote the probabilities of $S$ and $F$, respectively, so that

$$P(S) = p$$
$$P(F) = 1 - p = q$$

$n$ will denote the fixed number of trials.

$x$ will denote a specific number of successes in $n$ trials so that $x$ can be any whole number between 0 and $n$, inclusive.

$p$ denotes the probability of success in *one* of the $n$ trials.

$q$ denotes the probability of failure in *one* of the $n$ trials.

$P(x)$ denotes the probability of getting exactly $x$ successes among the $n$ trials.

The word *success* as used here does not necessarily correspond to a desired result. Selecting a defective parachute may be classified a success, even though the results of such a selection may be less than pleasant. Either of the two possible categories may be called the success $S$ as long as the corresponding probability is identified as $p$. The value of $q$ can always be found by subtracting $p$ from 1. If $p = 0.95$, then $q = 1 - 0.95 = 0.05$.

---

### Example

Again assume that thermostats are manufactured in batches of 6 with a 5% overall rate of defects. Also assume that we want to find the probability of getting 4 acceptable thermostats in a batch. Identify the values of $n$, $x$, $p$, and $q$.

### Solution

1. With 6 thermostats in each batch, we have $n = 6$.
2. We want 4 acceptable (successful) thermostats, so $x = 4$.
3. The probability of an acceptable (successful) thermostat is 0.95, so $p = 0.95$.
4. The probability of failure is 0.05, so $q = 0.05$.

---

In this section we present two methods for finding probabilities in a binomial experiment. The first method involves the use of Table A–1, while the second involves calculations using a formula. We will first describe the mechanics of both methods, then give a rationale for them.

### Method 1: Use Table A–1 in Appendix A.

To use Table A–1 for finding probabilities in a binomial experiment, first locate $n$ and the corresponding value of $x$ that is desired. At this stage, one row of numbers should be isolated. Now align that row with the proper probability of $p$ by using the column across the top. The isolated number represents the desired probability (missing its decimal point at the beginning). A very small probability such as 0.000000345 is indicated by $0+$. For example, the table indicates that for $n = 10$ trials, the probability of $x = 2$ successes when $P(S) = p = 0.05$ is 0.075. (A more precise value is 0.0746348, but the table values are approximate.)

Although the values in Table A–1 are approximate and only selected values of $p$ are included, the use of such a table often provides quick and easy results.

## Example

Using the thermostats from the preceding example, use Table A–1 to find the probability of getting 4 acceptable thermostats in a randomly selected batch.

## Solution

In the preceding example we established that

$$n = 6 \quad x = 4 \quad p = 0.95 \quad q = 0.05$$

We can now refer to Table A–1. On the first page of that table we locate $n = 6$ in the far left column. We then locate $x = 4$ and find the row entry directly below $p = 0.95$. The table shows that $P(4) = 0.031$. That is, there is a 0.031 probability that exactly 4 thermostats will be acceptable in a given batch.

## Method 2: Use the Binomial Probability Formula.

After identifying the values of $n$, $x$, $p$, and $q$, calculate $P(x)$ by using Formula 4–5.

**Formula 4–5**
**Binomial probability formula**

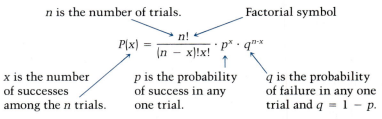

$n$ is the number of trials.    Factorial symbol

$$P(x) = \frac{n!}{(n - x)!x!} \cdot p^x \cdot q^{n-x}$$

$x$ is the number of successes among the $n$ trials.

$p$ is the probability of success in any one trial.

$q$ is the probability of failure in any one trial and $q = 1 - p$.

The factorial symbol, introduced in Section 3–6, denotes the product of decreasing factors. Many calculators have a factorial key. Two examples of factorials are $3! = 3 \cdot 2 \cdot 1 = 6$ and $0! = 1$ (by definition).

## Example

Use the binomial probability formula to find the probability described in the preceding example. That is, find $P(4)$ given that $n = 6$, $x = 4$, $p = 0.95$, and $q = 0.05$.

*continued*

## SHE WON THE LOTTERY TWICE!

■ Evelyn Marie Adams won the New Jersey Lottery twice in four months. This happy event was reported as an incredible coincidence with a likelihood of only 1 chance in 17 trillion. But Harvard mathematicians Persi Diaconis and Frederick Mosteller noted that there is 1 chance in 17 trillion that a particular person with one ticket in each of two New Jersey lotteries will win both times. However, there is about 1 chance in 30 that someone in the United States will win a lottery twice in a four-month period. Diaconis and Mosteller analyze coincidences and conclude that "with a large enough sample, any outrageous thing is apt to happen." ■

### Solution

Using the given values of $n$, $x$, $p$, and $q$ in Formula 4–5, we get

$$P(4) = \frac{6!}{(6-4)! \, 4!} \cdot 0.95^4 \cdot 0.05^{6-4}$$

$$= \frac{720}{2 \cdot 24} \cdot 0.95^4 \cdot 0.05^2$$

$$= (15)(0.81450625)(0.0025) = 0.0305$$

Here's a reasonable strategy for choosing between the use of Table A–1 and a calculation with the binomial probability formula: Try using Table A–1 first. If you can't get the desired probability directly from that table, then use the formula. The table is quick and easy to use and it sometimes replaces several computations with the formula, as the next example illustrates.

### Example

The operation of a component in a computer system is so critical that 3 of these components are built in for backup protection. The latest quality control tests indicate that this type of component has a 10% failure rate. (For optimists, that's a 90% success rate.) Assume that this is a binomial experiment and use Table A–1 to find the probabilities of the following events.

a. Among the 3 components, at least 1 continues to work.

b. Among the 3 components, none continue to work.

### Solution

For both parts of this problem, we stipulate that a success corresponds to a component that continues to work. With 3 components and a 10% failure rate, we have

$$n = 3 \qquad p = 0.9 \qquad q = 0.1$$

On the following page we show the portion of Table A–1 that relates to this problem. We have also constructed the table describing the probability distribution.

*continued*

**Solution** *continued*

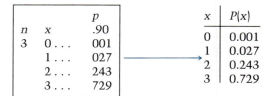

| n | x | p .90 |
|---|---|---|
| 3 | 0 . . . | 001 |
| | 1 . . . | 027 |
| | 2 . . . | 243 |
| | 3 . . . | 729 |

From Table A–1

| x | P(x) |
|---|---|
| 0 | 0.001 |
| 1 | 0.027 |
| 2 | 0.243 |
| 3 | 0.729 |

We can now proceed to find the indicated probabilities.

a. If at least 1 component continues to work, we have $x = 1, 2$ or 3. We get

$$P(\text{at least } 1) = P(1 \text{ or } 2 \text{ or } 3)$$
$$= P(1) + P(2) + P(3) \quad \text{(addition rule)}$$
$$= 0.027 + 0.243 + 0.729 \quad \text{(from preceding table)}$$
$$= 0.999$$

The probability of at least 1 working component is 0.999.

b. The probability of no component working is represented by $P(0)$. From the preceding table we see that $P(0) = 0.001$. There is a 0.001 probability that none of the 3 components will continue to work.

To really appreciate the ease of using Table A–1, let $n = 15$ and $p = 0.6$, and find the probability of at least 8 successes using Table A–1 and the binomial probability formula. The use of Table A–1 involves looking up the 8 probabilities (for $x = 8, 9, 10, \ldots, 15$) and adding them. The use of the binomial probability formula involves using that formula 8 times, computing the 8 different probabilities, and then adding them. Given the choice in this case, most rational people would choose the table. However, we must use the formula when the values of $n$ or $p$ do not allow us to use Table A–1, as in the next example.

**Example**

According to one study conducted at the University of Texas at Austin, 2/3 of all Americans can do routine computations. If an employer were to hire 8 randomly selected Americans, what is the probability that exactly 5 of them can do routine computations?

*continued*

---

## Solution

The experiment is binomial because of the following

1. We have a fixed number of trials (8).
2. The trials are independent since the employees are randomly selected.
3. There are 2 categories, since each employee either can or cannot do routine computations.
4. The probability of 2/3 remains constant from trial to trial.

We begin by identifying the values of $n$, $p$, $q$, and $x$ so that we can apply the binomial probability formula. We have

$$n = 8 \text{ (number of trials)}$$

$$p = 2/3 \text{ (probability of success)}$$

$$q = 1/3 \text{ (probability of failure)}$$

$$x = 5 \text{ (desired number of successes)}$$

We should check for consistency by verifying that what we call a success, as counted by $x$, is the same success with probability $p$. That is, we must be sure that $x$ and $p$ refer to the same concept of success. Using the values for $n$, $p$, $q$, and $x$ in the binomial probability formula, we get

$$P(5) = \frac{8!}{(8-5)!5!} \cdot \left(\frac{2}{3}\right)^5 \cdot \left(\frac{1}{3}\right)^{8-5}$$

$$= \frac{40,320}{(6)(120)} \cdot \frac{32}{243} \cdot \frac{1}{27}$$

$$= 0.273$$

There is a probability of 0.273 that 5 of the 8 employees can do routine computations.

**TABLE 4–4**

| $x$ | $P(x)$ |
|-----|--------|
| 0 | 0.000 |
| 1 | 0.002 |
| 2 | 0.017 |
| 3 | 0.068 |
| 4 | 0.171 |
| 5 | 0.273 |
| 6 | 0.273 |
| 7 | 0.156 |
| 8 | 0.039 |

In the last example we found $P(5)$. We could also use the binomial probability formula to find $P(0)$, $P(1)$, $P(2)$, $P(3)$, $P(4)$, $P(6)$, $P(7)$, and $P(8)$ so that the complete probability distribution for this case will be known. The results are shown in Table 4–4, where $x$ denotes the number of employees who can do routine computations. We can depict Table 4–4 in the form of a probability histogram, as in Figure 4–4. The general nature of the distribution can be seen from the probability histogram.

The entries in Table A–1 are calculated by using the binomial probability formula. The following is a rationale for that formula.

In the preceding example, we wanted the probability of getting 5 successes among the 8 trials, given a $\frac{2}{3}$ probability of success in any 1 trial.

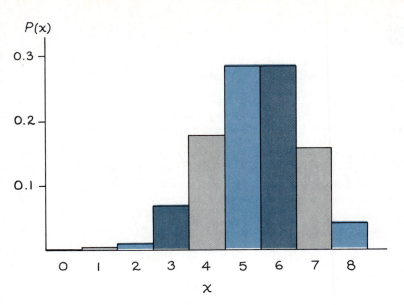

**Figure 4–4**

A common *error* is to compute

$$\overbrace{\frac{2}{3} \times \frac{2}{3} \times \frac{2}{3} \times \frac{2}{3} \times \frac{2}{3}}^{5 \text{ successes}} \times \overbrace{\frac{1}{3} \times \frac{1}{3} \times \frac{1}{3}}^{3 \text{ failures}} = \frac{32}{6561} = 0.00488$$

This incorrect solution uses the multiplication rule from Section 3–4, but it is incomplete because it contains an implicit assumption that the *first* 5 employees are "successes" while the last 3 are "failures." Yet we want the probability of exactly 5 successes with no stipulation about the order in which they occur. If we represent success by $S$ and failure by $F$, the product of 0.00488 corresponds to the specific event *SSSSSFFF*. However, there are other ways or orders of listing 5 successes and 3 failures. In fact, there are 56 possible arrangements and each has the same probability of 0.00488. The total probability is therefore $56 \times 0.00488 = 0.273$. We can generalize this specific result by stating that with $n$ independent trials in which $P(S) = p$ and $P(F) = q$, the multiplication rule indicates that the probability of the first $x$ cases being successes while the remaining cases are failures is

$$\underbrace{p \cdot p \cdots p}_{x \text{ times}} \cdot \underbrace{q \cdot q \cdots q}_{(n - x) \text{ times}}$$

or

$$p^x \cdot q^{n-x}$$

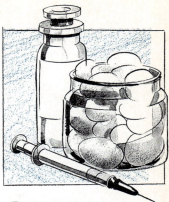

## COMPOSITE SAMPLING

■ The U.S. Army once tested for syphilis by giving each inductee an individual blood test that was analyzed separately. One researcher suggested mixing pairs of blood samples. After testing the mixed pairs, syphilitic inductees could be identified by retesting the few blood samples that were in the pairs that tested positive. The total number of analyses was reduced by pairing blood specimens, so why not put them in groups of three of four or more? Probability theory was used to find the most efficient group size, and a general theory was developed for detecting the defects in any population. This technique is known as *composite sampling.* ■

But $P(x)$ denotes the probability of $x$ successes among $n$ trials in *any* order, so that $p^x \cdot q^{n-x}$ must be multiplied by the number of ways the $x$ successes and $n - x$ failures can be arranged. The number of ways in which it is possible to arrange $x$ successes and $n - x$ failures is shown in Formula 4–6.

**Formula 4–6**    $$\frac{n!}{(n-x)!x!}$$    Number of outcomes with exactly $x$ successes among $n$ trials

The expression given in Formula 4–6 does correspond to $_nC_r$ as introduced in Section 3–6. (Coverage of Section 3–6 is not required for this chapter.) We won't derive Formula 4–6, but its role should be clear: It counts the number of ways you can arrange $x$ successes and $n - x$ failures. Combining this counting device (Formula 4–6) with the direct application of the multiplication rule for independent events results in the binomial probability formula.

$$P(x) = \underbrace{\frac{n!}{(n-x)!x!}}_{\substack{\text{The number of} \\ \text{outcomes with} \\ \text{exactly } x \text{ successes} \\ \text{among } n \text{ trials}}} \cdot \underbrace{p^x \cdot q^{n-x}}_{\substack{\text{The probability} \\ \text{of } x \text{ successes} \\ \text{among } n \text{ trials} \\ \text{for any one particular} \\ \text{order}}}$$

Many computer statistics packages include an option for generating binomial probabilities. The following are samples of output from STATDISK and Minitab obtained from a binomial experiment in which $n = 3$ and $p = 0.9$. The user simply enters those values and the entire probability distribution is displayed. The column labeled "Cum Prob" represents cumulative probabilities obtained by adding the values of $P(x)$ as you go down the column.

STATDISK DISPLAY

This program uses the BINOMIAL PROBABILITY FORMULA to find probabilities and cumulative probabilities
You must enter . . .
the number of trials or sample size . . . . . . . . . . . . . . . . . . . . . . . . . $n = 3$
the probability of success on a single trial . . . . . . . . . . . . . . . . . $p = 0.9$
Mean = 2.7                St. Dev. = .519615                Variance = .27

| X | P(X) | Cum Prob |
|---|------|----------|
| 0 | .00100 | .00100 |
| 1 | .02700 | .02800 |
| 2 | .24300 | .27100 |
| 3 | .72900 | 1.00000 |

MINITAB DISPLAY

```
MTB  > PDF;
SUBC > BINOMIAL n = 3 p = 0.9.
            BINOMIAL WITH N = 3 P = 0.900000
                K          P( X = K)
                0          0.0010
                1          0.0270
                2          0.2430
                3          0.7290
```

At the beginning of this chapter, we briefly discussed the Federal Aviation Administration's consideration of allowing twin-engine jets to make transatlantic flights. (Existing regulations required at least three engines.) A realistic estimate for the probability of an engine failing on a transatlantic flight is 1/14,100. Using that probability and the binomial probability formula, we can develop the probability distributions summarized in Tables 4–5 and 4–6, where $x$ represents the number of engines that fail on a transatlantic flight.

| TABLE 4–5 Three Engines | |
|---|---|
| $x$ | $P(x)$ |
| 0 | 0.9997872491 |
| 1 | 0.0002127358 |
| 2 | 0.0000000151 |
| 3 | 0.0000000000 |

| TABLE 4–6 Two Engines | |
|---|---|
| $x$ | $P(x)$ |
| 0 | 0.9998581611 |
| 1 | 0.0001418339 |
| 2 | 0.0000000050 |

Using those tables and assuming that a flight will be completed if at least one engine works, we get

$P$(safe flight with three-engine jet) = $P$(0, 1, or 2 engine failures)

$$= 0.9997872491 + 0.0002127358 + 0.0000000151$$
$$= 1.0000000000 \text{ (rounded to 10 decimal places)}$$

$P$(safe flight with twin-engine jet) = $P$(0 or 1 engine failures)

$$= 0.9998581611 + 0.0001418339 = 0.9999999950$$

# DISEASE CLUSTERS

■ Periodically, much media attention is given to a cluster of disease cases in a given community. One New Jersey community had 13 leukemia cases in 5 years, while the normal rate would have been 1 case in 10 years. Research of such clusters can be revealing. A study of a cluster of cancer cases near African asbestos mines led to the discovery that asbestos fibers can be carcinogenic.

When doing such an analysis, we should avoid the mistake of artificially creating a cluster by locating its outer boundary so that it just barely includes cases of disease or death. That would be like gerrymandering, even though it might be unintentional. ■

The three-engine result of 1.0000000000 is actually a rounded-off form of the more precise result of 0.9999999999996433. Comparing the two resulting probabilities, we see that the three-engine jet would be safer, as expected, but the difference doesn't seem to be too significant. All of those leading nines in the results suggest that both configurations are quite safe. The Federal Aviation Administration used this type of reasoning when it changed its regulations to allow transatlantic flights by jets with only two engines. We could analyze the significance of those final probabilities further, but this illustration does show a very useful application of the binomial probability formula.

To keep this section in perspective, remember that the binomial probability formula is only one of many probability formulas that can be used for different situations. It is, however, among the most important and most useful of all discrete probability distributions. It is often used in applications such as quality control, voter analysis, medical research, military intelligence, and advertising.

## 4–4 Exercises A

1. Which of the following can be treated as binomial experiments?
   a. Testing a sample of 5 capacitors (with replacement) from a population of 20 capacitors, of which 40% are defective
   b. Testing a sample of 5 capacitors (without replacement) from a population of 20 capacitors, of which 40% are defective
   c. Tossing an unbiased coin 500 times
   d. Tossing a biased coin 500 times
   e. Surveying 1700 television viewers to determine whether or not they watched a particular show
2. Which of the following can be treated as binomial experiments?
   a. Surveying 1200 registered voters to determine their preferences for the next president
   b. Surveying 1200 registered voters to determine whether or not they would again vote for the current president
   c. Sampling a randomly selected group of 500 prisoners to determine whether or not they have been in prison before
   d. Sampling a randomly selected group of 500 prisoners to determine the lengths of their current sentences
   e. Testing 500 randomly selected drivers to determine whether or not their blood alcohol content levels are over 0.10%

3.  In a binomial experiment, a trial is repeated $n$ times. Find the probability of $x$ successes given the probability $p$ of success on a given trial. (Use the given values of $n$, $x$, and $p$ and Table A–1.)
    a.   $n = 10, x = 3, p = 0.5$
    b.   $n = 10, x = 3, p = 0.4$
    c.   $n = 7, x = 0, p = 0.1$
    d.   $n = 7, x = 0, p = 0.99$
    e.   $n = 7, x = 7, p = 0.01$

4.  In a binomial experiment, a trial is repeated $n$ times. Find the probability of $x$ successes given the probability $p$ of success on a given trial. (Use the given values of $n$, $x$, and $p$ and Table A–1.)
    a.   $n = 15, x = 5, p = 0.7$
    b.   $n = 12, x = 11, p = 0.6$
    c.   $n = 9, x = 6, p = 0.1$
    d.   $n = 8, x = 5, p = 0.95$
    e.   $n = 14, x = 14, p = 0.9$

5.  For each of the following, use Formula 4–6 to find the number of ways you can arrange $x$ successes and $n - x$ failures.
    a.   $n = 5, x = 3$
    b.   $n = 5, x = 0$
    c.   $n = 8, x = 7$
    d.   $n = 8, x = 3$
    e.   $n = 8, x = 8$

6.  For each of the following, use Formula 4–6 to find the number of ways you can arrange $x$ successes and $n - x$ failures.
    a.   $n = 6, x = 2$
    b.   $n = 6, x = 6$
    c.   $n = 6, x = 0$
    d.   $n = 10, x = 3$
    e.   $n = 20, x = 18$

7.  In a binomial experiment, a trial is repeated $n$ times. Find the probability of $x$ successes given the probability $p$ of success on a single trial. Use the given values of $n$, $x$, and $p$ and the binomial probability formula. Leave answers in the form of fractions.
    a.   $n = 5, x = 3, p = 1/4$
    b.   $n = 4, x = 2, p = 1/3$
    c.   $n = 5, x = 1, p = 2/3$

8.  In a binomial experiment, a trial is repeated $n$ times. Find the probability of $x$ successes given the probability $p$ of success on a single trial. Use the given values of $n$, $x$, and $p$ and the binomial probability formula. Leave answers in the form of fractions.
    a.   $n = 6, x = 2, p = 1/2$
    b.   $n = 3, x = 1, p = 3/7$
    c.   $n = 4, x = 4, p = 2/3$

In Exercises 9–32, identify the values of $n$, $x$, $p$, and $q$, and find the value requested.

9. Find the probability of getting exactly 4 girls in 10 births. (Assume that male and female births are equally likely.)
10. Find the probability of getting exactly 6 girls in 7 births. (Assume that male and female births are equally likely.)
11. A Gallup poll showed that among convenience store shoppers, 60% gave closeness of location as their primary reason for shopping there. Find the probability that among 5 randomly selected convenience store customers, 3 of them give closeness of location as their primary reason for choosing that store.
12. In a study of brand recognition, 95% of consumers recognized Coke (based on data from Total Research Corporation). Find the probability that among 12 randomly selected consumers, exactly 11 will recognize Coke.
13. According to *Discover* magazine, 95% of airline passengers will survive a crash under certain conditions. Given those conditions, what is the probability that at least 8 of 10 passengers will survive a crash?
14. According to the U.S. Department of Justice, 5% of all U.S. households experienced at least one burglary last year. If you randomly select 15 households, what is the probability that fewer than 3 of them experienced at least one burglary last year?
15. According to the Labor Department, 40% of adult workers have a high school diploma and did not attend college. If 15 adult workers are randomly selected, find the probability that at least 10 of them have a high school diploma and did not attend college.
16. A Media General-AP poll showed that 20% of adult Americans are opposed to strict pollution controls on power plants that burn coal and oil. Assume that an environmental group launches a campaign to lower that number, and a post-campaign study begins with 15 randomly selected adults. If the 20% level of opposition hasn't changed, find the probability that among the 15 adults, fewer than 3 are opposed to the strict controls.
17. According to the Department of Defense, 93% of Air Force recruits have graduated from high school. If 15 Air Force recruits are randomly selected, find the probability that 13 of them have graduated from high school.
18. According to FBI data, 44% of those murdered are killed with handguns. If 30 murder cases are randomly selected, find the probability that 12 of the victims were killed with handguns.
19. A study conducted by the Centers for Disease Control found that 15% of New Yorkers average two or more alcoholic drinks per day. If 60 New Yorkers are randomly selected and surveyed, find the probability that 12 of them average two or more alcoholic drinks per day.

20. A multiple-choice test has 30 questions, and each one has 5 possible answers, of which 1 is correct. If all answers are guesses, find the probability of getting exactly 4 correct answers.

21. A study by the EPA (Environmental Protection Agency) showed that of the cars built with catalytic converters, 4.4% have them removed. If 50 cars built with catalytic converters are randomly selected and checked, find the probability that
    a. All cars continue to have their catalytic converters.
    b. One car has the catalytic converter removed.
    c. Two cars have the catalytic converter removed.

22. The National Coffee Association reports that among individuals in the 20 to 29 age bracket, 41% drink coffee. If 5 people in that age bracket are randomly selected, find the probability of the following.
    a. They all drink coffee.
    b. There is exactly 1 coffee drinker.
    c. There are 2 coffee drinkers.

23. In Table A–1, the probability corresponding to $n = 3$, $x = 2$, and $p = 0.01$ is shown as $0+$. Find the exact probability represented by this $0+$.

24. In Table A–1, the probability corresponding to $n = 5$, $x = 4$, and $p = 0.10$ is shown as $0+$. Find the exact probability represented by this $0+$.

25. A binomial experiment consists of 3 trials with a 1/3 probability of success in each trial. Construct the probability distribution table for this experiment. (Use the same format as Table 4–4.)

26. A binomial experiment consists of 4 trials with a 1/4 probability of success in each trial. Construct the probability distribution table for this experiment. (Use the same format as Table 4–4.)

27. Data from the U.S. Bureau of the Census show that among those in the 18–24 age bracket, 40.8% vote. If 16 individuals from that age bracket are randomly selected, find the probability that fewer than 2 of them vote.

28. A recent study by the A.C. Nielsen Company showed that 57% of all homes have cable TV. If 10 homes are randomly selected to test an experimental metering device, you would expect that about 6 of them have cable. Find the probability that exactly 6 of these homes have cable TV.

29. United Airlines Flight 470 from Denver to St. Louis has an on-time performance of 60% (based on data from the EAASY SABRE reservation system).
    a. Find the probability that among 12 such flights, at least 9 arrive on time.
    b. Find the probability that among 30 such flights, exactly 20 arrive on time.

30. Data from Survey Sampling, Inc., show that in Las Vegas, 46.4% of the telephone numbers are unlisted. If 10 telephone numbers are randomly selected, find the probability that more than 8 are unlisted numbers.

31. Ten percent of the general population is left-handed and, based on a study by psychologist Stanley Coren, 22% of children born to mothers over the age of 40 are left-handed. Find the probability of getting fewer than two left-handed people when 10 are randomly selected from the following groups.
   a. the general population
   b. children born to mothers over the age of 40

32. An auto marketing team finds that among New York State licensed drivers, there are 5,385,438 males and 4,643,599 females (based on data from the N.Y. State Department of Motor Vehicles).
   a. If 10 of these drivers are randomly selected, find the probability of getting exactly 6 males.
   b. What is the probability of getting exactly 6 males among 10 randomly selected subjects if you assume that males and females are equally likely?

# 4–4  Exercises B

33. Suppose that an experiment meets all conditions to be binomial except that the number of trials is not fixed. Then the **geometric distribution,** which gives us the probability of getting the first success on the $x$th trial, is described by $P(x) = p(1 - p)^{x-1}$, where $p$ is the probability of success on any one trial. Assume that the probability of a defective computer component is 0.2. Find the probability that the first defect is in the seventh component tested.

34. The **Poisson distribution** is used as a mathematical model describing the probability distribution for the arrivals of entities requiring service (such as cars arriving at a gas station, planes arriving at an airport, or people arriving at a ride in Disney World). The Poisson distribution is defined by the equation

$$P(x) = \frac{\mu^x \cdot e^{-\mu}}{x!}$$

where $x$ represents the number of arrivals during a given time interval (such as 1 hour), $\mu$ is the mean number of arrivals during the same time interval, and $e$ is a constant approximately equal to 2.718. Assume that $\mu = 15$ cars per hour for a gas station and find the probability of each number of arrivals in an hour.
   a. 0        b. 1        c. 10        d. 15        e. 20        f. 30

35. If we sample from a small finite population without replacement, the binomial distribution should not be used because the events are not independent. If sampling is done without replacement and the outcomes belong to 1 of 2 types, we can use the **hypergeometric distribution.** If a population has $A$ objects of one type, while the remaining $B$ objects are of the other type, and if $n$ objects are sampled without replacement, then the probability of getting $x$ objects of type $A$ and $n - x$ objects of type $B$ is

$$P(x) = \frac{A!}{(A - x)!x!} \cdot \frac{B!}{(B - n + x)!(n - x)!} \div \frac{(A + B)!}{(A + B - n)!n!}$$

Five people are randomly selected (without replacement) from a population of 7 men and 3 women. Find the probability of getting 4 men and 1 woman.

36. The binomial distribution applies only to cases involving 2 types of outcomes, whereas the **multinomial distribution** involves more than 2 categories. Suppose we have 3 types of mutually exclusive outcomes denoted by $A$, $B$, and $C$. Let $P(A) = p_1$, $P(B) = p_2$, and $P(C) = p_3$. In $n$ independent trials, the probability of $x_1$ outcomes of type $A$, $x_2$ outcomes of type $B$, and $x_3$ outcomes of type $C$ is given by

$$\frac{n!}{(x_1!)(x_2!)(x_3!)} \cdot p_1^{x_1} \cdot p_2^{x_2} \cdot p_3^{x_3}$$

a. Extend the result to cover 6 types of outcomes.
b. A genetics experiment involves 6 mutually exclusive genotypes identified as $A$, $B$, $C$, $D$, $E$, and $F$, and they are all equally likely. If 20 offspring are tested, find the probability of getting exactly 5 $A$'s, 4 $B$'s, 3 $C$'s, 2 $D$'s, 3 $E$'s, and 3 $F$'s.

## 4–5 Mean and Standard Deviation for the Binomial Distribution

The binomial distribution is a probability distribution, so the mean, variance, and standard deviation for the appropriate random variable can be found from the formulas presented in Section 4–3.

**Formula 4–1**         $\mu = \Sigma x \cdot P(x)$

**Formula 4–3**         $\sigma^2 = [\Sigma x^2 \cdot P(x)] - \mu^2$

**Formula 4–4**         $\sigma = \sqrt{[\Sigma x^2 \cdot P(x)] - \mu^2}$

However, these formulas, which apply to all probability distributions, can be made much simpler for the special case of binomial distributions. Given

## WHO IS SHAKESPEARE?

■ A poll of 1553 randomly selected adult Americans revealed the following:

· 89% could identify Shakespeare
· 58% could identify Napoleon
· 47% could identify Freud
· 92% could identify Columbus
· 71% knew what happened in 1776

Ronald Berman, a past director of the National Endowment for the Humanities, said, "I don't worry about them [those who don't know what happened in 1776]. I worry about the people who take polls." He goes on to say that the poll should ask substantive questions, such as, "What is the difference between democracy and totalitarianism?" ■

the binomial probability formula and the above general formulas for $\mu$, $\sigma$, and $\sigma^2$, we can pursue a series of somewhat complicated algebraic manipulations that ultimately lead to the following simple result.

For a binomial experiment,

**Formula 4–7** $\qquad\qquad\qquad\qquad \mu = n \cdot p$

**Formula 4–8** $\qquad\qquad\qquad\qquad \sigma^2 = n \cdot p \cdot q$

**Formula 4–9** $\qquad\qquad\qquad\qquad \sigma = \sqrt{n \cdot p \cdot q}$

The formula for the mean does make sense intuitively. If we were to analyze 100 births, we would expect to get about 50 girls, and $np$ in this experiment becomes $100 \cdot \frac{1}{2}$, or 50. In general, if we consider $p$ to be the proportion of successes, then the product $np$ will give us the actual number of expected successes among $n$ trials.

The variance and standard deviation are not so easily justified, and we prefer to omit the complicated algebraic manipulations that lead to the second formula. Instead, we will show that these simplified formulas (Formulas 4–7, 4–8, and 4–9) do lead to the same results as the more general formulas (Formulas 4–1, 4–3, and 4–4).

### Example

According to an analyst from Paine Webber, the Intel Corporation has a 35% yield for the computer chips it produces—35% of the chips are good and 65% are defective. Let's assume that this estimate is correct and that we have randomly selected a sample of 4 different chips. Find the mean, variance, and standard deviation for the numbers of good chips in such groups of 4.

### Solution

In this binomial experiment we have $n = 4$ and $P(\text{good chip}) = 0.35$. It follows that $P(\text{defective chip}) = q = 0.65$. We will find the mean and standard deviation by using two methods.

*Method 1:* Use Formulas 4–7, 4–8, and 4–9, which apply to binomial experiments only.

$$\mu = n \cdot p = 4 \cdot 0.35 = 1.4 \qquad \text{(Formula 4–7)}$$
$$\sigma^2 = n \cdot p \cdot q = 4 \cdot 0.35 \cdot 0.65 = 0.91 \qquad \text{(Formula 4–8)}$$
$$\sigma = \sqrt{n \cdot p \cdot q} = \sqrt{0.91} = 0.95 \qquad \text{(Formula 4–9)}$$

*continued*

**Solution** *continued*

*Method 2:* Use Formulas 4–1, 4–3, and 4–4, which apply to all discrete probability distributions. (*Note:* Method 1 provided us with the solutions we sought, but we want to show that these same values will result from the use of the more general formulas from Section 4–3.) We begin by computing the mean using Formula 4–1: $\mu = \Sigma x \cdot P(x)$. The possible values of $x$ are 0, 1, 2, 3, 4, but we also need the values of $P(0)$, $P(1)$, $P(2)$, $P(3)$, and $P(4)$. We use the binomial probability formula to find those values and enter them in Table 4–7.

**TABLE 4–7**

| $x$ | $P(x)$ | $x \cdot P(x)$ | $x^2$ | $x^2 \cdot P(x)$ |
|---|---|---|---|---|
| 0 | 0.179 | 0 | 0 | 0 |
| 1 | 0.384 | 0.384 | 1 | 0.384 |
| 2 | 0.311 | 0.622 | 4 | 1.244 |
| 3 | 0.111 | 0.333 | 9 | 0.999 |
| 4 | 0.015 | 0.060 | 16 | 0.240 |
| Total | | 1.399 | | 2.867 |

We now use results from Table 4–7 to apply the general formulas from Section 4–3 as follows.

$$\mu = \Sigma x \cdot P(x) = 1.4 \quad \text{(rounded off)}$$
$$\sigma^2 = [\Sigma x^2 \cdot P(x)] - \mu^2$$
$$= 2.867 - 1.4^2$$
$$= 0.91 \quad \text{(rounded off)}$$
$$\sigma = \sqrt{0.91} = 0.95 \quad \text{(rounded off)}$$

The two methods produced the same results, except for minor discrepancies due to rounding. There are two important points that we should recognize. First, the simplified binomial formulas (Formulas 4–7, 4–8, and 4–9) do lead to the same results as the more general formulas that apply to all discrete probability distributions. Second, the binomial formulas are much simpler, they provide fewer opportunities for arithmetic errors, and they are generally more conducive to a positive outlook on life. If we know an experiment is binomial, we should use the simplified formulas.

In the following example, we use only the simplified binomial formulas.

---

**Example**

A test consists of 100 multiple-choice questions with possible answers of *a, b, c, d,* and *e.* For people who know nothing and guess the answer to each question, find the mean and standard deviation for the number of correct answers per person.

**Solution**

For each person, the number of trials is $n = 100$ and the probability of correctly guessing an answer is $p = 1/5$, so that $q = 4/5$. (We get $p = 1/5$ because there is 1 correct answer among the 5 possible answers.) We now proceed to find the mean and standard deviation.

$$\mu = n \cdot p$$
$$= 100 \cdot \tfrac{1}{5}$$
$$= 20.0$$
$$\sigma = \sqrt{n \cdot p \cdot q}$$
$$= \sqrt{100 \cdot \tfrac{1}{5} \cdot \tfrac{4}{5}}$$
$$= 4.00$$

---

For the preceding example, the mean number of correct guesses is 20.0, so that a score of 20.0% on the test is actually an indication of no knowledge. The value of the standard deviation can be used to determine a reasonable range of scores for those who know nothing and guess. For example, using the empirical rule from Chapter 3, we can conclude that about 95% of all scores for guessers should be between 12 and 28. Using the same empirical rule, we can conclude that about 99.7% of all scores for guessers should be between 8 and 32. If someone gets more than 32 correct answers, it is very unlikely that they are guessing at all questions. A more likely explanation is that they probably know some of the answers.

When *n* is large and *p* is close to 0.5, the binomial distribution tends to resemble the smooth curve that approximates the histograms in Figures 4–5 and 4–6. Note that the data tend to form a bell-shaped curve. In Chapter 5, we will use this property for solving certain applied problems.

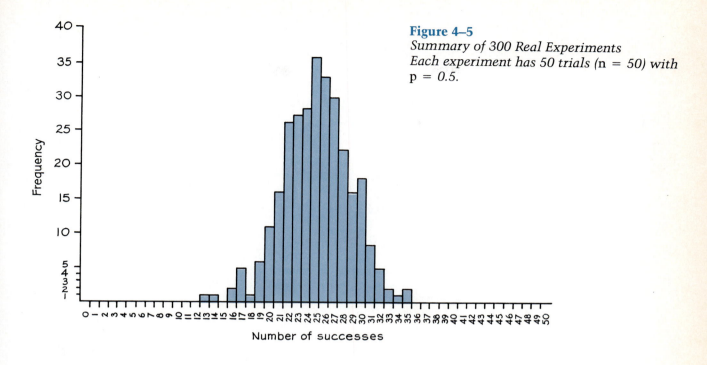

**Figure 4–5**
*Summary of 300 Real Experiments*
*Each experiment has 50 trials (n = 50) with*
*p = 0.5.*

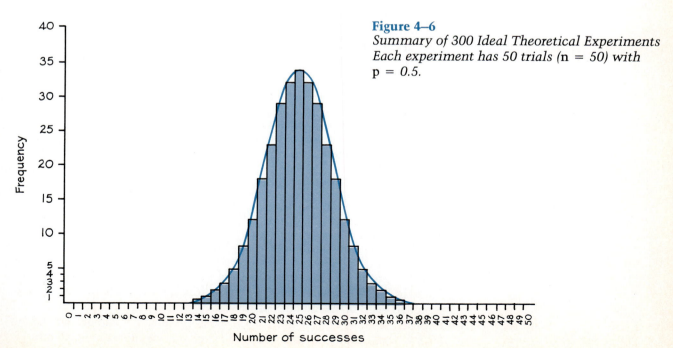

**Figure 4–6**
*Summary of 300 Ideal Theoretical Experiments*
*Each experiment has 50 trials (n = 50) with*
*p = 0.5.*

# 4–5 | Exercises A

In Exercises 1–12, find the mean $\mu$, variance $\sigma^2$, and standard deviation $\sigma$ for the given values of $n$ and $p$. Assume that the binomial conditions are satisfied in each case.

1. $n = 64, p = 0.5$
2. $n = 100, p = 0.5$
3. $n = 8, p = 0.6$
4. $n = 6, p = 0.3$
5. $n = 36, p = 0.25$
6. $n = 40, p = 0.85$
7. $n = 534, p = 0.173$
8. $n = 898, p = 0.392$
9. $n = 16, p = 1/5$
10. $n = 27, p = 1/4$
11. $n = 253, p = 2/3$
12. $n = 652, p = 3/8$

In Exercises 13–32, find the indicated values.

13. For a true-false test with 50 questions, several students are unprepared and all of their answers are guesses. Find the mean, variance, and standard deviation for the numbers of correct answers for such students.
14. For a multiple-choice test with 30 questions, each question has possible answers of *a*, *b*, *c*, and *d*, one of which is correct. For people who guess at all answers, find the mean, variance, and standard deviation for the number of correct answers.
15. Find the mean, variance, and standard deviation for the numbers of girls in families with 4 children. Assume that boys and girls are equally likely and that the sex of any child is independent of any brothers or sisters.
16. Find the mean, variance, and standard deviation for the numbers of girls in families with 6 children. Assume that boys and girls are equally likely and that the sex of any child is independent of any brothers or sisters.
17. According to U.S. Department of Justice reports, 54% of drivers arrested for DWI had been drinking beer only. A study of DWI arrests involves groups of randomly selected DWI arrests, with 25 per group. Find the mean and standard deviation for the numbers in each group who had been drinking beer only.
18. Recently, the highest scorer for the Boston Celtics made an average of 55.2% of his shots, not counting free throws. Assume that he shoots 15

times each game and the shots are independent. Find the mean and standard deviation for the number of shots he hits per game.

19. Among Americans aged 20 years or over, 12.5% sleep at least nine hours each night. A study of dreams requires 36 volunteers who are at least 20 years old. Find the mean and standard deviation for the number of such people in groups of 36 who sleep at least nine hours each night.

20. According to a survey of adults by the Roper Organization, 64% of adults have money in regular savings accounts. If we plan to conduct a survey with groups of 50 randomly selected adults, find the mean, variance, and standard deviation for the numbers who have regular savings accounts.

21. Among the 6665 films rated by the Motion Picture Association of America, 2945 have been rated R. Twenty rated films are randomly selected for a study. Find the mean, variance, and standard deviation for the number of R films in randomly selected groups of 20.

22. According to data from the U.S. Bureau of Labor Statistics, 70.4% of women in the 20 to 24 age bracket are working. Find the mean, variance, and standard deviation for the numbers of working women in randomly selected groups of 150 women between 20 and 24 years of age.

23. A study conducted by the National Transportation and Safety Board showed that among injured airline passengers, 47% of the injuries were caused by failure of the plane's seat. Two hundred different airline passenger injuries are to be randomly selected for a study. Find the mean, variance, and standard deviation for the number of injuries caused by seat failure in such groups of 200.

24. One test of extrasensory perception involves the determination of a color. Fifty blindfolded subjects are asked to identify the one color selected from the possibilities of red, yellow, green, blue, black, and white. Assuming that all 50 subjects make random guesses, find the mean, variance, and standard deviation for the number of correct responses in such groups of 50.

25. Of all individual tax returns, 37% include errors made by the taxpayer. If IRS examiners are assigned randomly selected returns in batches of 12, find the mean and standard deviation for the number of erroneous returns per batch.

26. A pathologist knows that 14.9% of all deaths are attributable to a myocardial infarction. Find the mean and standard deviation for the number of such deaths that will occur in a typical region with 5000 deaths.

27. Letter frequencies are analyzed in attempts to decipher intercepted messages. In standard English text, the letter *e* occurs with a relative frequency of 0.130. Find the mean and standard deviation for the number of times *e* will be found on standard pages of 2600 characters.

28. According to an Environmental Protection Agency study of cars originally built with catalytic converters, 4.4% have had them removed. One point of a certain highway is passed by 12,600 cars per hour. Assume that all of those cars were originally equipped with catalytic converters. What is the mean and standard deviation for the hourly numbers of cars with their catalytic converters removed?

29. A Roper survey showed that 73% of adult Americans do not feel that the government is doing enough to regulate toxic waste disposal. One thousand subjects are to be randomly selected for a related survey. Find the mean and standard deviation for the numbers in such groups of 1000 who share that same belief.

30. A survey has shown that 43% of all unregistered voters prefer the Democratic party. A follow-up study is to be conducted with 1200 randomly selected unregistered voters. Find the mean and standard deviation for the numbers of Democrats in groups of 1200.

31. Among all military personnel, 13.8% are officers. A study of military personnel involves random selections in groups of 50. Find the mean and standard deviation for the number of officers per group.

32. In a recent year, there were 68,593 motor vehicle accidents in New York State, and 42,000 of them involved injuries (based on data from the N.Y. State Department of Motor Vehicles). An insurance analyst will randomly select different groups of accidents, with 20 in each group. Use the sample data to estimate the probability that an accident involves an injury, then find the mean and standard deviation for the number of injury accidents in such groups of 20.

## 4–5 Exercises B

In Exercises 33–35, consider as unusual anything that differs from the mean by more than twice the standard deviation. That is, unusual values are either less than $\mu - 2\sigma$ or greater than $\mu + 2\sigma$.

33. a. Is it unusual to get 450 girls and 550 boys in 1000 independent births?
    b. Is it unusual to find 5 defective transistors in a sample of 20 if the defective rate is 20%?

34. A company manufactures an appliance, gives a warranty, and 95% of its appliances do not require repair before the warranty expires. Is it unusual for a buyer of 10 such appliances to require warranty repairs on 2 of the items?

35. A candidate is favored by 616 voters in a poll of 1100 randomly selected voters. If this candidate is actually favored by 50% of all voters, find

the lowest and highest *usual* numbers of supporters in groups of 1100. Does the poll result of 616 supporters seem to indicate a chance sample fluctuation?

36. a. If a company makes a product with an 80% yield (meaning that 80% are good), what is the minimum number of items that must be produced in order to be at least 99% sure that they have at least 5 good items?

    b. If the company produces batches of items, each with the minimum number determined in part *a*, find the mean and standard deviation for the number of good items in such batches.

 *Vocabulary List*

Define and give an example of each term.

random variable                  expected value
discrete random variable         fair game
continuous random variable       binomial experiment
probability distribution         binomial probability formula
probability histogram

 *Review*

The central concern of this chapter was the concept of a probability distribution. Here we dealt mostly with **discrete** probability distributions, while successive chapters deal with continuous probability distributions.

In an experiment yielding numerical results, the **random variable** can take on those different numerical values. A **probability distribution** consists of all values of a random variable, along with their corresponding probabilities. By constructing a probability histogram, we can see a useful correspondence between those probabilities and the areas of the rectangles in the histogram.

Of the infinite number of different probability distributions, special attention is given to the important and useful **binomial probability distribution,** which is characterized by these properties:

1. There is a fixed number of trials (denoted by *n*).

2. The trials must be independent.

3. Each trial must have outcomes that can be classified in *two* categories.

4. The probabilities involved must remain constant for each trial.

We saw that probabilities for the binomial distribution can be computed by using Table A–1 or by using the binomial probability formula, where $n$ is the number of trials, $x$ is the number of successes, $p$ is the probability of a success, and $q$ is the probability of a failure.

For the special case of the binomial probability distribution, the mean, variance, and standard deviations of the random variable can be easily computed by using the formulas given in the summary below.

---

### Important Formulas

Requirements for a discrete probability distribution:

1. $\Sigma\ P(x) = 1$ for all possible values of $x$.
2. $0 \leq P(x) \leq 1$ for any particular value of $x$.

For *any* discrete probability distribution:

$$\text{mean } \mu = \Sigma x \cdot P(x) \qquad \text{variance } \sigma^2 = \Sigma(x - \mu)^2 \cdot P(x)$$

$$\text{or} \quad \text{variance } \sigma^2 = [\Sigma x^2 \cdot P(x)] - \mu^2$$

$$\text{standard deviation } \sigma = \sqrt{[\Sigma x^2 \cdot P(x)] - \mu^2}$$

Expected value of discrete random variable: $E = \Sigma x \cdot P(x)$

Binomial probability formula: $P(x) = \dfrac{n!}{(n - x)!x!} \cdot p^x \cdot q^{n-x}$

For *binomial* probability distributions:

$$\text{mean } \mu = n \cdot p \qquad \text{variance } \sigma^2 = n \cdot p \cdot q$$

$$\text{standard deviation } \sigma = \sqrt{n \cdot p \cdot q}$$

---

## 2 *Review Exercises*

1. According to the National Highway Traffic Safety Administration, 66% of California motorists use seat belts. A California Highway Patrol study involves the random selection of groups with 12 motorists each.
   a. Find the mean and standard deviation for the numbers of motorists per group who wear seat belts.
   b. Find the probability that in one group of 12 California motorists, exactly 8 wear seat belts.
2. Does $P(x) = x/12$ (for $x = 1, 2, 3, 4$) determine a probability distribution? Explain.

3.  In a quality control study, it was found that 5% of all auto frames manufactured by one company had at least one defective weld.
    a.  Find the probability of getting 3 frames with defective welds when 20 frames are randomly selected.
    b.  If frames are examined in groups of 20, find the mean number of defective frames in each group.
    c.  Find the standard deviation for the number of defective frames in groups of 20.

4.  In a study of sleep patterns of adults, it has been found that 23% of the sleep time is spent in the REM (rapid eye movement) stage.
    a.  If a sleeping adult is observed at 5 randomly selected times, find the probability that exactly 1 of the 5 observations will be made during a REM stage.
    b.  Find the standard deviation for the number of REM stages observed in groups of 5 observations.

5.  Does $P(x) = (x + 1)/5$ (for $x = -1, 0, 1, 2$) determine a probability distribution? Explain.

6.  It has been found that the probabilities of 0, 1, 2, 3, and 4 wrong answers on a quiz are 0.20, 0.35, 0.30, 0.10, and 0.05, respectively.
    a.  Summarize the corresponding probability distribution.
    b.  Find the mean of the random variable $x$.
    c.  Find the standard deviation of the random variable $x$.

7.  An experiment in parapsychology involves the selection of one of the numbers 1, 2, 3, 4, and 5, and they are all equally likely.
    a.  Summarize the corresponding probability distribution.
    b.  Find the mean of the random variable $x$.
    c.  Find the variance of the random variable $x$.

8.  A study of enrollments at one college shows that 40% of all full-time undergraduates are under 20 years of age.
    a.  If 10 full-time undergraduates at this college are randomly selected, find the probability that at least half of them are under 20.
    b.  Find the mean number of full-time undergraduates under 20 years of age that would be found in groups of 10.
    c.  Find the standard deviation for the number of full-time undergraduates under 20 years of age that would be found in groups of 10.

9.  Among the voters in a certain region, 40% are Democrats, and 5 voters are randomly selected.
    a.  Find the probability that all 5 are Democrats.
    b.  Find the probability that exactly 3 of the 5 are Democrats.
    c.  Find the mean number of Democrats that would be selected in such groups of 5.
    d.  Find the variance of the number of Democrats that would be selected in such groups of 5.

*continued*

e.   Find the standard deviation for the number of Democrats that would be selected in such groups of 5.

10.   A probability distribution $P(x)$ is described by the accompanying table.
   a.   Complete the table.
   b.   Find the mean of the random variable $x$.
   c.   Find the standard deviation of the random variable $x$.

| $x$ | $P(x)$ |
|---|---|
| 0 | 0.20 |
| 1 | 0.70 |
| 5 | |

11.   If $P(x)$ is described by the accompanying table, does $P(x)$ form a probability distribution? Explain.

| $x$ | $P(x)$ |
|---|---|
| 0 | 0.4 |
| 1 | 0.4 |
| 2 | 0.4 |

12.   In a study of middle-aged adults (40 to 65 years), it was found that 7.8% suffer from hypertension. A follow-up study begins with the random selection of 20 middle-aged adults.
   a.   Find the probability that exactly one-fourth of the selected subjects suffer from hypertension.
   b.   Find the mean number of hypertension cases found in such groups of 20.
   c.   Find the standard deviation for the numbers of hypertension cases found in groups of 20.

13.   A solid-state device requires 15 electronic components, 2 of which cost less than 50¢ each. One component is randomly selected from each of 8 such devices.
   a.   Find the probability of getting at least 1 of the components costing less than 50¢.
   b.   Find the mean number of components costing under 50¢ in such groups of 8.
   c.   Find the standard deviation for the numbers of components costing under 50¢ in such groups of 8.

14.   If $P(x)$ is described by the accompanying table, does $P(x)$ form a probability distribution? Explain.

| $x$ | $P(x)$ |
|---|---|
| −1 | 0.35 |
| 0 | 0.15 |
| 1 | 0.40 |
| 2 | 0.10 |

15.   A recessive genetic trait is known to occur in 50% of all offspring born within a certain population. A study involves the random selection of offspring in groups of 16.

a.  Find the probability of selecting a group and getting exactly 8 offspring with the recessive trait.
b.  Find the mean number of offspring having the recessive trait in such groups of 16.
c.  Find the variance for the number of offspring having the recessive trait in such groups of 16.

16. A probability distribution $P(x)$ is described by the accompanying table.
a.  Fill in the missing probability.
b.  Find the mean of the random variable $x$.
c.  Find the variance of the random variable $x$.

| $x$ | $P(x)$ |
|-----|--------|
| 5   | 0.6    |
| 10  |        |
| 20  | 0.1    |

# Computer Projects

1.  Use a software package such as STATDISK or Minitab to find the probability that among 50 births, there are at least 30 girls.
2.  Develop your own program to compute probabilities using the binomial probability formula. The program should take the values of $n$, $x$, and $p$ as input, and output should be the corresponding probability of $x$ successes in $n$ trials of a binomial experiment. Run the program to generate all of the probabilities represented by $0+$ in the section of Table A–1 for which $n = 10$.

# Applied Projects

1.  Standardized tests are now used extensively in a wide variety of applications. Colleges use SAT and ACT tests among their admissions criteria, and many employers use standardized tests to assess suitability of job applicants. This project analyzes the effects of guessing. Assume that one particular test has 10 multiple choice questions, each with the 5 possible answers $a$, $b$, $c$, $d$, $e$.
a.  Assuming that everyone makes random guesses for each answer, construct a table that summarizes the probability distribution, and construct a probability histogram. Find the mean and standard deviation for the number of correct responses on a test.
b.  Simulate a test taken by someone who makes random guesses for each answer. Use a computer, telephone book, table of random

*continued*

numbers (found in reference books), or any other suitable method. Describe the process used to obtain the 10 guesses.

c. Construct a key consisting of 1 correct response for each question, then find the actual number of correct responses for the simulated test from part *b*.

d. Repeat parts *b* and *c* until you have 20 scores. Each score should be a number between 0 and 10. Construct a frequency table and histogram. Find the sample mean and standard deviation. Compare these results to the theoretical results found in part *a*.

2. When purchasing desks for classrooms, some consideration is being given to left-handed students. Refer to Data Set II in Appendix B to answer the following.

a. Estimate the probability that a randomly selected person is left-handed.

b. Using the result from part *a*, find the probability of getting exactly 2 left-handed students in a class of 10.

c. Using the result from part *a*, find the probability that when 25 students register for a class, the number of left-handed students is below 4.

d. For classes of 25 students, find the mean, variance, and standard deviation for the number of left-handed students. (Use the result from part *a*.)

3. For families with 4 children, assume that girls and boys are equally likely. Find the probabilities of 0, 1, 2, 3, and 4 girls, respectively, and summarize the results in the form of a probability distribution table. Calculate the mean and standard deviation for the numbers of girls in families with 4 children. Now proceed to collect data from 100 simulated families as follows. Refer to a telephone book and record the last 4 digits of 100 telephone numbers. Letting the digits 0, 1, 2, 3, 4 represent girls and letting 5, 6, 7, 8, 9 represent boys, find the number of girls in each of the 100 "families." Summarize the results in a frequency table and find the mean and standard deviation. Compare the results from the observed data to the theoretical results.

 *Writing Projects*

1. After reading the Chapter Problem given at the beginning of this chapter and its solution given at the end of Section 4–4, write a report for the Federal Aviation Administration. Outline the key issues and present your recommendations. Support your recommendations with specific results.

2. Write a report summarizing one of the videotape programs listed.

# *Videotapes*

Programs 16 and 17 from the series *Against All Odds: Inside Statistics* are recommended as supplements to this chapter.

# Chapter Five

## In This Chapter

**5–1** **Overview**
We identify chapter **objectives** and describe the **normal distribution.**

**5–2** **The Standard Normal Distribution**
We define the **standard normal distribution** and describe methods for determining probabilities by using that distribution.

**5–3** **Nonstandard Normal Distributions**
We use the $z$ score (or standard score) to work with normal distributions in which the mean is not 0 or the standard deviation is not 1.

**5–4** **Finding Scores When Given Probabilities**
With normal distributions, we determine the values of scores that correspond to various given probabilities.

**5–5** **Normal as Approximation to Binomial**
We can sometimes use the **normal** distribution to determine probabilities in a **binomial** experiment.

**5–6** **The Central Limit Theorem**
The sampling distribution of sample means tends to be a normal distribution with mean $\mu$ and standard deviation $\sigma/\sqrt{n}$.

# 5 Normal Probability Distributions

## Chapter Problem

It's not practical to design all products to be convenient for all humans. The standard doorway has a height of 6' 8", but this would not accommodate the 8' 2" height of Don Koehler (listed in the *Guinness Book of World Records*) or the extreme heights of many basketball players. Car designers use human engineering to create practical seating areas for most people, but not everyone.

After considering such factors as marketing reports, car size constraints, and the impact on passenger and trunk areas, an engineer designs a driver's seat so that it will comfortably fit women taller than 159.0 cm (or 62.6 in.). What is the percentage of women who are shorter than 159.0 cm? (According to data from the National Health Survey, adult women have heights with a mean of 161.5 cm and a standard deviation of 6.3 cm.)

This is one of the many applications to be considered in this chapter.

# RELIABILITY AND VALIDITY

■The reliability of data refers to the consistency with which results occur, whereas the validity of data refers to how well the data measure what they are supposed to measure. The reliability of an IQ test can be judged by comparing scores for the test given on one date to scores for the same test given at another time. To test the validity of an IQ test, we might compare the test scores to another indicator of intelligence, such as academic performance. Many critics charge that IQ tests are reliable, but not valid; they provide consistent results, but don't really measure intelligence. ■

# 5–1   Overview

Chapter 4 introduced the concept of a probability distribution, and considered only *discrete* types. There are also many different *continuous* probability distributions. This chapter focuses on the normal distribution, which is extremely important to the study of statistics. However, before formally considering the normal probability distribution, let's briefly examine a simpler continuous distribution called the uniform distribution, as illustrated in Figure 5–1. A **uniform distribution** has equally likely values over the range of possible outcomes so that its graph (called a *density curve* for continuous random variables) is always a rectangle with an enclosed area equal to 1. This property makes it very easy for us to solve probability problems. For example, given the uniform distribution of Figure 5–1, the probability of randomly selecting a voltage level between 7.00 V and 8.50 V is 0.375, which is the area of the shaded region. Here's an important point: **For a density curve depicting the distribution of a continuous random variable, there is a correspondence between area and probability.** In Figure 5–1, finding the probability of a value between 7.00 V and 8.50 V can be accomplished by finding the area of the corresponding shaded rectangular region with dimensions 0.25 by 1.50. Normal distributions involve a more complicated curve, so it's more difficult to find areas. But the basic principle is the same: There is a correspondence between area and probability.

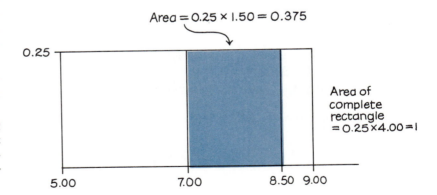

**Figure 5–1**

The smooth bell-shaped density curve shown in Figure 5–2 depicts the **normal distribution** that can be described by the equation

$$y = \frac{e^{-(x-\mu)^2/2\sigma^2}}{\sigma\sqrt{2\pi}}$$

(5–1)

Here $\mu$ represents the mean score of the entire population, $\sigma$ is the standard deviation of the population, $\pi$ is approximately 3.142, and $e$ is approximately 2.718. Equation 5–1 relates the horizontal scale of $x$ values to the vertical scale of $y$ values. And now for some really good news: We will not use this equation in actual computations because we would need some knowledge of calculus to do so.

In reality, we don't usually get scores that conform to the precise relationship expressed in Equation 5–1. In a theoretically ideal normal distribution, the tails extend infinitely far in both directions as they get closer to the horizontal axis. But this property is usually inconsistent with the limitations of reality. For example, IQ scores cannot be less than zero, so no distribution of IQ scores can extend infinitely far to the left. Similarly, objects cannot have negative weights nor can they have negative distances. Limitations such as these require that frequency distributions of real data can only approximate a normal distribution.

This chapter presents the standard methods used to work with normally distributed scores, and it includes applications. In addition to the importance of the normal distribution itself, the methods are important in establishing basic patterns and concepts that will apply to other continuous probability distributions.

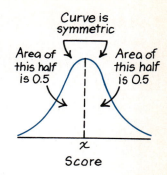

**Figure 5–2**
*The Normal Distribution*

## 5–2 The Standard Normal Distribution

There are actually many different normal probability distributions, each dependent on only two parameters: the population mean $\mu$ and the population standard deviation $\sigma$. Figure 5–3 shows three different normal distributions of IQ scores, with the differences due to changes in the mean and standard deviation. A change in the value of the population mean $\mu$ causes the curve to be shifted to the right or left. A change in the value of $\sigma$ causes

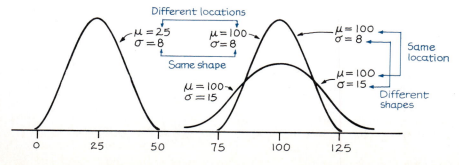

**Figure 5–3**

a change in the shape of the curve; the basic bell shape remains, but the curve becomes fatter or skinnier, depending on $\sigma$. Among the infinite possibilities, one particular normal distribution is of special interest.

---

### Definition

The **standard normal distribution** is a normal probability distribution that has a mean of 0 and a standard deviation of 1.

---

If we had to perform calculations with Equation 5–1 and we could choose any values for $\mu$ and $\sigma$, we would soon recognize that $\mu = 0$ and $\sigma = 1$ lead to the simplest form of that equation. Working with this simplified form, mathematicians are able to perform various analyses and computations. Figure 5–4 shows a graph of the standard normal distribution with some of the computed results. For example, the area under the curve and bounded by scores of 0 and 1 is 0.3413. The area under the curve and bounded by scores of $-1$ and $-2$ is 0.1359. The sum of the six known areas in Figure 5–4 is 0.9974, but if we include the small areas in the two tails we get a total of 1. (In any probability distribution, the total area under the curve is 1.) Nearly all (99.7%) of the values lie within three standard deviations from the mean.

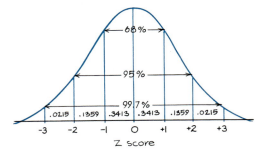

**Figure 5–4**
*Standard Normal Distribution*

Figure 5–4 illustrates only six probability values, but a more complete table has been compiled to provide more precise data. Table A–2 (in Appendix A) gives the probability corresponding to the area under the curve bounded on the left by a vertical line above the mean of zero and bounded on the right by a vertical line above any specific positive score denoted by $z$ (see Figure 5–5). Note that when you use Table A–2, the hundredths part of the $z$ score is found across the top row. To find the probability associated with a score between 0 and 1.23, for example, begin with the $z$ score of 1.23 by locating 1.2 in the left column. Then find the value in the adjoining row of probabilities that is directly below 0.03. There is a probability of 0.3907 of randomly selecting a score between 0 and 1.23. It is essential to remember that this table is designed only for the standard normal distribution, which has a mean of 0 and a standard deviation of 1. Nonstandard cases will be considered in the next section.

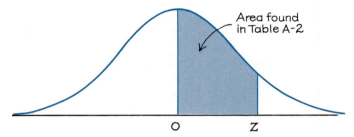

**Figure 5–5**
*The standard normal distribution. The area of the shaded region bounded by the mean of zero and the positive number z can be found in Table A–2.*

Normal distributions originally resulted from studies of experimental errors, and the following examples also deal with errors in measurements.

---

### Example

A manufacturer of scientific instruments produces thermometers that are supposed to give readings of 0° C at the freezing point of water. Tests on a large sample of these instruments reveal that some readings are too low (denoted by negative numbers) and some readings are too high (denoted by positive numbers). Assume that the mean reading is 0° C while the standard deviation of the readings is 1.00° C. Also assume that the frequency distribution of errors closely resembles the normal distribution. If one thermometer is randomly selected, find the probability that, at the freezing level of water, the reading is between 0° and +1.58°.

### Solution

We are dealing with a standard normal distribution and we are looking for the area of the shaded region in Figure 5–5 with $z = 1.58$. We find from Table A–2 that the shaded area is 0.4429. The probability of randomly selecting a thermometer with an error between 0° and +1.58° is therefore 0.4429.

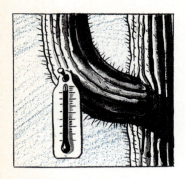

# HOTTEST SPOT

■ A firefighter in Bullhead City, Arizona, provided daily weather statistics. One day, a Weather Service representative demanded that the thermometer be moved from the firehouse lawn to a more natural setting. The move to a drier area 100 yards away led to readings about 5° higher, which often made Bullhead City the hottest spot in the United States. As Bullhead City gained prominence in many television weather reports, some residents denounced the notoriety as a handicap to business, while others felt that it helped. Under more standardized conditions, measuring instruments, such as thermometers, tend to produce errors that are normally distributed. ■

The solutions to this and the following examples are contingent on the values listed in Table A–2. But these values did not appear spontaneously. They were arrived at through calculations that relate directly to Equation 5–1. Table A–2 serves as a convenient means of circumventing difficult computations with that equation.

## Example

With the thermometers from the preceding example, find the probability of randomly selecting one thermometer that reads (at the freezing point of water) between 0° and −2.43°.

## Solution

We are looking for the region shaded in Figure 5–6(a), but Table A–2 is designed to apply only to regions to the right of the mean (zero) as in Figure 5–6(b). However, by observing that the normal probability distribution possesses symmetry about zero, we see that the shaded regions in parts (a) and (b) of Figure 5–6 have the same area. Referring to Table A–2, we can easily determine that the shaded area of Figure 5–6(b) is 0.4925, so the shaded area of Figure 5–6(a) must also be 0.4925. That is, the probability of randomly selecting a thermometer with an error between 0° and −2.43° is 0.4925.

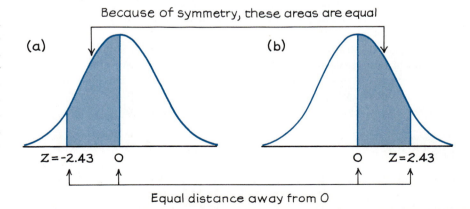

**Figure 5–6**

Section 2–5 presented the empirical rule, which states that for bell-shaped distributions:

- About 68% of all scores fall within *one* standard deviation of the mean.

- About 95% of all scores fall within *two* standard deviations of the mean.

- About 99.7% of all scores fall within *three* standard deviations of the mean.

If we refer to Figure 5–5 and let $z = 1$, Table A–2 shows us that the shaded area is 0.3413, therefore the proportion of scores between $z = -1$ and $z = 1$ will be $0.3413 + 0.3413 = 0.6826$. That is, about 68% of all scores fall within one standard deviation of the mean. Similar calculations with $z = 2$ and then $z = 3$ result in the values given above in the empirical rule. These values relate directly to Table A–2.

We incorporate an obvious but useful observation in the following example, but first go back for a minute to Figure 5–5. A vertical line directly above the mean of zero divides the area under the curve into two equal parts, each containing an area of 0.5. Since we are dealing with a probability distribution, the total area under the curve must be 1.

## Example

With these same thermometers, we again make a random selection. Find the probability that the chosen thermometer reads (at the freezing point of water) greater than $+1.27°$.

## Solution

We are again dealing with normally distributed values having a mean of 0° and a standard deviation of 1°. The probability of selecting a thermometer that reads above $+1.27°$ corresponds to the shaded area of Figure 5–7. Table A–2 cannot be used to find that area directly, but we can use the table to find the adjacent area of 0.3980. We can now reason that since the total area to the right of zero is 0.5, the shaded area is $0.5 - 0.3980$, or 0.1020. We conclude that there is a 0.1020 probability of randomly selecting one of the thermometers with a reading greater than $+1.27°$.

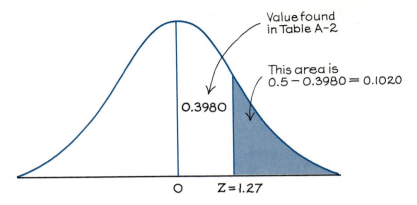

**Figure 5–7**

We are able to determine the area of the shaded region in Figure 5–7 by an *indirect* application of Table A–2. The following example illustrates yet another indirect use.

---

**Example**

Back to the same thermometers: Assuming that one thermometer is randomly selected, find the probability that it reads (at the freezing point of water) between 1.20° and 2.30°.

**Solution**

The probability of selecting a thermometer that reads between $+1.20°$ and $+2.30°$ corresponds to the shaded area of Figure 5–8. However, Table A–2 is designed to provide only for regions bounded on the left by the vertical line above zero. We can use the table to find the areas of 0.3849 and 0.4893 as shown in this figure. If we denote the area of the shaded region by $A$, we can see from the figure that

$$0.3849 + A = 0.4893$$

so that

$$A = 0.4893 - 0.3849$$
$$= 0.1044$$

The probability we seek is therefore 0.1044.

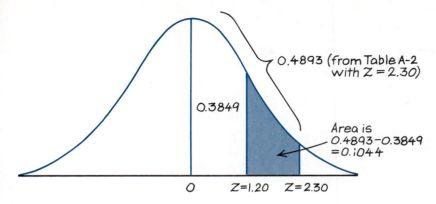

0.4893 (from Table A-2 with Z = 2.30)

0.3849

Area is
0.4893−0.3849
=0.1044

O    Z=1.20    Z=2.30

**Figure 5–8**

| Notation | |
|---|---|
| $P(a < z < b)$ | denotes the probability that the $z$ score is between $a$ and $b$. |
| $P(z > a)$ | denotes the probability that the $z$ score is greater than $a$. |
| $P(z < a)$ | denotes the probability that the $z$ score is less than $a$. |

Using this notation, we can express the result of the last example as $P(1.20 < z < 2.30) = 0.1044$. With a continuous probability distribution, such as the normal distribution, $P(z = a) = 0$. That is, the probability of getting any one precise value is 0. From this we can conclude that $P(a \leq z \leq b) = P(a < z < b)$.

The examples of this section were contrived so that the mean of 0 and the standard deviation of 1 coincided exactly with the values of the standard normal distribution described in Table A–2. In reality, it would be unusual to find such a nice relationship, since typical normal distributions involve means different from 0 and standard deviations different from 1.

These nonstandard normal distributions introduce another problem. What table of probabilities can be used, since Table A–2 is designed around a mean of 0 and a standard deviation of 1? For example, IQ scores are normally distributed with a mean of 100 and a standard deviation of 15. Scores in this range are far beyond the scope of Table A–2. Section 5–3 examines these nonstandard normal distributions and the methods used in dealing with them. We will see that Table A–2 can be used with nonstandard normal distributions after performing some very simple calculations. Learning how to use Table A–2 in this section prepares us for more practical and realistic circumstances that will be found in later sections.

# 5–2 Exercises A

In Exercises 1–36, assume that the readings on the thermometers are normally distributed with a mean of 0° and a standard deviation of 1.00°. A thermometer is randomly selected and tested. In each case, draw a sketch and find the probability of each reading in degrees.

1.  Between 0 and 0.25
2.  Between 0 and 1.00
3.  Between 0 and 1.50
4.  Between 0 and 1.96
5.  Between − 1.00 and 0
6.  Between − 0.75 and 0
7.  Between 0 and − 1.75
8.  Between 0 and − 2.33
9.  Greater than 1.00
10.  Greater than 0.37
11.  Greater than 1.83
12.  Greater than 2.05
13.  Less than − 1.00
14.  Less than − 2.17
15.  Less than − 0.91
16.  Less than − 1.37
17.  Greater than − 1.00
18.  Greater than − 0.09
19.  Less than 3.05
20.  Less than 0.42
21.  Between − 1.00 and 2.00
22.  Between − 0.25 and 0.75
23.  Between − 2.00 and 1.50
24.  Between − 1.96 and 1.96
25.  Between 1.00 and 2.00
26.  Between 1.96 and 2.33
27.  Between 1.28 and 2.58
28.  Between 0.27 and 2.27
29.  Between − 0.83 and − 0.51
30.  Between − 2.00 and − 1.50
31.  Between − 0.25 and − 1.35
32.  Between − 1.07 and − 2.11
33.  Greater than 0
34.  Less than 0
35.  Less than − 0.50 or greater than 1.50
36.  Less than − 1.96 or greater than 1.96

In Exercises 37–48, assume that the readings on the thermometers are normally distributed with a mean of 0° and a standard deviation of 1.00°. Find the indicated probability where $z$ is the reading in degrees.

37.  $P(z > 2.58)$
38.  $P(-1.36 < z < 1.36)$
39.  $P(z < 1.28)$
40.  $P(1.25 < z < 1.68)$
41.  $P(z < -0.57)$
42.  $P(0 < z < 1.68)$
43.  $P(-2.80 < z < -1.36)$
44.  $P(z > -0.50)$
45.  $P(-1.09 < z < 0)$
46.  $P(-2.73 < z < 2.51)$
47.  $P(z < 2.45)$
48.  $P(-0.81 < z < 0.63)$

# 5–2 Exercises B

49.  Assume that $\mu = 0$ and $\sigma = 1$ for a normally distributed population. Find the percentage of data that are
     a.  Within 1 standard deviation of the mean

  b.   Within 1.96 standard deviations of the mean
  c.   Between $\mu - 3\sigma$ and $\mu + 3\sigma$
  d.   Between 1 standard deviation below the mean and 2 standard deviations above the mean
  e.   More than 2 standard deviations away from the mean

50.   Assume that we have the same normally distributed thermometer readings with a mean of 0° and a standard deviation of 1.00°. If 5% of the thermometers are rejected because they read too high and another 5% are rejected because they read too low, what is the maximum error that will not lead to rejection?

51.   In Equation 5–1, if we let $\mu = 0$ and $\sigma = 1$ we get

$$y = \frac{e^{-x^2/2}}{\sqrt{2\pi}}$$

which can be approximated by

$$y = \frac{2.7^{-x^2/2}}{2.5}$$

Graph the last equation after finding the $y$ coordinates that correspond to the following $x$ coordinates: $-4$, $-3$, $-2$, $-1$, 0, 1, 2, 3, and 4. (A calculator capable of dealing with exponents will be helpful.) Attempt to determine the approximate area bounded by the curve, the $x$-axis, the vertical line passing through 0 on the $x$-axis, and the vertical line passing through 1 on the $x$-axis. Compare this result to Table A–2.

52.   Assume that $z$ scores are normally distributed with a mean of 0 and a standard deviation of 1.
  a.   If $P(0 < z < a) = 0.4778$, find $a$.
  b.   If $P(-b < z < b) = 0.7814$, find $b$.
  c.   If $P(z > c) = 0.0329$, find $c$.
  d.   If $P(z > d) = 0.8508$, find $d$.
  e.   If $P(z < e) = 0.0062$, find $e$.

# 5–3 | Nonstandard Normal Distributions

Section 5–2 considered only the standard normal distribution, but this section extends the same basic concepts to include nonstandard normal distributions. This inclusion will greatly expand the variety of practical applications we can make since, in reality, most normally distributed populations will have either a nonzero mean and/or a standard deviation different than 1.

  We continue to use Table A–2, but we require a way of standardizing these nonstandard cases. This is done by letting

$$z = \frac{x - \mu}{\sigma} \qquad\qquad (5\text{--}2)$$

Here $z$ is the number of standard deviations that a particular score $x$ is away from the mean. We call $z$ the **z score** or **standard score** and it is used in Table A–2 (see Figure 5–9). (This same definition was presented earlier in Section 2–6.)

(a)                                                                (b)

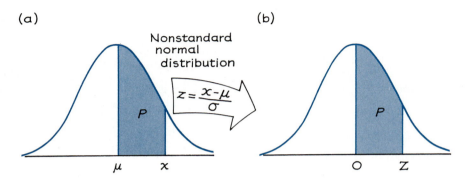

**Figure 5–9**

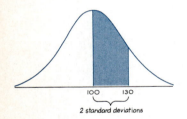

**Figure 5–10**

Suppose, for example, that we are considering a normally distributed collection of IQ scores known to have a mean of 100 and a standard deviation of 15. If we seek the probability of randomly selecting one IQ score that is between 100 and 130, we are concerned with the area shown in Figure 5–10. The difference between 130 and the mean of 100 is 30 IQ points, or exactly 2 standard deviations. The shaded area in Figure 5–10 will therefore correspond to the shaded area of Figure 5–5, where $z = 2$. We get $z = 2$ either by reasoning that 130 is 2 standard deviations above the mean of 100 or by computing

$$z = \frac{x - \mu}{\sigma} = \frac{130 - 100}{15} = \frac{30}{15} = 2$$

With $z = 2$, Table A–2 indicates that the shaded region we seek has an area of 0.4772, so that the probability of randomly selecting an IQ score between 100 and 130 is 0.4772. Thus Table A–2 can be indirectly applied to any normal probability distribution if we use Equation 5–2 as the algebraic way of recognizing that the $z$ score is actually the number of standard deviations that $x$ is away from the mean. The following examples illustrate that observation. Before considering the next example we should note that IQ and many other types of test scores are discrete whole numbers, while the normal distribution is continuous. We ignore that conflict in this and the following section since the results are minimally affected, but we will introduce a "continuity correction factor" in Section 5–5 when it becomes necessary. However, we might note for the present that finding the probability of getting an IQ score of *exactly* 130 would require that the discrete value of 130 be represented by the continuous interval from 129.5 to 130.5. This illustrates how the continuity correction factor can be used to deal with discrete data.

---

**Example**

If IQ scores are normally distributed with a mean of 100 and a standard deviation of 15, find the probability of randomly selecting a subject with an IQ between 100 and 133.

**Solution**

Referring to Figure 5–11, we seek the probability associated with the shaded region. In order to use Table A–2, the nonstandard data must be standardized by applying Equation 5–2.

$$z = \frac{x - \mu}{\sigma} = \frac{133 - 100}{15} = \frac{33}{15} = 2.20$$

The score of 133 therefore differs from the mean of 100 by 2.20 standard deviations. Corresponding to a $z$ score of 2.20, Table A–2 indicates a probability of 0.4861. There is therefore a probability of 0.4861 of randomly selecting a subject having an IQ between 100 and 133. We could also express this result as $P(100 < x < 133) = 0.4861$ by using the same notation introduced in Section 5–2. Note that in this nonstandard normal distribution, we represent the score in its original units by $x$, not $z$. Note also that $P(100 < x < 133) = P(0 < z < 2.20) = 0.4861$.

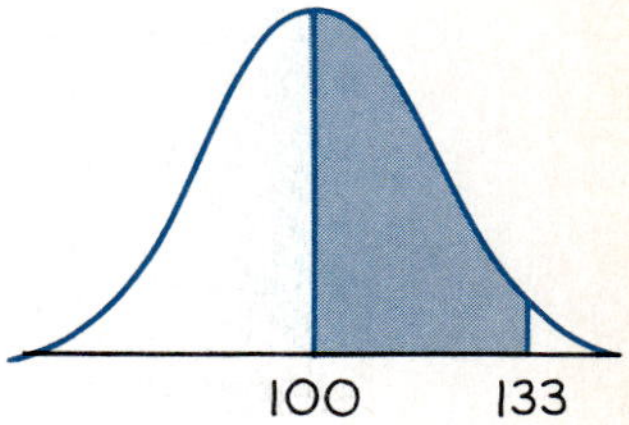

**Figure 5–11**

When Table A–2 is used in conjunction with Equation 5–2, the nonstandard population mean corresponds to the standard mean of 0. As a result, probabilities extracted directly from Table A–2 must represent regions whose left boundary is the line above the mean.

---

**Example**

In the beginning of this chapter we described a car designer's need to find the percentage of women who would not fit comfortably in a driver's seat. For the car design being considered, women are uncomfortable if they are shorter than 159.0 cm (or 62.6 in.). Based on data from the National Health Survey, we know that women's heights are normally distributed with a mean of 161.5 cm and a standard deviation of 6.3 cm. Find the percentage of women shorter than 159.0 cm.

*continued*

> ### Solution
>
> In Figure 5–12, the shaded region corresponds to the women who are too short for this car's design. We can't find the area of the shaded region directly, but we can use Equation 5–2 to find the area of the adjacent region $A$.
>
> $$z = \frac{x - \mu}{\sigma} = \frac{159.0 - 161.5}{6.3} = -0.40$$
>
> We now use Table A–2 to find that $z = -0.40$ corresponds to an area of 0.1554. Since $A = 0.1554$, the shaded area must be $0.5 - 0.1554 = 0.3446$. Since the shaded area is 0.3446, the probability of randomly selecting a woman less than 159.0 cm tall is also 0.3446, so the percentage of women less than 159.0 cm tall is 34.46%. Because this percentage is so high, the designers might do well to reconsider their plans, which would result in cars that would be uncomfortable for such a large market share.

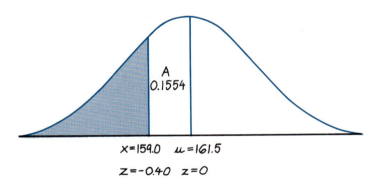

**Figure 5–12**

> ### Example
>
> A banker studying customer needs finds that the numbers of times people use automated-teller machines in a year are normally distributed with a mean of 30.0 and a standard deviation of 11.4 (based on data from Maritz Marketing Research, Inc.). Find the percentage of customers who
>
> *continued*

**Example** *continued*

use them between 40 and 50 times. Among 5000 customers, how many are expected to have between 40 and 50 uses in a year?

## Solution

Figure 5–13 shows the shaded region *B* corresponding to values between 40 and 50. We can find that shaded area indirectly as follows.

For areas *A* and *B* combined:

$$z = \frac{x - \mu}{\sigma} = \frac{50.0 - 30.0}{11.4} = 1.75$$

Table A–2 shows that $z = 1.75$ corresponds to 0.4599 so that the area of *A* and *B* combined is 0.4599.

For area *A* only: $z = \dfrac{x - \mu}{\sigma} = \dfrac{40.0 - 30.0}{11.4} = 0.88$

Table A–2 shows that $z = 0.88$ corresponds to 0.3106 so that area *A* alone is 0.3106.

Area *B* = (areas *A* and *B* combined) − (area *A*)
= 0.4599 − 0.3106 = 0.1493

The percentage of customers who use the automated-teller machine between 40 and 50 times is 14.93%. Among 5000 customers, we expect 14.93% of them will be in that category. The actual number expected is 5000 × 0.1493 = 746.5.

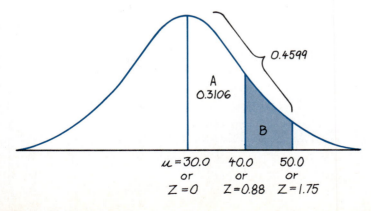

**Figure 5–13**

# POTATO CHIP CONTROL CHARTS

■ In the past, quality control was simply a matter of inspecting products after they were completed. Now, Statistical Process Control (SPC) is used to analyze samples at different points in the production process so that we know when to make adjustments and when to leave things alone. Frito Lay's bags of potato chips now have much less variability due to the use of SPC and control charts. Frito Lay aims for 1.6% salt content and continually monitors that level by using control charts. (See Program 18 from the series *Against All Odds: Inside Statistics* for a discussion of Frito Lay's use of SPC.)■

# Statistical Process Control (SPC)

Manufacturers and other segments of industry now use statistics to monitor and control the quality of the products they produce and distribute, and this text includes many of the basic tools required. In preceding chapters we discussed the sample mean, standard deviation, and range. This chapter presents concepts of the normal distribution, which plays an important role in the construction of a **control chart** used to depict some characteristic of a process in order to detect a trend over time. In Figure 5–14 we give an example of a control chart reflecting a production process that has gone out of control. (See Exercises 35 and 36.) Other types of control charts are $\bar{x}$-**charts** (with sample means plotted over time), **R-charts** (with sample ranges plotted over time), **p-charts** (with proportions of defects plotted over time), and **c-charts** (with numbers of defects per item plotted over time).

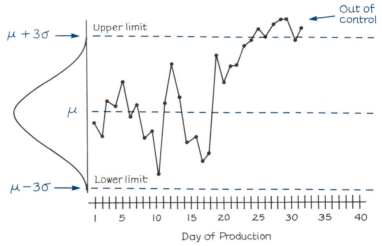

**Figure 5–14**

Such quality-control devices are often grouped under the heading of **statistical process control (SPC)**, which is defined in a Ford Motor Company publication as follows:

## Statistical Process Control

The use of statistical techniques such as control charts to analyze a process or its outputs so as to take appropriate actions to achieve and maintain a state of statistical control and to improve the process capability.

In this section we extended the concept of Section 5–2 to include more realistic nonstandard normal probability distributions. We noted that the formula $z = (x - \mu)/\sigma$ algebraically represents the number of standard deviations that a particular score $x$ is away from the mean. However, all the examples we have considered so far are of the same general type: A probability (or percentage) is determined by using the normal distribution (described in Table A–2) when given the values of the mean, standard deviation, and relevant score(s). In many practical and real cases, the probability (or percentage) is known and we must determine the relevant score(s). Problems of this type are considered in the following section.

## 5–3 Exercises A

In Exercises 1–8, assume that IQ scores are normally distributed with a mean of 100 and a standard deviation of 15. An IQ score is randomly selected from this population. Draw a graph and find the indicated probability.

1.  $P(100 < x < 145)$
2.  $P(x < 127)$
3.  $P(x > 140)$
4.  $P(88 < x < 112)$
5.  $P(110 < x < 120)$
6.  $P(120 < x < 130)$
7.  $P(x < 100)$
8.  $P(85 < x < 95)$

In Exercises 9–32, answer the given questions. In each case, draw a graph.

9.  One classic use of the normal distribution is inspired by a letter to Dear Abby in which a wife claimed to give birth 308 days after a brief visit from her husband, who was serving in the Navy. The lengths of pregnancies are normally distributed with a mean of 268 days and a standard deviation of 15 days. Given this information, find the probability of a pregnancy lasting 308 days or longer. Does this wife have a problem?

10. According to Nielsen Media Research, people watch television an average of 6.98 hours per day. Assume that these times are normally distributed with a standard deviation of 3.80 hours. Find the probability that a randomly selected person watches television more than 8.00 hours in a day.

11. According to the Federal Highway Administration, males aged 16–24 drive an average of 10,718 mi each year. Assume that the annual mileage totals are normally distributed with a standard deviation of 3573 mi. For a randomly selected male between the ages of 16 and 24, find the probability that he drives less than 12,000 mi in a year.

12. A study of VCR owners found that their annual household incomes are normally distributed with a mean of $41,182 and a standard deviation

of $19,990 (based on data from Nielsen Media Research). For households with VCRs, what is the percentage having incomes between $30,000 and $50,000?

13. In a Dutchess County manufacturing plant, the times required for employees to inspect silicon wafers are normally distributed with a mean of 9.98 min and a standard deviation of 3.74 min. If an employee is randomly selected to inspect a silicon wafer, what is the probability it will be done in less than 5 min?

14. In a study of facial behavior, people in a control group are timed for eye contact in a 5-minute period. Their times are normally distributed with a mean of 184.0 s and a standard deviation of 55.0 s (based on data from "Ethological Study of Facial Behavior in Nonparanoid and Paranoid Schizophrenic Patients," by Pitman, Kolb, Orr, and Singh, *Psychiatry*, 144:1). For a randomly selected person from the control group, find the probability that the eye contact time is greater than 230.0 s, which is the mean for paranoid schizophrenics.

15. The heights of six-year-old girls are normally distributed with a mean of 117.80 cm and a standard deviation of 5.52 cm (based on data from the National Health Survey, USDHEW publication 73-1605). Find the probability that a randomly selected six-year-old girl has a height between 117.80 cm and 120.56 cm.

16. For a certain population, scores on the Thematic Apperception Test are normally distributed with a mean of 22.83 and a standard deviation of 8.55 (based on "Relationships Between Achievement-Related Motives, Extrinsic Conditions, and Task Performance," by Schroth, *Journal of Social Psychology*, Vol. 127, No. 1). For a randomly selected subject, find the probability that the score is between 4.02 and 22.83.

17. A study shows that Michigan teachers have measures of job dissatisfaction that are normally distributed with a mean of 3.80 and a standard deviation of 0.95 (based on "Stress and Strain from Family Roles and Work-Role Expectations," by Cooke and Rousseau, *Journal of Applied Psychology*, Vol. 69, No. 2). If subjects with scores above 4.00 are to be given additional tests, what percentage will fall into that category?

18. For a certain population, scores on the Miller Analogies Test are normally distributed with a mean of 58.84 and a standard deviation of 15.94 (based on "Equivalencing MAT and GRE Scores Using Simple Linear Transformation and Regression Methods," by Kagan and Stock, *Journal of Experimental Education*, Vol. 49, No. 1). If subjects who score below 27.00 are to be given special training, what is the percentage of subjects who will be given the special training?

19. Using one measure of attractiveness, scores are normally distributed with a mean of 3.93 and a standard deviation of 0.75 (based on "Physical Attractiveness and Self Perception of Mental Disorder," by Burns and Farina, *Journal of Abnormal Psychology*, Vol. 96, No. 2). Find the probability of randomly selecting a subject with a measure of attractiveness that is less than 2.55.

20. The serum cholesterol levels in men aged 18 to 74 are normally distributed with a mean of 178.1 and a standard deviation of 40.7. All units are in mg/100 ml and the data are based on the National Health Survey (USDHEW publication 78-1654). If a man aged 18 to 24 is randomly selected, find the probability that his serum cholesterol level is between 100 and 200.

21. For a certain group, scores on the Mathematics Usage Test are normally distributed with a mean of 23.9 and a standard deviation of 8.7 (based on "Study of the Measurement Bias of Two Standardized Psychological Tests," by Drasgow, *Journal of Applied Psychology*, Vol. 72, No. 1). If a subject is randomly selected from this group, find the probability of a score between 25.0 and 30.0.

22. Scores on an antiaircraft artillery exam are normally distributed with a mean of 99.56 and a standard deviation of 25.84 (based on "Routinization of Mental Training in Organizations: Effects on Performance and Well Being," by Larson, *Journal of Applied Psychology*, Vol. 72, No. 1). For a randomly selected subject, find the probability of a score between 110.00 and 150.00.

23. The Beanstalk Club has a minimum height requirement of 5' 10" for women. If women have heights with a mean of 5' 5.5" and a standard deviation of 2.5", what percentage of women are eligible? (See Program 4 of *Against All Odds: Inside Statistics*.)

24. A standard IQ test produces normally distributed results with a mean of 100 and a standard deviation of 15. If an average IQ is defined to be any IQ between 90 and 109, find the probability of randomly selecting an IQ that is average.

25. Scores on the numeric part of the Minnesota Clerical Test are normally distributed with a mean of 119.3 and a standard deviation of 32.4. This test is used for selecting clerical employees. (The data are based on "Modification of the Minnesota Clerical Test to Predict Performance on Video Display Terminals," by Silver and Bennett, *Journal of Applied Psychology*, Vol. 72, No. 1.) If a firm requires scores above 172, find the percentage of subjects who don't qualify.

26. In a study of employee stock ownership plans, satisfaction by employees is measured and found to be normally distributed with a mean of 4.89 and a standard deviation of 0.63 (based on "Employee Stock Ownership and Employee Attitudes: A Test of Three Models," by Klein, *Journal of Applied Psychology*, Vol. 72, No. 2). If a subject from this population is randomly selected, find the probability of a job satisfaction score less than 6.78.

27. For males born in upstate New York, the gestation times are normally distributed with a mean of 39.4 weeks and a standard deviation of 2.43 weeks (based on data from the New York State Department of Health, Monograph No. 11). For a randomly selected male born in upstate New York, find the probability that the gestation time differs from the mean by more than 3.00 weeks.

28. In a study of the coliform contamination in streams, an environmentalist finds that one region has normally distributed coliform levels (number of cells per 100 ml) with a mean of 122 and a standard deviation of 14. For a randomly selected sample, find the probability that the coliform level differs from the mean by more than 25.

29. For a certain group of students, scores on an algebra placement test are normally distributed with a mean of 18.4 and a standard deviation of 5.1 (based on data from "Factors Affecting Achievement in the First Course in Calculus," by Edge and Friedberg, *Journal of Experimental Education*, Vol. 52, No. 3). If 50 different students are randomly selected from this population, how many of them are expected to score above 16.0?

30. The weights of women aged 18 to 24 are normally distributed with a mean of 132 lb and a standard deviation of 27.4 lb (based on data from the National Health Survey, USDHEW publication 79-1659). If 150 women 18 to 24 years old are randomly selected, how many of them are expected to weigh between 100 lb and 150 lb?

31. The weights of men aged 18 to 74 are normally distributed with a mean of 173 lb and a standard deviation of 30 lb (based on data from the National Health Survey, USDHEW publication 79-1659). Find the percentage of such weights between 190 lb and 225 lb. Among 400 men aged 18 to 74 years, how many are expected to weigh between 190 lb and 225 lb?

32. Scores on the biology portion of the Medical College Aptitude Test are normally distributed with a mean of 8.0 and a standard deviation of 2.6. Among 600 individuals taking this test, how many are expected to score between 6.0 and 7.0?

## 5–3 Exercises B

33. The following sample scores are times (in milliseconds) it took the author's disk drive to make one revolution. The times were recorded by a diagnostic software program.

| | | | | | | | | | |
|---|---|---|---|---|---|---|---|---|---|
| 200.5 | 199.7 | 201.1 | 200.4 | 200.3 | 200.1 | 200.4 | 200.4 | 200.4 | 200.5 |
| 200.1 | 200.1 | 200.3 | 200.5 | 200.3 | 200.3 | 200.6 | 200.5 | 200.4 | 200.5 |
| 200.3 | 201.2 | 200.5 | 200.6 | 200.4 | 200.5 | 200.3 | 200.7 | 200.6 | 200.5 |
| 200.4 | 200.0 | 201.2 | 200.6 | 200.4 | 200.8 | 200.6 | 200.3 | 200.6 | 200.5 |

a. Find the mean $\bar{x}$ of this sample.
b. Find the standard deviation $s$ of this sample.
c. Find the actual percentage of these sample scores that are greater than 201.0 milliseconds (ms).

d. Assuming a normal distribution, find the percentage of *population* scores greater than 201.0 ms. Use the sample values of $\bar{x}$ and $s$ as estimates of $\mu$ and $\sigma$.

e. The specifications require times between 198.0 ms and 202.0 ms. Based on these sample results, does the disk drive seem to be rotating at acceptable speeds?

34. a. The accompanying frequency table summarizes the age distribution for a number of students randomly selected from the population of students at a large university. Although the data do not appear to be normally distributed, assume that they are and find $\bar{x}$ and $s$. Then use those statistics as estimates of $\mu$ and $\sigma$ in order to find the proportion of students over 21 years of age. How does the result compare to the sample statistics summarized in the table?

| Age | f |
|-----|-----|
| 15–19 | 220 |
| 20–22 | 173 |
| 23–24 | 55 |
| 25–29 | 98 |
| 30–34 | 62 |
| 35–44 | 85 |
| 45–59 | 36 |
| 60–65 | 4 |

b. A population has a mean of 100 and a standard deviation of 15 but is uniformly distributed. Such a distribution has a minimum of 74 and a maximum of 126. Find the probability of randomly selecting a value between 80 and 110 with this uniform distribution, and compare it to the area between 80 and 110 for a normal distribution with the same mean of 100 and standard deviation of 15.

35. Given a production process designed for $\mu = 50.0$ and $\sigma = 4.0$, construct the control chart (see Figure 5–14) for the production values given below for 32 consecutive work days. The values are listed by rows so that 51.2 is the first day, 50.8 is the second day, and so on. What is happening to production?

$$
\begin{array}{ccccccc}
51.2 & 50.8 & 52.0 & 49.8 & 49.7 & 53.4 & 46.2 \\
49.3 & 55.0 & 48.3 & 50.4 & 42.6 & 58.3 & 58.0 \\
47.6 & 43.5 & 42.9 & 45.6 & 52.7 & 59.4 & 55.3 \\
41.0 & 40.8 & 38.0 & 56.5 & 60.4 & 62.5 & 57.3 \\
37.8 & 48.6 & 62.8 & 60.7 & & &
\end{array}
$$

36. A popular quality control tool is the **CUSUM** (cumulative sum) chart, which is very sensitive to changes in a mean. For the data in Exercise 35, construct a CUSUM chart as follows.

a. Using $\mu = 50.0$ as a reference value, let

$$
\begin{aligned}
S_1 &= 51.2 - 50.0 \\
S_2 &= S_1 + (50.8 - 50.0) \\
S_3 &= S_2 + (52.0 - 50.0)
\end{aligned}
$$

and so on.

These values of $S_1, S_2, S_3, \ldots, S_{32}$ are accumulating sums of differences between the scores and the reference value of 50.0.

*continued*

b. Use the same horizontal scale as in Figure 5–14, and use a vertical scale that ranges from −30 to 30. Plot the values $S_1, S_2, S_3, \ldots, S_{32}$. Does the CUSUM chart indicate a shift in the mean?

# 5–4 Finding Scores When Given Probabilities

All examples and exercises from Sections 5–2 and 5–3 involved a normal distribution in this format: Given some score, we used Table A–2 to find a probability. In this section that format is changed: Given a probability, we will find the score.

In the problem described at the beginning of this chapter, we saw a car design that was uncomfortable for women with heights under 159.0 cm (or 62.6 in.). In Section 5–3 we determined that this constitutes 34.46% of all women. Let's assume that after learning this, the engineers decide to make changes so that only 5% of all women would be uncomfortable. Here's our new problem: What height separates the shortest 5% of women from the tallest 95%? Recall that National Health Survey data reveal that women have normally distributed heights with a mean of 161.5 cm and a standard deviation of 6.3 cm. Figure 5–15 depicts this situation.

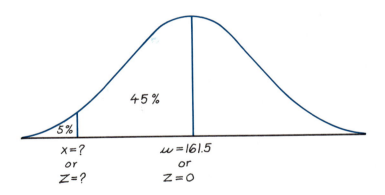

**Figure 5–15**

We can find the $z$ score corresponding to the $x$ value we seek after first noting that the region containing 45% corresponds to an area of 0.4500. Referring to Table A–2 we find that 0.4500 corresponds to $z = 1.645$. Since the $z$ score is negative whenever it is below the mean, we set $z = -1.645$. That is, the score $x$ is 1.645 standard deviations *below* the mean of 161.5.

Since the standard deviation is 6.3, we conclude that 1.645 standard deviations is $1.645 \times 6.3 = 10.3635$. Our $x$ score is therefore 10.3635 below 161.5 and we get $x = 161.5 - 10.3635 = 151.1$ (rounded off). That is, the height of 151.1 cm separates the shortest 5% of women from the tallest 95%. We could also state that 151.1 cm is the 5th percentile, or $P_5$.

We could have achieved the same result by noting that

$$z = \frac{x - \mu}{\sigma} \quad \text{becomes} \quad -1.645 = \frac{x - 161.5}{6.3}$$

when we substitute the given values for the mean $\mu$, the standard deviation $\sigma$, and the $z$ score corresponding to an area of 0.4500 to the left of the mean. We solve this equation by multiplying both sides by 6.3 and then adding 161.5 to both sides.

In the preceding solution, it is often too easy to make the mistake of forgetting the negative sign in $-1.645$. Omission of this negative sign would have led to an answer of 171.9 cm, but Figure 5–15 should show that $x$ can't possibly be 171.9 cm. This illustrates the importance of drawing a graph when working with the normal distribution. Always draw the graph with the relevant labels and use common sense to check that the results are reasonable.

Also note that in the preceding solution, Table A–2 led to a $z$ score of 1.645, which is midway between 1.64 and 1.65. When using Table A–2, we can usually avoid interpolation by simply selecting the closest value. However, there are two special cases involving important values that are commonly used. These are summarized in the table shown in the margin. Except for these two special cases, we can select the closest value in the table. (If a desired value is midway between two table values, select the larger value.) Also, for $z$ scores above 3.09, we can use 0.4999 as an approximation of the corresponding area.

| $z$ score | Area |
|-----------|--------|
| 1.645 | 0.4500 |
| 2.575 | 0.4950 |

In the preceding illustration, some women were made uncomfortable by cost-effective product design. The next example affects men.

---

**Example**

A clothing manufacturer finds it unprofitable to make clothes for very tall or very short adult males. The executives decide to discontinue production of goods for the tallest 7.5% and the shortest 7.5% of the adult male population. Find the minimum and maximum heights they will continue to serve. The heights of adult males are normally distributed with a mean of 69.0 in. and a standard deviation of 2.8 in. (The figures are based on data from the National Health Survey, USDHEW publication 79–1659.)

*continued*

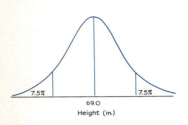

7.5%        7.5%

69.0
Height (in.)

**Figure 5–16**

### Solution

See Figure 5–16. Note that the two outer regions total 15%, so the two equal inner regions must comprise the remaining 85%. This implies that each of the two inner regions must represent 42.5%, which is equivalent to a probability of 0.425. We can use Table A–2 to find the $z$ score that yields a probability of 0.425. With $z = 1.44$, the corresponding probability of 0.4251 is close enough so we conclude that the upper and lower cutoff scores will correspond to $z = 1.44$ and $z = -1.44$. With $\mu = 69.0$ and $\sigma = 2.8$, we can use both values of $z$ in $z = (x - \mu)/\sigma$ to get

$$1.44 = \frac{x - 69.0}{2.8} \text{ and } -1.44 = \frac{x - 69.0}{2.8}$$

Solving both of these equations results in the values of 73.032 and 64.968. That is, the company will make clothing only for adult males between 65.0 in. and 73.0 in. tall.

## 5–4  Exercises A

In Exercises 1–12, assume that the errors on a scale (in meters) are normally distributed with a mean of 0 m and a standard deviation of 1 m. (The errors can be positive or negative.)

1. Ninety-five percent of the errors are below what value?
2. Ninety-nine percent of the errors are below what value?
3. Ninety-five percent of the errors are above what value?
4. Ninety-nine percent of the errors are above what value?
5. If the top 5% and the bottom 5% of all errors are unacceptable, find the minimum and maximum acceptable errors.
6. If the top 0.5% and the bottom 0.5% of all errors are unacceptable, find the minimum and maximum acceptable errors.
7. If the top 10% and the bottom 5% of all errors are unacceptable, find the minimum and maximum acceptable errors.
8. If the top 15% and the bottom 20% of all errors are unacceptable, find the minimum and maximum acceptable errors.
9. Find the value that separates the top 40% of all errors from the bottom 60%.
10. Find the value that separates the top 82% of all errors from the bottom 18%.
11. Find the value of the third quartile ($Q_3$), which separates the top 25% of all errors from the bottom 75%.
12. Find the value of $P_{18}$ (18th percentile), which separates the bottom 18% of all errors from the top 82%.

In Exercises 13–28, answer the given question. In each case, draw a graph.

13. Heights of women are normally distributed with a mean of 161.5 cm and a standard deviation of 6.3 cm (based on data from the National Health Survey). If a car design makes the driver's seat uncomfortable for the shortest 10% of women, find the height that separates the shortest 10% from the tallest 90%.

14. Adult males have normally distributed heights with a mean of 69.0 in. and a standard deviation of 2.8 in. (based on data from the National Health Survey). If a clothing manufacturer decides to produce goods that exclude the shortest 9% and the tallest 9% of adult males, find the minimum and maximum heights that they will fit.

15. A psychologist wants to interview subjects with IQ scores in the top 1%. IQ scores are normally distributed with a mean of 100 and a standard deviation of 15. Find the score separating the top 1% of IQs from the bottom 99%.

16. On the Graduate Record Exam in economics, scores are normally distributed with a mean of 615 and a standard deviation of 107. If a college admissions office decides to require scores above the 70th percentile, find the cutoff point.

17. A study of VCR owners found that their annual household incomes are normally distributed with a mean of $41,182 and a standard deviation of $19,990 (based on data from Nielsen Media Research). If an advertising campaign is to be targeted at those VCR owners whose household incomes are in the top 90%, find the minimum income level for this target group.

18. A manufacturer of car tires finds that the tires last distances that are normally distributed with a mean of 35,600 mi and a standard deviation of 4275 mi. The manufacturer wants to guarantee the tires so that only 3% will be replaced because of failure before the guaranteed number of miles. For how many miles should the tires be guaranteed?

19. An insurance researcher learns that for males aged 16–24, the mean number of miles driven each year is 10,718 (based on data from the Federal Highway Administration). Assume that the annual mileage totals are normally distributed with a standard deviation of 3573 mi and the company will impose a surcharge for those in the top 30%. Find the mileage total that separates those who will be surcharged.

20. In a study of facial behavior, people in a control group are timed for eye contact in a 5-minute period. Their times are normally distributed with a mean of 184 s and a standard deviation of 55.0 s (based on data from "Ethological Study of Facial Behavior in Nonparanoid and Paranoid Schizophrenic Patients," by Pitman, Kolb, Orr, and Singh, *Psychiatry*, 144:1). Since results showed that nonparanoid schizophrenic patients had much lower eye contact times, assume that you want to further analyze people in the control group who are in the bottom 15%. For the control group, find the eye contact time separating the bottom 15% from the rest.

21. A standard IQ test produces normally distributed results with a mean of 100 and a standard deviation of 15. A class of high school science students is grouped homogeneously by excluding students with IQ scores in either the top 20% or the bottom 20%. Find the lowest and highest possible IQ scores of students remaining in the class.

22. A machine fills sugar boxes in such a way that the weights (in grams) of the contents are normally distributed with a mean of 2260 g and a standard deviation of 20 g. Another machine checks the weights and rejects packages in the top 1% or bottom 1%. Find the minimum and maximum acceptable weights.

23. A study shows that Michigan teachers have measures of job dissatisfaction that are normally distributed with a mean of 3.80 and a standard deviation of 0.95. (See "Stress and Strain from Family Roles and Work-Role Expectations," by Cooke and Rousseau, *Journal of Applied Psychology*, Vol. 69, No. 2.) Find the value of the 10–90 percentile range. That is, find the difference between the 10th percentile and the 90th percentile.

24. For a certain population, scores on the Miller Analogies Test are normally distributed with a mean of 58.84 and a standard deviation of 15.94. (The data are based on "Equivalencing MAT and GRE Scores Using Simple Linear Transformation and Regression Methods," by Kagan and Stock, *Journal of Experimental Education*, Vol. 49, No. 1.) Find the interquartile range. That is, find the value of $Q_3 - Q_1$ where $Q_3$ is the third quartile and $Q_1$ is the first quartile.

25. A manufacturer has contracted to supply ball bearings. Product analysis reveals that the diameters are normally distributed with a mean of 25.1 mm and a standard deviation of 0.2 mm. The largest 7% of the diameters and the smallest 13% of the diameters are unacceptable. Find the limits for the diameters of the acceptable ball bearings.

26. A manufacturer of color television sets tests competing brands and finds that the amounts of energy they require are normally distributed with a mean of 320 kWh and a standard deviation of 7.5 kWh. If the lowest 30% and the highest 20% are not included in a second round of tests, what are the limits for the energy amounts of the remaining sets?

27. A particular X-ray machine gives radiation dosages (in milliroentgens) that are normally distributed with a mean of 4.13 and a standard deviation of 1.27. A dosimeter is set so that it displays yellow for radiation levels that are not in the top 10% or bottom 30%. Find the lowest and highest "yellow" radiation levels.

28. In a study of coliform contamination in streams, an environmentalist finds that one region has normally distributed coliform levels (number of cells per 100 ml) with a mean of 122 and a standard deviation of 14. The contamination levels are classified into three equal groups of low, medium, and high. Find the minimum and maximum levels for the medium category.

# 5–4 Exercises B

29. A teacher gives a test and gets normally distributed results with a mean of 50 and a standard deviation of 10. Grades are to be assigned according to the following scheme. Find the numerical limits for each letter grade.

  A: Top 10%
  B: Scores above the bottom 70% and below the top 10%
  C: Scores above the bottom 30% and below the top 30%
  D: Scores above the bottom 10% and below the top 70%
  F: Bottom 10%

30. Using recent data from the College Entrance Examination Board, the mean math SAT score is 475 and 17.0% of the scores are above 600. Find the standard deviation and then use that result to find the 99th percentile.

31. A city sponsored cross-country race has 4830 applicants, but only 200 are allowed to run in the final race. A qualifying run was held two weeks before the final, and the times are normally distributed with a mean of 36.2 min and a standard deviation of 3.8 min. If the 200 fastest times qualify, what is the cutoff time?

32.  a. Assume that the following scores in the given stem-and-leaf plot are representative of a normally distributed list of test scores.

  i. Find the mean of $\bar{x}$ of this sample
  ii. Find the standard deviation $s$ of this sample
  iii. If a grade of A is given to the top 5%, find the minimum numerical score that corresponds to A in this sample.
  iv. Find the theoretical score that separates the top 5% by using the sample mean and standard deviation as estimates for the population mean and standard deviation.

  | 4 | 7 |
  |---|---|
  | 5 | 0  3  4  6  8  9  9  9  9 |
  | 6 | 0  0  1  4  4  4  6  6  9  9  9  9 |
  | 7 | 2  2  2  2  4  4  6  6  6  7  7  7  8  9 |
  | 8 | 0  2  2 |
  | 9 | 2 |

  b. In constructing a boxplot for exploring a set of data, the median is found to be 650 while the hinges are 572 and 728. If the data appear to be normally distributed, use this information to find the standard deviation.

# 5–5    Normal as Approximation to Binomial

In Section 4–4 we learned (we hope) how to solve binomial problems by using Table A–1 or the binomial probability formula. However, there are many important binomial problems that are not practical to solve by such methods. As an example, consider this problem:

> The Intel Corporation makes computer chips with a 35% yield, meaning that 35% of the chips are good while 65% are not (based on data from a Paine Webber analyst). What is the probability that in a batch of 150 chips, there are at least 60 good chips that are needed to fill a special order?

This is a binomial problem with $n = 150$, $p = 0.35$, $q = 0.65$, and $x = 60$, $61, 62, \ldots, 150$. Table A–1 doesn't go up to $n = 150$ so it can't be used. In theory, we could use the binomial probability formula 91 times (Yikes!) beginning with

$$P(60) = \frac{150!}{(150 - 60)!60!} \cdot 0.35^{60} \cdot 0.65^{150 - 60}$$

The resulting 91 probabilities can be added to produce the correct result. However, these calculations would require days of work and tons of patience. Fortunately, this section introduces a simple and practical alternative: Under certain circumstances, we can approximate the binomial probability distribution by the normal distribution. The following summarizes the key point of this section.

> **If $np \geq 5$ and $nq \geq 5$, then the binomial random variable is approximately normally distributed with the mean and standard deviation given as**
>
> $$\mu = np$$
> $$\sigma = \sqrt{npq}$$

If you go back and review Figure 4–6 on page 211, you will see that a particular binomial distribution does have a probability histogram that has roughly the same shape as a normal distribution. There are other distributions (such as the $t$ distribution to be examined later) with the same basic bell shape, yet they cannot be approximated by the normal distribution since unacceptable errors result. The justification that allows us to use the normal distribution as an approximation to the binomial distribution results from more advanced mathematics. Unfortunately (or fortunately, depending on your perspective), it is not practical to outline here the details of the formal proof of the above result. For now, try to accept the intuitive evidence of the strong resemblance between the binomial and normal distributions and take the more rigorous evidence on faith. Relative to binomial experiments, past experience has shown that the normal distribution is a reasonable approximation to the binomial distribution as long as $np \geq 5$ and $nq \geq 5$.

We will now use the normal approximation approach to solve our computer chip problem. (For that binomial problem we have already ruled out the use of Table A–1 or the binomial probability formula.) We first verify that $np \geq 5$ and $nq \geq 5$.

$$np = 150 \times 0.35 = 52.5 \qquad \text{(Therefore } np \geq 5.\text{)}$$
$$nq = 150 \times 0.65 = 97.5 \qquad \text{(Therefore } nq \geq 5.\text{)}$$

As a result of satisfying these two requirements, we now know that it is reasonable to approximate the binomial distribution by a normal distribution and we therefore proceed to find values for $\mu$ and $\sigma$ that are needed. We get the following. (The exact values can be stored in your calculator for later use.)

$$\mu = np = 150 \times 0.35 = 52.5$$
$$\sigma = \sqrt{npq} = \sqrt{(150)(0.35)(0.65)} = 5.84$$

## Continuity Correction

We want the probability of getting *at least* 60 good chips in a batch of 150, so we include the probability of getting *exactly* 60. But the discrete value of 60 is approximated in the continuous normal distribution by the interval from 59.5 to 60.5. Such conversions from a discrete to a continuous distribution are called **continuity corrections.** (Figure 5–17 illustrates how the discrete value of 60 is corrected for continuity when represented in the continuous normal distribution.) In the problem we are considering, "at

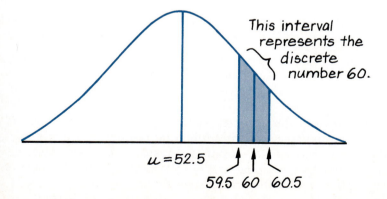

This interval represents the discrete number 60.

$\mu = 52.5$

59.5   60   60.5

**Figure 5–17**

## Survey Medium Can Affect Results

■ In a survey of Catholics in Boston, the subjects were asked if contraceptives should be made available to unmarried women. In personal interviews, 44% of the respondents said yes. But for a similar group contacted by mail or telephone, 75% of the respondents answered yes to the same question. ■

least 60″ means that we include the entire interval representing 60, so 59.5 becomes the actual boundary that we use. If we ignore or forget the continuity correction, the additional error will be relatively small as long as $n$ is large. However, the continuity correction should always be used with a normal distribution to approximate a binomial distribution.

Now let's get back to our problem of finding the probability of getting at least 60 good chips. Figure 5–18 illustrates our problem and it includes the continuity correction. We need to find the shaded area. Using our usual procedures associated with normal distributions, we must first find the area bound by 52.5 and 59.5. We get

$$z = \frac{x - \mu}{\sigma} = \frac{59.5 - 52.5}{5.84} = 1.20$$

Using Table A–2 we find that $z = 1.20$ corresponds to an area of 0.3849, so the shaded region has an area of $0.5 - 0.3849 = 0.1151$. The probability of getting at least 60 good chips is 0.1151.

In the preceding example, if we neglected to correct for continuity by using 60 instead of 59.5, our answer would have been 0.1003. The answer of 0.1151 is a better result.

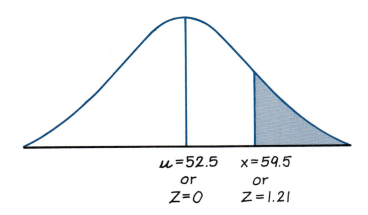

**Figure 5–18**

A computer printout from the statistics package STATDISK is given below. Our program in that package computes binomial probabilities, and the sample display shows one page of the output that results when we request the binomial probabilities corresponding to $n = 150$ and $p = 0.35$. The table shows that the cumulative probability corresponding to $x = 59$ is 0.88390, so that the probability of getting any value 60 and above is $1 - 0.88390 = 0.1161$, which is very close to our result of 0.1151. For this particular exam-

STATDISK DISPLAY

| | | | | | | | | |
|---|---|---|---|---|---|---|---|---|
| Mean = 52.5 | | | St. Dev. = 5.84166 | | | Variance = 34.125 | | |
| X | P(X) | Cum Prob | X | P(X) | Cum Prob | X | P(X) | Cum Prob |
| 39 | .00450 | .01170 | 52 | .06806 | .50339 | 65 | .00712 | .98594 |
| 40 | .00672 | .01842 | 53 | .06776 | .57115 | 66 | .00494 | .99088 |
| 41 | .00971 | .02813 | 54 | .06554 | .63670 | 67 | .00334 | .99422 |
| 42 | .01357 | .04170 | 55 | .06160 | .69830 | 68 | .00219 | .99641 |
| 43 | .01835 | .06005 | 56 | .05627 | .75457 | 69 | .00140 | .99781 |
| 44 | .02403 | .08409 | 57 | .04997 | .80454 | 70 | .00087 | .99869 |
| 45 | .03048 | .11457 | 58 | .04314 | .84768 | 71 | .00053 | .99922 |
| 46 | .03746 | .15203 | 59 | .03622 | .88390 | 72 | .00031 | .99953 |
| 47 | .04464 | .19667 | 60 | .02958 | .91349 | 73 | .00018 | .99971 |
| 48 | .05158 | .24825 | 61 | .02350 | .93699 | 74 | .00010 | .99981 |
| 49 | .05781 | .30606 | 62 | .01817 | .95515 | 75 | .00006 | .99987 |
| 50 | .06288 | .36894 | 63 | .01366 | .96882 | 76 | .00003 | .99990 |
| 51 | .06639 | .43533 | 64 | .01000 | .97882 | 77 | .00002 | .99991 |

ple, the STATDISK program is a good alternative to the normal approximation method, but every program will have some limitations that do not apply to the approximation method.

Figure 5–19 summarizes the procedure for using the normal distribution as an approximation to the binomial distribution. The following example follows the procedure outlined in Figure 5–19.

## Example

According to data from the Hertz Corporation, 80% of commuters use their own vehicle. Find the probability that of 100 randomly selected commuters, *exactly* 85 use their own vehicle.

## Solution

Refer to Figure 5–19. In step 1 we verify that the conditions described do satisfy the criteria for the binomial distribution and $n = 100$, $p = 0.80$, $q = 0.20$, and $x = 85$. Proceeding to step 2, we see that Table A–1 cannot

*continued*

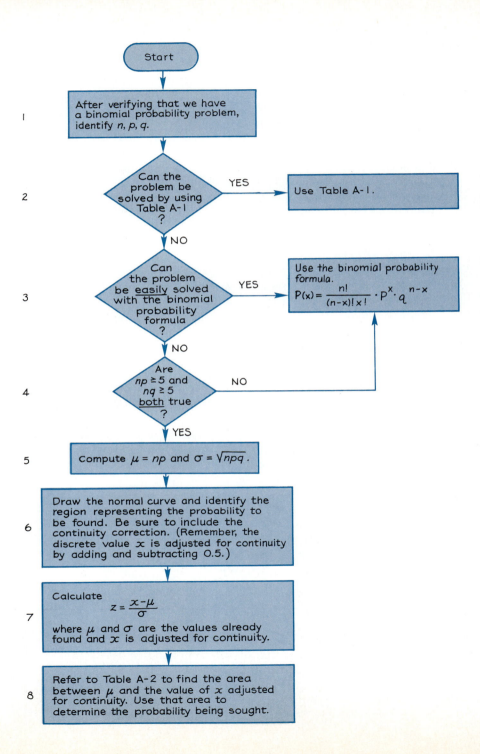

**Figure 5–19**

*Solving Binomial Probability Problems*

**Solution** *continued*

be used because $n$ is too large. In step 3, the binomial probability formula applies, but

$$P(85) = \frac{100!}{(100 - 85)! \, 85!} \cdot 0.80^{85} \cdot 0.20^{100 - 85}$$

is too difficult to compute. Many calculators cannot evaluate anything above 70!, but the availability of calculators or a computer would simplify this approach. In step 4 we get

$$np = 100 \cdot 0.80 = 80 \geq 5$$
$$nq = 100 \cdot 0.20 = 20 \geq 5$$

and since $np$ and $nq$ are both at least 5, we conclude that the normal approximation to the binomial is satisfactory. We now go on to step 5, where we obtain the values of $\mu$ and $\sigma$ as follows.

$$\mu = np = 100 \cdot 0.80 = 80.0$$
$$\sigma = \sqrt{npq} = \sqrt{100 \cdot 0.80 \cdot 0.20} = 4.0$$

Now we go to step 6, where we draw the normal curve shown in Figure 5–20. The shaded region of Figure 5–20 represents the probability we want. Use of the continuity correction results in the representation of 85 by the region extending from 84.5 to 85.5. We now proceed to step 7.

The format of Table A–2 requires that we first find the probability corresponding to the region bounded on the left by the vertical line through the mean of 80.0 and on the right by the vertical line through 85.5, so that one of the calculations required in step 7 is as follows.

$$z_2 = \frac{x - \mu}{\sigma} = \frac{85.5 - 80.0}{4.0} = 1.38$$

We also need the probability corresponding to the region bounded by 80.0 and 84.5, so we calculate

$$z_1 = \frac{x - \mu}{\sigma} = \frac{84.5 - 80.0}{4.0} = 1.13$$

Finally, in step 8 we use Table A–2 to find that a probability of 0.4162 corresponds to $z_2 = 1.38$ and 0.3708 corresponds to $z_1 = 1.13$. Consequently, the entire shaded region of Figure 5–20 depicts a probability of $0.4162 - 0.3708 = 0.0454$.

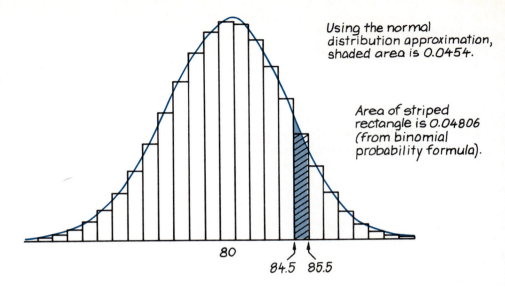

Using the normal distribution approximation, shaded area is 0.0454.

Area of striped rectangle is 0.04806 (from binomial probability formula).

80

84.5  85.5

**Figure 5–20**

In the preceding example, STATDISK or a calculator will result in a probability of 0.04806, while the normal approximation method resulted in a value of 0.0454. The discrepancy of 0.00266 is very small. The discrepancy occurs because we are finding the area of the shaded region in Figure 5–20, but the actual area would be a *rectangle* centered above 85. (Figure 5–20 illustrates this discrepancy.) The area of the rectangle is 0.04806, but the area of the approximating shaded region is 0.0454.

We now have three methods for determining probabilities in binomial experiments, and they are all summarized in Figure 5–19.

## 5–5  Exercises A

In Exercises 1–4, check that $np \geq 5$ and $nq \geq 5$ in order to determine whether the normal distribution is a suitable approximation. In each case, also find the values of $\mu$ and $\sigma$.

1.  $n = 25, p = 0.250$
2.  $n = 50, p = 0.333$
3.  $n = 84, p = 0.950$
4.  $n = 125, p = 0.961$

In Exercises 5–8, find the indicated binomial probabilities by using (a) Table A–1 in Appendix A and (b) the normal distribution as an approximation to the binomial probability distribution.

5.  With $n = 12$ and $p = 0.50$, find $P(8)$.
6.  With $n = 15$ and $p = 0.40$, find $P(7)$.
7.  With $n = 12$ and $p = 0.50$, find $P$(at least 8).
8.  With $n = 20$ and $p = 0.70$, find $P$(at most 12).
9.  Find the probability of getting at least 60 girls in 100 births.
10. Find the probability of getting exactly 40 girls in 80 births.
11. Find the probability of passing a true-false test of 100 questions if 65% is passing and all responses are random guesses.
12. A multiple-choice test consists of 40 questions with possible answers of $a, b, c, d, e$. Find the probability of getting at most 30% correct if all answers are random guesses.
13. Continental Airlines recently reported that its on-time arrival rate is 80%. Find the probability that among 100 randomly selected flights, fewer than 70 arrive on time.
14. Among U.S. households, 24% have telephone answering machines (based on data from the U.S. Consumer Electronics Industry). If a tele-marketing campaign involves 2500 households, find the probability that more than 650 have answering machines.
15. In a study by United Group Information Services, 66% of small businesses are based in homes. If a marketing study involves the random selection of 375 small businesses, find the probability that fewer than 250 are based in homes.
16. Based on U.S. Bureau of Justice data, 16% of those arrested are women. If 250 arrested people are randomly selected, find the probability that the number of women is at least 35.
17. A survey conducted by the U.S. Department of Transportation showed that 25% of New York City drivers wear seat belts. If 400 New York City drivers are randomly selected, find the probability that at least 125 of them wear seat belts.
18. Based on U.S. Bureau of the Census data, 12% of the men in the United States have earned bachelor's degrees. If 140 U.S. men are randomly selected, find the probability that at least 20 of them have a bachelor's degree.
19. *Popular Science* magazine reported that a study of chain saw accidents found that 25% were due to kickback. Find the probability of getting exactly 21 kickback accidents in 84 randomly selected chain saw accidents.
20. Among teenagers old enough to drive, 35% have their own cars (based on data from a Rand Youth Poll). If a marketing research team randomly selects 600 teenagers of driving age, find the probability that at least 210 of them have their own cars.

21. According to Bureau of the Census data, among men aged 18 to 24, 60% live at home with their parents. If 500 men aged 18 to 24 are randomly selected, find the probability that more than 325 of them live at home with their parents.

22. According to Helen Fisher of the American Museum of Natural History, among couples who divorce, 40% have no children. If 250 divorce cases are randomly selected, find the probability that more than 80 involve couples with no children.

23. Among women aged 18 to 24, 75% are more than 159 cm tall (based on data from the National Health Survey, USDHEW publication 79–1659). If 320 women aged 18 to 24 are randomly selected, find the probability that more than 250 of them are more than 159 cm tall.

24. An airline company experiences a 7% rate of no-shows on advance reservations. Find the probability that of 250 randomly selected advance reservations, there will be at least 10 no-shows.

25. A certain genetic characteristic appears in one-quarter of all offspring. Find the probability that of 40 randomly selected offspring, fewer than 5 exhibit the characteristic in question.

26. The IRS finds that of all taxpayers whose returns are audited, 70% end up paying additional taxes. Find the probability that of 500 randomly selected returns, at least 400 end up paying additional taxes.

27. According to the American Medical Association, 18.4% of college graduates smoke. If a health study begins with the random selection of 280 college graduates, find the probability that more than one-fifth of them smoke.

28. Twenty-five percent of doctors are under 35 years of age (data from Health Care Market Research). Find the probability that among 40 randomly selected doctors, exactly 10 are under 35 years of age.

29. Of those who commute to southern Manhattan, 6.5% use commuter railroads (based on data from the New York Metropolitan Transportation Council). If we randomly select 175 people who commute to southern Manhattan, find the probability that the number who use railroads is between 10 and 15 inclusive.

30. In Illinois, 17% of men surveyed were found to have at least two alcoholic drinks per day (based on data from the National Centers for Disease Control). Assuming that this rate is correct, find the probability that among 125 randomly selected Illinois men, the number who average at least two alcoholic drinks per day is between 20 and 25 inclusive.

31. Forty-five percent of us have group O blood, according to data provided by the Greater New York Blood Program. If 400 subjects are randomly selected, find the probability that the number with group O blood is between 200 and 205 inclusive.

32. Among workers aged 20 to 24, 26% work more than 40 hours per week (based on data from the U.S. Department of Labor). If we randomly

select 350 workers aged 20 to 24, find the probability that the number who work more than 40 hours per week is between 80 and 90 inclusive.

# 5–5 Exercises B

33. In a binomial experiment with $n = 15$ and $p = 0.4$, find $P$(at least 5) using the following.
    a.  The table of binomial probabilities (Table A–1)
    b.  The binomial probability formula
    c.  The normal distribution approximation

34. Assume that a baseball player hits .350 so that his probability of a hit is 0.350. Also assume that his hitting attempts are independent of each other.
    a.  Find the probability of at least 1 hit in 4 tries in 1 game.
    b.  Assuming that this batter gets up 4 times each game, find the probability of getting a total of at least 56 hits in 56 games.
    c.  Assuming that this batter gets up 4 times each game, find the probability of at least 1 hit in each of 56 consecutive games (Joe DiMaggio's 1941 record).
    d.  What minimum batting average would be required for the probability in part c to be greater than 0.1?

35. An airline company works only with advance reservations and experiences a 7% rate of no-shows. How many reservations could be accepted for an airliner with a capacity of 250 if there is at least a 0.95 probability that all reservation holders who show will be accommodated?

36. A company manufactures integrated circuit chips with a 23% rate of defects. What is the minimum number of chips that must be manufactured if there must be at least a 90% chance that 5000 good chips can be supplied?

# 5–6 The Central Limit Theorem

The central limit theorem is one of the most important and useful concepts in statistics. Before considering this theorem, we will first try to develop an intuitive understanding of one of its most important consequences:

**The sampling distribution of sample means tends to be a normal distribution.**

This implies that if we collect samples all of the same size, compute their means, and then develop a histogram of those means, it will tend to assume the bell shape of a normal distribution. This is true regardless of the shape of the distribution of the original population. The central limit theorem qualifies the preceding remarks and includes additional aspects, but stop and try to understand the thrust of these remarks before continuing.

Let's begin with some concrete numbers. Table 5–1 contains a block of data consisting of 300 sample scores. These scores were generated through a computer simulation, but they could have been extracted from a telephone directory or a book of random numbers. (Now *there's* exciting reading, although the plot is a little thin.) Figure 5–21 illustrates the histogram of the 300 sample scores, showing that their distribution is essentially uniform. Now consider the 300 scores to be 30 samples, with 10 scores in each sample. The resulting 30 sample means are listed in Table 5–1 and illustrated in the histogram shown in Figure 5–22. Note that the shape of Figure 5–22 is roughly that of a normal distribution. It is important to observe that even though the original population has a uniform distribution, the sample means seem to have a normal distribution. It was observations exactly like this that led to the formulation of the central limit theorem. If our sample means

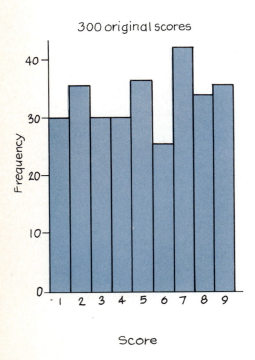

**Figure 5–21**
*Histogram of the 300 Original Scores Randomly Selected (Between 1 and 9)*

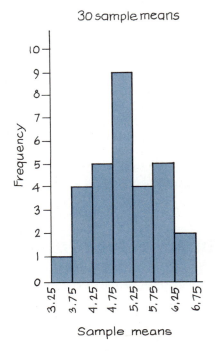

**Figure 5–22**
*Histogram of the 30 Sample Means*
*Each sample mean is based on 10 raw scores randomly selected between 1 and 9, inclusive.*

were based on samples larger than 10, Figure 5–22 would more closely resemble a normal distribution. We are now ready to consider the central limit theorem.

Let's assume that the variable $x$ represents scores that may or may not be normally distributed, and that the mean of the $x$ values is $\mu$ while the

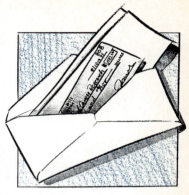

| Sample | Data | | | | | | | | | | Sample Mean |
|---|---|---|---|---|---|---|---|---|---|---|---|
| 1 | 2 | 7 | 5 | 5 | 2 | 1 | 7 | 7 | 9 | 4 | 4.9 |
| 2 | 5 | 8 | 1 | 1 | 5 | 7 | 1 | 4 | 1 | 4 | 3.7 |
| 3 | 7 | 6 | 9 | 8 | 5 | 1 | 6 | 4 | 7 | 9 | 6.2 |
| 4 | 7 | 3 | 1 | 7 | 3 | 6 | 7 | 9 | 4 | 3 | 5.0 |
| 5 | 9 | 7 | 7 | 6 | 1 | 6 | 8 | 3 | 4 | 7 | 5.8 |
| 6 | 5 | 3 | 3 | 4 | 2 | 5 | 9 | 9 | 1 | 9 | 5.0 |
| 7 | 5 | 5 | 3 | 9 | 5 | 3 | 1 | 9 | 1 | 5 | 4.6 |
| 8 | 4 | 3 | 9 | 5 | 5 | 9 | 1 | 7 | 7 | 8 | 5.8 |
| 9 | 2 | 1 | 7 | 8 | 6 | 7 | 7 | 9 | 8 | 3 | 5.8 |
| 10 | 3 | 4 | 5 | 6 | 8 | 4 | 8 | 3 | 4 | 5 | 5.0 |
| 11 | 5 | 3 | 2 | 2 | 6 | 8 | 1 | 5 | 5 | 9 | 4.6 |
| 12 | 7 | 5 | 9 | 6 | 8 | 2 | 2 | 7 | 2 | 1 | 4.9 |
| 13 | 3 | 1 | 4 | 1 | 7 | 9 | 3 | 2 | 3 | 8 | 4.1 |
| 14 | 6 | 2 | 7 | 4 | 4 | 5 | 2 | 6 | 8 | 6 | 5.0 |
| 15 | 9 | 6 | 2 | 9 | 4 | 2 | 6 | 3 | 5 | 5 | 5.1 |
| 16 | 9 | 2 | 2 | 3 | 6 | 2 | 6 | 6 | 8 | 3 | 4.7 |
| 17 | 5 | 4 | 2 | 1 | 9 | 4 | 2 | 9 | 4 | 2 | 4.2 |
| 18 | 8 | 1 | 2 | 1 | 4 | 3 | 2 | 8 | 5 | 4 | 3.8 |
| 19 | 5 | 8 | 9 | 6 | 2 | 7 | 9 | 3 | 8 | 5 | 6.2 |
| 20 | 5 | 6 | 8 | 7 | 5 | 9 | 6 | 4 | 8 | 7 | 6.5 |
| 21 | 7 | 9 | 9 | 8 | 3 | 5 | 5 | 1 | 4 | 6 | 5.7 |
| 22 | 8 | 4 | 7 | 8 | 7 | 8 | 7 | 7 | 1 | 8 | 6.5 |
| 23 | 5 | 5 | 1 | 7 | 5 | 7 | 7 | 2 | 9 | 8 | 5.6 |
| 24 | 9 | 5 | 2 | 5 | 9 | 2 | 5 | 3 | 5 | 8 | 5.3 |
| 25 | 4 | 5 | 8 | 4 | 2 | 9 | 2 | 6 | 6 | 1 | 4.7 |
| 26 | 1 | 7 | 7 | 3 | 4 | 7 | 7 | 2 | 8 | 7 | 5.3 |
| 27 | 8 | 1 | 1 | 7 | 6 | 2 | 2 | 1 | 4 | 9 | 4.1 |
| 28 | 9 | 4 | 3 | 7 | 3 | 7 | 8 | 4 | 3 | 2 | 5.0 |
| 29 | 1 | 2 | 9 | 3 | 8 | 2 | 4 | 6 | 2 | 8 | 4.5 |
| 30 | 2 | 9 | 3 | 3 | 1 | 2 | 6 | 7 | 8 | 7 | 4.8 |

**TABLE 5–1** Thirty samples are given here, each consisting of ten random numbers between 1 and 9. The right column consists of the corresponding sample means.

↑
See Figure 5–21.

↑
See Figure 5–22.

## SURVEY SOLICITS CONTRIBUTIONS

■ The American Institute for Cancer Research distributed a survey on diet and breast cancer. The stated purpose of the survey was to develop a "statistical profile of the eating habits of American women—and to help find the link between diet and breast cancer." However, the survey ended with an option for enclosing a gift ranging from $5 to $500. Two fundamental questions arise. What was real purpose of the survey? Does the contribution request affect the validity of the results? ■

standard deviation is $\sigma$. Suppose we collect a sample of size $n$ and calculate the sample mean $\bar{x}$. What do we know about the collection of all sample means that we produce by repeating this experiment, collecting a sample of size $n$ to get the sample mean? The central limit theorem tells us that as the sample size $n$ increases, the sample means will tend to approach a normal distribution with mean $\mu$ and standard deviation $\sigma/\sqrt{n}$ (see Figure 5–23). The distribution of sample means *tends* to be a normal distribution in the sense that as $n$ becomes larger, the distribution of sample means gets closer to a normal distribution. This conclusion is not intuitively obvious, and it was arrived at through extensive research and analysis. The formal rigorous proof requires advanced mathematics and is beyond the scope of this text. We will illustrate the theorem and give examples of its use.

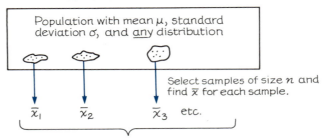

Central Limit Theorem

Population with mean $\mu$, standard deviation $\sigma$, and <u>any</u> distribution

Select samples of size $n$ and find $\bar{x}$ for each sample.

$\bar{x}_1$    $\bar{x}_2$    $\bar{x}_3$    etc.

These sample means will, as $n$ increases, approach a <u>normal</u> distribution with mean $\mu$ and standard deviation $\frac{\sigma}{\sqrt{n}}$.

**Figure 5–23**

---

**Central Limit Theorem**

Given:

1. The random variable $x$ has any distribution with mean $\mu$ and standard deviation $\sigma$.

2. Samples of size $n$ are randomly selected from this population.

Conclusions:

1. The distribution of all possible sample means $\bar{x}$ will approach a *normal* distribution.

2. The *mean* of the sample means will be $\mu$.

3. The *standard deviation* of the sample means will be $\sigma/\sqrt{n}$.

In the first conclusion above, when we say that the sample means *approach* a normal distribution, we are employing these commonly used rules:

1.  For samples of size $n$ larger than 30, the sample means can be approximated reasonably well by a normal distribution. The approximation gets better as the sample size $n$ becomes larger.

2.  If the original population is itself normally distributed, then the sample means will be normally distributed for *any* sample size $n$.

---

## Notation

If all possible random samples of size $n$ are selected from a population with mean $\mu$ and standard deviation $\sigma$, the mean of the sample means is denoted by $\mu_{\bar{x}}$ so that

$$\mu_{\bar{x}} = \mu$$

Also, the standard deviation of the sample means is denoted by $\sigma_{\bar{x}}$ so that

$$\sigma_{\bar{x}} = \frac{\sigma}{\sqrt{n}}$$

$\sigma_{\bar{x}}$ is often called the **standard error of the mean.**

---

Comparison of Figures 5–21 and 5–22 should confirm that the original numbers have a nonnormal distribution, while the sample means approximate a normal distribution. The central limit theorem also indicates that the mean of *all* such sample means should be $\mu$ (the mean of the original population) and the standard deviation of all such sample means should be $\sigma/\sqrt{n}$ (where $\sigma$ is the standard deviation of the original population and $n$ is the sample size of 10). We can find $\mu$ and $\sigma$ for the original population of numbers between 1 and 9 by noting that, if those numbers occur with equal frequency as they should, then the population mean $\mu$ is given by

$$\mu = \frac{1 + 2 + 3 + 4 + 5 + 6 + 7 + 8 + 9}{9} = 5.0$$

Similarly, we can find $\sigma$ by again using 1, 2, 3, 4, 5, 6, 7, 8, 9 as an ideal or theoretical representation of the population. Following this course, $\sigma$ is computed to be 2.58. The mean and standard deviation of the sample means can now be found as follows.

$$\mu_{\bar{x}} = \mu = 5.0$$

$$\sigma_{\bar{x}} = \frac{\sigma}{\sqrt{n}} = \frac{2.58}{\sqrt{10}} = \frac{2.58}{3.16} = 0.82$$

The preceding results represent *all* sample means of size $n = 10$. For the 30 sample means shown in Table 5–1, we have a mean of 5.08 and a standard deviation of 0.75. We can see that our real data conform quite well to the theoretically predicted values for $\mu_{\bar{x}}$ and $\sigma_{\bar{x}}$. Many important and practical problems can be solved with the central limit theorem. In the following example we use the same basic methods introduced earlier in Section 5–3, but we make some important adjustments required because our problem deals with the mean for a group of scores (instead of an individual score). Consequently, this example illustrates a method for dealing with the sampling distribution of means, instead of the sampling distribution of individual scores.

### Example

In a study of work patterns in one population, the mean family time devoted to child care is 23.08 h with a standard deviation of 15.58 h (based on data from "Nonstandard Work Schedules and Family Life," by Staines and Pleck, *Journal of Applied Psychology*, Vol. 69, No. 3). If 55 subjects are randomly selected from this population, find the probability that the mean of this sample group is between 23.08 h and 25.00 h.

### Solution

We weren't given the distribution of the original population, but the sample size $n = 55$ exceeds 30, so we use the central limit theorem and conclude that the distribution of sample means is the normal distribution with these parameters:

$$\mu_{\bar{x}} = \mu = 23.08$$

$$\sigma_{\bar{x}} = \frac{\sigma}{\sqrt{n}} = \frac{15.58}{\sqrt{55}} = 2.10$$

Figure 5–24 shows the shaded area corresponding to the probability we seek. We find that area by first determining the value of the $z$ score.

$$z = \frac{\bar{x} - \mu_{\bar{x}}}{\sigma_{\bar{x}}} = \frac{\bar{x} - \mu}{\dfrac{\sigma}{\sqrt{n}}} = \frac{25.00 - 23.08}{2.10} = 0.91$$

From Table A–2 we find that $z = 0.91$ corresponds to an area of 0.3186, so the probability we seek is 0.3186. That is, $P(23.08 < \bar{x} < 25.00) = P(0 < z < 0.91) = 0.3186$.

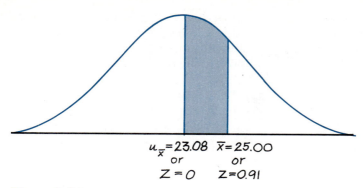

$$u_{\bar{x}} = 23.08 \quad \bar{x} = 25.00$$
$$\text{or} \qquad \text{or}$$
$$Z = 0 \qquad Z = 0.91$$

**Figure 5–24**

It is interesting to note that, as the sample size increases, the sample means tend to vary less, since $\sigma_{\bar{x}} = \sigma/\sqrt{n}$ gets smaller as $n$ gets larger. For example, IQ scores have a mean of 100 and a standard deviation of 15. Samples of 36 will produce means with $\sigma_{\bar{x}} = 15/\sqrt{36} = 2.5$, so that 99% of all such samples will have means between 93.6 and 106.4. If the sample size is increased to 100, $\sigma_{\bar{x}}$ becomes $15/\sqrt{100}$, or 1.5, so that 99% of the samples will have means between 96.1 and 103.9.

These results are supported by common sense: As the sample size increases, the corresponding sample mean will tend to be closer to the true population mean. The effect of an unusual or outstanding score tends to be dampened as it is averaged in as part of a sample.

## Finite Population Correction Factor

Our use of $\sigma_{\bar{x}} = \sigma/\sqrt{n}$ assumes that the population is infinite. When we sample with replacement of selected data, for example, the population is effectively infinite. Yet realistic applications involve sampling without replacement, so successive samples depend on previous outcomes. In manufacturing, for example, quality control inspectors typically sample items without replacement from a finite production run. For such finite populations, we may need to adjust $\sigma_{\bar{x}}$.

---

**Notation**

Just as $n$ denotes the *sample* size, $N$ denotes the size of a *population*. For finite populations of size $N$, we should incorporate the **finite population correction factor** $\sqrt{(N - n) \div (N - 1)}$ so that $\sigma_{\bar{x}}$ is found as follows.

$$\sigma_{\bar{x}} = \frac{\sigma}{\sqrt{n}} \sqrt{\frac{N - n}{N - 1}}$$

---

If the sample size $n$ is small in comparison to the population size $N$, the finite population correction factor will be close to 1. Consequently its impact will be negligible and it can therefore be ignored.

Statisticians have devised the following rule of thumb.

### Rule

Use the finite population correction factor when computing $\sigma_{\bar{x}}$ if the population is finite and $n > 0.05N$. That is, use the correction factor only if the sample size is greater than 5% of the population size.

We now have some important questions that should be answered when considering the use of the central limit theorem: Is the parent (original) population normally distributed? If not, are the sample sizes greater than 30? Is the population finite? If so, then are the sample sizes more than 5% of the population size? The answers to these questions affect our calculations. Figure 5–25 organizes our methods into one coherent scheme, summarizing the key points of this section.

The next example illustrates three different cases. The first case involves a simple and direct use of the normal distribution, the second case includes use of the central limit theorem, while the third case requires the central limit theorem along with the finite population correction factor.

### Example

In human engineering and product design, it is often important to consider weights of people so that airplanes or elevators aren't overloaded, chairs don't break, and other such unpleasantries don't occur. Assume that a population of 750 men has normally distributed weights with a mean of 173 pounds and a standard deviation of 30 pounds (based on data from the National Health Survey).

a. If 1 of these men is randomly selected, find the probability that his weight is greater than 180 pounds.
b. If 36 different men are randomly selected from this population, find the probability that their mean weight is greater than 180 pounds.
c. If 49 different men are randomly selected from this population, find the probability that their mean weight is greater than 180 pounds.

*continued*

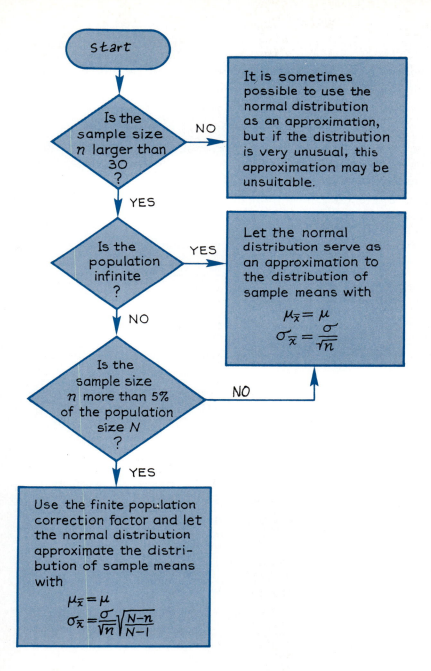

**Figure 5–25**
*Flowchart Summarizing the Decisions to Be Made
When Considering a Distribution of Sample Means*

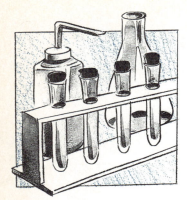

## ETHICS IN EXPERIMENTS

■ Sample data can often be obtained by simply observing or surveying members selected from the population. Many other situations require that we somehow manipulate circumstances to obtain sample data. In both cases, ethical questions may arise. Researchers in Tuskegee, Alabama, withheld the effective penicillin treatment to syphilis victims so that the disease could be studied. This continued for a period of 27 years! ■

### Solution

a. *Approach: Use the same methods presented in Section 5–3.*
   We seek the area of the shaded region in Figure 5–26(a).

$$z = \frac{x - \mu}{\sigma} = \frac{180 - 173}{30} = 0.23$$

We now refer to Table A–2 to find that region $A$ is 0.0910. The shaded region is therefore $0.5 - 0.0910 = 0.4090$. The probability of the man weighing more than 180 pounds is 0.4090.

b. *Approach: Use the central limit theorem without the finite population correction factor (since the sample size of 36 does not exceed 5% of 750).*
   We want the shaded area shown in Figure 5–26(b).

$$z = \frac{\overline{x} - \mu_{\overline{x}}}{\sigma_{\overline{x}}} = \frac{\overline{x} - \mu}{\dfrac{\sigma}{\sqrt{n}}} = \frac{180 - 173}{\dfrac{30}{\sqrt{36}}} = 1.40$$

Referring to Table A–2, we find that $z = 1.40$ corresponds to an area of 0.4192, so the shaded region is $0.5 - 0.4192 = 0.0808$. The probability that the 36 men have a mean weight greater than 180 pounds is 0.0808.

c. *Approach: Use the central limit theorem with the finite population correction factor (since the sample size of 49 does exceed 5% of 750).*
   With $\mu = 173$, $\sigma = 30$, $N = 750$, and $n = 49$, we find that

$$\mu_{\overline{x}} = \mu = 173$$

$$\sigma_{\overline{x}} = \frac{\sigma}{\sqrt{n}} \sqrt{\frac{N - n}{N - 1}} = \frac{30}{\sqrt{49}} \sqrt{\frac{750 - 49}{750 - 1}} = 4.146$$

We use the finite population correction factor because 5% of 750 is 37.5 and $49 > 37.5$. This indicates that the sample is large in comparison to the population size. We now proceed to find the shaded region of Figure 5–26(c).

$$z = \frac{\overline{x} - \mu_{\overline{x}}}{\sigma_{\overline{x}}} = \frac{180 - 173}{4.146} = 1.69$$

Referring to Table A–2, we find that $z = 1.69$ corresponds to an area of 0.4545, so the shaded region is $0.5 - 0.4545 = 0.0455$. The probability that the 49 men have a mean weight greater than 180 pounds is 0.0455.

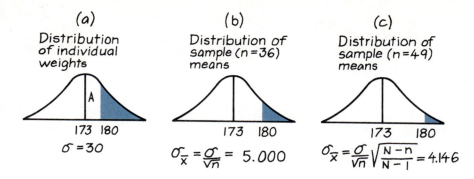

Figure 5–26

The central limit theorem is one of the most important and useful concepts in statistics because it allows us to use the basic normal distribution methods in a wide variety of different circumstances. Many of the topics and applications in the following chapters will depend on the central limit theorem.

# 5–6 Exercises A

1.  A large normally distributed population has a mean of 50 and a standard deviation of 10.
    a.  Find the probability that a randomly selected score is between 50 and 53.
    b.  If a sample of size $n = 36$ is randomly selected, find the probability that the sample mean $\bar{x}$ will be between 50 and 53.
2.  A large normally distributed population has a mean of 150 and a standard deviation of 20.
    a.  Find the probability that a randomly selected score is between 150 and 155.
    b.  If a sample of size $n = 100$ is randomly selected, find the probability that the sample mean $\bar{x}$ will be between 150 and 155.
3.  A large normally distributed population has a mean of 4.50 and a standard deviation of 1.05.
    a.  Find the probability that a randomly selected score is less than 5.00.
    b.  If a sample of size 40 is randomly selected, find the probability that the sample mean is less than 5.00.

4.  A large normally distributed population has a mean of 640 and a standard deviation of 53.
    a.  Find the probability that a randomly selected score is greater than 630.
    b.  If a sample of size 65 is randomly selected, find the probability that the sample mean is greater than 630.

5.  Scores on IQ tests are normally distributed with a mean of 100 and a standard deviation of 15.
    a.  If someone is randomly selected, find the probability that his or her IQ score will be between 100 and 105.
    b.  If 36 people are randomly selected, find the probability that their mean will be between 100 and 105.

6.  For college-bound high school juniors, math scores on the PSAT are normally distributed with a mean of 45.6 and a standard deviation of 11.1 (based on data from Educational Testing Service).
    a.  Find the probability of a randomly selected subject from this population scoring between 45.6 and 48.0.
    b.  If 100 subjects are randomly selected from this population, find the probability that their mean score is between 45.6 and 48.0.

7.  The Goodenough-Harris Drawing Test is used to measure the intellectual maturity of young people. For twelve-year-old girls, the scores are normally distributed with a mean of 34.8 and a standard deviation of 7.02. (The figures are based on data from the National Health Survey.) If 36 twelve-year-old girls are randomly selected, find the probability that their mean score is between 34.8 and 37.0.

8.  A study of the time (in hours) college freshmen use to study each week found that the mean is 7.06 h and the standard deviation is 5.32 h (based on data from *The American Freshman*). If 55 freshmen are randomly selected, find the probability that their mean weekly study time exceeds 7.00 h.

9.  A study of the time high school students spend working each week at a job found that the mean is 10.7 h and the standard deviation is 11.2 h (based on data from the National Federation of State High School Associations). If 42 high school students are randomly selected, find the probability that their mean weekly work time is less than 12.0 h.

10. The typical computer random number generator yields numbers in a uniform distribution between 0 and 1 with a mean of 0.500 and a standard deviation of 0.289. If 45 random numbers are generated, find the probability that their mean is below 0.565.

11. The ages of U.S. commercial aircraft have a mean of 13.0 years and a standard deviation of 7.9 years (based on data from Aviation Data Services). If the Federation Aviation Administration randomly selects 35 commercial aircraft for special stress tests, find the probability that the mean age of this sample group is greater than 15.0 years.

12.  An aircraft strobe light is designed so that the times between flashes have a mean of 10.15 s and a standard deviation of 0.40 s. A sample of 50 times is randomly selected. Find the probability that the sample mean is greater than 10.00 s.

13.  The times that managers spend on paperwork per day have a mean of 2.7 h and a standard deviation of 1.4 h (data from Adia Personnel Services). If a computer sales team randomly selects 75 managers, find the probability that this sample group has a mean less than 2.5 h.

14.  A study of 218,344 births in upstate New York found that the gestation times had a mean of 39.4 weeks and a standard deviation of 2.43 weeks, and those times had a distribution that is approximately normal (based on data from the New York State Department of Health). If 1000 of these births are randomly selected, find the probability that the mean gestation time exceeds 39.6 weeks.

15.  A total of 630 students were measured for "burnout." The resulting scores have a mean of 2.97 and a standard deviation of 0.60 (based on data from "Moderating Effects of Social Support on the Stress-Burnout Relationship," by Etzion, *Journal of Applied Psychology*, Vol. 69, No. 4). If 31 of these subjects are randomly selected, find the probability that their mean burnout score is between 3.00 and 3.10.

16.  In a study of work patterns, data from 669 subjects were collected. For the time spent in child care, the mean is 23.08 h and the standard deviation is 15.58 h (based on data from "Nonstandard Work Schedules and Family Life," by Staines and Pleck, *Journal of Applied Psychology*, Vol. 69, No. 3). If 32 of these subjects are randomly selected, find the probability that their mean time in child care is between 20.00 h and 22.00 h.

17.  A population has a standard deviation of 20. Samples of size $n$ are taken randomly and the means of the samples are computed. What happens to the standard error of the mean if the sample size is increased from 100 to 400?

18.  A population has a standard deviation of 20. Samples of size $n$ are randomly selected and the means of the samples are computed. What happens to the standard error of the mean if the sample size is decreased from 64 to 16?

In Exercises 19–20, assume that samples of size $n$ are randomly selected from a finite population of size $N$. Also assume that $\sigma = 15$. Find the value of $\sigma_{\bar{x}}$. (Be sure to include the finite population correction factor whenever $n > 0.05N$.)

19.  a.  $N = 5000, n = 200$
     b.  $N = 12,000, n = 1000$
     c.  $N = 4000, n = 500$
     d.  $N = 8000, n = 3000$
     e.  $N = 1500, n = 50$

20.  a.  $N = 750, n = 50$
     b.  $N = 673, n = 32$
     c.  $N = 866, n = 73$
     d.  $N = 50,000, n = 10,000$
     e.  $N = 8362, n = 935$

In Exercises 21–24, be sure to check for the use of the finite population correction factor and use it whenever necessary.

21. In a study of Reye's syndrome (by Holtzhauer and others, *American Journal of Diseases of Children*, Vol. 140), 160 children had a mean age of 8.5 years, standard deviation of 3.96 years, and their ages approximated a normal distribution. If 36 of those children are randomly selected, find the probability that their mean age is between 7.0 years and 10.0 years.

22. A study involves a population of 300 women who are 6 ft tall and are between 18 and 24 years of age. This population has a mean weight of 121.5 lb and a standard deviation of 6.5 lb. If 50 members of this population are randomly selected, find the probability that the mean weight of this sample group is greater than 120.0 lb.

23. A psychologist collects a population of 250 volunteers who will participate in a behavior modification experiment. A pretest of this population produces behavior indices with a mean of 436 and a standard deviation of 24. If 20% of the population is randomly selected for one phase of treatment, find the probability that the mean for this sample group is between the desired limits of 430 and 440.

24. In doing an economic impact study, a sociologist identifies a population of 1200 households with a mean annual income of $23,460 and a standard deviation of $3750. If 10% of these households are randomly selected for a more detailed survey, find the probability that the mean for this sample group will fall between the acceptable limits of $23,000 and $24,000.

# 5–6 | Exercises B

25. A population consists of these scores:

$$2 \quad 3 \quad 6 \quad 8 \quad 11 \quad 18$$

a. Find $\mu$ and $\sigma$.
b. List all samples of size $n = 2$.
c. Find the population of all values of $\bar{x}$ by finding the mean of each sample from part *b*.
d. Find the mean $\mu_{\bar{x}}$ and standard deviation $\sigma_{\bar{x}}$ for the population of sample means found in part *c*.
e. Verify that

$$\mu_{\bar{x}} = \mu \text{ and } \sigma_{\bar{x}} = \frac{\sigma}{\sqrt{n}} \sqrt{\frac{N-n}{N-1}}$$

26. The value of $\sigma_{\bar{x}}$ can be used as a measure of how close sample means will be to the population mean $\mu$. Assume that samples of size 36 are randomly selected from the population of IQ scores with $\mu = 100$ and $\sigma = 15$.
    a. What percentage of these sample means will fall within 2.5 of the mean?
    b. Between what two values (with the mean at the center) will 95% of these sample means fall?

27. The accompanying frequency table summarizes the number of defective units produced by a machine on 87 different days. Find $\mu$ and $\sigma$ for this population. If 32 of the 87 days are randomly selected, use the central limit theorem to find the probability that the mean for the 32 days is greater than 30.0.

| $x$ | $f$ |
|---|---|
| 0–4 | 2 |
| 5–9 | 0 |
| 10–14 | 5 |
| 15–19 | 8 |
| 20–24 | 12 |
| 25–29 | 17 |
| 30–34 | 20 |
| 35–39 | 14 |
| 40–44 | 6 |
| 45–49 | 3 |

28. a. Assume that a population is infinite. Find the probability that the mean of a sample of 100 differs from the population mean by more than $\sigma/4$.
    b. A sample of size 50 is randomly selected from a population of size $N$, with the result that the standard error of the mean is one-tenth the value of the population standard deviation. Find the size of the population.

## Vocabulary List

Define and give an example of each term.

normal distribution
standard normal distribution
z score
standard score
control chart

statistical process control
continuity correction
central limit theorem
standard error of the mean
finite population correction factor

## Review

The main concern of this chapter is the concept of a **normal distribution,** the most important of all continuous probability distributions. Many real and natural occurrences yield data that are normally distributed or can be approximated by a normal distribution. The

### Important Formulas

Standard normal distribution has $\mu = 0$ and $\sigma = 1$. Standard score or $z$ score:

$$z = \frac{x - \mu}{\sigma}$$

Prerequisites for approximating binomial by normal:

$$np \geq 5 \qquad nq \geq 5$$

Parameters used when approximating binomial by normal:

$$\mu = np \qquad \sigma = \sqrt{npq}$$

Parameters used when applying central limit theorem:

$$\mu_{\overline{x}} = \mu$$

$$\sigma_{\overline{x}} = \frac{\sigma}{\sqrt{n}} \qquad \text{(standard error of the mean)}$$

$$\sigma_{\overline{x}} = \frac{\sigma}{\sqrt{n}} \sqrt{\frac{N - n}{N - 1}} \qquad \text{(used when } n > 0.05N\text{)}$$

$$z = \frac{\overline{x} - \mu_{\overline{x}}}{\sigma_{\overline{x}}}$$

normal distribution, which appears bell-shaped when graphed, can be described algebraically by an equation, but the complexity of that equation usually forces us to use a table of values instead.

Table A–2 represents the **standard normal distribution,** which has a mean of 0 and a standard deviation of 1. This table relates deviations away from the mean with areas under the curve. Since the total area under the curve is 1, those areas correspond to probability values.

In the early sections of this chapter, we worked with the standard procedures used in applying Table A–2 to a variety of different situations. We saw that Table A–2 can be applied indirectly to normal distributions that are nonstandard. (That is, $\mu$ and $\sigma$ are not 0 and 1, respectively.) We were able to find the number of standard deviations that a score $x$ is away from the mean $\mu$ by computing $z = (x - \mu)/\sigma$.

In Sections 5–3 and 5–4 we considered real and practical examples as we converted from a nonstandard to a standard normal distribution. In Section 5–5 we saw that we can sometimes approximate a binomial probability distribution by a normal distribution. If both $np \geq 5$ and $nq \geq 5$, the binomial random variable $x$ is approximately normally distributed with the mean and standard deviation given as $\mu = np$ and $\sigma = \sqrt{npq}$. Since the binomial probability distribution deals with discrete data while the normal distribution deals with continuous data, we introduced the **continuity correction,** which should be used in normal approximations to binomial distributions. Finally, in Section 5–6, we considered the distribution of sample means that can come from normal or nonnormal populations. The **central limit theorem** asserts that the distribution of sample means $\bar{x}$ (based on random samples of size $n$) will, as $n$ increases, approach a normal distribution with mean $\mu$ and standard deviation $\sigma/\sqrt{n}$. This means that if samples are of size $n$ where $n > 30$, we can approximate the distribution of those sample means by a normal distribution. The **standard error of the mean** is $\sigma/\sqrt{n}$ as long as the population is infinite or the sample size is not more than 5% of the population. But if the sample $n$ exceeds 5% of the population $N$, then the standard error of the mean must be adjusted by the **finite population correction factor** with $\sigma/\sqrt{n}$ multiplied by $\sqrt{(N - n)/(N - 1)}$. Figure 5–25 summarizes these concepts.

Since basic concepts of this chapter serve as critical prerequisites for the following chapters, it would be wise to master these ideas and methods now.

# ② *Review Exercises*

1. Household incomes of VCR owners have a mean of $41,182 and a standard deviation of $19,990 (based on data from Nielsen Media Research). If 125 households with VCRs are randomly selected, find the probability that the mean income is between $40,000 and $45,000.
2. The heights of adult males are normally distributed with a mean of 69.0 in. and a standard deviation of 2.8 in. (based on data from the National Health Survey, USDHEW publication 79–1659). If 95% of all males satisfy a minimum height requirement for police officers, what is that minimum height requirement?
3. An insurance company finds that the ages of motorcyclists killed in crashes are normally distributed with a mean of 26.9 years and a standard deviation of 8.4 years (based on data from the U.S. Department of Transportation).
   a. If we randomly select one such motorcyclist, find the probability that he or she was under 25 years of age.
   b. If we randomly select 40 such motorcyclists, find the probability that their mean age was under 25 years.
4. Delta Airlines recently reported that its on-time rate was 82% (based on data from the U.S. Department of Transportation). Find the probability that among 500 randomly selected Delta flights, fewer than 400 arrive on time.

5. A population of 700 scores has a mean of 5.40, a standard deviation of 1.20, and its distribution is approximately normal.
   a. If a score is randomly selected, find the probability that it is greater than 5.00.
   b. If 32 different scores are randomly selected without replacement, find the probability that their mean is greater than 5.00.
   c. If 36 different scores are randomly selected without replacement, find the probability that their mean is greater than 5.00.

6. Errors from meter readings are normally distributed with a mean of 0 V and a standard deviation of 1 V. (The errors can be positive or negative.) One reading is randomly selected. Find the probability that the error is
   a. Between 0 V and 1.42 V
   b. Greater than $-1.05$ V
   c. Between 0.50 V and 1.50 V

7. Among Americans aged 18 and older, 8% are divorced (based on U.S. Bureau of the Census data). If 225 Americans aged 18 or older are randomly selected, find the probability that at least 20 of them are divorced.

8. A sociologist finds that for a certain segment of the population, the numbers of years of formal education are normally distributed with a mean of 13.20 years and a standard deviation of 2.95 years.
   a. For a person randomly selected from this group, find the probability that he or she has between 13.20 and 13.50 years of education.
   b. For a person randomly selected from this group, find the probability that he or she has at least 12.00 years of education.
   c. Find the first quartile, $Q_1$. That is, find the value separating the lowest 25% from the highest 75%.
   d. If an employer wants to establish a minimum education requirement, how many years of education would be required if only the top 5% of this group would qualify?
   e. If 35 people are randomly selected from this group, find the probability that their mean years of education is at least 12.00 years.

9. The systolic blood pressures of adults are normally distributed with a mean of 129.8 and a standard deviation of 21.9. (Units are in mm of Hg and the data are based on the National Health Survey, USDHEW publication 78-1648.) If 500 adults are randomly selected for a medical research project, how many of them are expected to have systolic blood pressures above 180.0?

10. A study has shown that among people without any preschool education, 32% were employed at age 19. If these figures are correct, find the probability that for a group of 100 people without any preschool education, 40 or fewer are employed at age 19.

11. Scores on a standard IQ test are normally distributed with a mean of 100 and a standard deviation of 15.
   a. Find the probability that a randomly selected subject will achieve a score between 90 and 120.
   b. Find the probability that a randomly selected subject will achieve a score above 105.
   c. If 30 subjects are randomly selected and tested, find the probability that their mean IQ score is above 105.
   d. Find $P_{95}$, the IQ score separating the top 5% from the lower 95%.
   e. Find $P_{15}$, the IQ score separating the bottom 15% from the top 85%.

12. Scores on a hearing test are normally distributed with a mean of 600 and a standard deviation of 100.
   a. If one subject is randomly selected, find the probability that the score is between 600 and 735.
   b. If one subject is randomly selected, find the probability that the score is more than 450.
   c. If one subject is randomly selected, find the probability that the score is between 500 and 800.
   d. If a job requires a score in the top 80%, find the lowest acceptable score.
   e. If 50 subjects are randomly selected, find the probability that their mean score is between 600 and 635.

13. Errors on a scale are normally distributed with a mean of 0 kg and a standard deviation of 1 kg. One item is randomly selected and weighed. (The errors can be positive or negative.)
   a. Find the probability that the error is between 0 and 0.74 kg.
   b. Find the probability that the error is greater than 1.76 kg.
   c. Find the probability that the error is greater than $-1.08$ kg.

14. The mean IQ of engineers is estimated to be 120. Among adults, IQ scores are normally distributed with a mean of 100 and a standard deviation of 15. If an adult is randomly selected, find the probability that his or her IQ is above the mean IQ of engineers.

15. In a study of 600 checkout times at a Caldor department store in upstate New York, the mean is found to be 1.80 min and the standard deviation is found to be 0.60 min. If 32 of these checkout times are randomly selected, find the probability that their mean is between 1.90 min and 2.00 min.

16. According to data from the American Medical Association, 10% of us are left-handed. In a freshman class of 200 students, find the probability that
   a. Exactly 9 are left-handed
   b. Fewer than 9 are left-handed

 *Computer Projects*

1. Listed below are two BASIC programs, which may require some minor modification in order to run on certain computers. The first program will produce 36 randomly generated numbers. The second program will produce a mean of 36 randomly generated numbers.

   a. Enter and run the first program and manually construct a histogram of these 36 values.

   b. Enter the second program and run it 36 times to get 36 sample means. Then manually construct a histogram of these 36 values. *Hint:* Instead of entering RUN 36 times, we can run the program 36 times by including these two lines:

      5 FOR J = 1 to 36
      65 NEXT J

   c. Compare the two resulting histograms.

   ```
   10 RANDOMIZE
   20 FOR I = 1 to 36
   30   PRINT INT(100*RND(X))
   40 NEXT I
   50 END


   10 RANDOMIZE
   20 LET T = 0
   30 FOR I = 1 to 36
   40   LET T = T + INT(100*RND(X))
   50 NEXT I
   60 PRINT T/36
   70 END
   ```

2. Listed below are the life spans of 50 randomly selected left-handed baseball players (based on data from the *Baseball Encyclopedia*). Use a software package such as STATDISK or Minitab to enter the data and find the mean and standard deviation. Also generate a histogram. Based on the histogram, do these scores appear to come from a normally distributed population?

   | | | | | | | | | | | | | | | |
   |---|---|---|---|---|---|---|---|---|---|---|---|---|---|---|
   | 78 | 59 | 74 | 75 | 81 | 68 | 69 | 60 | 81 | 64 | 50 | 39 | 70 | 57 | 55 |
   | 43 | 47 | 69 | 52 | 63 | 62 | 74 | 58 | 79 | 55 | 41 | 62 | 77 | 54 | 46 |
   | 94 | 81 | 62 | 53 | 77 | 59 | 60 | 45 | 60 | 67 | 52 | 58 | 57 | 68 | 52 |
   | 85 | 68 | 41 | 76 | 82 | | | | | | | | | | |

# *Applied Projects*

1.  Refer to the 50 life spans listed in the second computer project given above. If that project wasn't completed, manually calculate the mean, standard deviation, and construct a histogram. For the purposes of this project, assume that the data come from a normally distributed population and use the sample mean and standard deviation as estimates of the population mean and standard deviation. Using the normal distribution techniques presented in this chapter, find the following.
    a.  The probability of a left-handed baseball player living more than 65 years
    b.  The probability of a left-handed baseball player living less than 50 years
    c.  The value of the first quartile $Q_1$, the third quartile $Q_3$, and these percentiles: $P_{10}$, $P_{20}$, $P_{30}$, $P_{40}$, $P_{60}$, $P_{70}$, $P_{80}$, $P_{90}$

2.  Table 5–1 listed 30 samples, each with ten random numbers between 1 and 9. Use a table of random digits, a telephone book, or a computer to observe or generate 50 samples with ten random numbers between 0 and 9. Construct a frequency table and histogram of the combined sample of 500 scores, and find the mean and standard deviation. Then find the mean of each sample and, using the 50 sample means, construct a frequency table and histogram and find the mean and standard deviation. Compare the results obtained from the combined sample of 500 scores to those obtained for the 50 sample means. Show how the comparison supports the conclusions of the central limit theorem.

# *Writing Projects*

1.  In your own words, write a description of the central limit theorem. Describe an original example that illustrates its use.
2.  Write a report summarizing one of the programs listed below.

# *Videotapes*

Programs 4, 5, and 18 from the series *Against All Odds: Inside Statistics* are recommended as supplements to this chapter.

# Chapter Six

## In This Chapter

# 6 Estimates and Sample Sizes

## Chapter Problem

A *Newsweek* article described the use of "people meters" as a way of determining how many people are watching different television programs. This article noted that some preliminary tests revealed a "fatigue factor" whereby the people using the meters get tired of pushing the buttons and tend to quit. The article also noted that the technology of this television rating device tends to favor younger urban residents who are generally more adept at using more sophisticated technological devices. The article ended by stating, "Statisticians have long argued that the household samples used by the rating services are simply too small to accurately determine what America is watching. In that light, it may be illuminating to note that the 4000 homes reached by the people meters constitute exactly 0.0045% of the wired nation."

Television ratings are important since they are used to determine which shows are canceled and what rates to charge for advertising. In this chapter we will analyze the implied claim that a sample size of 4000 homes is too small.

# 6–1   Overview

We use descriptive statistics when we attempt to simply describe and better understand known data. We use inferential statistics when we attempt to make inferences (generalizations) about an unknown population by analyzing known sample data. In this chapter we will be working with inferential statistics as we use sample data to make **estimates of values of population parameters.** Section 6–2 begins by using the sample statistic $\bar{x}$ in estimating the value of the population parameter $\mu$. In subsequent sections we use sample proportions and sample variances to estimate the values of population proportions and population variances. We also introduce some of the common methods used to decide how large samples should be in order to obtain specific levels of precision.

While this chapter presents the use of inferential statistics for estimating population parameters, the following chapter presents another important aspect of inferential statistics: hypothesis testing.

# 6–2   Estimates and Sample Sizes of Means

When 40 U.S. commercial aircraft are randomly selected, they are found to have the following ages in years (based on data from Aviation Data Services).

| | | | | | | | |
|---|---|---|---|---|---|---|---|
| 3.2 | 22.6 | 23.1 | 16.9 | 0.4 | 6.6 | 12.5 | 22.8 |
| 26.3 | 8.1 | 13.6 | 17.0 | 21.3 | 15.2 | 18.7 | 11.5 |
| 4.9 | 5.3 | 5.8 | 20.6 | 23.1 | 24.7 | 3.6 | 12.4 |
| 27.3 | 22.5 | 3.9 | 7.0 | 16.2 | 24.1 | 0.1 | 2.1 |
| 7.7 | 10.5 | 23.4 | 0.7 | 15.8 | 6.3 | 11.9 | 16.8 |

Using only these sample scores, we want to estimate the mean age of *all* U.S. commercial aircraft. We could use statistics such as the sample median, midrange, or mode as estimates of $\mu$, but the sample mean $\bar{x}$ usually provides the best estimate of $\mu$. This is not simply an intuitive conclusion. It is based on careful study and analysis of the distributions of various estimators. For many populations, the distribution of sample means $\bar{x}$ has a smaller variance than the distribution of the other possible estimators, so $\bar{x}$ tends to be more consistent. For all populations we say that $\bar{x}$ is an **unbiased** estimator, meaning that the distribution of $\bar{x}$ values tends to center about the value of $\mu$. For these reasons, we will use $\bar{x}$ as the best estimate of $\mu$. Because $\bar{x}$ is a single number that corresponds to a point on the number scale, we call it a point estimate.

## Definition

The sample mean $\bar{x}$ is the best **point estimate of the population mean** $\mu$.

Computing the sample mean of the preceding 40 commercial aircraft ages, we get $\bar{x} = 13.41$, which becomes our point estimate of the mean age of all U.S. commercial aircraft. But even though 13.41 years is our *best* estimate of $\mu$, we have no indication of just how good that estimate is. Sometimes even the best is very poor. Suppose that we had only the first two ages of 3.2 years and 22.6 years. Their mean of 12.90 years would be the best point estimate of $\mu$, but we should not expect this best estimate to be very good since it is based on such a small sample of only two scores. Statisticians have developed another estimator that does reveal how good it is. We will present this estimator, illustrate its use, and then explain the underlying rationale. We begin with a new notation that uses the Greek letter $\alpha$ (alpha) to represent the total combined area of two tails in a normal distribution.

## Notation

$z_{\alpha/2}$ is the positive standard $z$ value that separates an area of $\alpha/2$ in the right tail of the standard normal distribution (see Figure 6–1.)

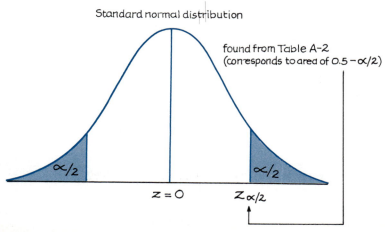

**Figure 6–1**

## CAPTURE-RECAPTURE

■ Ecologists need to determine population sizes of endangered species. One method is to capture a sample of some species, mark each member of this sample, and then free them. Later, another sample is captured and the ratio of marked subjects, coupled with the size of the first sample, can be used to estimate the population size. This capture-recapture method was used with other methods to estimate the blue whale population, and the result was alarming: The population was as small as 1000. That led the International Whaling Commission to ban the killing of blue whales to prevent their extinction. ■

---

### Example

If $\alpha = 0.05$, then $z_{\alpha/2} = 1.96$. That is, 1.96 is the standard $z$ value that separates a right-tail region with an area of $0.05 \div 2$ or 0.025. We find $z_{\alpha/2}$ by noting that the region to its left (and bounded by the mean of $z = 0$) must be $0.5 - 0.025$, or 0.475. In Table A–2, an area of 0.4750 corresponds exactly to a $z$ score of 1.96.

---

### Definition

The **maximum error of the estimate of** $\mu$ is given by

$$E = z_{\alpha/2} \cdot \frac{\sigma}{\sqrt{n}}$$

and there is a probability of $1 - \alpha$ that $\bar{x}$ differs from $\mu$ by less than $E$.

---

The maximum error of estimate $E$ is sometimes called the **bound on the error estimate** or the **margin of error.** For a given $\alpha$, there is a probability of $1 - \alpha$ that the point estimate $\bar{x}$ will miss $\mu$ by no more than the value of $E$. If $\alpha = 0.05$, for example, there is a 0.95 probability that $\bar{x}$ is in error (from $\mu$) by at most $E$. In this sense, $E$ is the maximum error of the point estimate $\bar{x}$.

---

### Definition

The **confidence interval** (or **interval estimate**) for the population mean is given by $\bar{x} - E < \mu < \bar{x} + E$.

We will use the preceding form of the confidence interval, but other equivalent forms are $\mu = \bar{x} \pm E$ and $(\bar{x} - E, \bar{x} + E)$.

The **degree of confidence** is the probability $1 - \alpha$ that the parameter $\mu$ is contained in the confidence interval. (The probability is often expressed as the equivalent percentage value.) The degree of confidence is also referred to as the **level of confidence** or the **confidence coefficient.**

---

Common choices for the degree of confidence are 95%, 99%, and 90%. The choice of 95% is most common, since it seems to represent a good balance between precision (as reflected in the width of the confidence interval) and reliability (as expressed by the degree of confidence).

The calculation of the value for the maximum error of estimate $E$ requires prior knowledge of the population standard deviation $\sigma$. In reality, it's rare that we know $\sigma$ for a population when we don't know $\mu$. The following is a common practice.

**In a normally distributed population with unknown $\sigma$, we can replace $\sigma$ by $s$ if $n > 30$.**

Small $(n \leq 30)$ sample cases will be discussed later in this section. The following example involves a large sample $(n > 30)$ so that we can use the sample deviation $s$ in place of the population standard deviation $\sigma$.

## Example

For the aircraft ages given earlier in this section, we have $n = 40$, $\bar{x} = 13.41$, and $s = 8.28$. For a 0.95 degree of confidence, use these statistics to find both of the following.

a. The maximum error of estimate $E$

b. The confidence interval for $\mu$

## Solution

a. The 0.95 degree of confidence implies that $\alpha = 0.05$, so that

$$E = z_{\alpha/2} \frac{\sigma}{\sqrt{n}} = 1.96 \cdot \frac{8.28}{\sqrt{40}} = 2.57$$

(Note that since $\sigma$ is unknown but $n > 30$, we used $s = 8.28$ for the value of $\sigma$. Also, recall from the previous example that with $\alpha = 0.05$, $z_{\alpha/2} = 1.96$.)

b. With $\bar{x} = 13.41$ and $E = 2.57$, we get

$$\bar{x} - E < \mu < \bar{x} + E$$

$$13.41 - 2.57 < \mu < 13.41 + 2.57$$

$$10.84 < \mu < 15.98$$

We interpret this result as follows: If we were to select different samples of size 40 from the given population and use the above method for finding the corresponding intervals, in the long run, 95% of those intervals would actually contain the value of $\mu$. Instead of expressing the result as $10.84 < \mu < 15.98$, two other equivalent forms are $\mu = 13.41 \pm 2.57$ and $(10.84, 15.98)$. We will not use these two forms in this book.

Assume that in the preceding example the aircraft ages really come from a population with a mean of 13.00 years. Then the confidence interval obtained from the given sample data contains the population mean, since 13.00 is between 10.84 and 15.98. This is illustrated in Figure 6–2 on page 286.

# Excerpts from a Department of Transportation Circular

■ The following excerpts from a Department of Transportation circular concern some of the accuracy requirements for navigation equipment used in aircraft. Note the use of the confidence interval.

"The total of the error contributions of the airborne equipment, when combined with the appropriate flight technical errors listed, should not exceed the following with a 95% confidence (2-sigma) over a period of time equal to the update cycle."

"The system of airways and routes in the United States has widths of route protection used on a VOR system use accuracy of ±4.5 degrees on a 95% probability basis." ■

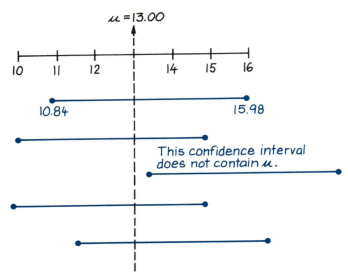

**Figure 6–2**

Note the wording of the interpretation of the confidence interval given in part *b* of the preceding example. We must be careful to interpret correctly the meaning of a confidence interval. Let's assume for the present discussion that we are using a 0.95 degree of confidence. Given the appropriate data, we can calculate the values of $\bar{x} - E$ and $\bar{x} + E$, which we will call the **confidence interval limits.** In general, these limits will have a 95% chance of enclosing $\mu$. That is, based on sample data to be collected, there will be a 0.95 probability that the confidence interval will contain $\mu$. If we recognize that different samples produce different confidence intervals, in the long run we will be correct 95% of the time when we say that $\mu$ is between the confidence interval limits. But once we use actual sample data to find specific limits, those limits either enclose $\mu$ or they do not, and we cannot determine if they do or don't without knowing the whole population. It is incorrect to state that $\mu$ has a 95% chance of falling within specific limits, since $\mu$ is a constant, not a random variable, and it will either fall within the interval or it won't—and there's no probability involved in that. Although it's wrong to say that $\mu$ has a 95% chance of falling between the confidence interval limits, it is correct to say that these methods will result in confidence limits that, in the long run, will contain $\mu$ in 95% of the random samples collected.

So far we have defined the confidence interval for the mean and illustrated its use. We now explain why the confidence interval has the form given in the definition.

The basic underlying idea relates to the central limit theorem, which indicates that the distribution of sample means is approximately normal as long as the samples are large ($n > 30$). The central limit theorem was also used to determine that sample means have a mean of $\mu$ while the standard deviation of means from samples of size $n$ is $\sigma/\sqrt{n}$. That is,

$$\mu_{\bar{x}} = \mu$$

$$\sigma_{\bar{x}} = \frac{\sigma}{\sqrt{n}}$$

Recall that $\sigma_{\bar{x}}$ is called the standard error of the mean. It is the standard deviation of means computed from samples of size $n$. Since a $z$ score is the number of standard deviations a value is away from the mean, we conclude that $z_{\alpha/2}\, \sigma/\sqrt{n}$ represents a number of standard deviations away from $\mu$. **There is a probability of $1 - \alpha$ that a sample mean will differ from $\mu$ by less than $z_{\alpha/2}\, \sigma/\sqrt{n}$.** See Figure 6–3 and note that the unshaded inner regions total $1 - \alpha$ and correspond to sample means that differ from $\mu$ by less than

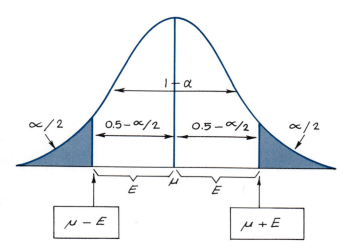

**Figure 6–3**
*(a) There is a $1 - \alpha$ probability that a sample mean will be in error by less than E or $z_{\alpha/2}\sigma/\sqrt{n}$.*
*(b) There is a probability of $\alpha$ that a sample mean will be in error by more than E (in one of the shaded tails).*

$z_{\alpha/2}\,\sigma/\sqrt{n}$. In other words, a sample mean error of $\mu - \bar{x}$ will be between $-z_{\alpha/2}\,\sigma/\sqrt{n}$ and $z_{\alpha/2}\,\sigma/\sqrt{n}$. This can be expressed as one inequality:

$$-z_{\alpha/2}\,\frac{\sigma}{\sqrt{n}} < \mu - \bar{x} < z_{\alpha/2}\,\frac{\sigma}{\sqrt{n}}$$

or

$$-E < \mu - \bar{x} < E$$

Adding $\bar{x}$ to each component of the inequality, we get the confidence interval $\bar{x} - E < \mu < \bar{x} + E$.

Applying these general concepts to the specific aircraft age data given at the beginning of this section, we can see that there is a 95% chance that a sample mean will be in error by less than

$$z_{\alpha/2}\,\frac{\sigma}{\sqrt{n}} = 1.96 \cdot \frac{8.28}{\sqrt{40}} = 2.57$$

Of the confidence intervals we construct by this method, in the long run, an average of 95% of them will involve sample means that are in error by less than 2.57. If $\bar{x} = 13.41$ is really in error by less than 2.57, then $\mu$ would be between 10.84 and 15.98. Another way of stating this is $10.84 < \mu < 15.98$, which corresponds to our confidence interval. It should be clear that this interval estimate of $\mu$ gives us much more insight than the point estimate of 13.41 alone. The interval estimate gives us some sense of how much the sample mean might be in error.

## Determining Sample Size

If we begin with the expression for $E$ and solve for $n$, we get

**Formula 6–1**
$$n = \left[\frac{z_{\alpha/2}\,\sigma}{E}\right]^2$$

This equation may be used to determine the sample size necessary to produce results accurate to a desired degree of confidence. It should be used when we know the value of $\sigma$ and want to determine the sample size necessary to establish, with a probability of $1 - \alpha$, the value of $\mu$ to within $\pm E$. The existence of such an equation is somewhat remarkable, since it implies that the sample size does not depend on the size of the population.

---

**Example**

We wish to be 99% sure that a random sample of IQ scores yields a mean that is within 2.0 of the true mean. How large should the sample be? Assume that $\sigma$ is 15.

## Solution

We seek $n$ given that $\alpha = 0.01$ (from 99% confidence) and the maximum allowable error is 2.0. Applying the equation for sample size $n$, we get

$$n = \left[\frac{z_{\alpha/2}\sigma}{E}\right]^2 = \left[\frac{2.575 \times 15}{2.0}\right]^2$$
$$= (19.3125)^2 = 373 \qquad \text{(rounded up)}$$

Therefore we should obtain at least 373 randomly selected IQ scores if we require 99% confidence that our sample mean is within 2.0 units of the true mean.

If we can settle for less accurate results and accept a maximum error of 4.0 instead of 2.0, we can see that the required sample size is reduced from 373 to 94. Direct application of the equation for $n$ produces a value of 93.24, which is *rounded up* to 94. (We always round up in sample size computations so that the required number is at least adequate instead of being slightly inadequate.)

Doubling the maximum error from 2.0 to 4.0 caused the required sample size to decrease to one-fourth of its original value. Conversely, if we want to halve the maximum error, we must quadruple the sample size. In the equation, $n$ is inversely proportional to the square of $E$ and directly proportional to the square of $z_{\alpha/2}$. All of this implies that we can obtain more accurate results with greater confidence, but the sample size will be substantially increased. Since larger samples generally require more time and money, there may be a need for a tradeoff between the sample size and the confidence level.

# Small Sample Cases and the Student *t* Distribution

Unfortunately, the application of the equation for sample size (Formula 6–1) requires prior knowledge of $\sigma$. Realistically, $\sigma$ is usually unknown unless previous research results are available. When we do sample from a population in which $\sigma$ is unknown, a preliminary study must be conducted so that $\sigma$ can be reasonably estimated. Only then can we determine the sample size required to meet our error tolerance and confidence demands.

If we intend to construct a confidence interval, do now know $\sigma$, and do not plan a preliminary study, we can use the **Student *t* distribution** developed by William S. Gosset (1876–1937). Gosset was a Guinness Brewery employee who needed a distribution that could be used with small samples. The Irish

## QUEUES

■ Queuing theory is a branch of mathematics that uses probability and statistics. The study of queues, or waiting lines, is important to businesses such as supermarkets, banks, fast food restaurants, airlines, and amusement parks. Grand Union supermarkets try to keep lines no longer than three shoppers. Wendy's introduced the "Express Pak" to expedite servicing its numerous drive-through customers. Disney conducts extensive studies of lines at its amusement parks so that it can keep patrons and plan for expansion. Bell Laboratories uses queuing theory to optimize telephone network usage, and factories use it to design efficient production lines. ■

brewery where he worked did not allow the publication of research results, so Gosset published under the pseudonym "Student." Factors such as cost and time often severely limit the size of a sample, so the normal distribution cannot be an appropriate approximation of the distribution of means from small samples. As a result of those earlier experiments and studies of small samples, we can now use the Student $t$ distribution instead.

---

### Definition

If a population is essentially normal, then the distribution of

$$t = \frac{\bar{x} - \mu}{s/\sqrt{n}}$$

is essentially a **Student $t$ distribution** for all samples of size $n$. (The Student $t$ distribution is often referred to as the **$t$ distribution.**)

---

Table A–3 lists **critical values** of $t$ along with corresponding areas. The critical $t$ value is obtained by locating the proper values for degrees of freedom in the left column and then proceeding across that corresponding row until reaching the number directly below the applicable value of $\alpha$. Roughly stated, degrees of freedom correspond to the number of values that may vary after certain restrictions have been imposed on all values. For example, if 10 scores must total 50, we can freely assign values to the first 9 scores, but the 10th score would then be determined so that there would be 9 degrees of freedom. For the applications of this section, the number of degrees of freedom is simply the sample size minus 1.

$$\textbf{degrees of freedom} = n - 1$$

As an example, suppose we have a sample of size $n = 10$. With this small ($n \leq 30$) sample size and unknown $\sigma$, we use $t_{\alpha/2}$ instead of $z_{\alpha/2}$. Just as $z_{\alpha/2}$ is the positive $z$ value that separates an area of $\alpha/2$ in the right tail of a normal distribution, the value of $t_{\alpha/2}$ separates an area of $\alpha/2$ in the right tail of a Student $t$ distribution. For a 95% confidence interval and a sample size of $n = 10$, the critical value of $t_{\alpha/2}$ is found from Table A–3 by noting the following:

1.   The area in the two tails is 0.05.
2.   With $n = 10$, the number of degrees of freedom is $10 - 1 = 9$.

Locating 9 at the left column and 0.05 (two tails) at the top row, Table A–3 shows that the critical score is $t_{\alpha/2} = 2.262$.

To use the Student $t$ distribution, the parent population must be essentially normal. The population does not have to be exactly normal, but if it has only one mode and is basically symmetric, we will generally get good results if we use the Student $t$ distribution. If there is strong evidence that the population has a very nonnormal distribution, then nonparametric methods (see Chapter 12) may apply.

# Important Properties of the Student $t$ Distribution

1.  The Student $t$ distribution is different for different sample sizes. (See Figure 6–4 for the cases $n = 3$ and $n = 12$.)

2.  The Student $t$ distribution has the same general bell shape of the normal distribution, but it reflects the greater variability that is expected with small samples.

3.  The Student $t$ distribution has a mean of $t = 0$ (just as the standard normal distribution has a mean of $z = 0$).

4.  The standard deviation of the Student $t$ distribution is greater than 1 (unlike the standard normal distribution, which has $\sigma = 1$).

5.  As the sample size $n$ gets larger, the Student $t$ distribution gets closer to the normal distribution. For values of $n > 30$, the differences are so small that we can use the critical $z$ values instead of developing a much larger table of critical $t$ values. (The values in the bottom row of Table A–3 are equal to the corresponding critical $z$ values from the normal distribution.)

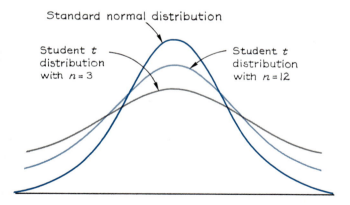

**Figure 6–4**
*The Student t distribution has the same general shape and symmetry as the normal distribution, but it reflects the greater variability that is expected with small samples.*

# How One Telephone Survey Was Conducted

■ A *New York Times*/CBS News survey was based on 1417 interviews of adult men and women in the United States. It reported a 95% certainty that the sample results differed by no more than three percentage points from the percentage that would have been obtained if *every* adult American had been interviewed. A computer was used to select telephone exchanges in proportion to the population distribution. After selecting an exchange number, the computer generated random numbers to develop a complete phone number so that listed and unlisted numbers could be included. ■

Given below is a summary of the conditions indicating use of a $t$ distribution instead of the normal distribution. These same conditions will also apply in Chapter 7.

---

### Conditions for Student $t$ Distribution

1. The sample is small $(n \le 30)$; and
2. $\sigma$ is unknown; and
3. the parent population is essentially normal.

---

We can now determine values for the maximum error $E$ when a $t$ distribution applies.

---

### Maximum Error

$$E = t_{\alpha/2} \frac{s}{\sqrt{n}}$$

when *all* the following conditions are met:

1. $n \le 30$; and
2. $\sigma$ is unknown; and
3. the population is normally distributed.

---

In the next example we use some of the same data from a previous example so that you can better compare and contrast similarities and differences.

---

### Example

Suppose we have only ten scores representing the ages (in years) of randomly selected U.S. commercial aircraft (based on data from Aviation Data Services).

3.2  22.6  23.1  16.9  0.4  6.6  12.5  22.8  26.3  8.1

For these scores, $n = 10$, $\bar{x} = 14.25$, and $s = 9.35$. Recognizing that this is a small sample $(n \le 30)$ and $\sigma$ is unknown, construct the 95% confidence interval for the mean age of all U.S. commercial aircraft. (Assume that other studies reveal that the distribution of such ages is approximately normal.)

## Solution

With a small sample and unknown $\sigma$, we know that the Student $t$ distribution applies and we begin by calculating the maximum error of estimate $E$.

$$E = t_{\alpha/2} \cdot \frac{s}{\sqrt{n}} = 2.262 \cdot \frac{9.35}{\sqrt{10}} = 6.69$$

We now substitute $E = 6.69$ and $\bar{x} = 14.25$ in

$$\bar{x} - E < \mu < \bar{x} + E$$

to get

$$14.25 - 6.69 < \mu < 14.25 + 6.69$$

or

$$7.56 < \mu < 20.94$$

For the given set of 10 scores, the point estimate of $\mu$ is $\bar{x} = 14.25$ years and the interval estimate is as given above. In the long run, 95% of such samples will lead to confidence limits that actually do contain the true value of the population mean $\mu$.

Shown below is the STATDISK computer display of the results from the data in the preceding example. The user enters the level of confidence, the sample size, sample mean, and sample standard deviation. (STATDISK allows entry of a population standard deviation if it is known, but this is not shown on the display below.) We also show the Minitab display that results from the data in the preceding example.

STATDISK DISPLAY

```
Confidence Level . . . . . . . . . . . . . . . . . . . . . . . . . . . . . . .  = .95
Sample size  . . . . . . . . . . . . . . . . . . . . . . . . . . . . . . n  = 10
Sample mean . . . . . . . . . . . . . . . . . . . . . . . . . . . . . .  = 14.25
Sample standard deviation . . . . . . . . . . . . . . . . . . . . .  = 9.35

                    The Confidence Interval is
           7.55632 < Population mean < 20.94368
```

## SHAKESPEARE'S VOCABULARY

■ According to Bradley Effron and Ronald Thisted, Shakespeare's writings included 31,534 different words. They used probability theory to conclude that Shakespeare probably knew at least another 35,000 words that he didn't use in his writings. The problem of estimating the size of a population is an important problem often encountered in ecology studies, but the result given here is another interesting application. (See "Estimating the Number of Unseen Species: How Many Words Did Shakespeare Know?" in *Biometrika*, Vol. 63, No. 3.) ■

MINITAB DISPLAY

```
MTB > SET C1
DATA > 3.2 22.6 23.1 16.9 0.4 6.6 12.5 22.8 26.3 8.1
DATA > ENDOFDATA
MTB > TINTERVAL with 95 percent for data in C1

        N      MEAN     STDEV     SE MEAN     95.0 PERCENT C.I.
C1     10      14.25     9.35       2.96      (  7.56,      20.94)
```

## The Bootstrap

In 1977, Stanford University's Bradley Efron introduced the **bootstrap** method, which can be used to construct confidence intervals even in situations in which traditional methods cannot (or should not) be used. For example, if we have a sample of size $n = 10$ randomly selected from a very nonnormal distribution, we can't use the methods discussed in this section, as they require a population that is at least approximately normal. The bootstrap method, which makes no assumptions about the original population, typically requires a computer to build a "bootstrap" population by replicating (duplicating) a sample many times. We can draw from the sample with replacement, thereby creating an approximation of the original population. In this way, we pull the sample up "by its own bootstraps" to simulate the original population. This is a relatively new technique that is now gaining much attention. (See Computer Projects 3 and 4 at the end of this chapter for more details about the actual process.)

## 6–2  Exercises A

1. a. If $\alpha = 0.05$, find $z_{\alpha/2}$.
   b. If $\alpha = 0.02$, find $z_{\alpha/2}$.
   c. Find $z_{\alpha/2}$ for the value of $\alpha$ corresponding to a confidence level of 96%.
   d. If $\alpha = 0.05$, find $t_{\alpha/2}$ for a sample of 20 scores.
   e. If $\alpha = 0.01$, find $t_{\alpha/2}$ for a sample of 15 scores.

2. a. If $\alpha = 0.10$, find $z_{\alpha/2}$.
   b. Find $z_{\alpha/2}$ for the value of $\alpha$ corresponding to a confidence level of 95%.
   c. Find $z_{\alpha/2}$ for the value of $\alpha$ corresponding to a confidence level of 80%.
   d. If $\alpha = 0.10$, find $t_{\alpha/2}$ for a sample of 10 scores.
   e. If $\alpha = 0.02$, find $t_{\alpha/2}$ for a sample of 25 scores.

In Exercises 3–8, use the given data to find the maximum error of estimate *E*. Be sure to use the correct expression for *E*, depending on whether the normal distribution or Student *t* distribution applies.

3.   $\alpha = 0.05$, $\sigma = 15$, $n = 100$
4.   $\alpha = 0.05$, $\sigma = 15$, $n = 64$
5.   $\alpha = 0.01$, $\sigma = 40$, $n = 25$
6.   $\alpha = 0.05$, $s = 30$, $n = 25$
7.   $\alpha = 0.01$, $s = 15$, $n = 100$
8.   $\alpha = 0.01$, $s = 30$, $n = 64$
9.   Find the 95% confidence interval for $\mu$ if $\sigma = 5$, $\bar{x} = 70.4$, and $n = 36$.
10.   Find the 95% confidence interval for $\mu$ if $\sigma = 7.3$, $\bar{x} = 84.2$, and $n = 40$.
11.   Find the 99% confidence interval for $\mu$ if $\sigma = 2$, $\bar{x} = 98.6$, and $n = 100$.
12.   Find the 90% confidence interval for $\mu$ if $\sigma = 5.5$, $\bar{x} = 123.6$, and $n = 75$.
13.   In an insurance company study of New York State licensed drivers' ages, 570 randomly selected ages have a mean of 41.8 years and a standard deviation of 16.7 years (based on data from the New York State Department of Motor Vehicles). Construct a 99% confidence interval for the mean age of all New York State licensed drivers.
14.   In a study at a Caldor department store in upstate New York, 40 randomly selected check-out times are found to have a mean of 1.80 min and a standard deviation of 0.60 min. Construct a 90% confidence interval for the mean of all such check-out times.
15.   A car repair business randomly selects receipts for parts purchased on 94 different jobs. The mean is $31.32 and the standard deviation is $30.42. Construct the 95% confidence interval for the mean cost of parts for all jobs.
16.   The U.S. Department of Health, Education, and Welfare collected sample data for 772 males between the ages of 18 and 24. That sample group has a mean height of 69.7 in. with a standard deviation of 2.8 in. (see USDHEW publication 79-1659). Use this sample data to find the 99% confidence interval for the mean height of all males between the ages of 18 and 24.
17.   In a time-use study, 20 randomly selected managers spend a mean of 2.40 h each day on paperwork. The standard deviation of the 20 scores is 1.30 h (based on data from Adia Personnel Services). Construct the 95% confidence interval for the mean paperwork time of all managers.
18.   A carpenter randomly selects 5 wood framing studs that are supposed to be 96 in. long. She finds that their lengths have a mean of 95.99 in. and a standard deviation of 0.15 in. Construct the 95% confidence interval for the mean length of all such framing studs.

19. In a study of times required for room service at a newly opened Radisson Hotel, 20 deliveries had a mean time of 24.2 min and a standard deviation of 8.7 min. Construct the 90% confidence interval for the mean of all deliveries.

20. The author's computer disk drive was timed (in milliseconds) for 25 different revolutions. The sample mean and standard deviation were found to be 200.45 ms and 0.24 ms, respectively. Construct the 99% confidence interval for the mean of all such times.

21. Find the sample size necessary to estimate a population mean to within three units if $\sigma = 16$ and we want 95% confidence in our results.

22. Find the sample size necessary to estimate a population mean. Assume that $\sigma = 20$, the maximum allowable error is 1.5, and we want 95% confidence in our results.

23. On a standard IQ test, $\sigma$ is 15. How many random IQ scores must be obtained if we want to find the true population mean (with an allowable error of 0.5) and we want 99% confidence in the results?

24. Do Exercise 23 assuming that the maximum error is 0.25 unit instead of 0.5 unit.

25. In a study of relationships between work schedules and family life, 29 subjects work at night. Their weekly times (in hours) spent in child care are measured and the mean and standard deviation are 26.84 and 17.66, respectively. (See "Nonstandard Work Schedules and Family Life," by Staines and Pleck, *Journal of Applied Psychology*, Vol. 69, No. 3.) Use this sample data to construct the 95% confidence interval for the mean time in child care for all night workers.

26. When 40 randomly selected customers purchased gas at one location, their mean amount was 8.03 gal while the standard deviation was 3.56 gal. Construct the 98% confidence interval for the mean amount purchased by all customers.

27. A sample of computer connect times (in hours) is obtained for 18 randomly selected students. For this sample, the mean is 16.2 while the standard deviation is 3.4. Construct the 99% confidence interval for the mean connect time of all such students.

28. In sampling the caloric content of 24 bottles of "light" beer, the mean and standard deviation are found to be 107.3 calories and 3.9 calories, respectively. Find the 90% confidence interval for the mean caloric content of all such bottles.

29. In a study of physical attractiveness and mental disorders, 231 subjects were rated for attractiveness and the resulting sample mean and standard deviation are 3.94 and 0.75, respectively. (See "Physical Attractiveness and Self-Perception of Mental Disorder," by Burns and Farina, *Journal of Abnormal Psychology*, Vol. 96, No. 2.) Use this sample data to construct the 95% confidence interval for the population mean.

30. The U.S. Department of Health, Education, and Welfare collected sample data for 1525 women aged 18 to 24. That sample group has a mean

serum cholesterol level (measured in mg/100 ml) of 191.7 with a standard deviation of 41.0 (see USDHEW publication 78-1652). Use this sample data to find the 97% confidence interval for the mean serum cholesterol level of all women in the 18–24 age bracket.

31. The National Center for Education Statistics surveyed 4400 college graduates about the lengths of time required to earn their bachelor's degrees. The mean is 5.15 years and the standard deviation is 1.68 years. Based on this sample data, construct the 99% confidence interval for the mean time required by all college graduates.

32. In a study of factors affecting soldiers' decisions to reenlist, 320 subjects were measured for an index of satisfaction and the sample mean and standard deviation are 28.8 and 7.3, respectively. (See "Affective and Cognitive Factors in Soldiers' Reenlistment Decisions," by Motowidlo and Lawton, *Journal of Applied Psychology*, Vol. 69, No. 1.) Use the given sample data to construct the 98% confidence interval for the population mean.

33. Assume that $\sigma = 2.8$ in. for the heights of adult males (based on data from the National Health Survey, publication 79-1659). Find the sample size necessary to estimate the mean height of all adult males to within 0.5 in. if we want 99% confidence in our results.

34. We want to estimate the mean weight of one type of coin minted by a certain country. How many coins must we sample if we want to be 99% confident that the sample mean is within 0.001 oz of the true mean? Assume that a pilot study has shown that $\sigma = 0.004$ oz can be used.

35. We want to determine the mean weight of all boxes of cereal labeled 400 grams. We need to be 98% confident that our sample mean is within 3 g of the population mean, and a pilot study suggests that $\sigma = 10$ g. How large must our sample be?

36. A psychologist has developed a new test of spatial perception, and she wants to estimate the mean score achieved by adult male pilots. How many people must she test if she wants the sample mean to be in error by no more than 2.0 points, with 95% confidence? A pilot study suggests that $\sigma = 21.2$.

37. In a study of the use of hypnosis to relieve pain, sensory ratings were measured for 16 subjects with the results given below. (See "An Analysis of Factors That Contribute to the Efficacy of Hypnotic Analgesia," by Price and Barber, *Journal of Abnormal Psychology*, Vol. 96, No. 1.) Use this sample data to construct the 95% confidence interval for the mean sensory rating for the population from which the sample was drawn.

| 8.8 | 6.6 | 8.4 | 6.5 | 8.4 | 7.0 | 9.0 | 10.3 |
| 8.7 | 11.3 | 8.1 | 5.2 | 6.3 | 11.6 | 6.2 | 10.9 |

38. The Internal Revenue Service conducts a study of estates valued at more than $300,000 and determines the value of bonds for a randomly se-

lected sample with the results (in dollars) given below. Find an interval estimate of the mean value of bonds for all such estates. Assume that the sample standard deviation can be used as an estimate of the population standard deviation.

| | | | | | |
|---|---|---|---|---|---|
| 45,300 | 36,200 | 72,500 | 50,500 | 15,300 | 58,500 |
| 26,200 | 97,100 | 74,200 | 83,700 | 72,000 | 10,000 |
| 63,000 | 15,000 | 49,200 | 37,500 | 81,000 | 24,000 |
| 145,000 | 27,900 | 53,100 | 27,500 | 94,000 | 23,800 |
| 74,600 | 36,800 | 65,900 | 29,400 | 86,300 | 25,600 |
| 53,200 | 47,200 | 61,800 | 33,200 | 18,200 | 75,000 |

39. In one region of a city, a random survey of households includes a question about the number of people in the household. The results are given in the accompanying frequency table. Construct the 90% confidence interval for the mean size of all such households. Assume that the sample standard deviation can be used as an estimate of the population standard deviation.

| Household Size | $f$ |
|---|---|
| 1 | 15 |
| 2 | 20 |
| 3 | 37 |
| 4 | 23 |
| 5 | 14 |
| 6 | 4 |
| 7 | 2 |

40. An aeronautical research team collects data on the stall speeds (in knots) of ultralight aircraft. The results are summarized in the accompanying stem-and-leaf plot. Construct the 95% confidence interval for the mean stall speed of all such aircraft.

| 21. | 7 8 |
|---|---|
| 22. | 3 4 4 6 |
| 23. | 2 2 5 8 9 9 |
| 24. | 0 1 3 |
| 25. | 2 |

## 6–2 Exercises B

41. A 95% confidence interval for the lives (in minutes) of Kodak AA batteries is $430 < \mu < 470$. (See Program 1 of *Against All Odds: Inside Statistics.*) Assume that this result is based on a sample of size $n = 100$.
   a. Construct the 99% confidence interval.
   b. What is the value of the sample mean?
   c. What is the value of the sample standard deviation?
   d. If the confidence interval $432 < \mu < 468$ is obtained from the same sample data, what is the degree of confidence?

42. a. Using the data from Exercise 38 as a pilot study, how many random estates must be surveyed in order to estimate the mean bond value? Assume that we want to be 95% confident that the sample mean is in error by at most $1000.

b. Using the data from Exercise 39 as a pilot study, how many random households must be surveyed if we want to estimate the mean household size? Assume that we want 99% confidence that the sample mean is in error by at most 0.1.

43. The development of Formula 6–1 assumes that the population is infinite, or we are sampling with replacement, or the population is very large. If we have a relatively small population and we sample without replacement, we should modify $E$ to include the finite population correction factor as follows:

$$E = z_{\alpha/2} \frac{\sigma}{\sqrt{n}} \sqrt{\frac{N - n}{N - 1}} \qquad \text{where } N \text{ is the population size}$$

a. Show that the preceding expression can be solved for $n$ to yield

$$n = \frac{N\sigma^2[z_{\alpha/2}]^2}{(N - 1)E^2 + \sigma^2[z_{\alpha/2}]^2}$$

b. Do Exercise 34 assuming that the coins are selected without replacement from a population of $N = 100$ coins.

44. The standard error of the mean is $\sigma/\sqrt{n}$ provided that the population size is infinite. If the population size is finite and is denoted by $N$, then the correction factor

$$\sqrt{\frac{N - n}{N - 1}}$$

should be used whenever $n > 0.05N$. This correction factor multiplies the standard error of the mean, as shown in Exercise 43. Find the 95% confidence interval for the mean of 100 IQ scores if a sample of 30 scores produces a mean and standard deviation of 132 and 10, respectively.

## 6–3 Estimates and Sample Sizes of Proportions

In this section we consider the same concepts of estimating and sample size determination that were discussed in Section 6–2, but we apply the concepts to proportions instead of means. We assume that the conditions given in Section 5–5 for the binomial distribution with $np \geq 5$ and $nq \geq 5$ are satisfied. These assumptions enable us to use the normal distribution as an approximation of the binomial distribution.

Although we make repeated references to proportions, keep in mind that the theory and procedures also apply to probabilities and percents. Proportions and probabilities are both expressed in decimal or fraction form. If we intend to deal with percents we can easily convert them to proportions by deleting the percent sign and dividing by 100. The symbol $p$ may therefore represent a proportion, a probability, or the decimal equivalent of a percent. We continue to use $p$ as the population proportion in the same way that we use $\mu$ to represent the population mean. We now introduce a new notation.

---

### Notation

$$\hat{p} = \frac{x}{n}$$

---

In this way, $p$ represents the population proportion while $\hat{p}$ (called "$p$ hat") represents the sample proportion. In previous chapters we stipulated that $q = 1 - p$, so it is natural to stipulate that $\hat{q} = 1 - \hat{p}$.

The term $\hat{p}$ denotes the sample proportion that is analogous to the relative frequency definition of a probability. As an example, suppose that a pollster is hired to determine the proportion of adult Americans who favor socialized medicine. Let's assume that 2000 adult Americans are surveyed with 1347 favorable reactions. The pollster seeks the value of $p$, the true proportion of all adult Americans favoring socialized medicine. Sample results indicate that $x = 1347$ and $n = 2000$, so that

$$\hat{p} = \frac{x}{n} = \frac{1347}{2000} = 0.6735$$

Just as $\bar{x}$ was selected as the point estimate of $\mu$, we now select $\hat{p}$ as the best point estimate of $p$.

---

### Definition

The sample proportion $\hat{p}$ is the best **point estimate of the population proportion $p$.**

---

Of the various estimators that could be used for $p$, $\hat{p}$ is deemed best because it is unbiased and the most consistent. It is unbiased in the sense that the distribution of sample proportions tends to center about the value of $p$. It is most consistent in the sense that the variance of sample proportions tends to be smaller than the variance of any other unbiased estimators.

We assume in this section that the binomial conditions are essentially satisfied and that the normal distribution can be used as an approximation to the distribution of sample proportions. This allows us to draw from results established in Section 5–5 and to conclude that the mean number of successes $\mu$ and the standard deviation of the number of successes $\sigma$ are given by

$$\mu = np$$

$$\sigma = \sqrt{npq}$$

where $p$ is the probability of a success. Both of these parameters pertain to $n$ trials, and we now convert them to a "per trial" basis simply by dividing by $n$:

$$\text{mean of sample proportions} = \frac{np}{n} = p$$

$$\text{standard deviation of sample proportions} = \frac{\sqrt{npq}}{n} = \sqrt{\frac{npq}{n^2}} = \sqrt{\frac{pq}{n}}$$

The first result may seem trivial since we have already stipulated that the true population proportion is $p$. The second result is nontrivial and very useful. In the last section, we saw that the sample mean $\overline{x}$ has a probability of $1 - \alpha$ of being within $z_{\alpha/2}\,\sigma/\sqrt{n}$ of $\mu$. Similar reasoning leads us to conclude that $\hat{p}$ has a probability of $1 - \alpha$ of being within $z_{\alpha/2}\sqrt{pq/n}$ of $p$. But if we already know the value of $p$ or $q$, we have no need for estimates or sample size determinations. Consequently, we must replace $p$ and $q$ by their point estimates of $\hat{p}$ and $\hat{q}$ so that an error factor can be computed in real situations. This leads to the following results.

<br>

### Definition

The **maximum error of the estimate of $p$** is given by

$$E = z_{\alpha/2}\,\sqrt{\frac{\hat{p}\hat{q}}{n}}$$

and the probability that $\hat{p}$ differs from $p$ by less than $E$ is $1 - \alpha$.

<br>

### Definition

The **confidence interval** (or interval estimate) for the population proportion $p$ is given by

$$\hat{p} - E < p < \hat{p} + E$$

The following example illustrates the construction of a confidence interval for a proportion.

## THE WISDOM OF HINDSIGHT

■ The Roper Organization erred by 10 percentage points when it predicted the outcome of the Mondale-Reagan presidential race. Roper noted that a sample error was caused when eight-year-old census data were not updated. This resulted in a sample with a disproportionate number of Democrats. Also, respondents were first asked to identify the candidate they favored and then asked if they planned to vote. This order of questions results in an inflated figure for the second question: Respondents who identify a favorite candidate are more inclined to say that they will vote, even if they are not really inclined to do so. The Roper Organization is usually quite accurate in its results, and steps have been taken to prevent such errors in the future. ■

---

### Example

In a Roper Organization poll of 2000 adults, 1280 have money in regular savings accounts. Find the 95% confidence interval for the true proportion of adults who have money in regular savings accounts.

### Solution

The sample results are $x = 1280$ and $n = 2000$, so that $\hat{p} = 1280/2000 = 0.640$ and $\hat{q} = 1 - 0.640 = 0.360$. A confidence level of 95% requires that $\alpha = 0.05$, so that $z_{\alpha/2} = 1.96$. We first calculate the maximum error of estimate $E$.

$$E = z_{\alpha/2} \sqrt{\frac{\hat{p}\hat{q}}{n}} = 1.96 \sqrt{\frac{(0.640)(0.360)}{2000}}$$
$$= 0.021$$

We can now find the confidence interval since we know that $\hat{p} = 0.640$ and $E = 0.021$.

$$\hat{p} - E < p < \hat{p} + E$$
$$0.640 - 0.021 < p < 0.640 + 0.021$$
$$0.619 < p < 0.661$$

If we wanted the 95% confidence interval for the true population *percent*, we could express the result as $61.9\% < p < 66.1\%$. This result is often reported as follows: "Among adults, the percent with money in regular savings accounts is estimated to be 64.0%, with a margin of error of plus or minus 2.1 percentage points." The level of confidence should also be reported.

## Sample Size

Having discussed point estimates and confidence intervals for $p$, we now consider the problem of determining how large a sample should be when we want to find the approximate value of a population proportion. In the previous section we started with the expression for the error $E$ and solved for $n$. Following that reasonable precedent, we begin with

$$E = z_{\alpha/2} \sqrt{\frac{\hat{p}\hat{q}}{n}}$$

and we solve for $n$ to get the sample size.

## Sample Size

$$n = \frac{[z_{\alpha/2}]^2 \hat{p}\hat{q}}{E^2}$$

| $\hat{p}$ | $\hat{q}$ | $\hat{p}\hat{q}$ |
|------|------|------|
| 0.1 | 0.9 | 0.09 |
| 0.2 | 0.8 | 0.16 |
| 0.3 | 0.7 | 0.21 |
| 0.4 | 0.6 | 0.24 |
| 0.5 | 0.5 | 0.25 |
| 0.6 | 0.4 | 0.24 |
| 0.7 | 0.3 | 0.21 |
| 0.8 | 0.2 | 0.16 |
| 0.9 | 0.1 | 0.09 |

But if we are going to determine the necessary sample size, we can assume that the sampling has not yet taken place, so $\hat{p}$ and $\hat{q}$ are not known. Mathematicians have cleverly circumvented this problem by showing that, in the absence of $\hat{p}$ and $\hat{q}$, we can assign the value of 0.5 to each of those statistics and the resulting sample size will be at least sufficient. The underlying reason for the assignment of 0.5 is found in the conclusion that the product $\hat{p} \cdot \hat{q}$ achieves a maximum possible value of 0.25 when $\hat{p} = 0.5$ and $\hat{q} = 0.5$. (See the accompanying table, which lists some values of $\hat{p}$ and $\hat{q}$.) In practice, this means that no knowledge of $\hat{p}$ or $\hat{q}$ requires that the preceding expression for $n$ be replaced as follows.

## Sample Size

$$n = \frac{[z_{\alpha/2}]^2 \cdot 0.25}{E^2}$$

Here the occurrence of 0.25 reflects the substitution of 0.5 for both $\hat{p}$ and $\hat{q}$. If we have evidence supporting specific known values of $\hat{p}$ or $\hat{q}$, we can substitute those values and thereby reduce the sample size accordingly. For example, if $\hat{p} = 0.6$, then $\hat{q} = 0.4$ and $\hat{p}\hat{q} = 0.24$, which is less than 0.25, so that the resulting value of $n$ will be smaller. Such evidence about $\hat{p}$ or $\hat{q}$ could come from a pilot study, previous experience, or other such sources.

### Example

We want to estimate, with a maximum error of 0.03, the true proportion of all TV households turned in to a particular show, and we want 95% confidence in our results. We have no prior information suggesting a possible value of $p$. How many TV households must we survey?

*continued*

## SMALL SAMPLE

■The Children's Defense Fund was organized to promote the welfare of children. The group published *Children Out of School in America*, which reported that in one area, 37.5% of the 16- and 17-year-old children were out of school. This statistic received much press coverage, but it was based on a sample of only 16 children. Another statistic was based on a sample size of only 3 students. (See "Firsthand Report: How Flawed Statistics Can Make an Ugly Picture Look Even Worse," *American School Board Journal*, Vol. 162.) ■

### Solution

With a confidence level of 95%, we have $\alpha = 0.05$, so that $z_{\alpha/2} = 1.96$. We are given $E = 0.03$, but in the absence of $\hat{p}$ or $\hat{q}$ we use the last expression for $n$. We get

$$n = \frac{[z_{\alpha/2}]^2 \cdot 0.25}{E^2} = \frac{[1.96]^2 \cdot 0.25}{0.03^2} = 1067.11 = 1068 \text{ (rounded up)}$$

To be 95% confident that we come within 0.03 of the true proportion of TV households who watch the show, we should poll 1068 randomly selected TV households.

At the beginning of this chapter we presented a brief excerpt from an article implying that a sample of 4000 may be too small, since it is only 0.0045% of all television households. We now know that this position is essentially incorrect, because the desired sample size doesn't depend on the population size; it depends on the levels of accuracy and confidence we desire. It is the absolute size of the sample that is important, not the size of the sample relative to the population. It is the sample size number that determines its credibility, not the percent of the population.

The previous example shows that with a sample of 1068 randomly selected TV households, we have 95% confidence that the sample proportion is within 0.03 of the true proportion. Based on Nielsen data, there are about 90 million TV households, so that a sample size of 1068 represents about 0.001% of the population. Although that percent is small, we are still about 95% confident that we are within 0.03 of the true value, so that the results will be quite good. Clearly, a sample of size 2000 or 4000 would be even better. In this sense, that article is misleading. Although Nielsen ratings might be criticized for being based on a sample that is not truly representative of the population, they should not be criticized because the sample is a small percent of the population.

### Example

You've been hired to conduct a study for the purpose of estimating the percentage of households in your state that have answering machines. You want an error of no more than three percentage points and a confidence level of 96%. A previous study by the U.S. Consumer Electronics Industry indicates that the percentage should be about 24%. How large should your sample be?

### Solution

With a 96% confidence level, we have $\alpha = 0.04$ and $z_{\alpha/2} = 2.05$. The error of three percentage points means that $E = 0.03$, since the actual calculations use proportions. The prior study gives us $\hat{p} = 0.24$ and $\hat{q} = 0.76$. We now calculate

$$n = \frac{[z_{\alpha/2}]^2 \hat{p}\hat{q}}{E^2} = \frac{[2.05]^2 (0.24)(0.76)}{(0.03)^2}$$

$$= 851.71 = 852 \text{ (rounded up)}$$

Rounding *up*, we find that the sample size should be 852. (If we had no prior knowledge of the percentage, we would have used 0.25 for $\hat{p}\hat{q}$ and our required sample size would have been the much larger value of 1168.)

Newspaper, magazine, television, and radio reports often feature results of polls. Reporters frequently provide percentages without any indication of the sample size or degree of confidence. For example, one national newspaper reported that "the American Lung Association says that 64 percent of smokers agree they shouldn't light up near nonsmokers." Without knowing the sample size and degree of confidence, we have no real sense for how good that statistic is. In contrast, the *New York Times* published an article giving results of a poll conducted on the popularity of the president. The *Times* included a five-paragraph insert explaining that the results were "based on telephone interviews conducted Nov. 20 through Nov. 24 with 1553 adults around the U.S., excluding Alaska and Hawaii." The *Times* also explained how the telephone numbers were selected to be representative of the population. They explained how results were weighted to be representative of region, race, sex, age, and education. They explained that "in theory, in 19 cases out of 20 the results based on such samples will differ by no more than 3 percentage points in either direction from what would have been obtained by interviewing all adult Americans." In general, the five-paragraph insert provided information that allows informed readers to recognize the quality of the poll.

Polling is an important and common practice in the United States. Polls can affect the television shows we watch, the leaders we elect, the legislation that governs us, and the products we consume. Understanding the concepts of this section should remove much of the mystery and misunderstanding created by polls.

## LARGE SAMPLE SIZE ISN'T GOOD ENOUGH

■Biased sample data should not be used for inferences, no matter how large the sample is. For example, in *Women and Love: A Cultural Revolution in Progress*, Shere Hite bases her conclusions on 4500 replies that she received after mailing 100,000 questionnaires to various women's groups. A *random* sample of 4500 subjects would usually provide good results, but Hite's sample is biased. It is criticized for overrepresenting women who join groups and women who feel strongly about the issues addressed. Because Hite's sample is biased, her inferences are not valid, even though the sample size of 4500 might seem to be sufficiently large. ■

## $6-3$ Exercises A

In Exercises 1–4, a trial is repeated $n$ times with $x$ successes. In each case find (a) $\hat{p}$; (b) $\hat{q}$; (c) the best point estimate for the value of $p$; (d) the maximum error of estimate $E$ (assuming that $\alpha = 0.05$).

1. $n = 500, x = 100$
2. $n = 2000, x = 300$
3. $n = 1068, x = 325$
4. $n = 1776, x = 50$

In Exercises 5–8, use the given data to find the appropriate confidence interval for the population proportion $p$.

5. $n = 400, x = 100$, 95% confidence
6. $n = 900, x = 400$, 95% confidence
7. $n = 512, x = 309$, 98% confidence
8. $n = 12,485, x = 3456$, 99% confidence
9. In considering the production of a new car accessory, a manufacturer wants to do a marketing study that includes determination of the proportion of cars with stereo tape decks. How many cars must be sampled if the manufacturer wants 92% confidence that the sample proportion is in error by no more than 0.035?
10. You want to estimate the percentage of employees who have been with their current employer for one year or less. Data from the U.S. Department of Labor suggest that this percentage is around 29%, but you must conduct a survey to be 99% confident that your randomly selected sample of employees leads to a sample percentage that is off by no more than two percentage points. How many employees must you survey?
11. A sports equipment supplier finds that when 1180 adults are surveyed, 79 play golf (based on data from a Roper Organization Poll). Construct the 95% interval estimate for the true proportion of all adults who play golf.
12. You want to estimate the proportion of home accident deaths that are caused by falls. How many home accident deaths must you survey in order to be 95% confident that your sample proportion is within 0.04 of the true population proportion?
13. You randomly select 650 home accident deaths and find that 180 of them are caused by falls (based on data from the National Safety Council). Construct the 95% confidence interval for the true population proportion of all home accident deaths caused by falls.
14. A survey of 1002 voters by Martilla and Kiley, Inc., showed that 13.0% of them felt that it is very likely that we will get into a nuclear war within the next 10 years. Construct the 95% confidence interval for the true percentage of all voters who feel that way.

15. In an Airport Transit Association poll of 4664 adults, 72% indicated that they have flown in an airplane. Find the 99% confidence interval for the percentage of all adults who have flown in an airplane.

16. In a survey of 1500 people, 63% indicated that they listened to their favorite radio station because of the music (based on data from a Strategic Radio Research poll). Construct the 98% interval estimate of the true population percentage of all people who listen to their favorite radio station because of the music.

17. You plan to conduct a poll to estimate the percentage of consumers who are satisfied with long-distance phone service. You want to be 90% confident that your sample percentage is within 2.5 percentage points of the true population value, and a Roper poll suggests that this percentage should be about 85%. How large must your sample be?

18. You want to estimate the percentage of small (10 or fewer employees) incorporated businesses that are owned by women. A previous study suggests that this percentage is 27%. How many of these small businesses must be surveyed if you want 94% confidence that your sample percentage is within three percentage points of the true population percentage?

19. A pollster is hired to determine the percentage of voters favoring the Republican presidential nominee. If we require 99% confidence that the estimated value is within two percentage points of the true value, how large should the random sample be?

20. A *New York Times* article about poll results states, "In theory, in 19 cases out of 20, the results from such a poll should differ by no more than one percentage point in either direction from what would have been obtained by interviewing all voters in the United States." Find the sample size suggested by this statement.

21. A coach observes a player trying out for a basketball team. The potential player makes 12 of 40 shots from three-point range. Construct the 95% interval estimate of the true proportion of such shots this player can make under similar conditions.

22. A Roper poll of 1998 adults resulted in 53% saying that security was an important aspect of money. Based on this sample data, construct the 95% confidence interval for the true percentage of the entire adult population.

23. Among 24,350 felons convicted in state courts, 67% were given jail sentences (based on data from the U.S. Department of Justice). Construct the 98% confidence interval for the proportion of all such felons who were given jail sentences.

24. According to Merrill Lynch, when 600 full-time workers in the 45–64 age bracket were surveyed, 75% of them planned to remain in their current homes after they retired. Construct the 90% confidence interval for the true proportion of all such workers with that intent.

25. A multiple-choice test question is considered easy if at least 80% of the responses are correct. A sample of 6503 responses to one question indicates that 5463 of those responses were correct. Construct the 99% confidence interval for the true proportion of correct responses. Is it likely that this question is really easy?

26. Of 600 people who completed the first item on a questionnaire, 24% responded "always," 60% responded "sometimes," and the others responded "never." Construct the 99% confidence interval for the true proportion of people who respond "never."

27. Among 785 randomly selected subjects who complete four years of college, 18.3% smoke (based on data from the American Medical Association). Construct the 98% confidence interval for the true percentage of smokers among all people who complete four years of college.

28. A marketing study found that 312 of the 650 buyers of compact cars were women (based on data from the Ford Motor Company). Construct the 95% interval estimate for the true percentage of all compact car buyers who are women.

## 6–3 | Exercises B

29. In this section we developed two formulas used for determining sample size, and in both cases we assume that the population is infinite, or we are sampling with replacement, or the population is very large. If we have a relatively small population and we sample without replacement, we should modify $E$ to include the finite population correction factor as follows:

$$E = z_{\alpha/2} \sqrt{\frac{\hat{p}\hat{q}}{n}} \sqrt{\frac{N-n}{N-1}}$$

Here $N$ is the size of the population.

a. Show that the above expression can be solved for $n$ to yield

$$n = \frac{N\hat{p}\hat{q}[z_{\alpha/2}]^2}{\hat{p}\hat{q}[z_{\alpha/2}]^2 + (N-1)E^2}$$

b. Do Exercise 18 assuming that there is a finite population of size $N = 500$ businesses with 10 or fewer employees.

30. Special tables are available for finding confidence intervals for proportions involving small numbers of cases where the normal distribution approximation cannot be used. For example, given 3 successes among 8 trials, the 95% confidence interval is $0.085 < p < 0.755$. Find the confidence interval that would result if you were to incorrectly use the normal distribution as an approximation to the binomial distribution. Are the results reasonably close?

31.  A newspaper article indicates that an estimate of the unemployment rate involves a sample of 47,000 people. If the reported unemployment rate must have an error no larger than 0.2 percentage point and the rate is known to be about 8%, find the corresponding confidence level.

32.  a.  If IQ scores of adults are normally distributed with a mean of 100 and a standard deviation of 15, use the methods presented in Chapter 5 to find the percentage of IQ scores above 130.

   b.  Now assume that you plan to test a sample of adults with the intention of estimating the percentage of IQ scores above 130. How many adults must you test if you want to be 98% confident that your error is no more than 2.5 percentage points? (Use the result from part *a*.)

33.  A **one-sided confidence interval** for $p$ can be written as

$$p < \hat{p} + E \quad \text{or} \quad p > \hat{p} - E$$

where $z_\alpha$ replaces $z_{\alpha/2}$ in the expression for $E$.

   If an airline company wants to report an on-time performance of "at least $x$ percent" with 95% confidence, construct the appropriate one-sided confidence interval and then find the percent in question. Assume that a random sample of 750 flights results in 630 that are on time.

 ## Estimates and Sample Sizes of Variances

Many real and practical situations, such as quality control in a manufacturing process, require that we estimate values of population variances or standard deviations. In addition to making products with good average quality, the manufacturer must make products of *consistent* quality that do not run the gamut from extremely good to extremely poor. This consistency can often be measured by the variance or standard deviation, so these become vital statistics in maintaining the quality of products. Also, there are many other circumstances in which variance and standard deviation are critically important.

   **In this section, we assume that the population has normally distributed values.** This assumption was made in earlier sections, but it is a more critical assumption here. In using the Student $t$ distribution in Section 6–2, for example, we require that the population of values be approximately normal, but we can accept deviations away from normality that are not too severe. However, when we deal with variances by using the chi-square distribution, to be introduced shortly, departures away from normality can lead to gross errors. Consequently, the assumption of normality must be adhered to much more strictly. We describe this sensitivity by saying that inferences about $\sigma^2$

(or $\sigma$) made on the basis of the chi-square distribution are not **robust** against departures from normality. In contrast, inferences made about $\mu$ based on the Student $t$ distribution are reasonably robust, since departures from normality that are not too extreme will not lead to gross errors.

# Chi-Square Distribution

In a normally distributed population with variance $\sigma^2$, we randomly select independent samples of size $n$ and compute the variance $s^2$ for each sample. The random variable $(n - 1)s^2/\sigma^2$ has a distribution called the **chi-square distribution.**

---

**Chi-Square Distribution**

$$\chi^2 = \frac{(n - 1)s^2}{\sigma^2}$$

$n$ = sample size

$s^2$ = sample variance

$\sigma^2$ = population variance

---

We denote chi-square by $\chi^2$ and we pronounce it "kigh square." The specific mathematical equations used to define this distribution are not given here since they are beyond the scope of this text. Instead, you can refer to Table A–4 for required values of the chi-square distribution. We should also note that the general form of the chi-square distribution is $(df)s^2/\sigma^2$, where $df$ represents degrees of freedom. In this chapter we have $n - 1$ degrees of freedom.

**degrees of freedom = $n - 1$**

In later chapters we will encounter situations in which the degrees of freedom are not $n - 1$. For that reason, we should not universally equate degrees of freedom with $n - 1$.

Here are some other important properties of the chi-square distribution:

1. The chi-square distribution is not symmetric, unlike the normal and Student $t$ distributions (see Figure 6–5).

2. The values of chi-square can be zero or positive, but they cannot be negative (see Figure 6–5).

3. The chi-square distribution is different for each number of degrees of freedom, which is $df = n - 1$ in this section (see Figure 6–6).

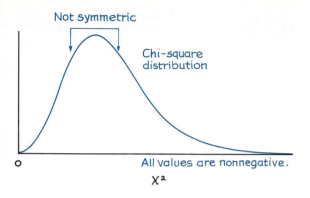

**Figure 6–5**

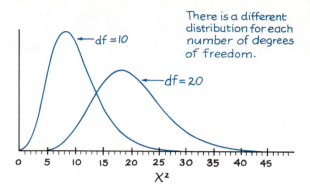

**Figure 6–6**

In previous sections of this chapter we focused on the topics of estimating population parameters and determining sample size. In this section we again consider those same topics as they relate to variances. However, due to the nature of the chi-square distribution, the techniques discussed in this section will not closely parallel those in the preceding two sections.

Since sample variances tend to center on the value of the population variance, we say that $s^2$ is an unbiased estimator of $\sigma^2$. Also, the variance of $s^2$ values tends to be smaller than the variance of the other unbiased estimators. For these reasons we decree that, among the various possible statistics we could use to estimate $\sigma^2$, the best is $s^2$.

---

### Definition

The sample variance $s^2$ is the best **point estimate of the population variance $\sigma^2$**.

---

Since $s^2$ is the best point estimate of $\sigma^2$, it would be natural to expect that $s$ is the best point estimate of $\sigma$, but this is not the case. For reasons we will not pursue, $s$ is a biased estimator of $\sigma$; if the sample size is large, however, the bias is small so that we can use $s$ as a reasonably good estimate of $\sigma$.

Although $s^2$ is the best point estimate of $\sigma^2$, there is no indication of how good this best estimate is. To compensate for that deficiency, we develop a more informative interval estimate (or confidence interval).

## HOW VALID ARE CRIME STATISTICS?

■ Police departments are often judged by the number of arrests because that statistic is readily available and clearly understood. However, this encourages police to concentrate on easily solved minor crimes (such as marijuana smoking) at the expense of more serious crimes, which are difficult to solve. A study of Washington, D.C., police records revealed that police can also manipulate statistics another way. The study showed that more than 1000 thefts in excess of $50 were intentionally valued at less than $50 so that they would be classified as petty larceny and not major crimes. ■

### Definition

The **confidence interval** (or **interval estimate**) for the population variance $\sigma^2$ is given by

$$\frac{(n-1)s^2}{\chi_R^2} < \sigma^2 < \frac{(n-1)s^2}{\chi_L^2}$$

The confidence interval (or interval estimate) for $\sigma$ is found by taking the square root of each component of the preceding inequality:

$$\sqrt{\frac{(n-1)s^2}{\chi_R^2}} < \sigma < \sqrt{\frac{(n-1)s^2}{\chi_L^2}}$$

In the above expressions, $\chi_R^2$ and $\chi_L^2$ are notations defined as follows. (Some books use $\chi_{\alpha/2}^2$ instead of $\chi_R^2$ and $\chi_{1-\alpha/2}^2$ for $\chi_L^2$.)

### Notation

With a total area of $\alpha$ divided equally between the two tails of a chi-square distribution, $\chi_L^2$ denotes the **left-tailed critical value** and $\chi_R^2$ denotes the **right-tailed critical value**. (See Figure 6–7.)

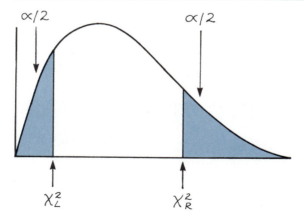

**Figure 6–7**

In constructing a confidence interval for $\sigma^2$ by using the above expression, we must determine the values of $\chi_R^2$ and $\chi_L^2$, which we refer to as critical values. In finding those values, we use Table A–4. An important feature of this table is that each critical value separates an area to the *right* that corresponds to the value given in the top row.

**Example**

Find the critical values of $\chi^2$ that determine critical regions containing areas of 0.025 in each tail. Assume that the relevant sample size is 10 so that the degrees of freedom are $10 - 1$, or 9.

**Solution**

See Figure 6–8 and refer to Table A–4. The critical value to the right (19.023) is obtained in a straightforward manner by locating 9 in the degrees-of-freedom column at the left and 0.025 across the top. The left critical value of 2.700 once again corresponds to 9 in the degrees-of-freedom column, but we must locate 0.975 across the top since the values in the top row are always *areas to the right* of the critical value.

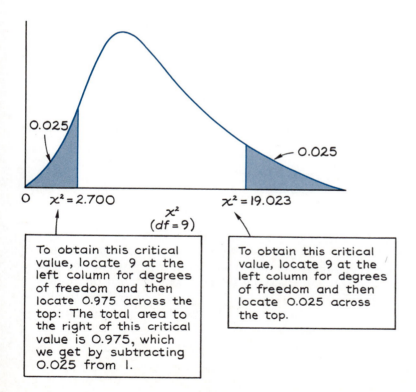

To obtain this critical value, locate 9 at the left column for degrees of freedom and then locate 0.975 across the top: The total area to the right of this critical value is 0.975, which we get by subtracting 0.025 from 1.

To obtain this critical value, locate 9 at the left column for degrees of freedom and then locate 0.025 across the top.

**Figure 6–8**
*Finding critical values of the chi-square distribution using Table A–4.*

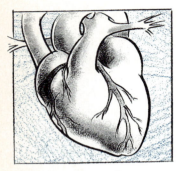

## GENDER GAP IN DRUG TESTING

■ A study of the relationships between heart attacks and doses of aspirin involved 22,000 male physicians. This study, like many others, excluded women. The General Accounting Office recently criticized the National Institutes of Health for not including both sexes in many studies because results of medical tests on males do not necessarily apply to females. For example, women's hearts are different than men's in many important ways. When forming conclusions based on sample results, we should be wary of an inference that extends to a population larger than the one from which the sample was drawn. ■

Figure 6–8 shows that, for a sample of 10 scores taken from a normally distributed population, the statistic $(n - 1)s^2/\sigma^2$ has a 0.95 probability of falling between 2.700 and 19.023. In general, there is a probability of $1 - \alpha$ that the statistic $(n - 1)s^2/\sigma^2$ will fall between $\chi_L^2$ and $\chi_R^2$. In other words (and symbols), there is a $1 - \alpha$ probability that both of the following are true:

$$\frac{(n - 1)s^2}{\sigma^2} < \chi_R^2 \quad \text{and} \quad \frac{(n - 1)s^2}{\sigma^2} > \chi_L^2$$

If we multiply both of the preceding inequalities by $\sigma^2$ and divide each inequality by the appropriate critical value of $\chi^2$, we see that the two inequalities can be expressed in the equivalent forms

$$\frac{(n - 1)s^2}{\chi_R^2} < \sigma^2 \quad \text{and} \quad \frac{(n - 1)s^2}{\chi_L^2} > \sigma^2$$

These last two inequalities can be combined into one inequality,

$$\frac{(n - 1)s^2}{\chi_R^2} < \sigma^2 < \frac{(n - 1)s^2}{\chi_L^2}$$

There is a probability of $1 - \alpha$ that the population variance $\sigma^2$ is contained in the above interval, and this result corresponds to the definition of the confidence interval for $\sigma^2$. We can now construct confidence intervals for $\sigma$ as illustrated in this next example.

### Example

A container of car antifreeze is supposed to hold 3785 mL of the liquid. Realizing that fluctuations are inevitable, the quality control manager wants to be quite sure that the standard deviation is less than 30 mL. Otherwise, some containers would overflow while others would not have enough of the coolant. She randomly selects a sample with the results given below. Use these sample results to construct the 99% confidence interval for the true value of $\sigma$. Does this confidence interval suggest that the fluctuations are at an acceptable level?

| 3761 | 3861 | 3769 | 3772 | 3675 | 3861 | $n = 18$ |
| 3888 | 3819 | 3788 | 3800 | 3720 | 3748 | $\bar{x} = 3787.0$ |
| 3753 | 3821 | 3811 | 3740 | 3740 | 3839 | $s = 55.4$ |

## Solution

Based on the sample data, the mean of $\bar{x} = 3787.0$ appears to be fine, but we will now construct the 99% confidence interval for $\sigma^2$ and then for $\sigma$. With a sample of size $n = 18$, we have $n - 1 = 17$ degrees of freedom. Since we want 99% confidence, we divide the 1% chance of error between the two tails to get an area of 0.005 in each tail. In the 17th row of Table A–4, we find that 0.005 in the left and right tails corresponds to $\chi_L^2 = 5.697$ and $\chi_R^2 = 35.718$. With these values and $n = 18$ and $s = 55.4$, we get

$$\frac{(18 - 1)(55.4)^2}{35.718} < \sigma^2 < \frac{(18 - 1)(55.4)^2}{5.697}$$

This then becomes $1460.8 < \sigma^2 < 9158.5$. Taking the square root of each part yields $38.2 < \sigma < 95.7$. It appears that the standard deviation is too large and corrective action must be taken to ensure more consistent container fillings.

# Sample Size

The problem of determining the sample size necessary to estimate $\sigma^2$ to within given tolerances and confidence levels becomes much more complex than it was in similar problems that dealt with means and proportions. Instead of developing very complicated procedures, we supply Table 6–1, which lists approximate sample sizes (see the following page).

## Example

You wish to estimate $\sigma^2$ to within 10% and you need 99% confidence in your results. How large should your sample be? Assume that the population is normally distributed.

## Solution

From Table 6–1, 99% confidence and an error of 10% for $\sigma^2$ correspond to a sample of size 1400. You should randomly select 1400 values from the population.

**TABLE 6–1**

| To be 95% confident that $s^2$ is within | of the value of $\sigma^2$, the sample size $n$ should be at least |
|---|---|
| 1% | 77,210 |
| 5% | 3,150 |
| 10% | 806 |
| 20% | 210 |
| 30% | 97 |
| 40% | 57 |
| 50% | 38 |

| To be 99% confident that $s^2$ is within | of the value of $\sigma^2$, the sample size $n$ should be at least |
|---|---|
| 1% | 133,362 |
| 5% | 5,454 |
| 10% | 1,400 |
| 20% | 368 |
| 30% | 172 |
| 40% | 101 |
| 50% | 67 |

| To be 95% confident that $s$ is within | of the value of $\sigma$, the sample size $n$ should be at least |
|---|---|
| 1% | 19,205 |
| 5% | 767 |
| 10% | 192 |
| 20% | 47 |
| 30% | 21 |
| 40% | 12 |
| 50% | 8 |

| To be 99% confident that $s$ is within | of the value of $\sigma$, the sample size $n$ should be at least |
|---|---|
| 1% | 33,196 |
| 5% | 1,335 |
| 10% | 336 |
| 20% | 85 |
| 30% | 38 |
| 40% | 22 |
| 50% | 14 |

In this section we saw that the best point estimate of $\sigma^2$ is $s^2$, but it does not follow that the best point estimate of $\sigma$ is $s$. We saw that when dealing with the distribution of sample variances, we use the chi-square distribution. We saw the method for constructing confidence intervals for $\sigma^2$ or $\sigma$, and we saw how to use Table 6–1 in determining sample size.

## 6–4 Exercises A

1. a. If a sample is described by the statistics $n = 100$, $\bar{x} = 146$, and $s^2 = 12$, find the best point estimate of $\sigma^2$.
   b. Use Table 6–1 to find the approximate minimum sample size necessary to estimate $\sigma^2$ with a 30% maximum error and 95% confidence.
   c. Find the $\chi_L^2$ and $\chi_R^2$ values for a sample of 25 scores and a confidence level of 99%.

2. a. If a sample is described by the statistics $n = 1087$, $\bar{x} = 77.3$, and $s = 4.0$, find the best point estimate of $\sigma^2$.
   b. Use Table 6–1 to find the approximate minimum sample size needed to estimate $\sigma$ with a 10% maximum error and 95% confidence.
   c. Find the $\chi_L^2$ and $\chi_R^2$ values for a sample of 15 scores and a confidence level of 95%.

3. a. Find the $\chi_L^2$ and $\chi_R^2$ values for a sample of 11 scores and a confidence level of 95%.
   b. Use Table 6–1 to find the approximate minimum sample size needed to estimate $\sigma$ with a 5% maximum error and 99% confidence.
   c. Find the best point estimate of $\sigma^2$ based on a sample for which $n = 17$, $\bar{x} = 69.2$, and $s = 1.2$.

4. a. Find the $\chi_L^2$ and $\chi_R^2$ values for a sample of 27 scores and a confidence level of 90%.
   b. Use Table 6–1 to find the approximate minimum sample size needed to estimate $\sigma$ with a 10% maximum error and 99% confidence.
   c. Find the best point estimate of $\sigma^2$ based on a sample for which $n = 6$, $\bar{x} = 428.2$, and $s = 1.9$.

5. The statistics $n = 30$, $\bar{x} = 16.4$, and $s = 2.5$ are obtained from a random sample drawn from a normally distributed population. Construct the 95% confidence interval about $\sigma^2$.

6. The statistics $n = 16$, $\bar{x} = 12.37$, and $s = 1.05$ are obtained from a random sample drawn from a normally distributed population. Construct the 99% confidence interval about $\sigma$.

7. Construct a 95% confidence interval about $\sigma^2$ if a random sample of 21 scores is selected from a normally distributed population and the sample variance is 100.0.

8. Construct a 95% confidence interval about $\sigma^2$ if a random sample of 10 scores is selected from a normally distributed population and the sample variance is 225.0.

9. In a study of the relationships between work schedules and family life, 29 subjects work at night. Their times (in hours) spent in child care are measured and the mean and standard deviation are 26.84 h and 17.66 h, respectively. (See "Nonstandard Work Schedules and Family Life," by Staines and Pleck, *Journal of Applied Psychology*, Vol. 69, No. 3.) Use this sample data to construct the 95% confidence interval for the standard deviation of the times in child care for all night workers.

10. Car owners were randomly selected and asked about the length of time they plan to keep their cars. The sample mean and standard deviation are 7.01 years and 3.74 years, respectively (based on data from a Roper poll). Assume that the sample size is 100 and construct the 95% confidence interval for the population standard deviation.

11. When working high school students are randomly selected and surveyed about the time they work at after-school jobs, the mean and standard deviation are found to be 17.6 h and 9.3 h, respectively (based on data from the National Federation of State High School Associations). Assume that this data is from a sample of 50 subjects and construct the 99% confidence interval about the standard deviation for all working high school students.

12. In a random sample of 40 college-bound high school juniors, their verbal PSAT scores have a mean of 40.7 and a standard deviation of 10.2 (based on data from Educational Testing Service). Construct the 99% confidence interval for the standard deviation of verbal PSAT scores for all college-bound juniors.

13. A researcher finds that the times required to fill 40 randomly selected room-service orders at a Radisson Hotel are normally distributed with mean of 24.2 min and a standard deviation of 8.7 min. Find the 99% confidence interval for the standard deviation of all room-service times at this hotel.

14. The author's computer disk drive was tested for times required to complete one revolution. The results for 30 randomly selected values are normally distributed with a mean of 200.45 ms and a standard deviation of 0.24 ms. Find the 98% confidence interval for the standard deviation of all such times.

15. In a Dutchess County silicon wafer manufacturing plant, a manager recorded the times required to inspect 40 randomly selected wafers. The results were normally distributed with a mean of 9.98 min and a standard deviation of 3.74 min. Construct the 90% confidence interval for the standard deviation of all such inspection times.

16. When 40 seventeen-year-old women are randomly selected and tested with the Goodenough-Harris Drawing Test, their mean and standard deviation are found to be 37.9 and 7.3, respectively (based on data from the National Health Survey, USDHEW publication 74–1620). Construct the 95% confidence interval for the standard deviation of the test scores for all seventeen-year-old women.

17. The following reaction times (in seconds) are randomly obtained from a normally distributed population: 0.60, 0.61, 0.63, 0.72, 0.91, 0.72. Find the 99% confidence interval for $\sigma$.

18. The following weights (in grams) are randomly obtained from a normally distributed population: 201, 203, 212, 222, 213, 215, 217, 230, 205, 208, 217, 225. Find the 95% confidence interval for $\sigma$.

19. A car designer studies variations in heights of women and begins with the following values (in inches) obtained from a random sample (based on data from the National Health Survey). Construct the 95% confidence interval about $\sigma$.

   60.8  63.8  64.8  64.3  62.7  68.7  62.9  61.8  66.0  63.1
   64.0  61.7  61.2  64.4  65.4  60.8  59.4  66.1  66.3  63.9

20. In a study of the use of hypnotism to relieve pain, sensory ratings were measured for 16 subjects with the results given below (based on data from "An Analysis of Factors that Contribute to the Efficacy of Hypnotic Analgesia," by Price and Barber, *Journal of Abnormal Psychology*, Vol. 96, No. 1). Use this set of sample data to construct the 95% confidence interval for the standard deviation of the sensory ratings for the population from which the sample was drawn.

   8.8   6.6  8.4  6.5  8.4   7.0  9.0  10.3
   8.7  11.3  8.1  5.2  6.3  11.6  6.2  10.9

# 6–4  Exercises B

21. A random sample is drawn from a normally distributed population and it is found that $n = 20$, $\bar{x} = 45.2$, and $s = 3.8$. Based on this sample, the following confidence interval is constructed.

$$2.8 < \sigma < 6.0$$

Find the degree of confidence.

22. A random sample of 12 scores is drawn from a normally distributed population and the 95% confidence interval is found to be

$$19.1 < \sigma < 45.8$$

Find the standard deviation of the sample.

23. In constructing confidence intervals for $\sigma$ or $\sigma^2$, we use Table A–4 to find $\chi_L^2$ and $\chi_R^2$, but that table applies only to cases in which $n \leq 101$ so that the number of degrees of freedom is 100 or fewer. For large numbers of degrees of freedom, we can approximate $\chi_L^2$ and $\chi_R^2$ by

$$\chi^2 = \frac{1}{2}[\pm z_{\alpha/2} + \sqrt{2k - 1}]^2$$

Here $k$ = number of degrees of freedom and $z_{\alpha/2}$ is as described in the preceding sections. Construct the 95% confidence interval about $\sigma$ by using the following sample data: For 772 males between the ages of 18 and 24, their measured heights have a mean of 69.7 in. and a standard deviation of 2.8 in. (based on data from the National Health Survey, USDHEW publication 79-1659).

24. When 500 items are randomly selected from a normally distributed population, the mean is 253.7 and the standard deviation is 4.8. Based on this data, the following confidence interval is obtained:

$$4.5459 < \sigma < 5.0788$$

What is the degree of confidence? (*Hint:* See Exercise 23.)

 ## *Vocabulary List*

Define and give an example of each term.

point estimate
maximum error of estimate
margin of error
confidence interval
interval estimate
degree of confidence
confidence interval limits
Student $t$ distribution
$t$ distribution
chi-square distribution

 ## *Review*

In this chapter we introduced important and fundamental concepts of inferential statistics. The main objective of this chapter was to develop procedures for estimating values of these population parameters: means (Section 6–2), proportions (Section 6–3), and variances (Section 6–4). We saw that these parameters have best **point estimates.** The best point estimate for a population mean is the value of the sample mean. The best point estimate of a population proportion is the value of the sample proportion. The best point estimate of a population variance is the sample variance. (However, the best point estimate of a population standard deviation is *not* the sample standard deviation.) As single values, the point estimates don't convey any real sense of how reliable they are, so we introduced **confidence intervals** (or **interval estimates**) as more informative estimates. We also considered ways of determining the **sample sizes** necessary to estimate parameters to within given tolerance factors.

This chapter also introduced the Student $t$ and chi-square distributions. We must be careful to use the correct distribution for each set of circumstances. The accompanying table summarizes the key concepts in this chapter.

## Important Formulas

| Parameter | Point Estimate | Confidence Interval | Sample Size |
|---|---|---|---|
| $\mu$ | $\bar{x}$ | $\bar{x} - E < \mu < \bar{x} + E$ <br><br> where $E = z_{\alpha/2}\dfrac{\sigma}{\sqrt{n}}$    (if $\sigma$ is known or if $n > 30$, in which case we use $s$ for $\sigma$) <br><br> or $E = t_{\alpha/2}\dfrac{s}{\sqrt{n}}$    (if $\sigma$ is unknown and $n \leq 30$) | $n = \left[\dfrac{z_{\alpha/2}\sigma}{E}\right]^2$ |
| $p$ | $\hat{p} = \dfrac{x}{n}$ | $\hat{p} - E < p < \hat{p} + E$ <br><br> where $E = z_{\alpha/2}\sqrt{\dfrac{\hat{p}\hat{q}}{n}}$ | $n = \dfrac{[z_{\alpha/2}]^2 \hat{p}\hat{q}}{E^2}$ <br><br> or <br><br> $n = \dfrac{[z_{\alpha/2}]^2 \cdot 0.25}{E^2}$ |
| $\sigma^2$ | $s^2$ | $\dfrac{(n-1)s^2}{\chi_R^2} < \sigma^2 < \dfrac{(n-1)s^2}{\chi_L^2}$ | See Table 6–1. |

# ? Review Exercises

1. Assume that the following statistics represent sample data randomly selected from a normally distributed population: $n = 60$, $\bar{x} = 83.2$ kg, $s = 4.1$ kg.
   a. What is the best point estimate of $\mu$?
   b. Construct the 95% confidence interval about $\mu$.

2. Use the sample data given in Exercise 1.
   a. What is the best point estimate of $\sigma^2$?
   b. Construct the 95% confidence interval about $\sigma$.

3. You want to determine the percentage of individual tax returns that include capital gains deductions. How many such returns must be randomly selected and checked? We want to be 90% confident that our sample percent is in error by no more than four percentage points.

4. Of 1475 transportation workers randomly selected, 32.0% belong to unions (based on data from the U.S. Bureau of Labor Statistics). Construct the 95% confidence interval for the true proportion of all transportation workers who belong to unions.

5. A medical researcher wishes to estimate the serum cholesterol level (in mg/100 ml) of all women aged 18 to 24. There is strong evidence suggesting that $\sigma = 41.0$ mg/100 ml (based on data from a survey of 1524 women aged 18 to 24, as part of the National Health Survey, USDHEW publication 78-1652). If the researcher wants to be 95% confident in obtaining a sample mean that is off by no more than four units, how large must the sample be?

6. A magazine reporter is conducting independent tests to determine the distance a certain car will travel while consuming only 1 gallon of gas. A sample of 5 cars is tested and a mean of 28.2 mi is obtained. Assuming that $\sigma = 2.7$ mi, find the 98% confidence interval for the mean distance traveled by all such cars using 1 gallon of gas.

7. Independent tests are conducted to determine the distance a car will travel while consuming only 1 gallon of gas. A sample of 5 cars is tested and the 5 distances have a mean of 28.2 mi and a standard deviation of 2.7 mi. Construct a 99% confidence interval for the standard deviation of distances traveled for all such cars.

8. A sociologist wants to determine the mean value of cars owned by retired people. If the sociologist wants to be 96% confident that the mean of the sample group is off by no more than $250, how many retired people must be sampled? A pilot study suggests that the standard deviation is $3050.

9. In clinical trials of the allergy medication Seldane, 70 subjects experienced drowsiness while 711 did not (based on data from Merrell Dow Pharmaceuticals). Construct the 95% confidence interval for the proportion of Seldane users who experience drowsiness.

10. A psychologist wants to determine the proportion of students in a large school district who have divorced parents. How many students must be surveyed if the psychologist wants 96% confidence that the sample proportion is in error by no more than 0.06?

11. a. Evaluate $z_{\alpha/2}$ for $\alpha = 0.10$.
    b. Evaluate $\chi^2_L$ and $\chi^2_R$ for $\alpha = 0.05$ and a sample of 10 scores.
    c. Evaluate $t_{\alpha/2}$ for $\alpha = 0.05$ and a sample of 10.
    d. What is the largest possible value of $p \cdot q$?

12. Assume that the following statistics represent sample data randomly selected from a normally distributed population: $n = 16$, $\bar{x} = 83.2$ kg, $s = 4.1$ kg.
    a. What is the best point estimate of $\mu$?
    b. Construct the 95% confidence interval about $\mu$.

13. Use the sample data given in Exercise 12.
    a. What is the best point estimate of $\sigma^2$?
    b. Construct the 95% confidence interval about $\sigma$.

14. Based on recent data from the U.S. Bureau of the Census, the proportion of Americans below the poverty level is 0.140. A researcher wants to verify that figure by conducting an independent survey. Assuming that 0.140 is approximately correct, how many randomly selected Americans must be surveyed? The researcher wants to be 96% confident that the sample proportion is within 0.015 of the true population proportion.

15. In a Roper survey of 1998 adults, 24% included loud commercials among the annoying aspects of television. Construct the 99% confidence interval for the proportion of all adults who are annoyed by loud commercials.

16. A botanist wants to determine the mean diameter of pine trees in a forest. She conducts a preliminary study to establish that the standard deviation of all such trees is about 6.35 cm. How many randomly selected trees must be measured if she wants 98% confidence that the sample mean is in error by no more than 0.5 cm?

17. The president of the student body at a very large university wants to determine the percent of students who are registered voters. How many students must be surveyed if we want 90% confidence that the sample is in error by no more than five percentage points?

18. a. Evaluate $z_{\alpha/2}$ for a confidence level of 96%.
    b. Evaluate $t_{\alpha/2}$ for a confidence level of 99% and a sample size of 16.
    c. Evaluate $\chi^2_L$ and $\chi^2_R$ for a confidence level of 99% and a sample size of 20.
    d. For the same set of data, confidence intervals are constructed for the 95% and 99% confidence levels. Which interval has limits that are farther apart?

19. In a random sample of 40 high school sophomores, scores on the math portion of the PSAT test are normally distributed with a mean of 34.5 and a standard deviation of 10.0 (based on data from Educational Testing Service). Construct the 95% confidence interval for the mean score for all high school sophomores.

20. Given the sample data in Exercise 19, construct the 95% confidence interval for the standard deviation of scores for all high school sophomores.

21. An advertising firm wants to estimate the mean time spent by preschool children watching television on Saturday morning. A pilot study suggests that $\sigma = 0.8$ h. How many subjects must be surveyed for 98% confidence that the sample mean is off by no more than 0.02 h?

22. In a Gallup poll of 1004 adults, 93% indicated that restaurants and bars should refuse service to patrons who have had too much to drink. Construct the 98% confidence interval for the proportion of all adults who feel the same way.

23. An employer measures arm lengths of a sample of male machine operators and the values given below (in centimeters) are obtained. Construct the 95% confidence interval for the mean arm length of all such employees.

| 76.8 | 75.6 | 69.3 | 75.7 | 75.5 | 71.2 | 72.5 | 71.9 |
| 70.9 | 69.4 | 71.7 | 72.5 | 72.2 | 68.5 | 75.9 | 73.0 |

24. Use the sample data given in Exercise 23 to find the 95% confidence interval for the standard deviation of the population from which the sample was drawn.

 ## Computer Projects

1. Computer software packages designed for statistics (such as STAT-DISK or Minitab) commonly provide programs for generating confidence intervals. Use such a program to find the confidence intervals referred to in the indicated exercises.

   Section 6–2: Exercises 37, 38, 39, 40
   Section 6–3: Exercises 13, 14, 15, 16
   Section 6–4: Exercises 17, 18, 19, 20

2. We know from this chapter that the formula

$$n = \frac{[z_{\alpha/2}]^2 \cdot 0.25}{E^2} = \frac{[1.96]^2 \cdot 0.25}{E^2} = \frac{0.9604}{E^2}$$

can be used to determine the sample size necessary to estimate a population proportion to within a maximum error $E$, and that result will correspond to a 95% confidence level. Develop a computer program that takes the value of $E$ as input. Output should consist of the sample size corresponding to a 95% confidence level. Run the program to determine the sample sizes corresponding to a variety of different values of $E$ ranging from 0.001 to 0.150.

3. The sample scores given below are randomly selected from a population that is very nonnormal, so the standard methods of this section don't apply. Instead, construct a 95% confidence interval about $\mu$ by using the **bootstrap** method as follows:

   a.  Create 1000 new samples, each of size 10, by selecting 10 scores (with replacement) from the sample scores given below.

   b.  Find the means of the 1000 bootstrap samples generated in part a.

   c.  Rank the 1000 means and then find the percentiles $P_{2.5}$ and $P_{97.5}$. These two values are the limits of the desired confidence interval. Identify the resulting confidence interval.

   2.9   564.2   1.4   4.7   67.6   4.8   51.3   3.6   18.0   3.6

4. Use the bootstrap method (see Computer Project 3) to find a 95% confidence interval about the population standard deviation $\sigma$. Compare your result to the interval

   $$318.4 < \sigma < 1079.6$$

   This interval was obtained by incorrectly using the standard methods described in Section 6–4; the true value of $\sigma$ is 232.1.

 # *Applied Projects*

1. Determine the sample size needed to estimate the mean weight of a passenger car in your region. The calculation of that number requires the value of the population standard deviation, which is unknown. Conduct a preliminary pilot study by finding the weights of at least 30 different passenger cars. Those weights can usually be found on vehicle registration documents. List the year and make of each vehicle and identify any factors suggesting that your sample is not representative of the population of all passenger cars in your region. Calculate the standard deviation of the sample weights and use the result as an estimate of the population standard deviation so that the required sample size can be determined. Assume that you want to be 98% confident that the sample mean is within 25 lb of the true population mean.

2. Refer to the data sets in Appendix B.
   a. Construct 95% and 99% confidence intervals for the mean home selling price.
   b. Construct 95% and 99% confidence intervals for the proportion of people who are left-handed.
   c. Construct 95% and 99% confidence intervals for the mean living area of homes recently sold in Dutchess County.
   d. You want to estimate the mean selling price of all homes in Dutchess County. Use the sample of 150 selling prices in order to estimate the value of the standard deviation. Then use that result to find the required sample size. Assume that you want to be 95% confident that you are within $5000 of the population mean.

# *Writing Projects*

1. The following excerpt is taken from an actual letter written by a corporation president and sent to the Associated Press.

   "When you or anyone else attempts to tell me and my associates that 1223 persons account for our opinions and tastes here in America, I get mad as hell! How dare you! When you or anyone else tells me that 1223 people represent America, it is astounding and unfair and should be outlawed."

   The writer goes on to state that since the sample size of 1223 represents 120 million people, then his letter represents 98,000 people (120 million divided by 1223) who share the same views. Write a response that addresses these points.

2. Using Data Set II in Appendix B, randomly select 36 females and construct a 95% confidence interval for the mean height of all females. Write a paragraph that explains how that confidence interval should be interpreted.

3. A cable news network conducts a survey by asking viewers to call a "900" phone number to register their opinions; each call costs 50¢. Assume that one survey results in 3000 responses, with 78% of the callers opposed to gun control. Write an argument supporting the correct position that it is not valid to use these results as being representative of the way all Americans feel.

4. Write a summary of the videotape program listed.

# *Videotapes*

Program 19 from the series *Against All Odds: Inside Statistics* is recommended as a supplement to this chapter.

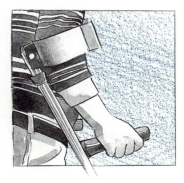

# POLIO EXPERIMENT

■ In 1954 a large-scale medical experiment was conducted to test the effectiveness of the Salk vaccine as a protection against the devastating effects of polio. Previously developed polio vaccines had been used, but it was discovered that some of those earlier vaccinations actually *caused* paralytic polio. Researchers justifiably developed a cautious and conservative approach to approving new vaccines for general use, and they decided to conduct a large-scale test of the Salk vaccine with volunteers.

Because of a variety of physiological factors, the vaccine could not be 100% effective, so its effectiveness had to be proved by a lowered incidence of polio among inoculated children. It was hoped that this experiment would result in a lower incidence of polio among vaccinated children that would be so significant as to be overwhelmingly convincing.

The number of children involved in this experiment was necessarily large because a small sample would not provide the conclusive evidence required by these cautious researchers. Approximately 200,000 children were injected with an ineffective salt solution, while 200,000 other children were injected with the Salk vaccine. Assignments of the real vaccine and the useless salt solution were made on a random basis. The children being injected did not know whether they were given the real vaccine or the salt solution. Even the doctors giving the injections and evaluating subsequent results did not know which injections contained the real Salk vaccine. Only 33 of the 200,000 vaccinated children later developed paralytic polio, while 115 of the 200,000 injected with the salt solution later developed paralytic polio. Statistical analysis of these and other results led to the conclusion that the Salk vaccine was indeed effective against paralytic polio.

The polio experiment is one example of a *double blind* experiment, which is characterized by the fact that neither the subjects nor those who evaluate results are aware of the true nature of the treatment. In a *single blind* experiment, only the subjects are not aware of the treatment they receive. Medical experiments are generally double blind because doctors are often more inclined to recognize "cures" when they know that the patient has been given the real medicine and not the placebo. (Please see Section 8-4.) ■

Photo Courtesy of Boeing Commercial Airplane Group

## AN INTERVIEW WITH
# DAVID HALL

### DIVISION STATISTICAL MANAGER AT THE BOEING COMMERCIAL AIRPLANE GROUP

*David Hall is Division Statistical Manager, Renton Division, Boeing Commercial Airplane Group. He manages the Statistical Methods Organization, which focuses on applying statistical and other quality technology techniques to continuous quality improvement. Before joining Boeing, he worked at Battelle Pacific Northwest Laboratories, where he was manager of a statistical applications group.*

**How extensive is the use of statistics at Boeing, and is it increasing, decreasing, or remaining about the same?**

The use of statistics is extensive and definitely increasing. I'm sure that this is true of the aircraft industry in general. People are becoming very aware of the need to improve, and to improve you must have data. Statistics is riding the wave of quality improvement. The use of control charts and, more generally, statistical process control, is increasing. Designed experiments are very common. In the beginning, 99% of the useful (statistical) tools are very, very simple. As the processes get refined and understanding increases, more sophisticated tools are required. Regression and correlation analysis, analysis of variance, contingency tables, hypothesis testing, confidence intervals, and time series analysis—virtually all techniques are used at some time. But given the state most American manufacturing is in now, incredible gains can be made with the simplest tools such as Pareto diagrams and run charts.

**How do you sample at Boeing?**

Our Quality Assurance Department uses many of the well-known sampling schemes for inspection. Currently, we check the daylights out of everything. However, sampling is also involved in statistical process control applications. The sampling can be quite complex and is handled on a case-by-case basis.

**Could you cite a specific example of how the use of statistics was helpful at Boeing?**

We were working with our Fabrication Division to produce more consistent hydropress formed parts. Through the use of designed experiments, we found that the type of rubber placed over the blanks during forming could drastically reduce the part-to-part variation. That's one simple example of what goes on all the time. An example of a more sophisticated application was the use of bootstrapping to estimate the variability in wind tunnel tests.

*"Statistics is riding the wave of quality improvement."*

**Do you feel job applicants are viewed more favorably if they have studied some statistics?**

We naturally have a great many engineers at Boeing and almost all of them have studied some statistics. We are now expecting our managers to have more familiarity with statistics than ever before. They are expected to understand variation and how to effectively use data.

**Do you have any advice for today's students?**

When I get a new statistician just out of school, whether they have a B.S., M.S., or PH.D., they still have a tremendous amount to learn before they are effective in our environment. Most of what they need involves people skills, team building, planning, and communication. Right now, American industry is crying out for people with an understanding of statistics and the ability to communicate its use. ■

# Chapter Seven

## *In This Chapter*

# 7  Testing Hypotheses

## Chapter Problem

Recently, an Aloha Airlines flight between two Hawaiian islands survived the disturbing incident of losing part of its fuselage. A 20-ft hole left several passengers exposed to open air. The cause of the structural failure of the Boeing 737 jet was apparently due to metal fatigue; the jet had been in service for 19 years. Subsequent investigations raised concerns about the effects of aircraft age on aviation safety.

Suppose an industry representative tries to argue that U.S. commercial aircraft really aren't too old by making this claim: "The mean age of U.S. commercial aircraft is only 10 years or less." Listed below are the ages (in years) of 40 randomly selected U.S. commercial aircraft (based on data from Aviation Data Services).

$$
\left.
\begin{array}{cccccccc}
3.2 & 22.6 & 23.1 & 16.9 & 0.4 & 6.6 & 12.5 & 22.8 \\
26.3 & 8.1 & 13.6 & 17.0 & 21.3 & 15.2 & 18.7 & 11.5 \\
4.9 & 5.3 & 5.8 & 20.6 & 23.1 & 24.7 & 3.6 & 12.4 \\
27.3 & 22.5 & 3.9 & 7.0 & 16.2 & 24.1 & 0.1 & 2.1 \\
7.7 & 10.5 & 23.4 & 0.7 & 15.8 & 6.3 & 11.9 & 16.8
\end{array}
\right\}
\begin{array}{l}
n = 40 \\
\bar{x} = 13.41 \\
s = 8.28
\end{array}
$$

In Chapter 6 we used this collection of sample data to *estimate* the mean age of U.S. commercial aircraft, but in this chapter we will use the sample data to *test the claim* made by the industry representative. The sample data have a mean of 13.41 years and this seems to contradict the claim that the mean age is 10 years or less. But is the sample mean of 13.41 years *significantly* greater than 10 years? Could it be that the population mean really is 10 years or less and the above sample results are due to chance fluctuations? This chapter provides us with the ability to answer such questions so that decisions can be made about a variety of different claims.

# $\boxed{7-1}$ Overview

In Chapter 6 we studied a major topic of inferential statistics—using sample statistics to *estimate* values of population parameters. In this chapter we study another major topic of inferential statistics as we use sample statistics to *test hypotheses* made about population parameters. In statistics, a **hypothesis** is a statement that something is true. The following statements are examples of hypotheses that will be tested by the procedures we develop in this chapter.

- An airline industry representative claims that the mean age of U.S. commercial aircraft is 10 years or less.

- A television executive claims that the majority of all adults are not annoyed by violence on television.

- A bank president claims that with a single line, customers have more consistent waiting times, with less than the 6.2 min standard deviation for multiple waiting lines.

Before beginning Section 7–2, it would be very helpful to have a general sense of the thinking used in **hypothesis tests,** also called **tests of significance.** Try to follow the reasoning behind the following example.

Suppose you take a dime from your pocket and claim that it favors heads when it is flipped. That claim is a hypothesis, and we can test it by flipping the dime 100 times. We would expect to get around 50 heads with a fair coin. If heads occur 94 times out of 100 tosses, most people would agree that the coin favors heads. If heads occur 51 times out of 100 tosses, we should not conclude that the dime favors heads since we could easily get 51 heads with a fair and unbiased coin. Here is the key point: We should conclude that the dime favors heads only if we get *significantly* more heads than we would expect with an unbiased dime.

This brief example illustrates the basic approach used in testing hypotheses. That approach involves a variety of standard terms and conditions in the context of an organized procedure. We suggest that you begin the study of this chapter by first reading Section 7–2 casually to obtain a general idea of its concepts. Then read the material more carefully to gain familiarity with the terminology. Subsequent readings should incorporate the details and refinements into the basic procedure. You are not expected to master the principles of hypothesis testing in one reading. It may take several readings for the material to become understandable.

# 7–2 Testing a Claim About a Mean

We begin with the problem described at the beginning of this chapter. Here are the key components:

- An airline industry representative claims that the ages of U.S. commercial aircraft have a mean of 10 years or less.

- A *sample* of 40 randomly selected U.S. commercial aircraft has a mean of $\bar{x} = 13.41$ years and a standard deviation of $s = 8.28$ years.

This is the key issue: Is the sample mean of 13.41 large enough for us to conclude that the population mean cannot be 10 or less as claimed? Or is the discrepancy more likely due to chance variations in samples?

Refer now to Figure 7–1 (page 334), which outlines the general procedure for testing hypotheses. Accompanying Figure 7–1 is the application of the general procedure to the specific problem we are considering. Carefully follow each step as we demonstrate how the given sample data lead us to reject the claim that the mean age of U.S. commercial aircraft is 10 years or less. Try to maintain a sense of the real issue: There is a claim that the population mean is 10 or less, but that claim is brought into question when a sample mean of $\bar{x} = 13.41$ is obtained. We need to determine if that discrepancy is *significant*, or if it could be explained away as a typical chance fluctuation. The critical value sets up a cutoff point for what is significant, while the test statistic allows us to see where $\bar{x} = 13.41$ lies relative to that cutoff point.

As outlined in Figure 7–1, the procedure of hypothesis testing uses these standard terms:

- **Null hypothesis** (denoted by $H_0$): The statement of a zero or null difference that is directly tested. This will correspond to the original claim if that claim includes the condition of no change or difference (such as $=, \leq, \geq$). Otherwise, the null hypothesis is the negation of the original claim. We test the null hypothesis directly in the sense that the final conclusion will be either rejection of $H_0$ or failure to reject $H_0$.

- **Alternative hypothesis** (denoted by $H_1$): The statement that must be true if the null hypothesis is false.

- **Type I error:** The mistake of rejecting the null hypothesis when it is true.

- **Type II error:** The mistake of failing to reject the null hypothesis when it is false.

- $\alpha$ **(alpha):** Symbol used to represent the probability of a type I error.

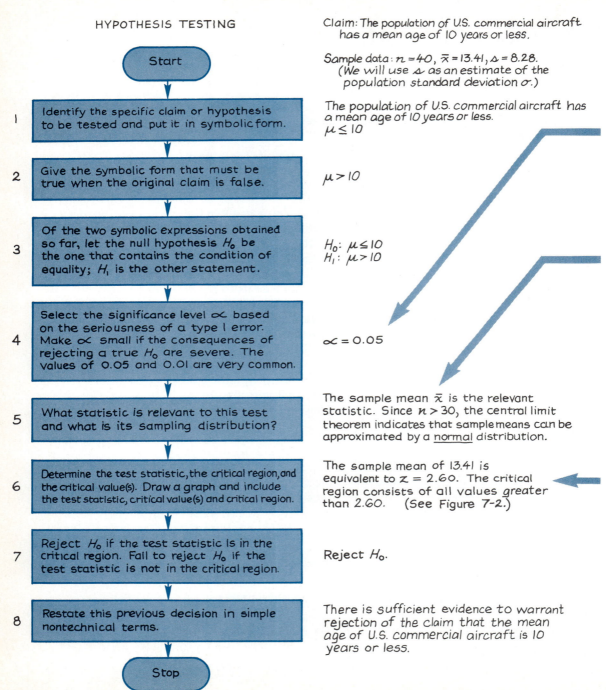

HYPOTHESIS TESTING

**Start**

1   Identify the specific claim or hypothesis to be tested and put it in symbolic form.

2   Give the symbolic form that must be true when the original claim is false.

3   Of the two symbolic expressions obtained so far, let the null hypothesis $H_0$ be the one that contains the condition of equality; $H_1$ is the other statement.

4   Select the significance level $\alpha$ based on the seriousness of a type I error. Make $\alpha$ small if the consequences of rejecting a true $H_0$ are severe. The values of 0.05 and 0.01 are very common.

5   What statistic is relevant to this test and what is its sampling distribution?

6   Determine the test statistic, the critical region, and the critical value(s). Draw a graph and include the test statistic, critical value(s) and critical region.

7   Reject $H_0$ if the test statistic is in the critical region. Fail to reject $H_0$ if the test statistic is not in the critical region.

8   Restate this previous decision in simple nontechnical terms.

**Stop**

Claim: The population of U.S. commercial aircraft has a mean age of 10 years or less.

Sample data: $n = 40$, $\bar{x} = 13.41$, $s = 8.28$. (We will use $s$ as an estimate of the population standard deviation $\sigma$.)

The population of U.S. commercial aircraft has a mean age of 10 years or less.
$\mu \leq 10$

$\mu > 10$

$H_0$: $\mu \leq 10$
$H_1$: $\mu > 10$

$\alpha = 0.05$

The sample mean $\bar{x}$ is the relevant statistic. Since $n > 30$, the central limit theorem indicates that sample means can be approximated by a <u>normal</u> distribution.

The sample mean of 13.41 is equivalent to $z = 2.60$. The critical region consists of all values *greater* than 2.60.   (See Figure 7-2.)

Reject $H_0$.

There is sufficient evidence to warrant rejection of the claim that the mean age of U.S. commercial aircraft is 10 years or less.

**Figure 7-1**

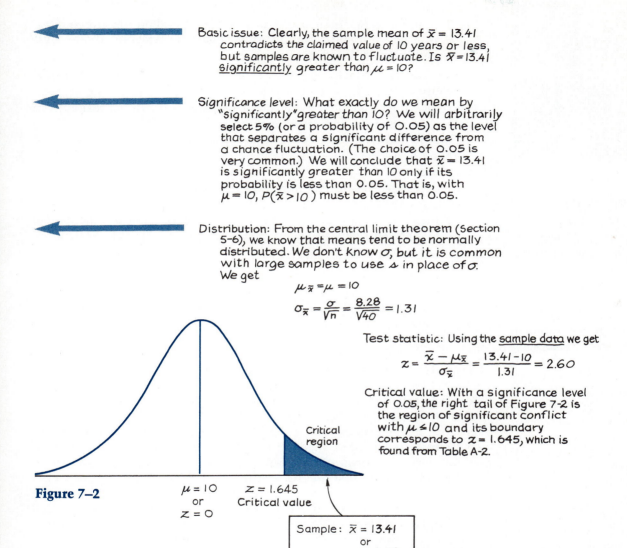

Basic issue: Clearly, the sample mean of $\bar{x} = 13.41$
contradicts the claimed value of 10 years or less,
but samples are known to fluctuate. Is $\bar{x} = 13.41$
<u>significantly</u> greater than $\mu = 10$?

Significance level: What exactly do we mean by
"significantly" greater than 10? We will arbitrarily
select 5% (or a probability of 0.05) as the level
that separates a significant difference from
a chance fluctuation. (The choice of 0.05 is
very common.) We will conclude that $\bar{x} = 13.41$
is significantly greater than 10 only if its
probability is less than 0.05. That is, with
$\mu = 10$, $P(\bar{x} > 10)$ must be less than 0.05.

Distribution: From the central limit theorem (Section
5-6), we know that means tend to be normally
distributed. We don't know $\sigma$, but it is common
with large samples to use $s$ in place of $\sigma$.
We get

$$\mu_{\bar{x}} = \mu = 10$$

$$\sigma_{\bar{x}} = \frac{\sigma}{\sqrt{n}} = \frac{8.28}{\sqrt{40}} = 1.31$$

Test statistic: Using the <u>sample data</u> we get

$$z = \frac{\bar{x} - \mu_{\bar{x}}}{\sigma_{\bar{x}}} = \frac{13.41 - 10}{1.31} = 2.60$$

Critical value: With a significance level
of 0.05, the right tail of Figure 7-2 is
the region of significant conflict
with $\mu \le 10$ and its boundary
corresponds to $z = 1.645$, which is
found from Table A-2.

Critical
region

**Figure 7–2**

$\mu = 10$
or
$z = 0$

$z = 1.645$
Critical value

Sample: $\bar{x} = 13.41$
or
$z = 2.60$    Test statistic

Conclusion: Figure 7-2 shows that $\bar{x} = 13.41$ is
significantly greater than 10. We reject the claim
that the mean age is 10 years or less. There is
sufficient sample evidence to warrant rejection
of the representative's claim that the mean age
is 10 years or less.

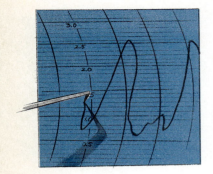

## LIE DETECTORS

■ Why not require all criminal suspects to take lie detector tests and dispense with trials by jury? The Council of Scientific Affairs of the American Medical Association states, "It is established that classification of guilty can be made with 75% to 97% accuracy, but the rate of false positives is often sufficiently high to preclude use of this (polygraph) test as the sole arbiter of guilt or innocence." A "false positive" is an indication of guilt when the subject is actually innocent. Even with accuracy as high as 97%, the percentage of false positive results can be 50%, so that half of the innocent subjects incorrectly appear to be guilty. ■

- **β (beta):** Symbol used to represent the probability of a type II error.
- **Test statistic:** A sample statistic or a value based on the sample data. It is used in making the decision about the rejection of the null hypothesis.
- **Critical region:** The set of all values of the test statistic that would cause us to reject the null hypothesis.
- **Critical value(s):** The value(s) that separates the critical region from the values of the test statistic that would not lead to rejection of the null hypothesis. The critical value(s) depends on the nature of the null hypothesis, the relevant sampling distribution, and the level of significance $\alpha$.
- **Significance level:** The probability of rejecting the null hypothesis when it is true. Typical values selected are 0.05 and 0.01. That is, the values of $\alpha = 0.05$ and $\alpha = 0.01$ are typically used. (We use the symbol $\alpha$ to represent the significance level.)
- **Elation:** The feeling experienced when the techniques of hypothesis testing are mastered.

In addition to knowing the general procedure outlined in Figure 7–1 and the preceding terms, you should be aware of some other details. We will first consider these topics and then present two additional examples.

## Null and Alternative Hypotheses

From steps 1, 2, and 3 of Figure 7–1 we see how to determine the null and alternative hypotheses. Note that the original claim may be the null or alternative hypothesis, depending on how it is stated. If we are making our own claims, we should arrange the null and alternative hypotheses so that the most serious error is a type I error (rejecting a true null hypothesis). In this text we assume that we are testing a claim made by someone else. Ideally, all claims would be made so that they would all be null hypotheses. Unfortunately, our real world is not ideal. There is poverty, war, crime, and people who make claims that are actually alternative hypotheses. This text was written with the understanding that not all original claims are as they should be. As a result, some examples and exercises involve claims that are null hypotheses whereas others involve claims that are alternative hypotheses.

In conducting a formal statistical hypothesis test, we are *always* testing the *null hypothesis*, whether it corresponds to the original claim or not. Sometimes the null hypothesis corresponds to the original claim and sometimes it corresponds to the opposite of the original claim. Since we always test the null hypothesis, we will be testing the original claim in some cases and the opposite of the original claim in other cases. Carefully examine the examples in the box below.

(Regarding notation: Even though we may write $H_0$ with the symbols $\leq$ or $\geq$ as in $H_0$: $\mu \leq 10$ or $H_0$: $\mu \geq 10$, we conduct the test by assuming that $H_0$: $\mu = 10$ is true. We must have a fixed and specific value for $\mu$ so that we can work with one particular distribution.)

| | **Original Claim** | | | |
|---|---|---|---|---|
| | The mean grade is 75. | The mean grade is not 75. | The mean grade is at least 75. | The mean grade is above 75. |
| Step 1: Symbolic form of original claim. | $\mu = 75$ | $\mu \neq 75$ | $\mu \geq 75$ | $\mu > 75$ |
| Step 2: Symbolic form that is true when original claim is false. | $\mu \neq 75$ | $\mu = 75$ | $\mu < 75$ | $\mu \leq 75$ |
| Step 3: Null hypothesis $H_0$ (must contain equality). | $H_0$: $\mu = 75$ | $H_0$: $\mu = 75$ | $H_0$: $\mu \geq 75$ | $H_0$:$\mu \leq 75$ |
| Alternative hypothesis $H_1$ (cannot contain equality). | $H_1$: $\mu \neq 75$ | $H_1$: $\mu \neq 75$ | $H_1$: $\mu < 75$ | $H_1$: $\mu > 75$ |

# Type I and Type II Errors

From Table 7–1 on page 338 we see that the conclusion in a hypothesis test may be correct or wrong. A type I error is the mistake of rejecting a true null hypothesis. A type II error is the mistake of failing to reject a false null hypothesis. The probability of a type I error is the significance level $\alpha$ and the probability of a type II error is denoted by $\beta$.

One step in our procedure for testing hypotheses involves the selection of $P(\text{type I error}) = \alpha$, but we don't select $P(\text{type II error}) = \beta$. We could select both $\alpha$ and $\beta$—the required sample size would then be determined—but the usual procedure used in research and industry is to determine in advance the values of $\alpha$ and $n$, so that the value of $\beta$ is determined. Based on the seriousness of a type I error, try to use the largest $\alpha$ that you can tolerate. The following practical considerations may be relevant to some experiments:

1. To decrease both $\alpha$ and $\beta$, increase the sample size $n$.
2. For any fixed sample size $n$, a decrease in $\alpha$ will cause an increase in $\beta$. Conversely, an increase in $\alpha$ will cause a decrease in $\beta$. To further consider $\beta$, see Exercise 32.

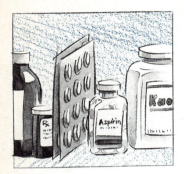

## DRUG APPROVAL

■ The Pharmaceutical Manufacturing Association has reported that the development and approval of a new drug costs around $87 million and takes about eight years. Extensive laboratory testing is followed by FDA approval for human testing, which is done in three phases. Phase I human testing involves about 80 people, while phase II involves about 250 people. In phase III, between 1000 and 3000 volunteers are used. Overseeing such a complex, extensive, and time-consuming process would be enough to give anyone a headache, but the process does protect us from dangerous or worthless drugs. ■

| TABLE 7–1 | | | |
|---|---|---|---|
| | | True State of Nature | |
| | | The null hypothesis is true. | The null hypothesis is false. |
| Decision | We decide to reject the null hypothesis. | Type I error | Correct decision |
| | We fail to reject the null hypothesis. | Correct decision | Type II error |

## Conclusions

We have already noted that the original claim sometimes becomes the null hypothesis and sometimes it becomes the alternative hypothesis. However, our procedure requires that we always test the null hypothesis. In Step 7 of Figure 7–1 we can see that our initial conclusion will always be one of the following:

1.   Fail to reject the null hypothesis, $H_0$.
2.   Reject the null hypothesis, $H_0$.

Some texts say that we "accept the null hypothesis" instead of "fail to reject the null hypothesis." Whether we use *accept* or *fail to reject*, we should recognize that *we are not proving the null hypothesis;* we are merely saying that the sample evidence is not strong enough to warrant rejection of the null hypothesis. It's like a jury saying that there is not enough evidence to convict a suspect. The term *accept* is somewhat misleading since it seems incorrectly to imply that the null hypothesis has been proved. The phrase *fail to reject* says, more correctly, "Let's withhold judgment because the available evidence isn't strong enough." In this text, we will use *fail to reject the null hypothesis* instead of *accept the null hypothesis*.

We either fail to reject the null hypothesis or we reject the null hypothesis. Such a conclusion is fine for those of us with the wisdom to take a statistics course, but it's usually necessary to use simple nontechnical terms in stating what the conclusion suggests. Figure 7–3 shows how to formulate the correct wording of the final conclusion. Note that only one case leads to wording indicating that the sample data actually *support* the conclusion. If you want to justify some claim, state it in such a way that it becomes the *alternative* hypothesis, and then hope that the null hypothesis gets rejected.

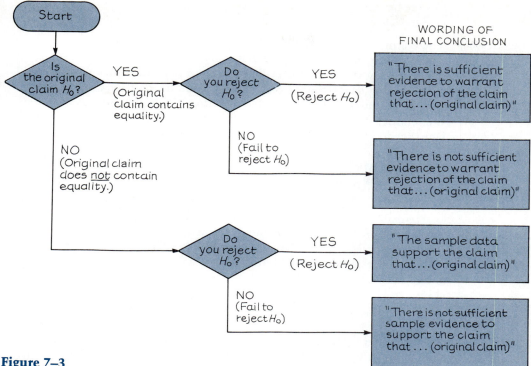

**Figure 7–3**

# Left-Tailed, Right-Tailed, Two-Tailed

Our first example of hypothesis testing involved a **right-tailed** test in the sense that the critical region of Figure 7–2 is in the extreme right region under the curve. We reject the null hypothesis $H_0$ if our test statistic is in the critical region, because that indicates a significant discrepancy between the null hypothesis and the sample data. Some tests will be **left-tailed,** with the critical region located in the extreme left region under the curve. Other tests may be **two-tailed,** because the critical region is comprised of two components located in the two extreme regions under the curve. *In the two-tailed case, $\alpha$ is divided equally between the two tails that constitute the critical region.*

By examining the null hypothesis $H_0$, we should be able to *deduce* whether a test is right-tailed, left-tailed, or two-tailed. The tail will correspond to the critical region where you have the values that would conflict significantly with the null hypothesis. A useful check is summarized in the accompanying box, which shows how the inequality sign in $H_1$ points in the direction of the critical region. The symbol $\neq$ is often expressed in programming languages as $<\,>$, and this reminds us that an alternative hypothesis such as $\mu \neq 10$ corresponds to a *two*-tailed test.

| Sign used in $H_1$ | Type of test |
|---|---|
| $>$ | Right-tailed |
| $<$ | Left-tailed |
| $\neq$ | Two-tailed |

### Example

The engineering department of a car manufacturer claims that the fuel consumption rate of one model is equal to 35 mi/gal. The advertising department wants to test this claim to see if the announced figure should be higher or lower than 35 mi/gal. The quality control group suggests that $\sigma = 4$ mi/gal, and a sample of 50 cars yields $\bar{x} = 33.6$ mi/gal. Test the claim of the engineering department by using a 0.05 level of significance.

### Solution

We outline our test of the manufacturer's claim by following the scheme in Figure 7–1. The result is shown in Figures 7–4 and 7–5. The test is two-tailed because a sample mean significantly greater than 35 mi/gal (right tail) or less than 35 mi/gal (left tail) is strong evidence against the null hypothesis that $\mu = 35$ mi/gal. Our sample mean of 33.6 mi/gal is found to be equivalent to $z = -2.47$ through the following computation:

$$z = \frac{\bar{x} - \mu_{\bar{x}}}{\sigma_{\bar{x}}} = \frac{33.6 - 35}{4/\sqrt{50}} = -2.47$$

The critical $z$ values are found by distributing $\alpha = 0.05$ equally between the two tails to get 0.025 in each tail. We then refer to Table A–2 (since we are assuming a normal distribution) to find the $z$ value corresponding to $0.5 - 0.025$ or 0.4750. After finding $z = 1.96$, we use the property of symmetry to conclude that the left critical value is $-1.96$. See Figure 7–5 for the conclusions.

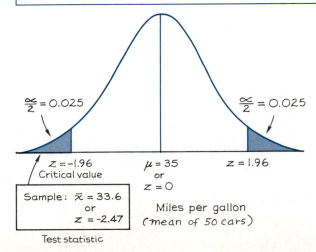

**Figure 7–4**

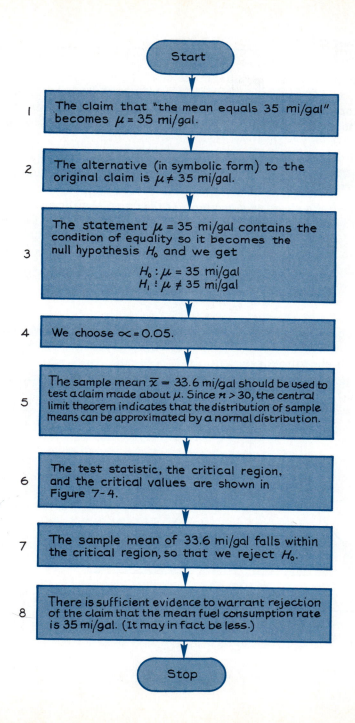

**Figure 7–5**

## Example

A brewery distributes beer in bottles labeled 32 oz. The local Bureau of Weights and Measures randomly selects 50 of these bottles, measures their contents, and obtains a sample mean of 31.0 oz. Assuming that $\sigma$ is known to be 0.75 oz, is it valid at the 0.01 significance level to conclude that the brewery is cheating consumers?

## Solution

The brewery is cheating consumers if they give significantly less than 32 oz of beer. We outline the test of the claim that the mean is less than 32 oz by again following the model of Figure 7–1. The results are presented in Figures 7–6 and 7–7. The $z$ value of $-9.43$ is computed as follows:

$$z = \frac{\bar{x} - \mu_{\bar{x}}}{\sigma_{\bar{x}}} = \frac{31 - 32}{0.75/\sqrt{50}} = -9.43$$

The critical $z$ value is found in Table A–2 as the $z$ value corresponding to an area of 0.4900. The conclusion in Figure 7–7 does suggest that the brewery is cheating consumers.

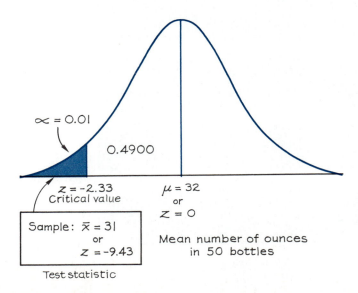

**Figure 7–6**

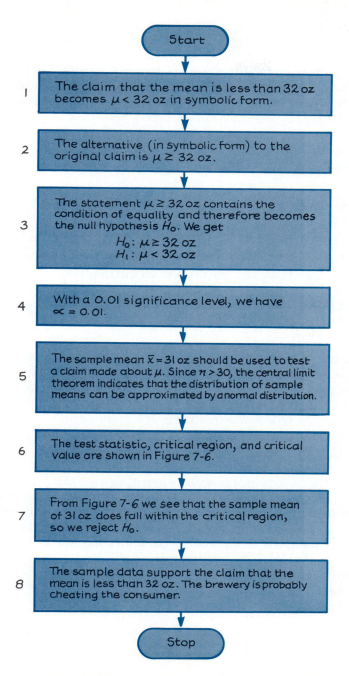

Start

1  The claim that the mean is less than 32 oz becomes $\mu < 32$ oz in symbolic form.

2  The alternative (in symbolic form) to the original claim is $\mu \geq 32$ oz.

3  The statement $\mu \geq 32$ oz contains the condition of equality and therefore becomes the null hypothesis $H_0$. We get
$$H_0: \mu \geq 32 \text{ oz}$$
$$H_1: \mu < 32 \text{ oz}$$

4  With a 0.01 significance level, we have $\alpha = 0.01$.

5  The sample mean $\bar{x} = 31$ oz should be used to test a claim made about $\mu$. Since $n > 30$, the central limit theorem indicates that the distribution of sample means can be approximated by a normal distribution.

6  The test statistic, critical region, and critical value are shown in Figure 7-6.

7  From Figure 7-6 we see that the sample mean of 31 oz does fall within the critical region, so we reject $H_0$.

8  The sample data support the claim that the mean is less than 32 oz. The brewery is probably cheating the consumer.

Stop

**Figure 7–7**

In presenting the results of a hypothesis test, it is not always necessary to show all of the steps included in Figure 7–7 on the previous page. However, the results should include the null hypothesis, the alternative hypothesis, the calculation of the test statistic, a graph such as Figure 7–6, and the initial conclusion (reject $H_0$ or fail to reject $H_0$) and the final conclusion stated in nontechnical terms. The graph should show the test statistic, critical value(s), critical region, and significance level.

## Assumptions

For the examples and exercises in this section, we are working with these assumptions.

1.    The claim is made about the mean of a single population.
2.    a.    The sample is large ($n > 30$), so that the central limit theorem applies and we can use the normal distribution.
      or
      b.    If the sample is small ($n \leq 30$), then the population is normally distributed and the value of the population standard deviation $\sigma$ is known.

A potentially unrealistic feature of some examples and exercises from this section is the assumption that $\sigma$ is known. Realistic tests of hypotheses must often be made without knowledge of the population standard deviation. *If the sample is large* ($n > 30$), *we can compute the sample standard deviation and we may be able to use that value of s as an estimate of* $\sigma$. When $\sigma$ is not known and the sample is small ($n \leq 30$), we may be able to use the $t$ statistic discussed in Section 7–4.

It is easy to become entangled in a complex web of steps without ever understanding the underlying rationale of hypothesis testing. The key to that understanding lies with recognition of this concept: **If an event can easily occur, we attribute it to chance, but if the event appears to be unusual, we attribute that significant departure to the presence of characteristics different from those assumed to be true.** If we keep this idea in mind as we examine various examples, hypothesis testing will become meaningful instead of a rote mechanical process.

There. Wasn't that easy?

# 7–2 Exercises A

In Exercises 1 and 2, read the given claim and identify $H_0$ and $H_1$ as in the following example: The mean IQ of doctors is greater than 110.

$$H_0: \mu \leq 110$$
$$H_1: \mu > 110$$

1. a. The mean age of professors is more than 30 years.
   b. The mean IQ of criminals is above 100.
   c. The mean IQ of college students is at least 100.
   d. The mean annual household income is at least $12,300.
   e. The mean monthly maintenance cost of an aircraft is $3271.
2. a. The mean height of females is 1.6 m.
   b. The mean annual salary of air traffic controllers is more than $26,000.
   c. The mean annual salary of college presidents is under $50,000.
   d. The mean weight of girls at birth is at most 3.2 kg.
   e. The mean life of a car battery is not more than 46 months.
3. Identify the type I error and the type II error corresponding to each claim in Exercise 1.
4. Identify the type I error and the type II error for each claim in Exercise 2.
5. For each claim in Exercise 1, categorize the hypothesis test as a right-tailed test, a left-tailed test, or a two-tailed test.
6. For each claim in Exercise 2, categorize the hypothesis test as a right-tailed test, a left-tailed test, or a two-tailed test.

In Exercises 7 and 8, find the critical $z$ value for the given conditions. In each case assume that the normal distribution applies, so that Table A–2 can be used. Also, draw a graph showing the critical value and critical region.

7. a. Right-tailed test; $\alpha = 0.05$
   b. Right-tailed test; $\alpha = 0.01$
   c. Two-tailed test; $\alpha = 0.05$
   d. Two-tailed test; $\alpha = 0.01$
   e. Left-tailed test; $\alpha = 0.05$

8. a. Left-tailed test; $\alpha = 0.02$
   b. Two-tailed test; $\alpha = 0.10$
   c. Right-tailed test; $\alpha = 0.005$
   d. Right-tailed test; $\alpha = 0.025$
   e. Left-tailed test; $\alpha = 0.025$

In each of the following exercises, test the given hypotheses by following the procedure suggested by Figure 7–1. Draw the appropriate graph, as in Figure 7–2.

9. Test the claim that $\mu \leq 100$ given a sample of $n = 81$ for which $\overline{x} = 100.8$. Assume that $\sigma = 5$, and test at the $\alpha = 0.01$ significance level.

10. Test the claim that $\mu \leq 40$ given a sample of $n = 150$ for which $\overline{x} = 41.6$. Assume that $\sigma = 9$, and test at the $\alpha = 0.01$ significance level.

11. Test the claim that $\mu \geq 20$ given a sample of $n = 100$ for which $\overline{x} = 18.7$. Assume that $\sigma = 3$, and test at the $\alpha = 0.05$ significance level.

12. Test the claim that $\mu \geq 15.5$ given a sample of $n = 45$ for which $\overline{x} = 14.3$. Assume that $\sigma = 5.5$, and test at the $\alpha = 0.05$ significance level.

13. Test the claim that a population mean equals 500. You have a sample of 300 items for which the sample mean is 510. Assume that $\sigma = 100$, and test at the $\alpha = 0.10$ significance level.

14. Test the claim that a population mean equals 65. You have a sample of 50 items for which the sample mean if 66.1 Assume that $\sigma = 4$, and test at the $\alpha = 0.05$ significance level.

15. Test the claim that a population mean exceeds 40. You have a sample of 50 items for which the sample mean is 42. Assume that $\sigma = 8$, and test at the $\alpha = 0.05$ significance level.

16. Test the claim that a population mean is less than 75.0. You have a sample of 32 items for which the sample mean is 73.8. Assume that $\sigma = 4.2$ and test at the $\alpha = 0.10$ significance level.

17. Use the aircraft age sample data ($n = 40$, $\overline{x} = 13.41$ years, $s = 8.28$ years) given at the beginning of the chapter. At the 0.05 level of significance, test the *Time* magazine claim that the mean age of aircraft in the U.S. fleet is 14 years. Assume that the sample standard deviation can be used for $\sigma$.

18. A brewery distributes beer in cans labeled 12 oz. The Bureau of Weights and Measures randomly selects 36 cans, measures their contents, and obtains a sample mean of 11.82 oz. Assuming that $\sigma$ is known to be 0.38 oz, is it valid at the 0.01 significance level to conclude that the brewery is cheating consumers?

19. In a study of distances traveled by buses before the first major engine failure, a sample of 191 buses results in a mean of 96,700 mi and a standard deviation of 37,500 mi (based on data in *Technometrics*, Vol. 22, No. 4). At the 0.05 level of significance, test the claim that

mean distance traveled before a major engine failure is more than 90,000 mi. (Assume that the sample standard deviation can be used for $\sigma$.)

20. When 150 randomly selected boys aged 6 to 11 are given the reading portion of the Wide Range Achievement Test, their mean score is 52.4 and the standard deviation is 13.14 (based on data from the National Health Survey, USDHEW publication 72-1011). At the 0.05 level of significance, test the claim that this sample is from a population with a mean greater than 51.0. (Assume that the sample standard deviation can be used for $\sigma$.)

21. A poll of 100 randomly selected car owners revealed that the mean length of time they plan to keep their car is 7.01 years and the standard deviation is 3.74 years (based on data from a Roper poll). Test the claim that the mean for all car owners is less than 7.5 years. (Assume that the sample standard deviation can be used for $\sigma$.)

22. When 200 convicted embezzlers are randomly selected, the mean length of prison sentence is found to be 22.1 months (based on data from the U.S. Department of Justice). Assuming that $\sigma$ is known to be 8.6 months, test the claim that prison terms for convicted embezzlers have a mean less than 2 years.

23. A paint is applied to tin panels and baked for 1 h so that the mean index of hardness is 35.2. Suppose 38 test panels are painted and baked for 3 h, producing a sample mean index of hardness equal to 35.9. Assuming that $\sigma = 2.7$, test (at the $\alpha = 0.05$ significance level) the claim that longer baking does not affect hardness of the paint.

24. A certain nighttime cold medicine bears a label indicating the presence of 600 mg of acetaminophen in each fluid ounce of the drug. The Food and Drug Administration randomly selects 65 1-oz samples and finds that the mean acetaminophen content is 589 mg, while the standard deviation is 21 mg. With $\alpha = 0.01$, test the claim that the population mean is equal to 600 mg. (Assume that the sample standard deviation can be used for $\sigma$.)

25. The mean time between failures (in hours) for a certain type of radio used in light aircraft is 420 h. Suppose 35 new radios have been modified for more reliability, and tests show that the mean time between failures for this sample is 385 h. Assume that $\sigma$ is known to be 24 h and let $\alpha = 0.05$. Test the claim that the modifications improved reliability. (Note that improved reliability should correspond to a *longer* mean time between failures.)

26. In an insurance study of driving habits, 750 female drivers aged 16–24 are randomly selected and their mean driving distance for one year is

6047 mi (based on data from the Federal Highway Administration). Assuming that $\sigma$ is known to be 2944 mi, use a 0.05 significance level to test the claim that the population mean for women in this age bracket is less than 7124 mi, which is the known mean for females in the 25–34 age bracket.

27. In the article "Multiple Spans in Transcription Typing" (*Journal of Applied Psychology*, Vol. 72, No. 2), data were given for a sample of 45 typists. Their mean normal typing score is 182, while the standard deviation is 52. Test the claim that the sample is from a population with a mean of 180. (Assume that the sample standard deviation can be used for $\sigma$.)

28. A late-night television show is seen by a relatively large percentage of household members who videotape the show for viewing at a more convenient time. The show's marketing manager claims that the mean income of households with VCRs is greater than $40,000. Test that claim. A sample of 1700 households with VCRs produces a sample mean of $41,182 (based on data from Nielsen Media Research). Assume that $\sigma$ is known to be $19,990.

## 7–2 Exercises B

29. A brewery claims that the consumers are getting a mean volume equal to 32 oz of beer in their quart bottles. The Bureau of Weights and Measures randomly selects 36 bottles and obtains the following measures in ounces:

| | | | | | | |
|---|---|---|---|---|---|---|
| 32.09 | 31.89 | 31.06 | 32.03 | 31.42 | 31.39 | 31.75 |
| 31.53 | 32.42 | 31.56 | 31.95 | 32.00 | 31.39 | 32.09 |
| 31.67 | 31.47 | 32.45 | 32.14 | 31.86 | 32.09 | 32.34 |
| 32.00 | 30.95 | 33.53 | 32.17 | 31.81 | 31.78 | 32.64 |
| 31.06 | 32.64 | 32.20 | 32.11 | 31.42 | 32.09 | 33.00 |
| 32.06 | | | | | | |

Using the sample standard deviation as an estimate for $\sigma$, test the claim of the brewery at the 0.05 significance level.

30.  The accompanying frequency table summarizes the num-
     bers of defective units produced by a machine for a sample
     of randomly selected days. Find $\bar{x}$ and $s$ for this sample and
     use $s$ as an estimate of $\sigma$. Then test the claim that the
     mean number of defective parts per day is equal to 30.0.
     Use a 0.04 level of significance.

| $x$ | $f$ |
|------|-----|
| 0–4 | 2 |
| 5–9 | 0 |
| 10–14 | 5 |
| 15–19 | 8 |
| 20–24 | 12 |
| 25–29 | 17 |
| 30–34 | 20 |
| 35–39 | 14 |
| 40–44 | 6 |
| 45–49 | 3 |

31.  When a sample of 93 IQ scores is randomly selected from a certain
     population, the results are described by the accompanying boxplot. At
     the 0.03 level of significance, test the claim that the population has a
     mean IQ score equal to 105.

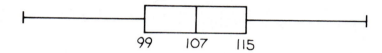

99      107      115

32.  **The probability $\beta$ of a type II error:** For a given hypothesis test, the
     probability $\alpha$ of a type I error is fixed, whereas the probability $\beta$ of a
     type II error depends on the particular value of $\mu$ that is used as an
     alternative to the null hypothesis. For hypothesis tests of the type found
     in this section, we can find $\beta$ as follows.

     1.  Find the value of $\bar{x}$ that corresponds to the critical value. In

     $$z = \frac{\bar{x} - \mu_{\bar{x}}}{\sigma_{\bar{x}}}$$

     substitute the critical score for $z$, enter the values for $\mu_{\bar{x}}$ and $\sigma_{\bar{x}}$, then
     solve for $\bar{x}$.
     2.  Given a particular value of $\mu$ that is an alternative to the null hy-
         pothesis $H_0$, draw the normal curve with this new value of $\mu$ at the
         center. Also plot the value of $\bar{x}$ found in step 1.
     3.  Refer to the graph from step 2 and find the area of the new critical
         region bounded by $\bar{x}$. This is the probability of rejecting the null
         hypothesis given that the new value of $\mu$ is correct.

*continued*

4. The value of $\beta$ is 1 minus the area from step 3. This is the probability of failing to reject the null hypothesis given that the new value of $\mu$ is correct.

The preceding steps allow you to find the probability of failing to reject $H_0$ when it is false. You are finding the area under the curve that *excludes* the critical region where you reject $H_0$; this area therefore corresponds to a failure to reject $H_0$ that is false, since we use a particular value of $\mu$ that goes against $H_0$. Refer to the aircraft age example discussed in this section (see Figures 7–1 and 7–2) and find the value of $\beta$ corresponding to the following.

  a.  $\mu = 10.50$
  b.  $\mu = 11.00$
  c.  $\mu = 14.00$

33. The **power** of a test is $1 - \beta$, the probability of rejecting a false null hypothesis. Refer to the aircraft age example discussed in this section. If that test has a power of 0.8, find the mean $\mu$ (see Example 32).

34.   a.  Using the sample data in Exercise 29, construct a 95% confidence interval about $\mu$.
  b.  Does that interval contain 32 oz?
  c.  Describe a rule for using confidence intervals in place of the traditional methods in two-tailed hypothesis tests. What feature of a confidence interval would cause you to reject a null hypothesis?

# $\boxed{7\text{–}3}$ *P*-Values

In Sections 7–1 and 7–2 we introduced the **classical**, or **traditional**, **approach** used to test a hypothesis or claim made about a population mean. We saw that the conclusion involved a decision either to reject or to fail to reject the null hypothesis, and that decision was determined by a comparison of the test statistic and the critical value.

In addition to the classical approach, there are other approaches to test claims made about population parameters, such as the mean. One approach involves the use of sample data in the construction of confidence intervals (or interval estimates). Consider this brief example: An airline industry representative claims that the mean age of U.S. commercial aircraft is equal to 10 years, but a random sample of 40 such aircraft yields a mean of 13.41 years. In Section 6–2 we used that sample data to conclude that there is a 95% chance that the following interval really does contain the true mean age of U.S. commercial aircraft.

$$10.84 \text{ years} < \mu < 15.98 \text{ years}$$

This implies that there is a relatively small chance of 5% or less that these limits don't contain the mean. Based on the above confidence interval, we can conclude that it is highly unlikely that the claimed mean of 10 years is correct. (In recent years, editors of medical journals have been campaigning to use confidence intervals in place of hypothesis tests, and the preceding example illustrates how this is done.)

Many professional articles and software packages use another approach to hypothesis testing that is based on the calculation of a **probability-value** or ***P*-value.** The *P*-value approach uses most of the same basic procedures as the classical approach, but these steps are different:

## *P*-Value Decision Criterion

1.  Find the *P*-value. This is the probability of getting a value of the sample mean $\bar{x}$ that is at least as extreme as the $\bar{x}$ found from the sample data (assuming that the hypothesized mean is correct).

2.  Some statisticians prefer to simply report the *P*-value and leave the conclusion to the reader. Others prefer to use this decision criterion:

*   *Reject* the null hypothesis if the *P*-value is less than or equal to the significance level $\alpha$.

*   *Fail to reject* the null hypothesis if the *P*-value is greater than the significance level $\alpha$.

In step 2 above, if the conclusion is based on the *P*-value alone, the following guide may be helpful.

| *P*-value | Interpretation |
|---|---|
| Less than 0.01 | Highly statistically significant<br>Very strong evidence against the null hypothesis |
| 0.01 to 0.05 | Statistically significant<br>Adequate evidence against the null hypothesis |
| Greater than 0.05 | Insufficient evidence against the null hypothesis |

Figure 7–8 (see the following page) outlines key steps and decisions that lead to the *P*-value. In a right-tailed test, the *P*-value is obtained by finding the area to the right of the test statistic. In a left-tailed test, the *P*-value is the area to the left of the test statistic. However, we must be careful to note that in a two-tailed test, the *P*-value is *twice* the area of the extreme region bounded by test statistic. This makes sense when we recognize that the *P*-value gives us the probability of getting a sample mean that is *at least as extreme* as the sample mean actually obtained, and the two-tailed case has critical or extreme regions in *both* tails.

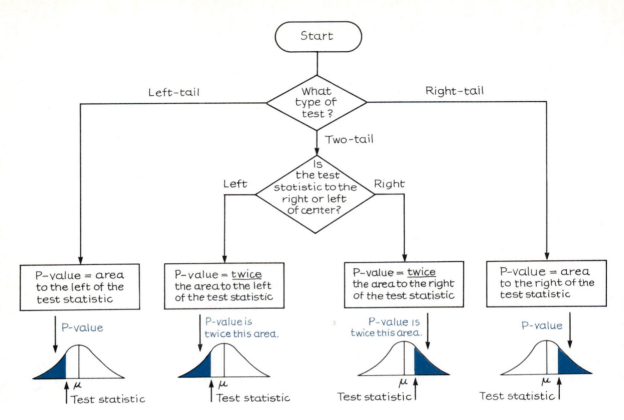

**Figure 7–8**

The following example uses the *P*-value approach for the same example presented in Section 7–2 (see Figures 7–1 and 7–2).

**Example**

An airline industry representative claims that the mean age of U.S. commercial aircraft is 10 years or less. A random sample of 40 such aircraft has a mean of 13.41 years and a standard deviation of 8.28 years. At the $\alpha = 0.05$ significance level, use the *P*-value approach to test the representative's claim.

## Solution

As in Section 7–2, we have

| | |
|---|---|
| Null hypothesis | $H_0: \mu \leq 10$ |
| Alternative hypothesis | $H_1: \mu > 10$ |
| Significance level | $\alpha = 0.05$ |
| Test statistic | $z = \dfrac{\bar{x} - \mu_{\bar{x}}}{\sigma_{\bar{x}}} = \dfrac{13.41 - 10}{\dfrac{8.28}{\sqrt{40}}} = 2.60$ |

We will now find the *P*-value. Refer to Figure 7–9 and observe that the area to the right of $\bar{x} = 13.41$ (or $z = 2.60$) can be found from Table A–2. Referring to Table A–2 with $z = 2.60$, find the area of 0.4953 as shown in Figure 7–9. That area is subtracted from 0.5 to yield the right-tail area of 0.0047. The *P*-value is 0.0047, since this is the probability of getting a value at least as extreme as $\bar{x} = 13.41$.

We now observe that since the *P*-value of 0.0047 is less than the significance level of $\alpha = 0.05$, we reject the null hypothesis. There is sufficient evidence to warrant rejection of the claim that the mean age is 10 years or less.

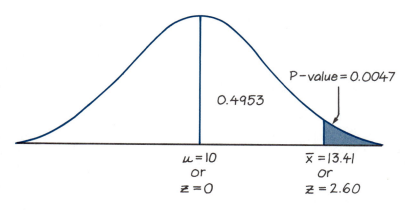

**Figure 7–9**

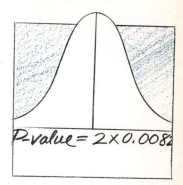

## BEWARE OF *P*-VALUE MISUSE

■John P. Campbell, editor of the *Journal of Applied Psychology*, wrote the following on the subject of *P*-values. "Books have been written to disuade people from the notion that smaller *P*-values mean more important results or that statistical significance has anything to do with substantive significance. It is almost impossible to drag authors away from their *P*-values, and the more zeros after the decimal point, the harder people cling to them." While it might be necessary to provide a statistical analysis of the results of a study, we should place strong emphasis on the significance of the results themselves. ■

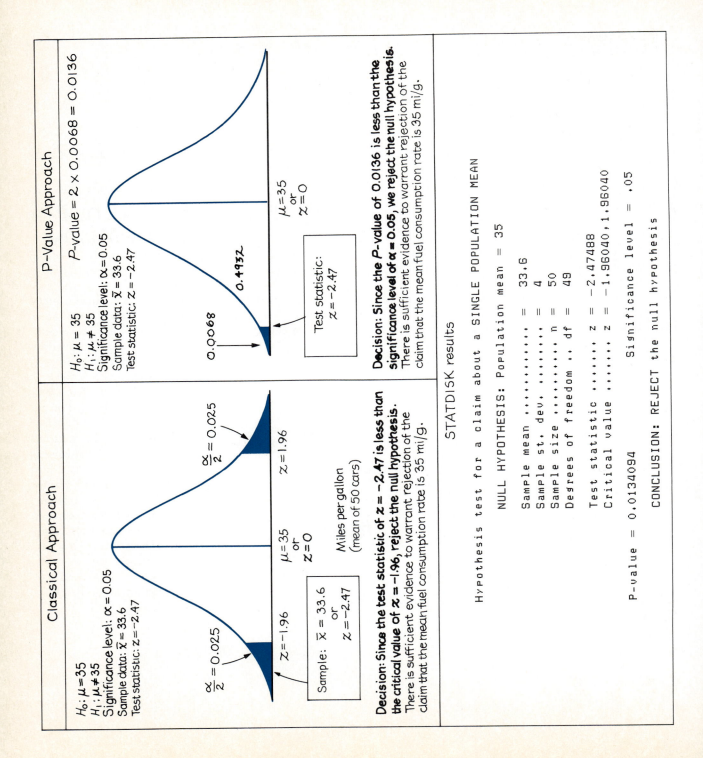

## Classical Approach

$H_0: \mu = 35$
$H_1: \mu \neq 35$
Significance level: $\alpha = 0.05$
Sample data: $\bar{x} = 33.6$
Test statistic: $z = -2.47$

$\frac{\alpha}{2} = 0.025$

$\frac{\alpha}{2} = 0.025$

$z = -1.96$    $\mu = 35$    $z = 1.96$
or
$z = 0$

Miles per gallon
(mean of 50 cars)

Sample: $\bar{x} = 33.6$
or
$z = -2.47$

**Decision: Since the test statistic of $z = -2.47$ is less than the critical value of $z = -1.96$, reject the null hypothesis.** There is sufficient evidence to warrant rejection of the claim that the mean fuel consumption rate is 35 mi/g.

## P-Value Approach

$P$-value $= 2 \times 0.0068 = 0.0136$

$H_0: \mu = 35$
$H_1: \mu \neq 35$
Significance level: $\alpha = 0.05$
Sample data: $\bar{x} = 33.6$
Test statistic: $z = -2.47$

0.4932

0.0068

$\mu = 35$
or
$z = 0$

Test statistic:
$z = -2.47$

**Decision: Since the $P$-value of 0.0136 is less than the significance level of $\alpha = 0.05$, we reject the null hypothesis.** There is sufficient evidence to warrant rejection of the claim that the mean fuel consumption rate is 35 mi/g.

STATDISK results

```
Hypothesis test for a claim about a SINGLE POPULATION MEAN

  NULL HYPOTHESIS: Population mean = 35

     Sample mean ............ = 33.6
     Sample st. dev. ........ = 4
     Sample size ............ n = 50
     Degrees of freedom .. df = 49

     Test statistic ........ z = -2.47488
     Critical value ........ z = -1.96040,1.96040

  P-value = 0.0134094        Significance level = .05

     CONCLUSION: REJECT the null hypothesis
```

In Section 7–2 we included an example of a two-tailed hypothesis test, and that example used the classical approach to hypothesis testing. In the figure on page 354 we have extracted the essential components of that example to compare them to the *P*-value approach. (We use the decision criterion that involves a comparison of the significance level $\alpha$ and the *P*-value.) Note that the only real difference is the decision criterion, which leads to the same conclusion in both cases.

In Section 7–2, we stated that the significance level $\alpha$ should be selected *before* a hypothesis test is conducted. Many statisticians consider this a good practice since it helps to prevent us from using the data to support subjective conclusions or beliefs. They feel that this practice becomes especially important with the *P*-value approach because we may be tempted to adjust the significance level based on the resulting *P*-value. With a 0.05 level of significance and a *P*-value of 0.06, we should fail to reject the null hypothesis, but it is sometimes tempting to say that a probability of 0.06 is small enough to warrant rejection of the null hypothesis. Consequently, we should always select the significance level first. Other statisticians feel that prior selection of a significance level reduces the usefulness of *P*-values. They contend that no significance level should be specified, and the conclusion should be left to the reader. We shall use the decision criterion that involves a comparison of a significance level and the *P*-value.

## 7–3  Exercises A

In Exercises 1–4, use the *P*-value and significance level to choose between rejecting the null hypothesis or failing to reject the null hypothesis.

1.  *P*-value: 0.03; significance level: $\alpha = 0.05$
2.  *P*-value: 0.04; significance level: $\alpha = 0.01$
3.  *P*-value: 0.405; significance level: $\alpha = 0.10$
4.  *P*-value: 0.09; significance level: $\alpha = 0.10$

In Exercises 5–12, first find the *P*-value. Then either reject or fail to reject the null hypothesis by assuming a significance level of $\alpha = 0.05$.

5.  $H_0: \mu \geq 152$; $H_1: \mu < 152$; test statistic: $z = -1.40$
6.  $H_0: \mu \geq 100$; $H_1: \mu < 100$; test statistic: $z = -0.46$
7.  $H_0: \mu \leq 15.7$; $H_1: \mu > 15.7$; test statistic: $z = 1.94$
8.  $H_0: \mu \leq 428$; $H_1: \mu > 428$; test statistic: $z = 1.66$
9.  $H_0: \mu = 75.0$; $H_1: \mu \neq 75.0$; test statistic: $z = 1.66$
10. $H_0: \mu = 1365$; $H_1: \mu \neq 1365$; test statistic: $z = -1.30$

11. $H_0: \mu = 2.53$; $H_1: \mu \neq 2.53$; test statistic: $z = -1.94$
12. $H_0: \mu = 12.8$; $H_1: \mu \neq 12.8$; test statistic: $z = 3.00$

In Exercises 13–20, use the $P$-value approach to test the given hypotheses.

13. Test the claim that $\mu \geq 100$, given a sample of $n = 45$ for which $\bar{x} = 95$. Assume that $\sigma = 15$, and test at the $\alpha = 0.05$ significance level.

14. Test the claim that $\mu \leq 500$, given a sample of $n = 35$ for which $\bar{x} = 508$. Assume that $\sigma = 90$, and test at the $\alpha = 0.01$ significance level.

15. Test the claim that $\mu = 75.6$, given a sample of $n = 81$ for which $\bar{x} = 78.8$. Assume that $\sigma = 12.0$, and test at the $\alpha = 0.01$ significance level.

16. Test the claim that $\mu = 98.6$, given a sample of $n = 200$ for which $\bar{x} = 97.1$. Assume that $\sigma = 2.35$, and test at the $\alpha = 0.10$ significance level.

17. Use the aircraft age sample data $(n = 40, \bar{x} = 13.41$ years, $s = 8.28$ years$)$ given at the beginning of this chapter. At the 0.05 level of significance, test the claim of *Time* magazine when it reported that the mean age of aircraft in the U.S. airline fleet is equal to 14 years. Assume that the sample standard deviation can be used for $\sigma$.

18. A brewery distributes beer in cans labeled 12 oz. The Bureau of Weights and Measures randomly selects 36 cans, measures their contents, and obtains a sample mean of 11.82 oz. Assuming that $\sigma$ is known to be 0.38 oz, is it valid at the 0.01 significance level to conclude that the brewery is cheating consumers?

19. In a study of distances traveled by buses before the first major engine failure, a sample of 191 buses results in a mean of 96,700 mi and a standard deviation of 37,500 mi (based on data in *Technometrics*, Vol. 22, No. 4). At the 0.05 level of significance, test the claim that mean distance traveled before a major engine failure is more than 90,000 mi. (Assume that the sample standard deviation can be used for $\sigma$.)

20. When 150 randomly selected boys aged 6 to 11 are given the reading portion of the Wide Range Achievement Test, their mean score is 52.4 and the standard deviation is 13.14 (based on data from the National Health Survey, USDHEW publication 72-1011). At the 0.05 level of significance, test the claim that this sample is from a population with a mean greater than 51.0. (Assume that the sample standard deviation can be used for $\sigma$.)

# 7–3 Exercises B

21. A journal article reports that a null hypothesis of $\mu = 100$ is rejected because the *P*-value is less than 0.01. The sample size is given as 62 and the sample mean is given as 103.6. Find the largest possible standard deviation.

22. For a random sample of 40 people, what is the mean IQ score if a *P*-value of 0.04 is obtained when testing the claim that $\mu \geq 100$? Assume that $\sigma = 15$.

23. When 93 IQ scores are randomly selected from a certain population, results are described by the accompanying boxplot. Find the *P*-value obtained when testing the claim that the mean IQ score of the population equals 105.

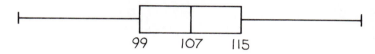

24. Find the smallest mean above $23,460 that leads to a rejection of the claim that the mean annual household income is $23,460. Assume a 0.02 significance level, and assume that $\sigma = \$3750$. The sample consists of 50 randomly selected households.

# 7–4 *t* Test

In Sections 7–2 and 7–3 we introduced the general method for testing hypotheses, but all the examples and exercises involved situations in which the central limit theorem applies so that the normal distribution can be used. The population standard deviations were given and/or the samples were large, and each hypothesis tested related to a population mean. In those cases, we can apply the central limit theorem and use the normal distribution as an approximation to the distribution of sample means. A very unrealistic feature of those examples and exercises is the assumption that $\sigma$ is known. If $\sigma$ is unknown and the sample is large, we can treat *s* as if it were $\sigma$ and proceed as in Section 7–2 or 7–3. This estimation of $\sigma$ by *s* is reasonable because large random samples tend to be representative of the population. But small random samples may exhibit unusual behavior, and they cannot be so trusted. **In this section we consider tests of hypotheses about a population mean, where the samples are small and $\sigma$ is unknown.** We begin by referring to Figure 7–10, which outlines the theory we are describing.

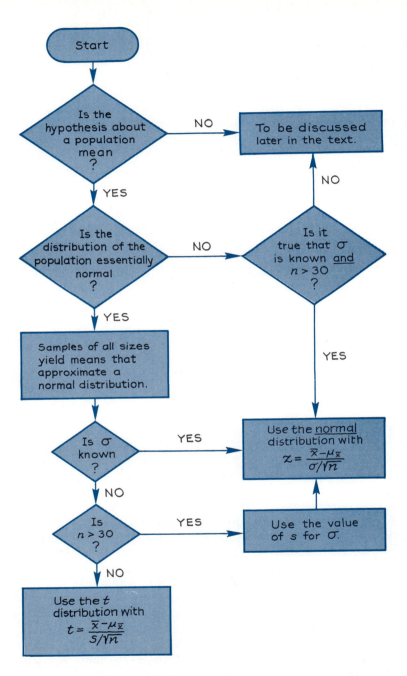

**Figure 7–10**

Starting at the top of Figure 7–10, we see that our immediate concerns lie only with the hypotheses made about one population mean. (In following sections we will consider hypotheses made about population parameters other than the mean.) Figure 7–10 summarizes the following observations:

1. In *any* population, the distribution of sample means can be approximated by the normal distribution as long as the random samples are large. This is justified by the central limit theorem.

2. In populations with distributions that are essentially normal, samples of *any* size will yield means having a distribution that is approximately normal. The value of $\mu$ would correspond to the null hypothesis, and the value of $\sigma$ must be known. If $\sigma$ is unknown and the samples are large, we can use $s$ as a substitute for $\sigma$, since large random samples tend to be representative of the populations from which they come.

3. In populations with distributions that are essentially normal, assume that we randomly select *small* samples and we do not know the value of $\sigma$. For this case, we can use the Student *t* distribution that was first introduced in Section 6–2.

4. If our random samples are small, $\sigma$ is unknown, and the population is grossly nonnormal, then we can use nonparametric methods, some of which are discussed in Chapter 12.

In Section 6–2 we introduced the Student *t* distribution and noted several important features. Those features particularly relevant to this section are given below.

## Conditions for using the Student *t* distribution:

1. The sample is small $(n \leq 30)$; and
2. The value of $\sigma$ is unknown; and
3. The parent population is essentially normal.

---

**Test Statistic**

If a population is essentially normal, then the distribution of

$$t = \frac{\bar{x} - \mu}{s/\sqrt{n}}$$

is essentially a **Student *t* distribution** for all samples of size $n$. (The Student *t* distribution is often referred to as the **t distribution.**)

## Critical Values

1. Critical values are found in Table A–3.
2. Degrees of freedom $= n - 1$.

## Important Properties

1. The Student $t$ distribution is different for different sample sizes (see Figure 6–4 in Section 6–2).
2. The Student $t$ distribution has the same general bell shape of the normal distribution, but it reflects the greater variability that is expected with small samples.
3. The Student $t$ distribution has a mean of $t = 0$ (just as the standard normal distribution has a mean of $z = 0$).
4. The standard deviation of the Student $t$ distribution is greater than 1 (unlike the standard normal distribution, which has $\sigma = 1$).
5. As the sample size $n$ gets larger, the Student $t$ distribution gets closer to the normal distribution. For values of $n > 30$, the differences are so small that we can use the critical $z$ values instead of developing a much larger table of critical $t$ values. (The values in the bottom row of Table A–3 are equal to the corresponding critical $z$ values from the normal distribution.)

### Example

A tobacco company claims that its best-selling cigarettes contain at most 40 mg of nicotine. Test this claim at the 1% significance level by using the results of 15 randomly selected cigarettes for which $\bar{x} = 42.6$ mg and $s = 3.7$ mg. Other evidence suggests that the distribution of nicotine contents is a normal distribution.

### Solution

We list the solution according to the steps outlined in Figure 7–1, which summarizes the classical procedure for testing hypotheses.

*Step 1:* The original claim expressed in symbolic form is $\mu \leq 40$.

*Step 2:* The opposite of the original claim is $\mu > 40$.

*Step 3:* $H_0$ must contain the condition of equality so we get

$H_0: \mu \leq 40$ (Null hypothesis)
$H_1: \mu > 40$ (Alternative hypothesis)

*continued*

**Solution** *continued*

*Step 4:* The significance level of 1% corresponds to $\alpha = 0.01$.

*Step 5:* The sample mean should be used in testing a claim about a population mean. Since the sample is small and $\sigma$ is unknown, we use the Student $t$ distribution.

*Step 6:* The test statistic is

$$t = \frac{\overline{x} - \mu}{s/\sqrt{n}} = \frac{42.6 - 40}{3.7/\sqrt{15}} = 2.722$$

The test statistic and critical value are shown in Figure 7–11.

*Step 7:* Since the test statistic of $t = 2.722$ does fall in the critical region, we reject $H_0$.

*Step 8:* There is sufficient evidence to warrant rejection of the tobacco company's claim. These cigarettes appear to contain significantly more than 40 mg of nicotine.

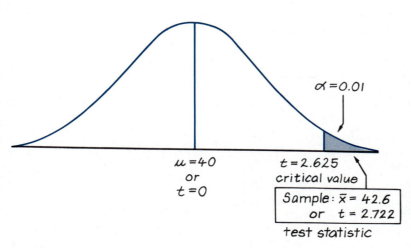

**Figure 7–11**

If this example had been two-tailed instead of right-tailed, there would have been two critical values ($t = -2.977$ and $t = 2.977$) corresponding to 14 degrees of freedom with $\alpha = 0.01$ divided equally between two tails. The test statistic of $t = 2.722$ would not fall within the critical region and we would fail to reject the claim that the mean equals 40 mg.

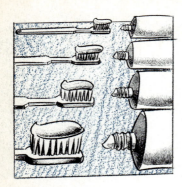

## PRODUCT TESTING

■ The United States Testing Company in Hoboken, New Jersey, is the world's largest independent product testing laboratory. It's often hired to verify advertising claims. A vice president has said that the most difficult part of his job is "telling a client when his product stinks. But if we didn't do that, we'd have no credibility." He says that there have been a few clients who wanted results fabricated, but most clients want honest results. United States Testing Laboratory evaluates laundry detergents, cosmetics, insulation materials, zippers, pantyhose, football helmets, toothpaste, fertilizer, and a wide variety of other products. ■

If the tobacco company had claimed that the mean nicotine content was below 50 mg, then we would have a left-tailed test with the test statistic

$$t = \frac{\overline{x} - \mu_{\overline{x}}}{s/\sqrt{n}} = \frac{42.6 - 50}{3.7/\sqrt{15}} = -7.746$$

and critical value $t = -2.625$ found from Table A–3. The test statistic would fall within the critical region and we would support the claim that the mean is below 50 mg.

## P-Values

In the example presented in this section, we used the classical approach to hypothesis testing. However, much of the literature and many computer packages will use the P-value approach. Shown here, for example, is the STATDISK display for the last example. The P-value of 8.27152E-03 is actually 0.00827152.

STATDISK DISPLAY

> Hypothesis test for a claim about a SINGLE POPULATION MEAN
>
> NULL HYPOTHESIS: Population mean < = 40
>
> Sample mean . . . . . . . . . . . . . . . . . . . . . . . . . . . . . . . . . . . . . . . . . . . . . . . . . . . . . . = 42.6
> Sample st. dev. . . . . . . . . . . . . . . . . . . . . . . . . . . . . . . . . . . . . . . . . . . . . . . . . . . . . = 3.7
> Sample size . . . . . . . . . . . . . . . . . . . . . . . . . . . . . . . . . . . . . . . . . . . . . . . . n = 15
> Degrees of freedom . . . . . . . . . . . . . . . . . . . . . . . . . . . . . . . . . . . . . . . . . df = 14
>
> Test statistic . . . . . . . . . . . . . . . . . . . . . . . . . . . . . . . . . . . . . . . . . . . . . . . . . . t = 2.72155
> Critical value . . . . . . . . . . . . . . . . . . . . . . . . . . . . . . . . . . . . . . . . . . . . . . . . . t = 2.62610
>
> P-value = 8.27152E-03                         Significance level = .01
>
> CONCLUSION: REJECT the null hypothesis

Since the $t$ distribution table (Table A–3) includes only selected values of $\alpha$, we cannot usually find the specific P-value from Table A–3. Instead, we can use that table to identify limits that contain the P-value. In the last example we found the test statistic to be $t = 2.722$, and we know that the test is one-tailed with 14 degrees of freedom. By examining the row of Table

A–3 corresponding to 14 degrees of freedom, we see that the test statistic of 2.722 falls between the table values of 2.977 and 2.625, which, in a one-tailed test, correspond to $\alpha = 0.005$ and $\alpha = 0.01$. While we cannot determine the exact *P*-value, we do know that it must fall between 0.005 and 0.01, so that

$$0.005 < P\text{-value} < 0.01$$

With a significance level of 0.01 and a *P*-value less than 0.01, we would reject the null hypothesis as we did in the classical approach. (Some calculators and computer programs allow us to find exact *P*-values.)

So far, we have discussed tests of hypotheses made about population means only. In the next section we will learn how to test hypotheses made about population proportions or percentages.

# 7–4 | Exercises A

In Exercises 1 and 2, find the critical *t* value suggested by the given data.

1.  a. $H_0: \mu = 12$
        $n = 27$
        $\alpha = 0.05$
    b. $H_0: \mu \leq 50$
        $n = 17$
        $\alpha = 0.10$
    c. $H_0: \mu \geq 1.36$
        $n = 6$
        $\alpha = 0.01$
    d. $H_0: \mu = 1.36$
        $n = 6$
        $\alpha = 0.01$
    e. $H_0: \mu \geq 10.75$
        $n = 29$
        $\alpha = 0.01$

2.  a. $H_0: \mu \leq 100$
        $n = 27$
        $\alpha = 0.10$
    b. $H_1: \mu \neq 500$
        $n = 16$
        $\alpha = 0.05$
    c. $H_1: \mu < 67.5$
        $n = 12$
        $\alpha = 0.05$
    d. $H_1: \mu > 98.4$
        $n = 7$
        $\alpha = 0.05$
    e. $H_1: \mu \neq 75$
        $n = 24$
        $\alpha = 0.05$

In Exercises 3–8, assume that the population is normally distributed.

3.  Test the claim that $\mu \leq 10$, given a sample of 9 for which $\bar{x} = 11$ and $s = 2$. Use a significance level of $\alpha = 0.05$.
4.  Test the claim that $\mu \leq 32$, given a sample of 27 for which $\bar{x} = 33.5$ and $s = 3$. Use a significance level of $\alpha = 0.10$.
5.  Test the claim that $\mu \geq 98.6$, given a sample of 18 for which $\bar{x} = 98.2$ and $s = 0.8$. Use a significance level of $\alpha = 0.025$.
6.  Test the claim that $\mu \geq 100$, given a sample of 22 for which $\bar{x} = 95$ and $s = 18$. Use a 5% level of significance.

7. Test the claim that $\mu = 75$, given a sample of 15 for which $\bar{x} = 77.6$ and $s = 5$. Use a significance level of $\alpha = 0.05$.

8. Test the claim that $\mu = 500$, given a sample of 20 for which $\bar{x} = 541$ and $s = 115$. Use a significance level of $\alpha = 0.10$.

In Exercises 9–28, test the given hypothesis by following the procedure suggested by Figure 7–1. Draw the appropriate graph. In each case, assume that the population is approximately normal.

9. A clothing manufacturer randomly selects 24 eleven-year-old girls and records various measurements. Their heights have a mean of 147.6 cm and the standard deviation is 7.85 cm (based on data from the National Health Survey). At the 0.02 significance level, test a clothing designer's claim that the mean height of all eleven-year-old girls is equal to 157.6 cm.

10. For each of 12 organizations, the cost of operation per client was found. The 12 scores have a mean of $2133 and a standard deviation of $345 (based on data from "Organizational Communication and Performance," by Snyder and Morris, *Journal of Applied Psychology*, Vol. 69, No. 3). At the 0.01 significance level, test the claim of a critic who complains that the mean for all such organizations exceeds $1800 as the cost per client.

11. The skid properties of a snow tire have been tested and the mean skid distance of 154 ft has been established for standardized conditions. A new, more expensive tire is developed, but tests on a sample of 20 new tires yield a mean skid distance of 141 ft with a standard deviation of 12 ft. Because of the cost involved, the new tires will be purchased only if they skid less at the $\alpha = 0.005$ significance level. Based on the sample, will the new tires be purchased?

12. An aircraft manufacturer randomly selects 12 planes of the same model and tests them to determine the distance (in meters) they require for takeoff. The sample mean and standard deviation are computed to be 524 m and 23 m, respectively. At the 5% level of significance, test the claim that the mean for all such planes is more than 500 m.

13. A pill is supposed to contain 20.0 mg of phenobarbitol. A random sample of 30 pills yields a mean and standard deviation of 20.5 mg and 1.5 mg, respectively. Are these sample pills acceptable at the $\alpha = 0.02$ significance level?

14. The Federal Aviation Administration randomly selects five light aircraft of the same type and tests the left wings for their loading capacities. The sample mean and the standard deviation are 16,735 lb and 978 lb, respectively. At the 5% level of significance, test the claim that the mean loading capacity for all such aircraft wings is equal to 17,850 lb.

15. In a study of consumer credit, 25 randomly selected credit card holders were surveyed, and the mean amount they charged in the past 12 months was found to be $1756, while the standard deviation was $843.

Use a 0.025 level of significance to test the claim that the mean amount charged by all credit card holders was greater than $1500.

16. A long-range missile misses its target by an average of 0.88 mi. A new steering device is supposed to increase accuracy, and a random sample of eight missiles is equipped with this new mechanism and tested. These eight missiles miss by distances with a mean of 0.76 mi and a standard deviation of 0.04 mi. At $\alpha = 0.01$, does the new steering mechanism lower the miss distance?

17. A high school principal is concerned with the amount of time her students devote to working at an after-school job. She randomly selects 25 students, obtains their working hours, and computes $\bar{x} = 12.3$ and $s = 11.2$. Both values are in hours for one week. She claims that this is significantly more than the mean of 10.7 h obtained from a study conducted by the National Federation of State High School Associations. Test her claim by using a 0.05 level of significance.

18. A study was conducted to determine whether or not a standard clerical test would need revision for use on video display terminals (VDT). The VDT scores of 22 subjects have a mean of 170.2 and a standard deviation of 35.3 (based on data from "Modification of the Minnesota Clerical Test to Predict Performance on Video Display Terminals," by Silver and Bennett, *Journal of Applied Psychology*, Vol. 72, No. 1). At the 0.05 level of significance, test the claim that the mean for all subjects differs from the mean of 243.5 for the standard printed version of the text.

19. A high school senior is concerned about attending college because she knows that many college students require more than four years to earn a bachelor's degree. At the 0.10 level of significance, test the claim of a guidance counselor who states that the mean time is greater than five years. Sample data consist of 28 randomly selected college graduates who had a mean of 5.15 years and a standard deviation of 1.68 years (based on data from the National Center for Education Statistics).

20. A study was conducted to determine the effects of mental training in organizations. (See "Routinization of Mental Training in Organizations: Effects on Performance and Well-Being," by Larsson, *Journal of Applied Psychology*, Vol. 72, No. 1.) For an experimental group of 20 subjects, a performance exam resulted in scores with a mean of 79.12 and a standard deviation of 17.49. At the 0.10 level of significance, test the claim that the experimental group comes from a population with a mean less than 85.70.

21. In a study of the long-term effects of promoting or holding back elementary school students, the following reading test results were obtained for a sample of 15 third grade students: $\bar{x} = 31.0$, $s = 10.5$. (The data are based on "A Longitudinal Study of the Effects of Retention/Promotion on Academic Achievement," by Peterson and others, *American Educational Research Journal*, Vol. 24, No. 1.) Does this third grade sample mean differ significantly from a first grade population mean of 41.9? Assume a 0.01 level of significance.

22. In a study of factors affecting hypnotism, visual analogue scale (VAS) sensory ratings were obtained for 16 subjects. For these sample ratings, the mean is 8.33 while the standard deviation is 1.96 (based on data from "An Analysis of Factors that Contribute to the Efficacy of Hypnotic Analgesia," by Price and Barber, *Journal of Abnormal Psychology*, Vol. 96, No. 1). At the 0.01 level of significance, test the claim that this sample comes from a population with a mean rating less than 10.00

23. A study of one effect of strenuous exercise obtained peak oxygen consumption levels for six subjects. The mean and standard deviation are 3.98 and 0.49, respectively; both values are in liters for one minute (based on data from "Supramaximal Exercise After Training-Induced Hypervolemia," by Green and others, *Journal of Applied Physiology*, Vol. 62, No. 5). Test the claim that this sample comes from a population with a mean that is less than 4.00 L for one minute.

24. At the 0.10 level of significance, test the claim that a brewery fills bottles with amounts having a mean greater than 32 oz. A sample of 27 bottles produces a mean of 32.2 oz and a standard deviation of 0.4 oz.

25. A standard final examination in an elementary statistics course produces a mean score of 75. At the 5% level of significance, test the claim that the following sample scores reflect an above-average class:

79  79  78  74  82  89  74  75  78  73
74  84  82  66  84  82  82  71  72  83

26. A sample of beer cans labeled 16 oz is randomly selected and the actual contents accurately measured. The results (in ounces) are as follows. Is the consumer being cheated?

15.8  16.2  16.3  15.9  15.5
15.9  16.0  15.6  15.8

27. Listed below are the total electric energy consumption amounts (in kWh) for the author's home during seven different years. Test the utility company's claim that the mean annual consumption amount is equal to 11,000 kWh.

11,943  11,463  10,789  9907  9012  9942  11,153

28. Given below are the birthweights (in kilograms) of male babies born to mothers on a special vitamin supplement (based on data from the New York State Department of Health). At the 0.05 level of significance, test the claim that the mean for all such (vitamin) male babies is equal to 3.39 kg, which is the mean for the population of all males.

3.73  4.37  3.73  4.33  3.39  3.68  4.68  3.52
3.02  4.09  2.47  4.13  4.47  3.22  3.43  2.54

# 7–4 Exercises B

29. For certain conditions, a hypothesis test requires the Student $t$ distribution, as described in this section. Assume that the standard normal distribution is incorrectly used instead. Using the standard normal distribution, are you more likely to reject the null hypothesis, less likely, or does it make no difference? Explain.

30. What do you know about the $P$-value in each of the following cases?
    a.  $H_0: \mu \leq 5.00$; $n = 10$; test statistic: $t = 2.205$
    b.  $H_0: \mu = 5.00$; $n = 20$; test statistic: $t = 2.678$
    c.  $H_0: \mu \geq 5.00$; $n = 16$; test statistic: $t = -1.234$

31. Some computer programs approximate critical $t$ values by

    $$t = \sqrt{DF \cdot (e^{A^2/DF} - 1)}$$

    where

    $$DF = n - 1$$

    $$e = 2.718$$

    $$A = z\left(\frac{8\ DF + 3}{8\ DF + 1}\right)$$

    and $z$ is the critical $z$ score.

    Use this approximation to find the critical $t$ score corresponding to $n = 10$ and a significance level of 0.05 in a right-tailed case. Compare the results to the critical $t$ found in Table A–3.

32. Refer to Exercise 24 and assume that you're testing the null hypothesis of $\mu \leq 32.0$ oz. Find $\beta$ (the probability of a type II error) given that $\mu = 32.3$. (See Exercise 32 from Section 7–2.)

# 7–5 Tests of Proportions

In Section 7–2 we learned the basic method for testing hypotheses. It is not difficult to modify that procedure for many other circumstances. In this section we consider a method for testing hypotheses made about a population proportion. The particular assumptions for this section are listed below.

## Assumptions

1.  We are testing a claim made about a population **proportion, probability,** or **percentage.**

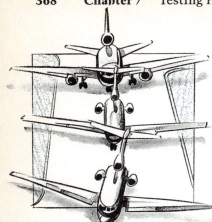

## MISLEADING STATISTICS

■ *Money* reported that among airlines, Air North had the second highest complaint rate in one particular year. But reporter George Bernstein investigated and found that Air North's high complaint rate of 38.6 per 100,000 passengers really represented only 22 complaints. (Air North flew 57,000 passengers that year.) "That many complaints could have come from a single delayed flight," wrote Bernstein. He went on to state, "If nothing else, this proves how dangerous statistics can be, and how easy it is for them to be blown out of proportion if you don't know exactly what's behind them." The following year, Air North had only three complaints. ■

2. The conditions for a **binomial experiment** are satisfied. (That is, we have a fixed number of independent trials having constant probabilities, and each trial has two outcome categories.)

3. The conditions $np \geq 5$ and $nq \geq 5$ are both satisfied so that **the binomial distribution of sample proportions can be approximated by a normal distribution** with $\mu = np$ and $\sigma = \sqrt{npq}$.

If $np \geq 5$ and $nq \geq 5$ are not both true, we may be able to use Table A–1 or the binomial probability formula described in Section 4–4, but this section deals only with situations in which the normal distribution is a suitable approximation for the distribution of sample proportions.

If the above three conditions are all satisfied, the value of the test statistic will be found by computing $z$, as follows.

---

### Test Statistic

$$z = \frac{\hat{p} - p}{\sqrt{\dfrac{pq}{n}}}$$

where $n$ = number of trials
$p$ = population proportion (given in the null hypothesis)
$q = 1 - p$
$\hat{p} = x/n$ (sample proportion)

---

The **critical value** is found from Table A–2 by using the same procedures described in Section 7–2.

The above test statistic is justified by noting that when using the normal distribution to approximate a binomial distribution, we substitute $\mu = np$ and $\sigma = \sqrt{npq}$ to get

$$z = \frac{x - \mu}{\sigma} = \frac{x - np}{\sqrt{npq}}$$

Here $x$ is the number of successes among $n$ trials. Divide the numerator and denominator of this last expression by $n$, then replace $x/n$ by the symbol $\hat{p}$, and you get the test statistic given above. In other words, the above test statistic is simply $z = (x - \mu)/\sigma$ modified for the binomial notation; the distribution of sample proportions $\hat{p}$ is a normal distribution with mean $p$ and standard deviation $\sqrt{pq/n}$.

We can now test hypotheses made about population proportions. Simply follow the same general steps listed in Figure 7–1 and use the test statistic given above.

### Example

A manufacturing plant manager defines production as out of control when the overall rate of defects exceeds 4%. A random sample of 150 items results in 9 defects. Use a 5% significance level to test the claim that production is out of control.

### Solution

We will follow the steps outlined in Figure 7–1.

*Step 1:*   The original claim is that production is out of control, or the rate of defects exceeds 4%. We express this in symbolic form as $p > 0.04$.

*Step 2:*   The opposite of the original claim is $p \leq 0.04$.

*Step 3:*   Since $p \leq 0.04$ contains equality, we have
$H_0$: $p \leq 0.04$      (Null hypothesis)
$H_1$: $p > 0.04$      (Alternative hypothesis)

*Step 4:*   The significance level is $\alpha = 0.05$.

*Step 5:*   The statistic relevant to this test is $\hat{p} = 9/150 = 0.06$. The sampling distribution of sample proportions is approximated by the normal distribution. (The requirements that $np \geq 5$ and $nq \geq 5$ are satisfied.)

*Step 6:*   The test statistic is

$$z = \frac{\hat{p} - p}{\sqrt{\dfrac{pq}{n}}} = \frac{0.06 - 0.04}{\sqrt{\dfrac{(0.04)(0.96)}{150}}} = 1.25$$

In Section 5–5 we included a correction for continuity, but we ignore it here since its effect is negligible. The critical value of $z = 1.645$ is found from Table A–2. The critical value corresponds to a table entry of 0.4500, which is equivalent to a right-tail area of $\alpha = 0.05$ (see Figure 7–12 on page 370).

*Step 7:*   Since the test statistic does not fall within the critical region, we fail to reject the null hypothesis.

*Step 8:*   There is not sufficient evidence to support the claim that the rate of defects exceeds 4%. We lack sufficient evidence to conclude that production is out of control.

The claim in the preceding example involved a percentage, but we must use the equivalent decimal form. The methods presented in this section can be used to test claims made about proportions, probabilities, or percentages.

## HAVE ATOMIC TESTS CAUSED CANCER?

■ *Conquerors* was a 1954 movie made in Utah. A few years ago a team of investigative reporters attempted to locate members of the cast and crew. Of the 79 people they found (some living and some dead), 27 had developed cancer, including John Wayne, Susan Hayward, and Dick Powell. Two key questions arise:

1. Is a cancer rate of 27 out of 79 significant, or could it be a coincidence?

2. Was the cancer caused by fallout from the nuclear tests previously conducted in the area?

We can use statistics to answer the first question, but not the second. ■

Whether we have a proportion, percentage, or probability, the value of $p$ must be between 0 and 1, and the sum of $p$ and $q$ must be exactly 1. (Yogi Berra revealed that he lacked formal training in statistics when he said, "Baseball is 90% mental; the other half is physical.")

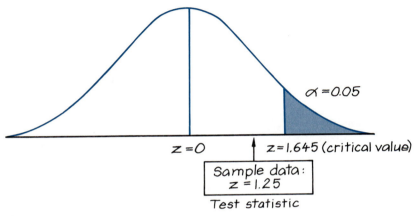

**Figure 7–12**

---

| **Example** |
| --- |

A television executive claims that "less than half of all adults are annoyed by the violence on television." (That is, violence in television shows, not atop television sets.) Test this claim using the sample data from a Roper poll in which 48% of 1998 surveyed adults indicated their annoyance with television violence. Use a 0.05 significance level.

| **Solution** |
| --- |

We summarize the key components of the hypothesis test.

$$H_0: p \geq 0.5$$

$$H_1: p < 0.5 \text{ (from the claim that "less than half are annoyed")}$$

Test statistic:  $z = \dfrac{\hat{p} - p}{\sqrt{\dfrac{pq}{n}}} = \dfrac{0.48 - 0.5}{\sqrt{\dfrac{(0.5)(0.5)}{1998}}} = -1.79$

*continued*

**Solution** *continued*

The test statistic, critical value, and critical region are shown in Figure 7–13. Since the test statistic is in the critical region, we reject the null hypothesis. There is sufficient sample evidence to support the claim that less than half of all adults are annoyed by the violence on television.

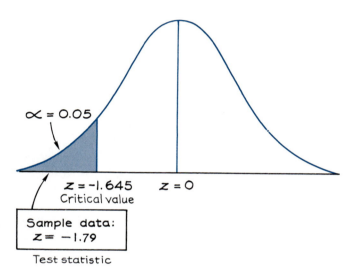

**Figure 7–13**

## THE YEAR WAS (SAFE) (UNSAFE)

■ Impressions can be manipulated by the statistics that are presented. For a recent year, there were 31 air traffic accidents among planes operated by scheduled airlines in the United States. That was the highest number of accidents since 1974. This all makes it sound like a bad year. But the Air Transport Association called this the seventh safest year in airline aviation history because the fatality rate was only 0.43 per 100,000 flights. ■

## *P*-Values

The examples in this section followed the traditional approach to hypothesis testing, but it would be easy to use the *P*-value approach, since the test statistic is a *z* score. The *P*-value is obtained by using the same procedure described in Section 7–3. In a right-tailed test, the *P*-value is the area to the right of the test statistic. In a left-tailed test, the *P*-value is the area to the left of the test statistic. In a two-tailed test, the *P*-value is twice the area of the extreme region bounded by the test statistic (see Figure 7–8). We reject the null hypothesis if the *P*-value is less than the significance level.

The last example was left-tailed, so the *P*-value is the area to the left of the test statistic $z = -1.79$. Table A–2 indicates that the area between $z = 0$ and $z = -1.79$ is 0.4633, so the *P*-value is $0.5 - 0.4633 = 0.0367$. Since the *P*-value of 0.0367 is less than the significance level of 0.05, we reject the null hypothesis and again conclude that there is sufficient sample evidence to support the claim that less than half of all adults are annoyed with television violence. Again, the *P*-value approach is another way of arriving at the same conclusion.

## 7–5  Exercises A

In Exercises 1–20, test the given hypotheses. Include the steps listed in Figure 7–1, and draw the appropriate graph.

1. At the 0.05 significance level, test the claim that the proportion of defects *p* for a certain product equals 0.3. Sample data consist of $n = 100$ randomly selected products, of which 45 are defective.

2. At the 0.05 significance level, test the claim that the proportion of females *p* at a given college equals 0.6. Sample data consist of $n = 80$ randomly selected students, of which 54 are females.

3. In a survey by Media General and the Associated Press, 813 of the 1084 respondents indicated support for a ban on household aerosols. At the 0.01 significance level, test the claim that more than 70% of the population supports the ban.

4. In a Roper Organization poll of 2000 adults, 1280 have money in regular savings accounts. Use this sample data to test the claim that less than 65% of all adults have money in regular savings accounts. Use a 0.05 level of significance.

5. According to a Harris Poll, 71% of Americans believe that the overall cost of lawsuits is too high. If a random sample of 500 people results in 74% who hold that belief, test the claim that the actual percentage is 71%. Use a 0.10 significance level.

6. In clinical studies of the allergy drug Seldane, 70 of the 781 subjects experienced drowsiness (based on data from Merrell Dow Pharmaceuticals, Inc.). Test the claim that more than 8% of Seldane users experience drowsiness. Use a 0.05 significance level.

7. Test the claim that more than one-fourth of all white collar criminals have attended college. Sample data (from U.S. Bureau of Justice statistics) consist of 1400 randomly selected white collar criminals, with 33% of them having attended college. Use a 0.02 level of significance.

8. A study of 500 aircraft accidents involving spatial disorientation of the pilot found that 91% of those accidents resulted in fatalities (based on data from the Department of Transportation). Test the claim that three-fourths of all such accidents will result in fatalities.

9. Test the claim that less than 10% of U.S. senior medical students prefer pediatrics. Sample data consist of 1068 randomly selected medical school seniors, with 64 of them choosing pediatrics (based on data

reported by the Association of American Medical Colleges). Use a 0.01 significance level.

10. Among 785 subjects who completed four years of college, 18.3% of them smoke (based on data from the American Medical Association). At the 0.04 significance level, test the claim that among those who complete four years of college, less than one-fifth smoke.

11. Recently, TWA reported an on-time arrival rate of 78.4%. Assume that a later random sample of 750 flights results in 630 that are on time. If TWA were to claim that its on-time arrival rate is now higher than 78.4%, would that claim be supported at the 0.01 level of significance?

12. A study by the Environmental Protection Agency led to the claim that 4.4% of catalytic converters are removed from cars originally installed with them. Test this claim if a study of 200 cars built with catalytic converters reveals that 8.0% of them were removed. Use a 0.01 level of significance.

13. An airline reservations system suffers from a 7% rate of no-shows. A new procedure is instituted whereby reservations are confirmed on the day preceding the actual flight, and a study is then made of 5218 randomly selected reservations made under the new system. If 333 no-shows are recorded, test the claim that the no-show rate is lower with the new system.

14. In a genetics experiment, the Mendelian law is followed as expected if one-eighth of the offspring exhibit a certain recessive trait. Analysis of 500 randomly selected offspring indicates that 83 exhibited the necessary recessive trait. Is the Mendelian law being followed as expected? Use a 2% level of significance.

15. Test the claim that fewer than one-half of San Francisco residential telephones have unlisted numbers. A random sample of 400 such phones results in an unlisted rate of 39%. Use a 0.01 level of significance.

16. In a randomly selected group of people who bought compact cars, 312 were women and 338 were men (based on data from the Ford Motor Company). Test the claim that men constitute more than half of the buyers of compact cars.

17. In a survey of randomly selected households, 288 had computers while 962 did not (based on data from the Electronic Industries Association). At the 0.02 level of significance, test the claim that computers are in 20% of all households.

18. A study of randomly selected loans revealed that 37 were defaulted while 1383 had all obligations satisfied. At the 0.01 level of significance, test the claim that the loan default rate is less than 4%.

19. In a study of brand recognition, 831 subjects recognized the Campbell's soup brand, while 18 did not. Use this sample data to test the claim that the recognition rate is equal to 98%. Use a 0.10 level of significance.

20. The Kennedy-Nixon presidential race was extremely close. Kennedy won with 34,227,000 votes to Nixon's 34,108,000 votes. At the 0.01 level of significance, test the claim that the true population proportion for Kennedy exceeded 0.5. Assume that the voters represent a random sampling of those eligible.

## 7–5  Exercises B

21. A supplier of chemical waste containers finds that 3% of a sample of 500 units are defective. Being somewhat devious, he wants to make a claim that the defective rate is no more than some specified percentage, and he doesn't want that claim rejected at the 0.05 level of significance if the sample data are used. What is the *lowest* defective rate he can claim under these conditions?

22. A reporter claims that 10% of the residents of her city feel that the mayor is doing a good job. Test her claim if it is known that, in a random sample of 15 residents, there are none who feel that the mayor is doing a good job. Use a 5% level of significance. Since $np = 1.5$ and is not at least 5, the normal distribution is not a suitable approximation of the distribution of sample proportions.

23. A study of 500 aircraft accidents involving spatial disorientation of the pilot found that 91% of those accidents resulted in fatalities. Someone with a vested interest wants to claim that the percentage is at least some particular value, and they don't want that claim rejected at the 0.01 level of significance if the sample data are used. What is the *highest* percentage that can be claimed under these conditions?

24. Refer to the example in this section that relates to television violence. If the true value of $p$ is 0.45, find $\beta$, the probability of a type II error (see Exercise 32 from Section 7–2). *Hint:* In step 3 use the values of $p = 0.45$ and $\sqrt{pq/n} = \sqrt{(0.45)(0.55)/1998}$.

## 7–6  Tests of Variances

In this section we discuss tests of hypotheses made about a population variance $\sigma^2$ or standard deviation $\sigma$. Since $\sigma$ is the square root of $\sigma^2$, if we know the value of one we also know the value of the other. As a result, we can use the same procedure for testing claims about $\sigma$ or $\sigma^2$. The preceding sections of this chapter used the normal distribution and the Student $t$ distribution. Tests of claims about $\sigma^2$ or $\sigma$ again require that the population have normally distributed values, so the discussions of this section are made with the following assumptions.

## Assumptions

1. We are testing a hypothesis made about a population variance $\sigma^2$ or standard deviation $\sigma$.
2. The population has values that are normally distributed.

**Given these assumptions, the following test statistic has a chi-square distribution with $n - 1$ degrees of freedom and critical values given in Table A–4.**

---

### Test Statistic

$$\chi^2 = \frac{(n - 1)s^2}{\sigma^2}$$

where $n$ = sample size

$s^2$ = sample variance

$\sigma^2$ = population variance (given in the null hypothesis)

---

The chi-square distribution was introduced in Section 6–4 where we noted these important properties:

## Properties of the Chi-Square Distribution

1. All values of $\chi^2$ are nonnegative and the distribution is not symmetric (see Figure 7–14).

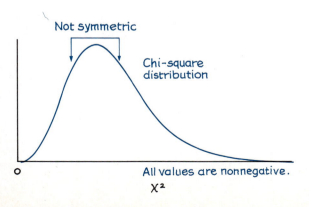

**Figure 7–14**

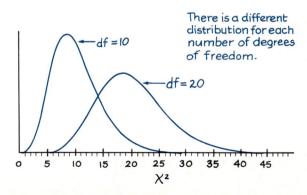

**Figure 7–15**

# AIRLINE CHILD SAFETY SEATS

■ Recently, a consumer's group lobbied for special safety seats to be made available for children on airliners. The Federal Aviation Administration responded with this argument: If babies are placed in their special safety seats, they would require their own seat on the plane, instead of being held by a parent. If babies require their own seats, they must be charged a fare. If you charge a fare for babies, the extra cost will force many parents to drive instead of fly. The fatality rate for cars is much higher than the rate for airliners. Therefore, in the long run, it's safer to let babies continue to fly in people's laps without requiring special safety seats. ■

2. There is a different distribution for each number of degrees of freedom (see Figure 7–15 on the previous page).
3. The **critical values** are found in Table A–4 where degrees of freedom $= n - 1$.

In using Table A–4, it is essential to note that each critical value separates an *area to the right* that corresponds to the value given in the top row.

As in Section 6–4, we should again note that later chapters will involve cases in which the degrees of freedom are not $n - 1$, so we should not universally equate degrees of freedom with $n - 1$. Once we have determined degrees of freedom, the significance level $\alpha$, and the type of test (left-tailed, right-tailed, or two-tailed), we can use Table A–4 to find the critical chi-square values.

In a right-tailed test, the value of $\alpha$ will correspond exactly to the areas given in the top row of Table A–4. In a left-tailed test, the value of $1 - \alpha$ will correspond exactly to the areas given in the top row of Table A–4. In a two-tailed test, the values of $\alpha/2$ and $1 - \alpha/2$ will correspond exactly to the areas given in the top row of Table A–4. (See Figure 6–8 and note the example on page 313.)

Many applications of statistics involve decisions or inferences about variances or standard deviations. In manufacturing, quality control engineers want to ensure that a product is, on the average, acceptable. But they also want to produce items of *consistent* quality so there will be few defective products. Consistency is often measured by variance or standard deviation. As a specific example, consider aircraft altimeters. Due to mass production techniques and various other factors, these altimeters don't give readings that are exactly correct; some errors are to be expected. Federal Aviation Regulation 91.36 requires that aircraft altimeters be tested and calibrated to give a reading "within 125 feet (on a 95-percent probability basis)." Even if the mean altitude reading is exactly correct, an excessively large standard deviation will result in individual readings that are dangerously low or high. Such a large standard deviation would mean that production is out of control and that unacceptable and dangerous altimeters are being manufactured.

## Example

A company has been successfully manufacturing aircraft altimeters with errors that have a mean of 0 ft (achieved by calibration) and a standard deviation of 43.7 ft. After installing new production equipment, 30 altimeters are randomly selected from the new line. This sample group has errors with a standard deviation of $s = 54.7$ ft. At the 0.05 level of significance, test the claim that the new population has errors with a standard deviation equal to 43.7 ft.

## Solution

We will follow the same general steps listed in Figure 7–1.

*Step 1:* The claim is that the new production method has resulted in a population with a standard deviation equal to 43.7 ft. In symbolic form we have $\sigma = 43.7$ ft.

*Step 2:* If the original claim is false, then $\sigma \neq 43.7$ ft.

*Step 3:* Since the null hypothesis must contain equality, we have
$H_0: \sigma = 43.7$
$H_1: \sigma \neq 43.7$

*Step 4:* The significance level is $\alpha = 0.05$.

*Step 5:* Because this claim is about $\sigma$, we will use the chi-square distribution.

*Step 6:* The test statistic is

$$\chi^2 = \frac{(n-1)s^2}{\sigma^2} = \frac{(30-1)(54.7)^2}{43.7^2} = 45.437$$

The critical values are 16.047 and 45.772. They are found in Table A–4, in the 29th row (degrees of freedom $= n - 1 = 29$) in the columns corresponding to 0.975 and 0.025. See Figure 7–16 on page 378 where the test statistic and critical values are shown.

*Step 7:* Because the test statistic is not in the critical region, we fail to reject the null hypothesis.

*Step 8:* There is not sufficient evidence to warrant rejection of the claim that the standard deviation is equal to 43.7 ft. However, it would be wise to continue monitoring and testing the new product line.

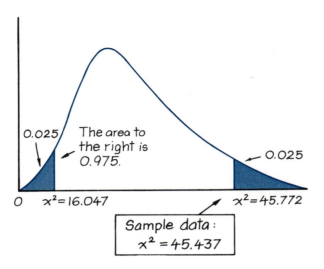

**Figure 7–16**

Another example in which variability is a major concern involves the waiting lines of banks. In the past, customers traditionally entered a bank and selected one of several lines formed at different windows. A different system growing in popularity involves one main waiting line, which feeds the various windows as vacancies occur. The mean waiting time isn't reduced, but the variability among waiting times is decreased and the irritation of being caught in a slow line is also diminished.

> ### Example
>
> With individual lines at its various windows, a bank finds that the standard deviation for normally distributed waiting times on Friday afternoons is 6.2 min. The bank experiments with a single main waiting line and finds that for a random sample of 25 customers, the waiting times have a standard deviation of 3.8 min. At the $\alpha = 0.05$ significance level, test the claim that a single line causes lower variation among the waiting times.

## Solution

We wish to test $\sigma < 6.2$ based on a sample of $n = 25$ for which $s = 3.8$. We begin by identifying the null and alternative hypotheses.

$$H_0: \sigma \geq 6.2$$

$$H_1: \sigma < 6.2$$

The significance level of $\alpha = 0.05$ has already been selected, so we proceed to compute the value of $\chi^2$ based on the given data:

$$\chi^2 = \frac{(n-1)s^2}{\sigma^2} = \frac{(25-1)(3.8)^2}{(6.2)^2} = 9.016$$

This test is left-tailed since $H_0$ will be rejected only for small values of $\chi^2$; with $\alpha = 0.05$ and $n = 25$, we go to Table A–4 and align 24 degrees of freedom with an area of 0.95 to obtain the critical $\chi^2$ value of 13.848 (see Figure 7–17). Since the test statistic falls within the critical region, we reject $H_0$ and conclude that the 3.8-min standard deviation is significantly less than the 6.2-min standard deviation that corresponds to multiple waiting lines. The sample data support the claim of lower variation. That is, the single main line does appear to lower the variation among waiting times.

# *P*-Values

The *P*-value approach to hypothesis testing can be used to test claims made about population standard deviations or variances. Since the chi-square distribution table (Table A–4) includes only selected values of $\alpha$, we cannot usually find the specific *P*-value from that table. Instead, we can use the table to identify limits that contain the *P*-value. In the last example we found the test statistic to be $\chi^2 = 9.016$ and we know that the test is left-tailed with 24 degrees of freedom. By examining the 24th row of Table A–4, we see that the test statistic of 9.016 is less than the lowest table value of 9.886, so the *P*-value must be less than 0.005. With a significance level of $\alpha = 0.05$, we reject the null hypothesis as we did in the classical approach used in the example given.

Although Table A–4 can be used to estimate *P*-values, we can find exact *P*-values using some calculators or statistical software, such as STATDISK or Minitab. The STATDISK display for the last example appears on the following page. We entered $6.2^2 = 38.44$ for the population variance, and $3.8^2 = 14.44$ for the sample variance. Figure 7–17 is also displayed by STATDISK (see the following page).

In this section we have illustrated one important use of the chi-square distribution. Other valuable uses, such as in analyzing the significance of differences between expected frequencies and the frequencies that actually occur, are considered in Chapter 10.

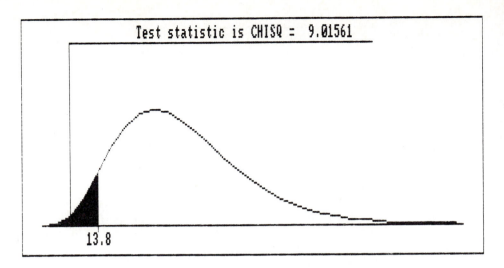

**Figure 7–17**

STATDISK DISPLAY

Hypothesis test for a claim about a SINGLE
POPULATION STANDARD DEVIATION or VARIANCE

NULL HYPOTHESIS: Population variance > = 38.44

Sample size . . . . . . . . . . . . . . . . . . . . . . . . . . . . . . . . . . . . . n = 25
Sample variance . . . . . . . . . . . . . . . . . . . . . . . . . . . . . . . . . . . . = 14.44
Degrees of freedom . . . . . . . . . . . . . . . . . . . . . . . . . . . . df = 24
Test statistic . . . . . . . . . . . . . . . . . . . . . . . . . . . . . . .CHISQ = 9.01561
Critical value . . . . . . . . . . . . . . . . . . . . . . . . . . . . . . . . .CHISQ = 13.8
P-value . . . . . . . . . . . . . . . . . . . . . . . . . . . . . . . . . . . . . . = .00257
Significance level . . . . . . . . . . . . . . . . . . . . . . . . . . . . . . . = .05

CONCLUSION: REJECT the null hypothesis

# 7–6  Exercises A

In Exercises 1 and 2, use Table A–4 to find the critical values of $\chi^2$ based on
the given data.

1.  a.  $\alpha = 0.05$
        $n = 20$
        $H_0: \sigma = 16$
    b.  $\alpha = 0.05$
        $n = 20$
        $H_0: \sigma^2 \geq 256$
    c.  $\alpha = 0.01$
        $n = 23$
        $H_0: \sigma^2 = 10$
    d.  $\alpha = 0.01$
        $n = 23$
        $H_0: \sigma^2 \leq 10$
    e.  $\alpha = 0.005$
        $n = 15$
        $H_1: \sigma < 4.83$

2.  a.  $\alpha = 0.10$
        $n = 6$
        $H_1: \sigma < 10$
    b.  $\sigma = 0.05$
        $n = 40$
        $H_1: \sigma^2 > 500$
    c.  $\alpha = 0.025$
        $n = 81$
        $H_0: \sigma^2 \geq 144$
    d.  $\alpha = 0.01$
        $n = 50$
        $H_0: \sigma = 15$
    e.  $\alpha = 0.05$
        $n = 75$
        $H_1: \sigma^2 \neq 31.5$

In Exercises 3–20, test the given hypotheses. Follow the pattern outlined in Figure 7–1 and draw the appropriate graph. In all cases, assume that the population is normally distributed.

3.  At the $\alpha = 0.05$ significance level, test the claim that $\sigma^2 > 100$ if a random sample of 27 yields $s^2 = 194$.

4.  At the $\alpha = 0.05$ significance level, test the claim that $\sigma^2 = 100$ if a random sample of 27 yields a variance of 57.

5.  At the $\alpha = 0.01$ significance level, test the claim that a population has a standard deviation of 10.0. A random sample of 18 items yields a standard deviation of 14.5.

6.  At the $\alpha = 0.05$ significance level, test the claim that a population has a standard deviation of 52.0. A random sample of 18 items yields a standard deviation of 71.2.

7.  At the $\alpha = 0.05$ significance level, test the claim that a population has a variance less than or equal to 9.00. A random sample of 81 items yields a variance of 12.25.

8.  At the $\alpha = 0.025$ level of significance, test the claim that a population has a standard deviation less than 98.6. A random sample of 51 items yields a standard deviation of 79.000.

9.  A new production method is used to manufacture aircraft altimeters. A random sample of 81 altimeters results in errors with a standard deviation of $s = 52.3$ ft. At the 0.05 level of significance, test the claim that the new production line has errors with a standard deviation equal to 43.7 ft., which was the standard deviation for the old production method.

10. If the standard deviation for the weekly downtimes of a computer is low, then availability of the computer is predictable and planning is facilitated. If 12 weekly downtimes for a computer are randomly selected and the standard deviation is computed to be 2.85 h, at the 0.025 significance level, test the claim that $\sigma > 2.00$ h.

11. A machine pours medicine into a bottle in such a way that the standard deviation of the weights is 0.15 oz. A new machine is tested on 71 bottles, and the standard deviation for this group is 0.12 oz. At the 0.05 significance level, test the claim that the new machine produces less variance.

12. Test the claim that scores on a standard IQ test have a standard deviation equal to 15 if a sample of 24 randomly selected subjects yields a standard deviation of 10. Use a significance level of $\alpha = 0.01$.

13. In comparing systolic blood pressure levels of men and women, a medical researcher obtains readings for a random sample of 50 women. The sample mean and standard deviation are found to be 130.7 and 23.4, respectively. If systolic blood pressure levels for men are known to have a mean and standard deviation of 133.4 and 19.7, respectively, test the claim that women have a larger standard deviation. Use a 0.05 level of significance. (All readings are in millimeters of mercury, and the data are based on the National Health Survey, USDHHS publication 81-1671.)

14. A software firm finds that the times required to run one particular computer program have a standard deviation of 52 h. A sample of 30 new computer programs produces a standard deviation of 60 h. At the 0.05 significance level, test the claim that the standard deviation for the new computer programs is equal to 52 h.

15. When 22 bolts are tested for hardness, their indices have a standard deviation of 65.0. Test the claim that the standard deviation of the hardness indices for all such bolts is greater than 50.0. Test at the 0.025 level of significance.

16. In a study of the wide ranges in the academic success of college freshmen, an obvious factor is the amount of time spent studying. At the 0.05 significance level, test the claim that the standard deviation is more than 4.00 h. Sample data consist of 70 randomly selected freshmen who have a standard deviation of 5.33 h (based on data reported by *USA Today*).

17. The caffeine contents (in mg) for a dozen randomly selected cans of a soft drink are given below. At the 0.025 level of significance, test the claim that the standard deviation for all such cans is less than 2.0 mg.

    34.2  33.7  31.9  34.3  31.6  32.7
    33.1  35.2  31.6  32.9  33.0  32.4

18. Based on data from the National Health Survey (USDHEW) publication 79-1659), men aged 25 to 34 have heights with a standard deviation of 2.9 in. Test the claim that men aged 45 to 54 have heights with a standard deviation less than 2.9 in. The heights of 25 randomly selected men in the 45 to 54 age bracket are listed at the top of the following page.

| | | | | | | |
|---|---|---|---|---|---|---|
| 66.80 | 71.22 | 65.80 | 66.24 | 69.62 | 70.49 | 70.00 |
| 71.46 | 65.72 | 68.10 | 72.14 | 71.58 | 66.85 | 69.88 |
| 68.69 | 72.77 | 67.34 | 68.40 | 68.96 | 68.70 | 72.69 |
| 68.67 | 67.79 | 63.97 | 67.19 | | | |

19. Given below are birthweights (in kilograms) of male babies born to mothers on a special vitamin supplement (based on data from the New York State Department of Health). Test the claim that this sample comes from a population with a standard deviation equal to 0.470 kg, which is the standard deviation for male birthweights in general.

| | | | | | | | |
|---|---|---|---|---|---|---|---|
| 3.73 | 4.37 | 3.73 | 4.33 | 3.39 | 3.68 | 4.68 | 3.52 |
| 3.02 | 4.09 | 2.47 | 4.13 | 4.47 | 3.22 | 3.43 | 2.54 |

20. Listed below are the total electric energy consumption amounts (in kWh) for the author's home during seven different years. At the 0.10 significance level, test the claim that the standard deviation for all such years is equal to 1000 kWh.

| | | | | | | |
|---|---|---|---|---|---|---|
| 11,943 | 11,463 | 10,789 | 9907 | 9012 | 9942 | 11,153 |

# 7–6 Exercises B

21. For large numbers of degrees of freedom, we can approximate values of $\chi^2$ as follows.

$$\chi^2 = \frac{1}{2}[z + \sqrt{2k - 1}]^2$$

Here $k$ = number of degrees of freedom and $z$ = the critical score, found in Table A–2.

For example, if we want to approximate the two critical values of $\chi^2$ in a two-tailed hypothesis test with $\alpha = 0.05$ and a sample size of 150, we let $k = 149$ with $z = -1.96$, followed by $k = 149$ and $z = 1.96$.

a. Use this approximation to estimate the critical values of $\chi^2$ in a two-tailed hypothesis test when $n = 101$ and $\alpha = 0.05$. Compare the results to those found in Table A–4.

b. Use this approximation to estimate the critical values of $\chi^2$ in a two-tailed hypothesis test when $n = 150$ and $\alpha = 0.05$.

22. Do Exercise 21 using the approximation

$$\chi^2 = k\left[1 - \frac{2}{9k} + z\sqrt{\frac{2}{9k}}\right]^3$$

Here $k$ and $z$ are as described in that exercise.

23. What do you know about the *P*-value in each of the following cases?
    a.   $H_1$: $\sigma > 15.0$; $n = 10$; test statistic is $\chi^2 = 19.735$.
    b.   $H_1$: $\sigma < 45.0$; $n = 20$; test statistic is $\chi^2 = 7.337$.
    c.   $H_0$: $\sigma = 1.52$; $n = 30$; test statistic is $\chi^2 = 54.603$.
24. Refer to the last example presented in this section. Assuming that $\sigma$ is actually 4.0, find $\beta$ (the probability of a type II error). See Exercise 32 from Section 7–2 and modify the procedure so that it applies to a hypothesis test involving $\sigma$ instead of $\mu$.

 *Vocabulary List*

Define and give an example of each term.

| | |
|---|---|
| hypothesis | significance level |
| hypothesis test | right-tailed test |
| test of significance | left-tailed test |
| null hypothesis | two-tailed test |
| alternative hypothesis | *P*-value |
| type I error | Student *t* distribution |
| type II error | *t* distribution |
| test statistic | degrees of freedom |
| critical region | chi-square distribution |
| critical value | |

 *Review*

Chapters 6 and 7 introduce two of the most important methods used in working with sample data to make inferences about a population. While the major objective of Chapter 6 was estimating the values of population parameters, this chapter focused on methods for testing hypotheses made about population parameters. We considered claims made about population means, proportions, variances, and standard deviations. Hypothesis tests are also called tests of significance, reflecting the fact that we decide whether sample differences are due to chance fluctuations or whether the differences are so dramatic that they are not likely to occur by chance. We are able to select exact **levels of significance**; 0.05 and 0.01 are common values. Sample results are said to reflect significant differences when their occurrences have probabilities less than the chosen level of significance.

Section 7–2 presented in detail the procedure for testing hypotheses. The essential steps are summarized in Figure 7–1. We defined **null hypothesis, alternative hypothesis, type I error, type II error, test statistic, critical region, critical value,** and **significance level.** All these standard terms are commonly used in discussing tests of hypotheses. We also identified the three basic types of tests: **right-tailed, left-tailed,** and **two-tailed.**

We introduced the method of testing hypotheses in Section 7–2 by using examples in which only the normal distribution applies, and we introduced other distributions in subsequent sections. The brief table that follows summarizes the hypothesis tests covered in this chapter.

Section 7–2 outlined the **classical approach** to testing hypotheses, while the **P-value** approach was presented in Section 7–3. In the classical approach we make a decision about the null hypothesis by comparing the test statistic and critical value. With the P-value approach we base that decision on a comparison of the significance level and the P-value, which represents the probability of getting a sample that is at least as extreme as the one obtained.

| Important Formulas | | | | |
|---|---|---|---|---|
| Parameter to which hypothesis refers | Applicable distribution | Assumption | Test statistic | Table of critical values |
| $\mu$ (population mean) | Normal | $\sigma$ is known and population is normally distributed | $z = \dfrac{\bar{x} - \mu}{\sigma/\sqrt{n}}$ | Table A–2 |
| | Normal | $n > 30$ (If $\sigma$ is not known, use $s$ for $\sigma$.) | $z = \dfrac{\bar{x} - \mu}{\sigma/\sqrt{n}}$ | Table A–2 |
| | Student $t$ | $\sigma$ is unknown and $n \le 30$ and population is normally distributed | $t = \dfrac{\bar{x} - \mu}{s/\sqrt{n}}$ | Table A–3 |
| $p$ (population proportion) | Normal | $np \ge 5$   and   $nq \ge 5$ | $z = \dfrac{\hat{p} - p}{\sqrt{pq/n}}$ where $\hat{p} = \dfrac{x}{n}$ | Table A–2 |
| $\sigma^2$ (population variance) $\sigma$ (population standard deviation) | Chi-square | Population is normally distributed | $\chi^2 = \dfrac{(n-1)s^2}{\sigma^2}$ | Table A–4 |

# 2 Review Exercises

In Exercises 1 and 2, find the appropriate critical values.

1.  a.  $\alpha = 0.05$
        $n = 160$
        $H_0: p \geq 0.5$
    b.  $\alpha = 0.01$
        $n = 35$
        $H_0: \mu \leq 16.5$
    c.  $\alpha = 0.01$
        $n = 12$
        $H_0: \mu = 38.4$
    d.  $\alpha = 0.01$
        $n = 25$
        $H_0: \sigma^2 \geq 225$
    e.  $\alpha = 0.05$
        $n = 30$
        $H_0: \sigma^2 = 84.3$

2.  a.  $\alpha = 0.10$
        $n = 15$
        $H_0: \mu = 1.23$
    b.  $\alpha = 0.10$
        $n = 15$
        $H_0: \sigma^2 = 123$
    c.  $\alpha = 0.06$
        $n = 100$
        $H_0: \mu = 72.3$
    d.  $\alpha = 0.05$
        $n = 10$
        $H_0: \sigma = 15$
    e.  $\alpha = 0.01$
        $n = 30$
        $H_1: \sigma < 5.8$

In Exercises 3 and 4, respond to each of the following:

a.  Give the null hypothesis in symbolic form.
b.  Is this test left-tailed, right-tailed, or two-tailed?
c.  In simple terms devoid of symbolism and technical language, describe the type I error.
d.  In simple terms devoid of symbolism and technical language, describe the type II error.
e.  What is the probability of making a type I error?

3.  At the 0.01 level of significance, the claim is that the mean treatment time for a dentist is at least 20.0 min.

4.  At the 5% level of significance, the claim is that the mean reading in a biofeedback experiment is 6.2 mV.

5.  At the 0.05 significance level, test the claim that 15% of U.S. families have incomes below the poverty level. A random sample of 600 families included exactly 81 below the poverty level (based on data from the U.S. Bureau of the Census).

6.  At the 0.05 level of significance, test the claim that $\mu = 10.0$ s. Sample data consist of 100 scores, for which $\bar{x} = 8.2$ s. Assume that $\sigma = 6.0$ s.

7.  At the 0.01 level of significance, test the claim that $\mu \leq 25.5$ ft. Sample data consist of 25 observations, for which $\bar{x} = 27.3$ ft and $s = 3.7$ ft.

8.  In a study of birth weights (in grams), a random sample of 30 baby girls produces a mean of 3264 g and a standard deviation of 485 g. Assume that the mean and standard deviation for birth weights of

boys are known to be 3393 g and 470 g, respectively (based on data from the New York State Department of Health, Monograph No. 11). At the 0.05 level of significance, test the claim that boys and girls have birth weights with the same standard deviation.

9.  Test the claim that a certain X-ray machine gives radiation dosages with a mean below 5.00 milliroentgens. Sample data consist of 36 observations with a mean of 4.13 milliroentgens and a standard deviation of 1.91 milliroentgens. Use a 0.01 level of significance.

10.  A manufacturer considers her production process to be out of control when defects exceed 3%. A random sample of 500 items includes exactly 22 defects, but the manager claims that this represents a chance fluctuation and that production is not really out of control. Test the manager's claim at the 5% level of significance.

11.  Use a 0.025 level of significance to test the claim that vehicle speeds at a certain location have a mean above 55.0 mi/h. A random sample of 50 vehicles produces a mean of 61.3 mi/h and a standard deviation of 3.3 mi/h.

12.  A standard test for braking reaction times (in seconds) has produced an average of 0.75 s for young females. A driving instructor claims that his class of young females exhibits an overall reaction time that is below the average. Test his claim if it is known that 13 of his students (randomly selected) produced a mean of 0.71 s and a standard deviation of 0.06 s. Test at the 1% level of significance.

13.  One large high school has found that students taking a standard college aptitude test earn scores with a variance of 6410. A counselor claims that the current group of test subjects includes a group with more varied aptitudes. Test the claim that the variance is larger than 6410 if a random sample of 60 students produces a variance of 8464. Use a 0.10 level of significance.

14.  A sociologist designs a test to measure prejudicial attitudes and claims that the mean population score is 60. The test is then administered to 28 randomly selected subjects and the results produce a mean and standard deviation of 69 and 12, respectively. At the 5% level of significance, test the sociologist's claim.

15.  In a Gallup poll of 1553 randomly selected adult Americans, 92% recognize the name Christopher Columbus. At the 0.01 significance level, test the claim that 95% of all adult Americans recognize the name Columbus.

16.  In a study of blood pressure levels, a sample of 75 randomly selected women aged 25 to 34 results in a mean systolic value of 116.7 and a standard deviation of 12.5, with both values measured in millimeters of mercury (based on data from the National Health Survey, USDHHS publications 81–1671). At the 0.05 level of significance, test the claim that this sample comes from a population with a mean equal to 120.0 mm of mercury.

17. The following sample scores have been randomly selected from a normally distributed population. At the 0.01 significance level, test the claim that the population has a mean of 100.

$$101 \quad 106 \quad 98 \quad 92 \quad 97 \quad 80 \quad 89 \quad 88$$
$$110 \quad 112 \quad 100 \quad 100 \quad 103 \quad 97 \quad 97$$

18. Using the same sample data from Exercise 17, test the claim that the standard deviation is less than 10.0. Use a 0.05 level of significance.

19. Test the claim that the mean female reaction time to a highway signal is less than 0.700 s. When 18 females are randomly selected and tested, their mean is 0.668 s. Assume that $\sigma = 0.100$ s, and use a 5% level of significance.

20. Test the claim that less than half of those earning a bachelor's degree in business are women. A random sample of 200 business graduates earning the bachelor's degree includes 80 women and 120 men.

 ## *Computer Projects*

1. Most statistics software packages include procedures for testing hypotheses. Use a software package such as STATDISK or Minitab for Exercise 27 from Section 7–4.

2. One advantage of computers is that they sometimes make possible approaches that are otherwise impractical. We will use a simulation technique for a hypothesis test. Assume that we have a sample of 10 IQ scores with a sample mean of 103. The population of all IQ scores has a standard deviation of 15. At the 5% significance level, we want to test the claim that the sample comes from a population with a mean greater than 100. Instead of the classical or *P*-value approach, we will use a computer simulation to decide whether a sample mean of 103 could easily occur by chance or whether it constitutes a significant difference. Proceed as follows:

   a. Use a software package such as STATDISK or Minitab to generate 10 sample score from a normally distributed population with a mean of 100 and a standard deviation of 15. Record the mean of these 10 scores.

   b. Repeat step *a* until you have 20 sample means.

   c. Examine the 20 sample means and decide whether or not a sample mean of 103 can easily occur by chance or whether it constitutes a significant difference from 100. (In this problem, "easily" means that you get 103 or above more than 5% of the time.)

   d. Form a conclusion.

*Applied Projects*

1. Using Data Set II in Appendix B, test the claim that the percentage of left-handed people is less than 15%.
2. Refer to the real estate data in Appendix B. For the same time period during which the 150 homes were sold, *U.S. Housing Markets* reported that the average price of a home in the United States was $121,000. Test the claim that the sample of the 150 homes comes from a population with a mean selling price that is greater than the national average of $121,000.
3. Use Data Set II in Appendix B to test the claim that for women who are at least 66 in. tall, the mean weight is greater than 120 lb.
4. Conduct a survey by asking the question "Do you favor or oppose the death penalty for people convicted of murder?" Survey at least 50 people and test the claim that the proportion in favor is less than or equal to 0.5. Identify the population from which you are sampling, and identify any factors that might suggest that your sample is not representative of the population. (If you record the response along with the sex of the respondent, you may be able to use the same data in Chapter 10.)
5. A 1921 study of 1 million draft recruits aged 21–30 years found that heights had a mean of 67.49 in. and a standard deviation of 2.71 in. (based on data from "Army Anthropology" by Davenport and Love, *American Journal of Anatomy*). Refer to Data Set II in Appendix B, find the heights of males in the same age bracket, and test the claim that this more recent sample comes from a population with the same mean. If the heights now appear to be different, how have they changed? Why?

 *Writing Projects*

1. In one full page or less, describe the basic rationale that underlies the hypothesis testing procedure.
2. In testing the claim that the mean IQ of statistics professors is greater than 100, a *P*-value of 0.003 is obtained. Interpret this result.
3. Describe how a confidence interval can be used as a substitute for a two-tailed hypothesis test.
4. Summarize one of the programs listed below.

 *Videotapes*

Programs 20 and 21 from the series *Against All Odds: Inside Statistics* are recommended as supplements to this chapter.

# Chapter Eight

## In This Chapter

# 8 Inferences from Two Samples

## Chapter Problem

In a recent study of 22,000 male physicians, half were given regular doses of aspirin while the other half were given placebos. The study ran for six years at a cost of $4.4 million. Among those who took the aspirin, 104 suffered heart attacks. Among those who took the placebos, 189 suffered heart attacks. (The figures are based on data from *Time* and the *New England Journal of Medicine*, Vol. 318. No. 4.) Criticism, disagreement, and many words of caution followed the report of these results, but we will focus on the central statistical issue: Do these results show a statistically significant decrease in heart attacks among the sample group who took aspirin? The issue is clearly quite important, since it can affect many lives, not to mention the sales of aspirin.

In this chapter we will test the claim that the proportion of heart attacks among the aspirin group is significantly lower than the proportion for the placebo group.

# 8–1 Overview

We use inferential statistics when we use sample data to form conclusions about populations. The last two chapters introduced two of the most important and practical topics in the field of inferential statistics. Chapter 6 introduced methods of estimating population parameters, while Chapter 7 introduced methods of testing hypotheses made about population parameters. Both chapters shared this feature: All examples and exercises involved the use of *one* sample to form an inference about *one* population. In reality, there are many cases in which the main objective is the comparison of *two* groups of data. The preceding chapter problem identifies a typical experiment based on the comparison of a control group and a treatment group. This chapter presents methods for using data from two samples so that inferences can be made about the populations from which they came.

Chapters 6 and 7 both began with means, followed by proportions, then variances. In this chapter we depart from that order because we will sometimes use results of tests comparing two variances as a prerequisite for tests comparing two means. Consequently, we begin this chapter with a discussion of the method for comparing two variances.

# 8–2 Comparing Two Variances

## Hypothesis Tests

This section presents a method for using two samples to compare the variances of the populations from which the samples are drawn. The method we use requires the following assumptions.

### Assumptions

1. Based on two samples drawn from two populations, we want to compare the **variances** of the two populations.
2. The two populations are **independent** of each other.
3. The two populations are each **normally distributed.**

The third assumption—that both populations have normal distributions—is critical for the $F$-test, which is presented in this section. The relevant test statistic is very sensitive to departures from normality, and this extreme sensitivity does not diminish with large samples. Other tests in the following sections of this chapter are not so sensitive to departures from normality.

| TABLE 8–1    Life of Car Batteries (in years) | |
|---|---|
| Production Method A | Production Method B |
| 2.0 | 3.6 |
| 2.1 | 3.7 |
| 2.5 | 3.9 |
| 3.0 | 3.9 |
| 3.3 | 3.9 |
| 4.2 | 4.0 |
| 4.2 | 4.0 |
| 4.3 | 4.0 |
| 6.8 | 4.1 |
| 7.6 | 4.2 |
| | 4.3 |
| | 4.4 |
| $n_1 = 10$ | $n_2 = 12$ |
| $\bar{x}_1 = 4.00$ | $\bar{x}_2 = 4.00$ |
| $s_1^2 = 3.59$ | $s_2^2 = 0.05$ |

For an example, see the two groups of sample data in Table 8–1. A manufacturer of car batteries wants to compare two production methods. From the statistics included in the table, we can see that both groups of batteries seem to last the same length of time, but the batteries produced by method A show much less consistency. The different degrees of consistency are reflected in the obviously different sample variances of 3.59 and 0.05. Since not all comparisons of variances are so obvious, we need more standardized procedures. Even for the data in Table 8–1, the difference between the sample variances of 3.59 and 0.05 must be weighed against the sample sizes to determine whether this "obvious" difference is statistically significant.

Our computations will be simplified if we stipulate that $s_1^2$ represents the larger of the two sample variances. This stipulation presents no logical difficulties because the identification of the samples through subscript notation is arbitrary. We choose the following notation.

## Notation

$s_1^2$ = **larger** of the two **sample variances**

$n_1$ = size of the sample with the **larger** variance

$\sigma_1^2$ = variance of the **population** from which the sample
     was drawn

The symbols $s_2^2$, $n_2$, $\sigma_2^2$ correspond to the other population and sample.

Extensive analyses have shown that **for two normally distributed populations with equal variances (that is, $\sigma_1^2 = \sigma_2^2$), the sampling distribution of the following test statistic is the $F$ distribution shown in Figure 8–1 with critical values listed in Table A–5.**

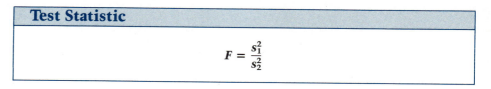

**Test Statistic**

$$F = \frac{s_1^2}{s_2^2}$$

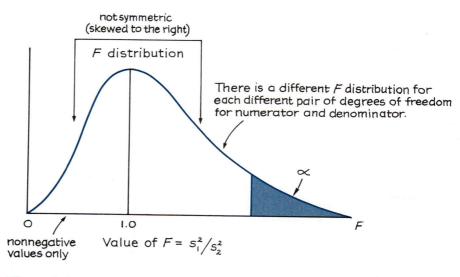

not symmetric
(skewed to the right)

$F$ distribution

There is a different $F$ distribution for each different pair of degrees of freedom for numerator and denominator.

$\alpha$

O

nonnegative
values only

1.0

Value of $F = s_1^2/s_2^2$

$F$

**Figure 8–1**

If the two populations really do have equal variances, then $F = s_1^2/s_2^2$ tends to be close to 1 since $s_1^2$ and $s_2^2$ tend to be close in value. But if the two populations have radically different variances, $s_1^2$ and $s_2^2$ tend to be very different numbers. Denoting the larger of the sample variances by $s_1^2$, we see that the ratio $s_1^2/s_2^2$ will be a large number whenever $s_1^2$ and $s_2^2$ are far apart in value. Consequently, a value of $F$ near 1 will be evidence in favor of the conclusion that $\sigma_1^2 = \sigma_2^2$. A large value of $F$ will be evidence against the conclusion of equality of the population variances.

When we use Table A–5 we obtain **critical $F$ values** that are determined by the following three values:

1.  The significance level $\alpha$.

2.  The **degrees of freedom for the numerator,** $(n_1 - 1)$

3.  The **degrees of freedom for the denominator,** $(n_2 - 1)$

When using Table A–5, be sure that $n_1$ corresponds to the sample having variance $s_1^2$, while $n_2$ is the size of the sample with variance $s_2^2$. Identify a level of significance and determine whether the test is one-tailed or two-tailed. For a one-tailed test use the significance level found in Table A–5. (Since we stipulate that the larger sample variance is $s_1^2$, all one-tailed tests will be right-tailed.) For a two-tailed test, first divide the area of the critical region (equal to the significance level) equally between the two tails and refer to that part of Table A–5 that represents *one-half* of the significance level. In that part of Table A–5, intersect the column representing the degrees of freedom for $s_1^2$ with the row representing the degrees of freedom for $s_2^2$. (Unlike the normal and Student $t$ distributions, the $F$ distribution is not symmetric and does not have 0 at its center. Consequently, left-tail critical values *cannot* be found by using the negative of the right-tail critical values. Instead, left-tail critical values can be found by using the reciprocal of the right-tail value with the numbers of degrees of freedom reversed. See Exercise 21.)

The following example illustrates the testing of hypotheses about two population variances.

---

**Example**

Table 8–1 lists the lives (in years) of car batteries produced by two different production methods. The statistics are given below with extra decimal places included. At the 0.05 significance level, test the claim that the variance of all battery lives from production method A is equal to the variance for production method B. It is reasonable to assume that each set of sample data comes from a normally distributed population.

| Production Method A | Production Method B |
|---|---|
| $n_1 = 10$ | $n_2 = 12$ |
| $\bar{x}_1 = 4.0000$ | $\bar{x}_2 = 4.0000$ |
| $s_1^2 = 3.5911$ | $s_2^2 = 0.0527$ |

**Solution**

Since the larger variance is already denoted by $s_1^2$, we use the same subscript notation given above. (If $s_2^2$ had been larger, we would have interchanged the subscripts so that $s_1^2$ would be the larger variance.) We now proceed to follow the steps for hypothesis testing as outlined in Figure 7–1.

*Step 1:* The claim of equal variances is expressed in symbolic form as
$$\sigma_1^2 = \sigma_2^2.$$

*continued*

## POLLSTER LOU HARRIS

■ Lou Harris has been in the polling business since 1947, and he has been involved with about 250 political campaigns. While George Gallup believes that pollsters should remain detached and objective, Lou Harris prefers a more personal involvement. He advised John F. Kennedy to openly attack the prejudice against a Catholic becoming president. Kennedy followed that advice and won the nomination. ■

**Solution** *continued*

*Step 2:* If the original claim is false, then $\sigma_1^2 \neq \sigma_2^2$.

*Step 3:* Since the null hypothesis must contain equality, we have
$$H_0: \sigma_1^2 = \sigma_2^2$$
$$H_1: \sigma_1^2 \neq \sigma_2^2$$

*Step 4:* The significance level is $\alpha = 0.05$.

*Step 5:* Because this test involves two population variances, we use the $F$ distribution.

*Step 6:* The test statistic is

$$F = \frac{s_1^2}{s_2^2} = \frac{3.5911}{0.0527} = 68.1423$$

For the critical values, we first note that this is a two-tailed test with 0.025 in each tail. As long as we are stipulating that the larger variance is placed in the numerator of the $F$ test statistic, we need to find only the right-tailed critical value. From Table A–5 we get 3.5879, which corresponds to 0.025 in the right tail, with 9 degrees of freedom for the numerator and 11 degrees of freedom for the denominator (see Figure 8–2).

*Step 7:* Figure 8–2 shows that the test statistic does fall well within the critical region, so we reject the null hypothesis.

*Step 8:* There is sufficient evidence to warrant rejection of the claim that the two variances are equal.

If the preceding test had been right-tailed instead of two-tailed, we would have obtained a critical value of $F = 2.8962$ for a right-tailed area of 0.05.

We can use the same procedures to test claims about two population standard deviations. Any claim about two population standard deviations can be restated in terms of the corresponding variances. For example, suppose we want to test the claim that $\sigma_1 = \sigma_2$, and we are given the following sample data taken from normal populations.

| Sample 1 | Sample 2 |
|---|---|
| $n_1 = 16$ | $n_2 = 10$ |
| $s_1 = 15$ | $s_2 = 10$ |

Restating the claim as $\sigma_1^2 = \sigma_2^2$, we get $s_1^2 = 15^2 = 225$ and $s_2^2 = 10^2 = 100$, and we can proceed with the $F$ test in the usual way. Rejection of $\sigma_1^2 = \sigma_2^2$ is

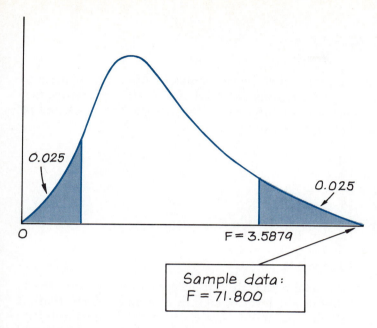

**Figure 8–2**

equivalent to rejection of $\sigma_1 = \sigma_2$. Failure to reject $\sigma_1^2 = \sigma_2^2$ implies failure to reject $\sigma_1 = \sigma_2$.

Note that in all tests of hypotheses made about population variances and standard deviations, the values of the means are irrelevant. In the next section we consider tests comparing two population means, and we will see that some of those tests will require the hypothesis tests described in this section.

## Confidence Intervals

When comparing two population variances, hypothesis tests are used much more often than confidence intervals. Confidence intervals (or interval estimates) of the ratio $\sigma_1^2/\sigma_2^2$ can be constructed by using

$$\frac{s_1^2}{s_2^2} \cdot \frac{1}{F_R} < \frac{\sigma_1^2}{\sigma_2^2} < \frac{s_1^2}{s_2^2} \cdot \frac{1}{F_L}$$

Here $F_L$ and $F_R$ are as described in Exercise 21 (see Exercises 21 and 22). The confidence intervals in the following sections are generally more important and we will examine them more closely.

# 8–2 | Exercises A

In Exercises 1–4, test the claim that the two samples come from populations having equal variances. Use a significance level of $\alpha = 0.05$ and assume that all populations are normally distributed. Follow the pattern suggested by Figure 7–1 and draw the appropriate graphs.

1. Sample A: $n = 10, s^2 = 50$
   Sample B: $n = 10, s^2 = 25$
2. Sample A: $n = 10, s^2 = 50$
   Sample B: $n = 15, s^2 = 25$
3. Sample A: $n = 5, \bar{x} = 372, s = 14.3$
   Sample B: $n = 15, \bar{x} = 298, s = 1.1$
4. Sample A: $n = 25, \bar{x} = 583, s = 3.9$
   Sample B: $n = 10, \bar{x} = 648, s = 6.3$

In Exercises 5–8, test the claim that the variance of population A exceeds that of population B. Use a 5% level of significance and assume that all populations are normally distributed. Follow the pattern outlined in Figure 7–1 and draw the appropriate graphs.

5. Sample A: $n = 10, \bar{x} = 200, s^2 = 48$
   Sample B: $n = 10, \bar{x} = 180, s^2 = 12$
6. Sample A: $n = 50, \bar{x} = 75.3, s^2 = 18.2$
   Sample B: $n = 20, \bar{x} = 75.9, s^2 = 8.7$
7. Sample A: $n = 16, \bar{x} = 124, s^2 = 225$
   Sample B: $n = 200, \bar{x} = 128, s^2 = 160$
8. Sample A: $n = 35, \bar{x} = 238, s^2 = 42.3$
   Sample B: $n = 25, \bar{x} = 254, s^2 = 16.2$
9. Car batteries are manufactured with two different production methods. The lives (in years) are found for a sample from each group with the following results.

   Production method A: $n = 25, \bar{x} = 4.00, s = 0.37$

   Production method B: $n = 30, \bar{x} = 4.00, s = 0.31$

   At the 0.05 significance level, test the claim that the two production methods yield batteries with the same variance.
10. When 15 randomly selected adult males are given a test on reaction times, their scores produce a variance of 1.04. When 17 other randomly selected adult males are given a double martini before taking the same test, their scores produce a variance of 3.26. At the 0.05 significance level, test the claim that the population of all drinkers will have a variance larger than the population of all nondrinkers.

11.  A department store manager experiments with two methods for checking out customers, and obtains the following sample data. At the 0.02 significance level, test the claim that $\sigma_1^2 = \sigma_2^2$. Assume that the sample data came from normally distributed populations.

| Sample A | Sample B |
|---|---|
| $n_1 = 16$ | $n_2 = 10$ |
| $s_1^2 = 225$ | $s_2^2 = 100$ |

12.  An experiment is devised to study the variability of grading procedures among college professors. Two different professors are asked to grade the same set of 25 exam solutions, and their grades have variances of 103.4 and 39.7, respectively. At the 0.05 significance level, test the claim that the first professor's grading exhibits greater variance.

13.  In a study of the effect of job previews on work expectation, subjects from different groups were tested. For 60 subjects given specific job previews, the mean promotion expectation score is 19.14 and the standard deviation is 6.56. For a "no booklet" sample group of 40 subjects, the mean is 20.81 and the standard deviation is 4.90. (The data are based on "Effects of Realistic Job Previews on Hiring Bank Tellers," by Dean and Wanous, *Journal of Applied Psychology*, Vol. 69, No. 1.) At the 0.10 level of significance, test the claim that the two sample groups come from populations with the same standard deviation.

14.  In an insurance study of pedestrian deaths in New York State, monthly fatalities are totaled for two different time periods. Sample data for the first time period are summarized by these statistics: $n = 12$, $\bar{x} = 46.42$, $s = 11.07$. Sample data for the second time period are summarized by these statistics: $n = 12$, $\bar{x} = 51.00$, $s = 10.39$. (Based on data from the New York State Department of Motor Vehicles.) At the 0.05 significance level, test the claim that both time periods have the same variance.

15.  A study was conducted to investigate relationships between different types of standard test scores. On the Graduate Record Examination Verbal test, 68 females had a mean of 538.82 and a standard deviation of 114.16, while 86 males had a mean of 525.23 and a standard deviation of 97.23. (See "Equivalencing MAT and GRE Scores using Simple Linear Transformation and Regression Methods," by Kagan and Stock, *Journal of Experimental Education*, Vol. 49, No. 1.) At the 0.02 significance level, test the claim that the two groups come from populations with the same standard deviation.

16.  As part of the National Health Survey, data were collected on the weights of men. For 804 men aged 25 to 34, the mean if 176 lb and the standard deviation is 35.0 lb. For 1657 men aged 65 to 74, the mean and standard deviation are 164 lb and 27.0 lb, respectively. (The data are based on the National Health Survey, USDHEW publication 79-1659.) At the 0.01 significance level, test the claim that the older men come

from a population with a standard deviation less than that for men in the 25 to 34 age bracket.

17. The effectiveness of a mental training program was tested in a military training program. In an antiaircraft artillery examination, scores for an experimental group and a control group were recorded. Use the given data to test the claim that both groups come from populations with the same variance. Use a 0.05 significance level. (The data are based on "Routinization of Mental Training in Organizations: Effects on Performance and Well-Being," by Larsson, *Journal of Applied Psychology*, Vol. 72, No. 1.)

| Experimental | | | | | Control | | | |
|---|---|---|---|---|---|---|---|---|
| 60.83 | 117.80 | 44.71 | 75.38 | | 122.80 | 70.02 | 119.89 | 138.27 |
| 73.46 | 34.26 | 82.25 | 59.77 | | 118.43 | 54.22 | 118.58 | 74.61 |
| 69.95 | 21.37 | 59.78 | 92.72 | | 121.70 | 70.70 | 99.08 | 120.76 |
| 72.14 | 57.29 | 64.05 | 44.09 | | 104.06 | 94.23 | 111.26 | 121.67 |
| 80.03 | 76.59 | 74.27 | 66.87 | | | | | |

18. The arrangement of test items was studied for its effect on anxiety. At the 0.05 significance level, test the claim that the two given samples come from populations with the same variance. (See "Item Arrangement, Cognitive Entry Characteristics, Sex and Test Anxiety as Predictors of Achievement Examination Performance," by Klimko, *Journal of Experimental Education*, Vol. 52, No. 4.)

| Easy to difficult | | | | | Difficult to easy | | | |
|---|---|---|---|---|---|---|---|---|
| 24.64 | 39.29 | 16.32 | 32.83 | | 33.62 | 34.02 | 26.63 | 30.26 |
| 28.02 | 33.31 | 20.60 | 21.13 | | 35.91 | 26.68 | 29.49 | 35.32 |
| 26.69 | 28.90 | 26.43 | 24.23 | | 27.24 | 32.34 | 29.34 | 33.53 |
| 7.10 | 32.86 | 21.06 | 28.89 | | 27.62 | 42.91 | 30.20 | 32.54 |
| 28.71 | 31.73 | 30.02 | 21.96 | | | | | |
| 25.49 | 38.81 | 27.85 | 30.29 | | | | | |
| 30.72 | | | | | | | | |

19. Sample data were collected in a study of calcium supplements and the effects on blood pressure. A placebo group and a calcium group began the study with measures of blood pressures. At the 0.05 significance level, test the claim that the two sample groups come from populations with the same standard deviation. (See "Blood Pressure and Metabolic Effects of Calcium Supplementation in Normotensive White and Black Men," by Lyle and others, *Journal of the American Medical Association*, Vol. 257, No. 13.)

| Placebo | | | | | Calcium | | | |
|---|---|---|---|---|---|---|---|---|
| 124.6 | 104.8 | 96.5 | 116.3 | | 129.1 | 123.4 | 102.7 | 118.1 |
| 106.1 | 128.8 | 107.2 | 123.1 | | 114.7 | 120.9 | 104.4 | 116.3 |
| 118.1 | 108.5 | 120.4 | 122.5 | | 109.6 | 127.7 | 108.0 | 124.3 |
| 113.6 | | | | | 106.6 | 121.4 | 113.2 | |

20. A prospective home buyer is investigating home selling price differences between zone 1 (southern Dutchess County) and zone 7 (northern Dutchess County). The given random sample data are in thousands of dollars. (This will not affect the value of the test statistic $F$, but it allows us to work with more manageable numbers.) At the 0.05 significance level, test the claim that both zones have the same variance. Assume that the sample data come from normally distributed populations. (In many cases, home selling prices might not be normally distributed. But histograms show that the assumption of normal distributions is reasonable here.)

| Zone 7 (north) | | Zone 1 (south) | |
|---|---|---|---|
| 270.000 | | 115.000 | |
| 107.000 | | 136.900 | |
| 148.000 | | 121.000 | |
| 125.000 | | 164.000 | |
| 127.500 | $n = 11$ | 175.000 | |
| 125.500 | $\overline{x} = 142.32$ | 128.500 | |
| 126.000 | $s^2 = 2122$ | 147.500 | $n = 14$ |
| 109.000 | | 147.000 | $\overline{x} = 138.24$ |
| 113.500 | | 105.000 | $s^2 = 455$ |
| 147.000 | | 163.750 | |
| 167.000 | | 115.000 | |
| | | 149.165 | |
| | | 120.500 | |
| | | 147.000 | |

# 8–2  Exercises B

21. For hypothesis tests in this section that were two-tailed, we found only the upper critical value. Let's denote that value by $F_R$, where the subscript suggests the right side. The lower critical value $F_L$ (for the left side) can be found by first interchanging the degrees of freedom and then taking the reciprocal of the resulting $F$ value found in Table A–5. ($F_R$ is often denoted by $F_{\alpha/2}$ while $F_L$ is often denoted by $F_{1-\alpha/2}$.) Find the critical values $F_L$ and $F_R$ for two-tailed hypothesis tests based on the following values.

a.  $n_1 = 10, n_2 = 10, \alpha = 0.05$
b.  $n_1 = 10, n_2 = 7, \alpha = 0.05$
c.  $n_1 = 7, n_2 = 10, \alpha = 0.05$
d.  $n_1 = 25, n_2 = 10, \alpha = 0.02$
e.  $n_1 = 10, n_2 = 25, \alpha = 0.02$

22. In addition to testing claims involving $\sigma_1^2$ and $\sigma_2^2$, we can also construct interval estimates of the ratio $\sigma_1^2/\sigma_2^2$ using the following.

$$\frac{s_1^2}{s_2^2} \cdot \frac{1}{F_R} < \frac{\sigma_1^2}{\sigma_2^2} < \frac{s_1^2}{s_2^2} \cdot \frac{1}{F_L}$$

    Here $F_L$ and $F_R$ are as described in Exercise 21. Construct the 95% interval estimate for the ratio of the experimental group variance to the control group variance for the data in Exercise 17.

23. Sample data consist of temperatures recorded for two different groups of items that were produced by two different production techniques. A quality control specialist plans to analyze the results. She begins by testing for equality of the two population standard deviations.
    a. If she adds the same constant to every temperature from both groups, is the value of the test statistic $F$ affected? Explain.
    b. If she uses the same constant to multiply every score from both groups, is the value of the test statistic $F$ affected? Explain.
    c. If she converts all temperatures from the Fahrenheit scale to the Celsius scale, is the value of the test statistic $F$ affected? Explain.

24. a. Two samples of equal size produce variances of 37 and 57. At the 0.05 significance level, we test the claim that the variance of the second population exceeds that of the first, and that claim is upheld by the data. What is the approximate minimum size of each sample?
    b. A sample of 21 scores produces a variance of 67.2 and another sample of 25 produces a variance that causes rejection of the claim that the two populations have equal variances. If this test is conducted at the 0.02 level of significance, find the maximum variance of the second sample if you know that it is smaller than that of the first sample.

## 8–3 Inferences from Two Means

In this section we consider methods for testing hypotheses about two population means as well as methods for constructing confidence intervals for the difference between two population means. The following assumptions apply.

### Assumptions

1. We intend to **test a hypothesis** made about the means of two populations, or we intend to construct an **interval estimate** of the difference between the means of two populations.

2. We have sample data from each of two **normally distributed** populations.

The particular methods we use will be affected by the presence or absence of a relationship between the two samples, so we begin by distinguishing between dependent and independent samples.

### Definition

Two samples are **independent** if the sample selected from one population has no effect on the sample selected from the other population. If the two samples are not independent, they are **dependent.**

Consider the sample data given below. We would expect the sample of pretraining weights and the sample of posttraining weights to be two *dependent* samples, since each pair is matched according to the person involved. Such "before-and-after" data are usually matched and are usually dependent. (The table is based on data from the *Journal of Applied Psychology*, Vol. 62, No. 1.)

| Pretraining weights (kg) | 99 | 62 | 74 | 59 | 70 |
|---|---|---|---|---|---|
| Posttraining weights (kg) | 94 | 62 | 66 | 58 | 70 |

However, for the data given below, the two samples are *independent* since the sample of females is completely independent of the sample of males. The data are not matched as they are in the table above.

| Weights of females (lb) | 115 | 107 | 110 | 128 | 130 | | |
|---|---|---|---|---|---|---|---|
| Weights of males (lb) | 128 | 150 | 160 | 140 | 163 | 155 | 175 |

When dealing with two dependent samples, it is very wasteful to reduce the sample data to $\bar{x}_1$, $s_1$, $n_1$, $\bar{x}_2$, $s_2$, and $n_2$ since the relationship between matched pairs of values would be completely lost. Instead, we compute the *differences* ($d$) between the pairs of data as follows.

| $x$ | 22 | 48 | 27 | 29 | 32 |
|---|---|---|---|---|---|
| $y$ | 23 | 51 | 25 | 29 | 32 |
| $d = x - y$ | $-1$ | $-3$ | 2 | 0 | 0 |

## STATISTICAL SIGNIFICANCE VERSUS PRACTICAL SIGNIFICANCE

■ The hypothesis tests described in this text address the issue of statistical significance. We try to determine if the observations are so unlikely that we are led to believe that differences are due to factors other than chance sample fluctuations. Experimental results can sometimes be statistically significant without being practically significant. A diet causing a weight loss of 1/2 lb might be statistically significant if the sample size is 10,000, but such a diet would not have practical significance. Nobody would bother with a diet that results in a loss of only 1/2 lb.■

---

**Notation**

Let $\bar{d}$ denote the mean value of $d$ or $x - y$ for the paired sample data.

Let $s_d$ denote the standard deviation of the $d$ values for the paired sample data.

Let $n$ denote the number of *pairs* of data.

---

**Example**

For the $d$ values of $-1, -3, 2, 0, 0$ taken from the preceding table, we get

$$\bar{d} = \frac{\Sigma d}{n} = \frac{(-1) + (-3) + 2 + 0 + 0}{5} = \frac{-2}{5} = -0.4$$

For the $d$ values of $-1, -3, 2, 0, 0$, we get $s_d = 1.8$ as follows.

$$s_d = \sqrt{\frac{\Sigma(d - \bar{d})^2}{n-1}} = \sqrt{\frac{13.2}{4}} = 1.8$$

For the data of the last table, $n = 5$.

---

# Hypothesis Testing

**In repeated random sampling from two normal and dependent populations in which the mean of the paired differences is $\mu_d$, the following test statistic possesses a Student $t$ distribution with $n - 1$ degrees of freedom:**

---

**Test Statistic**

$$t = \frac{\bar{d} - \mu_d}{s_d/\sqrt{n}}$$

---

Note that the involved populations must be normally distributed. If the populations depart radically from normal distributions, we should not use the methods given in this section. Instead, we may be able to apply the sign test (Section 12–2) or the Wilcoxon signed-ranks test (Section 12–3).

If we claim that there is no difference between the two population means, then we are claiming that $\mu_d = 0$. This makes sense if we recognize that $\bar{d}$ should be around 0 if there is no difference between the two population means.

In the following example we illustrate a complete hypothesis test for a situation involving dependent samples.

## Example

A study was conducted to investigate the effectiveness of hypnotism in reducing pain. Results for randomly selected subjects are given below. At the 0.05 significance level, test the claim that the sensory measurements are lower after hypnotism. (The values are before and after hypnosis. The measurements are in centimeters on the mean visual analog scale, and the data are based on "An Analysis of Factors That Contribute to the Efficacy of Hypnotic Analgesia," by Price and Barber, *Journal of Abnormal Psychology*, Vol. 96, No. 1.)

| Subject | A | B | C | D | E | F | G | H |
|---|---|---|---|---|---|---|---|---|
| Before | 6.6 | 6.5 | 9.0 | 10.3 | 11.3 | 8.1 | 6.3 | 11.6 |
| After | 6.8 | 2.4 | 7.4 | 8.5 | 8.1 | 6.1 | 3.4 | 2.0 |
| Difference | −0.2 | 4.1 | 1.6 | 1.8 | 3.2 | 2.0 | 2.9 | 9.6 |

## Solution

Since each pair of scores is matched for one particular person, we can conclude that the values are dependent. Each difference is the "before" score minus the "after" score. If the hypnotism is effective, we would expect the after scores to be lower so that $\bar{d}$ would be positive and *significantly* greater than 0. That is, the claim of lower after scores is equivalent to $\mu_d > 0$. We therefore have the null hypothesis $H_0$: $\mu_d \leq 0$ and the alternative hypothesis $H_1$: $\mu_d > 0$.

The mean and standard deviation of the differences are found as follows.

$$\bar{d} = \frac{\Sigma d}{n} = \frac{-0.2 + 4.1 + \cdots + 9.6}{8}$$

$$= \frac{25}{8} = 3.125$$

$$s_d = \sqrt{\frac{\Sigma(d - \bar{d})^2}{n - 1}}$$

$$= \sqrt{\frac{(-0.2 - 3.125)^2 + (4.1 - 3.125)^2 + \cdots + (9.6 - 3.125)^2}{7}}$$

$$= \sqrt{\frac{59.335}{7}} = 2.911$$

*continued*

**Solution** *continued*

With $\bar{d} = 3.125$, $s_d = 2.911$, and $n = 8$, we calculate the value of the test statistic as

$$t = \frac{\bar{d} - \mu_d}{s_d/\sqrt{n}} = \frac{3.125 - 0}{2.911/\sqrt{8}} = 3.036$$

The critical value of $t = 1.895$ can be found in Table A–3; this is a right-tailed test, $\alpha = 0.05$, and there are $8 - 1 = 7$ degrees of freedom.

Since the test statistic of $t = 3.036$ falls within the critical region (see Figure 8–3), we reject the null hypothesis. There is sufficient evidence to support the claim that the after scores are significantly lower than the before scores. It appears that hypnosis does have a significant effect on pain, as measured by the sensory scores.

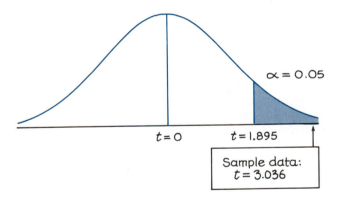

**Figure 8–3**

This test could also be conducted by using the STATDISK software package or any other software package designed to treat the case of two dependent samples. After the user selects the appropriate case from the menu of options, STATDISK prompts him or her to enter the necessary data, and the results are then displayed as follows.

This test could also be run as a $t$ test for a single population mean. Enter the data as differences and use 0 as the value of the claimed population mean. Shown below is the Minitab display that begins with the original before-after data. The differences are calculated and assigned to the variable C3.

STATDISK DISPLAY

> Hypothesis test for a claim about two DEPENDENT populations
>
> NULL HYPOTHESIS: Mean 1 <= Mean 2
>
> Mean of differences . . . . . . . . . . . . . .  =  3.125
> St. Dev. of differences . . . . . . . . . . . .  =  2.91143
> Degrees of freedom . . . . . . . . . . . df  =  7
> Test statistic . . . . . . . . . . . . . . . . . . .t  =  3.03591
> Critical value . . . . . . . . . . . . . . . . . . .t  =  1.89554
> P-value . . . . . . . . . . . . . . . . . . . . . . .  =  .00948
> Significance level . . . . . . . . . . . . . . .  =  .05
>
> CONCLUSION: REJECT the null hypothesis

MINITAB DISPLAY

```
MTB > SET C1
DATA> 6.6  6.5  9.0  10.3  11.3  8.1  6.3  11.6
DATA> ENDOFDATA
MTB > SET C2
DATA> 6.8  2.4  7.4  8.5  8.1  6.1  3.4  2.0
DATA> ENDOFDATA
MTB > LET C3 = C1 − C2
MTB > TTEST of mu = 0 for data in C3;
SUBC> ALTERNATIVE = +1 .

TEST OF mu = 0.000 VS mu G.T. 0.000
```

|     | N | MEAN  | STDEV | SE MEAN | T    | P VALUE |
|-----|---|-------|-------|---------|------|---------|
| C3  | 8 | 3.125 | 2.911 | 1.029   | 3.04 | 0.0095  |

The test is then run with the option "ALTERNATIVE = +1"—the Minitab code for a hypothesis test that is right-tailed.

Note that the *P*-value of 0.0095 is less than the significance level of 0.05. This indicates that such sample results are not likely to occur by chance, assuming that the null hypothesis is true.

As we consider other tests of hypotheses made about two population means, we begin to encounter a maze, which can easily lead to confusion. Questions must be answered regarding the independence of samples, knowledge of $\sigma_1$ and $\sigma_2$, and sample size before the correct procedure can be selected. Most of the confusion can be avoided by referring to Figure 8–4, which summarizes the procedures discussed in this section. We illustrate the use of Figure 8–4 through specific examples and then present the underlying theory that led to its development. Since we have already presented an example involving dependent populations, our next examples will involve independent populations.

### Example

Two machines fill packages, and samples are selected from each machine. Denoting the results from machine A as group 1 and the results from machine B as group 2, we have

| Machine A | Machine B |
|---|---|
| $n_1 = 50$ | $n_2 = 100$ |
| $\bar{x}_1 = 4.53$ kg | $\bar{x}_2 = 4.01$ kg |

If the standard deviations of the contents filled by machine A and machine B are 0.80 kg and 0.60 kg, respectively, test the claim that the mean contents produced by machine A equal the mean for machine B. Assume a 0.05 significance level.

### Solution

The two means are independent and $\sigma_1$ and $\sigma_2$ are known. Referring to Figure 8–4, we see that we should use a normal distribution test with

$$z = \frac{(\bar{x}_1 - \bar{x}_2) - (\mu_1 - \mu_2)}{\sqrt{\dfrac{\sigma_1^2}{n_1} + \dfrac{\sigma_2^2}{n_2}}} = \frac{(4.53 - 4.01) - 0}{\sqrt{\dfrac{0.80^2}{50} + \dfrac{0.60^2}{100}}} = 4.06$$

With the null and alternative hypotheses described as

$$H_0: \mu_1 = \mu_2 \qquad (\text{or } \mu_1 - \mu_2 = 0)$$
$$H_1: \mu_1 \neq \mu_2 \qquad (\text{or } \mu_1 - \mu_2 \neq 0)$$

and $\alpha = 0.05$, we conclude that the test involves two tails. From Table A–2 we extract the critical $z$ values of 1.96 and $-1.96$. The test statistic of 4.06 is well into the critical region, and we therefore reject $H_0$ and conclude that the population means corresponding to the two machines are not equal. It appears that machine A fills with amounts that are significantly greater than those of machine B.

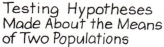

Testing Hypotheses
Made About the Means
of Two Populations

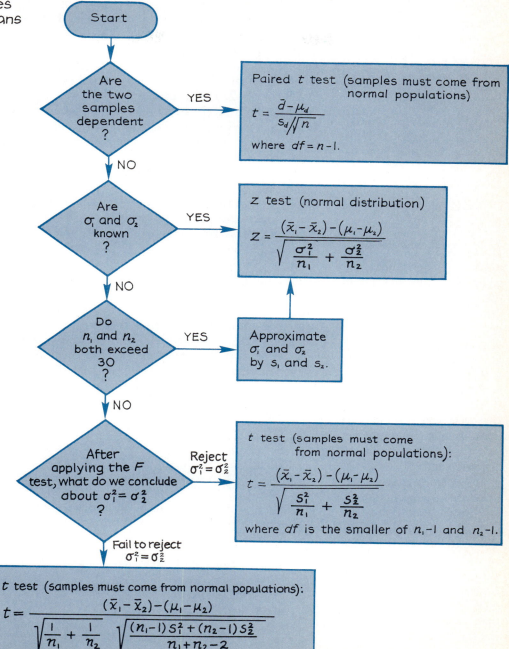

**Figure 8–4**

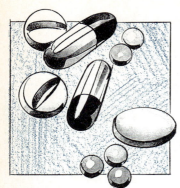

## DRUG SCREENING: FALSE POSITIVES

■ For a job applicant undergoing drug screening, a *false positive* is an indication of drug use when he or she does not use them. A *false negative* is an indication of no drug use when he or she is a user. The *test sensitivity* is the probability of a positive indication for a drug user; the *test specificity* is the probability of a negative indication for a nonuser. Suppose that 3% of job applicants use drugs, and a test has a sensitivity of 0.99 and a specificity of 0.98. The high probabilities make the test seem reliable, but 40% of the positive indications will be false positives. The American Management Association reports that more than half of all companies test for drugs.■

In the preceding example, we arbitrarily identified the machine A results as group 1, while those of machine B are identified as group 2. For this example, the two large sample sizes allow us to avoid the $F$ test of $\sigma_1^2 = \sigma_2^2$. Although not always necessary, it is often a good strategy to **identify the data set with the larger sample variance as group 1,** and identify the data set with the smaller sample variance as group 2. This is the same procedure followed in Section 8–2. If it becomes necessary to use the $F$ test of $\sigma_1^2 = \sigma_2^2$, this identification will be helpful.

The preceding example is somewhat contrived because we seldom know the values of $\sigma_1$ and $\sigma_2$. It is rare to sample from two populations with unknown means but known standard deviations. To cover more realistic cases involving independent samples with unknown standard deviations or variances, we next examine the sizes of the two samples, as suggested by Figure 8–4. In this next example and the remaining examples of this section, the samples must come from normally distributed populations. If this condition is not satisfied, we may be able to use the Wilcoxon rank-sum test described in Section 12–4.

If both samples are large (greater than 30), we can estimate $\sigma_1$ and $\sigma_2$ by $s_1$ and $s_2$. We can then proceed as in the last example. However, if either sample is small, we must apply the $F$ test to determine whether the two sample variances are significantly different. The next example illustrates these points since the populations are independent, both population standard deviations are unknown, and both samples are small (less than or equal to 30). We see that the $F$ test suggests that the two population variances are not equal and, from Figure 8–4, we see that the circumstances of this next example cause us to turn right at the last diamond.

### Example

Random samples of home selling prices are obtained from two zones in Dutchess County. Results are summarized below.

| Zone 7 (north) | Zone 1 (south) |
| --- | --- |
| $n_1 = 11$ | $n_2 = 14$ |
| $\bar{x}_1 = \$142{,}318$ | $\bar{x}_2 = \$138{,}237$ |
| $s_1 = \$46{,}068$ | $s_2 = \$21{,}336$ |

At the 0.05 significance level, test the claim that the two zones have the same mean selling price.

## Solution

Since the Zone 7 results have a larger variance, we made them the first sample. Since the samples are independent, neither standard deviation is known, and both samples are small, Figure 8–4 indicates that we should begin by applying the $F$ test discussed in Section 8–2. (See Section 8–2, where we presented a complete hypothesis test of the claim that $\sigma_1^2 = \sigma_2^2$.) We want to decide whether $\sigma_1^2 = \sigma_2^2$, so we formulate the following null and alternative hypotheses:

$$H_0: \sigma_1^2 = \sigma_2^2$$
$$H_1: \sigma_1^2 \neq \sigma_2^2$$

With $\alpha = 0.05$, we do a complete hypothesis test to decide whether or not $\sigma_1^2 = \sigma_2^2$. We then do another complete hypothesis test of the claim that $\mu_1 = \mu_2$ using the appropriate Student $t$ distribution. For the preliminary test we get

$$F = \frac{s_1^2}{s_2^2} = \frac{(46,068)^2}{(21,336)^2} = 4.6620$$

The critical $F$ value obtained from Table A–5 is 3.2497. (The test involves two tails with $\alpha = 0.05$, and the degrees of freedom for the numerator and the denominator are 10 and 13, respectively.) These results cause us to reject the null hypothesis of equal variances and we reject $\sigma_1^2 = \sigma_2^2$. In Figure 8–4 we test the claim that $\mu_1 = \mu_2$ by using the Student $t$ distribution and the test statistic given in the box to the *right* of the fourth diamond. With

$$H_0: \mu_1 = \mu_2 \qquad (\text{or } \mu_1 - \mu_2 = 0)$$
$$H_1: \mu_1 \neq \mu_2 \qquad (\text{or } \mu_1 - \mu_2 \neq 0)$$
$$\alpha = 0.05$$

we compute the test statistic based on the sample data.

$$t = \frac{(\overline{x}_1 - \overline{x}_2) - (\mu_1 - \mu_2)}{\sqrt{\dfrac{s_1^2}{n_1} + \dfrac{s_2^2}{n_2}}} = \frac{(142,318 - 138,237) - 0}{\sqrt{\dfrac{46,068^2}{11} + \dfrac{21,336^2}{14}}}$$
$$= 0.272$$

This is a two-tailed test with $\alpha = 0.05$ and 10 degrees of freedom, so the critical $t$ values obtained from Table A–3 are $t = 2.228$ and $t = -2.228$. The computed $t$ value of 0.272 does not fall within the critical region and we fail to reject the null hypothesis of equal means. Based on the available sample data, we cannot reject the claim that the two zones have the same mean selling price.

# POLL RESISTANCE

■ Surveys based on relatively small samples can be quite accurate, provided the sample is random or representative of the population. However, increasing survey refusal rates are now making it more difficult to obtain random samples. The Council of American Survey Research Organizations reported that in a recent year, 38% of consumers refused to respond to surveys. The head of one market research company said, "Everyone is fearful of self-selection and worried that generalizations you make are based on cooperators only." Results from the multibillion-dollar market research industry affect the products we buy, the television shows we watch, and many other facets of our lives. ■

The next example illustrates a hypothesis test comparing two means for a situation featuring the following characteristics:

- The two samples come from independent and normal populations.
- $\sigma_1$ and $\sigma_2$ are unknown.
- Both sample sizes are small ($\leq 30$).
- The sample variances suggest, through the $F$ test, that $\sigma_1^2 = \sigma_2^2$.

The conditions inherent in this next example cause us to follow the path leading to the bottom of the flowchart in Figure 8–4.

### Example

Samples of girls aged six and seven are given the Wide Range Achievement Test with the results summarized below (based on data from the National Health Survey, USDHEW publication 72-1011). At the 0.05 level of significance, test the claim that the mean score for seven-year-old girls is greater than the mean for six-year-old girls. (We identify the age seven group as group 1 because the variance is larger.)

| Age 7 | Age 6 |
|---|---|
| $n_1 = 16$ | $n_2 = 12$ |
| $\overline{x}_1 = 44.0$ | $\overline{x}_2 = 27.5$ |
| $s_1 = 13.2$ | $s_2 = 10.2$ |

### Solution

Referring to Figure 8–4, we begin by questioning the independence of the two populations and conclude that they are independent because separate groups of subjects are used. We continue with the flowchart by noting that $\sigma_1$ and $\sigma_2$ are not known and that neither $n_1$ nor $n_2$ exceeds 30. At this stage, the flowchart brings us to the last diamond, which requires application of the $F$ test. With $H_0$: $\sigma_1^2 = \sigma_2^2$, $H_1$: $\sigma_1^2 \neq \sigma_2^2$, and $\alpha = 0.05$, we compute

$$F = \frac{s_1^2}{s_2^2} = \frac{13.2^2}{10.2^2} = 1.6747$$

With $\alpha = 0.05$ in a two-tailed $F$ test and with 15 and 11 as the degrees of freedom for the numerator and denominator, respectively, we use Table A–5 to obtain the critical $F$ value of 3.3299. Since the computed test statistic of $F = 1.6747$ is not within the critical region, we fail to reject

*continued*

**Solution** *continued*

the null hypothesis of equal variances. Leaving the last diamond in Figure 8–4, we proceed downward to the bottom and apply the required $t$ test as follows.

$$H_0: \mu_1 \leq \mu_2 \quad (\text{or } \mu_1 - \mu_2 \leq 0)$$
$$H_1: \mu_1 > \mu_2 \quad (\text{or } \mu_1 - \mu_2 > 0)$$
$$\alpha = 0.05$$

$$t = \frac{(\bar{x}_1 - \bar{x}_2) - (\mu_1 - \mu_2)}{\sqrt{\dfrac{1}{n_1} + \dfrac{1}{n_2}} \sqrt{\dfrac{(n_1 - 1)s_1^2 + (n_2 - 1)s_2^2}{n_1 + n_2 - 2}}}$$

$$= \frac{(44.0 - 27.5) - 0}{\sqrt{\dfrac{1}{16} + \dfrac{1}{12}} \sqrt{\dfrac{(16 - 1)(13.2)^2 + (12 - 1)(10.2)^2}{16 + 12 - 2}}}$$

$$= 3.594$$

Noting that $\mu_1 > \mu_2$ is equivalent to $\mu_1 - \mu_2 > 0$, we can better see that the test is right-tailed. With $\alpha = 0.05$ in this right-tailed $t$ test, and with $n_1 + n_2 - 2 = 16 + 12 - 2 = 26$ degrees of freedom, we obtain a critical $t$ value of 1.706. The computed test statistic of $t = 3.594$ falls within the critical region, so we reject the null hypothesis. The given sample data support the claim that the mean for seven-year-old girls is greater than that for six-year-old girls.

The calculations in this last example might seem somewhat complex, but we should recognize that calculators and computers can be used to ease that burden. STATDISK, for example, can be used to conduct this test. The user selects the option indicating a hypothesis test involving the means of two independent samples, and the user is then prompted for the necessary data. The resulting display is shown on page 414. Note that the results of the prerequisite $F$ test are included in the display. Also note the inclusion of the $P$-values in both tests. Using the given sample statistics, it takes about 60 seconds to run this test on STATDISK.

One of the most difficult aspects of tests comparing two means is the determination of the correct test to be used. Careful and consistent use of Figure 8–4 should help us avoid using the wrong procedures for a situation involving a hypothesis test. There is a danger of being overwhelmed by the work involved in the five different cases considered here. However, we can use Figure 8–4 to decompose a complex problem into simpler components that can be treated individually.

STATDISK DISPLAY

Hypothesis test for a claim about two INDEPENDENT populations

NULL HYPOTHESIS: Mean 1 $<=$ Mean 2

CONCLUSION: REJECT the null hypothesis

------------------------------------------------------------

Test statistic ................. F = 1.67474
P-value ....................... = .39218

F Test CONCLUSION: FAIL TO REJECT equality of variances

------------------------------------------------------------

Degrees of freedom ........... df = 26
Test statistic ................. t = 3.59386
Critical value ................. t = 1.70601
P-value ....................... = .00067
Significance level .............. = .05

T Test CONCLUSION: REJECT the null hypothesis

# *P*-Values

The comments made in Section 7–3 about *P*-values apply to this section as well. (See Figure 7–8, which summarizes key decisions to be made in determining *P*-values.) The preceding example is right-tailed with a test statistic of $t = 3.594$ and 26 degrees of freedom. Refer to the 26th row of Table A–3, where we can see that a test statistic of $t = 3.594$ corresponds to a *P*-value less than 0.005.

It is not practical to outline in full detail the derivations leading to the general test statistics given in Figure 8–4, but we can give some reasons for their existence. We have already discussed the case for dependent populations having normal distributions. Actual experiments and mathematical derivations show that, in repeated random samplings from two normal and dependent populations, the values of $\overline{d}$ possess a Student $t$ distribution with mean $\mu_d$ and standard deviation $\sigma_d/\sqrt{n}$.

Two cases of Figure 8–4 lead to the normal distribution with the test statistic given by

$$z = \frac{(\overline{x}_1 - \overline{x}_2) - (\mu_1 - \mu_2)}{\sqrt{\sigma_1^2/n_1 + \sigma_2^2/n_2}}$$

This expression is essentially an application of the central limit theorem, which tells us that sample means $\bar{x}$ are normally distributed with mean $\mu$ and standard deviation $\sigma/\sqrt{n}$.

In Section 5–6 we saw that, when samples are size 31 or larger, the normal distribution serves as a reasonable approximation to the distribution of sample means. By similar reasoning, the values of $\bar{x}_1 - \bar{x}_2$ also tend to approach a normal distribution with mean $\mu_1 - \mu_2$. When both samples are large, we conclude that the values of $\bar{x}_1 - \bar{x}_2$ will have a standard deviation of

$$\sqrt{\frac{\sigma_1^2}{n_1} + \frac{\sigma_2^2}{n_2}}$$

by using a property of variances: **The variance of the differences between two independent random variables equals the variance of the first random variable plus the variance of the second random variable.** That is, the variance of values $\bar{x}_1 - \bar{x}_2$ will tend to equal $\sigma_{\bar{x}_1}^2 + \sigma_{\bar{x}_2}^2$ provided that $\bar{x}_1$ and $\bar{x}_2$ are independent. This is a difficult concept, and we therefore illustrate it by the specific data given in Table 8–2. The $x$ and $y$ scores were independently and randomly selected as the last digits of numbers in a telephone book. The variance of the $x$ values is 7.68, the variance of the $y$ values is 8.48, and the variance of the $x - y$ values is 17.22

$$\left.\begin{array}{rl} s_x^2 &= 7.68 \\ s_y^2 &= 8.48 \\ s_{x-y}^2 &= 17.22 \end{array}\right] \quad s_{x-y}^2 \approx s_x^2 + s_y^2$$

We can see that $s_x^2 + s_y^2$ is roughly equal to $s_{x-y}^2$. By comparing the values of $x$, $y$, and $x - y$, we see that the variance is largest for the $x - y$ values. The $x$ values range from 0 to 9, the $y$ values range from 0 to 9, but the $x - y$ values exhibit greater variation by ranging from $-8$ to 9. If our $x$, $y$, and $x - y$ sample sizes were much larger than the sample sizes of 36 for Table 8–2, we would approach these theoretical values: $\sigma_x^2 = 8.25$, $\sigma_y^2 = 8.25$, and $\sigma_{x-y}^2 = 16.50$.

These theoretical values illustrate that $\sigma_{x-y}^2 = \sigma_x^2 + \sigma_y^2$ when $x$ and $y$ are independent random variables. When we deal with means of large random sample sizes, the standard deviation of those sample means is $\sigma/\sqrt{n}$, so the variance is $\sigma^2/n$.

We can now combine our additive property of variances with the central limit theorem's expression for variance of sample means to obtain the following.

$$\sigma_{\bar{x}_1 - \bar{x}_2}^2 = \sigma_{\bar{x}_1}^2 + \sigma_{\bar{x}_2}^2 = \frac{\sigma_1^2}{n_1} + \frac{\sigma_2^2}{n_2}$$

In this expression, we assume that we have population 1 with variance $\sigma_1^2$ and population 2 with variance $\sigma_2^2$. Samples of size $n_1$ are randomly drawn

| TABLE 8–2 | | |
|---|---|---|
| $x$ | $y$ | $x - y$ |
| 1 | 7 | -6 |
| 3 | 0 | 3 |
| 1 | 0 | 1 |
| 3 | 4 | -1 |
| 2 | 6 | -4 |
| 8 | 3 | 5 |
| 3 | 9 | -6 |
| 3 | 7 | -4 |
| 9 | 0 | 9 |
| 1 | 1 | 0 |
| 4 | 2 | 2 |
| 9 | 4 | 5 |
| 4 | 9 | -5 |
| 9 | 6 | 3 |
| 5 | 2 | 3 |
| 6 | 7 | -1 |
| 5 | 6 | -1 |
| 5 | 8 | -3 |
| 3 | 3 | 0 |
| 3 | 6 | -3 |
| 1 | 3 | -2 |
| 0 | 2 | -2 |
| 4 | 4 | 0 |
| 3 | 1 | 2 |
| 3 | 9 | -6 |
| 6 | 4 | 2 |
| 9 | 5 | 4 |
| 5 | 1 | 4 |
| 2 | 2 | 0 |
| 9 | 6 | 3 |
| 2 | 2 | 0 |
| 6 | 1 | 5 |
| 1 | 9 | -8 |
| 7 | 0 | 7 |
| 9 | 1 | 8 |
| 6 | 5 | 1 |

from population 1 and the mean $\bar{x}$ is computed. The same is done for population 2. Here, $\sigma^2_{\bar{x}_1 - \bar{x}_2}$ denotes the variance of $\bar{x}_1 - \bar{x}_2$ values. This result shows that the standard deviation of $\bar{x}_1 - \bar{x}_2$ values is

$$\sqrt{\frac{\sigma^2_1}{n_1} + \frac{\sigma^2_2}{n_2}}$$

Since $z$ is a standard score that corresponds in general to

$$z = \frac{(\text{sample statistic}) - (\text{population mean})}{(\text{population standard deviation})}$$

we get

$$z = \frac{(\bar{x}_1 - \bar{x}_2) - (\mu_1 - \mu_2)}{\sqrt{\sigma^2_1/n_1 + \sigma^2_2/n_2}}$$

by noting that the sample values of $\bar{x}_1 - \bar{x}_2$ will have a mean of $\mu_1 - \mu_2$ and the standard deviation just given.

The test statistic located at the bottom of Figure 8–4 is appropriate for tests of hypotheses about two means from independent and normal populations when the samples are small and the population variances appear to be equal. The numerator $(\bar{x}_1 - \bar{x}_2) - (\mu_1 - \mu_2)$ again describes the difference between the sample statistic $(\bar{x}_1 - \bar{x}_2)$ and the mean of all such values $(\mu_1 - \mu_2)$. The denominator represents the standard deviation of $\bar{x}_1 - \bar{x}_2$ values, which come from repeated sampling when the stated assumptions are satisfied. Since these assumptions include equality of population variances, the denominator should be an estimate of

$$\sqrt{\frac{\sigma^2}{n_1} + \frac{\sigma^2}{n_2}} = \sigma\sqrt{\frac{1}{n_1} + \frac{1}{n_2}}$$

Both $n_1$ and $n_2$ will be known, so we need to estimate only $\sigma$, and we pool (combine) the sample variances to get the best possible estimate. The pooled estimate of $\sigma$ is

$$\sqrt{\frac{(n_1 - 1)s^2_1 + (n_2 - 1)s^2_2}{n_1 + n_2 - 2}}$$

Here the expression under the square root sign is just a weighted average of $s^2_1$ and $s^2_2$. (Weights of $n_1 - 1$ and $n_2 - 1$ are used.)

We use the test statistic

$$t = \frac{(\bar{x}_1 - \bar{x}_2) - (\mu_1 - \mu_2)}{\sqrt{\frac{s^2_1}{n_1} + \frac{s^2_2}{n_2}}}$$

in tests of hypotheses about two means coming from independent and normal populations when the samples are small and the two population variances appear to be different. The form of that test statistic seems to follow from similar reasoning used in the other situations, but that is not the case. In fact, no exact test has been found for testing the equality of two means when all the following conditions are met:

1. The two population variances are unknown.
2. The two population variances are unequal.
3. At least one of the samples is small $(n \le 30)$.
4. The two populations are independent and normal.

The approach for this case given in Figure 8–4 is only an approximate test, though it is widely used. Also, in Figure 8–4 we indicate that the number of degrees of freedom for this case is found by selecting the smaller of $n_1 - 1$ and $n_2 - 1$. This is a more conservative and simplified alternative to computing the number of degrees of freedom by using Formula 8–1. If the result of Formula 8–1 is not an integer, round off to the nearest integer.

$$df = \frac{(A + B)^2}{\dfrac{A^2}{n_1 - 1} + \dfrac{B^2}{n_2 - 1}} \qquad \text{where } A = \frac{s_1^2}{n_1} \text{ and } B = \frac{s_2^2}{n_2} \qquad \textbf{(8–1)}$$

Better results are obtained by using Formula 8–1. In either case, the results are only approximate.

The two bottom cases of Figure 8–4 begin with a preliminary $F$ test. If we apply the $F$ test with a certain level of significance and then do a $t$ test at that same level of significance, the overall result will not be at that same level of significance. Also, in addition to being sensitive to differences in population variances, the $F$ statistic is also sensitive to departures from normal distributions, so it is possible to reject a null hypothesis for the wrong reason. Because of these factors, some statisticians do not recommend a preliminary $F$ test, while others consider this the better approach. In any event, there is no universal agreement.

## Confidence Intervals

Figure 8–5 summarizes the decisions and choices to be made when constructing confidence intervals for the difference between two population means. If you compare the confidence interval flowchart (Figure 8–5 on page 418) to the hypothesis testing flowchart (Figure 8–4), you will see very strong similarities. The same underlying theory described for hypothesis tests also applies to confidence intervals.

Confidence
Intervals
for the
Difference
Between
Two
Population
Means

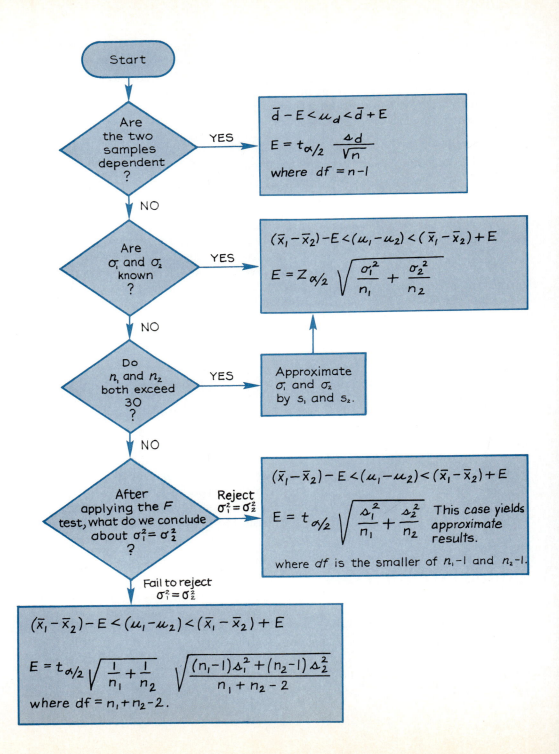

Figure 8–5

The following four examples use the same data as the corresponding four examples of hypothesis testing given earlier in this section. They illustrate the four cases included in Figure 8–5.

## Example

When 8 subjects were tested for sensory measurements before and after hypnosis, the following results were obtained. (See the example on page 405.) Construct a 95% confidence interval for the mean of all such differences.

| Subject | A | B | C | D | E | F | G | H |
|---|---|---|---|---|---|---|---|---|
| Before | 6.6 | 6.5 | 9.0 | 10.3 | 11.3 | 8.1 | 6.3 | 11.6 |
| After | 6.8 | 2.4 | 7.4 | 8.5 | 8.1 | 6.1 | 3.4 | 2.0 |
| Difference | −0.2 | 4.1 | 1.6 | 1.8 | 3.2 | 2.0 | 2.9 | 9.6 |

## Solution

We are dealing with dependent samples and Figure 8–5 reminds us of the format that is appropriate for these circumstances. Our previous hypothesis test of this same data included these statistics: $n = 8$, $\bar{d} = 3.125$, $s_d = 2.911$. From Table A–3 we get $t_{\alpha/2} = 2.365$, which corresponds to $n - 1 = 7$ degrees of freedom and area 0.05 in two tails (corresponding to 95% confidence). Using these results, we first evaluate $E$:

$$E = t_{\alpha/2} \frac{s_d}{\sqrt{n}} = 2.365 \cdot \frac{2.911}{\sqrt{8}} = 2.434$$

With $\bar{d} = 3.125$ and $E = 2.434$,

$$\bar{d} - E < \mu_d < \bar{d} + E$$

becomes

$$3.125 - 2.434 < \mu_d < 3.125 + 2.434$$

or

$$0.691 < \mu_d < 5.559$$

In the long run, 95% of such samples will lead to confidence limits that actually do contain the true population mean of the differences.

**Example**

On page 408 we conducted a test of the claim that two machines provide the same mean amount of package contents. The sample data shown below were given along with these values for the population standard deviations: $\sigma = 0.80$ for machine A; $\sigma = 0.60$ for machine B. Use the same data to construct a 95% confidence interval for the difference between the two population means.

| Machine A | Machine B |
|-----------|-----------|
| $n_1 = 50$ | $n_2 = 100$ |
| $\bar{x}_1 = 4.53$ kg | $\bar{x}_2 = 4.01$ kg |

**Solution**

With these independent samples and known population standard deviations, Figure 8–5 leads us to the form of the confidence interval and the maximum error of estimate $E$. We first evaluate $E$.

$$E = z_{\alpha/2} \sqrt{\frac{\sigma_1^2}{n_1} + \frac{\sigma_2^2}{n_2}} = 1.96 \sqrt{\frac{0.80^2}{50} + \frac{0.60^2}{100}} = 0.25$$

With $\bar{x}_1 = 4.53$, $\bar{x}_2 = 4.01$, and $E = 0.25$, the confidence interval

$$(\bar{x}_1 - \bar{x}_2) - E < (\mu_1 - \mu_2) < (\bar{x}_1 - \bar{x}_2) + E$$

becomes

$$(4.53 - 4.01) - 0.25 < (\mu_1 - \mu_2) < (4.53 - 4.01) + 0.25$$

or

$$0.27 < (\mu_1 - \mu_2) < 0.77$$

Note that the above confidence interval for $\mu_1 - \mu_2$ does not contain 0, suggesting that $\mu_1 - \mu_2 \neq 0$. This is consistent with the previous hypothesis test based on the same data; there we concluded that the two means are not equal, so it follows that $\mu_1 - \mu_2 \neq 0$, and this is affirmed by the fact that the above confidence interval does not contain 0. Here we have a direct correspondence between the confidence interval and the two-tailed hypothesis test.

## Example

Random samples of home selling prices are obtained from two zones in Dutchess County. Results are summarized below.

| Zone 7 (north) | Zone 1 (south) |
|---|---|
| $n_1 = 11$ | $n_2 = 14$ |
| $\bar{x}_1 = \$142,318$ | $\bar{x}_2 = \$138,237$ |
| $s_1 = \$46,068$ | $s_2 = \$21,336$ |

Construct a 95% confidence interval for the difference between the two population means.

## Solution

This same collection of data was considered in the hypothesis test illustrated on page 410. The samples are independent, neither population standard deviation is known, and the preliminary $F$ test caused us to reject $\sigma_1^2 = \sigma_2^2$. (Figure 8–5 indicates that we turn right at the fourth diamond.) We now calculate

$$E = t_{\alpha/2} \sqrt{\frac{s_1^2}{n_1} + \frac{s_2^2}{n_2}} = 2.228 \sqrt{\frac{46,068^2}{11} + \frac{21,336^2}{14}} = 33,453$$

In the above calculation, $t_{\alpha/2} = 2.228$ corresponds to 0.05 in two tails (95% confidence) and 10 degrees of freedom (the smaller of 10 and 13). With $\bar{x}_1 = 142,318$, $\bar{x}_2 = 138,237$, and $E = 33,453$,

$$(\bar{x}_1 - \bar{x}_2) - E < (\mu_1 - \mu_2) < (\bar{x}_1 - \bar{x}_2) + E$$
$$\text{becomes} \quad -\$29,372 < (\mu_1 - \mu_2) < \$37,534$$

Note that the above confidence interval for $\mu_1 - \mu_2$ does contain 0. This is consistent with the previous hypothesis test based on the same data; we failed to reject equality of the two population means. If $\mu_1 = \mu_2$, then $\mu_1 - \mu_2 = 0$ and the above confidence interval does contain 0. Again we have a direct correspondence between the confidence interval and the two-tailed hypothesis test.

In the preceding example, we could have obtained better results by calculating the number of degrees of freedom by using Formula 8–1 instead of simply selecting the smaller of $n_1 - 1$ and $n_2 - 1$. (The confidence limits would be $-\$28,351$ and $\$36,513$.) Even if we use Formula 8–1 to calculate the number of degrees of freedom, this case continues to yield only approximate results.

## Example

On page 412 we used the sample data shown below to test a claim about the two population means.

| Age 7 | Age 6 |
|-------|-------|
| $n_1 = 16$ | $n_2 = 12$ |
| $\bar{x}_1 = 44.0$ | $\bar{x}_2 = 27.5$ |
| $s_1 = 13.2$ | $s_2 = 10.2$ |

Construct a 95% confidence interval for the difference between the two population means.

## Solution

The samples are independent, the population standard deviations are not known, and the samples are small, so Figure 8–5 directs us to the $F$ test, leading to this conclusion: Fail to reject $\sigma_1^2 = \sigma_2^2$. We now have the case located at the bottom of Figure 8–5 and we first calculate

$$E = t_{\alpha/2} \sqrt{\frac{1}{n_1} + \frac{1}{n_2}} \sqrt{\frac{(n_1 - 1)s_1^2 + (n_2 - 1)s_2^2}{n_1 + n_2 - 2}}$$

$$= 2.056 \sqrt{\frac{1}{16} + \frac{1}{12}} \sqrt{\frac{(16 - 1)(13.2^2) + (12 - 1)(10.2)^2}{16 + 12 - 2}} = 9.4$$

with $\bar{x}_1 = 44.0$, $\bar{x}_2 = 27.5$, and $E = 9.4$, the confidence interval

$$(\bar{x}_1 - \bar{x}_2) - E < (\mu_1 - \mu_2) < (\bar{x}_1 - \bar{x}_2) + E$$

becomes

$$7.1 < (\mu_1 - \mu_2) < 25.9$$

The corresponding hypothesis test that used the same data was a one-tailed test, so a direct comparison between this confidence interval and that test is not appropriate. We can make a direct comparison between the confidence interval and the hypothesis test only when that test is two-tailed. However, a one-tailed hypothesis test with significance level $\alpha$ can be compared to a confidence interval with degree of confidence $1 - 2\alpha$. For example, a right-tailed hypothesis test with a 0.05 significance level can be compared to a 0.90 confidence interval. The data in the last example will result in this 0.90 confidence interval:

$$8.7 < \mu_1 - \mu_2 < 24.3$$

In testing the hypothesis (at the 0.05 significance level) that $\mu_1 > \mu_2$ (or $\mu_1 - \mu_2 > 0$), the above 0.90 confidence interval shows that this claim of

$\mu_1 - \mu_2 > 0$ is supported. That is, there is sufficient evidence to support the claim that the mean for seven-year-old girls is greater than that for six-year-old girls. Although it's a bit trickier, we can make correspondences between confidence intervals and one-tailed hypothesis tests.

STATDISK users: You can get confidence interval limits for the problems of this section. From the main STATDISK menu, select "Hypothesis Testing" and proceed as if you were testing the claim that $\mu_1 = \mu_2$. If you choose $\mu_1 = \mu_2$ for your null hypothesis, the confidence interval limits will be included in the output.

# 8–3 | Exercises A

In Exercises 1–4, use a 0.05 significance level to test the claim that $\mu_1 = \mu_2$. In each case, the two samples are independent and are randomly selected from populations with normal distributions.

1.  | Control Group | Experimental Group |
    |---|---|
    | $n_1 = 40$ | $n_2 = 40$ |
    | $\bar{x}_1 = 79.6$ | $\bar{x}_2 = 84.2$ |
    | $s_1 = 12.4$ | $s_2 = 12.2$ |

2.  | Brand X | Brand Y |
    |---|---|
    | $n_1 = 16$ | $n_2 = 14$ |
    | $\bar{x}_1 = 64.3$ | $\bar{x}_2 = 65.1$ |
    | $s_1 = 2.50$ | $s_2 = 2.50$ |

3.  | Treated | Untreated |
    |---|---|
    | $n_1 = 16$ | $n_2 = 16$ |
    | $\bar{x}_1 = 98.6$ | $\bar{x}_2 = 97.8$ |
    | $s_1 = 8.60$ | $s_2 = 4.20$ |

4.  | Production Method A | Production Method B |
    |---|---|
    | $n_1 = 20$ | $n_2 = 25$ |
    | $\bar{x}_1 = 127.4$ | $\bar{x}_2 = 108.3$ |
    | $s_1 = 15.6$ | $s_2 = 14.3$ |

In Exercises 5–8, use the data in the indicated exercise to construct a 95% confidence interval for the difference $\mu_1 - \mu_2$. In each case, the two samples are independent and are randomly selected from populations with normal distributions.

5.  Exercise 1                6.  Exercise 2
7.  Exercise 3                8.  Exercise 4

9. A study was conducted to investigate the effectiveness of hypnotism in reducing pain. Results for randomly selected subjects are given below. At the 0.05 significance level, test the claim that the affective responses to pain are the same before and after hypnosis. (The data are based on "An analysis of Factors That Contribute to the Efficacy of Hypnotic Analgesia," by Price and Barber, *Journal of Abnormal Psychology*, Vol. 96, No. 1.)

| Before | − 5.5 | − 5.0 | − 6.6 | − 9.7 | − 4.0 | − 7.0 | − 7.0 | − 8.4 |
|--------|-------|-------|-------|-------|-------|-------|-------|-------|
| After  | − 1.4 | − 0.5 | 0.7   | 1.0   | 2.0   | 0.0   | − 0.6 | − 1.8 |

10. Use the sample data from Exercise 9 to construct a 95% confidence interval for the mean of the differences $(\mu_d)$.

11. In a study of the effect of job previews on work expectation, subjects from different groups were tested. For 60 subjects given specific job previews, the mean promotion expectation score is 19.14 and the standard deviation is 6.56. For a "no booklet" sample group of 40 subjects, the mean is 20.81 and the standard deviation is 4.90. (See "Effects of Realistic Job Previews on Hiring Bank Tellers," by Dean and Wanous, *Journal of Applied Psychology*, Vol. 69, No. 1.) At the 0.10 level of significance, test the claim that the two sample groups come from populations with the same mean.

12. Use the sample data from Exercise 11 to construct a 90% confidence interval for the difference between the population means $(\mu_1 - \mu_2)$, where $\mu_1$ is the mean score for all subjects given specific job previews and $\mu_2$ is the mean for the "no booklet" population.

13. An investor considering two possible locations for a new restaurant commissions a study of the pedestrian traffic at both sites. At each location, the pedestrians are observed in 1-hour units and, for each hour, an index of desirable characteristics is compiled. The sample results are given below. Construct a 95% confidence interval for the difference between the two mean indices.

| East | West |
|------|------|
| $n = 35$ | $n = 50$ |
| $\bar{x} = 421$ | $\bar{x} = 347$ |
| $s = 122$ | $s = 85$ |

14. Use the sample data given in Exercise 13 to test the claim that both sites have the same mean index. Use a 0.05 level of significance.

15. A large firm collects sample data on the lengths of telephone calls (in minutes) made by employees in two different divisions, and the results are as shown on the following page. At the 0.02 level of significance, test the claim that there is no difference between the mean times of all long distance calls made in the two divisions.

| Sales Division | Customer Service Division |
|---|---|
| $n_1 = 40$ | $n_2 = 20$ |
| $\bar{x}_1 = 10.26$ | $\bar{x}_2 = 6.93$ |
| $s_1 = 8.65$ | $s_2 = 4.93$ |

16. Use the sample data in Exercise 15 to construct a 98% confidence interval for the difference between the two population means.

17. Twelve different and independent samples from each of two competing cold medicines are tested for the amount of acetaminophen, and the results (in milligrams) are given below. At the 0.05 significance level, test the claim that the mean amount of acetaminophen is the same in each brand.

| Brand X | 472 | 487 | 506 | 512 | 489 | 503 | 511 | 501 | 495 | 504 | 494 | 462 |
|---|---|---|---|---|---|---|---|---|---|---|---|---|
| Brand Y | 562 | 512 | 523 | 528 | 554 | 513 | 516 | 510 | 524 | 510 | 524 | 508 |

18. Using the data from Exercise 17, construct a 95% confidence interval for this difference: The mean of the Brand Y levels minus the mean of the Brand X levels.

19. A dose of the drug captopril, designed to lower systolic blood pressure, is administered to 10 randomly selected volunteers, with the following results. Construct the 95% confidence interval for $\mu_d$, the mean of the differences between the before and after scores.

| Before pill | 120 | 136 | 160 | 98 | 115 | 110 | 180 | 190 | 138 | 128 |
|---|---|---|---|---|---|---|---|---|---|---|
| After pill | 118 | 122 | 143 | 105 | 98 | 98 | 180 | 175 | 105 | 112 |

20. Using the sample data from Exercise 19, test the claim that systolic blood pressure is not affected by the pill. Use a 0.05 significance level.

21. Two ambulance services are tested for response times. A sample of 50 responses from the first firm produces a mean of 12.2 min and a standard deviation of 1.5 min. A sample of 50 responses from the second firm produces a mean of 14.0 min with a standard deviation of 2.1 min. Construct a 95% confidence interval for the difference between the two means.

22. Using the data from Exercise 21, test the claim that the two ambulance services have the same mean response time. Use a 0.05 level of significance.

23. Two different and independent procedures are used to control air traffic at an airport. The numbers of operations for 30 different randomly selected hours are listed below. At the 0.05 significance level, test the claim that the use of System 2 results in a mean number of operations per hour exceeding the mean for System 1.

| System 1 | 63 | 62 | 67 | 66 | 53 | 72 | 62 | 57 | 49 | 57 | 60 | 68 | 58 | 64 | 61 |
|---|---|---|---|---|---|---|---|---|---|---|---|---|---|---|---|
| System 2 | 62 | 63 | 67 | 61 | 68 | 64 | 66 | 62 | 67 | 66 | 62 | 62 | 65 | 65 | 66 |

24. Using the data found in Exercise 23, construct a 95% confidence interval for the difference between the means of the two systems.

25. A course is designed to increase reading speed and comprehension. To evaluate the effectiveness of this course, students are given a test both before and after this course, and the sample results follow. Construct a 95% confidence interval for the mean of the differences between the before and after scores.

| Before | 100 | 110 | 135 | 167 | 200 | 118 | 127 | 95 | 112 | 116 |
|--------|-----|-----|-----|-----|-----|-----|-----|-----|-----|-----|
| After | 136 | 160 | 120 | 169 | 200 | 140 | 163 | 101 | 138 | 129 |

26. Use the data from Exercise 25 to test the claim that the scores are higher after the course. Use a 0.05 level of significance.

27. As part of the National Health Survey, data were collected on the weights of men. For 804 men aged 25 to 34, the mean is 176 lb and the standard deviation is 35.0 lb. For 1657 men aged 65 to 74, the mean and standard deviation are 164 lb and 27.0 lb, respectively. (See the National Health Survey, USDHEW publication 79-1659.) Construct a 99% confidence interval for the difference between the means of the men in the two age brackets.

28. Use the data from Exercise 27 to test the claim that the older men come from a population with a mean less than that for men in the 25–34 age bracket. Use a 0.01 level of significance.

29. A study was conducted to investigate some effects of physical training. Sample data are listed at the right. (See "Effect of Endurance Training on Possible Determinants of $VO_2$ During Heavy Exercise," by Casaburi and others, *Journal of Applied Physiology*, Vol. 62, No. 1.) At the 0.05 level of significance, test the claim that mean pretraining weight equals mean posttraining weight. All weights are given in kilograms.

| Pre-training | Post-training |
|--------------|---------------|
| 99 | 94 |
| 57 | 57 |
| 62 | 62 |
| 69 | 69 |
| 74 | 66 |
| 77 | 76 |
| 59 | 58 |
| 92 | 88 |
| 70 | 70 |
| 85 | 84 |

30. Use the data from Exercise 29 to construct a 95% confidence interval for the mean of the differences between pretraining and posttraining weights.

31. The effectiveness of a mental training program was tested in a military training program. In an antiaircraft artillery examination, scores for an experimental group and a control group were recorded. Use the given data to test the claim that both groups come from populations with the same mean. Use a 0.05 significance level. (See "Routinization of Mental

Training in Organizations: Effects on Performance and Well-Being,"
by Larsson, *Journal of Applied Psychology*, Vol. 72, No. 1.)

| Experimental | | | | | Control | | | |
|---|---|---|---|---|---|---|---|---|
| 60.83 | 117.80 | 44.71 | 75.38 | | 122.80 | 70.02 | 119.89 | 138.27 |
| 73.46 | 34.26 | 82.25 | 59.77 | | 118.43 | 54.22 | 118.58 | 74.61 |
| 69.95 | 21.37 | 59.78 | 92.72 | | 121.70 | 70.70 | 99.08 | 120.76 |
| 72.14 | 57.29 | 64.05 | 44.09 | | 104.06 | 94.23 | 111.26 | 121.67 |
| 80.03 | 76.59 | 74.27 | 66.87 | | | | | |

32. Using the sample data from Exercise 31, construct a 95% confidence interval for the difference between the two population means.

# 8–3 Exercises B

33. In constructing the confidence interval for Exercise 7, the value of $t_{\alpha/2}$ was found by setting the degrees of freedom equal to the smaller of $n_1 - 1$ and $n_2 - 1$. A better result is obtained by using Formula 8–1 to find the number of degrees of freedom. Use Formula 8–1 to determine the number of degrees of freedom, then use it to find the confidence interval requested in Exercise 7. Compare the results to the confidence interval originally obtained in Exercise 7.

34. Refer to Exercise 28. If the actual difference between the population means is 6.0, find $\beta$, the probability of a type II error. (See Exercise 32 in Section 7–2.) *Hint:* In step 1, replace $\bar{x}$ by $(\bar{x}_1 - \bar{x}_2)$, replace $\mu_{\bar{x}}$ by 0, and replace $\sigma_{\bar{x}}$ by

$$\sqrt{\frac{\sigma_1^2}{n_1} + \frac{\sigma_2^2}{n_2}}$$

35. a. Find the variance for this *population* of x scores: 5, 10, 15.
    b. Find the variance for this *population* of y scores: 1, 2, 3.
    c. List the population of all possible $x - y$ scores, then find the variance of this population.
    d. Use the results from parts a, b, c to verify that

    $$\sigma_{x-y}^2 = \sigma_x^2 + \sigma_y^2$$

    (This principle is used to derive the test statistic and confidence interval for several cases in this section.)

36. Refer to the indicated exercise and find the *P*-value.
    a. Exercise 1
    b. Exercise 2
    c. Exercise 3
    d. Exercise 9

# $\boxed{8\text{--}4}$ Inferences from Two Proportions

In this section we consider inferences made about two population proportions. The concepts and procedures we develop can be used to answer questions such as the following.

- When one group is given an experimental drug and another group is given a placebo, is there a difference between the cure rates in the two groups?
- When one group is given an experimental drug and another group is given a placebo, what is an estimate of the difference between the two cure rates?

In this section we make the following assumptions.

### Assumptions

1. We intend to *test a hypothesis* made about two population proportions or we intend to construct a *confidence interval* for the difference between two population proportions.
2. We have two sets of sample data that come from *independent* populations.
3. For both samples, the conditions $np \geq 5$ and $nq \geq 5$ are satisfied.

We will also use the following notation.

---

**Notation**

For Population 1 we let:

$p_1$ denote the population proportion
$n_1$ denote the size of the sample
$x_1$ denote the number of successes

$$\hat{p}_1 = \frac{x_1}{n_1}$$

The corresponding meanings are attached to $p_2$, $n_2$, $x_2$, and $\hat{p}_2$, which come from Population 2.

---

# Hypothesis Tests

We know from Section 5–5 that if $n_1 p_1 \geq 5$ and $n_1 q_1 \geq 5$, then we can approximate the binomial distribution by the normal distribution. This applies to the second population as well. As a result, our test statistic will be approximately normally distributed. **We will be testing only claims including the assumption that $p_1 = p_2$**, and we will use the following pooled estimate of their common value.

## Definition

The **pooled estimate of $p_1$ and $p_2$** is denoted by $\bar{p}$ and is given by

$$\bar{p} = \frac{x_1 + x_2}{n_1 + n_2}$$

$$\bar{q} = 1 - \bar{p}$$

With a null hypothesis of $p_1 = p_2$, or $p_1 \geq p_2$, or $p_1 \leq p_2$, we can use the following test statistic.

## Test Statistic

(For $H_0: p_1 = p_2$, $H_0: p_1 \geq p_2$, or $H_0: p_1 \leq p_2$)

$$z = \frac{(\hat{p}_1 - \hat{p}_2) - (p_1 - p_2)}{\sqrt{\bar{p}\bar{q}\left(\dfrac{1}{n_1} + \dfrac{1}{n_2}\right)}}$$

where $(p_1 - p_2) = 0$
and

$$\hat{p}_1 = \frac{x_1}{n_1} \qquad \hat{p}_2 = \frac{x_2}{n_2}$$

$$\bar{p} = \frac{x_1 + x_2}{n_1 + n_2}$$

$$\bar{q} = 1 - \bar{p}$$

We will first illustrate the procedure for testing hypotheses, then we will justify the test statistic given above.

## Example

In a study of 22,000 male physicians, half were given regular doses of aspirin while the other half were given placebos. (The study was reported in *Time* and the *New England Journal of Medicine*, Vol. 318, No. 4.) Among those who took the aspirin doses, 104 suffered heart attacks. Among those who took the placebos, 189 suffered heart attacks. At the 0.01 significance level, test the claim that the aspirin group has a significantly lower heart attack rate.

*continued*

# BETTER RESULTS WITH SMALLER CLASS SIZE

■ An experiment at the State University of New York at Stony Brook found that students did significantly better in classes limited to 35 students than in large classes with 150 to 200 students. For a calculus course, failure rates were 19% for the small classes compared to 50% for the large classes. The percentages of A's were 24% for the small classes and 3% for the large classes. These results suggest that students benefit from smaller classes, which allow for more direct interaction between students and teachers. ■

**Example** *continued*

$n_1 = 11,000$
Heart attacks: $x_1 = 104$

0.945%

$n_2 = 11,000$
Heart attacks: $x_2 = 189$

1.72%

The fundamental question is really this: Under the given circumstances, is the 0.945% rate significantly lower than the 1.72% rate? The claim of a lower heart attack rate for the aspirin group can be represented by $p_1 < p_2$. We therefore have

$$H_0: p_1 \geq p_2$$
$$H_1: p_1 < p_2$$

With $\alpha = 0.01$, we use the given sample data to get

$$\hat{p}_1 = \frac{x_1}{n_1} = \frac{104}{11,000} = 0.00945$$

$$\hat{p}_2 = \frac{x_2}{n_2} = \frac{189}{11,000} = 0.0172$$

$$\bar{p} = \frac{x_1 + x_2}{n_1 + n_2} = \frac{104 + 189}{11,000 + 11,000} = \frac{293}{22,000} = 0.0133$$

$$\bar{q} = 1 - \bar{p} = 1 - 0.0133 = 0.9867$$

$$z = \frac{(\hat{p}_1 - \hat{p}_2) - 0}{\sqrt{\bar{p}\bar{q}\left(\frac{1}{n_1} + \frac{1}{n_2}\right)}} = \frac{(0.00945 - 0.0172) - 0}{\sqrt{(0.0133)(0.9867)\left(\frac{1}{11,000} + \frac{1}{11,000}\right)}}$$

$$= -5.02$$

Figure 8–6 shows the test statistic of $z = -5.02$ along with the critical value of $z = -2.33$. The test statistic is well within the critical region so we reject the null hypothesis. The sample data support the claim that the heart attack rate is lower for the aspirin group.

Also, we can determine the *P*-value by using the procedures from Section 7–3. With a test statistic of $z = -5.02$ in a left-tailed test, Table A–2 shows that the *P*-value is less than 0.0001. Since the *P*-value is less than the significance level of $\alpha = 0.01$, we reject the null hypothesis and again conclude that the sample data support the claim of a lower heart attack rate for the aspirin group.

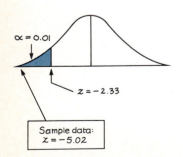

$\alpha = 0.01$

$z = -2.33$

Sample data:
$z = -5.02$

**Figure 8–6**

Does aspirin really prevent heart attacks? The preceding results show that the rate of 104 heart attacks among 11,000 subjects is significantly less than the rate of 189 heart attacks among 11,000 subjects. The media reports of these results included many words of caution and criticism. At the very least, these results suggest that medical researchers should continue with further research. While we cannot use statistics to *prove* that aspirin *causes* fewer heart attacks, we do have strong evidence that helps to guide future medical research.

The symbols $x_1$, $x_2$, $n_1$, $n_2$, $\hat{p}_1$, $\hat{p}_2$, $\bar{p}$, and $\bar{q}$ should have become more meaningful through this example. In particular, you should recognize that, under the assumption of equal proportions, the best estimate of the common proportion is obtained by pooling both samples into one larger sample. Then

$$\bar{p} = \frac{x_1 + x_2}{n_1 + n_2}$$

becomes a more obvious estimate of the common population proportion.

So far we have discussed only proportions in this section, but probabilities are already in decimal or fractional form, so they can directly replace proportions in the preceding discussion. Percentages can also be dealt with by using the corresponding decimal equivalents, as illustrated in the following example.

## Example

In a study of people who stop to help drivers with disabled cars, researchers hypothesized that more people would stop if they first saw someone else getting help. In one experiment, 2000 drivers first saw a woman being helped with a flat tire and 2.90% of them stopped to help a second woman with a flat tire. Among 2000 drivers who did not see the first helper, only 1.75% stopped to help a woman with a flat tire. (See "Help On the Highway," by McCarthy, *Psychology Today*, July 1987.) At the 0.05 significance level, test the claim that the two proportions are equal.

## Solution

The claim of equal proportions leads to $H_0$: $p_1 = p_2$ and $H_1$: $p_1 \neq p_2$. We can summarize the given data as follows.

| Saw earlier helper | Saw no earlier helper |
|---|---|
| $n_1 = 2000$ | $n_2 = 2000$ |
| $x_1 = 2.90\%$ of $2000 = 58$ | $x_2 = 1.75\%$ of $2000 = 35$ |

*continued*

**Solution** *continued*

With $\hat{p}_1 = x_1/n_1 = 0.0290$ and $\hat{p}_2 = x_2/n_2 = 0.0175$, we can now find $\bar{p}$ and $\bar{q}$.

$$\bar{p} = \frac{x_1 + x_2}{n_1 + n_2} = \frac{58 + 35}{2000 + 2000} = \frac{93}{4000} = 0.02325$$
$$\bar{q} = 1 - \bar{p} = 0.97675$$

We continue by computing the value of the test statistic.

$$z = \frac{(\hat{p}_1 - \hat{p}_2) - 0}{\sqrt{\bar{p}\bar{q}\left(\dfrac{1}{n_1} + \dfrac{1}{n_2}\right)}} = \frac{(0.0290 - 0.0175) - 0}{\sqrt{(0.02325)(0.97675)\left(\dfrac{1}{2000} + \dfrac{1}{2000}\right)}}$$
$$= 2.41$$

With $\alpha = 0.05$ in this two-tailed test, we use Table A–2 to obtain the critical $z$ values of $-1.96$ and $1.96$ (see Figure 8–7). Since the test statistic falls within the critical region, we reject the null hypothesis. We conclude that there is sufficient sample evidence to warrant rejection of the claim that the two proportions are equal.

The $P$-value can be found by using Table A–2. The test statistic of $z = 2.41$ corresponds to an area of 0.4920. Since this test is two-tailed, the $P$-value is $2 \times (0.5000 - 0.4920) = 0.016$. Since the $P$-value of 0.016 is less than the significance level of 0.05, we should reject the null hypothesis of equal proportions.

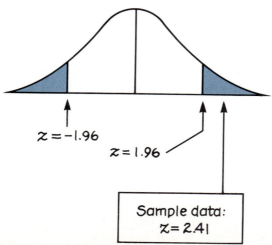

$z = -1.96$

$z = 1.96$

Sample data:
$z = 2.41$

**Figure 8–7**

Shown below is the STATDISK display for the data in the preceding example.

STATDISK DISPLAY

Hypothesis test for a claim about TWO POPULATION PROPORTIONS

NULL HYPOTHESIS: p1 = p2

| Sample 1: | x1 = 58 | n1 = 2000 | x1/n1 = .02900 |
|---|---|---|---|
| Sample 2: | x2 = 35 | n2 = 2000 | x2/n2 = .01750 |

Test statistic . . . . . . . . . . . . . . . . . . .z = 2.41321
Critical value . . . . . . . . . . . . . . . . . .z = −1.96039, 1.96039
P-value . . . . . . . . . . . . . . . . . . . . . . . . . = .01591
Conf. interval limits . . . . . . . . . . . . . . = .00216, .02084
Significance level . . . . . . . . . . . . . . . = .05

CONCLUSION: REJECT the null hypothesis

The test statistic we are using can be justified as follows.

1. With $n_1 p_1 \geq 5$ and $n_1 q_1 \geq 5$, the distribution of $\hat{p}_1$ can be approximated by a normal distribution with mean $p_1$, standard deviation $\sqrt{p_1 q_1 / n_1}$, and variance $p_1 q_1 / n_1$. These conclusions follow from Sections 5–5 and 6–3. They also apply to the second sample and population.

2. Since $\hat{p}_1$ and $\hat{p}_2$ are each approximated by a normal distribution, $\hat{p}_1 - \hat{p}_2$ will also be approximated by a normal distribution with mean $p_1 - p_2$ and variance

$$\sigma^2_{(\hat{p}_1 - \hat{p}_2)} = \sigma^2_{\hat{p}_1} + \sigma^2_{\hat{p}_2} = \frac{p_1 q_1}{n_1} + \frac{p_2 q_2}{n_2}$$

(In Section 8–3 we establish that the variance of the differences between two independent random variables is the sum of their individual variances.)

3. Since the values of $p_1$, $p_2$, $q_1$, and $q_2$ are typically unknown, and from the null hypothesis we assume that $p_1 = p_2$, we can pool the sample data. The pooled estimate of the common value of $p_1$ and $p_2$ is

$$\bar{p} = \frac{x_1 + x_2}{n_1 + n_2}$$

## MORE POLICE, FEWER CRIMES?

■ Does an increase in the number of police officers result in lower crime rates? The question was studied in a New York City experiment that involved a 40% increase in police officers in one precinct while adjacent precincts maintained a constant level of officers. Statistical analysis of the crime records showed that crimes visible from the street (such as auto thefts) did decrease, but crimes not visible from the street (such as burglaries) were not significantly affected. ■

If we replace $p_1$ and $p_2$ by $\overline{p}$, and replace $q_1$ and $q_2$ by $\overline{q} = 1 - \overline{p}$, the variance from step 2 above leads to this standard deviation:

$$\sigma_{(\hat{p}_1 - \hat{p}_2)} = \sqrt{\frac{\overline{p}\,\overline{q}}{n_1} + \frac{\overline{p}\,\overline{q}}{n_2}} = \sqrt{\overline{p}\,\overline{q}\left(\frac{1}{n_1} + \frac{1}{n_2}\right)}$$

4. We now know that the distribution of $\hat{p}_1 - \hat{p}_2$ is approximately normal, with mean $p_1 - p_2$ and standard deviation as given above. This corresponds to the test statistic

$$z = \frac{\text{(sample statistic)} - \text{(population mean)}}{\text{(population standard deviation)}}$$

or

$$z = \frac{(\hat{p}_1 - \hat{p}_2) - (p_1 - p_2)}{\sqrt{\overline{p}\,\overline{q}\left(\dfrac{1}{n_1} + \dfrac{1}{n_2}\right)}}.$$

which is the test statistic given earlier.

Remember, in order to use this test statistic, we must assume that $p_1 = p_2$, $p_1 \le p_2$, or $p_1 \ge p_2$. With this assumption, $p_1 - p_2$ in the test statistic will always be 0. For testing claims that the difference $p_1 - p_2$ is equal to a nonzero constant, a different test statistic is used (see Exercise 25).

## Confidence Intervals

With the same assumptions given at the beginning of this section, a confidence interval for the difference between population proportions $p_1$ and $p_2$ can be constructed by evaluating

$$(\hat{p}_1 - \hat{p}_2) - E < (p_1 - p_2) < (\hat{p}_1 - \hat{p}_2) + E$$

where

$$E = z_{\alpha/2}\sqrt{\frac{\hat{p}_1\hat{q}_1}{n_1} + \frac{\hat{p}_2\hat{q}_2}{n_2}}$$

We will first give an example illustrating the construction of a confidence interval, then we will justify the format given above.

**Example**

Use the sample data given in the preceding example to construct the 95% confidence interval for the difference between the two population proportions.

| Saw earlier help | Saw no earlier help |
|---|---|
| $n_1 = 2000$ | $n_2 = 2000$ |
| $x_1 = 58$ | $x_2 = 35$ |

**Solution**

With a 95% degree of confidence, $z_{\alpha/2} = 1.96$ (from Table A–2). With $\hat{p}_1 = 58/2000 = 0.0290$ and $\hat{p}_2 = 35/2000 = 0.0175$, we first evaluate the maximum error of estimate $E$.

$$E = z_{\alpha/2} \sqrt{\frac{\hat{p}_1 \hat{q}_1}{n_1} + \frac{\hat{p}_2 \hat{q}_2}{n_2}} = 1.96 \sqrt{\frac{(0.0290)(0.9710)}{2000} + \frac{(0.0175)(0.9825)}{2000}}$$

$$= 0.0093$$

With $\hat{p}_1 = 0.0290$, $\hat{p}_2 = 0.0175$, and $E = 0.0093$, the confidence interval

$$(\hat{p}_1 - \hat{p}_2) - E < (p_1 - p_2) < (\hat{p}_1 - \hat{p}_2) + E$$

becomes

$$(0.0290 - 0.0175) - 0.0093 < (\hat{p}_1 - \hat{p}_2) < (0.0290 - 0.0175) + 0.0093$$

or

$$0.0022 < (p_1 - p_2) < 0.0208$$

We should be careful when interpreting confidence intervals. Because $p_1$ and $p_2$ have fixed values and are not variables, it is wrong to state that there is a 95% chance that the value of $p_1 - p_2$ falls between 0.0022 and 0.0208. It is correct to state that if we repeat the same sampling process and construct 95% confidence intervals, in the long run, 95% of the intervals will actually contain the value of $p_1 - p_2$.

In the preceding example, if we had reversed the order of the samples, the result would have been

$$-0.0208 < (p_1 - p_2) < -0.0022$$

When there does appear to be a difference, be sure that you know which proportion is larger.

You can see from this example that once you have identified the values of $n_1$, $\hat{p}_1$, $\hat{q}_1$, $n_2$, $\hat{p}_2$, and $\hat{q}_2$, the construction of the confidence interval is not very difficult.

The form of the confidence interval comes directly from the test statistic if we use the variance

$$\sigma^2_{(\hat{p}_1 - \hat{p}_2)} = \sigma^2_{\hat{p}_1} + \sigma^2_{\hat{p}_2} = \frac{p_1 q_1}{n_1} + \frac{p_2 q_2}{n_2}$$

to estimate the standard deviation as

$$\sqrt{\frac{\hat{p}_1 \hat{q}_1}{n_1} + \frac{\hat{p}_2 \hat{q}_2}{n_2}}$$

(We don't use pooled estimates of proportions because we're not assuming from a null hypothesis that $p_1 = p_2$.) In the test statistic

$$z = \frac{(\hat{p}_1 - \hat{p}_2) - (p_1 - p_2)}{\sqrt{\frac{\hat{p}_1 \hat{q}_1}{n_1} + \frac{\hat{p}_2 \hat{q}_2}{n_2}}}$$

let $z$ be positive and negative (for two tails) and solve for $p_1 - p_2$. The result is the confidence interval given earlier.

Users of STATDISK can get confidence interval limits for the problems in this section. From the main STATDISK menu, select "Hypothesis Testing" and proceed as if you were testing the claim $p_1 = p_2$. If you choose that null hypothesis, the confidence interval limits will be included in the output.

# 8–4 Exercises A

1. Let samples from two populations be such that $x_1 = 45$, $n_1 = 100$, $x_2 = 115$, and $n_2 = 200$.
   a. Compute the $z$ test statistic based on the given data.
   b. If the significance level is 0.05 and the test is two-tailed, find the critical $z$ values.
   c. Test the claim that the two populations have equal proportions using the significance level of $\alpha = 0.05$.
   d. Find the $P$-value.
   e. Find the 95% confidence interval for $p_1 - p_2$.
2. Samples taken from two populations yield the data $x_1 = 30$, $n_1 = 250$, $x_2 = 44$, and $n_2 = 800$.
   a. Compute the $z$ test statistic based on the given data.
   b. If the significance level is 0.01 and the test is two-tailed, find the critical $z$ values.
   c. Test the claim that the two populations have equal proportions using the significance level of $\alpha = 0.01$.
   d. Find the $P$-value.
   e. Find the 99% confidence interval for $p_1 - p_2$.

3. In a *New York Times*/CBS News survey, 35% of 552 Democrats felt that the government should regulate airline prices, compared to 41% of the 417 Republicans surveyed. At the 0.05 significance level, test the claim that there is no difference between the proportions of Democrats and Republicans who feel that way.

4. Use the sample data from Exercise 3 to construct a 95% confidence interval for the difference between the proportions of Democrats and Republicans who feel that the government should regulate airline prices.

5. After a recent national election, surveys showed that among 200 randomly selected people aged 18–24, 36.0% voted. Among 250 people in the 25–44 age bracket, 54.0% voted (based on data from the U.S. Bureau of the Census). Construct a 95% confidence interval for the difference between the proportions of voters in the two different age brackets.

6. Using the sample data given in Exercise 5, test the claim that the proportions of voters in the two age brackets are the same.

7. The *New York Times* ran an article about a study in which Professor Denise Korniewicz and other Johns Hopkins researchers subjected laboratory gloves to stress. Among 240 vinyl gloves, 63% leaked viruses. Among 240 latex gloves, 7% leaked viruses. At the 0.01 significance level, test the claim that vinyl and latex gloves have the same virus leak rates.

8. Using the sample data from Exercise 7, construct the 99% confidence interval for the difference between the two proportions of gloves that leak viruses.

9. When 294 central city residents were surveyed, 28.9% refused to respond. A survey of 1015 residents not in a central city resulted in a 17.1% refusal rate (based on data from "I Hear You Knocking But You Can't Come In," by Fitzgerald and Fuller, *Sociological Methods and Research*, Vol. 11, No. 1). At the 0.01 significance level, test the claim that the central city refusal rate is the same as the refusal rate in other areas.

10. Using the sample data in Exercise 9, construct a 99% confidence interval for the difference between the proportions of refusals in a central city and elsewhere.

11. For 2750 randomly selected arrests of criminals under 21 years of age, 4.25% involve violent crimes. For 2200 randomly selected arrests of criminals 21 years of age or older, 4.55% involve violent crimes (based on data from the Uniform Crime Reports). Construct a 95% confidence interval for the difference between the two proportions of violent crimes.

12. Use the data from Exercise 11 to test, at the 0.05 significance level, the claim that both age groups have the same rate of violent crimes.

13. In a recent survey of 500 males aged 14–24, 3.6% of them were living alone. In a 1960 survey of 750 males aged 14–24, 1.6% were living alone (based on data from the U.S. Bureau of the Census). Construct a 99% confidence interval for the difference between the two proportions of males living alone.

14. Refer to the sample data in Exercise 13. At the 0.01 level of significance, test the claim that the more recent rate is greater than the rate in 1960.

15. A study was conducted to investigate the use of seat belts in taxicabs. Among 72 taxis observed in Pittsburgh, 36 had seat belts at least partially visible. Among 129 taxis observed in Chicago, 77 had seat belts at least partially visible. (See "The Phantom Taxi Seat Belt," by Welkon and Reisinger, *American Journal of Public Health*, Vol. 67, No. 11.) At the 0.05 level of significance, test the claim that Pittsburgh and Chicago have the same proportion of taxis with seat belts at least partially visible.

16. Refer to the sample data in Exercise 15 and construct a 95% interval estimate of the difference between the two proportions of cabs with seat belts at least partially visible.

17. The American College Testing Program provides data showing that 30% of four-year public college freshmen drop out, while the dropout rate at four-year private colleges is 26%. Assume that these results are based on observations of 1000 four-year public college freshmen and 500 four-year private college freshmen. At the 0.05 significance level, test the claim that four-year public and private colleges have the same freshman dropout rate.

18. Use the data from Exercise 17 to construct a 95% confidence interval for the difference between the proportion of dropout rates for four-year public and private colleges.

19. In initial tests of the Salk vaccine, 33 of 200,000 vaccinated children later developed polio. Of 200,000 children vaccinated with a placebo, 115 later developed polio. At the 1% level of significance, test the claim that the Salk vaccine is effective in lowering the polio rate.

20. The New York State Department of Motor Vehicles provided the following motor vehicle conviction data for a recent year.

|  | Albany County | Queens County |
|---|---|---|
| Total convictions | 24,384 | 166,197 |
| DWI convictions | 558 | 1,214 |

At the 0.01 level of significance, test the claim that the proportion of DWI (driving while intoxicated) convictions is lower in Queens County. Assume that the given data represent random samples drawn from a larger population.

21. About 15 years ago, a survey of 2000 adults showed that 65% were concerned about nuclear power plants. A recent survey of 2000 adults showed that 82% expressed that same concern (based on data from a Roper poll). Construct a 99% confidence interval for the difference between the proportions of concerned adults shown in these two surveys.

22. Using the data from Exercise 21, test the claim that the proportion of concerned adults was lower 15 years ago. Use a 0.01 significance level.

23. An advertiser studies the proportion of radio listeners who prefer country music. In Region A, 38% of the 250 listeners surveyed indicated a preference for country music. In Region B, country music was preferred by 14% of the 400 listeners surveyed. Construct a 98% confidence interval for the difference between the proportions of listeners who prefer country music.

24. Using the data from Exercise 23, and using a 0.02 significance level, test the claim that region A has a greater proportion of listeners who prefer country music.

## 8–4 Exercises B

25. To test the null hypothesis that the difference between two population proportions is equal to a nonzero constant $c$, use

$$z = \frac{(\hat{p}_1 - \hat{p}_2) - c}{\sqrt{\dfrac{\hat{p}_1(1 - \hat{p}_1)}{n_1} + \dfrac{\hat{p}_2(1 - \hat{p}_2)}{n_2}}}$$

As long as $n_1$ and $n_2$ are both large, the sampling distribution of the above test statistic $z$ will be approximately the standard normal distribution. Suppose a winery is conducting market research in New York and California. In a sample of 500 New Yorkers, 120 like the wine, while a sample of 500 Californians shows that 210 like the wine. Use a 0.05 level of significance to test the claim that the percentage of Californians who like the wine is 25% more than the percentage of New Yorkers who like it.

26. Sample data are randomly drawn from three independent populations. The sample sizes and the numbers of successes follow.

| Population 1 | Population 2 | Population 3 |
|---|---|---|
| $n = 100$ | $n = 100$ | $n = 100$ |
| $x = 40$ | $x = 30$ | $x = 20$ |

a. At the 0.05 significance level, test the claim that $p_1 = p_2$.
b. At the 0.05 significance level, test the claim that $p_2 = p_3$.

*continued*

c. At the 0.05 significance level, test the claim that $p_1 = p_3$.
d. In general, if hypothesis tests lead to the decisions that $p_1 = p_2$ and $p_2 = p_3$, does it follow that the decision $p_1 = p_3$ will be reached under the same conditions?

27. The **sample size** needed to estimate the difference between two population proportions to within $E$ with a confidence level of $1 - \alpha$ can be found as follows. In the expression

$$E = z_{\alpha/2} \sqrt{\frac{p_1 q_1}{n_1} + \frac{p_2 q_2}{n_2}}$$

replace $n_1$ and $n_2$ by $n$ (assuming that both samples have the same size) and replace each of $p_1$, $q_1$, $p_2$, and $q_2$ by 0.5 (since their values are not known). Then solve for $n$.

Use this approach to find the size of each sample if you want to estimate the difference between the proportions of men and women who own cars. Assume that you want 95% confidence that your error is no more than 0.03.

28. Refer to Exercise 20 and assume that you want to test the given claim by constructing an appropriate confidence interval. Find that confidence interval, identify its degree of confidence, and state the conclusion.

 ## *Vocabulary List*

Define and give an example of each term.

*F* distribution
numerator degrees of freedom
denominator degrees of freedom
independent samples
dependent samples
pooled estimate of $p_1$ and $p_2$

 ## *Review*

In Chapters 6 and 7 we introduced two major concepts of inferential statistics: the estimation of population parameters and the methods of testing hypotheses made about population parameters. The examples and exercises in Chapters 6 and 7 were restricted to cases involving a single sample drawn from a single population. This chapter extended our coverage to include two samples drawn from two populations.

We began by presenting a test for comparing two population variances (or standard deviations) that come from two independent populations having normal distributions. We began with a test for comparing two variances or standard deviations, since such a test is sometimes used as part of an overall test for comparing two population means. In Section 8–2 we saw that when $\sigma_1^2 = \sigma_2^2$ the sampling distribution of $F = s_1^2/s_2^2$ is the $F$-distribution, for which Table A–5 was computed. We briefly discussed the construction of confidence intervals for the ratio of two population variances.

In Section 8–3 we considered various situations that can occur when we want to use a hypothesis test for comparing two means. We should begin such a test by determining whether or not the two populations are **dependent** in the sense that they are related in some way. When comparing population means that come from two dependent and normal populations, we compute the differences between corresponding pairs of values. Those differences have a mean and standard deviation denoted by $\bar{d}$ and $s_d$, respectively. In repeated random samplings, the values of $\bar{d}$ possess a Student $t$ distribution with mean $\mu_d$ and standard deviation $\sigma_d/\sqrt{n}$.

When using hypothesis tests to compare two population means from independent populations, we encounter four situations that can be summarized best by Figure 8–4. These cases incorporate standard deviations reflecting the property that, if one random variable $x$ has variance $\sigma_x^2$ and another independent random variable $y$ has variance $\sigma_y^2$, the random variable $x - y$ will have variance $\sigma_x^2 + \sigma_y^2$.

Section 8–3 also included confidence intervals that serve as estimates of differences between population means. Figure 8–5 can be used to determine the appropriate form of the confidence interval for the four different cases that occur.

In Section 8–4 we considered hypothesis tests and confidence intervals for proportions, probabilities, or percentages that come from two independent populations. We saw that the sample proportions have differences

$$\frac{x_1}{n_1} - \frac{x_2}{n_2} \quad \text{or} \quad \hat{p}_1 - \hat{p}_2$$

that tend to have a distribution that is approximately normal, with mean $p_1 - p_2$ and a standard deviation estimated by

$$\sqrt{\bar{p}\bar{q}\left(\frac{1}{n_1} + \frac{1}{n_2}\right)}$$

when $p_1 = p_2$. Also, $\bar{q} = 1 - \bar{p}$ and $\bar{p}$ is the pooled proportion $(x_1 + x_2)/(n_1 + n_2)$.

Figure 8–8 provides a reference chart for locating the appropriate material. One of the most difficult aspects of inferential statistics is the identification of the most appropriate distribution and the selection of the proper test statistic or confidence interval. Figure 8–8 should help in that determination.

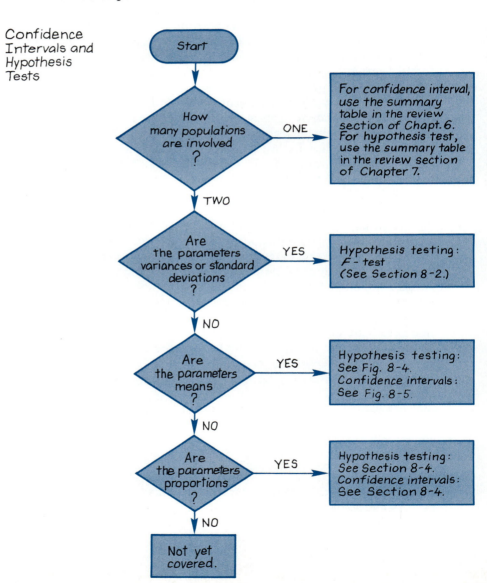

**Figure 8–8**

## Important Formulas

| Parameters | Applicable Distribution | Testing Hypotheses (Test statistic) | Confidence Intervals | Table of Critical Values |
|---|---|---|---|---|
| $\sigma_1, \sigma_2$ (two standard deviations) or $\sigma_1^2, \sigma_2^2$ (two variances) | $F$ | $F = \dfrac{s_1^2}{s_2^2}$ where $s_1^2 \geq s_2^2$ | See Exercises 21 and 22 in Section 8–2. | Table A–5 |
| $\mu_1, \mu_2$ (two means): dependent samples | Student $t$ | $t = \dfrac{\bar{d} - \mu_d}{s_d/\sqrt{n}}$ | $\bar{d} - E < \mu_d < \bar{d} + E$ where $E = t_{\alpha/2}\dfrac{s_d}{\sqrt{n}}$ | Table A–3 |
| independent samples (Hypothesis tests: see Figure 8–4; confidence intervals: see Figure 8–5) | Normal or Student $t$ | $z = \dfrac{(\bar{x}_1 - \bar{x}_2) - (\mu_1 - \mu_2)}{\sqrt{\dfrac{\sigma_1^2}{n_1} + \dfrac{\sigma_2^2}{n_2}}}$ | $(\bar{x}_1 - \bar{x}_2) - E < (\mu_1 - \mu_2) < (\bar{x}_1 - \bar{x}_2) + E$ where $E = z_{\alpha/2}\sqrt{\dfrac{\sigma_1^2}{n_1} + \dfrac{\sigma_2^2}{n_2}}$ | Table A-2 |
| | | $t = \dfrac{(\bar{x}_1 - \bar{x}_2) - (\mu_1 - \mu_2)}{\sqrt{\dfrac{s_1^2}{n_1} + \dfrac{s_2^2}{n_2}}}$ | or $E = t_{\alpha/2}\sqrt{\dfrac{s_1^2}{n_1} + \dfrac{s_2^2}{n_2}}$ | Table A-3 |
| | | $t = \dfrac{(\bar{x}_1 - \bar{x}_2) - (\mu_1 - \mu_2)}{\sqrt{\dfrac{1}{n_1} + \dfrac{1}{n_2}}\sqrt{\dfrac{(n_1 - 1)s_1^2 + (n_2 - 1)s_2^2}{n_1 + n_2 - 2}}}$ | or $E = t_{\alpha/2}\sqrt{\dfrac{1}{n_1} + \dfrac{1}{n_2}}\sqrt{\dfrac{(n_1 - 1)s_1^2 + (n_2 - 1)s_2^2}{n_1 + n_2 - 2}}$ | Table A-3 |
| $p_1, p_2$ (two proportions) | Normal | $z = \dfrac{(\hat{p}_1 - \hat{p}_2) - (p_1 - p_2)}{\sqrt{\bar{p}\bar{q}\left(\dfrac{1}{n_1} + \dfrac{1}{n_2}\right)}}$ | $(\hat{p}_1 - \hat{p}_2) - E < (p_1 - p_2) < (\hat{p}_1 - \hat{p}_2) + E$ where $E = z_{\alpha/2}\sqrt{\dfrac{\hat{p}_1\hat{q}_1}{n_1} + \dfrac{\hat{p}_2\hat{q}_2}{n_2}}$ | Table A-2 |

## Review Exercises

1. The New York State Department of Motor Vehicles provided the following motor vehicle conviction data for a recent year.

|  | Albany County | Monroe County |
|---|---|---|
| Total convictions | 24,384 | 60,961 |
| Speeding convictions | 10,292 | 26,074 |

   a. At the 0.10 significance level, test the claim that the proportion of speeding convictions is the same for both counties. Assume that the given data represent random samples drawn from a larger population.
   b. Construct a 90% confidence interval for the difference between the proportions of speeding tickets in both counties.

2. Two separate counties use different procedures for selecting jurors. The 40 randomly selected subjects from one county produce a mean waiting time of 183.0 min and a standard deviation of 21.0 min. The 50 randomly selected subjects from the other county produce a mean waiting time of 253.1 min and a standard deviation of 29.2 min.
   a. At the 0.05 significance level, test the claim that the mean waiting time of prospective jurors is the same for both counties.
   b. Construct a 95% interval estimate of the difference between the two mean waiting times.

3. In a study of techniques used to measure lung volumes, physiological data were collected for 10 subjects. The values given in the table are in liters, representing the measured forced vital capacities of the 10 subjects in a sitting position and in a supine (lying) position. (Table is based on data from "Validation of Esophageal Balloon Technique at Different Lung Volumes and Postures," by Baydur, Cha, and Sassoon, *Journal of Applied Physiology*, Vol. 62, No. 1.) At the 0.05 significance level, test the claim that both positions have the same mean.

| Sitting | 4.66 | 5.70 | 5.37 | 3.34 | 3.77 | 7.43 | 4.15 | 6.21 | 5.90 | 5.77 |
|---|---|---|---|---|---|---|---|---|---|---|
| Supine | 4.63 | 6.34 | 5.72 | 3.23 | 3.60 | 6.96 | 3.66 | 5.81 | 5.61 | 5.33 |

4. The arrangement of test items was studied for its effect on anxiety. (See "Item Arrangement, Cognitive Entry Characteristics, Sex and Test Anxiety as Predictors of Achievement Examination Performance," by Klimko, *Journal of Experimental Education*, Vol. 52, No. 4.) Sample results are as follows.

| Easy to difficult | | | | Difficult to easy | | | |
|---|---|---|---|---|---|---|---|
| 24.64 | 39.29 | 16.32 | 32.83 | 33.62 | 34.02 | 26.63 | 30.26 |
| 28.02 | 33.31 | 20.60 | 21.13 | 35.91 | 26.68 | 29.49 | 35.32 |
| 26.69 | 28.90 | 26.43 | 24.23 | 27.24 | 32.34 | 29.34 | 33.53 |
| 7.10 | 32.86 | 21.06 | 28.89 | 27.62 | 42.91 | 30.20 | 32.54 |
| 28.71 | 31.73 | 30.02 | 21.96 | | | | |
| 25.49 | 38.81 | 27.85 | 30.29 | | | | |
| 30.72 | | | | | | | |

a. At the 0.05 significance level, test the claim that the two given samples come from populations with the same mean.

b. Construct a 95% confidence interval for the difference between the two population means.

5. In a study involving motivation and test scores, data were obtained for females and males. Use the data given below to test the claim that the two groups come from populations with the same standard deviation. Use a 0.02 significance level. (See "Relationships Between Achievement-Related Motives, Extrinsic Conditions, and Task Performance," by Schroth, *Journal of Social Psychology*, Vol. 127, No. 1.)

| Female | | | | Male | | | |
|---|---|---|---|---|---|---|---|
| 31.13 | 18.71 | 14.34 | 23.90 | 12.27 | 39.53 | 32.56 | 23.93 |
| 13.96 | 13.88 | 29.85 | 20.15 | 19.54 | 25.73 | 32.20 | 19.84 |
| 6.66 | 19.20 | 15.89 | | 20.20 | 23.01 | 25.63 | 17.98 |
| | | | | 22.99 | 22.12 | 12.63 | 18.06 |

6. Among 200 randomly selected female physicians, 2.5% are surgeons. Among 250 randomly selected male physicians, 19.2% are surgeons. (The data are based on information provided by the American Medical Association.)

a. At the 0.05 significance level, test the claim that the percentage of male surgeons is greater than the percentage of female surgeons.

b. Construct a 95% confidence interval for the difference between the two proportions.

7. Two different firms manufacture garage door springs that are designed to produce a tension of 68 kg. Random samples are selected from each of these two suppliers and tension test results are as follows:

| Firm A | Firm B |
|---|---|
| $n = 20$ | $n = 32$ |
| $\bar{x} = 66.0$ kg | $\bar{x} = 68.3$ kg |
| $s = 2.1$ kg | $s = 0.4$ kg |

a. At the 5% level of significance, test the claim that both firms produce the same standard deviation.

*continued*

    b.    Test the claim that both firms' springs produce the same mean tension. Use a 0.05 level of significance.

    c.    Construct a 95% confidence interval for the difference between the mean for Firm A and the mean for Firm B.

8.    Automobiles are selected at random and tested for fuel economy with each of two different carburetors. The following results show the distance traveled on 1 gallon of gas.

    a.    At the 5% level of significance, test the claim that both carburetors produce the same mean mileage.

    b.    Construct a 95% confidence interval for the mean of the differences.

| Car | Distance with Carburetors A and B | | | | | | | | |
|-----|------|------|------|------|------|------|------|------|------|
|     | 1    | 2    | 3    | 4    | 5    | 6    | 7    | 8    | 9    |
| A   | 16.1 | 21.3 | 19.2 | 14.8 | 29.3 | 20.2 | 18.6 | 19.7 | 16.4 |
| B   | 18.2 | 23.4 | 19.7 | 14.7 | 28.7 | 23.4 | 19.0 | 21.2 | 18.2 |

9.    Stores and theaters are randomly selected and their inside temperatures are measured in degrees Celsius. The results are as follows.

| Stores | Theaters |
|--------|----------|
| $n = 40$ | $n = 32$ |
| $\bar{x} = 18.3$ | $\bar{x} = 22.2$ |
| $s = 0.8$ | $s = 0.9$ |

    a.    At the 5% level of significance, test the claim that theaters are warmer than stores.

    b.    Construct a 95% confidence interval for the difference between the mean store temperature and the mean theater temperature.

10.    In a survey, 500 men and 500 women were randomly selected. Among the men, 52 were ticketed for speeding within the last year, while 27 of the women were ticketed for speeding during the same period (based on data from R. H. Bruskin Associates).

    a.    At the 0.01 significance level, test the claim that women have a lower proportion of speeding tickets.

    b.    Construct a 99% confidence interval for the difference between the two proportions.

11.    Researchers studying commercial air-filtering systems for noise pollution reported the following sample results. At the 5% level of significance, test the claim that there is no difference in the mean noise levels.

| Unit A | Unit B |
|--------|--------|
| $n = 8$ | $n = 6$ |
| $\bar{x} = 87.5$ | $\bar{x} = 91.3$ |
| $s = 0.8$ | $s = 1.1$ |

12. A bank with branches in two different cities uses a standard credit-rating system for all loan applicants. Randomly selected applicants are chosen from each branch and the results are summarized below.
    a. At the 0.05 level of significance, test the claim that both populations have the same mean.
    b. Construct a 95% confidence interval for the difference between the mean for city A and the mean for city B.

    | City A | City B |
    |--------|--------|
    | $n = 40$ | $n = 60$ |
    | $\bar{x} = 43.7$ | $\bar{x} = 48.2$ |
    | $s = 16.2$ | $s = 16.5$ |

13. A test question is considered good if it discriminates between good and poor students. The first question on a test is answered correctly by 62 of 80 good students, while 23 of 50 poor students give correct answers. At the 5% level of significance, test the claim that this question is answered correctly by a greater proportion of good students.

14. In order to test the effectiveness of a lesson, a teacher gives randomly selected students a pretest and a follow-up test. The results are given below. At the 0.025 level of significance, test the claim that the lesson was effective.

    | Student | A | B | C | D | E | F | G | H |
    |---------|---|---|---|---|---|---|---|---|
    | Before | 6 | 8 | 5 | 4 | 3 | 5 | 4 | 7 |
    | After | 9 | 10 | 8 | 7 | 6 | 8 | 7 | 10 |

15. The manager of a movie theater conducts a study of the ages of those who view two different movies. The study yields the following sample results.

    | Movie X | Movie Y |
    |---------|---------|
    | $n = 45$ | $n = 65$ |
    | $\bar{x} = 22.6$ years | $\bar{x} = 31.0$ years |
    | $s = 5.8$ years | $s = 4.7$ years |

    a. At the 0.025 level of significance, test the claim that there is no difference between the two population means.
    b. Construct a 97.5% confidence interval for the difference between the mean age for movie X and the mean age for movie Y.

16. Sample data were collected in a study of calcium supplements and the effects on blood pressure. A placebo group and a calcium group began the study with measures of blood pressures. (See "Blood Pressure and Metabolic Effects of Calcium Supplementation in Normotensive

White and Black Men," by Lyle and others, *Journal of the American Medical Association*, Vol. 257, No. 13.)

| Placebo | | | | Calcium | | | |
|---|---|---|---|---|---|---|---|
| 124.6 | 104.8 | 96.5 | 116.3 | 129.1 | 123.4 | 102.7 | 118.1 |
| 106.1 | 128.8 | 107.2 | 123.1 | 114.7 | 120.9 | 104.4 | 116.3 |
| 118.1 | 108.5 | 120.4 | 122.5 | 109.6 | 127.7 | 108.0 | 124.3 |
| 113.6 | | | | 106.6 | 121.4 | 113.2 | |

a. At the 0.05 significance level, test the claim that the two sample groups come from populations with the same mean.

b. Construct a 95% confidence interval for the difference between the mean for the placebo population and the mean for the calcium population.

 ## *Computer Projects*

1. Use an existing software package, such as STATDISK or Minitab, to solve the following exercises.
   a. Exercise 20 from Section 8–2
   b. Exercises 25, 26, 31, and 32 from Section 8–3
   c. Exercises 23 and 24 from Section 8–4
2. The following 25 scores between 0 and 1 are taken from a table of uniformly distributed random numbers.

   | | | | | | | | | |
   |---|---|---|---|---|---|---|---|---|
   | 0.482 | 0.694 | 0.521 | 0.905 | 0.337 | 0.915 | 0.106 | 0.296 | 0.943 |
   | 0.715 | 0.424 | 0.279 | 0.491 | 0.599 | 0.205 | 0.020 | 0.835 | 0.361 |
   | 0.907 | 0.522 | 0.073 | 0.229 | 0.399 | 0.411 | 0.164 | | |

   a. Using the BASIC programming language, enter

   $$? \text{RND (X)}$$

   25 times to get another sample of 25 scores between 0 and 1 that come from a uniformly distributed population.

   b. Why can't we use the methods given in this chapter to test the hypothesis that both samples come from populations with the same mean?

   c. Although we can't use the methods given in this chapter to test the hypothesis of equal means, we can use other tests for agreement between the table values given above and the computer-generated values. Identify and execute another approach.

# Applied Projects

1. Use Data Set II in Appendix B to test the claim that men and women have the same employment rate.
2. Use Data Set II in Appendix B to test the claim that right-handed people have the same mean weight as left-handed people.
3. Using the sample data for males in Appendix B, test the claim that males under 70 in. tall have a mean weight equal to the mean for males at least 70 in. tall.
4. In designing a car seat, an engineer needs to study sitting heights of men and women. Using the sample results in Data Set II of Appendix B, construct a 95% confidence interval for $\mu_M - \mu_W$, where $\mu_M$ and $\mu_W$ are the mean sitting heights of all men and women, respectively.

# Writing Projects

1. In a television commercial, Chrysler Chairman Lee Iacocca said, "Two cars come off the assembly line in the same American plant. A Japanese nameplate goes on one, an American nameplate goes on the other, and the people prefer the Japanese version! We've got to get people to wake up to the truth!" Describe an experiment that would test the claim included in this statement.
2. Do Applied Project 4 above and explain how to interpret the resulting confidence interval.
3. Summarize one of the programs listed below.

# Videotapes

Programs 22 and 23 from the series *Against All Odds: Inside Statistics* are recommended as supplements to this chapter.

# Chapter Nine

## In This Chapter

# 9 Correlation and Regression

## Chapter Problem

You might expect to find a relationship between the selling price of a home and its size, as measured by the living area. Table 9–1 lists the actual living areas and selling prices for eight homes recently sold in Dutchess County, New York. The living areas are given in hundreds of square feet, so that the first value of 15 indicates a living area of 1500 sq ft. The selling prices are in thousands of dollars, and so the first value of 145 indicates a selling price of $145,000. Using the given data, can we conclude that there is a relationship between selling price and living area? If so, what is the relationship? An objective of this chapter is to analyze such relationships. The data given in Table 9–1 will be considered in the following sections.

**TABLE 9–1**

| Living area   | 15  | 38  | 23  | 16  | 16  | 13  | 20  | 24  |
|---------------|-----|-----|-----|-----|-----|-----|-----|-----|
| Selling price | 145 | 228 | 150 | 130 | 160 | 114 | 142 | 265 |

# 9–1 | Overview

In using statistical methods, it is extremely important to understand that the arrangement and nature of data can dramatically affect the particular methods that should be used. Examine Table 9–2. It consists of paired data, but the actual nature of the data determines the statistical approach that should be used. Consider the following two scenarios.

| TABLE 9–2 | | | | | | |
|---|---|---|---|---|---|---|
| x | 110 | 103 | 105 | 98 | 140 | 112 |
| y | 108 | 104 | 101 | 96 | 135 | 107 |

*Scenario 1:*  In Table 9–2, the *x-y* pairs represent "before and after" weights of subjects who were weighed before and after an experimental diet. An effective diet should result in lower weights. The reasonable and sensible question is this: "Are the after weights *significantly less* than the before weights?" A relevant method of analysis is the test of the hypothesis $\mu_x > \mu_y$ where the two populations are dependent. (See Section 8–3.)

*Scenario 2:*  In Table 9–2, the *x-y* pairs represent the adult weights of pairs of identical twins. A reasonable and sensible question is this: "Is there a *relationship between* the weights of the twins?" The relevant method of analysis will be presented in this chapter.

For Scenario 1, we are interested in the differences between the numbers in each pair. For Scenario 2, we are interested in the presence or absence of a relationship between the two variables. These are fundamentally different issues, and the statistical methods used to consider them will be very different.

In this chapter we investigate ways of analyzing the relationship between two or more variables. Sample data will be paired as in Tables 9–1 and 9–2. Such paired data is sometimes referred to as **bivariate** data. We begin Section 9–2 by describing the **scatter diagram**, which serves as a graph of the sample data. We then investigate the concept of **correlation**, which is used to decide whether there is a statistically significant relationship between two variables. Section 9–3 investigates regression analysis as we attempt to identify the exact nature of the relationship between two variables. Specifically, we show how to determine an equation that relates the two variables. In Section 9–4 we analyze the **variation** between predicted and observed values. In Section 9–5 we use concepts of **multiple regression** to describe the relationship among three or more variables.

Throughout this chapter we deal only with **linear** relationships between two or more variables. (Advanced texts consider more variables and nonlinear relationships.)

# 9–2 | Correlation

As we mentioned in the overview, this section deals with correlation and scatter diagrams as tools that help us decide whether a linear relationship exists between two variables. To use these tools, therefore, the sample data must be collected as paired data. We start with an example.

Using the sample data in Table 9–1, is there a linear relationship between the selling price and living area?

We can often form intuitive and qualitative conclusions about paired data by constructing a scatter diagram similar to the one in Figure 9–1, which represents the data in Table 9–1. The points in the figure seem to follow an upward pattern, so we might conclude that there is a relationship between selling price and living area.

Other examples of scatter diagrams are shown in Figure 9–2 on page 455. In general, scatter diagrams often reveal patterns and are easy to plot, but conclusions drawn from them tend to be subjective. More precise and objec-

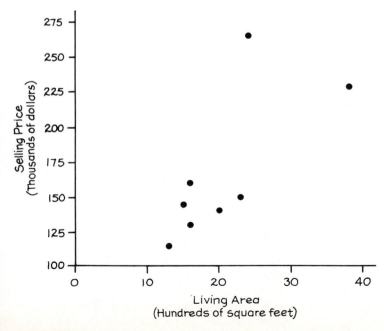

**Figure 9–1**

tive analyses accompany the computation of the **linear correlation coefficient,** which is denoted by $r$ and is given in Formula 9–1.

**Formula 9–1**
$$r = \frac{n\Sigma xy - (\Sigma x)(\Sigma y)}{\sqrt{n(\Sigma x^2) - (\Sigma x)^2}\ \sqrt{n(\Sigma y^2) - (\Sigma y)^2}}$$

Since $r$ is calculated using sample data, it is a sample statistic. We might think of $r$ as a point estimate of the population parameter $\rho$ (rho), which is the linear correlation coefficient for all pairs of data in a population.

When calculating $r$ and other statistics in this chapter, rounding errors can sometimes wreak havoc with your results. Try using your calculator's memory to store intermediate results, and round off only at the end. Don't round off intermediate results. Many inexpensive calculators have Formula 9–1 built in so that you can automatically evaluate $r$ after entering the sample data.

We now describe the way to compute and interpret the linear correlation coefficient $r$ given a list of paired data. Later in this section we present the underlying theory that led to the development of this formula. Before computing the correlation coefficient $r$ for the data in Table 9–1, we explain the notation relevant to Formula 9–1.

---

### Notation

| | |
|---|---|
| $n$ | denotes the **number of pairs** of data present. In Table 9–1 $n = 8$. |
| $\Sigma$ | denotes the addition of the items indicated. |
| $\Sigma x$ | denotes the sum of all $x$ scores. |
| $\Sigma x^2$ | indicates that each $x$ score should be squared, and then those squares added. |
| $(\Sigma x)^2$ | indicates that the $x$ scores should be added and the total then squared. It is extremely important to avoid confusion between $\Sigma x^2$ and $(\Sigma x)^2$. |
| $\Sigma xy$ | indicates that each $x$ score should be multiplied by its corresponding $y$ score. After obtaining all such products, find their sum. |
| $r$ | is the linear correlation coefficient, which measures the strength of the relationship between the paired $x$ and $y$ values in a *sample*. $r$ is a sample statistic. |
| $\rho$ | (rho) is the linear correlation coefficient, which measures the strength of the relationship between all paired $x$ and $y$ values in a *population*. $\rho$ is a population parameter. |

---

For the data in Table 9–1, we compute the individual components and then use the results to determine the value of $r$.

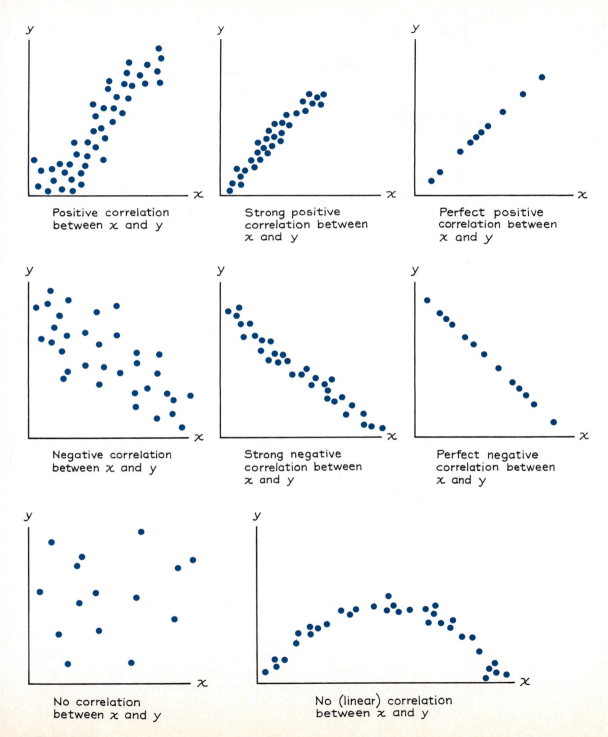

Positive correlation
between $x$ and $y$

Strong positive
correlation between
$x$ and $y$

Perfect positive
correlation between
$x$ and $y$

Negative correlation
between $x$ and $y$

Strong negative
correlation between
$x$ and $y$

Perfect negative
correlation between
$x$ and $y$

No correlation
between $x$ and $y$

No (linear) correlation
between $x$ and $y$

**Figure 9–2**

## Example

Using the data in Table 9–1, find the value of the linear correlation coefficient $r$.

## Solution

For the sample paired data in Table 9–1 we get $n = 8$ because there are 8 pairs of data. The other components required in Formula 9–1 are found from the calculations in the table below. Note how this vertical format makes the calculations easier.

| Living Area $x$ | Selling Price $y$ | $x \cdot y$ | $x^2$ | $y^2$ |
|---|---|---|---|---|
| 15 | 145 | 2,175 | 225 | 21,025 |
| 38 | 228 | 8,664 | 1,444 | 51,984 |
| 23 | 150 | 3,450 | 529 | 22,500 |
| 16 | 130 | 2,080 | 256 | 16,900 |
| 16 | 160 | 2,560 | 256 | 25,600 |
| 13 | 114 | 1,482 | 169 | 12,996 |
| 20 | 142 | 2,840 | 400 | 20,164 |
| 24 | 265 | 6,360 | 576 | 70,225 |
| Total: 165 | 1,334 | 29,611 | 3,855 | 241,394 |
| ↑ $\Sigma x$ | ↑ $\Sigma y$ | ↑ $\Sigma xy$ | ↑ $\Sigma x^2$ | ↑ $\Sigma y^2$ |

Using the calculated values, we can now evaluate $r$ as follows.

$$r = \frac{n(\Sigma xy) - (\Sigma x)(\Sigma y)}{\sqrt{n(\Sigma x^2) - (\Sigma x)^2}\sqrt{n(\Sigma y^2) - (\Sigma y)^2}}$$

$$= \frac{8(29,611) - (165)(1334)}{\sqrt{8(3855) - (165)^2}\sqrt{8(241,394) - (1334)^2}}$$

$$= \frac{16,778}{\sqrt{3615}\sqrt{151,596}} = 0.717$$

After calculating $r$, how do we interpret the result? Given the way in which Formula 9–1 was derived, it can be shown that the computed value of $r$ must always fall between $-1$ and $+1$ inclusive. A strong positive linear correlation between $x$ and $y$ is reflected by a value of $r$ near $+1$, while a strong negative linear correlation is indicated by a value of $r$ near $-1$. If $r$ is close to 0, we conclude that there is no significant linear correlation between $x$ and $y$. We can make this decision process more objective through a formal hypothesis test by using one of two equivalent procedures described below and in Figure 9–3.

## Method 1

This method follows the format presented in earlier chapters. It uses the Student $t$ distribution.

Null hypothesis: $H_0$: $\rho = 0$

Test statistic: $t = \dfrac{r - \mu_r}{s_r} = \dfrac{r}{\sqrt{\dfrac{1 - r^2}{n - 2}}}$

Critical value: $t$ (from Table A–3), with $n - 2$ degrees of freedom

In this test statistic, we expressed $t$ as $(r - \mu_r)/s_r$, following the format of earlier chapters, but $\mu_r$ is actually $\rho$, which is assumed to be 0.

## Method 2

This method uses fewer calculations. Instead of calculating the test statistic given above, use the computed value of $r$ as the test statistic. With $r$ as the test statistic, we can find the critical value from Table A–6.

Null hypothesis: $\rho = 0$

Test statistic: $r$ (from Formula 9–1)

Critical value: $r$ (from Table A–6)

The critical values of $r$ in Table A–6 are found by solving

$$t = \frac{r}{\sqrt{\dfrac{1 - r^2}{n - 2}}}$$

for $r$ to get

$$r = \frac{t}{\sqrt{t^2 + n - 2}}$$

Here the $t$ value is found from Table A–3 by assuming a two-tailed case with $n - 2$ degrees of freedom. Table A–6 lists the results for selected values of $n$ and $\alpha$.

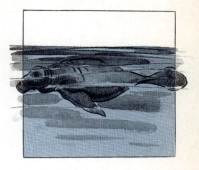

## MANATEES SAVED

■ Manatees are large mammals that like to float just below the water's surface, where they are in danger from powerboat propellers. A Florida study of the number of powerboat registrations and the numbers of accidental manatee deaths confirmed that there was a significant positive correlation. As a result, Florida created coastal sanctuaries where powerboats are prohibited so that manatees could thrive. (See Program 1 from the series *Against All Odds: Inside Statistics* for a discussion of this case.) This is one of many examples of the beneficial use of statistics. ■

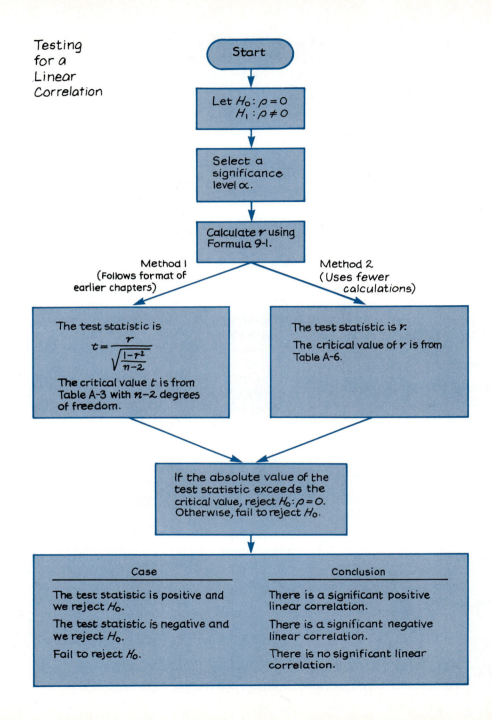

Testing
for a
Linear
Correlation

**Figure 9–3**

---

**Example**

Use Table A–6 to find the critical values of the linear correlation coefficient $r$ if we have 8 pairs of data (as in Table 9–1), the significance level is $\alpha = 0.05$, and the hypothesis test is two-tailed.

**Solution**

Refer to Table A–6 and locate the critical $r$ value of 0.707 corresponding to $n = 8$ and $\alpha = 0.05$. Since the hypothesis test is two-tailed, we have the critical values of 0.707 and $-0.707$.

---

If we reject $H_0$: $\rho = 0$ and $r$ is positive, we conclude that there is a significant positive linear correlation. If we reject $H_0$: $\rho = 0$ and $r$ is negative, we conclude that there is a significant negative linear correlation. If we fail to reject $H_0$: $\rho = 0$, we conclude that there is no significant linear correlation.

Figure 9–3 summarizes the two methods we have described. Some instructors prefer the first method because it reinforces concepts introduced in earlier chapters. Others prefer the second method because it eliminates the step of calculating the test statistic in terms of $t$.

**Example**

From 8 pairs of data, $r$ is computed to be 0.717. When testing the claim that $\rho = 0$, what can you conclude at the significance level of $\alpha = 0.05$?

**Solution**

With $H_0$: $\rho = 0$ and $H_1$: $\rho \neq 0$ and $\alpha = 0.05$, we proceed to obtain the value of the test statistic and the critical value.

*Method 1:* The test statistic is

$$t = \frac{r}{\sqrt{\dfrac{1 - r^2}{n - 2}}} = \frac{0.717}{\sqrt{\dfrac{1 - 0.717^2}{8 - 2}}} = 2.520$$

The critical value of $t = 2.447$ is found from Table A–3. It corresponds to $n - 2 = 8 - 2 = 6$ degrees of freedom and $\alpha = 0.05$ in two tails. Since the test statistic falls in the critical region, we reject $H_0$. (See Figure 9–4.) We conclude that there is a significant positive linear correlation.

*continued*

## PALM READING

■ Some people believe that the length of their palm's life line can be used to predict longevity. In a letter published in the *Journal of the American Medical Association*, authors M.E. Wilson and L.E. Mather refuted that belief with a study of cadavers. Ages at death were recorded along with the lengths of palm life lines. The authors concluded that there is no significant correlation between age at death and length of life line. Palmistry lost, hands down. ■

**Solution** *continued*

*Method 2:* The test statistic is $r = 0.717$. The critical value of $r = 0.707$ is found from Table A–6. (See Figure 9–5.) Since the test statistic falls within the critical region, we reject $H_0$. We conclude that there is a significant positive linear correlation.

Based on the results of either of the two equivalent methods, there appears to be a significant positive linear correlation between selling price and living area. Higher selling prices seem to correspond to larger living areas.

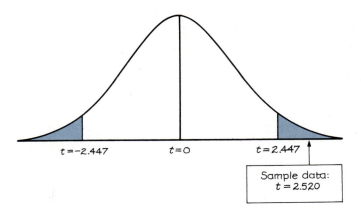

**Figure 9–4**

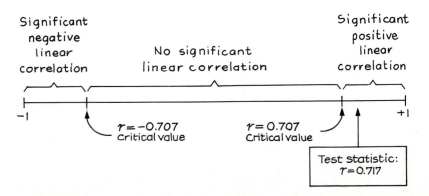

**Figure 9–5**

The preceding example and Figures 9–4 and 9–5 correspond to a two-tailed hypothesis test. The examples and exercises in this section will generally involve only two-tailed tests. One-tailed tests can occur with a claim of a positive correlation or a claim of a negative correlation. In such cases, the hypotheses will be as shown below.

| Left-tailed Test | Right-tailed Test |
|---|---|
| $H_0$: $\rho \geq 0$ | $H_0$: $\rho \leq 0$ |
| $H_1$: $\rho < 0$ | $H_1$: $\rho > 0$ |

Method 1 can be handled as in earlier chapters. For Method 2, either double the significance level given in Table A–6 or calculate the critical value as in Exercise 30.

# Common Errors

We now identify three of the most common errors made in interpreting results involving correlation.

1. *We must be careful to avoid concluding that a significant linear correlation between two variables is proof that there is a cause-effect relationship between them.* The statistical correlation between smoking and cancer is not proof that smoking *causes* cancer. The techniques in this chapter can be used only to establish a linear relationship. We cannot establish the existence or absence of any inherent cause-effect relationship between the variables. This problem is for the various professionals, such as medical researchers, psychologists, sociologists, biologists, and others. Mark Twain satirized the issue of causality when he commented on a cold winter by saying, "Cold! If the thermometer had been an inch longer, we'd all have frozen to death."

2. *Another source of potential error arises with data based on rates or averages.* When we use rates or averages for data, we suppress the variation among the individuals or items, and this may easily lead to an inflated correlation coefficient. One study produced a 0.4 linear correlation coefficient for paired data relating income and education among *individuals*, but the correlation coefficient became 0.7 when regional *averages* were used.

3. *A third error involves the property of linearity.* The conclusion that there is no significant linear correlation does not mean that x and y are not related in any way. The data depicted in Figure 9–6 on the following page result in a value of $r = 0$, an indication of no *linear* relationship between the two variables. However, we can easily see

from Figure 9–6 that there is a pattern reflecting a very strong (non-linear) relationship. (Figure 9–6 depicts the relationship between distance above ground and time elapsed for an object thrown upward.)

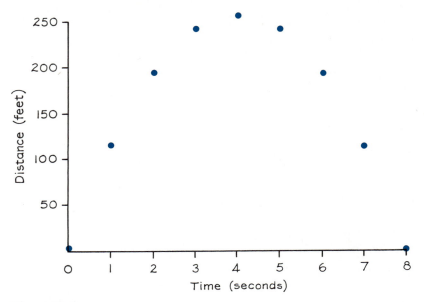

**Figure 9–6**

We have presented Formula 9–1 for calculating $r$, illustrated its use, and will now give a justification for it. Formula 9–1 simplifies the calculations used in this equivalent formula:

$$r = \frac{\Sigma(x - \bar{x})(y - \bar{y})}{(n - 1)s_x s_y}$$

We will temporarily use this latter version since its form relates more directly to the underlying theory. We will consider the following paired data, which is depicted in the scatter diagram shown in Figure 9–7. Figure 9–7 includes the point $(\bar{x}, \bar{y})$, which is called the **centroid** of the sample points.

| $x$ | 1 | 1 | 2 | 4 | 7 |
|---|---|---|---|---|---|
| $y$ | 4 | 5 | 8 | 15 | 23 |

Sometimes $r$ is called **Pearson's product moment.** This title reflects both the fact that it was first developed by Karl Pearson (1857–1936) and that it is based on the product of the moments $(x - \bar{x})$ and $(y - \bar{y})$. That is, Pearson based the measure of scattering on the statistic $\Sigma(x - \bar{x})(y - \bar{y})$. In any scatter diagram, vertical and horizontal lines through the centroid $(\bar{x}, \bar{y})$ divide the diagram into four quadrants (see Figure 9–7). If the points of the scatter diagram tend to approximate an uphill line (as in the figure), individual values of $(x - \bar{x})(y - \bar{y})$ tend to be positive, since most of the points are found in the first and third quadrants, where the products of $(x - \bar{x})$ and $(y - \bar{y})$ are positive. If the points of the scatter diagram approximate a downhill line, most of the points are in the second and fourth quadrants where $(x - \bar{x})$ and $(y - \bar{y})$ are opposite in sign, so $\Sigma(x - \bar{x})(y - \bar{y})$ tends to be negative. If the points follow no linear pattern, they tend to be scattered among the four quadrants, so $\Sigma(x - \bar{x})(y - \bar{y})$ tends to be close to zero.

The sum $\Sigma(x - \bar{x})(y - \bar{y})$ depends on the magnitude of the numbers used, yet $r$ should not be affected by the particular scale used. For example, $r$ should not change whether heights are measured in meters, centimeters, inches, or feet. We can make $r$ independent of the scale used by incorporating the sample standard deviations as follows.

$$r = \frac{\Sigma(x - \bar{x})(y - \bar{y})}{(n - 1)s_x s_y}$$

This expression can be algebraically manipulated into the equivalent form of Formula 9–1 (see Exercise 34, part a).

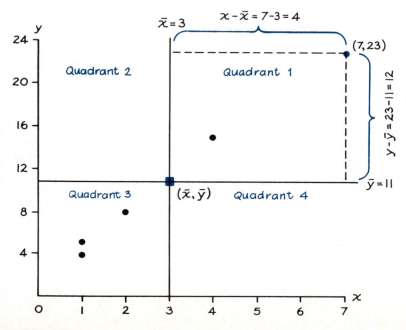

**Figure 9–7**

## STUDENT RATINGS OF TEACHERS

■ Many colleges equate high student ratings with good teaching—an equation often fostered by the fact that student evaluations are easy to administer and measure.

However, one study that compared student evaluations of teachers with the amount of material learned found a strong *negative* correlation between the two factors. Teachers rated highly by students seemed to induce less learning.

In a related study, an audience gave a high rating to a lecturer who conveyed very little information but was interesting and entertaining. ■

The format of Formula 9–1 or any equivalent formula leads to the following properties of the linear correlation coefficient *r*.

## Properties of *r*

1. *The value of r is always between* −1 *and 1*. That is, −1 ≤ *r* ≤ 1.
2. *The value of r does not change if all values of either variable are converted to a different scale*. For example, if the units of *x* are so large that they cause calculator errors, you can divide them all by a constant number, such as 1000.
3. *The value of r is not affected by the choice of* x *or* y. Interchange all *x* and *y* values and the value of *r* will not change.
4. *r measures the strength of a linear relationship*. It is not designed to measure the strength of a relationship that is not linear.

For the methods presented in this section, these assumptions apply.

## Assumptions

1. Both *x* and *y* are random variables.
2. The pairs of (*x*, *y*) data have a **bivariate normal distribution.** The key feature of such a distribution is that for any fixed value of *x*, the values of *y* have a normal distribution, and for any fixed value of *y*, the values of *x* have a normal distribution.

The second assumption is usually difficult to check, but a partial check can be made by determining whether the values of *x* are normally distributed and the values of *y* are also normally distributed. If one of the variables consists of the *ranks* 1, 2, 3, . . . , 10, for example, then the second assumption is clearly violated and we shouldn't use the techniques discussed in this section. Instead, you should consider *rank correlation*, which will be covered in Section 12–6. In general, if either variable has a distribution that is very nonnormal, then you know that the second assumption is not satisfied.

The construction of **confidence intervals** about the population parameter *ρ* involves somewhat complicated transformations, so we present that process as a *B* exercise. (See Exercise 36.)

We can use the linear correlation coefficient to decide whether there is a linear relationship between two variables. After deciding that a relationship exists, we can then proceed to determine what that relationship is. This next step is pursued in the following section.

## 9–2 Exercises A

For Exercises 1 and 2, would you expect positive correlation, negative correlation, or no correlation for each of the given sets of paired data?

1. a. The ages of cars and their resale values
   b. The years of education and the incomes of taxpayers
   c. The amounts of rainfall and vegetation growth
   d. The weights of cars and fuel consumption as measured in miles per gallon
   e. The hat sizes of adults and their IQ scores
2. a. Golf tournament scores and the prize money won by players
   b. Hours spent studying for tests and the resulting test scores
   c. The amount of china made in England and the price of tea in China
   d. The number of absences in a course and the grades in that course
   e. Annual per capita income for different nations and per capita entertainment expenditures for those nations

For each part of Exercises 3 and 4, a sample of paired data produces a linear correlation coefficient $r$. What do you conclude in each case? Assume a significance level of $\alpha = 0.05$.

3. a. $n = 20, r = 0.5$
   b. $n = 20, r = -0.5$
   c. $n = 50, r = 0.2$
   d. $n = 50, r = -0.2$
   e. $n = 37, r = 0.25$
4. a. $n = 77, r = 0.35$
   b. $n = 22, r = 0.37$
   c. $n = 22, r = 0.40$
   d. $n = 22, r = -0.5$
   e. $n = 6, r = -0.8$

In Exercises 5–8, use the given list of paired data.

   a. Construct the scatter diagram.
   b. Determine $n$.
   c. Find $\Sigma x$.
   d. Find $\Sigma x^2$.
   e. Find $(\Sigma x)^2$.
   f. Find $\Sigma xy$.
   g. Find $r$.

5.

| $x$ | 1 | 1 | 2 | 3 |
|-----|---|---|---|---|
| $y$ | 1 | 5 | 4 | 2 |

6.

| $x$ | 1 | 2 | 2 | 3 |
|-----|---|---|---|---|
| $y$ | 5 | 4 | 3 | 1 |

7.

| $x$ | 0 | 1 | 1 | 2 | 5 |
|-----|---|---|---|---|---|
| $y$ | 3 | 3 | 4 | 5 | 6 |

8.

| $x$ | 1 | 3 | 3 | 4 | 5 | 5 |
|-----|---|---|---|---|---|---|
| $y$ | 5 | 3 | 2 | 2 | 0 | 1 |

In Exercises 9–24:

    a.  Construct the scatter diagram.

    b.  Compute the linear correlation coefficient $r$.

    c.  Assume that $\alpha = 0.05$ and find the critical value of $r$ from Table A–6.

    d.  Based on the results of parts $b$ and $c$, decide whether there is a significant positive linear correlation, a significant negative linear correlation, or no significant linear correlation. In each case, assume a significance level of $\alpha = 0.05$.

    e.  Save your work because the same data will be used in the next section.

9.  A study was conducted to investigate any relationship between age (in years) and the BAC (blood alcohol concentration) measured when convicted DWI jail inmates were first arrested. Sample data are given below for randomly selected subjects (based on data from the Dutchess County STOP-DWI Program).

| Age | 17.2 | 43.5 | 30.7 | 53.1 | 37.2 | 21.0 | 27.6 | 46.3 |
|-----|------|------|------|------|------|------|------|------|
| BAC | 0.19 | 0.20 | 0.26 | 0.16 | 0.24 | 0.20 | 0.18 | 0.23 |

10.  Randomly selected subjects ride a bicycle at 5.5 mi/h for one minute. Their weights (in pounds) are given with the numbers of calories used (based on data from *Diet Free* by Kuntzlemann).

| Weight | 167 | 191 | 112 | 129 | 140 | 173 | 119 |
|--------|-----|-----|-----|-----|-----|-----|-----|
| Calories | 4.23 | 4.69 | 3.21 | 3.47 | 3.72 | 4.45 | 3.36 |

11.  The table below lists the value of exports (in billions of dollars) and the value of imports (in billions of dollars) for several different years (based on data from the U.S. Department of Commerce).

| Exports | 10 | 20 | 43 | 221 | 218 | 218 |
|---------|-----|-----|-----|-----|-----|-----|
| Imports | 9 | 15 | 40 | 245 | 326 | 370 |

12.  In a study of the factors that affect success in a calculus course, data were collected for 10 different people. Scores on an algebra placement test are given along with calculus achievement scores. (See "Factors Affecting Achievement in the First Course in Calculus" by Edge and Friedberg, *Journal of Experimental Education*, Vol. 52, No. 3.)

| Algebra | 17 | 21 | 11 | 16 | 15 | 11 | 24 | 27 | 19 | 8 |
|---------|-----|-----|-----|-----|-----|-----|-----|-----|-----|-----|
| Calculus | 73 | 66 | 64 | 61 | 70 | 71 | 90 | 68 | 84 | 52 |

13.  The accompanying table lists the number of registered automatic weapons (in thousands) along with the murder rate (in murders per 100,000)

for randomly selected states. (The data are provided by the FBI and the Bureau of Alcohol, Tobacco, and Firearms.)

| Automatic weapons | 11.6 | 8.3 | 3.6 | 0.6 | 6.9 | 2.5 | 2.4 | 2.6 |
|---|---|---|---|---|---|---|---|---|
| Murder rate | 13.1 | 10.6 | 10.1 | 4.4 | 11.5 | 6.6 | 3.6 | 5.3 |

14. When loads were added to a hanging copper wire, the wire stretched. The loads (in Newtons) and increases in length (in centimeters) are given below. (The table is based on data from *College Physics* by Sears, Zemansky, and Young.)

| Added load | 0 | 10 | 20 | 30 | 40 | 50 | 60 | 70 |
|---|---|---|---|---|---|---|---|---|
| Increase in length | 0 | 0.05 | 0.10 | 0.15 | 0.20 | 0.25 | 0.30 | 1.25 |

15. Emissions data are given (in grams per meter) for a sample of different vehicles. (See "Determining Statistical Characteristics of a Vehicle Emissions Audit Procedure" by Lorenzen, *Technometrics*, Vol. 22, No. 4.)

| HC | 0.65 | 0.55 | 0.72 | 0.83 | 0.57 | 0.51 | 0.43 | 0.37 |
|---|---|---|---|---|---|---|---|---|
| CO | 14.7 | 12.3 | 14.6 | 15.1 | 5.0 | 4.1 | 3.8 | 4.1 |

16. Randomly selected girls are given the Wide Range Achievement Test. Their ages are listed along with their scores on the reading part of that test (based on data from the National Health Survey, USDHEW publication 72-1011).

| Age | 6.1 | 7.2 | 5.9 | 6.3 | 10.5 | 11.0 |
|---|---|---|---|---|---|---|
| Score | 17.8 | 47.4 | 25.8 | 24.3 | 66.6 | 91.4 |

17. For randomly selected homes recently sold in Dutchess County, New York, the annual tax amounts (in thousands of dollars) are listed along with the selling prices (in thousands of dollars).

| Taxes | 1.9 | 3.0 | 1.4 | 1.4 | 1.5 | 1.8 | 2.4 | 4.0 |
|---|---|---|---|---|---|---|---|---|
| Selling price | 145 | 228 | 150 | 130 | 160 | 114 | 142 | 265 |

18. For randomly selected homes recently sold in Dutchess County, New York, the living areas (in hundreds of square feet) are listed along with the annual tax amounts (in thousands of dollars).

| Living area | 15 | 38 | 23 | 16 | 16 | 13 | 20 | 24 |
|---|---|---|---|---|---|---|---|---|
| Taxes | 1.9 | 3.0 | 1.4 | 1.4 | 1.5 | 1.8 | 2.4 | 4.0 |

19. Ten subjects underwent a physical training program. After the program, their weights (in kilograms) and anaerobic thresholds (in liters/min) were measured, with the results below. (Based on data from "Effect of Endurance Training On Possible Determinants of $\dot{V}O_2$ During Heavy

Exercise," by Casaburi, Storer, Ben-Dov, and Wasserman, *Journal of Applied Physiology*, Volume 62, No. 1.)

| Weight | 62 | 57 | 94 | 69 | 66 | 76 | 58 | 88 | 70 | 84 |
|---|---|---|---|---|---|---|---|---|---|---|
| Anaerobic threshold | 1.37 | 1.34 | 1.93 | 1.92 | 2.24 | 2.02 | 1.35 | 2.21 | 1.79 | 1.74 |

20. In a study of employee stock ownership plans, data on satisfaction with the plan and the amount of organizational commitment were collected at 8 companies. Results are given in the accompanying table, which is based on "Employee Stock Ownership and Employee Attitudes: A Test of Three Models" by Klein, *Journal of Applied Psychology*, Vol. 72, No. 2.

| Satisfaction | 5.05 | 4.12 | 5.39 | 4.17 | 4.00 | 4.49 | 5.40 | 4.86 |
|---|---|---|---|---|---|---|---|---|
| Commitment | 5.37 | 4.49 | 5.42 | 4.45 | 4.24 | 5.34 | 5.62 | 4.90 |

21. There are many regions where the winter accumulation of snowfall is a primary source of water. Several investigations of snowpack characteristics have used satellite observations from the Landsat series along with measurements taken on Earth. Given here are ground measurements of snow depth (in centimeters) along with the corresponding temperatures (in degrees Celsius). (The data are based on information in Kastner's *Space Mathematics*, published by NASA.)

| Temperature (°C) | −62 | −41 | −36 | −26 | −33 | −56 | −50 | −66 |
|---|---|---|---|---|---|---|---|---|
| Snow depth (cm) | 21 | 13 | 12 | 3 | 6 | 22 | 14 | 19 |

22. The following table lists per capita cigarette consumption in the United States for various years, along with the percentage of the population admitted to mental institutions as psychiatric cases.

| Cigarette consumption | 3522 | 3597 | 4171 | 4258 | 3993 | 3971 | 4042 | 4053 |
|---|---|---|---|---|---|---|---|---|
| Percentage of psychiatric admissions (in percentage points) | 0.20 | 0.22 | 0.23 | 0.29 | 0.31 | 0.33 | 0.33 | 0.32 |

23. The PSAT and SAT scores for 12 Dutchess County high school students are randomly selected from the population of applicants for a summer enrichment program.

| PSAT | 144 | 138 | 136 | 111 | 129 | 122 | 116 | 136 | 118 | 112 | 137 | 124 |
|---|---|---|---|---|---|---|---|---|---|---|---|---|
| SAT | 1290 | 1310 | 1340 | 1130 | 1170 | 1180 | 1220 | 1340 | 1210 | 1270 | 1160 | 1130 |

24. At one point in a recent season of the National Basketball Association, *USA Today* reported the current statistics. Given below are the total minutes played and the total points scored for 9 randomly selected players.

| Minutes | 1364 | 53 | 457 | 717 | 384 | 1432 | 365 | 1626 | 840 |
|---|---|---|---|---|---|---|---|---|---|
| Points | 652 | 20 | 163 | 210 | 175 | 821 | 143 | 1098 | 459 |

In Exercises 25–28, identify the error in the stated conclusion.

25.  *Given:* The paired sample data of age and score result in a linear correlation coefficient very close to 0.
*Conclusion:* Older people tend to get lower scores.

26.  *Given:* There is a significant positive linear correlation between per
capita income and per capita spending.
*Conclusion:* Increased spending is caused by increased income.

27.  *Given:* The paired sample data result in a linear correlation coefficient
very close to 0.
*Conclusion:* The two variables are not related in any way.

28.  *Given:* There is a significant linear correlation between state average
tax burdens and state average incomes.
*Conclusion:* There is a significant linear correlation between individual tax burdens and individual incomes.

## 9–2  Exercises B

29.  In addition to testing for a linear correlation between $x$ and $y$, we can
often use **transformations** of data to explore for other relationships.
Given the paired data below, construct the scatter diagram, then test
for a linear correlation between $y$ and each of the following.
a.  $x$
b.  $x^2$
c.  $\log x$
d.  $\sqrt{x}$
e.  $\dfrac{1}{x}$

Which case results in the largest value of $r$?

| $x$ | 1.3 | 2.4 | 2.6 | 2.8 | 2.4 | 3.0 | 4.1 |
|---|---|---|---|---|---|---|---|
| $y$ | 0.11 | 0.38 | 0.41 | 0.45 | 0.39 | 0.48 | 0.61 |

30.  Use the following formula to find the critical values of $r$ for the indicated cases.

$$r = \frac{t}{\sqrt{t^2 + n - 2}}$$

Be sure to use $n - 2$ degrees of freedom when referring to Table A–3.
a.  $H_0: \rho = 0, n = 50, \alpha = 0.05$
b.  $H_1: \rho \neq 0, n = 75, \alpha = 0.10$
c.  $H_0: \rho \geq 0, n = 20, \alpha = 0.05$
d.  $H_0: \rho \leq 0, n = 10, \alpha = 0.05$
e.  $H_1: \rho > 0, n = 12, \alpha = 0.01$

31. First plot the scatter diagram for the data in the accompanying table, then try to compute the linear correlation coefficient $r$. Comment on the results.

| $x$ | 0 | 3 | 5 | 5 | 6 |
|---|---|---|---|---|---|
| $y$ | 2 | 2 | 2 | 2 | 2 |

32. Do Exercise 21 after interchanging each $x$ value with the corresponding $y$ value and then changing all the depths from centimeters to millimeters by multiplying each depth by 10. How is the value of the linear correlation coefficient affected? How is it affected if we change from degrees Celsius to degrees Fahrenheit?

33. Do Exercise 21 after changing the depth of 19 cm to 190 cm. How much effect does an extreme value have on the value of the linear correlation coefficient?

34. a. Show that $\dfrac{\Sigma(x - \bar{x})(y - \bar{y})}{(n - 1)s_x s_y} = \dfrac{n\Sigma xy - (\Sigma x)(\Sigma y)}{\sqrt{n(\Sigma x^2) - (\Sigma x)^2}\sqrt{n(\Sigma y^2) - (\Sigma y)^2}}$

   b. Show that Formula 9–1 is equivalent to

   $$r = \frac{(\overline{xy}) - \bar{x} \cdot \bar{y}}{\sqrt{[(\overline{x^2}) - (\bar{x})^2][(\overline{y^2}) - (\bar{y})^2]}}$$

   where $\overline{xy} = \Sigma xy/n$ and $(\overline{x^2}) = \Sigma x^2/n$.

35. The graph of $y = x^2$ is a parabola, not a straight line, so we might expect that the value of $r$ would not reflect a linear correlation between $x$ and $y$. Using $y = x^2$, make a table of $x$ and $y$ values for $x = 0, 1, 2, \ldots, 10$ and calculate the value of $r$. What do you conclude? How do you explain the result?

36. Given $n$ pairs of data from which the linear correlation coefficient $r$ can be found, use the following procedure to construct a **confidence interval** about the population parameter $\rho$.

   i. Use Table A–2 to find $z_{\alpha/2}$ that corresponds to the desired degree of confidence.

   ii. Evaluate the interval limits $w_L$ and $w_R$:

   $$w_L = \frac{1}{2}\ln\left(\frac{1 + r}{1 - r}\right) - z_{\alpha/2} \cdot \frac{1}{\sqrt{n - 3}}$$

   $$w_R = \frac{1}{2}\ln\left(\frac{1 + r}{1 - r}\right) + z_{\alpha/2} \cdot \frac{1}{\sqrt{n - 3}}$$

   iii. Now evaluate the confidence interval limits in the expression below.

   $$\frac{e^{2w_L} - 1}{e^{2w_L} + 1} < \rho < \frac{e^{2w_R} - 1}{e^{2w_R} + 1}$$

   Use this procedure to construct a 95% confidence interval for $\rho$, given 50 pairs of data for which $r = 0.600$.

# 9–3  Regression

Sir Francis Galton (1822–1911), a cousin of Charles Darwin, studied the phenomenon of heredity in which certain characteristics regress or revert to more typical values. Galton noted, for example, that children of tall parents tend to be shorter than their parents, while short parents tend to have children taller than themselves (when fully grown, of course). These original studies of regression evolved into a fairly sophisticated branch of mathematics called *regression analysis*, which includes the consideration of linear and nonlinear relationships.

While Section 9–2 used the correlation statistic $r$ to test for a linear relationship between two variables, this section presents a method for identifying the particular **regression line** (or *line of best fit* or *least squares line*) that describes the relationship. As in Section 9–2, we consider only *linear* relationships. The actual relationship between the two variables is often linear, or can be effectively approximated by a linear relationship, and nonlinear or curvilinear cases involve difficulties beyond the scope of this text.

We now want to find an equation of the form $y = Mx + B$, where $M$ and $B$ are the **slope** and **y-intercept** of the true regression line. Using only sample data, we can't find the exact values of the population parameters $M$ and $B$, but we can find their point estimates $m$ and $b$ by using Formulas 9–2 and 9–3.

**Formula 9–2**
$$m = \frac{n(\Sigma xy) - (\Sigma x)(\Sigma y)}{n(\Sigma x^2) - (\Sigma x)^2}$$

**Formula 9–3**
$$b = \frac{(\Sigma y)(\Sigma x^2) - (\Sigma x)(\Sigma xy)}{n(\Sigma x^2) - (\Sigma x)^2}$$

Some inexpensive calculators accept entries of paired data and provide the $m$ and $b$ values directly. These formulas appear formidable, but three observations make the required computations easier. First, if the correlation coefficient $r$ has been computed using Formula 9–1, the values of $\Sigma x$, $\Sigma y$, $\Sigma x^2$, $(\Sigma x)^2$, and $\Sigma xy$ have already been computed. These values can now be used again in Formulas 9–2 and 9–3. (Note that the numerator for $r$ in Formula 9–1 is identical to the numerator for $m$ in Formula 9–2.) Second, examine the denominators of the formulas for $m$ and $b$ and note that they are identical. This means that the computation of $n(\Sigma x^2) - (\Sigma x)^2$ need be done only once, and the result can be used in both formulas. Third, the regression line always passes through the centroid $(\bar{x}, \bar{y})$, so the equation $\bar{y} = m\bar{x} + b$ must be true. This implies that $b = \bar{y} - m\bar{x}$, and it is usually easier to evaluate $b$ by computing $\bar{y} - m\bar{x}$ than by using Formula 9–3.

**Formula 9–4**
$$b = \bar{y} - m\bar{x}$$

## WHAT SAT SCORES MEASURE

■ According to Harvard psychologist David McClellan, many statistical surveys "have shown that no consistent relationship exists between SAT scores in college students and their actual college accomplishments in social leadership, the arts, science, music, writing, speech, and drama." There does appear to be a significant correlation between SAT scores and income levels of the tested students' families. Higher scores tend to come from high-income families. Some studies show that SAT scores have limited value as predictors of college grades and *no* significant relationship to career success. Motivation seems to be a major success factor not measured by the SAT scores. ■

### Example

In Section 9–2 we used the Table 9–1 data ($x$ = living area, $y$ = home selling price) to find that the linear correlation coefficient of $r = 0.717$ indicates that at the 0.05 significance level, there is a significant positive linear correlation. Now find the regression equation of the straight line that relates $x$ and $y$. Recall from Section 9–2 that the following statistics were found.

| TABLE 9–1 | | | | | | | | |
|---|---|---|---|---|---|---|---|---|
| Living area | 15 | 38 | 23 | 16 | 16 | 13 | 20 | 24 |
| Selling price | 145 | 228 | 150 | 130 | 160 | 114 | 142 | 265 |

$n = 8$      $\Sigma x^2 = 3855$

$\Sigma x = 165$      $\Sigma y^2 = 241,394$

$\Sigma y = 1334$      $\Sigma xy = 29,611$

### Solution

Having determined the values of the individual components, we can now compute $m$ by using Formula 9–2.

$$m = \frac{n(\Sigma xy) - (\Sigma x)(\Sigma y)}{n(\Sigma x^2) - (\Sigma x)^2}$$
$$= \frac{8(29,611) - (165)(1334)}{8(3855) - (165)^2}$$
$$= \frac{16,778}{3615} = 4.64$$

We now use Formula 9–4 to find the value of $b$.

$$b = \bar{y} - m\bar{x} = \frac{1334}{8} - \left(\frac{16,778}{3615}\right)\left(\frac{165}{8}\right) = 71.0$$

We will use $y'$ to represent the **predicted value** of $y$ so that the preceding results allow us to express $y' = mx + b$ as $y' = 4.64x + 71.0$. Figure 9–8 shows the graph of this line in the scatter diagram of the original sample data.

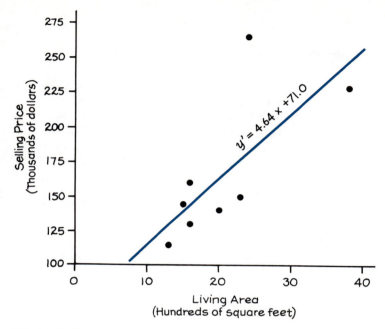

**Figure 9–8**

We should realize that $y' = 4.64x + 71.0$ is an *estimate* of the true straight-line equation $y = Mx + B$. This estimate is based on one particular set of sample data. Another sample drawn from the same population will probably lead to a slightly different equation.

Let's review the notation we are using.

| Notation | | |
|---|:---:|:---:|
| | **Population Parameter** | **Point Estimate** |
| Linear correlation coefficient | $\rho$ | $r$ |
| Slope of regression equation | $M$ | $m$ |
| $y$-intercept of regression equation | $B$ | $b$ |
| Equation of the regression line | $y = Mx + B$ | $y' = mx + b$ |

# Predictions

Regression equations can be helpful when used in *predicting* the value of one variable when given some particular value for the other variable. If the regression line fits the data quite well, then it makes sense to use its equation for predictions, provided that we don't go beyond the scope of the available scores. However, **we should use the equation of the regression line only if $r$ indicates that there is a significant linear correlation. In the absence of a significant linear correlation, we should not use the regression equation for projecting or predicting; instead, our best estimate of the second variable is simply its sample mean.** (See Figure 9–9, which summarizes this process.) This makes sense if we think of $r$ as a measure of how well the regression line fits the sample data. If $r$ is near $+1$ or $-1$, then the regression line fits the data well, but if $r$ is near 0, then the regression line fits poorly.

---

### Example

For a particular collection of sample data, we find the following.

$$n = 25 \qquad \bar{x} = 4.50 \qquad \bar{y} = 12.00$$
$$\text{Regression line: } y' = 2.00x + 3.00$$

Assuming a significance level of $\alpha = 0.05$ with $n = 25$, Table A–6 shows that the critical value is $r = 0.396$. Find the predicted point estimate of $y$ when $x = 5.00$, given the following computed values of the linear correlation coefficient.

a. $r = 0.987$
b. $r = 0.013$

### Solution

a. The test statistic is $r = 0.987$ and the critical value is $r = 0.396$, so we conclude that there is a significant positive linear correlation. *Because there is a significant positive linear correlation, we use the regression equation to find predicted values.* We therefore find the predicted point estimate of $y$ when $x = 5.00$ by substitution in the regression equation as follows:

$$y' = 2.00x + 3.00 = 2.00(5.00) + 3.00 = 13.00$$

When $x = 5.00$, the best predicted point estimate of $y$ is $y' = 13.00$.

b. The test statistic is $r = 0.013$ and the critical value is $r = 0.396$, so we conclude that there is no significant linear correlation. *Because there*

*continued*

**Solution** *continued*

*is not a significant linear correlation, we should not use the regression equation for predictions; instead, our best estimate of* y *is simply the mean* ȳ. We therefore conclude that the best predicted point estimate of y is $\bar{y} = 12.00$.

In part *a*, the presence of a significant linear correlation indicates that the regression equation $y' = 2.00x + 3.00$ fits the data well, so it makes sense to use it for predictions. In part *b*, the absence of a significant linear correlation indicates that the regression equation fits the data poorly, so we shouldn't use it for predictions. Instead, we use ȳ, which is the best point estimate of *y* under these circumstances.

Predicting
the Value
of a
Variable

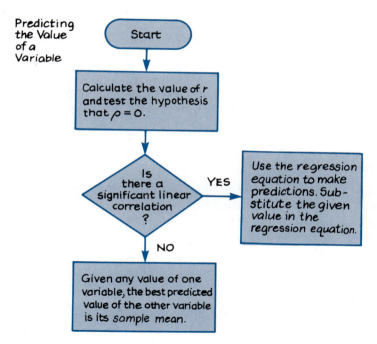

**Figure 9–9**

In part *a* of the preceding example, we were able to use the regression equation for predictions because there is a significant linear correlation; a common error is to use the regression equation when there is no significant linear correlation. That error is now listed along with three other common errors.

## Common Errors

1. *If there is no significant linear correlation, don't use the regression equation to make predictions.*

2. *When using the regression equation for predictions, stay within the scope of the available sample data.* If you find a regression equation that relates women's heights and shoe sizes, it's absurd to predict the shoe size of a woman who is 10 ft tall.

3. *A regression equation based on old data is not necessarily valid now.* The regression equation relating used car prices and the ages of cars is no longer usable if it's based on data from the 1950s.

4. *Don't make predictions about a population that is different from the population from which the sample data were drawn.* If we collect sample data from males and develop a regression equation relating SAT math scores and SAT verbal scores, the results don't necessarily apply to females. If we use *state averages* to develop a regression equation relating SAT math scores and SAT verbal scores, the results don't necessarily apply to *individuals*.

We have noted a qualitative relationship between the value of $r$ and how well the regression line fits the data. Another, more exact, relationship between $r$ and the regression line is the following formula, which relates $r$ to $m$ (the slope of the regression line).

**Formula 9–5**
$$r = \frac{ms_x}{s_y}$$

Here $s_x$ is the standard deviation of the $x$ values and $s_y$ is the standard deviation of the $y$ values. In the preceding expression, we can easily solve for $m$ to get $m = rs_y/s_x$, and this expression may simplify the calculation of $m$ when $r$, $s_y$, and $s_x$ have already been found.

## Least-Squares Property

Formulas 9–2 and 9–3 or 9–4 describe the computations necessary to obtain the regression line equation $y' = mx + b$, and the criterion used to arrive at these particular formulas is the **least-squares property: The sum of the squares of the vertical deviations of the sample points from the regression line is the smallest sum possible.** This least-squares property can be understood (believe it or not!) by examining Figure 9–10 and the text that follows. Figure 9–10 shows the paired data contained in the following table.

| $x$ | 1 | 2 | 4 | 5 |
|---|---|---|---|---|
| $y$ | 4 | 24 | 8 | 32 |

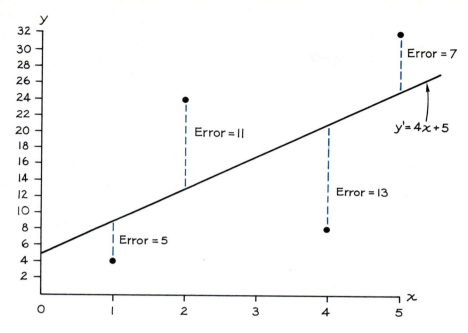

**Figure 9–10**

Figure 9–10 also shows the regression line, with equation $y' = 4x + 5$ found by using the formulas for $m$ and $b$. The distances identified as errors in Figure 9–10 are the differences between the actual *observed* y values and the *predicted* $y'$ values on the regression line. The least squares property dictates that the sum of the squares of those vertical errors ($5^2 + 11^2 + 13^2 + 7^2 = 364$) is the *minimum* sum that is possible with the given data. Any other line will yield a sum of squares larger than 364. It is in this sense that the line $y' = 4x + 5$ fits the data best. (For example, $y' = 3x + 8$ will have vertical errors of 7, 10, 12, and 9 and the sum of their squares is 374, which exceeds the minimum of 364.)

Fortunately, we need not deal directly with the least-squares property when we want to find the equation of the regression line. Calculus has been used to build the least-squares property into our formulas for $m$ and $b$. Because the derivations of those formulas require calculus, we don't include them in this text.

When dealing with larger collections of paired data, a calculator or computer is required to find the values of $r$, $m$, and $b$. We have already mentioned that some calculators allow the entry of paired data and provide the values of $r$, $m$, and $b$. Many computer software packages take paired data as input and produce more complete results, such as the STATDISK and Minitab output shown on the next page. The displayed output corresponds to the home selling price and living area data (Table 9–1) introduced in Section 9–1 and used in Sections 9–2 and 9–3. (The coefficient of determination, the standard error of estimate, and the variation statistics will be discussed in the following section.)

STATDISK DISPLAY

Linear correlation coefficient . . . . . . r = .716708
Coefficient of determination . . . . . . . = .51367
Standard error of estimate . . . . . . . . = 39.1912

Explained variation . . . . . . . . . . . . . . = 9733.79
Unexplained variation . . . . . . . . . . . . = 9215.71
Total variation . . . . . . . . . . . . . . . . . . = 18949.5

Equation of regression line . . Y = 4.64122 X + 71.0249

Test for linear correlation
Level of significance . . . . . . . . . . . . . = .05
Test statistic is . . . . . . . . . . . . . . . . r = .716708
Critical value is . . . . . . . . . . . . . . . . r = .707365
REJECT the claim of no significant linear correlation

MINITAB DISPLAY

```
MTB > READ C1 C2
DATA> 15 145
DATA> 38 228
DATA> 23 150
DATA> 16 130
DATA> 16 160
DATA> 13 114
DATA> 20 142
DATA> 24 265
DATA> ENDOFDATA
      8 ROWS READ
MTB > NAME C1 'AREA' C2 'PRICE'
MTB > CORRELATION C1 and C2

Correlation of AREA and PRICE = 0.717

MTB > REGRESSION C2 1 C1

The regression equation is
PRICE = 71.0 + 4.64 AREA
```

For the regression methods given in this section, the following assumptions apply.

## Assumptions

1. We are investigating only *linear* relationships.
2. For each $x$ value, $y$ is a random variable having a normal distribution. All of these $y$ distributions have the same variance. Also, for a given value of $x$, the distribution of $y$ values has mean $Mx + B$, which lies on the regression line. (Results are not seriously affected if departures from the normal distributions and equal variances are not too extreme.)

## Confidence Intervals

Although not easy, it is possible to construct **confidence intervals** about the population parameters $M$ and $B$, which represent the slope and $y$-intercept of the regression line for the entire population of data. The difficult computations can be simplified somewhat by using the standard error of estimate $s_e$, which will be introduced in the next section. (See Exercise 17 in Section 9–4 for a procedure that allows you to construct confidence intervals about the slope $M$ and the $y$-intercept $B$ of the regression line.)

## 9–3 Exercises A

In Exercises 1–8, use the given data to obtain the equation of the regression line.

1. 

| $x$ | 1 | 1 | 2 | 3 |
|-----|---|---|---|---|
| $y$ | 1 | 5 | 4 | 2 |

2. 

| $x$ | 1 | 2 | 2 | 3 |
|-----|---|---|---|---|
| $y$ | 5 | 4 | 3 | 1 |

3. 

| $x$ | 0 | 1 | 1 | 2 | 5 |
|-----|---|---|---|---|---|
| $y$ | 3 | 3 | 4 | 5 | 6 |

4. 

| $x$ | 1 | 3 | 3 | 4 | 5 | 5 |
|-----|---|---|---|---|---|---|
| $y$ | 5 | 3 | 2 | 2 | 0 | 1 |

5. 

| $x$ | 1 | 3 | 4 | 6 | 8 |
|-----|-----|-----|------|------|------|
| $y$ | 4.9 | 7.2 | 11.1 | 14.8 | 19.0 |

6. 

| $x$ | 1.4 | 7.2 | 3.9 | 4.5 | 6.2 |
|-----|-----|-----|-----|-----|-----|
| $y$ | 6 | 7 | 7 | 3 | 8 |

7. 

| $x$ | 15.4 | 17.3 | 19.7 | 14.9 | 18.5 |
|-----|------|------|------|------|------|
| $y$ | 7.71 | 8.65 | 9.32 | 7.51 | 9.22 |

8. 

| $x$ | 134 | 256 | 179 | 385 | 217 |
|-----|------|-------|------|-------|------|
| $y$ | 52.1 | 113.4 | 75.2 | 177.4 | 91.6 |

In Exercises 9–24, find the equation of the regression line, then plot it on the scatter diagram. (The data are taken from exercises in Section 9–2.)

9.

| Age | 17.2 | 43.5 | 30.7 | 53.1 | 37.2 | 21.0 | 27.6 | 46.3 |
|-----|------|------|------|------|------|------|------|------|
| BAC | 0.19 | 0.20 | 0.26 | 0.16 | 0.24 | 0.20 | 0.18 | 0.23 |

10.

| Weight | 167 | 191 | 112 | 129 | 140 | 173 | 119 |
|--------|-----|-----|-----|-----|-----|-----|-----|
| Calories | 4.23 | 4.69 | 3.21 | 3.47 | 3.72 | 4.45 | 3.36 |

11.

| Exports | 10 | 20 | 43 | 221 | 218 | 218 |
|---------|----|----|----|-----|-----|-----|
| Imports | 9 | 15 | 40 | 245 | 326 | 370 |

12.

| Algebra | 17 | 21 | 11 | 16 | 15 | 11 | 24 | 27 | 19 | 8 |
|---------|----|----|----|----|----|----|----|----|----|---|
| Calculus | 73 | 66 | 64 | 61 | 70 | 71 | 90 | 68 | 84 | 52 |

13.

| Automatic weapons | 11.6 | 8.3 | 3.6 | 0.6 | 6.9 | 2.5 | 2.4 | 2.6 |
|-------------------|------|-----|-----|-----|-----|-----|-----|-----|
| Murder rate | 13.1 | 10.6 | 10.1 | 4.4 | 11.5 | 6.6 | 3.6 | 5.3 |

14.

| Added load | 0 | 10 | 20 | 30 | 40 | 50 | 60 | 70 |
|------------|---|----|----|----|----|----|----|----|
| Increase in length | 0 | 0.05 | 0.10 | 0.15 | 0.20 | 0.25 | 0.30 | 1.25 |

15.

| HC | 0.65 | 0.55 | 0.72 | 0.83 | 0.57 | 0.51 | 0.43 | 0.37 |
|----|------|------|------|------|------|------|------|------|
| CO | 14.7 | 12.3 | 14.6 | 15.1 | 5.0 | 4.1 | 3.8 | 4.1 |

16.

| Age | 6.1 | 7.2 | 5.9 | 6.3 | 10.5 | 11.0 |
|-----|-----|-----|-----|-----|------|------|
| Score | 17.8 | 47.4 | 25.8 | 24.3 | 66.6 | 91.4 |

17.

| Taxes | 1.9 | 3.0 | 1.4 | 1.4 | 1.5 | 1.8 | 2.4 | 4.0 |
|-------|-----|-----|-----|-----|-----|-----|-----|-----|
| Selling price | 145 | 228 | 150 | 130 | 160 | 114 | 142 | 265 |

18.

| Living area | 15 | 38 | 23 | 16 | 16 | 13 | 20 | 24 |
|-------------|----|----|----|----|----|----|----|----|
| Taxes | 1.9 | 3.0 | 1.4 | 1.4 | 1.5 | 1.8 | 2.4 | 4.0 |

19.

| Weight | 62 | 57 | 94 | 69 | 66 | 76 | 58 | 88 | 70 | 84 |
|--------|----|----|----|----|----|----|----|----|----|----|
| Anaerobic threshold | 1.37 | 1.34 | 1.93 | 1.92 | 2.24 | 2.02 | 1.35 | 2.21 | 1.79 | 1.74 |

20.

| Satisfaction | 5.05 | 4.12 | 5.39 | 4.17 | 4.00 | 4.49 | 5.40 | 4.86 |
|--------------|------|------|------|------|------|------|------|------|
| Commitment | 5.37 | 4.49 | 5.42 | 4.45 | 4.24 | 5.34 | 5.62 | 4.90 |

21.

| Temperature (°C) | −62 | −41 | −36 | −26 | −33 | −56 | −50 | −66 |
|------------------|-----|-----|-----|-----|-----|-----|-----|-----|
| Snow depth (cm) | 21 | 13 | 12 | 3 | 6 | 22 | 14 | 19 |

22.

| Cigarette consumption | 3522 | 3597 | 4171 | 4258 | 3993 | 3971 | 4042 | 4053 |
|-----------------------|------|------|------|------|------|------|------|------|
| Percentage of psychiatric admissions (in percentage points) | 0.20 | 0.22 | 0.23 | 0.29 | 0.31 | 0.33 | 0.33 | 0.32 |

23.

| PSAT | 144 | 138 | 136 | 111 | 129 | 122 | 116 | 136 | 118 | 112 | 137 | 124 |
|------|-----|-----|-----|-----|-----|-----|-----|-----|-----|-----|-----|-----|
| SAT | 1290 | 1310 | 1340 | 1130 | 1170 | 1180 | 1220 | 1340 | 1210 | 1270 | 1160 | 1130 |

24.

| Minutes | 1364 | 53 | 457 | 717 | 384 | 1432 | 365 | 1626 | 840 |
|---------|------|----|-----|-----|-----|------|-----|------|-----|
| Points | 652 | 20 | 163 | 210 | 175 | 821 | 143 | 1098 | 459 |

25. Using a collection of paired sample data, the regression equation is found to be $y' = 50.0x + 10.0$ and the sample means are $\bar{x} = 0.30$ and $\bar{y} = 25.0$. In each of the following cases, use the additional information and find the best predicted point estimate of $y$ when $x = 2.0$. Assume a significance level of $\alpha = 0.05$.
   a. $n = 100$; $r = 0.999$
   b. $n = 10$; $r = 0.005$
   c. $n = 15$; $r = 0.519$
   d. $n = 25$; $r = 0.393$
   e. $n = 22$; $r = 0.567$

26. Using a collection of paired sample data, the regression equation is found to be $y' = -20.0x + 50.0$ and the sample means are $\bar{x} = 0.50$ and $\bar{y} = 40.0$. In each of the following cases, use the additional information and find the best predicted point estimate of $y$ when $x = 1.00$. Assume a significance level of $\alpha = 0.05$.
   a. $n = 5$; $r = -0.102$
   b. $n = 50$; $r = -0.997$
   c. $n = 20$; $r = -0.403$
   d. $n = 20$; $r = -0.449$
   e. $n = 65$; $r = -0.229$

27. In each of the following cases, find the best predicted point estimate of $y$ when $x = 5$. The given statistics are summarized from paired sample data. Assume a significance level of $\alpha = 0.01$.
   a. $n = 40$, $\bar{y} = 6$, $r = 0.01$, and the equation of the regression line is $y' = 3x + 2$.
   b. $n = 40$, $\bar{y} = 6$, $r = 0.93$, and the equation of regression line is $y' = 3x + 2$.
   c. $n = 20$, $\bar{y} = 6$, $r = -0.654$, and the equation of the regression line is $y' = -3x + 2$.
   d. $n = 20$, $\bar{y} = 6$, $r = 0.432$, and the equation of the regression line is $y' = 1.2x + 3.7$.
   e. $n = 100$, $\bar{y} = 6$, $r = -0.175$, and the equation of the regression line is $y' = -2.4x + 16.7$.

28. In each of the following cases, find the best predicted point estimate of $y$ when $x = 8.0$. The given statistics are summarized from paired sample data. Assume a significance level of $\alpha = 0.05$.
   a. $n = 10$, $\bar{y} = 8.40$, $r = -0.236$, and the equation of the regression line is $y' = -2.0x + 3.5$.
   b. $n = 10$, $\bar{y} = 8.40$, $r = -0.654$, and the equation of the regression line is $y' = -2.0x + 3.5$.

*continued*

c.  $n = 10, \bar{y} = 8.40, r = 0.602$, and the equation of the regression line
is $y' = 2.0x + 3.5$.

d.  $n = 32, \bar{y} = 8.40, r = -0.304$, and the equation of the regression
line is $y' = -2.0x + 3.5$.

e.  $n = 75, \bar{y} = 8.40, r = 0.257$, and the equation of the regression line
is $y' = 2.0x + 3.5$.

# 9–3 | Exercises B

29.  Large numbers, such as those in the table below, often cause computational problems. Find the equation of the regression line. Now divide
each $x$ entry by 1000 and find the equation of the regression line. How
are the results affected by the change in $x$? Also, how would the results
be affected if each $y$ entry is divided by 1000?

| $x$ | 924,736 | 832,985 | 825,664 | 793,427 | 857,366 |
|---|---|---|---|---|---|
| $y$ | 142 | 111 | 109 | 95 | 119 |

30.  Use the paired data in Exercise 2 to verify that $r = ms_x/s_y$ where $m$, $s_x$,
and $s_y$ are as defined in this section and $r$ is computed by using Formula
9–1.

31.  Using the data in Exercise 2 and the equation of the regression line, find
the sum of the squares of the vertical deviations for the given points.
Show that this sum is less than the corresponding sum obtained by
replacing the regression equation with $y' = -x + 6$.

32.  Prove that the point $(\bar{x}, \bar{y})$ will always lie on the regression line.

33.  Prove that $r$ and $m$ have the same sign.

34.  Verify Formula 9–5 by showing that

$$\frac{ms_x}{s_y} = \frac{n\Sigma xy - (\Sigma x)(\Sigma y)}{\sqrt{n(\Sigma x^2) - (\Sigma x)^2}\sqrt{n(\Sigma y^2) - (\Sigma y)^2}}$$

35.  We have noted that $\rho$ is the linear correlation coefficient for a population, while the regression equation for the population is $y = Mx + B$.
Explain why a test of the null hypothesis $H_0: \rho = 0$ is *equivalent* to a
test of the null hypothesis $H_0: M = 0$.

36.  If the scatter diagram reveals a nonlinear pattern that we recognize as
another type of curve, we may be able to apply the methods of this
section. For the data given in the table below, find an equation of the
form $y' = mx^2 + b$ by using the values of $x^2$ and $y$ instead of the values
of $x$ and $y$. Use Formulas 9–2 and 9–3 or 9–4, but enter the values of $x^2$
wherever $x$ occurs.

| $x$ | 4 | 5 | 6 | 7 |
|---|---|---|---|---|
| $y$ | 3 | 7 | 11 | 20 |

# 9–4 Variation and Prediction Intervals

In the preceding two sections we introduced the fundamental concepts of linear correlation and regression. We saw that the linear correlation coefficient $r$ could be used to determine whether or not there is a significant statistical relationship between two variables. We interpret $r$ in a very limited way when we make one of the following three conclusions:

1. There is a significant positive linear correlation.
2. There is a significant negative linear correlation.
3. There is not a significant linear correlation.

The actual values of $r$ can provide us with more information. We begin with a sample case, which leads to an important definition.

Let's assume that we have a large collection of paired data, which yields these results:

1. There is a significant positive linear correlation.
2. The equation of the regression line is $y' = 2x + 3$.
3. $\bar{y} = 9$.
4. The scatter diagram contains the point $(5, 19)$, which comes from the original set of observations.

When $x = 5$ we can find the predicted value $y'$ as follows.

$$y' = 2x + 3 = 2(5) + 3 = 13$$

Note that the point $(5, 13)$ is on the regression line, while the point $(5, 19)$ is not. Take the time to carefully examine Figure 9–11 on the following page.

---

### Definition

Given a collection of paired data, the **total deviation** (from the mean) of a particular point $(x, y)$ is $y - \bar{y}$, the **explained deviation** is $y' - \bar{y}$, and the **unexplained deviation** is $y - y'$.

---

For the specific data under consideration, the total deviation of $(5, 19)$ is $y - \bar{y} = 19 - 9 = 10$.

If we were totally ignorant of correlation and regression concepts and wanted to predict a value of $y$ given a value of $x$ and a collection of paired $(x, y)$ data, our best guess would be $\bar{y}$. But we are not totally ignorant of correlation and regression concepts: We know that in this case the way to predict the value of $y$ when $x = 5$ is to use the regression equation, which

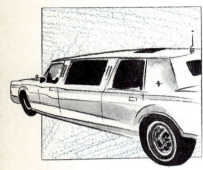

## UNUSUAL ECONOMIC INDICATORS

■ Forecasting and predicting are important goals of statistics. Investors seek indicators that can be used to forecast stock market behavior. Some of them are quite colorful. The hemline index is based on heights of women's skirts; rising hemlines supposedly precede a rise in the Dow Jones Industrial Average. According to the Super Bowl omen, a Super Bowl victory by a team with NFL origins is followed by a year in which the New York Stock Exchange index rises; otherwise, it falls. This indicator has been correct in 21 of the past 23 years. Other indicators: aspirin sales, limousines on Wall Street, and elevator traffic at the New York Stock Exchange. ■

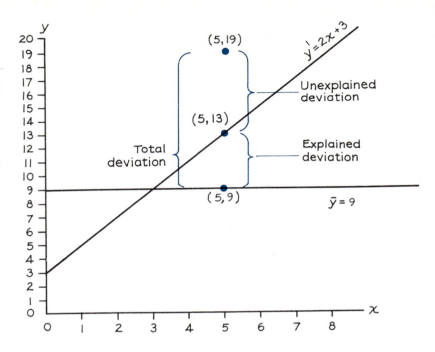

**Figure 9–11**

yields $y' = 13$. We can explain the discrepancy between $\bar{y} = 9$ and $y' = 13$ by simply noting that there is a significant positive linear correlation best described by the regression line. Consequently, when $x = 5$, $y$ *should* be 13, and not 9. But while $y$ *should* be 13, *it is* 19, and that discrepancy between 13 and 19 cannot be explained by the regression line and is called an unexplained deviation or a residual. This specific case illustrated in Figure 9–11 can be generalized as follows.

(total deviation) = (explained deviation) + (unexplained deviation)

or   $(y - \bar{y})$   $= (y' - \bar{y})$   $+ (y - y')$

This last expression can be further generalized and modified to include all of the pairs of sample data as follows. (By popular demand, we omit the intermediate algebraic manipulations.)

**Formula 9–6**
**(total variation)** = **(explained variation)** + **(unexplained variation)**
  or   $\Sigma(y - \bar{y})^2$   $= \Sigma(y' - \bar{y})^2$   $+ \Sigma(y - y')^2$

The components of this last expression are used in the next important definition.

## Definition

The amount of the variation in $y$ that is explained by the regression line is indicated by the **coefficient of determination,** which is given by

$$r^2 = \frac{\text{explained variation}}{\text{total variation}}$$

We can compute $r^2$ by using this definition with Formula 9–6 above, or we can simply square the linear correlation coefficient $r$, which is found by using the methods given in Section 9–2. To make some sense of this last definition, consider the following example.

## Example

If $r = 0.8$, then the coefficient of determination is $r^2 = 0.8^2 = 0.64$, which means that 64% of the total variation can be explained by the regression line. Thus 36% of the total variation remains unexplained.

Table 9–3 illustrates that the linear correlation coefficient $r$ can be computed with the total variation and explained variation. This procedure for

**TABLE 9–3**

| $x$ | $y$ | $y'$ | $y - \bar{y}$ | $(y - \bar{y})^2$ | $y' - \bar{y}$ | $(y' - \bar{y})^2$ |
|---|---|---|---|---|---|---|
| 1 | 3 | 3.651 | −4 | 16 | −3.349 | 11.2158 |
| 1 | 4 | 3.651 | −3 | 9 | −3.349 | 11.2158 |
| 2 | 6 | 5.744 | −1 | 1 | −1.256 | 1.5775 |
| 3 | 8 | 7.837 | 1 | 1 | 0.837 | 0.7006 |
| 6 | 14 | 14.116 | 7 | 49 | 7.116 | 50.6375 |
| | | | Totals: | 76 | | 75.3472 |

                ↑                            ↑

     Total variation          Explained variation

$$r^2 = \frac{\text{explained variation}}{\text{total variation}} = \frac{\Sigma(y' - \bar{y})^2}{\Sigma(y - \bar{y})^2} = \frac{75.3472}{76} = 0.9914$$

so that

$$r = \sqrt{0.9914} = 0.996$$

determining $r$ is not recommended since it is longer, less direct, and more likely to produce errors than the method given in Section 9–2. The paired $xy$ data are in the first two columns. The third column of $y'$ values is determined by substituting each value of $x$ into the equation of the regression line $(y' = 2.093x + 1.558)$. Also, $\bar{y} = 7.00$. The computed value of $r = 0.996$ agrees with the value found by using Formula 9–1. (If fewer decimal places are used, there will be a small discrepancy due to rounding errors.)

While the computations shown in the table don't seem too involved, we first had to compute the equation of the regression line, which is why this method for finding $r$ is inferior. Also observe that $r$ is not always the *positive* square root of $r^2$ since we can have negative values for the linear correlation coefficient. Examination of the scatter diagram should reveal whether $r$ is negative or positive. If the pattern of points is predominantly uphill (from left to right), then $r$ should be positive, but if there is a downward pattern, $r$ should be negative. That is, the linear correlation coefficient $r$ and the slope $m$ must have the same sign.

## Prediction Intervals

Prediction intervals are confidence intervals about a predicted value of $y$. Suppose we use the regression equation to predict a value of $y$ from a given value of $x$. For example, we have seen that with a significant positive linear correlation, the regression equation $y' = 2x + 3$ can be used to predict $y$ when $x = 5$; the result is $y' = 13$. Having made that prediction, we should consider how dependable it is. Intuition suggests that if $r = 1$, all sample points lie exactly on the regression line and the regression equation should therefore provide very dependable results, as long as we don't try to project beyond reasonable limits. If $r$ is very close to 1, the sample points are close to the regression line, and again our $y'$ values should be very dependable. What we need is a less subjective measure of the spread of the sample points about the regression line. The next definition provides such a measure.

---

**Definition**

The **standard error of estimate** is given by

$$s_e = \sqrt{\frac{\Sigma(y - y')^2}{n - 2}}$$

where $y'$ is the predicted $y$ value.

---

**Example**

Find the standard error of estimate $s_e$ for the 5 pairs of data given in the first two columns of Table 9–4.

**Solution**

We extend the table to include the necessary components.

**TABLE 9–4**

| $x$ | $y$ | $y'$ | $y - y'$ | $(y - y')^2$ |
|---|---|---|---|---|
| 1 | 3 | 3.651 | $-0.651$ | 0.4238 |
| 1 | 4 | 3.651 | 0.349 | 0.1218 |
| 2 | 6 | 5.744 | 0.256 | 0.0655 |
| 3 | 8 | 7.837 | 0.163 | 0.0266 |
| 6 | 14 | 14.116 | $-0.116$ | 0.0135 |
| | | | | 0.6512 |

$\uparrow$
unexplained variation

We can now evaluate $s_e$:

$$s_e = \sqrt{\frac{\Sigma(y - y')^2}{n - 2}} = \sqrt{\frac{0.6512}{3}} = 0.466$$

The development of the standard error of estimate closely parallels that of the ordinary standard deviation introduced in Chapter 2. Just as the standard deviation is a measure of how scores deviate from their mean, the standard error $s_e$ is a measure of how points deviate from their regression line. The reasoning behind dividing by $n - 2$ is similar to the reasoning that led to division by $n - 1$ for the ordinary standard deviation, and we will not pursue the complex details. Do understand that smaller values of $s_e$ reflect points that stay close to the regression line, while larger values indicate greater dispersion of points away from the regression line.

Formula 9–7 can also be used to compute the standard error of estimate (see top of following page). It is algebraically equivalent to the expression in the definition, but this form is generally easier to work with since it doesn't require that we compute each of the $y'$ values.

**Formula 9–7**
$$s_e = \sqrt{\frac{\Sigma y^2 - b\Sigma y - m\Sigma xy}{n - 2}}$$

Here $m$ and $b$ are the slope and $y$-intercept of the regression equation.

We can use the standard error of estimate, $s_e$, to construct interval estimates that will help us to see how dependable our point estimates of $y$ really are. Assume that for each fixed value of $x$, the corresponding sample values of $y$ are normally distributed about the regression line, and those normal distributions have the same variance. The following interval estimate applies to an individual $y$. (For a confidence interval about the *mean* value of $y$, see Exercise 20.)

---

### Definition

Given the fixed value $x_0$, the **prediction interval for an individual $y$** is
$$y' - E < y < y' + E$$
where
$$E = t_{\alpha/2} s_e \sqrt{1 + \frac{1}{n} + \frac{n(x_0 - \overline{x})^2}{n(\Sigma x^2) - (\Sigma x)^2}}$$

$x_0$     represents the given value of $x$.
$t_{\alpha/2}$    has $n - 2$ degrees of freedom.
$s_e$     is found from Formula 9–7.

---

The regression equation for the data given in Table 9–4 is $y' = 2.093x + 1.558$, so that when $x = 4$ we find the best point estimate for the predicted value of $y$ to be $y' = 9.930$. Using the same data, we will construct the 95% prediction interval for the individual $y$ value corresponding to the given value of $x = 4$. We have already found that $s_e = 0.466$. With $n = 5$ we have 3 degrees of freedom, so that $t_{\alpha/2} = 3.182$ for $\alpha = 0.05$ in two tails. With $x_0 = 4$, $\overline{x} = \frac{13}{5} = 2.6$, $\Sigma x^2 = 51$, and $\Sigma x = 13$, we get

$$E = (3.182)(0.466) \sqrt{1 + \frac{1}{5} + \frac{5(4 - 2.6)^2}{5(51) - 169}}$$
$$= (3.182)(0.466)(1.146) = 1.699$$

so that $y' - E < y < y' + E$ becomes

$$9.930 - 1.699 < y < 9.930 + 1.699$$

Given that $x = 4$, we therefore get the following 95% prediction interval for an individual $y$ value.

$$8.231 < y < 11.629$$

## Example

Refer to the Table 9–1 sample data that lists home selling prices along with living area. In previous sections we have shown the following.

a. There is a significant linear correlation (at the 0.05 significance level).

b. The regression equation is $y' = 4.64x + 71.0$.

c. When $x = 20$, the predicted $y$ value is 164.

Recall that $x$ is the living area (in hundreds of square feet) and $y$ is the selling price (in thousands of dollars), so that result c shows that for a living area of 2000 square feet, the predicted selling price is \$164,000. Construct a 95% prediction interval for an individual selling price, given that the living area is 2000 sq ft. This will give us a sense of how reliable that \$164,000 estimate really is.

## Solution

We have used the Table 9–1 sample data in previous sections to find the following.

$$n = 8 \qquad \Sigma y = 1334 \qquad \bar{x} = 20.625$$
$$\Sigma x = 165 \qquad \Sigma y^2 = 241{,}394$$
$$\Sigma x^2 = 3855 \qquad \Sigma xy = 29{,}611$$

From the regression equation (with extra digits carried):

$$m = 4.6412 \quad \text{and} \quad b = 71.025$$

When $x = 20$, the predicted $y$ is $y' = 164$.
   Using the preceding values, we first find the standard error $s_e$.

$$s_e = \sqrt{\frac{\Sigma y^2 - b\Sigma y - m\Sigma xy}{n - 2}}$$

$$= \sqrt{\frac{241{,}394 - (71.025)(1334) - (4.6412)(29{,}611)}{8 - 2}}$$

$$= \sqrt{\frac{9216.1}{6}} = 39.192$$

From Table A–3 we find $t_{\alpha/2} = 2.447$. (We used $8 - 2 = 6$ degrees of freedom with $\alpha = 0.05$ in two tails.) We can now calculate $E$ by letting $x_0 = 20$, since we want the prediction interval of $y$ for $x = 20$.

*continued*

---

**Solution** *continued*

$$E = t_{\alpha/2} s_e \sqrt{1 + \frac{1}{n} + \frac{n(x_0 - \bar{x})^2}{n(\Sigma x^2) - (\Sigma x)^2}}$$

$$= (2.447)(39.192) \sqrt{1 + \frac{1}{8} + \frac{8(20 - 20.625)^2}{8(3855) - (165)^2}}$$

$$= (2.447)(39.192)(1.0611) = 102 \text{ (rounded off)}$$

With $y' = 164$ and $E = 102$, we get the prediction interval

$$y' - E < y < y' + E$$

$$164 - 102 < y < 164 + 102$$

$$62 < y < 266$$

Recognizing that the selling price $y$ is in thousands of dollars, we see that for a 2000-sq-ft house $(x = 20)$, the predicted selling price is between $62,000 and $266,000. That's quite a range. This interval is so wide mainly because the sample is so small $(n = 8)$. The correlation coefficient of 0.717 barely made it into the critical region bounded by the critical value of 0.707. This shows that the regression equation used for the prediction does fit the sample data well, but the fit could be much better.

In addition to knowing that the predicted selling price is $164,000, we now have a sense for how reliable that estimate really is. The 95% prediction interval of

$$\$62,000 < \text{selling price} < \$266,000$$

shows that the $164,000 estimate can vary substantially.

---

# 9–4  Exercises A

1.  If $r = 0.333$, find the coefficient of determination and the percentage of the total variation that can be explained by the regression line.
2.  If $r = -0.444$, find the coefficient of determination and the percentage of the total variation that can be explained by the regression line.
3.  If $r = 0.800$, find the coefficient of determination and the percentage of the total variation that can be explained by the regression line.
4.  If $r = -0.700$, find the coefficient of determination and the percentage of the total variation that can be explained by the regression line.

In Exercises 5–8, find the (a) explained variation, (b) unexplained variation, (c) total variation, (d) coefficient of determination, (e) standard error of estimate $s_e$.

5.
| x | 1 | 2 | 3 | 5 | 6 |
|---|---|---|---|---|---|
| y | 5 | 8 | 11 | 17 | 20 |

(The equation of the regression line is $y' = 3x + 2$.)

6.
| x | 1 | 2 | 3 | 5 | 6 |
|---|---|---|---|---|---|
| y | 5 | 4 | 11 | 13 | 20 |

(The equation of the regression line is $y' = 2.9535x + 0.55814$.)

7.
| x | 4 | 7 | 5 | 2 | 3 |
|---|---|---|---|---|---|
| y | 13 | 22 | 16 | 7 | 9 |

(The equation of the regression line is $y' = 3.0811x + 0.45946$.)

8.
| x | 2 | 5 | 7 | 10 | 15 |
|---|---|---|---|---|---|
| y | 20 | 15 | 14 | 12 | 10 |

(The equation of the regression line is $y' = -0.71660x + 19.789$.)

9. Refer to the data given in Exercise 5, and assume that the necessary conditions of normality and variance are met.
   a. For $x = 4$, find $y'$, the predicted value of $y$.
   b. Find $s_e$. How does this value of $s_e$ affect the construction of the 95% prediction interval of $y$ for $x = 4$?

10. Refer to Exercise 6, and assume that the necessary conditions of normality and variance are met.
    a. For $x = 4$, find $y'$, the predicted value of $y$.
    b. Find the 99% prediction interval of $y$ for $x = 4$.

11. Refer to the data given in Exercise 7, and assume that the necessary conditions of normality and variance are met.
    a. For $x = 6$, find $y'$, the predicted value of $y$.
    b. Find the 95% prediction interval of $y$ for $x = 6$.

12. Refer to the data given in Exercise 8, and assume that the necessary conditions of normality and variance are met.
    a. For $x = 1$, find $y'$, the predicted value of $y$.
    b. Find the 99% prediction interval of $y$ for $x = 1$.

In Exercises 13–16, refer to the Table 9–1 sample data. Let $x$ represent the living area (in hundreds of feet) and let $y$ represent the selling price (in thousands of dollars). Construct a prediction interval for an individual selling price that corresponds to the living area and degree of confidence given below. (*Hint:* See the example in this section.)

13. 2500 sq ft; 95% confidence
14. 1800 sq ft; 90% confidence
15. 3000 sq ft; 90% confidence
16. 1600 sq ft; 99% confidence

# 9–4 Exercises B

17. **Confidence intervals** for the slope $M$ and $y$-intercept $B$ for a regression line $(y = Mx + B)$ can be found by evaluating the limits in the intervals below.

$$m - E < M < m + E \text{ where } E = t_{\alpha/2} \cdot \frac{s_e}{\sqrt{\Sigma x^2 - \frac{(\Sigma x)^2}{n}}}$$

$$b - E < B < b + E \text{ where } E = t_{\alpha/2} \cdot s_e \cdot \sqrt{\frac{1}{n} + \frac{\bar{x}^2}{\Sigma x^2 - \frac{(\Sigma x)^2}{n}}}$$

In these expressions, $m$ and $b$ are found from the sample data and $t_{\alpha/2}$ is found from Table A–3 by using $n - 2$ degrees of freedom. Using the data from Table 9–4, find the 95% confidence intervals for $M$ and $B$.

18. a. Assuming that a collection of paired data includes at least 3 pairs of values, what do you know about the linear correlation coefficient if $s_e = 0$?
    b. If a collection of paired data is such that the total explained variation is 0, what can be deduced about the slope of the regression line?

19. a. Find an expression for the unexplained variation in terms of the sample size $n$ and the standard error of estimate $s_e$.
    b. Find an expression for the explained variation in terms of the coefficient of determination $r^2$ and the unexplained variation.
    c. Suppose we have a collection of paired data for which $r^2 = 0.900$ and the regression equation is $y' = -2x + 3$. Find the linear correlation coefficient.

20. The formula

$$s_{y'} = s_e \sqrt{1 + \frac{1}{n} + \frac{n(x_0 - \bar{x})^2}{n(\Sigma x^2) - (\Sigma x)^2}}$$

gives the **standard error of the prediction** when predicting for a *single* $y$ given that $x = x_0$. When predicting for the *mean* of *all* values of $y$ for which $x = x_0$, the point estimate $y'$ is the same, but $s_{y'}$ is described as follows.

$$s_{y'} = s_e \sqrt{\frac{1}{n} + \frac{n(x_0 - \bar{x})^2}{n(\Sigma x^2) - (\Sigma x)^2}}$$

Extend the last example of this section to find the point estimate and a 95% confidence interval for the *mean* selling price of *all* homes with a living area of 2000 sq ft.

# 9–5 Multiple Regression

We began this chapter by considering the relationship between home selling price and living area. In Section 9–2 we used the linear correlation coefficient as a measure of the strength of the relationship. Based on the sample data in Table 9–1, we concluded that there is a significant positive linear correlation between the variable of home selling price and the variable of living area. In Section 9–3 we proceeded to describe that relationship by the regression equation $y' = 4.64x + 71.0$. In this equation, $x$ is the living area (in hundreds of square feet) and $y$ is the selling price (in thousands of dollars). For a living area of 2000 sq ft we have a predicted selling price of $164,000, and we know from Section 9–4 that this estimate is not very reliable. (It can easily vary between $62,000 and $266,000.)

It was not too surprising to find a significant linear correlation between home selling price and living area. After all, larger homes cost more to build and that increased cost is reflected in increased value. But we should also recognize that in addition to the living area of a home, there are other important factors that should be considered when trying to predict a home's value and selling price. The number of rooms (or bathrooms!) is an important factor for buyers with children who need their own space. The size of the lot can affect the selling price of a home. There are other less tangible factors such as the type of neighborhood, the state of the economy, the interest rate on mortgages, and so on.

We began with the simple regression equation

$$y' = 4.64x + 71.0$$

$$\underset{\substack{\text{selling price} \\ \text{(in thousands} \\ \text{of dollars)}}}{\uparrow} \qquad \underset{\substack{\text{living area} \\ \text{(in hundreds of} \\ \text{square feet)}}}{\uparrow}$$

We will now expand Table 9–1 to include the annual tax bill (in thousands of dollars) corresponding to each home. The result is Table 9–5.

| TABLE 9–5 | | | | | | | | |
|---|---|---|---|---|---|---|---|---|
| Tax bill ($x_2$) | 1.9 | 3.0 | 1.4 | 1.4 | 1.5 | 1.8 | 2.4 | 4.0 |
| Living area ($x_1$) | 15 | 38 | 23 | 16 | 16 | 13 | 20 | 24 |
| Selling price ($y$) | 145 | 228 | 150 | 130 | 160 | 114 | 142 | 265 |

## PREDICTING WINE BEFORE ITS TIME

■ Princeton University economist Orley Ashenfelter applies regression analysis to use weather as a predictor of the quality and price of vintage wines. He includes these variables: rainfall preceding the growing season, average growing season temperature, and rainfall during harvest. Ashenfelter states "Predicting the quality and price of a wine could be like predicting any other market item. All you need is the right equation and the right values for your variables." He successfully tested his multiple regression equation on past results, finding that wine auction prices confirmed his prediction of quality. ■

If we calculate the linear correlation coefficient for the paired data of taxes and selling prices, we get $r = 0.883$, which is significant at the $\alpha = 0.05$ level. This suggests that a predicted selling price should take the living area *and* the tax bill into account. That is, we need a regression equation of this type:

$$y' = b + m_1(\text{living area}) + m_2(\text{tax bill})$$

or

$$y' = b + m_1x_1 + m_2x_2$$

where $b$, $m_1$, $m_2$ are constants and $x_1$, $x_2$ are variables representing the living area and the tax bill. Such equations can be found. The result here is

$$y' = 38.4 + 2.04x_1 + 39.7x_2$$

A regression equation that includes more than two variables is a **multiple regression** equation. A multiple regression equation is determined according to the same "least-squares" criterion introduced in Section 9–3. It minimizes the sum of squares $\Sigma(y - y')^2$, where $y'$ represents the predicted value of $y$ that is found through substitution in the regression equation.

The actual process of finding constants $b$, $m_1$, $m_2$ for a multiple regression equation is extremely complex. It involves calculations that are extensive, time consuming, and error-prone. In reality, almost everyone uses a computer software package such as STATDISK, Minitab, SPSS, or SAS. To get some sense for the magnitude of the problem, consider the sample data in Table 9–5. To find the values of the constants $b$, $m_1$, $m_2$ in the multiple regression equation, we must solve these equations.

$$\Sigma y = bn + m_1\Sigma x_1 + m_2\Sigma x_2$$

$$\Sigma x_1 y = b\Sigma x_1 + m_1\Sigma x_1^2 + m_2\Sigma x_1 x_2$$

$$\Sigma x_2 y = b\Sigma x_2 + m_1\Sigma x_1 x_2 + m_2\Sigma x_2^2$$

For the data in Table 9–5 we get

$$1334 = 8b + 165m_1 + 17.4m_2$$

$$29{,}611 = 165b + 3855m_1 + 388.5m_2$$

$$3197.5 = 17.4b + 388.5m_1 + 43.78m_2$$

Solving the preceding system of equations, we get $b = 38.4$, $m_1 = 2.04$, $m_2 = 39.7$. If our regression equation involves a total of four variables, we need to find the constants $b$, $m_1$, $m_2$, $m_3$ in the equation

$$y' = b + m_1x_1 + m_2x_2 + m_3x_3$$

and we would get four equations with four unknowns. Get the point? This is *messy* arithmetic and algebra. Now let's proceed on the much more reasonable assumption that we can use a computer. Shown below are the displays from STATDISK and Minitab that result from the data in Table 9–5.

## STATDISK DISPLAY

> $b = 38.35$
> $m1 = + 2.037$
> $m2 = + 39.71$
> $Y = 38.35 + 2.037\ X1 + 39.71\ X2$
> (Results may be unreliable if there is
> a high correlation between variables.)
>
> Multiple coefficient of determination (R-sq) ............... = .845933
> Adjusted R-sq ........................................ = .784306
> Standard error of estimate ........................... = 24.164

## MINITAB DISPLAY

```
MTB > READ C1 C2 C3
DATA> 1.9 15 145
DATA> 3.0 38 228
DATA> 1.4 23 150
DATA> 1.4 16 130
DATA> 1.5 16 160
DATA> 1.8 13 114
DATA> 2.4 20 142
DATA> 4.0 24 265
DATA> ENDOFDATA
        8 ROWS READ
MTB > NAME C1 'TAXES' C2 'AREA' C3 'PRICE'
MTB > REGRESSION C3 2 C2 C1
The regression equation is
PRICE = 38.4 + 2.04 AREA + 39.7 TAXES
```

| Predictor | Coef | Stdev | t-ratio | p |
|---|---|---|---|---|
| Constant | 38.35 | 26.86 | 1.43 | 0.213 |
| AREA | 2.038 | 1.386 | 1.47 | 0.201 |
| TAXES | 39.71 | 12.09 | 3.28 | 0.022 |

s = 24.16     R-sq = 84.6%     R-sq(adj) = 78.4%

Data entered here

Multiple regression ← equation

← $R^2 = 0.846$

Once a multiple regression equation has been determined, it can be used to make predictions, as in the following example.

---

**Example**

Using the sample data from Table 9–5, the multiple regression equation is found to be

$$y' = 38.4 + 2.04x_1 + 39.7x_2$$

Here $y'$ is the predicted home selling price (in thousands of dollars).
    $x_1$ is the living area (in hundreds of square feet).
    $x_2$ is the tax bill (in thousands of dollars).
Find the predicted selling price for a home with a living area of 2000 sq ft $(x_1 = 20)$ and a tax bill of $2800 $(x_2 = 2.8)$.

**Solution**

Since $x_1$ is the living area in hundreds of square feet, we represent the 2000-sq-ft area by $x_1 = 20$. Since $x_2$ is the tax bill in thousands of dollars, we represent the $2800 tax bill by $x_2 = 2.8$. We now calculate the predicted value of $y$.

$$y' = 38.4 + 2.04(20) + 39.7(2.8) = 190$$

The predicted selling price is $190,000.

---

We will not consider exact procedures for determining how much a predicted value is likely to vary. However, we can use the **multiple coefficient of determination** $R^2$ as a measure of how well the multiple regression equation fits the available data. A perfect fit would result in $R^2 = 1$. A very good fit results in a value near 1. A very poor fit will result in a value of $R^2$ close to 0. The actual value of $R^2$ can be found by calculating

$$R^2 = 1 - \frac{\Sigma(y - y')^2}{\Sigma(y - \bar{y})^2}$$

This is also a messy calculation, but the value of $R^2$ can usually be found by using the same computer software that gives us the constants in the multiple regression equation. For the data in Table 9–5, $R^2 = 0.846$. (See the STAT-DISK and Minitab displays.) This value shows that the multiple regression equation $y' = 38.4 + 2.04x_1 + 39.7x_2$ fits the sample data quite well. It also indicates that 84.6% of the variation in the selling price can be explained by the living area $x_1$ and the tax bill $x_2$. This is a substantial increase over the 51.4% of the variation explained by using the living area alone. (In Section 9–2, we used the selling price/living area pairs to get $r = 0.717$ and 0.514 is the square of 0.717.)

Let's briefly review what we have done so far. Using the sample data in Table 9–1, we began by considering the relationship between home selling price and living area. Having found a significant positive linear correlation, we proceeded to find the regression equation $y' = 4.64x + 71.0$, which relates those two variables. In this section we then proceeded to find the multiple regression equation $y' = 38.4 + 2.04x_1 + 39.7x_2$, which expresses the selling price $y$ in terms of living area $x_1$ and the annual tax bill $x_2$.

We could go on to include other relevant factors, such as the size of the lot and the number of rooms. For the homes represented in Table 9–1, the inclusion of the lot size as a third variable $x_3$ will lead to the following multiple regression equation.

$$y' = 46.6 + 1.21x_1 + 39.0x_2 + 7.22x_3$$

selling price (in thousands of dollars)    living area (in hundreds of square feet)    taxes (in thousands of dollars)    lot size (in acres)

However, the inclusion of this third variable of lot size leads to a multiple coefficient of determination given by $R^2 = 0.852$. This is only slightly better than the previous value of 0.846, and it suggests that we really didn't gain much by including the acreage. (See the *adjusted* $R^2$ in Exercise 20.) We might normally expect that larger lot sizes lead to increased values and selling prices, but this is not true here. For the region being considered, some large lots are in rural areas where homes are less expensive. Other large lots are in more populated areas where the prices are higher. These conflicting patterns produce a lower correlation between the lot size and the selling price. Because of the negligible increase in $R^2$, we should not include the lot size variable $x_3$ in our multiple regression equations. In general, the best multiple regression equation has the following properties:

1. It includes relatively few independent variables.

2. $R^2$ is largest for the given number of independent variables. That is, no other combination of the same number of independent variables will yield a larger value of $R^2$.

3. The inclusion of another independent variable will not lead to a substantial increase in $R^2$.

For cases involving a large number of independent variables, many statistical software packages include a program for performing **stepwise regression,** whereby different combinations are tried until the best model is obtained.

Here is another example of the use of multiple regression: Illinois State University conducted a study to identify the factors that affect the academic successes of students in their first calculus course. (See "Factors Affecting Achievement in the First Course in Calculus" by Edge and Friedberg, *Journal of Experimental Education*, Vol. 52, No. 3.) The calculus course had high

# RISING TO NEW HEIGHTS

■ The concept of regression was introduced in 1877 by Sir Francis Galton. His research showed that when tall parents have children, those children tend to have heights that "regress," or revert to the mean height of the population.

Stanford University's Dr. Darrell Wilson reported on a correlation between height and intelligence. Taller children tend to earn higher scores on intelligence tests as well as on achievement tests. The correlation was found to be significant, but the test scores of tall people were not very much higher than those of shorter people. ■

failure and dropout rates. It was hoped that the study would identify the most important characteristics for success so that better placement would improve the situation. The study included factors such as American College Test (ACT) scores, high school rank, high school average, high school algebra grades, scores on an algebra placement test, sex, birth order, family size, and size of high school. The best multiple regression equation was

$$y' = 34.8 + 1.21x_1 + 0.23x_2$$

| | | |
|---|---|---|
| ↑ | ↑ | ↑ |
| calculus grade | score on algebra placement test | high school rank (percentile) |

This multiple regression equation shows that the algebra placement test score and high school rank can be used to predict the calculus grade, and these two variables became factors in placing students. The algebra score appears to measure algebra skills, while the high school rank is a measure of long-term perseverance and competitiveness. The end result is an improved placement procedure that helps students as well as faculty. This is an ideal use of statistics, whereby people are helping people. (Does that sound too much like a commercial for General Electric?)

When we discussed regression in Section 9–3, we listed four common errors that should be avoided when using regression equations to make predictions. These same errors should be avoided when using multiple regression equations. We should be especially careful about ascribing causal relationships. In this section we expressed the selling price in terms of the living area and taxes. It's reasonable to claim that a home's selling price will be increased if its living area is increased. But a homeowner would be foolish to think that a higher selling price can be obtained by getting the taxes raised. Here, the annual tax bill may be an important characteristic that's helpful in *predicting* a home's selling price, but a change in taxes will not necessarily *cause* a change in the selling price. Be very wary of claiming that a variable has a cause-effect relationship with another variable.

# 9–5 Exercises A

In Exercises 1–4, use the following regression equation.

$$y' = 34.8 + 1.21x_1 + 0.23x_2$$

Here $y'$ is the predicted calculus grade, $x_1$ is the score on an algebra placement test, and $x_2$ is the high school rank expressed as a percentile. (The equation is based on data from "Factors Affecting Achievement in the First Course in Calculus" by Edge and Friedberg, *Journal of Experimental Education*, Vol. 52, No. 3.)

1. Find the predicted calculus grade if the score on the algebra pretest $(x_1)$ is 24 and the high school rank is the 92nd percentile.
2. Find the predicted calculus grade if the score on the algebra pretest $(x_1)$ is 12 and the high school rank is the 71st percentile.
3. Find the predicted calculus grade if the score on the algebra pretest $(x_1)$ is 18 and the high school rank is the 81st percentile.
4. Find the predicted calculus grade if the score on the algebra pretest $(x_1)$ is 31 and the high school rank is the 99th percentile.

In Exercises 5–8, refer to the accompanying Minitab display to answer the given questions.

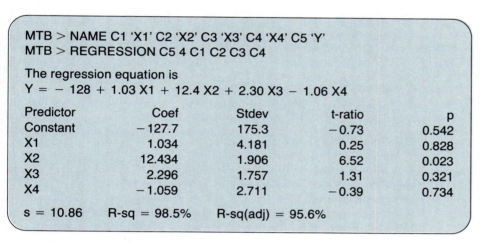

```
MTB > NAME C1 'X1' C2 'X2' C3 'X3' C4 'X4' C5 'Y'
MTB > REGRESSION C5 4 C1 C2 C3 C4

The regression equation is
Y = - 128 + 1.03 X1 + 12.4 X2 + 2.30 X3 - 1.06 X4

Predictor          Coef          Stdev        t-ratio            p
Constant          - 127.7         175.3        - 0.73          0.542
X1                 1.034          4.181          0.25          0.828
X2                12.434          1.906          6.52          0.023
X3                 2.296          1.757          1.31          0.321
X4                - 1.059         2.711        - 0.39          0.734

s = 10.86      R-sq = 98.5%     R-sq(adj) = 95.6%
```

5. What multiple regression equation is suggested by the Minitab display?
6. What is the value of the multiple coefficient of determination?
7. Find the predicted value of $y$ given that $x_1 = 7$, $x_2 = 20$, $x_3 = 75$, and $x_4 = 42$.
8. Find the predicted value of $y$ given that $x_1 = 9$, $x_2 = 29$, $x_3 = 80$, and $x_4 = 40$.

In Exercises 9–12, use Table 9–6 and software such as STATDISK or Minitab to find the indicated multiple regression equation and multiple coefficient of determination. Also, find the predicted $y$ value for $x_1 = 2$, $x_2 = 9$, $x_3 = 1$.

9. Express the variable $y$ in terms of $x_1$ and $x_2$.
10. Express the variable $y$ in terms of $x_1$ and $x_3$.
11. Express the variable $y$ in terms of $x_2$ and $x_3$.
12. Express the variable $y$ in terms of $x_1$, $x_2$, and $x_3$.

| TABLE 9–6 | | | | |
|---|---|---|---|---|
| $y$ | 25 | 13 | 16 | 13 | 10 |
| $x_1$ | 5 | 2 | 1 | 3 | 4 |
| $x_2$ | 10 | 8 | 7 | 11 | 7 |
| $x_3$ | 2 | 3 | 4 | 3 | 1 |

In Exercises 13–16, use the sample data given in Table 9–7 and software such as STATDISK or Minitab. The data are based on recent sales of homes in Dutchess County, New York. Selling prices and taxes are in thousands of dollars. The living areas are in hundreds of square feet and the acreage amounts are in acres.

| TABLE 9–7 | | | | | | | | |
|---|---|---|---|---|---|---|---|---|
| Selling price | 145 | 228 | 150 | 130 | 160 | 114 | 142 | 265 |
| Living area | 15 | 38 | 23 | 16 | 16 | 13 | 20 | 24 |
| Taxes | 1.9 | 3.0 | 1.4 | 1.4 | 1.5 | 1.8 | 2.4 | 4.0 |
| Acreage | 2.0 | 3.6 | 1.8 | 0.53 | 0.50 | 0.31 | 0.75 | 2.0 |
| Rooms | 5 | 11 | 9 | 7 | 7 | 7 | 9 | 7 |

13. a. Find the multiple regression equation that expresses the tax bill in terms of selling price and living area.
    b. Find the value of the multiple coefficient of determination obtained by using the tax, selling price, and living area data.
    c. Compare the results from parts $a$ and $b$ to those found by expressing the selling price in terms of living area and tax bill. ($R^2 = 0.846$ and $y' = 38.4 + 2.04x_1 + 39.7x_2$.)

14. a. Find the multiple regression equation that expresses the selling price in terms of the living area, acreage, and number of rooms.
    b. Find the value of the multiple coefficient of determination obtained by using the same four variables listed in part $a$.

15. a. Find the multiple regression equation that expresses the taxes in terms of the living area, acreage, and number of rooms.
    b. Find the value of the multiple coefficient of determination obtained by using the same four variables listed in part $a$.

16. a. Find the multiple regression equation that expresses the selling price in terms of living area, taxes, acreage, and number of rooms.
    b. Find the value of the multiple coefficient of determination obtained by using the same five variables listed in part $a$.

# 9–5 Exercises B

17. In some cases, the best-fitting multiple regression equation is of the form $y = b + m_1x + m_2x^2$. The graph of such an equation is a parabola. Let $x_1 = x$ and let $x_2 = x^2$. Use the values of $y$, $x_1$, and $x_2$ to find the

multiple regression equation for the parabola that best fits the data given below. Based on the value of the multiple coefficient of determination, how well does this equation fit the given data?

| $x$ | 1 | 3 | 4 | 7 | 5 |
|---|---|---|---|---|---|
| $y$ | 5 | 14 | 19 | 42 | 26 |

18. a. Given the paired data below, find the linear correlation coefficient and the equation of the regression line.

    b. Use the paired data below to find the multiple regression equation

$$y = b + m_1 x + m_2 x^2$$

(*Hint:* Let $x_1 = x$ and let $x_2 = x^2$.) Also find the value of the multiple coefficient of determination.

    c. Use the paired data below to find the multiple regression equation

$$y = b + m_1 x + m_2 x^2 + m_3 x^3$$

(Let $x_1 = x$, $x_2 = x^2$, $x_3 = x^3$.) Also find the value of the multiple coefficient of determination.

    d. Use the paired data below to find the multiple regression equation

$$y = b + m_1 x + m_2 x^2 + m_3 x^3 + m_4 x^4$$

Also find the value of the multiple coefficient of determination.

    e. Based on the preceding results, which equation best fits the given data?

| $x$ | $-2.0$ | $-1.0$ | 0.0 | 1.0 | 2.0 | 3.0 |
|---|---|---|---|---|---|---|
| $y$ | 13 | 4.0 | 5.0 | 4.0 | 13 | 68 |

19. We noted that with three variables $(y, x_1, x_2)$, solution of the equations

$$\Sigma y = bn + m_1 \Sigma x_1 + m_2 \Sigma x_2$$

$$\Sigma x_1 y = b\Sigma x_1 + m_1 \Sigma x_1^2 + m_2 \Sigma x_1 x_2$$

$$\Sigma x_2 y = b\Sigma x_2 + m_1 \Sigma x_1 x_2 + m_2 \Sigma x_2^2$$

will lead to the values of $b$, $m_1$, $m_2$ in the multiple regression equation

$$y = b + m_1 x_1 + m_2 x_2$$

Show that if we have only the two variables $y$ and $x_1$, the solution of the preceding system of equations will lead to the formulas for the slope and $y$-intercept of the regression line discussed in Section 9–3.

20. When we are trying to use different combinations of independent variables for finding the best multiple regression equation, the value of $R^2$ alone is inadequate because the largest $R^2$ is achieved by using all of the variables. The **adjusted $R^2$** is often used instead.

*continued*

$$\text{Adjusted } R^2 = 1 - \frac{(n-1)}{[n-(k+1)]}(1 - R^2)$$

where $k$ = number of independent variables included
$n$ = number of observations

For the real estate data considered in this section, we noted that the inclusion of lot size as a third independent variable $x_3$ increases $R^2$ from 0.846 to 0.852. Find the adjusted $R^2$ values corresponding to those two $R^2$ values that are based on 8 observations. Based on the adjusted $R^2$ values, should lot size be included in the multiple regression equation?

 ## *Vocabulary List*

Define and give an example of each term.

| | |
|---|---|
| bivariate data | total deviation |
| scatter diagram | explained deviation |
| correlation | unexplained deviation |
| linear correlation coefficient | total variation |
| centroid | explained variation |
| Pearson's product moment | unexplained variation |
| bivariate normal distribution | coefficient of determination |
| regression line | standard error of estimate |
| slope | multiple regression |
| y-intercept | multiple coefficient of |
| predicted value | determination |
| least-squares property | stepwise regression |

 ## *Review*

In this chapter we studied the concepts of **linear correlation** and **regression** so that we could analyze paired sample data. We limited our discussion to linear relationships because consideration of nonlinear relationships requires more advanced mathematics. With correlation, we attempted to decide whether there is a significant linear relationship between the two variables. With regression, we attempted to specify what that relationship is. While a scatter diagram provides a graphic display of the paired data, the linear correlation coefficient $r$ and the equation of the regression line serve as more precise and objective tools for analysis.

Given a list of paired data, we can compute the linear correlation coefficient $r$ by using Formula 9–1. We can use the Student $t$ distribution or Table A–6 to decide whether there is a significant linear relationship. The presence of a significant linear correlation does not necessarily mean that there is a direct cause-and-effect relationship between the two variables.

In Section 9–3 we developed procedures for obtaining the equation of the regression line which, by the least-squares criterion, is the straight line that best fits the paired data. When there is a significant linear correlation, the regression line can be used to predict the value of one variable when given some value of the other variable. The regression line has the form $y' = mx + b$, where the constants $m$ and $b$ can be found by using the formulas given in this section.

In Section 9–4 we introduced the concept of **total variation,** with components of explained and unexplained variation. We defined the coefficient of determination $r^2$ to be the quotient of explained variation by total variation. We saw that we could measure the amount of spread of the sample points about the regression line by the standard error of estimate, $s_e$. We developed **prediction intervals** for estimated values of $y$.

In Section 9–5 we considered **multiple regression,** which allows us to investigate relationships among several variables. We discussed procedures for obtaining a multiple regression equation as well as the value of the multiple coefficient of determination $R^2$. That value gives us an indication of how well the multiple regression equation actually fits the available sample data. Due to the nature of the calculations involved, finding the multiple regression equation and the multiple coefficient of determination usually requires the use of computer software, such as STATDISK or Minitab.

## Important Formulas

$$r = \frac{n\Sigma xy - (\Sigma x)(\Sigma y)}{\sqrt{n(\Sigma x^2) - (\Sigma x)^2}\sqrt{n(\Sigma y^2) - (\Sigma y)^2}} \quad \text{or} \quad r = \frac{ms_x}{s_y}$$

$$m = \frac{n\Sigma xy - (\Sigma x)(\Sigma y)}{n(\Sigma x^2) - (\Sigma x)^2} \quad \text{or} \quad m = r\frac{s_y}{s_x}$$

$$b = \frac{(\Sigma y)(\Sigma x^2) - (\Sigma x)(\Sigma xy)}{n(\Sigma x^2) - (\Sigma x)^2} \quad \text{or} \quad b = \bar{y} - m\bar{x}$$

$$y' = mx + b$$

$$r^2 = \frac{\text{explained variation}}{\text{total variation}}$$

$$s_e = \sqrt{\frac{\Sigma(y - y')^2}{n - 2}} \quad \text{or} \quad \sqrt{\frac{\Sigma y^2 - b\Sigma y - m\Sigma xy}{n - 2}}$$

$$y' - E < y < y' + E$$

$$\text{where } E = t_{\alpha/2}s_e\sqrt{1 + \frac{1}{n} + \frac{n(x_0 - \bar{x})^2}{n(\Sigma x^2) - (\Sigma x)^2}}$$

# ? *Review Exercises*

1. In each of the following, determine whether correlation or regression analysis is more appropriate.
   a. Is the value of a car related to its age?
   b. What is the relationship between the age of a car and annual repair costs?
   c. How are Celsius and Fahrenheit temperatures related?
   d. Is the age of a car related to annual repair costs?
   e. Is there a relationship between cigarette smoking and lung cancer?

2. a. What should you conclude if 40 pairs of data produce a linear correlation coefficient of $r = -0.508$?
   b. What should you conclude if 10 pairs of data produce a linear correlation coefficient of $r = 0.608$?

In Exercises 3–8, use the given paired data.

   a. Find the value of the linear correlation coefficient $r$.
   b. Assuming a 5% level of significance, find the critical value of $r$ from Table A–6.
   c. Use the results from parts $a$ and $b$ to decide whether there is a significant linear correlation.
   d. Find the equation of the regression line.
   e. Plot the regression line on the scatter diagram.

3. Some high school students from Dutchess County applied for a summer enrichment program. The PSAT verbal and math scores for a dozen randomly selected male applicants are given below.

| Verbal | 43 | 68 | 55 | 57 | 50 | 55 | 55 | 57 | 75 | 53 | 43 | 45 |
|--------|----|----|----|----|----|----|----|----|----|----|----|----|
| Math   | 59 | 76 | 58 | 54 | 61 | 63 | 58 | 60 | 67 | 76 | 57 | 63 |

4. A manager in a factory randomly selects 15 assembly-line workers and develops scales to measure their dexterity and productivity levels. The results are listed in the following table.

| Productivity | 63 | 67 | 88 | 44 | 52 | 106 | 99 | 110 | 75 | 58 | 77 | 91 | 101 | 51 | 86 |
|--------------|----|----|----|----|----|-----|----|-----|----|----|----|----|-----|----|----|
| Dexterity    | 2  | 9  | 4  | 5  | 8  | 6   | 9  | 8   | 9  | 7  | 4  | 10 | 7   | 4  | 6  |

5. Randomly selected subjects are given a standard IQ test and then tested for their receptivity to hypnosis. The results are listed below.

| IQ | 103 | 113 | 119 | 107 | 78 | 153 | 114 | 101 | 103 | 111 | 105 | 82 | 110 | 90 | 92 |
|----|-----|-----|-----|-----|----|-----|-----|-----|-----|-----|-----|----|-----|----|----|
| Receptivity to hypnosis | 55 | 55 | 59 | 64 | 45 | 72 | 42 | 63 | 62 | 46 | 41 | 49 | 57 | 52 | 41 |

6. The table below lists the values (in billions of dollars) of U.S. exports and incomes on foreign investments for various sample years.

| Exports | 16 | 20 | 27 | 39 | 56 | 63 |
|---|---|---|---|---|---|---|
| Incomes on foreign investments | 2 | 3 | 4 | 7 | 11 | 11 |

7. The given paired data lists the heights (in inches) of various males who were measured on their eighth and sixteenth birthdays.

| Height (8 yrs) | 52 | 50 | 48 | 51 | 51 | 48 |
|---|---|---|---|---|---|---|
| Height (16 yrs) | 69 | 66 | 64 | 67 | 69 | 65 |

8. A test designer develops two separate tests, each intended to measure a person's level of creative thinking. Randomly selected subjects were given both versions and their results follow.

| Test A | 85 | 97 | 100 | 76 | 80 | 116 | 120 | 105 |
|---|---|---|---|---|---|---|---|---|
| Test B | 92 | 109 | 100 | 74 | 85 | 118 | 125 | 90 |

In Exercises 9–16, use Table 9–8. The sample data in the table have been obtained from the New York State Department of Motor Vehicles and they represent randomly selected counties for a one-year period.

| **TABLE 9–8** | | | | | | | | | |
|---|---|---|---|---|---|---|---|---|---|
| County | A | B | C | D | E | F | G | H | I |
| Fatalities | 71 | 37 | 52 | 212 | 167 | 27 | 15 | 169 | 31 |
| Property accidents | 3,988 | 2,042 | 1,117 | 11,217 | 12,587 | 895 | 372 | 7,536 | 570 |
| Number of convictions | 71,676 | 30,361 | 22,497 | 130,771 | 116,937 | 23,357 | 9,698 | 192,951 | 9,826 |
| Registered cars | 410,938 | 167,562 | 102,307 | 927,462 | 947,765 | 96,135 | 47,143 | 435,456 | 56,731 |
| Registered motorcycles | 12,108 | 5,382 | 4,227 | 17,274 | 11,608 | 4,393 | 3,167 | 3,791 | 3,399 |
| Population | 719,949 | 261,317 | 165,269 | 1,351,094 | 1,335,697 | 164,060 | 89,486 | 2,238,961 | 122,685 |
| Licensed drivers | 479,423 | 176,622 | 114,850 | 947,886 | 982,936 | 116,193 | 62,068 | 733,557 | 77,214 |

9. a. Using the fatalities and the numbers of licensed drivers, find the value of the linear correlation coefficient $r$.

*continued*

    b.   Assuming a 0.05 level of significance, find the critical value of $r$ from Table A–6.

    c.   Use the results from parts $a$ and $b$ to decide whether there is a significant linear correlation.

    d.   Find the equation of the regression line. (Let $y$ = number of fatalities and $x$ = number of licensed drivers.)

10.  a.   Using the fatalities and the number of registered cars, find the value of the linear correlation coefficient $r$.

    b.   Assuming a 0.05 level of significance, find the critical value of $r$ from Table A–6.

    c.   Use the results from parts $a$ and $b$ to decide whether there is a significant linear correlation.

    d.   Find the equation of the regression line. (Let $y$ = number of fatalities and $x$ = number of registered cars.)

11.  a.   Using the fatalities and the number of property damage accidents, find the value of the linear correlation coefficient $r$.

    b.   Assuming a 0.05 level of significance, find the critical value of $r$ from Table A–6.

    c.   Use the results from parts $a$ and $b$ to decide whether there is a significant linear correlation.

    d.   Find the equation of the regression line. (Let $y$ = number of fatalities and $x$ = number of property damage accidents.)

12.  a.   Using the fatalities and the number of motor vehicle convictions, find the value of the linear correlation coefficient $r$.

    b.   Assuming a 0.05 level of significance, find the critical value of $r$ from Table A–6.

    c.   Use the results from parts $a$ and $b$ to decide whether there is a significant linear correlation.

    d.   Find the equation of the regression line. (Let $y$ = number of fatalities and $x$ = number of motor vehicle convictions.)

13.  a.   Let $y$ = fatalities, $x_1$ = number of property damage accidents, and $x_2$ = number of registered motorcycles. Use software such as STATDISK or Minitab to find the multiple regression equation of the form $y' = b + m_1 x_1 + m_2 x_2$.

    b.   Find the multiple coefficient of determination.

    c.   Based on the result of part $b$, how well does the multiple regression equation fit the sample data?

14.  a.   Let $y$ = number of property damage accidents, $x_1$ = population, and $x_2$ = number of registered cars. Use software such as STATDISK or Minitab to find the multiple regression equation of the form $y' = b + m_1 x_1 + m_2 x_2$.

    b.   Find the multiple coefficient of determination.

    c.   Based on the result of part $b$, how well does the multiple regression equation fit the sample data?

15. a.  Let $y$ = number of property damage accidents, $x_1$ = number of licensed drivers, and $x_2$ = number of motor vehicle convictions. Use software such as STATDISK or Minitab to find the multiple regression equation of the form $y' = b + m_1x_1 + m_2x_2$.
    b.  Find the multiple coefficient of determination.
    c.  Based on the result of part $b$, how well does the multiple regression equation fit the sample data?

16. a.  Let $y$ = number of property damage accidents, $x_1$ = number of licensed drivers, and $x_2$ = number of fatalities. Use software such as STATDISK or Minitab to find the multiple regression equation of the form $y' = b + m_1x_1 + m_2x_2$.
    b.  Find the multiple coefficient of determination.
    c.  Based on the result of part $b$, how well does the multiple regression equation fit the sample data?

In Exercises 17–20, use the given data set to find (a) explained variation, (b) unexplained variation, (c) total variation, (d) coefficient of determination, and (e) standard error of estimate $s_e$.

17. 

| $x$ | 1 | 2 | 4 | 6 |
|---|---|---|---|---|
| $y$ | 3 | 0 | 1 | 5 |

18. 

| $x$ | 2 | 5 | 4 | 3 | 7 |
|---|---|---|---|---|---|
| $y$ | 5 | 11 | 9 | 7 | 15 |

19. 

| $x$ | 4 | 2 | 1 | 6 |
|---|---|---|---|---|
| $y$ | 2 | 4 | 5 | 0 |

20. 

| $x$ | 8 | 4 | 2 | 9 | 6 |
|---|---|---|---|---|---|
| $y$ | 2 | 7 | 7 | 1 | 4 |

## Computer Projects

1.  Refer to the Appendix B data on homes recently sold in Dutchess County, New York. Find the multiple regression equation that relates the selling price $(y)$ to the living area $(x_1)$, the acreage $(x_2)$, the number of rooms $(x_3)$, and the number of baths $(x_4)$.

2.  Refer to the sample data in Table 9–8 found among the review exercises for this chapter. Assume that you want a multiple regression equation with $y'$ corresponding to the number of fatalities.
    a.  Exclude the numbers of property accidents for this reason: There is an extremely high $(r = 0.989)$ correlation between the numbers of property accidents and the numbers of licenses, so only

one of these variables should be included. (We chose the numbers of licenses because the fatalities/license correlation is $r = 0.964$, which is slightly better than the fatalities/property accidents correlation of $r = 0.951$.)

b.  Exclude any other variables that should be excluded by the same reasoning used in part *a*.

c.  After excluding the variables from parts *a* and *b*, find the multiple regression equation that results in the largest value of *adjusted* $R^2$. (See Exercise 20 in Section 9–5.) Identify that equation along with the values of $R^2$ and adjusted $R^2$.

3.  Refer to Data Set II in Appendix B and randomly select 20 right-handed females. Assume that you want to predict sitting heights based on these other measurements: weight (Wt), height (Ht), bitrochanteric breadth (Bitro), elbow length (Elbow), and upper arm girth (UAG). Experiment with different multiple regression equations and find the one that results in the largest value of adjusted $R^2$. (See Computer Project 2 above.)

 # *Applied Projects*

1.  Refer to Data Set II in Appendix B. Use the methods presented in this chapter to investigate relationships between bitrochanteric breadth (Bitro) and upper arm girth (UAG) among right-handed males.

2.  Refer to Data Set II in Appendix B. Use the methods in this chapter to investigate relationships between height (Ht) and sitting height (Sit) among right-handed females.

3.  Conduct a study to determine whether there is a correlation between two variables. Collect your own sample paired data consisting of at least 15 different pairs. Identify the population and the method used to obtain the data, and identify any factors suggesting that the sample data are not representative. Calculate the linear correlation coefficient and determine the equation of the regression line. Graph the regression line on the scatter diagram. Assuming a 0.05 level of significance, find the critical value and form a conclusion about the correlation between the two variables.

 # *Writing Projects*

1.  In studying the effects of heredity and environment on intelligence, it has been helpful to analyze IQs of identical twins that were separated soon after birth. Identical twins share identical genes inherited from the same fertilized egg. By studying identical twins raised apart, we

can eliminate the variable of heredity and better isolate the effects of environment. Given below are the IQs of identical twins (older twins are $x$) raised apart (based on data from "IQs of Identical Twins Reared Apart" by Arthur Jensen, *Behavioral Genetics*). Find the linear correlation coefficient $r$, the coefficient of determination $r^2$, and the equation of the regression line. Then construct a scatter diagram. Based on the results, write a report about the effect of heredity and environment on intelligence. Note that your conclusions are based on this relatively small sample of 12 pairs of identical twins.

| $x$ | 107 | 96 | 103 | 90 | 96 | 113 | 86 | 99 | 109 | 105 | 96 | 89 |
|---|---|---|---|---|---|---|---|---|---|---|---|---|
| $y$ | 111 | 97 | 116 | 107 | 99 | 111 | 85 | 108 | 102 | 105 | 100 | 93 |

2.  Write a report summarizing one of the programs listed below.

# Videotapes

Programs 8, 9, 11, and 25 from the series *Against All Odds: Inside Statistics* are recommended as supplements to this chapter.

Photo courtesy of Young & Rubicam Advertising

## AN INTERVIEW WITH
## JAY DEAN

### SENIOR VICE PRESIDENT AT THE SAN FRANCISCO OFFICE OF YOUNG & RUBICAM ADVERTISING

*Jay Dean is Director of the Consumer Insights Department at the San Francisco office of Young and Rubicam. He has worked with many well-known advertisers, including AT&T, Chevron, Clorox, Coors, Dr. Pepper, Ford Motor Co., General Foods, Gillette, Gulf Oil, H.J. Heinz, Kentucky Fried Chicken, and Warner-Lambert.*

**How extensive is your use of statistics at Young & Rubicam?**

We use statistics every day. I'm now working on a typical design, an advertising test for Take Heart salad dressing, one of the brands we handle for The Clorox Company. Two commercials were shown to independent samples of consumers. We then asked questions about what was being communicated by each commercial, perceptions of the brand, likes and dislikes, and so on. Significance testing will be used to compare the results and to choose the best commercial.

On larger surveys we use multivariate techniques such as factor analysis and multiple regression. Marketing is becoming tougher and tougher today. There are more brands, increasing price competition, and more sophisticated consumers. A tougher marketing environment requires more marketing research, and that means we're using statistics more often as well.

**Could you cite a case in which the use of statistics was instrumental in determining a successful strategy?**

We recently conducted a major creative exploratory for Pine-Sol, another Clorox brand. About half a dozen commercials were produced in rough form and shown to independent samples of consumers. Statistics helped us to identify the best commercial for the new Pine-Sol advertising campaign. Early marketplace results are very encouraging. At Y&R we believe consumer research helps us to produce the most effective advertising. Statistics help us to make informed decisions based upon the results of the research.

### How do you typically collect data for statistical analysis?

There are many ways to collect data, but typically it's either random telephone sampling for survey work, or if you want to show people something like a commercial, we use a central location interview. For example, respondents are intercepted in a shopping mall by an interviewer, screened for eligibility, asked to go into a testing facility to be shown a commercial, and so on.

> *"If I could go back to school, I would certainly study more math, statistics, and computer science."*

### Do you find it difficult to obtain representative and unbiased samples?

Yes. The marketing research industry now is very concerned about the growing refusal rate among the general public. That is driven in part by salespeople who use marketing research as a guise for selling. Mail surveys are subject to a huge self-selection bias and response rates are typically quite low. Shopping mall interviews have other problems. Not everyone goes to malls, and there is a certain degree of interviewer bias in approaching prospective respondents.

### What is your typical sample size?

For a national survey, the rule of thumb is about a thousand people, although you might go with as few as 600 in some cases. For an advertising test, a sample on the order of 200 would be a pretty healthy sample size—sometimes as few as a hundred.

### Do you feel that job applicants in your field are viewed more favorably if they have studied some statistics?

Yes, absolutely. Everyone has to understand and use statistics at some level. It's very important for entry-level people to know statistics well because they do much of our research project work.

### Do you have any advice for today's students?

If I could go back to school, I would certainly study more math, statistics, and computer science. I studied a lot of it, but I would like to know even more. There's an enormous data explosion in business these days. All businesses are becoming much more quantitative than they ever were in the past. Today, there is more information than you know what to do with, and you've got to have the analytical tools and knowledge to deal with this information if you want to be successful. ■

# Chapter Ten

## In This Chapter

# 10 Multinomial Experiments and Contingency Tables

## Chapter Problem

Many people (especially nonsmokers) believe that smoking is un-healthy. In a study of 1000 randomly selected deaths of males aged 45–64, the causes of death are listed along with their smoking habits (see Table 10–1, which is based on data from "Chartbook on Smoking, Tobacco, and Health," USDHEW publication CDC75-7511).

| TABLE 10–1 | | | |
|---|---|---|---|
| | Cause of Death | | |
| | Cancer | Heart disease | Other |
| Smoker | 135 | 310 | 205 |
| Nonsmoker | 55 | 155 | 140 |

What can we conclude from this data? We cannot conclude that smoking causes deaths, because *every one* of the 1000 subjects died. There are two issues we might explore using the sample data.

*continued*

1.  *Smokers have a disproportionately large representation among the 1000 deaths.* Table 10–1 shows that among 1000 randomly selected deaths of males aged 45–64, a total of 650 (or 65%) were smokers. Other data from the U.S. National Center for Health Statistics show that only 45% of males in the 45–64 age bracket are smokers. If tobacco lobbyists are correct when they claim that smoking doesn't affect health, we would expect 45% of the deceased to be smokers, not the 65% that the table shows. (This discrepancy from what we expect can be shown to be statistically significant. We use the hypothesis testing procedure given in Section 7–5 to test the claim that the proportion of deaths by smokers is $p = 0.45$; the sample proportion of $\hat{p} = 0.65$ leads to a test statistic of $z = 12.71$, which is well into the critical region for usual significance levels, so we reject the claim that the proportion is 0.45.)

2.  *The cause of death seems to be related to whether or not the men smoked.* In this chapter we will analyze tables such as Table 10-1 as we test the claim that smoking is independent of the cause of death. In general, we will test the claim that the row variable is independent of the column variable.

    Note that we can use one method from Chapter 7 to investigate one issue (whether smoking is unhealthy) and we can use another method from this chapter to investigate a different issue (whether smoking is related to the cause of death). Given sample data, there are often different questions that can be considered using different statistical methods.

## 10–1 Overview

The nature and configuration of data dramatically affect the type of statistical analyses that can and should be used. This chapter considers categorical data consisting of frequency counts for different categories. The configuration of those frequency counts will either be a single row (or column) or a table.

In Section 10–2 we begin by considering a method for testing a hypothesis made about several population proportions. Instead of working with the sample proportions, we deal directly with the frequencies with which the events occur. Our objective is to test for the significance of the differences between observed frequencies and the frequencies we would expect in theory. Since we test for how well an observed frequency distribution conforms to (or "fits") some theoretical frequency distribution, this procedure is often referred to as a **goodness-of-fit test.** We introduce the test statistic that measures the differences between observed frequencies and expected frequencies.

In Section 10–3 we analyze tables of frequencies called **contingency tables** or **two-way tables.** In these tables the rows represent categories of one variable, while the columns represent categories of another variable. We test the hypothesis that the two classification variables are independent. We will determine whether there is a statistically significant difference between the observed sample frequencies and the frequencies we would expect in theory if the two variables are independent. Contingency tables are *extremely* important in many different fields, especially the social sciences.

The two major topics (multinomial experiments and contingency tables) of this chapter share some common elements. They both have test statistics that are approximated by the chi-square distribution used earlier in Chapters 6 and 7. We should recall these important properties of the chi-square distribution:

1. Unlike the normal and Student $t$ distributions, the chi-square distribution is not symmetric. (See Figure 10–1.)

2. The values of the chi-square distribution can be 0 or positive, but they cannot be negative. (See Figure 10–1.)

3. The chi-square distribution is different for each number of degrees of freedom. (See Figure 10–2.)

Critical values of the chi-square distribution are found in Table A–4.

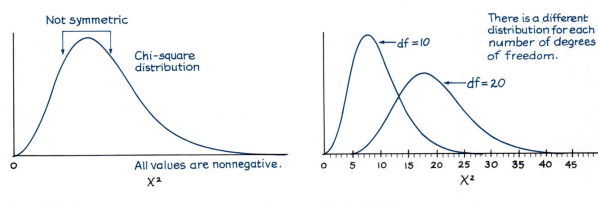

**Figure 10–1**                    **Figure 10–2**

# 10–2  Multinomial Experiments

In Chapter 4 we introduced the binomial probability distribution and indicated that each trial must have all outcomes classified into exactly one of two categories. This requirement for two categories is reflected in the prefix *bi*, which begins the term *binomial*. In this section we consider **multinomial experiments,** which require each trial to yield outcomes belonging to one of several categories. Except for this difference, binomial and multinomial experiments are essentially the same.

## DANGEROUS TO YOUR HEALTH

■ In 1965, the first warning labels were put on cigarette packs, and, in 1969, cigarette commercials were banned from television and radio. Reports of the surgeon general include many convincing statistics showing that smoking greatly increases the danger of cancer. The Tobacco Institute contests these reports. Horace Kornegay, the Tobacco Institute's chairman, says, "While many people believe a causal link between smoking and cancer is a given, scientific research has not been able to establish that link." Critics note that not a single case has shown smoking to be the cause of cancer, but the evidence clearly indicates a significant correlation. ■

### Definition

A **multinomial experiment** is one that meets these conditions.

1. There is a fixed number of trials.
2. The trials are independent.
3. Each trial must have all outcomes classified into exactly one of several categories.
4. The probabilities remain constant for each trial.

We have already discussed methods for testing hypotheses made about one population proportion (Section 7–5) and two population proportions (Section 8–4). However, there is often a need to deal with cases involving a population with more than two proportions. Consider the case involving a study of 147 industrial accidents that required medical attention. The sample data are summarized in Table 10–2, and we will test the claim that accidents occur on the five days with equal frequencies. (The data are based on results from "Counted Data CUSUM's" by Lucas, *Technometrics*, Vol. 27, No. 2.)

| TABLE 10–2 | | | | | |
|---|---|---|---|---|---|
| Day | Mon | Tues | Wed | Thurs | Fri |
| Observed accidents | 31 | 42 | 18 | 25 | 31 |

If accidents occur with equal frequencies on the five different days, the 147 accidents would average out to 29.4 per day. Table 10–3 lists the observed frequencies along with the theoretically expected values.

| TABLE 10–3 | | | | | |
|---|---|---|---|---|---|
| Day | Mon | Tues | Wed | Thurs | Fri |
| Observed accidents | 31 | 42 | 18 | 25 | 31 |
| Expected accidents | 29.4 | 29.4 | 29.4 | 29.4 | 29.4 |

We know that samples deviate from what we theoretically expect, so we now present the key question: Are the differences between the actual *observed* values and the theoretically *expected* values differences that occur just by chance, or are the differences statistically significant? To answer this question we need some way of measuring the significance of the differences between the observed values and the theoretical values. **The expected frequency of an outcome is the product of the probability of that outcome and the total number of trials.** If there are 5 possible outcomes that are supposed to be equally likely, the probability of each outcome is $\frac{1}{5}$. For 147 trials and 5 equally likely outcomes, the expected frequency of each outcome is $\frac{1}{5} \times 147 = 29.4$.

In testing for the differences among the five sample proportions, one approach might be to compare them two at a time by using the methods given in Section 8–4, but that would involve ten different pairings, ten different hypothesis tests, and a serious distortion of the significance level. Instead of pairing off samples, we will develop one comprehensive test that uses all of the data. We will use a test statistic that measures the disagreement between the observed frequencies and the frequencies we theoretically expect. This test statistic, based on observed and expected frequencies (instead of proportions), will use the following notation.

---

### Notation

$O$ represents the **observed frequency** of an outcome.

$E$ represents the theoretical or **expected frequency** of an outcome.

$k$ represents the **number of different categories** or outcomes.

---

From Table 10–3, we see that the $O$ values are 31, 42, 18, 25, and 31, and the corresponding values of $E$ are 29.4, 29.4, 29.4, 29.4, and 29.4. Since there are 5 categories, $k = 5$. The method generally used in testing for agreement between $O$ and $E$ values is based on the following test statistic.

---

### Test Statistic

$$\chi^2 = \sum \frac{(O - E)^2}{E}$$

---

Simply summing the differences between observed and expected frequencies would not lead to a good measure, since that sum is always 0.

$$\Sigma(O - E) = \Sigma O - \Sigma E = n - n = 0$$

Squaring the $O - E$ values provides a better statistic, which does reflect the differences between observed and expected frequencies, but $\Sigma(O - E)^2$ reflects only the magnitude of the differences; we need the magnitude of the differences relative to what was expected. We get it through division by the expected frequencies. For our accident data, Table 10–3 shows that for Monday, the observed frequency is 31 while the expected frequency is 29.4, so that

$$\frac{(O - E)^2}{E} = \frac{(31 - 29.4)^2}{29.4} = \frac{1.6^2}{29.4} = 0.0871$$

Proceeding in a similar manner with the remaining data, we get

$$\chi^2 = \Sigma \frac{(O - E)^2}{E}$$

$$= \frac{(31 - 29.4)^2}{29.4} + \frac{(42 - 29.4)^2}{29.4} + \frac{(18 - 29.4)^2}{29.4}$$

$$+ \frac{(25 - 29.4)^2}{29.4} + \frac{(31 - 29.4)^2}{29.4}$$

$$= 0.0871 + 5.4000 + 4.4204 + 0.6585 + 0.0871$$

$$= 10.653$$

The theoretical distribution of $\Sigma(O - E)^2/E$ is a discrete distribution, since there are a limited number of possible values. However, this distribution can be approximated by a chi-square distribution. This approximation is generally considered acceptable, provided that all values of $E$ are at least 5. We include this requirement with the assumptions that apply to this section.

## Assumptions

1.  We intend to test a hypothesis that for the $k$ categories of outcomes in a multinomial experiment, the population proportion for each of the $k$ categories is as claimed.

2.  The sample data consist of frequency counts for the $k$ different categories, and the data constitute a random sample.

3.  For every one of the $k$ categories, the expected frequency is at least 5.

In section 5–5 we saw that the continuous normal probability distribution can reasonably approximate the discrete binomial probability distribution, provided that $np$ and $nq$ are both at least 5. We now see that the continuous chi-square distribution can reasonably approximate the discrete distribution of $\Sigma(O - E)^2/E$, provided that all values of $E$ are at least 5. There are ways of circumventing the problem of an expected frequency that is less than 5. One procedure requires us to combine categories so that all expected frequencies are at least 5.

When we use the chi-square distribution as an approximation, we obtain the critical value from Table A–4 after determining the level of significance $\alpha$ and the number of degrees of freedom. In a multinomial experiment with $k$ possible outcomes, the number of degrees of freedom is $k - 1$.

**degrees of freedom $= k - 1$**

This reflects the fact that, for $n$ trials, the frequencies of $k - 1$ outcomes can be freely varied, but the frequency of the last outcome is determined. Our 147 accidents are distributed among 5 categories or cells, but we can freely vary the frequencies of only 4 cells, since the last cell would be 147 minus the total of the first 4 cell frequencies. In this case, we say that the number of degrees of freedom is 4.

Note that close *agreement* between observed and expected values will lead to a *small* value of $\chi^2$. A large value of $\chi^2$ will indicate strong disagreement between observed and expected values. A significantly large value of $\chi^2$ will therefore cause rejection of the null hypothesis of no difference between observed and expected frequencies. Our test is therefore right-tailed, since the critical value and critical region are located at the extreme right of the distribution. Unlike previous hypothesis tests, we had to determine whether the test was left-tailed, right-tailed, or two-tailed, **these multinomial tests are all right-tailed.**

For the sample accident data, we have determined the value of the test statistic ($\chi^2 = 10.653$) and the number of degrees of freedom (4). Let's assume a 5% level of significance so that $\alpha = 0.05$. With 4 degrees of freedom and $\alpha = 0.05$, Table A–4 indicates a critical value of 9.488.

Figure 10–3 indicates that our test statistic of 10.653 falls within the critical region, so we reject the null hypothesis that accidents occur on the different weekdays with equal frequencies. (While it appears that Wednesday has a lower accident rate, such a conclusion would require other methods of analysis.)

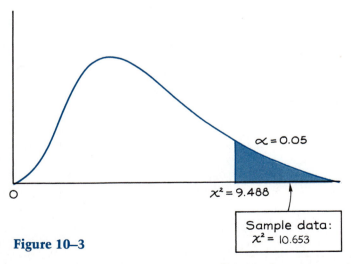

**Figure 10–3**

$\alpha = 0.05$

$\chi^2 = 9.488$

Sample data:
$\chi^2 = 10.653$

# DID MENDEL FUDGE HIS DATA?

■ R. A. Fisher analyzed the results of Mendel's experiments in hybridization. Fisher noted that the data were unusually close to theoretically expected outcomes. He says, "The data have evidently been sophisticated systematically, and after examining various possibilities, I have no doubt that Mendel was deceived by a gardening assistant, who knew only too well what his principal expected from each trial made." Fisher used chi-square tests and concluded that only about a 0.00004 probability exists of such close agreement between expected and reported observations. ■

The first example of this section dealt with the null hypothesis that the frequencies of accidents on the five working days were all equal. This can be represented as $H_0$: $p_1 = p_2 = p_3 = p_4 = p_5$ with the alternative hypothesis

$H_1$ being the statement that at least one of the proportions is different. The theory and methods we present here can also be used in cases where the claimed frequencies are different, as in the next example.

### Example

Use a 0.05 significance level and the same industrial accident data in Table 10–2 to test the claim that the accidents are distributed on the workdays as follows: 30% on Monday, 15% on Tuesday, 15% on Wednesday, 20% on Thursday, and 20% on Friday.

### Solution

The null hypothesis is the claim that the stated percentages are correct. This leads to the following null and alternative hypotheses:

$H_0$: $p_1 = 0.30$ and $p_2 = 0.15$ and $p_3 = 0.15$ and $p_4 = 0.20$ and $p_5 = 0.20$.
$H_1$: At least one of the preceding proportions is not equal to the value claimed.

Before computing the test statistic, we must first determine the values of expected frequency $E$. According to the claim, we expect 30% of the accidents to occur on Monday. Since 30% of 147 is $0.30 \times 147 = 44.1$, we have $E = 44.1$ for Monday. The expected value $E$ for Tuesday is found by taking 15% of 147, and so on.

| TABLE 10–4 | | | | | |
|---|---|---|---|---|---|
| Day | Mon | Tues | Wed | Thurs | Fri |
| Observed accidents | 31 | 42 | 18 | 25 | 31 |
| Expected accidents | 44.1 | 22.05 | 22.05 | 29.4 | 29.4 |

Using the observed and expected frequencies from Table 10–4, we now compute the test statistic.

$$
\begin{aligned}
\chi^2 &= \sum \frac{(O - E)^2}{E} \\
&= \frac{(31 - 44.1)^2}{44.1} + \frac{(42 - 22.05)^2}{22.05} + \frac{(18 - 22.05)^2}{22.05} \\
&\quad + \frac{(25 - 29.4)^2}{29.4} + \frac{(31 - 29.4)^2}{29.4} \\
&= 3.8914 + 18.0500 + 0.7439 + 0.6585 + 0.0871 \\
&= 23.431
\end{aligned}
$$

*continued*

**Solution** *continued*

This is a right-tailed test with $\alpha = 0.05$ and $5 - 1 = 4$ degrees of freedom, so the critical value from Table A–4 is 9.488. Since the test statistic of $\chi^2 = 23.431$ falls within the critical region, we reject the null hypothesis. There is sufficient evidence to warrant rejection of the claim that the accidents are distributed according to the given percentages.

The techniques in this section can be used to test for how well an observed frequency distribution conforms to some theoretical frequency distribution. For the employee accident data considered in this section, we used a goodness-of-fit test to decide whether the observed accidents conformed to a uniform distribution, and we found that the differences were significant. It appears that the observed frequencies do not make a good fit with a uniform distribution. Since many statistical analyses require a normally distributed population, we may use the chi-square test in this section to determine whether given samples are drawn from normally distributed populations. (See Exercise 26.)

## CHEATING SUCCESS

■ Eighteen students from a high school in a depressed area of Los Angeles took an Advanced Placement test in calculus and they all passed. Representatives of the Educational Testing Service challenged the validity of 14 students' results. A retest of 12 of the students confirmed the original scores; the students knew their calculus. What appeared to be cheating was actually the result of hard work by a group of dedicated students and an exceptional teacher, Jaime Escalante. His accomplishments were later dramatized in the movie *Stand and Deliver.* ■

## 10–2 Exercises A

1. The following table is obtained from a random sample of 100 absences. At the $\alpha = 0.01$ significance level, test the claim that absences occur on the 5 days with equal frequency.

   | Day | Mon | Tues | Wed | Thurs | Fri |
   |---|---|---|---|---|---|
   | Number absent | 27 | 19 | 22 | 20 | 12 |

   a. Find the $\chi^2$ value based on the sample data.
   b. Find the critical value of $\chi^2$.
   c. What conclusion can you draw?

2. A sample of motor vehicle deaths in Montana is randomly selected for a recent year. The numbers of fatalities are listed below for the different days of the week (based on data from the Insurance Institute for Highway Safety). At the 0.05 significance level, test the claim that accidents occur on the different days with equal frequency.

   | Day | Sun | Mon | Tues | Wed | Thurs | Fri | Sat |
   |---|---|---|---|---|---|---|---|
   | Number of fatalities | 31 | 20 | 20 | 22 | 22 | 29 | 36 |

3. In the decimal representation of $\pi$, the first 100 digits occur with the frequencies described in the table below. At the 0.05 significance level, test the claim that the digits are uniformly distributed.

| Digit | 0 | 1 | 2 | 3 | 4 | 5 | 6 | 7 | 8 | 9 |
|---|---|---|---|---|---|---|---|---|---|---|
| Frequency | 8 | 8 | 12 | 11 | 10 | 8 | 9 | 8 | 12 | 14 |

4. In the decimal representation of $\frac{5}{71}$, the first 100 digits occur with the frequencies described in the table below. At the 0.05 significance level, test the claim that the digits are uniformly distributed.

| Digit | 0 | 1 | 2 | 3 | 4 | 5 | 6 | 7 | 8 | 9 |
|---|---|---|---|---|---|---|---|---|---|---|
| Frequency | 17 | 12 | 15 | 9 | 7 | 11 | 12 | 6 | 8 | 3 |

5. One common test for authenticity of data is to analyze the frequencies of digits. When people are weighed and their weights are rounded to the nearest pound, we expect that the last digits (0, 1, 2, . . . , 9) occur with about the same frequency. In contrast, if we *ask* people how much they weigh, the digits 0 and 5 tend to occur at higher rates. When the author randomly selected 80 students, asked them for their weights, and recorded only the last digits, the results below were obtained. At the 0.01 significance level, test the claim that the last digits occur with the same frequency.

| Last digit | 0 | 1 | 2 | 3 | 4 | 5 | 6 | 7 | 8 | 9 |
|---|---|---|---|---|---|---|---|---|---|---|
| Frequency | 35 | 0 | 2 | 1 | 4 | 24 | 1 | 4 | 7 | 2 |

6. In a study of fatal car crashes, 216 cases are randomly selected from the pool in which the driver was found to have a blood alcohol content over 0.10. These cases are broken down according to the day of the week, with the results listed in the accompanying table (based on data from the Dutchess County STOP-DWI Program). At the 0.05 significance level, test the claim that such fatal crashes occur on the days of the week with equal frequency.

| Day | Sun | Mon | Tues | Wed | Thurs | Fri | Sat |
|---|---|---|---|---|---|---|---|
| Number | 40 | 24 | 25 | 28 | 29 | 32 | 38 |

7. An ice cream supplier claims that among the four most popular flavors, customers have these preference rates: 62% prefer vanilla, 18% prefer chocolate, 12% prefer neapolitan, and 8% prefer vanilla fudge. (The data are based on results from the International Association of Ice Cream Manufacturers.) A random sample of 200 customers produces the results below. At the $\alpha = 0.05$ significance level, test the claim that the percentages given by the supplier are correct.

| Flavor | Vanilla | Chocolate | Neapolitan | Vanilla fudge |
|---|---|---|---|---|
| Number prefer | 120 | 40 | 18 | 22 |

8. A genetics experiment involves 320 peas and is designed to determine whether Mendelian principles hold for a certain list of characteristics. The following table summarizes the actual experimental results and the expected Mendelian results for the five characteristics being considered. At the $\alpha = 0.01$ significance level, test the claim that Mendelian principles hold.

| Characteristic | A | B | C | D | E |
|---|---|---|---|---|---|
| Observed frequency | 30 | 15 | 58 | 83 | 134 |
| Expected (Mendelian) frequency | 20 | 20 | 40 | 120 | 120 |

9. Golf pros hit different types of balls and tried to identify the type based on their "feel." Fifty-three of the correct identifications are randomly selected and categorized by type, with the results given below (based on "The Fallacy of the Feel" by Pelz, *Golf* magazine). At the 0.05 significance level, test the claim that the correct identifications are equally distributed among the three types.

| Type | Three-piece balsata | Three-piece surlyn | Two-piece surlyn |
|---|---|---|---|
| Number | 18 | 19 | 16 |

10. Among the defective items a company produces, 300 are randomly selected and identified according to the production line that manufactured them. The results are given below. At the 0.05 significance level, test the claim that defects are equally distributed among the different production lines.

| Production line | A | B | C | D | E |
|---|---|---|---|---|---|
| Number of defects | 68 | 62 | 57 | 49 | 64 |

11. A marketing specialist claims that among supermarket shoppers who prefer a particular day, these rates apply: 7% prefer Sunday, 5% prefer Monday, 9% prefer Tuesday, 11% prefer Wednesday, 19% prefer Thursday, 24% prefer Friday, and 25% prefer Saturday. (The data are based on results from the Food Marketing Institute.) Sample results are given here. At the 0.05 significance level, test the claim that the given percentages are correct.

| Day | Sun | Mon | Tues | Wed | Thurs | Fri | Sat |
|---|---|---|---|---|---|---|---|
| Number | 9 | 6 | 10 | 8 | 19 | 23 | 28 |

12. Nicorette, a chewing gum designed to help people stop smoking cigarettes, was tested for adverse reactions. Among subjects with mouth or throat soreness, 129 are randomly selected and categorized as shown in the table below (based on data from Merrell Dow Pharmaceuticals, Inc.). If the drug and placebo produce the same effect, we would expect (because of different sample sizes) that the given categories have 30.9%, 19.1%, 30.9%, and 19.1% of the subjects, respectively. At the 0.05 significance level, test the claim that the actual frequencies agree with the expected rates.

| Subject | Drug (U.S.) | Drug (British) | Placebo (U.S.) | Placebo (British) |
|---|---|---|---|---|
| Number with mouth or throat soreness | 35 | 33 | 30 | 31 |

13. In an experiment on extrasensory perception, subjects were asked to identify the month showing on a calendar in the next room. If the results are as shown, test the claim that months were selected with equal frequencies. Assume a significance level of 0.05.

| Month | Jan | Feb | Mar | Apr | May | June | July | Aug | Sept | Oct | Nov | Dec |
|---|---|---|---|---|---|---|---|---|---|---|---|---|
| Number selected | 8 | 12 | 9 | 15 | 6 | 12 | 4 | 7 | 11 | 11 | 5 | 20 |

14. Among drivers who have had an accident in the last year, 88 are randomly selected and categorized by age, with the results listed below (based on data from the Insurance Information Institute). If all ages have the same accident rate, we would expect (because of the age distribution of licensed drivers) that the given categories have 16%, 44%, 27%, and 13% of the subjects, respectively. At the 0.05 significance level, test the claim that the actual frequencies agree with the expected rates.

| Age | Under 25 | 25–44 | 45–64 | Over 64 |
|---|---|---|---|---|
| Number | 36 | 21 | 12 | 19 |

15. A study of the color choices for buyers of compact cars claims that among the five most frequent choices, these preference rates apply: 22% prefer light red/brown, 22% prefer white, 20% prefer light blue, 18% prefer dark blue, and 18% prefer red. (The data are based on results from the Automotive Information Center.) When 270 compact cars are randomly selected, the following results are found. At the 0.05 level of significance, test the claim that the given percentages are correct.

| Color | Lt. red/brown | White | Lt. blue | Dk. blue | Red |
|---|---|---|---|---|---|
| Frequency | 60 | 61 | 43 | 41 | 65 |

16. A television company is told by a consulting firm that its eight leading shows are favored according to the percentages given in the following table. A separate and independent sample is obtained by another consulting firm. Do the figures agree? Assume a significance level of 0.05.

| Show | A | B | C | D | E | F | G | H |
|---|---|---|---|---|---|---|---|---|
| First consultant | 22% | 18% | 12% | 12% | 10% | 9% | 9% | 8% |
| Second consultant (number of respondents favoring show) | 29 | 30 | 20 | 16 | 9 | 17 | 10 | 19 |

17. In an insurance study of motor vehicle accidents in New York City, randomly selected fatal accidents are categorized according to time of day, with the results given in the table (based on data from the New York State Department of Motor Vehicles). Test the claim that fatal accidents occur at the different times of day with the same rate.

| | AM | | | | PM | | | |
|---|---|---|---|---|---|---|---|---|
| Time of day | 1–4 | 4–7 | 7–10 | 10–1 | 1–4 | 4–7 | 7–10 | 10–1 |
| Number of fatal accidents | 73 | 60 | 53 | 68 | 80 | 67 | 87 | 81 |

18. A car manufacturer tells a new dealer that among the new cars sold, 14% will be full-size, 30% will be mid-size, 37% will be compact, and 19% will be subcompact cars (based on data from Wards Automotive Reports). The dealer later checks records and finds that sales consisted of 18 full-size, 47 mid-size, 53 compact, and 21 subcompact new cars. Test the claim that actual sales agree with the manufacturer's claimed percentages.

19. A pair of dice has 36 possible outcomes that are supposed to be equally likely. If we plan to test a certain pair of dice for fairness using the chi-square distribution, what is the minimum number of rolls necessary if each outcome is to have an expected frequency of at least 5? Suppose we conduct the minimum number of rolls and obtain a $\chi^2$ value of 67.2. What do we conclude at the 0.01 significance level?

20. A roulette wheel has 38 possible outcomes that are supposed to be equally likely. If we plan to test a roulette wheel for fairness using the chi-square distribution, what is the minimum number of trials we must make if we are to satisfy the requirement that each expected frequency must be at least 5? Suppose we conduct the minimum number of trials and compute $\chi^2$ to be 49.6. What do we conclude at the 0.01 significance level?

# 10–2 | Exercises B

21. What do you know about the P-value for the hypothesis test in Exercise 18?

22. In executing a chi-square test in this section, suppose we multiply each observed frequency by the same positive integer greater than 1. How is the critical value affected? How is the test statistic affected?

23. In this exercise we will show that a hypothesis test involving a multinomial experiment with only two categories is equivalent to a hypothesis test for a proportion (Section 7–5). Assume that a particular multinomial experiment has only two possible outcomes $A$ and $B$ with observed frequencies of $f_1$ and $f_2$ respectively.
    a. Find an expression for the $\chi^2$ test statistic and find the critical value for a 0.05 significance level. Assume that we are testing the claim that both categories have the same frequency $(f_1 + f_2)/2$.
    b. The test statistic

$$z = \frac{\hat{p} - p}{\sqrt{\dfrac{pq}{n}}}$$

    is used to test the claim that a population proportion is equal to some value $p$. With the claim that $p = 0.5$, with $\alpha = 0.05$, and with

$$\hat{p} = \frac{f_1}{f_1 + f_2}$$

    show that $z^2$ is equivalent to $\chi^2$ (from part $a$). Also show that the square of the critical $z$ score is equal to the critical $\chi^2$ value from part $a$.

24. An observed frequency distribution is given below.
    a. Assuming a binomial distribution with $n = 3$ and $p = \frac{1}{3}$, use the binomial probability formula to find the probability corresponding to each category of the table.
    b. Using the probabilities found in part $a$, find the expected frequency for each category.
    c. Use a 0.05 level of significance to test the claim that the observed frequencies fit a binomial distribution for which $n = 3$ and $p = \frac{1}{3}$.

| Number of successes | 0 | 1 | 2 | 3 |
|---|---|---|---|---|
| Frequency | 89 | 133 | 52 | 26 |

    d. Show that

$$\sum \frac{(O - E)^2}{E} = \left( \sum \frac{O^2}{E} \right) - n$$

    where $n$ is the sum of all frequencies, then repeat part $c$ using this latter form of the test statistic.

25. In a survey of radio listeners, data are collected for different time slots, beginning at 1:00 PM The results are listed along with the expected frequencies based on past surveys. We cannot use the chi-square distribution, since all expected values are not at least 5. However, we can combine some columns so that all expected values do equal or exceed 5. Use this suggestion to test the claim that the observed and expected frequencies are compatible. Try to combine categories in a meaningful way.

| Time slot | 1 | 2 | 3 | 4 | 5 | 6 | 7 | 8 | 9 | 10 |
|---|---|---|---|---|---|---|---|---|---|---|
| Observed frequency | 2 | 8 | 8 | 9 | 3 | 5 | 3 | 0 | 12 | 3 |
| Expected frequency | 4 | 5 | 8 | 7 | 4 | 6 | 5 | 2 | 9 | 3 |

26. An observed frequency distribution of IQ scores is given below.
   a. Assuming a normal distribution with $\mu = 100$ and $\sigma = 15$, use the methods given in Chapter 5 to find the probability of a randomly selected subject belonging to each class. (The class boundaries are 79.5, 95.5, 110.5, 120.5.)
   b. Using the probabilities found in part *a*, find the expected frequency for each category.
   c. Use a 0.01 level of significance to test the claim that the IQ scores were randomly selected from a normally distributed population with $\mu = 100$ and $\sigma = 15$.

| IQ score | Under 80 | 80–95 | 96–110 | 111–120 | Above 120 |
|---|---|---|---|---|---|
| Frequency | 20 | 20 | 80 | 40 | 40 |

# 10–3  Contingency Tables

Many people believe that smoking is unhealthy. In a study of 1000 deaths of males aged 45 to 64, the causes of death are listed along with their smoking habits (see Table 10–5). (The data are based on results given in "Chartbook on Smoking, Tobacco, and Health," USDHEW publication CDC75-7511.)

**TABLE 10–5**

| | Cause of Death | | |
|---|---|---|---|
| | Cancer | Heart disease | Other |
| Smoker | 135 | 310 | 205 |
| Nonsmoker | 55 | 155 | 140 |

In the introduction of this problem at the beginning of the chapter, we noted that smokers are disproportionately represented among the sample of 1000 deaths. That suggests that smoking is unhealthy. We will now consider a different issue: Is smoking related to the cause of death? We will use the sample data in Table 10–5 to test the claim that *the cause of death is independent of smoking.* Having identified the key issue to be considered here, we should also remember that our analysis applies only to the population from which the sample was drawn. That is, we are considering only males who died between the ages of 45 and 64, and we are considering only the three categories of cause of death as given in Table 10–5.

Tables similar to Table 10–5 are generally called contingency tables, or two-way tables. In this context, the word *contingency* refers to dependence, and the contingency table serves as a useful medium for analyzing the dependence of one variable on another. This is only a statistical dependence that cannot be used to establish an inherent cause-and-effect relationship.

We need to test the *null hypothesis that the two variables in question are independent.* That is, we will test the claim that smoking habits and cause of death are independent. Let's select a significance level of $\alpha = 0.05$. We can now compute a test statistic based on the data and then compare that test statistic to the appropriate critical test value. As in the previous section, we use the chi-square distribution, with the test statistic given by $\chi^2$.

---

**Test Statistic**

$$\chi^2 = \sum \frac{(O - E)^2}{E}$$

---

This test statistic allows us to measure the degree of disagreement between the frequencies actually observed and those that we would theoretically expect when the two variables are independent. The reasons underlying the development of the $\chi^2$ statistic in the previous section also apply here. In repeated large samplings, *the distribution of the test statistic $\chi^2$ can be approximated by the chi-square distribution provided that all expected frequencies are at least 5.* We include this requirement with the assumptions that apply to this section.

## Assumptions

1.  We intend to test the hypothesis that for a contingency table, the row variable and column variable are *independent.*

2.  The sample data are randomly selected.

3.  For every cell in the contingency table, the expected frequency $E$ is at least 5.

In the preceding section we knew the corresponding probabilities and could easily determine the expected values, but the typical contingency table does not come with the relevant probabilities. Consequently, we need to devise a method for obtaining the corresponding expected values. We will first describe the procedure for finding the values of the expected frequencies, and then proceed to justify that procedure. For each cell in the frequency table, the expected frequency $E$ can be calculated by using the following.

## Expected Frequency

$$\text{Expected frequency } E = \frac{(\text{row total}) \cdot (\text{column total})}{(\text{grand total})}$$

Here *grand total* refers to the total number of observations in the table. For example, in the lower right cell of Table 10–5, we see the observed frequency of 140. The total of all frequencies for that row is 350, the total of the column frequencies is 345, and the total of all frequencies in the table is 1000, so we get an expected frequency of

$$E = \frac{(350)(345)}{1000} = 120.75$$

In the lower right cell, the *observed* frequency is 140, while the *expected* frequency is 120.75. Table 10–6 reproduces Table 10–5 with the expected frequencies inserted in parentheses. As in Section 10–2, we require that all expected frequencies be at least 5 before we can conclude that the chi-square distribution serves as a suitable approximation to the distribution of $\chi^2$ values.

| **TABLE 10–6** | | | | |
|---|---|---|---|---|
| | Cause of Death | | | |
| | Cancer | Heart disease | Other | |
| Smoker | 135 (123.50) | 310 (302.25) | 205 (224.25) | 650 |
| Nonsmoker | 55 (66.50) | 155 (162.75) | 140 (120.75) | 350 |
| | 190 | 465 | 345 | (Grand total: 1000) |

## MAGAZINE SURVEYS

■ Magazines often boost sales through reader surveys, but such surveys are usually biased and reflect only the views of the respondents. A *Time* article on magazine surveys noted that when wives were asked if they ever had an extramarital affair, the results were 21% yes for *Ladies Home Journal*, 34% yes for *Playboy*, and 54% yes for *Cosmopolitan*. One pollster suggested that a *Reader's Digest* survey "would probably find that *nobody* had any extramarital affairs." When people decide whether or not to include themselves in a survey, the survey is "self-selected" and cannot be considered representative of a larger population. ■

To better understand the rationale for this procedure, let's pretend that we know only the row and column totals and that we must fill in the cell frequencies by assuming that there is no relationship between the two variables involved (see Table 10–7).

| TABLE 10–7 | | | | |
|---|---|---|---|---|
| | Cause of Death | | | |
| | Cancer | Heart disease | Other | |
| Smoker | | | | 650 |
| Nonsmoker | | | | 350 |
| | 190 | 465 | 345 | (Grand total: 1000) |

We begin with the cell in the upper left corner that corresponds to smokers who died of cancer. Since 650 of the 1000 subjects are smokers, we have $P(\text{smoker}) = 650/1000$. Similarly, 190 of the 1000 subjects died from cancer, so that $P(\text{death by cancer}) = 190/1000$. Since we assume that smoking and cause of death are independent, conclude that

$$P(\text{smoker and death by cancer}) = \frac{650}{1000} \times \frac{190}{1000}$$

This follows from the multiplication rule of probability whereby $P(A \text{ and } B) = P(A) \times P(B)$ if $A$ and $B$ are independent events. To obtain the expected value for the upper left cell, we simply mulitply the probability for that cell by the total number of subjects available to get

$$\frac{650}{1000} \times \frac{190}{1000} \times 1000 = 123.50$$

The form of this product suggests a general way to obtain the expected frequency of a cell:

$$\text{expected frequency } E = \frac{(\text{row total})}{(\text{grand total})} \cdot \frac{(\text{column total})}{(\text{grand total})} \cdot (\text{grand total})$$

This expression can be simplified to

$$E = \frac{(\text{row total}) \cdot (\text{column total})}{(\text{grand total})}$$

Using the observed and expected frequencies shown in Table 10–6, we can now compute the $\chi^2$ test statistic based on the sample data.

$$\chi^2 = \sum \frac{(O - E)^2}{E}$$

$$= \frac{(135 - 123.50)^2}{123.50} + \frac{(310 - 302.25)^2}{302.25} + \frac{(205 - 224.25)^2}{224.25}$$

$$+ \frac{(55 - 66.50)^2}{66.50} + \frac{(155 - 162.75)^2}{162.75} + \frac{(140 - 120.75)^2}{120.75}$$

$$= 1.0709 + 0.1987 + 1.6525$$
$$+ 1.9887 + 0.3690 + 3.0688$$

$$= 8.349$$

With $\alpha = 0.05$ we proceed to Table A–4 to obtain the critical value of $\chi^2$, but we must first know where the critical region lies and the number of degrees of freedom. **Tests of independence with contingency tables involve only right-tailed critical regions,** since small values of $\chi^2$ support the claimed independence of the two variables. That is, $\chi^2$ is small if observed and expected frequencies are close. Large values of $\chi^2$ are to the right of the chi-square distribution, and they reflect significant differences between observed and expected frequencies.

In a contingency table with $r$ rows and $c$ columns, the number of degrees of freedom is given by the following.

**degrees of freedom $= (r - 1)(c - 1)$**

Thus Table 10–5 has $(2 - 1)(3 - 1) = 2$ degrees of freedom. In a right-tailed test with $\alpha = 0.05$ and with 2 degrees of freedom, we refer to Table A–4 to get a critical $\chi^2$ value of 5.991. Since the calculated $\chi^2$ value of 8.349 falls in the critical region bounded by $\chi^2 = 5.991$, we reject the null hypothesis of independence between the two variables. (See Figure 10–4.) It appears that smoking and cause of death are dependent.

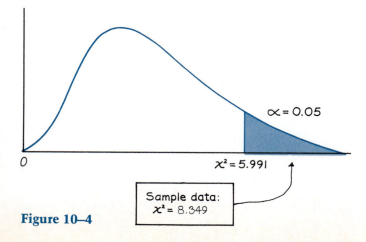

$\alpha = 0.05$

$\chi^2 = 5.991$

Sample data:
$\chi^2 = 8.349$

**Figure 10–4**

## ILLITERACY

We now summarize the important components of this hypothesis test. We used the sample data in Table 10–5 with these hypotheses.

$H_0$: Smoking is independent of the cause of death (cancer, heart disease, other).

$H_1$: Smoking and the cause of death are dependent.

Significance level: $\alpha = 0.05$

Test statistic: $\chi^2 = 8.349$

Critical value: $\chi^2 = 5.991$

Conclusion: Reject the null hypothesis. There is sufficient sample evidence to warrant rejection of the claim that smoking and cause of death are independent. It appears that they are dependent.

Note that this hypothesis test doesn't show that smoking is unhealthy, as we discussed at the beginning of the chapter. Rather, it suggests that smoking and cause of death are dependent. We should always be careful about jumping to a conclusion that there is a cause-effect relationship. We have not established that smoking has a causal effect on the cause of death. We have simply shown that smoking and cause of death appear to be dependent—we have not identified the true causal factors.

We might note that the original study involved a sample size of 43,221 deaths of males aged 45 to 64, and the results are even more dramatic than for the sample of 1000 included here. We might also comment that smoking really *is* unhealthy. No ifs, ands, or . . .

Because of the nature of the calculations, it is often helpful to use a statistics software package for tests of the type discussed in this section. The STATDISK and Minitab displays show the results obtained for the data in Table 10–5.

**STATDISK DISPLAY**

```
A [ 135 ]  [ 310 ]  [ 205 ]
  (123.5)  (302.3)  (224.3)

B [  55 ]  [ 155 ]  [ 140 ]
  ( 66.5)  (162.8)  (120.8)

Test statistic . . . . . . . . . Chi Square  = 8.3
Critical value . . . . . . . . . Chi Square  = 5.9
P-value . . . . . . . . . . . . . . . . . . . . . = 0.0156547
Significance level . . . . . . . . . . . . . . = .05
Degrees of freedom . . . . . . . . . . . . = 2

CONCLUSION: REJECT the null hypothesis that row and column
variables are independent.
```

MINITAB DISPLAY

```
MTB > READ C1 C2 C3
DATA> 135 310 205
DATA>  55 155 140
DATA> ENDOFDATA
       2 ROWS READ
MTB > CHISQUARE analysis for C1 C2 C3
Expected counts are printed below observed counts
```

|       | C1     | C2     | C3     | Total |
|-------|--------|--------|--------|-------|
| 1     | 135    | 310    | 205    | 650   |
|       | 123.50 | 302.25 | 224.25 |       |
| 2     | 55     | 155    | 140    | 350   |
|       | 66.50  | 162.75 | 120.75 |       |
| Total | 190    | 465    | 345    | 1000  |

ChiSq = 1.071 + 0.199 + 1.652 +
        1.989 + 0.369 + 3.069 = 8.349

df = 2

# 10–3 | Exercises B

1. Nicorette is a chewing gum designed to help people stop smoking cigarettes. Tests for adverse reactions yield the results given in the table (based on data from Merrell Dow Pharmaceuticals, Inc.). At the 0.05 significance level, test the claim that the treatment (drug or placebo) is independent of the reaction (whether or not mouth or throat soreness was experienced).

|                               | Drug | Placebo |
|-------------------------------|------|---------|
| Mouth or throat soreness      | 43   | 35      |
| No mouth or throat soreness   | 109  | 118     |

2. Many people believe that criminals who plead guilty tend to get lighter sentences than those who are convicted in trials. The table below summarizes sample data for San Francisco defendants in burglary cases. All of the subjects had prior prison sentences. (See "Does It Pay to Plead Guilty? Differential Sentencing and the Functioning of the Criminal Courts" by Brereton and Casper, *Law and Society Review*,

Vol. 16, No. 1.) At the 0.05 significance level, test the claim that the sentence (sent to prison or not sent to prison) is independent of the plea.

|  | Guilty plea | Plea of not guilty |
|---|---|---|
| Sent to prison | 392 | 58 |
| Not sent to prison | 564 | 14 |

3. In the judicial case *United States v. City of Chicago*, fair employment practices were challenged. A minority group (group A) and a majority group (group B) took the Fire Captain Examination. At the 0.05 significance level, use the given results and test the claim that success on the test is independent of the group.

|  | Pass | Fail |
|---|---|---|
| Group A | 10 | 14 |
| Group B | 417 | 145 |

4. A study of jail inmates charged with DWI provided the sample data in the table (based on data from the U.S. Department of Justice). At the 0.05 significance level, test the claim that the unconvicted/convicted status is independent of the prior DWI sentences.

|  | Unconvicted | Convicted |
|---|---|---|
| No prior DWI sentence | 100 | 672 |
| 1 prior DWI sentence | 56 | 394 |
| 2 prior DWI sentences | 16 | 173 |
| 3 or more prior DWI sentences | 11 | 72 |

5. A study of medical practices involved subjects from different groups. Sample data are given in the accompanying table (based on data from "A Randomized Controlled Trial of Academic Group Practice" by Goldberg and others, *Journal of the American Medical Association*, Vol. 257, No. 15). At the 0.05 level of significance, test the claim that the payer is independent of the group.

|  |  | Control Group | Experimental Group |
|---|---|---|---|
|  | Medicare | 38 | 43 |
| Payer | Medicaid | 16 | 17 |
|  | County welfare | 11 | 10 |

6. The following table provides sample data from a study of dropout rates among college freshmen. At the 0.05 significance level, test the claim

that the type of college is independent of the dropout rate. (The sample data are based on results from the American College Testing Program.)

|  | Four-year Public | Four-year Private | Two-year Public | Two-year Private |
|---|---|---|---|---|
| Freshmen dropouts | 10 | 9 | 15 | 9 |
| Freshmen who stay | 26 | 28 | 18 | 27 |

7. A study of car accidents and drivers who use cellular phones provided the following sample data. (The data are based on results from AT&T and the Automobile Association of America.) At the 0.05 level of significance, test the claim that the occurrence of accidents is independent of the use of cellular phones.

|  | Had accident in last year | Had no accident in last year |
|---|---|---|
| Cellular phone user | 23 | 282 |
| Not cellular phone user | 46 | 407 |

8. The following table summarizes results from randomly selected drivers convicted of motor vehicle violations in New York State. (The data are based on results from the New York State Department of Motor Vehicles.) At the 0.05 significance level, test the claim that the county is independent of the type of violation. (DWI is driving while intoxicated and DTD is driving with a disabled traffic device.)

|  | Albany | Monroe | Orange | Westchester |
|---|---|---|---|---|
| DWI | 6 | 19 | 6 | 10 |
| Speeding | 103 | 261 | 160 | 226 |
| DTD | 60 | 152 | 25 | 174 |

9. A study of seat belt users and non-users yielded the sample data summarized in the table (based on data from "What Kinds of People Do Not Use Seat Belts?" by Helsing and Comstock, *American Journal of Public Health*, Vol. 67, No. 11). Test the claim that the amount of smoking is independent of seat belt use.

|  | Number of cigarettes smoked per day | | | |
|---|---|---|---|---|
|  | 0 | 1–14 | 15–34 | 35 and over |
| Wear seat belts | 175 | 20 | 42 | 6 |
| Don't wear seat belts | 149 | 17 | 41 | 9 |

10. The following table summarizes sample data collected in a study of seat belt use in taxi cabs (based on "The Phantom Taxi Seat Belt" by Welkon and Reisinger, *American Journal of Public Health*, Vol. 67, No. 11). At the 0.05 level of significance, test the claim that the occurrence of usable seat belts is independent of the city.

| | | New York | Chicago | Pittsburgh |
|---|---|---|---|---|
| Taxi has usable seat belt? | Yes | 3 | 42 | 2 |
| | No | 74 | 87 | 70 |

11. For a random sample of deaths of U.S. Army veterans, a medical panel established the cause of death by reviewing records. Among deaths considered to be alcohol-related, the cause of death and the results are included in the accompanying table. (See "Underreporting of Alcohol-Related Mortality on Death Certificates of Young U.S. Army Veterans" by Pollock and others, *Journal of the American Medical Association*, Vol. 258, No. 3.) At the 0.05 significance level, test the claim that the cause of death is independent of the determining source.

| | Natural | Motor Vehicle | Suicide/Homicide |
|---|---|---|---|
| Original death certificate | 9 | 7 | 4 |
| Medical panel | 30 | 53 | 35 |

12. Colleges often compile data describing the distribution of grades. The table lists sample data taken from the author's college. At the 0.10 significance level, test the claim that these two departments have the same grade distribution.

| | A | B | C | D | F |
|---|---|---|---|---|---|
| Math | 49 | 69 | 64 | 42 | 61 |
| Physical Sciences | 52 | 73 | 78 | 23 | 18 |

13. A study conducted to determine the rate of smoking among people from different age groups provided the sample data summarized in the accompanying table (based on data from the National Center for Health Statistics). At the 0.05 significance level, test the claim that smoking is independent of the four listed age groups.

| | Age (years) | | | |
|---|---|---|---|---|
| | 20–24 | 25–34 | 35–44 | 45–64 |
| Smoke | 18 | 15 | 17 | 15 |
| Don't smoke | 32 | 35 | 33 | 35 |

14. A study of people who refused to answer survey questions provided the sample data in the table (based on data from "I Hear You Knocking But You Can't Come In," by Fitzgerald and Fuller, *Sociological Methods and Research*, Vol. 11, No. 1). At the 0.01 significance level, test the claim that the cooperation of the subject (response/refuse) is independent of the age category.

| | Age | | | | | |
|---|---|---|---|---|---|---|
| | 18–21 | 22–29 | 30–39 | 40–49 | 50–59 | 60 and over |
| Responded | 73 | 255 | 245 | 136 | 138 | 202 |
| Refused | 11 | 20 | 33 | 16 | 27 | 49 |

15. Sample data collected by the New York State Department of Motor Vehicles provided the results summarized in the table. At the 0.10 significance level, test the claim that the time of day of fatal accidents is the same in New York City as in all other New York State locations.

| | Time of Day of Fatal Accidents | | | | | | | |
|---|---|---|---|---|---|---|---|---|
| | AM | | | | PM | | | |
| | 1–4 | 4–7 | 7–10 | 10–1 | 1–4 | 4–7 | 7–10 | 10–1 |
| New York City | 73 | 60 | 53 | 68 | 80 | 67 | 87 | 81 |
| Other New York State | 187 | 115 | 136 | 161 | 196 | 257 | 237 | 235 |

16. In an experiment designed to test golfers' "feel" for different ball types, professional golfers hit equal numbers of three different types. They hit normally, while blinded, with hearing suppressed, and with both sight and sound suppressed. The number of correct identifications for each case are listed in the table (based on data from "The Fallacy of Feel," by Pelz, *Golf* magazine). Test the claim that when identifying the type of golf ball used, the sensory state of the golfer is independent of the ball type.

| | Ball Type | | |
|---|---|---|---|
| | Three-piece balata | Three-piece surlyn | Two-piece surlyn |
| Normal | 18 | 19 | 16 |
| Deaf | 19 | 18 | 9 |
| Blind | 19 | 12 | 19 |
| Blind and deaf | 16 | 11 | 9 |

# $\boxed{10\text{–}3}$ Exercises B

17. What do you know about the $P$-value for the hypothesis test in Exercise 16?

18. The chi-square distribution is continuous while the test statistic used in this section is actually discrete. Some statisticians use **Yates' correction for continuity** in cells with an expected frequency less than 10 or in all cells of a contingency table with two rows and two columns. With Yates' correction, we replace

$$\sum \frac{(O-E)^2}{E} \text{ with } \sum \frac{(|O-E|-0.5)^2}{E}$$

|   | X  | Y  |
|---|----|----|
| A | 5  | 25 |
| B | 65 | 5  |

Given the accompanying contingency table, find the value of the $\chi^2$ test statistic with and without Yates' correction. In general, what effect does Yates' correction have on the value of the test statistic?

19. If each observed frequency in a contingency table is multiplied by a positive integer $K$ (where $K \geq 2$), how is the value of the test statistic affected? How is the critical value affected?

20. a. For the contingency table given below, verify that the test statistic becomes

$$\chi^2 = \frac{(a+b+c+d)(ad-bc)^2}{(a+b)(c+d)(b+d)(a+c)}$$

Column

|        | 1 | 2 |
|--------|---|---|
| Row 1  | a | b |
| Row 2  | c | d |

b. Let $\hat{p}_1 = \frac{a}{a+c}$ and let $\hat{p}_2 = \frac{b}{b+d}$ and show that the test statistic

$$z = \frac{(\hat{p}_1 - \hat{p}_2) - 0}{\sqrt{\bar{p}\bar{q}\left(\frac{1}{n_1} + \frac{1}{n_2}\right)}}$$

is such that $z^2 = \chi^2$ (the same result from part $a$). This shows that the chi-square test involving a $2 \times 2$ table is equivalent to the test for the difference between two proportions, as described in Section 8–4.

# Vocabulary List

Define and give an example of each term.

goodness-of-fit test                    multinomial experiment
contingency table                       observed frequency
two-way table                           expected frequency

# Review

We began this chapter by developing methods for testing hypotheses made about more than two population proportions. For **multinomial experiments** we tested for goodness-of-fit or agreement between observed and expected frequencies by using the chi-square test statistic given in the table that follows. In repeated large samplings, the distribution of the $\chi^2$ test statistic can be approximated by the chi-square distribution. This approximation is generally considered acceptable as long as all expected frequencies are at least 5. In a multinomial experiment with $k$ cells or categories, the number of degrees of freedom is $k - 1$.

   In Section 10–3 we used the sample $\chi^2$ test statistic to measure disagreement between observed and expected frequencies in **contingency tables.** A contingency table contains frequencies; the rows correspond to categories of one variable while the columns correspond to categories of another variable. With contingency tables, we test the hypothesis that the two variables of classification are independent. In this test of independence, we can again approximate the sampling distribution of that statistic by the chi-square distribution as long as all expected frequencies are at least 5. In a contingency table with $r$ rows and $c$ columns, the number of degrees of freedom is $(r - 1)(c - 1)$.

| Important Formulas | | | | |
|---|---|---|---|---|
| Application | Applicable Distribution | Test Statistic | Degrees of Freedom | Table of Critical Values |
| Multinomial | chi-square | $\chi^2 = \sum \dfrac{(O - E)^2}{E}$ | $k - 1$ | Table A–4 |
| Contingency table | chi-square | $\chi^2 = \sum \dfrac{(O - E)^2}{E}$ <br> where $E = \dfrac{\text{(row total)(column total)}}{\text{(grand total)}}$ | $(r - 1)(c - 1)$ | Table A–4 |

# ? *Review Exercises*

1. Clinical tests of the allergy drug Seldane were conducted. Among subjects who experienced drowsiness, 212 were randomly selected and categorized as shown in the table below (based on data from Merrell Dow Pharmaceuticals, Inc.). At the 0.05 significance level, test the claim that those who experience drowsiness are equally distributed among the three categories.

| Group | Seldane Users | Placebo Users | Control |
|---|---|---|---|
| Experienced drowsiness | 54 | 49 | 109 |

2. Clinical tests of the allergy drug Seldane yielded results summarized in the table (based on data from Merrell Dow Pharmaceuticals, Inc.). At the 0.05 significance level, test the claim that the occurrence of headaches is independent of the group (Seldane/placebo/control).

| | Seldane Users | Placebo Users | Control |
|---|---|---|---|
| Headache | 49 | 49 | 24 |
| No headache | 732 | 616 | 602 |

3. Many people believe that criminals who plead guilty tend to get lighter sentences than those who are convicted in trials. In the following table we summarize sample data for 434 San Francisco defendants in robbery cases. All of the subjects had prior prison sentences. (See "Does It Pay to Plead Guilty? Differential Sentencing and the Functioning of Criminal Courts" by Brereton and Casper, *Law and Society Review*, Vol. 16, No. 1.) At the 0.01 level of significance, test the claim that the sentence (sent to prison or not sent to prison) is independent of the plea.

| | Guilty Plea | Not Guilty Plea |
|---|---|---|
| Sent to prison | 191 | 64 |
| Not sent to prison | 169 | 10 |

4. In a certain county, 80% of the drivers have no accidents in a given year, 16% have one accident, and 4% have more than one accident. A survey of 200 randomly selected teachers from the county produced 172 with no accidents, 23 with one accident, and 5 with more than one accident. At the 5% level of significance, test the claim that the teachers exhibit the same accident rate as the countywide population.

5. When *Time* magazine tracked U.S. deaths by gunfire in one week, the results in the table were obtained. At the 0.05 significance level, test the claim that gunfire death rates are the same for the different days of the week.

| Weekday | Mon | Tues | Wed | Thurs | Fri | Sat | Sun |
|---|---|---|---|---|---|---|---|
| Number of deaths by gunfire | 74 | 60 | 66 | 71 | 51 | 66 | 76 |

6. The manager of an assembly operation wants to determine whether product quality is dependent on the day of the week. She develops the following sample data. Test the claim of the union representative that the day of the week makes no difference in product quality.

|  | Mon | Tues | Wed | Thurs | Fri |
|---|---|---|---|---|---|
| Acceptable products | 80 | 100 | 95 | 93 | 82 |
| Defective products | 15 | 5 | 5 | 7 | 12 |

7. In a certain region, a survey of companies that officially declared bankruptcy during the past year yielded the following sample data. Of the 120 small bankrupt businesses, 72 advertised in weekly newspapers. Of the 65 medium-sized bankrupt businesses, 25 advertised in weekly newspapers, while 8 of the 15 large bankrupt businesses did so. At the 5% level of significance, test the claim that the three proportions of bankrupt businesses that used weekly newspaper ads are equal.

8. A marketing study is conducted in order to determine if a product's appeal is affected by geographical region. Given the sample data in the following table, use a 0.01 significance level to test the claim that the consumer's opinion is independent of region.

|  | Like | Dislike | Uncertain |
|---|---|---|---|
| Northeast | 30 | 15 | 15 |
| Southeast | 10 | 30 | 20 |
| West | 40 | 60 | 15 |

*Computer Projects*

1. Because of the nature of the calculations required, statistical software packages become very useful in testing hypotheses involving multinomial experiments and contingency tables.
   a. Use existing software such as STATDISK or Minitab for Exercise 15 from Section 10–3.
   b. For the sample data used in part *a*, multiply each frequency by 10 and note the effect on the test statistic, critical value, and conclusion. In general, what is the effect of multiplying each frequency in a contingency table by a positive integer greater than 1?
   c. For the same sample data used in part *a*, add 100 to each frequency and note the effects on the test statistic, critical value, and conclusion. In general, what is the effect of adding a positive integer to each entry in a contingency table?
   d. For the same sample date used in part *a*, interchange the first row (New York City) frequencies with those of the second row (Other New York State) and note the effect on the test statistic, critical value, and conclusion. In general, what is the effect of interchanging any two rows in a contingency table? What is the effect of interchanging any two columns in a contingency table?

 *Applied Projects*

1. Using the Appendix B data for homes sold, enter the frequencies in the table below.

|  | Number of Baths | |
| --- | --- | --- |
|  | 1 or 1.5 | 2 or more |
| Under 2000 sq ft |  |  |
| 2000 sq ft or more |  |  |

   Use this data to test the claim that the number of baths is independent of the living area.

2. Using Data Set II in Appendix B, construct a contingency table with sex (female, male) as one variable and employment (employed, unemployed) as the other variable. Test the claim that sex and employment are independent.

3.  Using Data Set II in Appendix B, construct a contingency table with sex (female, male) as one variable and "handedness" (right, left) as the other variable. Test the claim that sex and right/left handedness are independent.

4.  Conduct a survey by asking the question "Do you favor or oppose the death penalty for people convicted of murder?" Record each response (yes, no, undecided) along with the sex (female, male) of the respondent. Include in your data set only responses of "yes" or "no." At the 0.05 level of significance, test the claim that the opinion is independent of the sex of the respondent. Be sure to survey enough people so that the expected frequency of each cell in the resulting contingency table is at least 5. Identify any factors suggesting that your sample is not representative of the people in your region.

 ## *Writing Projects*

1.  In random sampling, each member of the population has the same chance of being selected. When the author asked 60 students to "randomly" select three digits each, the results listed below were obtained. Use this sample of 180 digits to test the claim that students select digits randomly. Were the students successful in choosing their own random numbers? Write a report summarizing your results and conclusions.

| 213 | 169 | 812 | 125 | 749 | 137 | 202 | 344 | 496 | 348 | 714 | 765 |
|-----|-----|-----|-----|-----|-----|-----|-----|-----|-----|-----|-----|
| 831 | 491 | 169 | 312 | 263 | 192 | 584 | 968 | 377 | 403 | 372 | 123 |
| 493 | 894 | 016 | 682 | 390 | 123 | 325 | 734 | 316 | 357 | 945 | 208 |
| 115 | 776 | 143 | 628 | 479 | 316 | 229 | 781 | 628 | 356 | 195 | 199 |
| 223 | 114 | 264 | 308 | 105 | 357 | 333 | 421 | 107 | 311 | 458 | 007 |

2.  Write a report summarizing the program listed below.

 ## *Videotapes*

Program 24 of the series *Against All Odds: Inside Statistics* is recommended as a supplement to this chapter.

# Chapter Eleven

## In This Chapter

# 11 Analysis of Variance

## Chapter Problem

Three different transmission types are installed on different cars, all having 6 cylinders and an engine size of 3.0 liters. The fuel consumption (in mi/gal) is measured under identical highway conditions and the results are given below (based on data from the Environmental Protection Agency). The letters $A$, $M$, $L$ represent automatic, manual, and lockup torque converter, respectively. Does the type of transmission have an effect on fuel consumption? The sample means of 21.8, 25.6, and 24.7 are different, but are those differences statistically significant?

| A | M | L |
|---|---|---|
| 23 | 27 | 24 |
| 23 | 29 | 26 |
| 20 | 25 | 24 |
| 21 | 23 | |
| | 24 | |
| $n_1 = 4$ | $n_2 = 5$ | $n_3 = 3$ |
| $\bar{x}_1 = 21.8$ | $\bar{x}_2 = 25.6$ | $\bar{x}_3 = 24.7$ |
| $s_1^2 = 2.3$ | $s_2^2 = 5.8$ | $s_3^2 = 1.3$ |

Such problems will be considered in this chapter. This particular problem will be addressed in Section 11–3.

# 11–1 | Overview

In Section 8–3 we developed procedures for testing the hypothesis that two population means are equal ($H_0$: $\mu_1 = \mu_2$). In this chapter we develop a procedure for testing the hypothesis that differences among three or more sample means are due to chance. A typical null hypothesis will be $H_0$: $\mu_1 = \mu_2 = \mu_3$. The following assumptions apply to the methods in this chapter.

## Assumptions

1. We intend to test the hypothesis that three or more samples come from populations with the same mean.

2. The populations being considered have normal distributions.

3. The populations being considered have the same variance (or standard deviation).

4. The samples are random and independent of each other.

The requirements of normality and equal variances are somewhat loose since the methods in this chapter work reasonably well unless there is a very nonnormal distribution or the population variances differ by large amounts. If the samples are independent but the distributions are very nonnormal, we can use the Kruskal-Wallis test presented in Section 12–5. The method we will describe is called analysis of variance (often referred to as ANOVA) because it is based on a comparison of two different estimates of the variance common to the different populations. Here's a helpful hint for the calculations in this chapter: Because the test statistics are based on estimates of variance, we can add (or subtract) any convenient constant to each score, and the results will not be affected. Adding or subtracting a constant throughout may simplify calculations.

The following sections use the **F distribution**, which was first presented in Section 8–2, where we noted these important properties (see Figure 11–1):

1. The $F$ distribution is not symmetric; it is skewed to the right.

2. The values of $F$ can be 0 or positive, but they cannot be negative. Critical values of $F$ are found in Table A–5.

3. There is a different $F$ distribution for each pair of degrees of freedom for numerator and denominator.

Section 11–2 begins with cases involving samples having the same number of scores. Section 11–3 will then proceed to consider cases in which the sample sizes are not all equal. Sections 11–2 and 11–3 apply to cases where there is only one way to classify the populations being considered; those sections use a method called *one-way analysis of variance*. In many other

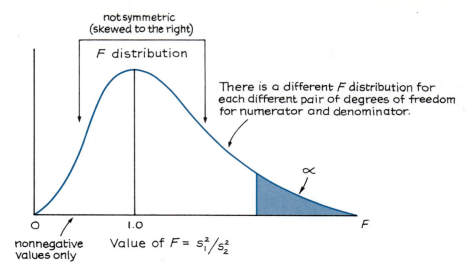

not symmetric
(skewed to the right)

F distribution

There is a different F distribution for each different pair of degrees of freedom for numerator and denominator.

$\alpha$

O

nonnegative values only

1.0

Value of $F = s_1^2 / s_2^2$

F

**Figure 11–1**

situations, there is more than one way to classify the populations. Section 11–4 introduces *two-way analysis of variance*, which allows us to deal with populations classified two ways.

## 11–2 ANOVA with Equal Sample Sizes

In this section we consider tests of hypotheses that more than two means are all equal, such as $H_0$: $\mu_1 = \mu_2 = \mu_3$; this section is restricted to cases in which the different samples are all of the same size and there is only one way to classify the populations involved. We will introduce the basic concepts underlying the method of *analysis of variance*, then we will use these same general concepts as we move on to consider cases with unequal sample sizes (Section 11–3) and cases with more than one way to classify the populations (Section 11–4). Let's begin with some sample data.

### Example

A pilot does extensive bad weather flying and decides to buy a battery-powered radio as an independent backup for her regular radios, which depend on the airplane's electrical system. She has a choice of three brands of rechargeable batteries that vary in cost. She obtains the sample

*continued*

## COMMERCIALS

■ Television networks have their own clearance departments for screening commercials and verifying claims. The National Advertising Division, a branch of the Council of Better Business Bureaus, investigates advertising claims. The Federal Trade Commission and local district attorneys also become involved. In the past, Firestone had to drop a claim that its tires resulted in 25% faster stops, and Warner Lambert had to spend $10 million informing customers that Listerine doesn't prevent or cure colds. Many deceptive ads are voluntarily dropped and many others escape scrutiny simply because the regulatory mechanisms can't keep up with the flood of commercials. ■

**Example** *continued*

data in the following table. She randomly selects five batteries for each brand, and tests them for the operating time (in hours) before recharging is necessary. Do the three brands have the same mean usable time before recharging is required?

| Brand X | Brand Y | Brand Z |
|---|---|---|
| 26.0 | 29.0 | 30.0 |
| 28.5 | 28.8 | 26.3 |
| 27.3 | 27.6 | 29.2 |
| 25.9 | 28.1 | 27.1 |
| 28.2 | 27.0 | 29.8 |
| $\bar{x}_1 = 27.18$ | $\bar{x}_2 = 28.10$ | $\bar{x}_3 = 28.48$ |
| $s_1^2 = 1.46$ | $s_2^2 = 0.69$ | $s_3^2 = 2.81$ |

In the above example, you might wonder why we don't test the claim that $\mu_x = \mu_y = \mu_z$ by using three individual tests with two means at a time ($\mu_x = \mu_y$ and $\mu_y = \mu_z$ and $\mu_x = \mu_z$). After all, Chapter 8 already covered procedures for testing equality of two means. That approach would lead to a significance level that is substantially changed because of the number of tests used. If we increase the number of individual tests of significance, we increase the chance of finding a difference by chance alone. There would be an excessively high risk that we would make the type I error of finding a difference in one of the pairs, when no such difference actually exists. Instead, we will use the analysis of variance method based on the following test statistic.

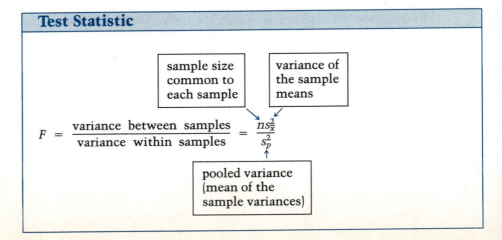

## Test Statistic

$$F = \frac{\text{variance between samples}}{\text{variance within samples}} = \frac{n s_{\bar{x}}^2}{s_p^2}$$

sample size common to each sample

variance of the sample means

pooled variance (mean of the sample variances)

Our analysis of variance method requires that we obtain two different estimates of the common population variance $\sigma^2$ and then compare those two estimates using the $F$ distribution. From the given test statistic we can see that $F$ is the ratio of two components. Those components are the different estimates of the common population variance $\sigma^2$.

# Variance Between Samples

The **variance between samples** (also called **variation due to treatment**) is an estimate of $\sigma^2$ based on the sample means. The intent of this estimate is to measure the variability caused by differences among the sample means that correspond to the different treatments or categories of classification. From the above test statistic, we see that with all samples of the same size $n$,

**variance between samples $= ns_{\bar{x}}^2$**

where $s_{\bar{x}}^2$ is the variance of the sample means.

Let's relate this to the sample data given above. The factor of brand has three "treatments" (Brand X, Brand Y, Brand Z) and the corresponding sample means (27.18, 28.10, 28.48) have a variance of 0.45; you can verify this on a calculator or by using Formula 2–5 or 2–6. We therefore have

variance between samples $= ns_{\bar{x}}^2 = (5)(0.45) = 2.25$

The expression $ns_{\bar{x}}^2$ can be justified as an estimate of $\sigma^2$ by first noting that $\sigma_{\bar{x}} = \sigma/\sqrt{n}$. Squaring both sides of this last expression and solving for $\sigma^2$, we get $\sigma^2 = n\sigma_{\bar{x}}^2$, which suggests that $\sigma^2$ can be estimated by $ns_{\bar{x}}^2$.

# Variance Within Samples

The **variance within samples** (also called **variation due to error**) is an estimate of $\sigma^2$ based on the sample variances. From the above test statistic, we see that with all samples of the same size $n$,

**variance within samples $= s_p^2$**

where $s_p^2$ is the *pooled variance* obtained by finding the mean of the sample variances.

For the sample data given above, we have

$$s_p^2 = \frac{s_1^2 + s_2^2 + s_3^2}{3} = \frac{1.46 + 0.69 + 2.81}{3} = 1.65$$

This approach seems reasonable since we assume that equal population variances imply that representative samples yield sample variances which, when pooled, provide a good estimate of the common population variance.

# Interpreting $F$

We can now proceed to include both estimates of $\sigma^2$ in the determination of the value of the $F$ statistic based on the data.

$$F = \frac{\text{variance between samples}}{\text{variance within samples}} = \frac{ns_{\bar{x}}^2}{s_p^2} = \frac{2.25}{1.65} = 1.36$$

If the two estimates of variance are close, the calculated value of $F$ will be close to 1 and we could conclude that there are no significant differences among the sample means. But if the value of $F$ is excessively *large*, then we would reject the claim of equal means. The estimate of variance in the denominator $(s_p^2)$ depends only on the sample variances and is not affected by differences among the sample means. However, if there are extreme differences among the sample means, the numerator will be larger so the value of $F$ will be larger. In general, as the sample means move farther apart, the value of $ns_{\bar{x}}^2$ grows larger and the value of $F$ itself grows larger. **Since excessively large values of $F$ reflect unequal means, the test is right tailed.**

The critical value of $F$ that separates excessive values from acceptable values is found in Table A–5, where $\alpha$ is again the level of significance and the numbers of degrees of freedom are as follows (assuming that there are $k$ sets of separate samples with $n$ scores in each set).

For cases with equal sample sizes:

$$\text{numerator degrees of freedom} = k - 1$$
$$\text{denominator degrees of freedom} = k(n - 1)$$

Our battery example involves $k = 3$ separate sets of samples and there are $n = 5$ scores in each set, so

$$\text{numerator degrees of freedom} = k - 1 = 3 - 1 = 2$$
$$\text{denominator degrees of freedom} = k(n - 1) = 3(5 - 1) = 12$$

With $\alpha = 0.05$, the critical $F$ value corresponding to these degrees of freedom is 3.8853. The computed test statistic of $F = 1.36$ does not fall within the right-tailed critical region bounded by $F = 3.8853$ (see Figure 11–2), so we fail to reject the null hypothesis of equal population means. That is, there are no significant differences among the means of the 3 given sets of samples. The pilot should purchase the least expensive battery.

The preceding analysis led to a right-tailed critical region, and every other similar situation will also involve a right-tailed critical region. Since the individual components of $n$, $s_{\bar{x}}^2$, and $s_p^2$ are all positive, $F$ is always positive, and we know that values of $F$ near 1 correspond to relatively close sample means. The only values of the $F$ test statistic that are indicative of significant differences among the sample means are those $F$ values that exceed 1 beyond the critical value obtained from Table A–5.

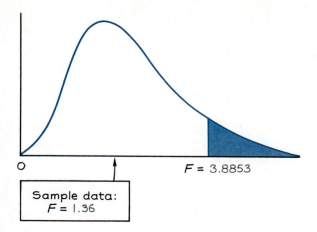

**Figure 11–2**

## DEATH PENALTY AS DETERRENT

■ A common argument supporting the death penalty is that it discourages others from committing murder. Jeffrey Grogger, of the University of California, analyzed daily homicide data in California during a four-year period with frequent executions. Among his conclusions published in the *Journal of the American Statistical Association* (Vol. 85, No. 410): "The analyses conducted consistently indicate that these data provide no support for the hypothesis that executions deter murder in the short term." This is a major social policy issue, and the efforts of people such as Professor Grogger help to dispel misconceptions so that we have accurate information with which to address such issues. ■

It may seem strange that we are testing equality of several means by analyzing only variances, but the sample means directly affect the value of the test statistic $F$. To illustrate this important point, we add 10 to each score listed under Brand X but we leave the Brand Y and Brand Z values unchanged. The revised sample statistics follow.

| Brand X | Brand Y | Brand Z |
|---|---|---|
| $\overline{x}_1 = 37.18$ | $\overline{x}_2 = 28.10$ | $\overline{x}_3 = 28.48$ |
| $s_1^2 = 1.46$ | $s_2^2 = 0.69$ | $s_3^2 = 2.81$ |

We again assume a significance level of $\alpha = 0.05$ and test the claim that the three brands have the same population mean. In this way, we can see the effect of increasing the Brand X scores by 10. The $F$ statistic based on the revised sample data follows.

$$F = \frac{\text{variance between samples}}{\text{variance within samples}} = \frac{ns_{\overline{x}}^2}{s_p^2} = \frac{(5)(26.38)}{(1.46 + 0.69 + 2.81)/3}$$

$$= \frac{131.90}{1.65} = 79.94$$

We can see the effect of adding 10 to each Brand X score. Note that the variance between samples changes from 2.25 to 131.90, indicating more variability among the three sample means. However, the variance within samples doesn't change at all, since the variability within a sample isn't affected when we add a constant to every score. The net effect is a change in the $F$ test statistic from 1.36 to 79.94. This difference in $F$ is attributable only to the change in $\overline{x}_1$. This illustrates that the $F$ test statistic is very sensitive to sample *means*, even though it is obtained through two different estimates of the common population *variance*.

We now summarize the key components of the hypothesis test for equality of means in three or more populations. Note that the following test statistic and numbers of degrees of freedom apply only to cases involving samples of the same size. That is, we have $k$ samples, each with $n$ scores.

---

### Notation

$k$ = number of samples
$n$ = size of each sample

#### Null and Alternative Hypotheses

$H_0$: $\mu_1 = \mu_2 = \ldots = \mu_k$
$H_1$: The preceding means are not all equal.

#### Test Statistic

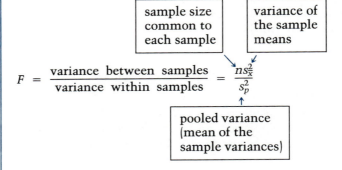

$$F = \frac{\text{variance between samples}}{\text{variance within samples}} = \frac{ns_{\bar{x}}^2}{s_p^2}$$

sample size common to each sample

variance of the sample means

pooled variance (mean of the sample variances)

#### Critical Value

- Test is always right-tailed.
- Critical values are found in Table A–5 with

   **numerator degrees of freedom = $k - 1$**

   **denominator degrees of freedom = $k(n - 1)$**

---

We must also keep in mind that we should satisfy the assumptions described in Section 11–1. That is, the populations should have nearly normal distributions with approximately the same variance, and the samples must be independent of each other.

# 11–2  Exercises A

1. The dean of a college wants to compare grade-point averages of resident, commuting, and part-time students. A random sample of each group is selected, and the results are as follows. At the $\alpha = 0.05$ level of significance, test the claim that the three populations have equal means.

| Residents | Commuters | Part-time |
|---|---|---|
| $n = 15$ | $n = 15$ | $n = 15$ |
| $\overline{x} = 2.60$ | $\overline{x} = 2.55$ | $\overline{x} = 2.30$ |
| $s^2 = 0.30$ | $s^2 = 0.25$ | $s^2 = 0.16$ |

2. Do Exercise 1 after changing the sample mean grade-point average for resident students from 2.60 to 3.00.

3. Five socioeconomic classes are being studied by a sociologist, and members from each class are rated for their adjustments to society. The sample data are summarized as follows. At the $\alpha = 0.05$ level of significance, test the claim that the five populations have equal means.

| A | B | C | D | E |
|---|---|---|---|---|
| $n_1 = 10$ | $n_2 = 10$ | $n_3 = 10$ | $n_4 = 10$ | $n_5 = 10$ |
| $\overline{x}_1 = 103$ | $\overline{x}_2 = 97$ | $\overline{x}_3 = 102$ | $\overline{x}_4 = 100$ | $\overline{x}_5 = 110$ |
| $s_1^2 = 230$ | $s_2^2 = 75$ | $s_3^2 = 200$ | $s_4^2 = 150$ | $s_5^2 = 100$ |

4. A unit on elementary algebra is taught to five different classes of randomly selected students with the same academic backgrounds. A different method of teaching is used in each class, and the final averages of the 20 students in each class are compiled. The results yield the following data. At the $\alpha = 0.05$ level of significance, test the claim that the five population means are equal.

| Traditional | Programmed | Audio | Audiovisual | Visual |
|---|---|---|---|---|
| $n = 20$ | $n = 20$ | $n = 20$ | $n = 20$ | $n = 20$ |
| $\overline{x} = 76$ | $\overline{x} = 74$ | $\overline{x} = 70$ | $\overline{x} = 75$ | $\overline{x} = 74$ |
| $s^2 = 60$ | $s^2 = 50$ | $s^2 = 100$ | $s^2 = 36$ | $s^2 = 40$ |

5. Five car models are studied in a test that involves four of each model. For each of the four cars in each of the five samples, exactly 1 gallon of gas is placed in the tank and the car is driven until the gas is used up. The results are given at the right. At the $\alpha = 0.05$ significance level, test the claim that the five population means are all equal.

| Distance Traveled in Miles | | | | |
|---|---|---|---|---|
| A | B | C | D | E |
| 16 | 18 | 18 | 19 | 15 |
| 22 | 23 | 18 | 21 | 16 |
| 17 | 15 | 20 | 22 | 20 |
| 17 | 20 | 20 | 22 | 17 |

| Stable | Divorced | Transition |
|--------|----------|------------|
| 110 | 115 | 90 |
| 105 | 105 | 120 |
| 100 | 110 | 125 |
| 95 | 130 | 100 |
| 120 | 105 | 105 |

6. A sociologist randomly selects subjects from three types of family structures: stable families, divorced families, and families in transition. The selected subjects are interviewed and rated for their social adjustment. The sample results are given at left. At the $\alpha = 0.05$ level of significance, test the claim that the three population means are equal.

7. Do Exercise 6 after adding 30 to each score in the stable group.

8. Readability studies are conducted to determine the clarity of four different texts, and the sample scores follow. At the $\alpha = 0.05$ level of significance, test the claim that the four texts produce the same mean readability score.

| Text A | Text B | Text C | Text D |
|--------|--------|--------|--------|
| 50 | 59 | 48 | 60 |
| 51 | 60 | 51 | 65 |
| 53 | 58 | 47 | 62 |
| 58 | 57 | 49 | 68 |
| 53 | 61 | 50 | 70 |

In Exercises 9–12, use the data given at the left. The data represent 30 different homes recently sold in Dutchess County, New York. The zones (1, 4, 7) correspond to different geographical regions of the county. The values of SP are the selling prices in thousands of dollars. The values of LA are the living areas in hundreds of square feet. The Acres values are the lot sizes in acres, and the Taxes values are the annual tax bills in thousands of dollars. For example, the first home is in zone 1, it sold for $147,000, it has a living area of 2000 square feet, it is on a 0.50-acre lot, and the annual taxes are $1900.

| Zone | SP | LA | Acres | Taxes |
|------|-----|----|-------|-------|
| 1 | 147 | 20 | 0.50 | 1.9 |
| 1 | 160 | 18 | 1.00 | 2.4 |
| 1 | 128 | 27 | 1.05 | 1.5 |
| 1 | 162 | 17 | 0.42 | 1.6 |
| 1 | 135 | 18 | 0.84 | 1.6 |
| 1 | 132 | 13 | 0.33 | 1.5 |
| 1 | 181 | 24 | 0.90 | 1.7 |
| 1 | 138 | 15 | 0.83 | 2.2 |
| 1 | 145 | 17 | 2.00 | 1.6 |
| 1 | 165 | 16 | 0.78 | 1.4 |
| 4 | 160 | 18 | 0.55 | 2.8 |
| 4 | 140 | 20 | 0.46 | 1.8 |
| 4 | 173 | 19 | 0.94 | 3.2 |
| 4 | 113 | 12 | 0.29 | 2.1 |
| 4 | 85 | 9 | 0.26 | 1.4 |
| 4 | 120 | 18 | 0.33 | 2.1 |
| 4 | 285 | 28 | 1.70 | 4.2 |
| 4 | 117 | 10 | 0.50 | 1.7 |
| 4 | 133 | 15 | 0.43 | 1.8 |
| 4 | 119 | 12 | 0.25 | 1.6 |
| 7 | 215 | 21 | 3.04 | 2.7 |
| 7 | 127 | 16 | 1.09 | 1.9 |
| 7 | 98 | 14 | 0.23 | 1.3 |
| 7 | 147 | 23 | 1.00 | 1.7 |
| 7 | 184 | 17 | 6.20 | 2.2 |
| 7 | 109 | 17 | 0.46 | 2.0 |
| 7 | 169 | 20 | 3.20 | 2.2 |
| 7 | 110 | 14 | 0.77 | 1.6 |
| 7 | 68 | 12 | 1.40 | 2.5 |
| 7 | 160 | 18 | 4.00 | 1.8 |

9. At the 0.05 significance level, test the claim that the means of the selling prices are the same in zones 1, 4, and 7.

10. At the 0.05 significance level, test the claim that the means of the living areas are the same in zones 1, 4, and 7.

11. At the 0.05 significance level, test the claim that the means of the lot sizes (in acres) are the same in zones 1, 4, and 7.

12. At the 0.05 significance level, test the claim that the means of the tax amounts are the same in zones 1, 4, and 7.

# 11–2 Exercises B

13. A study is made of three police precincts to determine the time (in seconds) required for a police car to be dispatched after a crime is reported. Sample results are given at right.

    a. At the 5% level of significance, test the claim that $\mu_1 = \mu_2$. Use the methods discussed in Chapter 8.

    b. At the 5% level of significance, test the claim that $\mu_2 = \mu_3$. Use the methods discussed in Chapter 8.

    c. At the 5% level of significance, test the claim that $\mu_1 = \mu_3$. Use the methods discussed in Chapter 8.

    d. At the 5% level of significance, test the claim that $\mu_1 = \mu_2 = \mu_3$. Use analysis of variance.

    e. Compare the methods and results of parts $a$, $b$, and $c$ to part $d$.

14. Five independent samples of 50 scores are randomly drawn from populations that are normally distributed with equal variances. We wish to test the claim that $\mu_1 = \mu_2 = \mu_3 = \mu_4 = \mu_5$.

    a. If we use only the methods given in Chapter 8, we would test the individual claims $\mu_1 = \mu_2$, $\mu_1 = \mu_3$, . . ., $\mu_4 = \mu_5$. What is the number of claims? That is, how many ways can we pair off five means?

    b. Assume that for each test of equality between two means, there is a 0.95 probability of not making a type I error. If all possible pairs of means are tested for equality, what is the probability of making no type I errors? (Although the tests are not actually independent, assume that they are.)

    c. If we use analysis of variance to test the claim that $\mu_1 = \mu_2 = \mu_3 = \mu_4 = \mu_5$ at the 5% level of significance, what is the probability of not making a type I error?

    d. Compare the results of parts $b$ and $c$.

15. Five independent samples of 50 scores are randomly drawn from populations that are normally distributed with equal variances, and the values of $n$, $\overline{x}$, and $s$ are obtained in each case. Analysis of variance is then used to test the claim that $\mu_1 = \mu_2 = \mu_3 = \mu_4 = \mu_5$.

    a. If a constant is added to each of the five sample means, how is the value of the test statistic affected?

    b. If each of the five means is multiplied by a constant, how is the value of the test statistic affected?

Precinct 1

$n_1 = 50$
$\overline{x}_1 = 170 \text{ s}$
$s_1 = 18 \text{ s}$

Precinct 2

$n_2 = 50$
$\overline{x}_2 = 202 \text{ s}$
$s_2 = 20 \text{ s}$

Precinct 3

$n_3 = 50$
$\overline{x}_3 = 165 \text{ s}$
$s_3 = 23 \text{ s}$

# 11–3  ANOVA with Unequal Sample Sizes

In Section 11–2 our discussion of analysis of variance involved only examples with the same number of scores in each sample. We will now proceed to consider the method for working with samples of unequal sizes. However, we continue to deal with a single factor; two factors will be considered in Section 11–4. We will again proceed by analyzing the significance of the difference between *variance between samples* (**variation due to treatment**) and *variance within samples* (**variation due to error**). As with the case of equal sample sizes, we again use the test statistic

$$F = \frac{\text{variance between samples}}{\text{variance within samples}}$$

However, for unequal sample sizes we weight the two estimates of variance to account for the different sample sizes. We get the following.

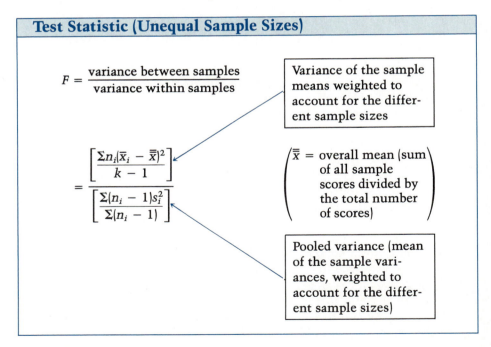

**Test Statistic (Unequal Sample Sizes)**

$$F = \frac{\text{variance between samples}}{\text{variance within samples}}$$

Variance of the sample means weighted to account for the different sample sizes

$$= \frac{\left[\dfrac{\Sigma n_i(\bar{x}_i - \bar{\bar{x}})^2}{k - 1}\right]}{\left[\dfrac{\Sigma(n_i - 1)s_i^2}{\Sigma(n_i - 1)}\right]}$$

$\left(\begin{array}{l}\bar{\bar{x}} = \text{overall mean (sum}\\ \text{of all sample}\\ \text{scores divided by}\\ \text{the total number}\\ \text{of scores)}\end{array}\right)$

Pooled variance (mean of the sample variances, weighted to account for the different sample sizes)

Note that the numerator is really a form of the formula

$$s^2 = \frac{\Sigma(x - \bar{x})^2}{n - 1}$$

for variance that was given in Chapter 2. The factor of $n_i$ is included so that larger samples carry more weight. The denominator of the test statistic is simply the mean of the sample variances, but it is a weighted mean with the weights based on the sample sizes. (If all samples have the same number of scores, the test statistic simplifies to $F = ns_{\bar{x}}^2/s_p^2$, as given earlier.)

Once the test statistic has been found, we need to find the critical value. As in Section 11–2, this test is also right-tailed and the critical values are found in Table A–5, but with unequal sample sizes we have

**degrees of freedom (numerator) = $k - 1$**     where $k$ is the number of samples

**degrees of freedom (denominator) = $N - k$**     where $N$ is the total number of scores in all samples combined

## DISCRIMINATION CASE USES STATISTICS

■ Statistics often play a key role in discrimination cases. One such case involved Matt Perez and more than 300 other FBI agents who won a class action suit charging that Hispanics in the FBI were discriminated against in the areas of promotions, assignments, and disciplinary actions. The plaintiff employed statistician Gary Lafree, who showed that the FBI's upper management positions had significantly low proportions of Hispanic employees. Statistics were instrumental in the plaintiff's victory in this case. (See Program 20 from the series *Against All Odds: Inside Statistics* for a discussion of this case.) ■

### Example

Three different transmission types are installed on different cars, all having 6 cylinder engines and an engine size of 3.0 liters. The fuel consumption (in mi/gal) is found for highway conditions with the results given below (based on data from the Environmental Protection Agency). The letters $A$, $M$, $L$ represent automatic, manual, and lockup torque converter, respectively. At the 0.05 significance level, test the claim that the mean fuel consumption values are the same for all three transmission types.

| A | M | L |
|---|---|---|
| 23 | 27 | 24 |
| 23 | 29 | 26 |
| 20 | 25 | 24 |
| 21 | 23 | |
| | 24 | |
| $n_1 = 4$ | $n_2 = 5$ | $n_3 = 3$ |
| $\bar{x}_1 = 21.8$ | $\bar{x}_2 = 25.6$ | $\bar{x}_3 = 24.7$ |
| $s_1^2 = 2.3$ | $s_2^2 = 5.8$ | $s_3^2 = 1.3$ |

### Solution

The null hypothesis is the claim that the samples come from populations with equal means.

We have

$H_0$: $\mu_A = \mu_M = \mu_L$
$H_1$: The preceding means are not all equal.

*continued*

> ### Solution *continued*
>
> With $\alpha = 0.05$, we proceed to calculate the test statistic as follows.
>
> $$\text{Number of samples: } k = 3 \qquad \bar{\bar{x}} = \text{mean of all scores} = \frac{289}{12} = 24.1$$
>
> variance between samples (numerator of $F$) =
>
> $$\frac{\Sigma n_i(\bar{x}_i - \bar{\bar{x}})^2}{k-1} = \frac{4(21.8 - 24.1)^2 + 5(25.6 - 24.1)^2 + 3(24.7 - 24.1)^2}{3-1}$$
>
> $$= \frac{33.5}{2} = 16.8$$
>
> variance within samples (denominator of $F$) =
>
> $$\frac{\Sigma(n_i - 1)s_i^2}{\Sigma(n_i - 1)} = \frac{(4-1)(2.3) + (5-1)(5.8) + (3-1)(1.3)}{(4-1) + (5-1) + (3-1)}$$
>
> $$= \frac{32.7}{9} = 3.6$$
>
> $$F = \frac{\text{variance between samples}}{\text{variance within samples}} = \frac{16.8}{3.6} = 4.7$$
>
> With a test statistic of $F = 4.7$ we now proceed to obtain the critical value. The significance level indicates that $\alpha = 0.05$. The degrees of freedom are as follows.
>
> $$\text{degrees of freedom (numerator)} = k - 1 = 3 - 1 = 2$$
>
> $$\text{degrees of freedom (denominator)} = N - k = 12 - 3 = 9$$
>
> With 2 degrees of freedom for the numerator and 9 degrees of freedom for the denominator, we refer to Table A–5 and find the critical value of $F = 4.2565$. Since the test statistic of $F = 4.7$ does exceed the critical value of $F = 4.2565$, we reject the null hypothesis that the means are equal. There is sufficient sample evidence to warrant rejection of the claim that the means are equal. It appears that the fuel consumption is not the same for all three transmission types.

The method illustrated in the preceding example is based on sample statistics $(n_i, \bar{x}_i, s_i^2)$ that would probably be found in the early stages of research. That method is a generalization of the same method used for the case of equal sample sizes; it reinforces earlier concepts and follows patterns already established. However, it often leads to larger errors from rounding

$k$ = number of different samples (or the number of columns of data)

$n_i$ = number of values in the $i$th sample

$c_i$ = total of the values in the $i$th sample

$N = \Sigma n_i$ = total number of values in all samples combined ($N$ has been used previously to represent the size of a population.)

$\Sigma x = \Sigma c_i$ = sum of the scores of all samples combined

$\Sigma x^2$ = sum of the squares of all scores from all samples combined

| | |
|---|---|
| $SS(\text{total}) = \Sigma x^2 - \dfrac{(\Sigma x)^2}{N}$ | Total sum of squares. This expression is algebraically the same as $\Sigma(x - \bar{x})^2$. |
| $SS(\text{treatment}) = \left(\Sigma \dfrac{c_i^2}{n_i}\right) - \dfrac{(\Sigma c_i)^2}{N}$ | Sum of squares that represents variation among the different samples (sometimes called *explained variation*). |
| $SS(\text{error}) =$ $SS(\text{total}) - SS(\text{treatment})$ | Sum of squares that represents the variation within samples that is due to chance (sometimes called *unexplained variation*.) |
| $df(\text{treatment}) = k - 1$ | Degrees of freedom associated with the different treatments. |
| $df(\text{error}) = N - k$ | Degrees of freedom associated with the errors within samples. |
| $MS(\text{treatment}) = \dfrac{SS(\text{treatment})}{df(\text{treatment})}$ | Mean square or variance estimate explained by the different treatments. |
| $MS(\text{error}) = \dfrac{SS(\text{error})}{df(\text{error})}$ | Mean square or variance estimate that is unexplained (due to chance). |
| $F = \dfrac{MS(\text{treatment})}{MS(\text{error})}$ | Test statistic representing the ratio of two estimates of variance. |

**TABLE 11–1    ANOVA Table**

| Source of Variation | Sum of Squares $SS$ | Degrees of Freedom | Mean Square $MS$ | Test Statistic |
|---|---|---|---|---|
| Treatments | $SS(\text{treatment})$ | $k - 1$ | $MS(\text{treatment}) = \dfrac{SS(\text{treatment})}{df(\text{treatment})}$ | $F = \dfrac{MS(\text{treatment})}{MS(\text{error})}$ |
| Error | $SS(\text{error})$ | $N - k$ | $MS(\text{error}) = \dfrac{SS(\text{error})}{df(\text{error})}$ | |
| Total | $SS(\text{total})$ | $N - 1$ | | |

than some other equivalent methods. Refer now to Table 11–1, which summarizes a second method that produces the same results, but usually with less error due to rounding.

Using Table 11–1 with the data from the preceding example, we get

$k = 3$           $N = 12$
$\Sigma x = 289$       $\Sigma x^2 = 7027$
$df(\text{treatment}) = 2$     $df(\text{error}) = 9$
$SS(\text{treatment}) = 34.3000$     $SS(\text{error}) = 32.6167$
$MS(\text{treatment}) = 17.1500$     $MS(\text{error}) = 3.6241$

$$F = \frac{MS(\text{treatment})}{MS(\text{error})} = \frac{17.1500}{3.6241} = 4.7322$$

These results show that the same test statistic is obtained with this second method. The first method is superior if you want to better understand the underlying concepts, and the second method is generally better if you want easier calculations. However, the "easier" calculations of the second method are complex enough that the use of computer software should be seriously considered for most real problems. The STATDISK and Minitab displays resulting from the data of the preceding problem are as follows.

STATDISK DISPLAY

```
D.F. (Treatment) . . . . . . . . . . . . . . . . . = 2
D.F. (Error) . . . . . . . . . . . . . . . . . . . . . = 9

Sum of Squares (Treatment) . . . . . . . = 34.2967
Sum of Squares (Error) . . . . . . . . . . = 32.62
Total Sum of Squares . . . . . . . . . . . . . = 66.9167

Mean Squares (Treatment) . . . . . . . . = 17.1483
Mean Squares (Error) . . . . . . . . . . . . = 3.62444

Significance level . . . . . . . . . . . . . . . . = .05
F. . . . . . . . . . . . . . . . . . . . . . . . . . . F = 4.7313
P-value . . . . . . . . . . . . . . . . . . . . . . . = .0394253
                    REJECT
        the null hypothesis of equal means
```

MINITAB DISPLAY

```
MTB > SET C1                        In the display below, Minitab
DATA> 23 23 20 21                   uses "FACTOR" instead of
DATA> ENDOFDATA                     "TREATMENT." That display
MTB > SET C2                        is one type of ANOVA table.
DATA> 27 29 25 23 24
DATA> ENDOFDATA
MTB > SET C3
DATA> 24 26 24
DATA> ENDOFDATA
MTB > AOVONEWAY C1 C2 C3


ANALYSIS OF VARIANCE
SOURCE    DF      SS      MS       F       P
FACTOR     2    34.30   17.15    4.73    0.039
ERROR      9    32.62    3.62
TOTAL     11    66.92
```

As efficient and reliable as such computer programs may be, they are totally worthless if we don't understand the relevant concepts. We should recognize that the methods in this section are used to test the claim that several samples come from populations with the same mean. **These methods require normally distributed populations with the same variance, and the samples must be independent.** We reject or fail to reject the null hypothesis of equal means by analyzing the two estimates of variance. The $MS$(treatment) is an estimate of the variation between samples, while the $MS$(error) is an estimate of the variation within samples. If $MS$(treatment) is significantly greater than $MS$(error), then we reject the claim of equal means, otherwise we fail to reject that claim.

When we use analysis of variance and conclude that three or more treatment groups have sample means that are not all equal, we do not identify the particular means that are different. There are several other tests that could be used to make such identifications. Such procedures for specifically identifying the means that are different are called **multiple comparison procedures.** The **extended Tukey test** and the **Bonferroni test** are two common multiple comparison procedures included in some of the more advanced texts.

# 11–3 Exercises A

1. An introductory calculus course is taken by students with varying high school records. The sample results from each of three groups follow. The values given are the final numerical averages in the calculus course. At the $\alpha = 0.05$ level of significance, test the claim that the mean scores are equal in the three groups.

| Good high school record | Fair high school record | Poor high school record |
|---|---|---|
| 90 | 80 | 60 |
| 86 | 70 | 60 |
| 88 | 61 | 55 |
| 93 | 52 | 62 |
| 80 | 73 | 50 |
|  | 65 | 70 |
|  | 83 |  |

2. A preliminary study is conducted to determine whether there is any relationship between education and income. The sample results are as follows. The figures represent, in thousands of dollars, the lifetime incomes of randomly selected workers from each category. At the $\alpha = 0.05$ level of significance, test the claim that the samples come from populations with equal means.

| Years of Education | | | | |
|---|---|---|---|---|
| 8 years or less | 9–11 years | 12 years | 13–15 years | 16 or more years |
| 300 | 270 | 400 | 420 | 570 |
| 210 | 330 | 430 | 480 | 640 |
| 260 | 380 | 370 | 510 | 590 |
| 330 | 310 | 390 | 390 | 700 |
|  |  | 420 | 470 | 620 |
|  |  |  |  | 660 |

3. Three car models are studied in tests that involve several cars of each model. In each case, the car is run on exactly 1 gallon of gas until the fuel supply is exhausted. The distances traveled are on the right. Test the claim that the population means are equal.

| A | B | C |
|---|---|---|
| 16 | 14 | 20 |
| 20 | 16 | 21 |
| 18 | 16 | 19 |
| 18 | 17 | 22 |
|  | 15 | 18 |
|  | 18 | 24 |
|  |  | 18 |
|  |  | 20 |

4. Three groups of adult men were selected for an experiment designed to measure their blood alcohol levels after consuming five drinks. Members of group A were tested after one hour, members of group B were tested after two hours, and members of group C were tested after four hours. The results are given at right. At the 0.05 level of significance, test the claim that the three groups have the same mean level.

| A | B | C |
|---|---|---|
| 0.11 | 0.08 | 0.04 |
| 0.10 | 0.09 | 0.04 |
| 0.09 | 0.07 | 0.05 |
| 0.09 | 0.07 | 0.05 |
| 0.10 | 0.06 | 0.06 |
| | | 0.04 |
| | | 0.05 |

5. Medical researchers used three different treatments in an experiment involving volunteer patients, with the recovery times (in days) given at right. Using a 0.01 significance level, test the claim that the different treatments result in the same means.

| A | B | C |
|---|---|---|
| 6 | 9 | 11 |
| 8 | 8 | 9 |
| 12 | 7 | 10 |
| 9 | 6 | 8 |
| 7 | 9 | 11 |
| 2 | | 9 |
| | | 12 |
| | | 14 |

6. A dental research team investigating a new tooth filling material experimented with four different hardening methods, and obtained the sample results given at right. At the 0.01 significance level, test the claim that the four methods yield the same mean index of hardness.

| A | B | C | D |
|---|---|---|---|
| 8.2 | 6.9 | 8.2 | 8.0 |
| 7.9 | 7.3 | 8.5 | 7.2 |
| 8.4 | 7.5 | 8.9 | 7.3 |
| 8.0 | 8.2 | 8.7 | 7.1 |
| 8.0 | 6.3 | 8.6 | 7.9 |
| | 6.8 | 8.4 | 7.3 |
| | 6.7 | 8.8 | 7.1 |
| | | | 7.4 |

7. The numbers of program errors are recorded for four different programmers on randomly selected days. Test the claim that they produce the same mean number of errors. Use a 0.01 level of significance.

| 1 | 2 | 3 | 4 |
|---|---|---|---|
| 14 | 3 | 17 | 16 |
| 16 | 5 | 20 | 18 |
| 18 | 12 | 22 | 20 |
| 14 | 8 | 24 | 17 |
| 22 | 7 | 26 | 21 |
| 9 | 6 | 18 | |
| | 6 | 9 | |
| | 4 | 11 | |
| | 7 | | |
| | 16 | | |

8. Five different machines are used to produce floppy disks and the number of defects are recorded for batches randomly selected on different days. Use a 0.01 significance level to test the claim that the machines produce the same mean number of defects.

| 1 | 2 | 3 | 4 | 5 |
|---|---|---|---|---|
| 8 | 11 | 14 | 32 | 10 |
| 9 | 11 | 13 | 33 | 8 |
| 6 | 8 | 9 | 26 | 11 |
| 10 | 10 | 10 | 15 | 14 |
| 12 | 13 | 12 | 18 | 22 |
| | 12 | | 25 | |
| | | | 31 | |
| | | | 40 | |

## 11–3   Exercises B

9. Complete the following ANOVA table if it is known that there are three samples with sizes of 5, 7, and 7, respectively.

| Source of Variation | Sum of Squares SS | Degrees of Freedom | Mean Square MS | Test Statistic |
|---|---|---|---|---|
| Treatments | ? | ? | ? | |
| Error | 112.57 | ? | ? | $F = ?$ |
| Total | 114.74 | ? | | |

10. Complete the following ANOVA table that resulted from samples of sizes 12, 25, 50, and 32.

| Source of Variation | Sum of Squares SS | Degrees of Freedom | Mean Square MS | Test Statistic |
|---|---|---|---|---|
| Treatments | 21.34 | ? | ? | |
| Error | 144.45 | ? | ? | $F = ?$ |
| Total | ? | ? | | |

11. Refer to the sample data given in Exercise 7. Assume that the first entry of 14 is incorrectly entered as 1400. How are the results affected by this outlier?

12. A researcher plans to conduct an analysis of variance on three samples of temperatures. Does it make any difference if the temperatures are entered in the Fahrenheit scale or the Celsius scale? In general, is the test statistic affected by the scale used?

13. Assuming that we have $k$ samples all of the same size $n$, verify the following:
   a. Variance between samples

$$\frac{\Sigma n_i(\overline{x}_i - \overline{\overline{x}})^2}{k - 1} \quad \text{reduces to } ns_{\overline{x}}^2$$

   b. Variance within samples

$$\frac{\Sigma(n_i - 1)s_i^2}{\Sigma(n_i - 1)} \quad \text{reduces to } \frac{\Sigma s_i^2}{k} \text{ or } s_p^2$$

14. Do Exercise 17 in Section 8–3 using the $t$-test described in that section and then solve it again using analysis of variance described in this section. Verify that the two tests are equivalent and the test statistics and critical values are related by $t^2 = F$.

# 11–4  Two-Way ANOVA

The preceding sections of this chapter illustrated the use of analysis of variance in deciding whether differences among sample means are statistically significant or attributable to chance. Those sections used procedures referred to as **one-way analysis of variance** or **single-factor analysis of variance.** These terms remind us that the data are classified into groups according to a single factor. (A **factor** is a property that is the basis for categorizing the different groups of data.) In the example illustrated in Section 11–3, the three samples were categorized according to the single factor of transmission type. In Table 11–2, three samples are categorized according to the single factor of engine size.

| TABLE 11–2 | Mi/gal (highway) of Twelve Different 4-Cylinder Cars | |
|---|---|---|
| | Engine Size (liters) | |
| 1.5 | 2.2 | 2.5 |
| 31 | 28 | 31 |
| 32 | 26 | 23 |
| 33 | 33 | 27 |
| 36 | 30 | 34 |

Since the three samples in Table 11–2 are all of the same size, $n = 4$, we could use the methods from Sections 11–2 or 11–3 to test the claim that $\mu_1 = \mu_2 = \mu_3$. Following is the Minitab display for the data in Table 11–2.

MINITAB DISPLAY

```
MTB  > SET C1
DATA > 31 32 33 36
DATA > ENDOFDATA
MTB  > SET C2
DATA > 28 26 33 30
DATA > ENDOFDATA
MTB  > SET C3
DATA > 31 23 27 34
DATA > ENDOFDATA
MTB  > AOVONEWAY C1 C2 C3

ANALYSIS OF VARIANCE
SOURCE    DF      SS      MS       F       p
FACTOR     2     43.2    21.6     1.77    0.224
ERROR      9    109.5    12.2
TOTAL     11    152.7
```

From this Minitab display we can see that the test statistic is

$$F = \frac{\text{variance between samples}}{\text{variance within samples}} = \frac{MS(\text{treatment})}{MS(\text{error})} = \frac{21.6}{12.2} = 1.77$$

(Note that Minitab uses "FACTOR" instead of "TREATMENT.") The $P$-value of 0.224 indicates that at the 0.05 significance level, we fail to reject the claim of equal means. Based on our limited sample data, it appears that the engine size does not have a significant effect on fuel consumption.

The preceding one-way analysis of variance involves the single factor of engine size. A more complex analysis might include additional factors, such as transmission type and weight of the car. In this section we consider **two-way analysis of variance,** which involves two factors.

Suppose we now have the sample data shown in Table 11–3 (based on data from the Environmental Protection Agency). Twelve different 4-cylinder

| TABLE 11–3 | Mi/gal (highway) of Different 4-Cylinder Compact Cars | | |
|---|---|---|---|
| | | Engine Size (liters) | |
| | | 1.5 | 2.2 | 2.5 |
| Transmission | Automatic | 31, 32 | 28, 26 | 31, 23 |
| | Manual | 33, 36 | 33, 30 | 27, 34 |

cars were tested for fuel consumption (in mi/gal) while driving under identical highway conditions. Tables 11–2 and 11–3 include the same numbers, but Table 11–2 classifies the data with the single factor of engine size, whereas Table 11–3 classifies the data with the two factors of engine size and transmission type.

Since we've already completed the one-way analysis of variance for the single factor of engine size, it might seem reasonable that we simply proceed with another one-way ANOVA for the factor of transmission type. Unfortunately, that approach wastes information and totally ignores any effect that might be due to an interaction between the two factors. Instead, we will proceed with two-way analysis of variance. The calculations are quite involved, so *we will assume that a software package is being used.* Shown below is the Minitab display for the data in Table 11–3.

# DELAYING DEATH

■ University of California sociologist David Phillips has studied the ability of people to postpone their death until after some important event. Analyzing death rates of Jewish men who died near Passover, he found that the death rate dropped dramatically in the week before Passover, but rose the week after. He found a similar phenomenon occurring among Chinese-American women; their death rate dropped the week before their important Harvest Moon Festival, then rose the week after. ■

MINITAB DISPLAY

```
MTB > READ MPG IN C1 TRANS IN C2 SIZE IN C3
DATA> 31 1 1
DATA> 32 1 1
DATA> 28 1 2
DATA> 26 1 2
DATA> 31 1 3
DATA> 23 1 3
DATA> 33 2 1
DATA> 36 2 1
DATA> 33 2 2
DATA> 30 2 2
DATA> 27 2 3
DATA> 34 2 3
DATA> ENDOFDATA
       12 ROWS READ
MTB > NAME C1 'MPG' C2 'TRANS' C3 'SIZE'
MTB > TWOWAY C1 C2 C3

ANALYSIS OF VARIANCE MPG
SOURCE          DF      SS      MS
TRANS            1     40.3    40.3
SIZE             2     43.2    21.6
INTERACTION      2      1.2     0.6
ERROR            6     68.0    11.3
TOTAL           11    152.7
```

Since the one-factor Table 11–2 and the two-factor Table 11–3 use the same data, we might expect some of the calculations to be the same. A comparison of the corresponding Minitab displays is given below.

| One-Way ANOVA | Two-Way ANOVA |
|---|---|
| Sample data: Table 11–2 | Sample data: Table 11–3 |
| One factor: engine size | Two factors: engine size and transmission type |
| $SS$(size) = 43.2 | Same |
| $MS$(size) = 21.6 | Same |
| $df$(size) = 2 | Same |
| $SS$(total) = 152.7 | Same |
| $df$(total) = 11 | Same |
| | $SS$(transmission) = 40.3 |
| $SS$(error) = 109.5 | $SS$(interaction) = 1.2 |
| | $SS$(error) = 68.0 |
| | $df$(transmission) = 1 |
| $df$(error) = 9 | $df$(interaction) = 2 |
| | $df$(error) = 6 |
| | $MS$(transmission) = 40.3 |
| $MS$(error) = 12.2 | $MS$(interaction) = 0.6 |
| | $MS$(error) = 11.3 |

Since the engine size numbers are the same in both cases, the calculated values of $SS$(size), $MS$(size), and $df$(size) are also the same in both cases. But note that as we go from the one-factor case (Table 11–2) to the two-factor case (Table 11–3) by partitioning the data according to transmission type, the value of $SS$(error) is partitioned into $SS$(transmission), $SS$(interaction), and $SS$(error). Also, $df$(error) is partitioned into $df$(transmission), $df$(interaction), and $df$(error). The term **interaction** refers to the effect due to the interaction of the two factors; there is an interaction effect if the *combination* of engine size and transmission type has an effect on mileage (see Figure 11–3). In executing a two-way analysis of variance, we consider three effects:

1. The effect due to the *interaction* between the two factors.
2. The effect due to the *row* factor.
3. The effect due to the *column* factor.

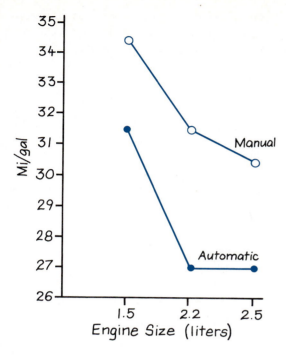

*This is a graph of the cell means from Table 11–3. When the lines are nearly parallel, as in this figure, there is no interaction between the factors. Here, the difference between the two lines is about the same for all three engine sizes. The presence of an interaction between factors would have been indicated by lines that are far from parallel, such as lines that cross to form an "x," or lines that go off in different directions with one rising while the other is falling.*

**Figure 11–3**

The following comments summarize the basic procedure for two-way analysis of variance.

## Procedure for Two-Way ANOVA

1. In two-way analysis of variance, begin by testing the null hypothesis that there is no interaction between the two factors. Using Minitab for the data in Table 11–3, we get the following test statistic.

$$F = \frac{MS(\text{interaction})}{MS(\text{error})} = \frac{0.6}{11.3} = 0.0531$$

With $df(\text{interaction}) = 2$ and $df(\text{error}) = 6$, we get a critical value of $F = 5.1433$ (assuming a 0.05 significance level). Since the test statistic does not exceed the critical value, we fail to reject the null hypothesis of no interaction between the two factors. There is not sufficient evidence to conclude that fuel consumption is affected by an interaction between engine size and transmission type.

## STATISTICS AND BASEBALL STRATEGY

■ Statisticians are using computers to develop very sophisticated measures of baseball performance and strategy. They have found, for example, that sacrifice bunts, sacrifice fly balls, and stolen bases rarely help win games and that it is seldom wise to have a pitcher intentionally walk the batter. They can identify the ballparks that favor pitchers and those that favor batters. Instead of simply comparing the batting averages of two different players, they can develop better measures of offensive strength by taking into account such factors as pitchers faced, position in the lineup, ballparks played, and weather conditions. ■

2. If we fail to reject the null hypothesis of no interaction between factors, then we should proceed to test the following two hypotheses:
   i. $H_0$: There are no main effects from the row factor. (The row means are equal.)
   ii. $H_0$: There are no main effects from the column factor. (The column means are equal.)

   If we do reject the null hypothesis of no interaction between factors, then we should not proceed with these two tests. (If there is an interaction between factors, we shouldn't consider the effects of either factor without the other.)

In step 1 above, we failed to reject the null hypothesis of no interaction between factors, so we proceed with the next two hypothesis tests identified in step 2.

i. Row factor:

$$F = \frac{MS(\text{transmission})}{MS(\text{error})} = \frac{40.3}{11.3} = 3.5664$$

- Not significant since the critical value is $F = 5.9874$. (Based on $df(\text{transmission}) = 1$, $df(\text{error}) = 6$, and a 0.05 significance level.)
- Conclusion: Fail to reject the null hypothesis of no main effects from transmission type. The type of transmission does not appear to have an effect on fuel consumption.

ii. Column factor:

$$F = \frac{MS(\text{size})}{MS(\text{error})} = \frac{21.6}{11.3} = 1.9115$$

- Not significant since the critical value is $F = 5.1433$. (Based on $df(\text{size}) = 2$, $df(\text{error}) = 6$, and a 0.05 significance level.)
- Conclusion: Fail to reject the null hypothesis of no main effects from engine size. The size of the engine does not appear to have an effect on the fuel consumption.

Some of these conclusions might not be consistent with our more general knowledge of cars, but these are the correct conclusions based on the very limited data included in Table 11–3. For such a small set of data to provide significant evidence of an effect, the effect would have to be very strong.

The procedures discussed here are actually quite similar to the procedures presented in Sections 11–2 and 11–3. We form conclusions about equal means by analyzing two estimates of variance, and the test statistic $F$ is the ratio of those two estimates. A significantly large value for $F$ indicates that there is a statistically significant difference in means.

# Special Case: One Observation Per Cell

Table 11–3 contains two observations per cell. If our sample data consist of only one observation per cell, we lose $MS$(interaction), $SS$(interaction), and $df$(interaction). If it seems reasonable to assume that there is no interaction between the two factors, make that assumption and then proceed as before to test these two hypotheses separately:

$H_0$: There are no main effects from the row factor.

$H_0$: There are no main effects from the column factor.

As an example, suppose we have only the first score in each cell of Table 11–3. Note that the two row means are 30.0 and 31.0. Is that difference significant, suggesting that there is an effect due to transmission type? The column means are 32.0, 30.5, and 29.0. Are those differences significant, suggesting that there is an effect due to engine size? It is reasonable to assume that there is no interaction between engine size and transmission type, so we make that assumption. Following is the Minitab display for the data in Table 11–3, with only the first score from each cell.

MINITAB DISPLAY

```
MTB > READ MPG IN C1 TRANS IN C2 SIZE IN C3
DATA> 31 1 1
DATA> 28 1 2
DATA> 31 1 3
DATA> 33 2 1
DATA> 33 2 2
DATA> 27 2 3
DATA> ENDOFDATA
        6 ROWS READ
MTB > NAME C1 'MPG' C2 'TRANS' C3 'SIZE'
MTB > TWOWAY C1 C2 C3
```

ANALYSIS OF VARIANCE MPG

| SOURCE | DF | SS | MS |
|--------|-----|------|------|
| TRANS  | 1  | 1.5  | 1.5  |
| SIZE   | 2  | 9.0  | 4.5  |
| ERROR  | 2  | 21.0 | 10.5 |
| TOTAL  | 5  | 31.5 |      |

We first use the above results to test the null hypothesis of no main effects from the row factor of transmission type. We get the test statistic

$$F = \frac{MS(\text{transmission})}{MS(\text{error})} = \frac{1.5}{10.5} = 0.1429$$

Assuming a 0.05 significance level, we find that the critical value is $F = 18.513$. (Refer to Table A–5 and note that the Minitab display provides degrees of freedom of 1 and 2 for numerator and denominator.) Because the test statistic of $F = 0.1429$ does not exceed the critical value of $F = 18.513$, we fail to reject the null hypothesis; it seems that the transmission type does not affect mileage.

We now use the Minitab display to test the null hypothesis of no main effect from the column factor of engine size. The test statistic is

$$F = \frac{MS(\text{size})}{MS(\text{error})} = \frac{4.5}{10.5} = 0.4286$$

which does not exceed the critical value of $F = 19.000$. (Refer to Table A–5 with $df(\text{size}) = 2$ and $df(\text{error}) = 2$.) We fail to reject the null hypothesis; it seems that the engine size does not affect mileage. Again, these conclusions are based on very limited data and they might not be valid when more sample data are acquired.

## Randomized Block Experiment

When we use the single-factor analysis of variance technique and conclude that the differences among the means are significant, we cannot necessarily conclude that the given factor is responsible for those differences. Perhaps there is some other unknown factor whose variation is responsible for the differences. One way to reduce the effect of the extraneous factors is to design the experiment so that it has a **completely randomized design** in which each element is given the same chance of belonging to the different categories or treatments. Another way to reduce the effect of extraneous factors is to use a **rigorously controlled design** in which elements are carefully chosen so that all other factors have no variability. That is, select elements that are the same in every characteristic except for the single factor being considered.

Another way to control for extraneous variation is to use a **randomized block experiment**, which we will now illustrate.

Suppose we want to use four different cars to test the mileage produced by three different grades (regular, extra, premium) of a gasoline. We want to control for the differences among the cars. Using a randomized block design, we burn each grade of gas in each of the four cars. We should randomly select the order in which this is done. We have three treatments corresponding to the three grades of gas. We have four blocks corresponding to the four cars used. Suppose the results are as given in Table 11–4.

## ZIP CODES REVEAL MUCH

■The Claritas Corporation has developed a way of obtaining considerable information about people from their zip codes. Zip code data are extracted from a variety of mailing lists, purchase orders, warranty cards, census data, and market research surveys. With people of the same social and economic levels tending to live in the same areas, it is possible to match zip codes with such factors as purchase patterns, types of cars driven, foods preferred, leisure time activities, and television viewing choices. The company helps clients target their advertising efforts to regions that are most likely to accept particular products. ■

| TABLE 11–4 | Mileage (mi/gal) | | | | |
|---|---|---|---|---|---|
| | | | | Block | |
| | | Car 1 | Car 2 | Car 3 | Car 4 |
| Treatment | Regular | 19 | 33 | 23 | 27 |
| | Extra | 19 | 34 | 26 | 29 |
| | Premium | 22 | 39 | 26 | 34 |

With this configuration, we can use the two-way analysis of variance software to test the claim that the three row-factor treatments (regular, extra, premium) produce the same mean. Shown below is the Minitab display for the data in Table 11–4.

MINITAB DISPLAY

```
ANALYSIS OF VARIANCE MPG

SOURCE       DF         SS         MS
GRADE         2       47.17      23.58
BLOCK         3      390.25     130.08
ERROR         6       11.50       1.92
TOTAL        11      448.92
```

Considering the treatment (row) effects, we calculate the test statistic

$$F = \frac{MS(\text{grade})}{MS(\text{error})} = \frac{23.58}{1.92} = 12.2813$$

From the display we get $df(\text{grade}) = 2$ and $df(\text{error}) = 6$ and we can now refer to Table A–5. Using 0.05 significance level, the critical value is $F = 5.1433$. Since the test statistic of $F = 12.2813$ exceeds the critical value of $F = 5.1433$, we reject the null hypothesis. The grade of gas does seem to have an effect on the mileage.

Considering the block (column) effects, the test statistic is

$$F = 130.08/1.92 = 67.7500$$

and the critical value is $F = 4.7571$. We reject the null hypothesis of equal block means and conclude that the different cars have different mileage values.

In this section we have briefly discussed an important branch of statistics. We have emphasized the interpretation of computer displays while omitting the manual calculations and formulas, which are quite formidable. More advanced texts typically discuss this topic in much greater detail, but our intent here is to give some general insight into the nature of two-way analysis of variance.

## 11–4 Exercises A

In Exercises 1–4, use the following Minitab display that corresponds to the data in Table 11–5.

| TABLE 11–5 Mi/gal (highway) | | | | |
|---|---|---|---|---|
| | | Engine Size (liters) | | |
| | | 1.5 | 2.2 | 2.5 |
| Transmission | Automatic | 34, 32, 32 | 30, 28, 28 | 24, 23, 24 |
| | Manual | 33, 35, 36 | 30, 31, 29 | 24, 27, 25 |

MINITAB DISPLAY

```
ANALYSIS OF VARIANCE MPG

SOURCE          DF        SS        MS
TRANS           1       12.50     12.50
SIZE            2      252.33    126.17
INTERACTION     2        0.33      0.17
ERROR          12       17.33      1.44
TOTAL          17      282.50
```

1. Identify the indicated values.
   a. $MS$(interaction)
   b. $MS$(error)
   c. $MS$(row treatments)
   d. $MS$(column treatments)

2. Find the test statistic and critical value for the null hypothesis of no interaction between engine size and type of transmission. What do you conclude?

3. Assume that mileage is not affected by an interaction between engine size and type of transmission. Find the test statistic and critical value for the null hypothesis that engine size has no effect on mileage. What do you conclude?

4. Assume that there is no interaction between engine size and type of transmission. Find the test statistic and critical value for the null hypothesis that the type of transmission has no effect on mileage. What do you conclude?

In Exercises 5 and 6, use only the first value from each cell in Table 11–5. Using only these first values, the Minitab display is as follows.

MINITAB DISPLAY

```
ANALYSIS OF VARIANCE MPG

SOURCE        DF        SS         MS
TRANS          1      0.167      0.167
SIZE           2     92.333     46.167
ERROR          2      0.333      0.167
TOTAL          5     92.833
```

5. Assuming that there is no effect on mileage from the interaction between engine size and transmission type, test the null hypothesis that transmission type has no effect on mileage. Identify the test statistic, critical value, and state the conclusion. Use a 0.05 significance level.

6. Assuming that there is no effect on mileage from the interaction between engine size and transmission type, test the null hypothesis that engine size has no effect on mileage. Identify the test statistic and critical value, and state the conclusion. Use a 0.05 significance level.

Exercises 7 and 8 refer to the sample data in Table 11–6 and the corresponding Minitab display. The table entries are the times (in minutes) required to complete a document. The same document was entered by four different typists using each of three different word processors.

| | | | Word Processor | |
|---|---|---|---|---|
| | | I | II | III |
| | A | 16 | 21 | 13 |
| Typist | B | 20 | 25 | 16 |
| | C | 18 | 21 | 14 |
| | D | 19 | 22 | 15 |

TABLE 11–6

MINITAB DISPLAY

```
ANALYSIS OF VARIANCE TIMES

SOURCE       DF          SS          MS
TYPIST        3      22.000       7.333
WORDPROC      2     120.167      60.083
ERROR         6       2.500       0.417
TOTAL        11     144.667
```

7. At the 0.05 significance level, test the claim that the choice of typist has no effect on the time. Identify the test statistic, critical value, and state the conclusion.

8. At the 0.05 significance level, test the claim that the choice of word processor has no effect on the time. Identify the test statistic, critical value, and state the conclusion.

# 11–4 Exercises B

9. Use a statistics software package that can produce results for two-way analysis of variance, such as Minitab or SPSS/PC+. First enter the data in Table 11–3 and verify that the results are as given in this section. Then transpose that table by making transmission type the column variable while making engine size the row variable. Obtain the computer display for the transposed table and compare the results to those found earlier in this section.

10. Refer to the data in Table 11–3 and subtract 10 from each table entry. Use a statistics software package with a two-way analysis of variance capability and determine the effects of subtracting 10 from each entry.

11. Refer to the data in Table 11–3 and multiply each table entry by 10. Use a statistics software package with a two-way analysis of variance capability and determine the effects of multiplying each entry by 10.

12. In analyzing Table 11–3, we concluded that fuel consumption was not affected by an interaction between engine size and transmission type, it was not affected by engine size, and it was not affected by transmission type.

   a. Change the table entries so that there is an effect from the interaction between engine size and transmission type.

   b. Change the table entries so that there is no effect from the interaction between engine size and transmission type, there is no effect from engine size, but there is an effect from transmission type.

   c. Change the table entries so that there is no effect from the interaction between engine size and transmission type, there is no effect from transmission type, but there is an effect from engine size.

## Vocabulary List

Define and give an example of each term.

| | |
|---|---|
| analysis of variance | single-factor analysis of variance |
| ANOVA | factor |
| variance between samples | two-way analysis of variance |
| variance within samples | interaction |
| variation due to treatment | completely randomized design |
| variation due to error | rigorously controlled design |
| multiple comparison procedures | randomized block experiment |
| one-way analysis of variance | |

## Review

In this chapter we used **analysis of variance** to determine whether differences among means are due to chance fluctuations or whether the differences are significant. This method requires (1) normally distributed populations, (2) populations with the same standard deviation (or variance), and (3) random samples that are independent of each other.

Our test statistics are based on the ratio of two different estimates of the common population variance. In repeated samplings, the distribution of the $F$ test statistic can be approximated by the $F$ distribution, which has critical values given in Table A–5.

In Section 11–2 we considered one-way analysis of variance for samples with the same number of scores. In Section 11–3 we extended that method to include samples of unequal sizes.

In Section 11–4 we considered two-way analysis of variance. The data were categorized by two factors instead of only one. We also considered two special cases: two-way analysis of variance with one observation per cell, and randomized block experiments. Due to the nature of the calculations required, this last section emphasized the interpretation of computer displays.

## Important Formulas

| Application | Distribution | Test Statistic | Degrees of Freedom | Critical Values |
|---|---|---|---|---|
| Analysis of variance (equal sample sizes only) | $F$ | $F = \dfrac{ns_{\bar{x}}^2}{s_p^2}$ | num: $k - 1$<br>den: $k(n - 1)$ | Table A–5 |
| (all cases) | $F$ | $F = \dfrac{\left[\dfrac{\Sigma n_i(\bar{x}_i - \bar{\bar{x}})^2}{k - 1}\right]}{\left[\dfrac{\Sigma(n_i - 1)s_i^2}{\Sigma(n_i - 1)}\right]}$ | num: $k - 1$<br>den: $N - k$ | Table A–5 |
| | | or $F = \dfrac{MS(\text{treatment})}{MS(\text{error})}$<br>(see below) | num: $k - 1$<br>den: $N - k$ | Table A–5 |

For analysis of variance:

$k$ = number of samples

$n_i$ = number of values in the $i$th sample

$\Sigma x$ = sum of all sample values

$SS(\text{total}) = \Sigma x^2 - \dfrac{(\Sigma x)^2}{N}$

$SS(\text{error}) = SS(\text{total}) - SS(\text{treatment})$

$df(\text{error}) = N - k$

$MS(\text{error}) = \dfrac{SS(\text{error})}{df(\text{error})}$

$N$ = total number of values

$c_i$ = total of values in the $i$th sample

$\Sigma x^2$ = sum of the squares of all sample values

$SS(\text{treatment}) = \left(\Sigma \dfrac{c_i^2}{n_i}\right) - \dfrac{(\Sigma c_i)^2}{N}$

$df(\text{treatment}) = k - 1$

$MS(\text{treatment}) = \dfrac{SS(\text{treatment})}{df(\text{treatment})}$

$F = \dfrac{MS(\text{treatment})}{MS(\text{error})}$

# 2 *Review Exercises*

1. A lawyer is studying punishments for a certain crime and wants to compare the sentences imposed by three different judges. Randomly selected results follow. At the $\alpha = 0.05$ level of significance, test the claim that the three judges impose sentences that have the same mean.

| Judge A | Judge B | Judge C |
|---|---|---|
| $n = 36$ | $n = 36$ | $n = 36$ |
| $\bar{x} = 5.2$ years | $\bar{x} = 4.1$ years | $\bar{x} = 5.5$ years |
| $s = 1.4$ years | $s = 1.1$ years | $s = 1.5$ years |

2. Two different research teams attempt to develop paint mixtures with more durability than the current product. Samples are tested and the resulting measures of durability are given below. At the 0.05 level of significance, test the claim that the three different mixtures have the same mean index of durability.

| Current Product | Team I Product | Team II Product |
|---|---|---|
| 12.5 | 12.1 | 14.0 |
| 12.3 | 12.6 | 14.2 |
| 11.8 | 12.9 | 12.6 |
| 12.4 | 13.5 | 14.8 |
| 12.9 | 12.7 | 15.1 |
|  | 12.7 | 13.9 |
|  |  | 14.3 |
|  |  | 14.4 |

3. In testing the effectiveness of four different diets, subjects with the same overweight characteristics are randomly selected for each diet. The weight losses are listed below. At the 0.01 level of significance, test the claim that the diets produce the same mean weight loss.

| Diet 1 | Diet 2 | Diet 3 | Diet 4 | |
|---|---|---|---|---|
| 8 | 10 | 6 | 21 | 35 |
| 12 | 14 | 24 | 23 | 12 |
| 14 | 14 | 12 | 16 | 19 |
| 16 | 21 | 10 | 19 | 15 |
| 3 | 5 | 10 | 27 | 40 |

4. Three teaching methods are used with three groups of randomly selected students with the following results. At the 0.05 level of signif-

icance, test the claim that the samples came from populations with equal means.

| Method A | Method B | Method C |
|---|---|---|
| $n = 20$ | $n = 20$ | $n = 20$ |
| $\bar{x} = 72.0$ | $\bar{x} = 76.0$ | $\bar{x} = 71.0$ |
| $s = 9.0$ | $s = 10.0$ | $s = 12$ |

In Exercises 5–7, use the following: A manager records the numbers of items produced by three employees who each work on three different machines for three different days. The sample results are given in Table 11–7 and the Minitab results follow.

**TABLE 11–7**

| | | Employee A | Employee B | Employee C |
|---|---|---|---|---|
| Machine | I | 15, 16, 19 | 12, 13, 8 | 12, 14, 11 |
| | II | 17, 14, 18 | 9, 7, 11 | 10, 12, 9 |
| | III | 20, 18, 17 | 8, 10, 11 | 13, 10, 11 |

MINITAB DISPLAY

ANALYSIS OF VARIANCE ITEMS

| SOURCE | DF | SS | MS |
|---|---|---|---|
| MACHINE | 2 | 10.89 | 5.44 |
| EMPLOYEE | 2 | 262.89 | 131.44 |
| INTERACTION | 4 | 8.22 | 2.06 |
| ERROR | 18 | 62.67 | 3.48 |
| TOTAL | 26 | 344.67 | |

5. Using a 0.05 significance level, test the claim that the interaction between employee and machine has no effect on the number of items produced. Identify the test statistic and critical value, and state the conclusion.

6. Using a 0.05 significance level, test the claim that the machine has no effect on the number of items produced. Identify the test statistic and critical value, and state the conclusion.

7. Using a 0.05 significance level, test the claim that the choice of employee has no effect on the number of items produced. Identify the test statistic and critical value, and state the conclusion.

8. Three different computer programming languages (A, B, and C) are used by different students to solve a problem; the times (in hours) required for the solution are listed below each language designation. At the 0.05 level of significance, test the claim that the mean times for all three languages are the same.

| A | B | C |
|---|---|---|
| 7 | 9 | 2 |
| 4 | 5 | 3 |
| 4 | 7 | 5 |
| 3 |   | 3 |
|   |   | 8 |
| $n_1 = 4$ | $n_2 = 3$ | $n_3 = 5$ |
| $\bar{x}_1 = 4.5$ | $\bar{x}_2 = 7.0$ | $\bar{x}_3 = 4.2$ |
| $s_1^2 = 3.0$ | $s_2^2 = 4.0$ | $s_3^2 = 5.7$ |

 *Computer Project*

1. Use software such as STATDISK or Minitab to solve Exercise 1 in Section 11–3. Repeat that exercise after adding 10 to each score and note the effect of that change. Then repeat that exercise a third time after multiplying each score by 5, and note the effect of this change. Also describe the effects caused by interchanging columns, and by changing one of the scores (such as 90) to an extreme score (such as 9000).

 *Applied Projects*

1. Using the Appendix B data for homes sold, enter the selling prices in the appropriate category below. Then test the claim that the samples in all three categories come from populations having the same mean.

| Number of baths | | |
|---|---|---|
| 1 or 1.5 | 2 or 2.5 | 3 or more |

2. Refer to Data Set II in Appendix B. Use the two factors of sex (male, female) and age (construct four age brackets). Randomly select an equal number of subjects from each of the eight categories and record their weights. Using two-way analysis of variance, what do you conclude?

*Writing Projects*

1. Your department conducts an experiment to determine the costs of manufacturing an airport security screening device. Three different production methods (denoted by A, B, C) are used. The Minitab analysis of variance display is shown below. Write a report summarizing your conclusions.

MINITAB DISPLAY

ANALYSIS OF VARIANCE

| SOURCE | DF | SS | MS | F | P |
|--------|-----|--------|--------|-------|-------|
| FACTOR | 2 | 210.53 | 105.27 | 27.22 | 0.000 |
| ERROR | 12 | 46.40 | 3.87 | | |
| TOTAL | 14 | 256.93 | | | |

2. Write a report summarizing the program listed below.

 *Videotapes*

Program 13 of the series *Against All Odds: Inside Statistics* is recommended as a supplement to this chapter.

# UNCLE SAM WANTS YOU, IF YOU'RE RANDOMLY SELECTED

■A considerable amount of controversy was created when a lottery was used to determine who would be drafted into the U.S. Army. In 1970, the lottery approach was instituted in an attempt to make the selection process random, but many claimed that the outcome was unfair because men born later in the year had a better chance of being drafted.

The 1970 lottery involved 366 capsules corresponding to the dates in a leap year. (The first dates selected would be the birthdays of the first men drafted.) First, the 31 January capsules were placed in a box. The 29 February capsules were added and the two months were mixed. Then the 31 March capsules were added and the three months were mixed. This process continued; one result was that January capsules were mixed 11 times, while December was mixed only once. Later arguments claimed that this process tended to place early dates closer to the bottom while dates later in the year tended to be near the top. The first ten dates selected, in order of priority, were September 14, April 24, December 30, February 14, October 18, September 6, October 26, September 7, November 22, and December 6. Although the runs test (in Section 12-7) indicated randomness, other statistical tests did not. There was enough criticism of the process to cause a revised procedure the following year.

In 1971, statisticians used two drums of capsules. One drum contained capsules with dates, which were deposited in a random order according to a table of random numbers. A second drum contained the draft priority numbers, which were also deposited randomly according to a table of random numbers. Both drums were rotated for an hour before the selection process was begun. September 16 was drawn from one drum and 139 was drawn from the other. This meant that men born on September 16 had draft priority number 139. This process continued until all dates had a priority number. This procedure was significantly more random, and less controversial.

See Computer Project 6 at the end of Chapter 12. ■

# Chapter Twelve

## In This Chapter

# 12 Nonparametric Statistics

## Chapter Problem

Employees like to feel that their salaries are fair compensations for their efforts, abilities, stress levels, and physical demands. In Table 12–1 we list randomly selected jobs along with rankings for salary and stress levels (based on data from *The Jobs Rated Almanac*). Among these ten jobs, stockbrokers have the second highest salary and the second highest stress level.

| TABLE 12–1 | | |
|---|---|---|
| Job | Salary Rank | Stress Rank |
| Stockbroker | 2 | 2 |
| Zoologist | 6 | 7 |
| Electrical engineer | 3 | 6 |
| School principal | 5 | 4 |
| Hotel manager | 7 | 5 |
| Bank officer | 10 | 8 |
| Occupational safety inspector | 9 | 9 |
| Home economist | 8 | 10 |
| Psychologist | 4 | 3 |
| Commercial airline pilot | 1 | 1 |

Is there a relationship between a job's salary and its level of stress? Because these values are ranks, we cannot use the linear correlation coefficient discussed in Chapter 9. The linear correlation coefficient requires that the two variables be normally distributed, and ranks don't satisfy that requirement. While we can't use the methods in Chapter 9, we can use an alternative method that can be applied to data in the form of ranks.

In Section 12–6 we will apply the concept of rank correlation to the data in Table 12–1 as we test for a relationship between job salary and stress level.

# 12–1 Overview

Most of the methods of inferential statistics covered before Chapter 10 can be called **parametric methods**, because their validity is based on sampling from a population with particular parameters such as the mean, standard deviation, or proportion. Parametric methods can usually be applied only to circumstances in which some fairly strict requirements are met. One typical requirement is that the sample data must come from a normally distributed population. What do we do when the necessary requirements are not satisfied? There may be an alternative approach among the many methods that are classified as **nonparametric.** In addition to being an alternative to parametric methods, nonparametric techniques are frequently valuable in their own right.

In this chapter, we introduce six of the more popular nonparametric methods currently used. These methods have advantages and disadvantages.

## Advantages of Nonparametric Methods

1. Nonparametric methods can be applied to a wider variety of situations, because they do not have the more rigid requirements of their parametric counterparts. In particular, nonparametric methods do not require normally distributed populations. For this reason, nonparametric tests of hypotheses are often called **distribution-free** tests. (Actually, some of these tests do depend on a parameter such as the median, but they don't require a particular distribution. While "distribution-free" is a more accurate description, the term *nonparametric* is commonly used. Sorry about that.)

2. Unlike the parametric methods, nonparametric methods can often be applied to nominal data that lack exact numerical values.

3. Nonparametric methods usually involve computations that are simpler than the corresponding parametric methods.

4. Since nonparametric methods tend to require simpler computations, they tend to be easier to understand.

If all of these terrific advantages could be accrued without any significant disadvantages, we could ignore the parametric methods and enjoy much simpler procedures. Unfortunately, there are some disadvantages.

# Disadvantages of Nonparametric Methods

1. Nonparametric methods tend to waste information, since exact numerical data are often reduced to a qualitative form.
2. Nonparametric methods are generally less sensitive than the corresponding parametric methods. This means that we need stronger evidence before we reject a null hypothesis.

As an example of the way that information is wasted, there is one nonparametric method in which weight losses by dieters are recorded simply as negative signs. With this particular method, a weight loss of only 1 pound receives the same representation as a weight loss of 50 pounds. This would not thrill dieters.

## Efficiency

Although nonparametric tests are less sensitive than their parametric counterparts, this can be compensated for by an increased sample size. The **efficiency** of a nonparametric method is one concrete measure of its sensitivity. Section 12–6 deals with a concept called the *rank correlation coefficient*, which has an efficiency rating of 0.91 when compared to the linear correlation coefficient of Chapter 9. This means that with all things being equal, this nonparametric approach would require 100 sample observations to achieve the same results as 91 sample observations analyzed through the parametric approach, assuming the stricter requirements for using the parametric method are met. Not bad! The point, though, is that an increased sample size can overcome lower sensitivity. Table 12–2 on the following page lists nonparametric methods covered in this chapter, along with the corresponding parametric approach and efficiency rating. You can see from this table that the lower efficiency might not be a critical factor.

In choosing between a parametric method and a nonparametric method, the key factors that should govern our decision are cost, time, efficiency, amount of data available, type of data available, method of sampling, nature of the population, and probabilities ($\alpha$ and $\beta$) of making type I and type II errors. In one experiment we might have abundant data with strong assurances that all of the requirements of a parametric test are satisfied, and we would probably be wise to choose that parametric test. But given another experiment with relatively few cases drawn from some mysterious population, we would probably fare better with a nonparametric test. Sometimes we don't really have a choice. Only nonparametric methods can be used on data consisting of observations that can only be ranked.

**TABLE 12–2**

| Application | Parametric Test | Nonparametric Test | Efficiency of Nonparametric Test with Normal Population |
|---|---|---|---|
| Two dependent samples | t test or z test | Sign test<br>Wilcoxon signed-ranks | 0.63<br>0.95 |
| Two independent samples | t test or z test | Wilcoxon rank-sum | 0.95 |
| Several independent samples | Analysis of variance (F test) | Kruskal-Wallis test | 0.95 |
| Correlation | Linear correlation | Rank correlation | 0.91 |
| Randomness | No parametric test | Runs test | No basis for comparison |

## Ranks

Some methods included in this chapter are based on ranks. Instead of describing ranks in each section or making some sections dependent on others, we will now discuss ranks so that we will be prepared to use them wherever they are required.

Data are **ranked** when they are arranged according to some criterion, such as smallest to largest or best to worst. The first item in the arrangement is given a rank of 1, the second item is given a rank of 2, and so on. For example, the numbers 5, 3, 40, 10, and 12 can be arranged from lowest to highest as 3, 5, 10, 12, and 40, and these numbers have ranks of 1, 2, 3, 4, 5, respectively (see the following illustration). If a tie in ranks should occur, the usual procedure is to find the mean of the ranks involved and then assign that mean rank to each of the tied items. The numbers 3, 5, 5, 10, and 12 would be given ranks of 1, 2.5, 2.5, 4, and 5, respectively. In this case, there is a tie for ranks 2 and 3, so we find the mean of 2 and 3 (which is 2.5) and assign it to the scores that created the tie. As another example, the scores 3, 5, 5, 7, 10, 10, 10, and 15 would be ranked 1, 2.5, 2.5, 4, 6, 6, 6, and 8, respectively. From these examples we can see how to convert numbers to ranks, but there are many situations in which the original data consist of ranks. If a judge ranks five piano contestants, we get ranks of 1, 2, 3, 4, 5 corresponding to five names; it's this type of data that precludes the use of parametric methods and demonstrates the importance of nonparametric methods.

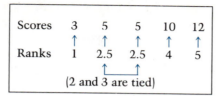

| Scores | 5 | 3 | 40 | 10 | 12 |
|---|---|---|---|---|---|
| Scores in order | 3 ↑ | 5 ↑ | 10 ↑ | 12 ↑ | 40 ↑ |
| Ranks | 1 | 2 | 3 | 4 | 5 |

| Scores | 3 ↑ | 5 ↑ | 5 ↑ | 10 ↑ | 12 ↑ |
|---|---|---|---|---|---|
| Ranks | 1 | 2.5 | 2.5 | 4 | 5 |

(2 and 3 are tied)

# 12–2  Sign Test

The **sign test** is one of the easiest nonparametric tests to use, and it is applicable to a few different situations. One application is to the paired data that form the basis for hypothesis tests involving two dependent samples. We considered such cases by using parametric tests in Section 8–3. One example from Section 8–3 refers to a study conducted to investigate the effectiveness of hypnotism in reducing pain. The sample results for randomly selected subjects are given in Table 12–3. (The values are before and after hypnosis. The measurements are in centimeters on the mean visual analogue scale, and the data are based on "An Analysis of Factors That Contribute to the Efficacy of Hypnotic Analgesia" by Price and Barber, *Journal of Abnormal Psychology*, Vol. 96, No. 1.) We will test the claim that the sensory measurements are lower after hypnotism. Note that Table 12–3 also includes the *signs* of the changes from the "before" values to the "after" values.

In Section 8–3 we used the parametric Student $t$ test, but in this section we apply the nonparametric sign test, which can be used to test for equality between two medians. The key concept underlying the sign test is this: **If the two sets of data have equal medians, the number of positive signs should be approximately equal to the number of negative signs.** For the data in Table 12–3, we can conclude that hypnotism is effective if there is an excess of negative signs and a deficiency of positive signs.

| TABLE 12–3 | | | | | | | | |
|---|---|---|---|---|---|---|---|---|
| Subject | A | B | C | D | E | F | G | H |
| Before | 6.6 | 6.5 | 9.0 | 10.3 | 11.3 | 8.1 | 6.3 | 11.6 |
| After | 6.8 | 2.4 | 7.4 | 8.5 | 8.1 | 6.1 | 3.4 | 2.0 |
| Sign of change from before to after | + | − | − | − | − | − | − | − |

## BUCKLE UP

■ Statistical analysis of data can lead to changes in public policy. The Highway Users Federation estimated that if we all used safety belts, each year there would be 1200 fewer highway deaths and 330,000 fewer disabling injuries. One study of 1126 accidents showed that riders wearing safety belts had 86% fewer life-threatening injuries. Some people don't use safety belts because they know of cases where a serious injury was avoided when an unbelted rider was thrown clear of the wreck. While there are cases where the safety belt had a negative effect, they are far outnumbered by cases in which it was clearly helpful. The wisest strategy is to buckle up. ■

In our sign test procedure, we exclude any ties (represented by zeros). (There are other ways to handle ties; see Exercise 22.) We now have this specific question: Do the seven negative signs in Table 12–3 *significantly* outnumber the single positive sign? Or, to put it another way, is the number of positive signs small enough to be significant? The answer to this question depends on the level of significance, so let's use $\alpha = 0.05$ as we did in Section 8–3. When we assume the null hypothesis of no decrease in sensory measurements, we assume that positive signs and negative signs occur with equal frequency, so $P(\text{positive sign}) = P(\text{negative sign}) = 0.5$. (The null hypothesis of no decrease also includes the possibility of an increase, but we continue to assume that positive signs and negative signs are equally likely.)

$H_0$: There is no decrease in sensory measurements.
$H_1$: There is a decrease in sensory measurements.

Since our results fall into two categories (positive or negative) and we have a fixed number of independent cases, we could use the binomial probability distribution (Section 4–4) to determine the likelihood of getting one or no positive signs among the eight subjects. Instead, we have used the binomial probability formula to construct a separate table (Table A–7) that lists critical values for the sign test. For consistency and ease, we will stipulate the following.

**The test statistic $x$ is the number of times the less frequent sign occurs.**

With seven negative signs and one positive sign, the test statistic $x$ is the lesser of 7 and 1, so $x = 1$. We have 8 sample cases so that $n = 8$. Our test is one-tailed with $\alpha = 0.05$, and Table A–7 indicates that the critical value is 1. We should therefore reject the null hypothesis only if the test statistic is less than or equal to 1. With a test statistic of $x = 1$ we do reject the null hypothesis of no decrease. It appears that hypnotism does result in lower sensory measurements.

Because of the way that we are determining the value of the test statistic, we should check to ensure that our conclusion is consistent with the circumstances. It is only when the sense of the sample data is *against* the null hypothesis that we should even consider rejecting it. If the sense of the data supports the null hypothesis, we should fail to reject it regardless of the test statistic and critical value. Figure 12–1 summarizes the procedure for the sign test and includes this check for consistency of results.

In the preceding example we arrived at the same conclusion obtained in Section 8–3. However, consider the new data set given in Table 12–4 (page 592).

For this data set there are two positive signs and six negative signs. Using the sign test, the test statistic is $x = 2$ and we fail to reject the null hypothesis of no decrease. But if we use the Student $t$ test from Section 8–3, we get

$$t = \frac{\overline{d} - 0}{s_d/\sqrt{n}} = \frac{-5.0875}{3.6725/\sqrt{8}} = -3.918$$

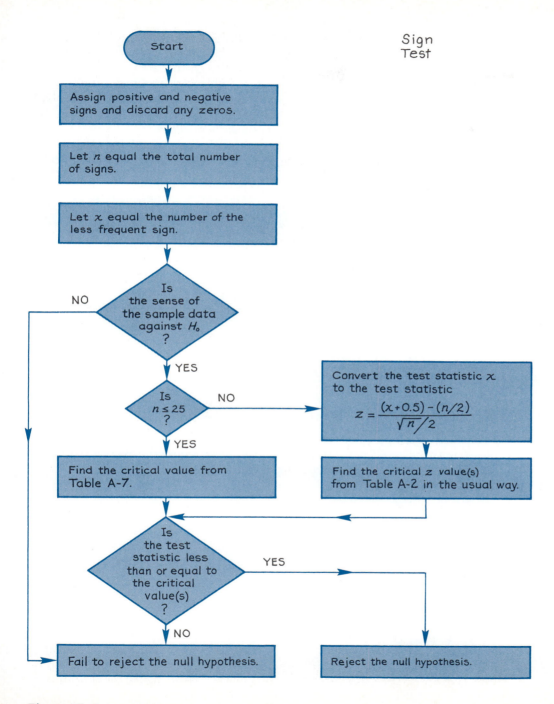

Sign
Test

**Figure 12–1**

This causes *rejection* of the null hypothesis because the test statistic of $t = -3.918$ is in the critical region bounded by the critical score of $t = -1.895$. An intuitive analysis of Table 12–4 suggests that the "after" scores are significantly lower, but the sign test is blind to the *magnitude* of the changes. This illustrates the previous assertion that nonparametric tests lack the sensitivity of parametric tests, with the resulting tendency that stronger evidence is required before a null hypothesis is rejected.

An examination of Figure 12–1 shows that when $n \leq 25$, we should use Table A–7 to find the critical value, but for $n > 25$, we use a normal approximation to obtain the critical values.

| **TABLE 12–4** | | | | | | | | |
|---|---|---|---|---|---|---|---|---|
| Subject | A | B | C | D | E | F | G | H |
| Before | 6.6 | 6.5 | 9.0 | 10.3 | 11.3 | 8.1 | 6.3 | 11.6 |
| After | 6.8 | 6.6 | 2.4 | 3.1 | 2.9 | 2.8 | 2.4 | 2.0 |
| Difference | 0.2 | 0.1 | −6.6 | −7.2 | −8.4 | −5.3 | −3.9 | −9.6 |
| Sign of change from before to after | + | + | − | − | − | − | − | − |

# Claims Involving Nominal Data

The next example illustrates the fact that nonparametric methods can be used with nominal data. Also, since this next example involves a sample size greater than 25, the normal approximation is used.

### Example

A company claims that its hiring practices are fair, it does not discriminate on the basis of sex, and the fact that 40 of the last 50 new employees are men is just a fluke. The company acknowledges that applicants are about half men and half women. Test the null hypothesis that men and women are equal in their ability to be employed by this company. Use a significance level of 0.05.

### Solution

$H_0: p_1 = p_2$ (the proportions of men and women are equal)
$H_1: p_1 \neq p_2$

*continued*

**Solution** *continued*

If we denote hired women by + and hired men by −, we have 10 positive signs and 40 negative signs. Refer now to the flowchart in Figure 12–1. The test statistic $x$ is the smaller of 10 and 40, so $x = 10$. This test involves two tails since a disproportionately low number of either sex will cause us to reject the claim of equality. The sense of the sample data is against the null hypothesis because 10 and 40 aren't exactly equal. Continuing with the procedure in Figure 12–1, we note that the value of $n = 50$ is above 25, so the test statistic $x$ is converted to the test statistic $z$ as follows.

$$
\begin{aligned}
z &= \frac{(x + 0.5) - (n/2)}{\sqrt{n}/2} \\
&= \frac{(10 + 0.5) - (50/2)}{\sqrt{50}/2} \\
&= -4.10
\end{aligned}
$$

With $\alpha = 0.05$ in a two-tailed test, the critical values are $z = -1.96$ and $1.96$. The test statistic $z = -4.10$ is less than these critical values (see Figure 12–2), so we reject the null hypothesis of equality. There is sufficient sample evidence to warrant rejection of the claim that the hiring practices are fair.

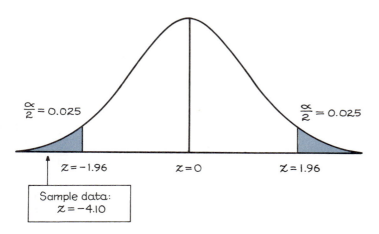

**Figure 12–2**

## SMOKING AIR

■ A consumer testing group studied tar and nicotine levels of cigarettes. The brand Now was advertised as having the lowest levels, but they seemed to burn more quickly than other brands. Samples of Now and Winston were randomly selected and weighed. Both brands were products of the R. J. Reynolds Tobacco Company, and Winston was their best-seller at the time. Results showed that the average weight of the tobacco in a Now cigarette was about two-thirds that of a Winston cigarette. With a third less tobacco, it isn't too difficult to get lower tar and nicotine levels. The study also showed that smokers tend to consume more cigarettes when they smoke those with lower levels of tar and nicotine. ■

When $n > 25$, the test statistic $z$ is based on a normal approximation to the binomial probability distribution with $p = q = \frac{1}{2}$. In Section 5–5 we saw that the normal approximation to the binomial distribution is acceptable when both $np \geq 5$ and $nq \geq 5$. Also, in Section 4–5 we saw that $\mu = np$ and $\sigma = \sqrt{n \cdot p \cdot q}$ for binomial experiments. Since this sign test assumes that $p = q = \frac{1}{2}$, we meet the $np \geq 5$ and $nq \geq 5$ prerequisites whenever $n \geq 10$; we have a table of critical values (Table A–7) for $n$ up to 25, so that we need the normal approximation only for values of $n$ above 25. Also, with the assumption that $p = q = \frac{1}{2}$, we get $\mu = np = n/2$ and $\sigma = \sqrt{n \cdot p \cdot q} = \sqrt{n/4} = \sqrt{n}/2$, so that

$$z = \frac{x - \mu}{\sigma} \text{ becomes } z = \frac{x - \left(\dfrac{n}{2}\right)}{\dfrac{\sqrt{n}}{2}}$$

Finally, we replace $x$ by $x + 0.5$ as a correction for continuity. That is, the values of $x$ are discrete, but since we are using a continuous probability distribution, a discrete value such as 10 is actually represented by the interval from 9.5 to 10.5. Because $x$ represents the less frequent sign, we need to concern ourselves only with $x + 0.5$; we thus get the test statistic $z$ as given above and in Figure 12–1.

## Claims About a Median

The previous examples involved application of the sign test to a comparison of *two* sets of data, but we can also use the sign test to investigate a claim made about the median of one set of data, as the next example shows.

### Example

Use the sign test to test the claim that the median IQ of pilots is at least 100 if a sample of 50 pilots contained exactly 22 members with IQs of 100 or higher.

### Solution

The null hypothesis is the claim that the median is equal to or greater than 100; the alternative hypothesis is the claim that the median is less than 100.

$H_0$: Median is at least 100. (Median $\geq$ 100)
$H_1$: Median is less than 100. (Median $<$ 100)

*continued*

**Solution** *continued*

We select a significance level of 0.05, and we use + to denote each IQ score that is at least 100. We therefore have 22 positive signs and 28 negative signs. We can now determine the significance of getting 22 positive signs out of a possible 50. Referring to Figure 12–1, we note that $n = 50$ and $x = 22$ (the smaller of 22 and 28). The sense of the data is against the null hypothesis, since a median of at least 100 would require at least 25 (half of 50) scores of 100 or higher. The value of $n$ exceeds 25, so we convert the test statistic $x$ to the test statistic $z$.

$$z = \frac{(x + 0.5) - (n/2)}{\sqrt{n}/2}$$
$$= \frac{(22 + 0.5) - (50/2)}{\sqrt{50}/2}$$
$$= \frac{22.5 - 25}{\sqrt{50}/2} = -0.71$$

In this one-tailed test with $\alpha = 0.05$, we use Table A–2 to get the critical $z$ value of $-1.645$. From Figure 12–3, we can see that the computed value of $-0.71$ does not fall within the critical region. We therefore fail to reject the null hypothesis. Based on the available sample evidence, we cannot reject the claim that the median IQ is at least 100. A corresponding parametric test may or may not lead to the same conclusion, depending on the specific values of the 50 sample scores.

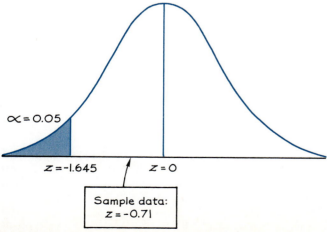

$\alpha = 0.05$

$z = -1.645$      $z = 0$

Sample data:
$z = -0.71$

**Figure 12–3**

We have shown that the sign test wastes information because it uses only information about the direction of the differences between pairs of data, while the magnitudes of those differences are ignored. The next section introduces the Wilcoxon signed-ranks test, which largely overcomes that disadvantage.

## 12–2 | Exercises A

In Exercises 1–20, use the sign test.

1. A study was conducted to investigate the effectiveness of hypnotism in reducing pain. At the 0.05 significance level, test the claim that the affective responses to pain are the same before and after hypnosis. Results for randomly selected subjects are given below. (The data are based on results from "An Analysis of Factors That Contribute to the Efficacy of Hypnotic Analgesia," by Price and Barber, *Journal of Abnormal Psychology*, Vol. 96, No. 1.)

| Before | −5.5 | −5.0 | −6.6 | −9.7 | −4.0 | −7.0 | −7.0 | −8.4 |
|--------|------|------|------|------|------|------|------|------|
| After  | −1.4 | −0.5 | 0.7  | 1.0  | 2.0  | 0.0  | −0.6 | −1.8 |

2. A study was conducted to investigate some effects of physical training. Sample data are listed in the accompanying table (based on data from "Effect of Endurance Training on Possible Determinants of $\dot{V}O_2$ During Heavy Exercise," by Casaburi and others, *Journal of Applied Physiology*, Vol. 62, No. 1). At the 0.05 level of significance, test the claim that pretraining weights equal posttraining weights (in kilograms).

| Pretraining  | 99 | 57 | 62 | 69 | 74 | 77 | 59 | 92 | 70 | 85 |
|--------------|----|----|----|----|----|----|----|----|----|----|
| Posttraining | 94 | 57 | 62 | 69 | 66 | 76 | 58 | 88 | 70 | 84 |

3. At the 0.05 level of significance, test the claim that the sample x and y values come from the same population. Assume that the x and y values are paired as shown.

| x | 1 | 2 | 2 | 3 | 5 | 6 | 8 | 4 | 6 | 7 | 2 | 5 | 3 | 2 |
|---|---|---|---|---|---|---|---|---|---|---|---|---|---|---|
| y | 2 | 0 | 1 | 4 | 4 | 4 | 9 | 3 | 5 | 7 | 2 | 6 | 2 | 1 |

4. Two different firms design their own IQ tests, and a psychologist administers both tests to randomly selected subjects. The results are given below. At the 0.05 level of significance, test the claim that there is no significant difference between the two tests.

| Subject | A | B | C | D | E | F | G | H | I | J |
|---------|-----|-----|-----|-----|-----|-----|-----|-----|-----|-----|
| Test I  | 98  | 94  | 111 | 102 | 108 | 105 | 92  | 88  | 100 | 99  |
| Test II | 105 | 103 | 113 | 98  | 112 | 109 | 97  | 95  | 107 | 103 |

5.  A pill designed to lower systolic blood pressure is administered to 10 randomly selected volunteers. The results follow. At the $\alpha = 0.05$ significance level, test the claim that systolic blood pressure is not affected by the pill. That is, test the claim that the before and after values are equal.

| Before pill | 120 | 136 | 160 | 98 | 115 | 110 | 180 | 190 | 138 | 128 |
|---|---|---|---|---|---|---|---|---|---|---|
| After pill | 118 | 122 | 143 | 105 | 98 | 98 | 180 | 175 | 105 | 112 |

6.  A test of driving ability is given to a random sample of 10 student drivers before and after they completed a formal driver education course. The results follow. At the $\alpha = 0.05$ significance level, test the claim that the course does not affect scores.

| Before course | 100 | 121 | 93 | 146 | 101 | 109 | 149 | 130 | 127 | 120 |
|---|---|---|---|---|---|---|---|---|---|---|
| After course | 136 | 129 | 125 | 150 | 110 | 138 | 136 | 130 | 125 | 129 |

7.  In a study of techniques used to measure lung volumes, physiological data were collected for 10 subjects. The values given in the table are in liters and they represent the measured functional residual capacities of the 10 subjects in a sitting position and in a supine (lying) position. (See "Validation of Esophageal Balloon Technique at Different Lung Volumes and Postures," by Baydur, Cha, and Sassoon, *Journal of Applied Physiology*, Vol. 62, No. 1.) At the 0.05 significance level, test the claim that there is no significant difference between the measurements from the two positions.

| Sitting | 2.96 | 4.65 | 3.27 | 2.50 | 2.59 | 5.97 | 1.74 | 3.51 | 4.37 | 4.02 |
|---|---|---|---|---|---|---|---|---|---|---|
| Supine | 1.97 | 3.05 | 2.29 | 1.68 | 1.58 | 4.43 | 1.53 | 2.81 | 2.70 | 2.70 |

8.  A course is designed to increase readers' speed and comprehension. To evaluate the effectiveness of this course, a test is given both before and after the course, and sample results follow. At the 0.05 significance level, test the claim that the scores are higher after the course.

| Before | 100 | 110 | 135 | 167 | 200 | 118 | 127 | 95 | 112 | 116 |
|---|---|---|---|---|---|---|---|---|---|---|
| After | 136 | 160 | 120 | 169 | 200 | 140 | 163 | 101 | 138 | 129 |

9.  The following chart lists a random sampling of the ages of married couples. The age of each husband is listed above the age of his wife. At the 0.01 significance level, test the claim that there is no difference between the ages of husbands and wives.

| Husband | 28.1 | 33.0 | 29.8 | 53.1 | 56.7 | 41.6 | 50.6 | 21.4 | 62.0 | 19.7 |
|---|---|---|---|---|---|---|---|---|---|---|
| Wife | 28.4 | 27.6 | 32.7 | 52.0 | 58.1 | 41.2 | 50.7 | 20.6 | 61.1 | 18.1 |

10.  Ten randomly selected volunteers test a new diet, with the following results. At the 0.01 level of significance, test the claim that the diet is effective—that is, that weights (in kilograms) are lower after the diet.

| Subject | A | B | C | D | E | F | G | H | I | J |
|---|---|---|---|---|---|---|---|---|---|---|
| Weight before diet | 68 | 54 | 59 | 60 | 57 | 62 | 62 | 65 | 88 | 76 |
| Weight after diet | 65 | 52 | 52 | 60 | 58 | 59 | 60 | 63 | 78 | 75 |

11.  A political party preference poll is taken among 20 randomly selected voters. If 7 prefer the Republican Party while 13 prefer the Democratic Party, apply the sign test to test the claim that both parties are preferred equally. Use a 0.05 level of significance.

12.  A television commercial advertises that 7 out of 10 dentists surveyed prefer Covariant toothpaste over the leading competitor. Assume that 10 dentists are surveyed and 7 do prefer Covariant, while 3 favor the other brand. Is this a reasonable basis for making the claim that most (more than half) dentists favor Covariant toothpaste? Use the sign test with a significance level of 0.05.

13.  Use the sign test to test the claim that the median life of a battery is at least 40 hours if a random sample of 75 includes exactly 32 that last 40 hours or more. Assume a significance level of 0.05.

14.  A college aptitude test is given to 100 randomly selected high school seniors. After a period of intensive training, another similar test is given to the same students, and 59 receive higher grades, 36 receive lower grades, and 5 students receive the same grades. At the 0.05 level of significance, use the sign test to test the claim that the training is effective.

15.  A company is experimenting with new fertilizer at 50 different locations. In 32 of the locations there is an increase in production, while in 18 locations there is a decrease. At the 0.05 level of significance, use the sign test to test the claim that production is increased by the new fertilizer.

16.  A new diet is designed to lower cholesterol levels. In six months, 36 of the 60 subjects on the diet have lower cholesterol levels, 22 have slightly higher levels, and 2 register no change. At the 0.01 level of significance, use the sign test to test the claim that the diet produces no change in cholesterol levels.

17.  After 30 drivers are tested for reaction times, they are given two drinks and tested again, with the result that 22 have slower reaction times, 6 have faster reaction times, and 2 receive the same scores as before the drinks. At the 0.01 significance level, use the sign test to test the claim that the drinks had no effect on the reaction times.

18.  In target practice, 40 police academy students use two different pistols. Analysis of the scores shows that 24 students get higher scores with

the more expensive pistol, while 16 students get better scores with the less expensive pistol. At the 0.05 level of significance, use the sign test to test the claim that both pistols are equally effective.

19.  Of 50 voters surveyed, 28 favor a tax revision bill before Congress, while all the others are opposed. At the 0.10 level of significance, use the sign test to test the claim that the majority (more than half) of voters favor the bill.

20.  A standardized aptitude test yields a mathematics score $M$ and a verbal score $V$ for each person. Among 15 male subjects, $M - V$ is positive in 12 cases, negative in 2 cases, and 0 in 1 case. At the 0.05 level of significance, use the sign test to test the claim that males do better on the mathematics portion of the test.

## 12–2  Exercises B

21.  Given $n$ sample scores sorted in ascending order $(x_1, x_2, \ldots, x_n)$, if we wish to find the approximate $1 - \alpha$ **confidence interval for the population median $M$,** we get

$$x_{k+1} < M < x_{n-k}$$

Here $k$ is the critical value (Table A–7) for the number of signs in a two-tailed hypothesis test conducted at the significance level of $\alpha$. Find the approximate 95% confidence interval for the sample scores listed below.

   3,  8,  6,  2,  1,  7,  9,  11,  17,  23,  25,  10,  14,  8,  30

22.  **Differences of zero:** In the sign test procedure described in this section, we excluded ties (represented by 0 instead of a sign of $+$ or $-$). A second approach is to treat half of the zeros as positive signs and half as negative signs. (If there's an odd number of zeros, exclude one so they can be divided equally.) With a third approach, in two-tailed tests we make half the zeros positive and half negative; in one-tailed tests make all zeros either positive or negative, whichever supports the null hypothesis. Assume that in using the sign test of a claim that the median score is at least 100, we get 60 scores below 100, 40 scores above 100, and 21 scores equal to 100. Identify the test statistic and conclusion for the three different ways of handling differences of zero. Assume a 0.05 significance level in all three cases.

23.  Of $n$ subjects tested for high blood pressure, a majority of exactly 50 provided negative results. (That is, their blood pressure is not high.) This is sufficient for us to apply the sign test and reject (at the 0.01 level of significance) the claim that the median blood pressure level is high. Find the largest value $n$ can assume.

24. Table A–7 lists critical values for limited choices of $\alpha$. Use Table A–1 to add a new column in Table A–7 (down to $n = 15$) that would represent a significance level of 0.03 in one tail or 0.06 in two tails. For any particular $n$ we use $p = 0.5$, since the sign test requires the assumption that

$$P(\text{positive sign}) = P(\text{negative sign}) = 0.5$$

The probability of $x$ or fewer like signs is the sum of the probabilities up to and including $x$.

## 12–3  Wilcoxon Signed-Ranks Test for Two Dependent Samples

In the preceding section, we used the sign test to analyze the differences between paired data. The sign test used only the signs of the differences, while ignoring their actual magnitudes. In this section we introduce the **Wilcoxon signed-ranks test,** which takes the magnitudes into account. Because this test incorporates and uses more information than the ordinary sign test, it tends to yield better results than the sign test. However, the Wilcoxon signed-ranks test requires a stronger assumption than the sign test. It requires that the two sets of data come from populations with a common distribution. Unlike the $t$ test for paired data (see Section 8–3), the Wilcoxon signed-ranks test does *not* require normal distributions.

Consider the data given in Table 12–5. The 13 subjects are given a test for logical thinking. They are then given a tranquilizer and retested. We will

**TABLE 12–5**

| Subject | Before | After | Difference | Ranks of Differences | Signed-Ranks |
|---------|--------|-------|------------|----------------------|--------------|
| A | 67 | 68 | −1 | 1 | −1 |
| B | 78 | 81 | −3 | 2 | −2 |
| C | 81 | 85 | −4 | 3 | −3 |
| D | 72 | 60 | +12 | 10 | +10 |
| E | 75 | 75 | 0 | — | — |
| F | 92 | 81 | +11 | 8.5 | +8.5 |
| G | 84 | 73 | +11 | 8.5 | +8.5 |
| H | 83 | 78 | +5 | 4 | +4 |
| I | 77 | 84 | −7 | 5 | −5 |
| J | 65 | 56 | +9 | 6 | +6 |
| K | 71 | 61 | +10 | 7 | +7 |
| L | 79 | 64 | +15 | 11 | +11 |
| M | 80 | 63 | +17 | 12 | +12 |

use the Wilcoxon signed-ranks test to test the claim that the tranquilizer has no effect, so that there is no significant difference between before and after scores. We will assume a 0.05 level of significance.

$H_0$: The tranquilizer has no effect on logical thinking.
$H_1$: The tranquilizer has an effect on logical thinking.

In general, the null hypothesis will be the claim that both samples come from the same population distribution. We summarize here the procedure for using the Wilcoxon signed-ranks test with paired data.

## Procedure

1. For each pair of data, find the difference $d$ by subtracting the second score from the first. Retain signs, but discard any pairs for which $d = 0$.

2. Ignoring the signs of the differences, rank them from lowest to highest. When differences have the same numerical value, assign to them the mean of the ranks involved in the tie. (See Section 12–1 for the method of ranking data.)

3. Assign to each rank the sign of the difference from which it came.

4. Find the sum of the absolute values of the negative ranks. Also find the sum of the positive ranks.

5. Let $T$ be the smaller of the two sums found in step 4.

6. Let $n$ be the number of pairs of data for which the difference $d$ is not zero.

7. If $n \leq 30$, use Table A–8 to find the critical value of $T$. Reject the null hypothesis if the sample data yield a value of $T$ less than or equal to the value in Table A–8. Otherwise, fail to reject the null hypothesis. If $n > 30$, compute the test statistic $z$ by using Formula 12–1.

**Formula 12–1**
$$z = \frac{T - \dfrac{n(n + 1)}{4}}{\sqrt{\dfrac{n(n + 1)(2n + 1)}{24}}}$$

When Formula 12–1 is used, the critical $z$ values are found using Table A–2 in the usual way. Again reject the null hypothesis if the test statistic $z$ is less than or equal to the critical value(s) of $z$. Otherwise, fail to reject the null hypothesis.

### Example

Use the Wilcoxon signed-ranks test to test the claim that the tranquilizer has no effect on logical thinking. The sample results are given in Table 12–5.

### Solution

We will follow the seven steps listed in the above procedure.

*Step 1.* In Table 12–5, the column of differences is obtained by subtracting each "after" score from the corresponding "before" score. Differences of zero are discarded. See the fourth column of Table 12–5.

*Step 2.* Ignoring their signs, the differences are then ranked from lowest to highest, with ties being treated in the manner described in Section 12–1. See the fifth column of Table 12–5.

*Step 3.* The signed-ranks column is then created by applying to each rank the sign of the corresponding difference. The results are listed in the last column of Table 12–5. If the tranquilizer really has no effect, we would expect the number of positive ranks to be approximately equal to the number of negative ranks. If the tranquilizer tends to lower scores, then ranks of positive sign would tend to outnumber ranks of negative sign. If the tranquilizer tends to raise scores, then ranks of negative sign would tend to outnumber ranks of positive sign. We can detect a domination by either sign through analysis of the rank sums.

*Step 4.* Now find the sum of the absolute values of the negative ranks and the sum of the positive ranks. For the data in Table 12–5, we get

sum of absolute values of negative ranks $= 1 + 2 + 3 + 5 = 11$

sum of positive ranks $= 10 + 8.5 + 8.5 + 4 + 6 + 7 + 11 + 12 = 67$

*Step 5.* We will base our test on the smaller of those two sums, denoted by $T$. For the given data we have $T = 11$. We use the following notation.

### Notation

| | |
|---|---|
| $T$ | Smaller of these two sums: |
| | 1. The sum of the absolute values of the negative ranks. |
| | 2. The sum of the positive ranks. |
| $n$ | Number of *pairs* of data after excluding any pairs in which both values are the same. |

*continued*

> **Solution** *continued*
>
> *Step 6.* $n = 12$ since there are 12 pairs of data with nonzero differences.
>
> *Step 7.* Whenever $n \leq 30$, we use Table A–8 to find the critical values. If $n > 30$ we can use a normal approximation with the test statistic given in Formula 12–1 and critical values given in Table A–2. As in Section 12–2, we would be justified in using the normal approximation whenever $n \geq 10$, but we have a table of critical values for values of $n$ up to 30 so that we really need the normal approximation only for $n > 30$. Because the data of Table 12–5 yield $n = 12$, we use Table A–8 to get the critical value of 14. ($\alpha = 0.05$ and the test is two-tailed since our null hypothesis is the claim that the scores have not changed significantly.) Since $T = 11$ is less than or equal to the critical value of 14, we reject the null hypothesis. It appears that the drug does affect scores.

In this last example, the unsigned ranks of 1 through 12 have a total of 78. If the two sets of data have no significant differences, each of the two signed-rank totals should be in the neighborhood of $78 \div 2$, or 39. However, for the given sample data, we got 11 for one total and 67 for the other; this 11-67 split was a significant departure from the 39-39 split expected with a true null hypothesis. The table of critical values shows that at the 0.05 level of significance with 12 pairs of data, a 14-64 split represents a significant departure from the null hypothesis, and any split farther apart (such as 13-65 or 12-66) will also represent a significant departure from the null hypothesis. Conversely, splits like 15-63, 16-62, or 38-40 do not represent significant departures away from a 39-39 split, and they would not be a basis for rejecting the null hypothesis. The Wilcoxon signed-ranks test is based on the lower rank total, so that instead of analyzing both numbers that constitute the split, it is necessary to analyze only the lower number.

In general, the sum $1 + 2 + 3 + \cdots + n$ is equal to $n(n + 1)/2$; if this is a rank sum to be divided equally between two categories (positive and negative), each of the two totals should be near $n(n + 1)/4$, which is $n(n + 1)/2$ after it is halved. Recognition of this principle helps us to understand Formula 12–1. The denominator in that formula represents a standard deviation of $T$ and is based on the principle that

$$1^2 + 2^2 + 3^2 + \cdots + n^2 = n(n + 1)(2n + 1)/6.$$

If we were to apply the ordinary sign test (Section 12–2) to the example given in this section, we would fail to reject the null hypothesis of no change in before and after scores. This is not the conclusion reached through the

Wilcoxon signed-ranks test, which is more sensitive to the magnitudes of the differences and is therefore more likely to be correct.

This section can be used for paired data only, but the next section involves a rank-sum test that can be applied to two sets of data that are not paired.

## 12–3 | Exercises A

In Exercises 1–8, first arrange the given data in order of lowest to highest and then find the rank of each entry.

1. 5, 8, 12, 15, 10
2. 1, 3, 6, 8, 99
3. 150, 600, 200, 100, 50, 400
4. 47, 53, 46, 57, 82, 63, 90, 55
5. 6, 8, 8, 9, 12, 20
6. 6, 8, 8, 8, 9, 12, 20
7. 16, 13, 16, 13, 13, 14, 15, 18, 12
8. 36, 27, 27, 27, 41, 39, 58, 63, 63

In Exercises 9–12, use the given before and after test scores in the Wilcoxon signed-ranks test procedure to do the following.

    a.   Find the differences $d$.
    b.   Rank the differences while ignoring their signs.
    c.   Find the signed ranks.
    d.   Find $T$.

9.
| Before | 103 | 98 | 112 | 94 | 118 | 99 | 90 | 101 |
|--------|-----|-----|-----|-----|-----|-----|-----|-----|
| After | 100 | 105 | 114 | 98 | 119 | 99 | 100 | 116 |

10.
| Before | 66 | 58 | 59 | 58 | 63 | 52 | 54 | 60 |
|--------|-----|-----|-----|-----|-----|-----|-----|-----|
| After | 58 | 51 | 56 | 53 | 53 | 51 | 45 | 64 |

11.
| Before | 83 | 76 | 91 | 59 | 62 | 75 | 80 | 66 | 73 |
|--------|-----|-----|-----|-----|-----|-----|-----|-----|-----|
| After | 82 | 77 | 89 | 62 | 68 | 70 | 90 | 86 | 73 |

12.
| Before | 52 | 49 | 37 | 45 | 50 | 48 | 39 | 49 | 55 | 42 | 40 |
|--------|-----|-----|-----|-----|-----|-----|-----|-----|-----|-----|-----|
| After | 44 | 46 | 40 | 35 | 41 | 43 | 41 | 34 | 35 | 35 | 40 |

In Exercises 13–16, assume a 0.05 level of significance in a two-tailed hypothesis test. Use the given statistics to find the critical score from Table A–8, then form a conclusion about the null hypothesis $H_0$.

13.  a.  $T = 24, n = 15$
     b.  $T = 25, n = 15$
14.  a.  $T = 26, n = 15$
     b.  $T = 81, n = 24$
15.  a.  $T = 25, n = 17$
     b.  $T = 8, n = 10$
16.  a.  $T = 5, n = 10$
     b.  $T = 15, n = 12$

In Exercises 17–28, use the Wilcoxon signed-ranks test.

17.  A psychologist wants to test the claim that two different IQ tests produce the same results. Both tests are given to a sample of nine randomly selected students, with the results given below. At the 0.05 level of significance, test the claim that both tests produce the same results.

| Test A | 100 | 111 | 93 | 92 | 99 | 85 | 117 | 110 | 98 |
|--------|-----|-----|----|----|----|----|-----|-----|-----|
| Test B | 106 | 112 | 95 | 90 | 107 | 100 | 126 | 105 | 110 |

18.  A biomedical researcher wants to test the effectiveness of a synthetic antitoxin. The 12 randomly selected subjects are tested for resistance to a particular poison. They are retested after receiving the antitoxin, with the results given below. At the 0.05 level of significance, test the claim that the antitoxin is not effective and produces no change.

| Before | 18.2 | 21.6 | 23.5 | 22.9 | 16.3 | 19.2 | 21.6 | 21.8 | 20.3 | 19.5 | 18.9 | 20.3 |
|--------|------|------|------|------|------|------|------|------|------|------|------|------|
| After  | 18.4 | 20.3 | 21.5 | 20.2 | 17.6 | 18.5 | 21.7 | 22.3 | 19.4 | 18.6 | 20.1 | 19.7 |

19.  In a study of techniques used to measure lung volumes, physiological data were collected for 10 subjects. The values given in the table are in liters and they represent the measured forced vital capacities of the 10 subjects in a sitting position and in a supine (lying) position. (See "Validation of Esophageal Balloon Technique at Different Lung Volumes and Postures," by Baydur, Cha, and Sassoon, *Journal of Applied Physiology*, Vol. 62, No. 1.) At the 0.05 significance level, test the claim that both positions have the same distribution.

| Sitting | 4.66 | 5.70 | 5.37 | 3.34 | 3.77 | 7.43 | 4.15 | 6.21 | 5.90 | 5.77 |
|---------|------|------|------|------|------|------|------|------|------|------|
| Supine  | 4.63 | 6.34 | 5.72 | 3.23 | 3.60 | 6.96 | 3.66 | 5.81 | 5.61 | 5.33 |

20.  In a study of techniques used to measure lung volumes, physiological data were collected for 10 subjects. The values given in the table are in liters and they represent the measured functional residual capacities of the 10 subjects in a sitting position and in a supine (lying) position. (See "Validation of Esophageal Balloon Technique at Different Lung Volumes and Postures," by Baydur, Cha, and Sassoon, *Journal of Applied Physiology*, Vol. 62, No. 1.) At the 0.05 significance level, test the

claim that both positions have the same distribution.

| Sitting | 2.96 | 4.65 | 3.27 | 2.50 | 2.59 | 5.97 | 1.74 | 3.51 | 4.37 | 4.02 |
|---------|------|------|------|------|------|------|------|------|------|------|
| Supine  | 1.97 | 3.05 | 2.29 | 1.68 | 1.58 | 4.43 | 1.53 | 2.81 | 2.70 | 2.70 |

21. A researcher devises a test of depth perception while the subject has one eye covered. The test is repeated with the other eye covered. Results are given below for 8 randomly selected subjects. At the 0.05 level of significance, test the claim that depth perception is the same for both eyes.

| Right eye | 14.7 | 16.3 | 12.4 | 8.1 | 21.6 | 13.9 | 14.2 | 15.8 |
|-----------|------|------|------|-----|------|------|------|------|
| Left eye  | 15.2 | 16.7 | 12.6 | 10.4 | 24.1 | 17.2 | 11.9 | 18.4 |

22. To test the effect of smoking on pulse rate, a researcher compiled data consisting of pulse rate before and after smoking. The results are given below. At the 0.05 level of significance, test the claim that smoking does not affect pulse rate.

| Before smoking | 68 | 72 | 69 | 70 | 70 | 74 | 66 | 71 |
|----------------|----|----|----|----|----|----|----|----|
| After smoking  | 69 | 76 | 68 | 73 | 72 | 76 | 66 | 71 |

23. An anxiety-level index is invented for third-grade students, and results for each of 14 randomly selected students are obtained in a classroom setting and a recess situation. The results follow. At the 0.05 level of significance, test the claim that anxiety levels are the same for both situations.

| Classroom | 7.2 | 7.8 | 6.0 | 5.1 | 3.9 | 4.7 | 8.2 | 9.1 | 8.7 | 4.1 | 3.8 | 6.7 | 5.8 | 4.9 |
|-----------|-----|-----|-----|-----|-----|-----|-----|-----|-----|-----|-----|-----|-----|-----|
| Recess    | 8.7 | 7.1 | 5.8 | 4.2 | 4.0 | 3.6 | 7.0 | 8.7 | 7.9 | 2.5 | 2.4 | 5.0 | 4.0 | 3.4 |

24. Two types of cooling systems are being tested in preparation for construction of a nuclear power plant. Eight different standard experimental situations yield the temperature in degrees Fahrenheit of water expelled by each of the cooling systems, and the results follow. At the 0.05 level of significance, test the claim that both cooling systems produce the same results.

| Type A | 72 | 78 | 81 | 77 | 84 | 76 | 79 | 74 |
|--------|----|----|----|----|----|----|----|----|
| Type B | 75 | 71 | 71 | 72 | 73 | 74 | 70 | 74 |

25. Randomly selected voters are given two different tests designed to measure their attitudes about conservatism. Both tests supposedly use the same rating scale and the same criteria. At the 0.05 level of significance, test the claim that both tests produce the same results. The sample data follow.

| Test A | 237 | 215 | 312 | 190 | 217 | 250 | 341 | 380 | 270 | 245 |
|--------|-----|-----|-----|-----|-----|-----|-----|-----|-----|-----|
| Test B | 217 | 190 | 307 | 192 | 220 | 233 | 314 | 367 | 249 | 238 |

26. Randomly selected executives are surveyed in an attempt to measure their attitudes toward two different minority groups, and the sample results follow. At the 0.05 level of significance, test the claim that there is no difference in their attitudes toward the two groups.

| Group A | 420 | 490 | 380 | 570 | 630 | 710 | 425 | 576 | 550 | 610 | 580 | 575 |
|---|---|---|---|---|---|---|---|---|---|---|---|---|
| Group B | 520 | 510 | 450 | 530 | 600 | 705 | 415 | 600 | 625 | 730 | 500 | 530 |

27. A study was conducted to investigate the effectiveness of hypnotism in reducing pain. At the 0.05 significance level, test the claim that the distribution of affective responses to pain is the same before and after hypnosis. Results for randomly selected subjects are given in the accompanying table, which is based on data from "An Analysis of Factors That Contribute to the Efficacy of Hypnotic Analgesia," by Price and Barber, *Journal of Abnormal Psychology*, Vol. 96, No. 1.

| Before | −5.5 | −5.0 | −6.6 | −9.7 | −4.0 | −7.0 | −7.0 | −8.4 |
|---|---|---|---|---|---|---|---|---|
| After | −1.4 | −0.5 | 0.7 | 1.0 | 2.0 | 0.0 | −0.6 | −1.8 |

28. A study was conducted to investigate some effects of physical training. Sample data are listed in the following table (based on data from "Effect of Endurance Training on Possible Determinants of $\dot{V}O_2$ During Heavy Exercise," by Casaburi and others, *Journal of Applied Physiology*, Vol. 62, No. 1). At the 0.05 level of significance, test the claim that the pretraining weights and the posttraining weights have the same distribution. All weights are given in kilograms.

| Pretraining | 99 | 57 | 62 | 69 | 74 | 77 | 59 | 92 | 70 | 85 |
|---|---|---|---|---|---|---|---|---|---|---|
| Posttraining | 94 | 57 | 62 | 69 | 66 | 76 | 58 | 88 | 70 | 84 |

## 12–3  Exercises B

29. a. Two checkout systems are being tested at a department store. One system uses an optical scanner to record prices while the other system has prices manually entered by the clerk. Randomly selected customers are paid to use both checkout systems, and their processing times are recorded. Listed here are the differences (in seconds) obtained when the times for the scanner system are subtracted from the corresponding times for the manual system. At the 0.01 significance level, use the Wilcoxon signed-ranks test to test the claim that both systems require the same times.

```
 30  33  27   0   −5   −3  18  10  16  12    3  52  14   −8
−27   0  42  26   19   35  72  14   5   1   12  −6  23   52
 47  33  19  16    0  −12  44  40  29  59   38
```

*continued*

b.   Part *a* is a two-tailed test. Now test the claim that the scanner times are lower. That is, the distribution of scanner times is shifted to the left of the distribution of times for the manual system. Use the same 0.01 significance level.

30.   a.   With $n = 8$ pairs of data, find the lowest and highest possible values of *T*.

b.   With $n = 10$ pairs of data, find the lowest and highest possible values of *T*.

c.   With $n = 50$ pairs of data, find the lowest and highest possible values of *T*.

31.   Use Formula 12–1 to find the critical value of *T* for a two-tailed hypothesis test with a significance level of 0.05. Assume that there are $n = 100$ pairs of data with no differences of zero.

32.   The Wilcoxon signed-ranks test can be used to test the claim that a sample comes from a population with a specified median. This use of the Wilcoxon signed-ranks test requires that the population be approximately symmetrical. That is, when the population distribution is separated in the middle, the left half approximates a mirror image of the right half. The procedure for testing hypotheses is the same as the one described in this section, except that the differences (step 1) are obtained by subtracting the value of the hypothesized median from each score. At the 0.05 level of significance, test the claim that the values below are drawn from a population with a median of 10,000 lb. The scores are the weights (in pounds) of 50 different loads handled by a moving company in Dutchess County, New York.

| | | | | |
|---|---|---|---|---|
| 8,090 | 9,110 | 17,810 | 12,350 | 3,670 |
| 14,800 | 10,100 | 26,580 | 17,330 | 15,970 |
| 8,800 | 11,860 | 7,770 | 8,450 | 12,430 |
| 10,780 | 13,260 | 5,030 | 10,220 | 11,430 |
| 13,490 | 11,600 | 13,520 | 7,470 | 4,510 |
| 14,310 | 14,760 | 13,410 | 4,480 | 7,450 |
| 7,540 | 3,250 | 10,630 | 6,400 | 10,330 |
| 8,160 | 10,510 | 9,310 | 12,700 | 9,900 |
| 7,200 | 6,170 | 12,010 | 16,200 | 11,450 |
| 8,770 | 9,140 | 6,820 | 7,280 | 6,390 |

## 12–4  Wilcoxon Rank-Sum Test for Two Independent Samples

While Section 12–3 used ranks to analyze dependent or paired data, this section introduces the **Wilcoxon rank-sum test,** which can be applied to situations involving two samples that are *independent* and not paired. This test is equivalent to the **Mann-Whitney U test** found in some other books (see Exercise 18). The Wilcoxon rank-sum test is used under these conditions.

## Assumptions

1.  We have two independent samples.

2.  We are testing the null hypothesis that the two independent samples come from the same distribution; the alternative hypothesis is the claim that the two distributions are different in some way.

3.  Each of the two samples must have more than 10 scores. For cases with samples having 10 or fewer values, special tables are available in other reference books.

4.  Unlike the corresponding hypothesis test in Section 8–3, the Wilcoxon rank-sum test does *not* require normally distributed populations.

5.  Unlike the corresponding hypothesis test in Section 8–3, the Wilcoxon rank-sum test *can* be used with data at the ordinal level of measurement, such as data consisting of ranks.

In Section 12–1 we noted that the Wilcoxon rank-sum test has a 0.95 efficiency rating when compared with the parametric $t$ test or $z$ test. Because this test has such a high efficiency rating and involves easier calculations, it is often preferred over the parametric tests presented in Section 8–3, even when the condition of normality is satisfied.

We will illustrate the procedure used for the Wilcoxon rank-sum test with the following example. The basis for the procedure is the principle that if two samples are drawn from identical populations and the individual scores are all ranked as one combined collection of values, then the high and low ranks should be dispersed evenly between the two samples. If we find that the low (or high) ranks are found predominantly in one of the samples, we suspect that the two populations are not identical.

## CLASS ATTENDANCE *DOES* HELP

■ In a study of 424 undergraduates at the University of Michigan, it was found that students with the worst attendance records tended to get the lowest grades. (Is anybody surprised?) Those who were absent less than 10% of the time tended to receive grades of B or above. The study also showed that students who sit in the front of the class tend to get significantly better grades. ■

### Example

Random samples of teachers' salaries from Massachusetts and Pennsylvania are as follows. (The data are based on a survey by the National Education Association.) At the 0.05 level of significance, test the claim that the salaries of teachers are the same in both states.

*continued*

**Example** *continued*

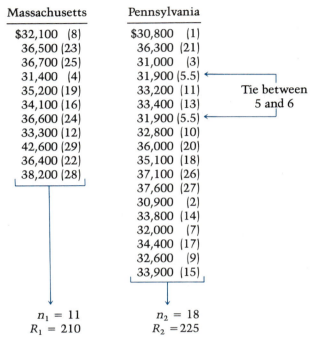

| Massachusetts | Pennsylvania | |
|---|---|---|
| \$32,100 (8) | \$30,800 (1) | |
| 36,500 (23) | 36,300 (21) | |
| 36,700 (25) | 31,000 (3) | |
| 31,400 (4) | 31,900 (5.5) ← | |
| 35,200 (19) | 33,200 (11) | Tie between |
| 34,100 (16) | 33,400 (13) | 5 and 6 |
| 36,600 (24) | 31,900 (5.5) ← | |
| 33,300 (12) | 32,800 (10) | |
| 42,600 (29) | 36,000 (20) | |
| 36,400 (22) | 35,100 (18) | |
| 38,200 (28) | 37,100 (26) | |
| | 37,600 (27) | |
| | 30,900 (2) | |
| | 33,800 (14) | |
| | 32,000 (7) | |
| | 34,400 (17) | |
| | 32,600 (9) | |
| | 33,900 (15) | |

$$n_1 = 11 \qquad n_2 = 18$$
$$R_1 = 210 \qquad R_2 = 225$$

### Solution

$H_0$: The populations of salaries are identical.
$H_1$: The populations are not identical.

We may be tempted to use the Student $t$ test to compare the means of two independent samples (as in Section 8–3), but there may be a question about the normality of the distribution of teachers' salaries. We therefore use the Wilcoxon rank-sum test, which does not require normal distributions.

We rank all 29 salaries, beginning with a rank of 1 (assigned to the lowest salary of \$30,800). The ranks corresponding to the various salaries are shown in parentheses in the above table. Note that the tie between the fifth and sixth scores results in assigning the rank of 5.5 to each of those two salaries. We denote by $R$ the sum of the ranks for one of the two samples. If we choose the Massachusetts salaries we get

$$R = 8 + 23 + 25 + 4 + 19 + 16 + 24 + 12 + 29 + 22 + 28$$
$$= 210$$

*continued*

**Solution** *continued*

**With a null hypothesis of identical populations and with both sample sizes greater than 10, the sampling distribution of $R$ is approximately normal.** Denoting the mean of the sample $R$ values by $\mu_R$ and the standard deviation of the sample $R$ values by $\sigma_R$, we are able to use the test statistic

**Formula 12–2**
$$z = \frac{R - \mu_R}{\sigma_R}$$

where $\mu_R = \dfrac{n_1(n_1 + n_2 + 1)}{2}$

$\sigma_R = \sqrt{\dfrac{n_1 n_2(n_1 + n_2 + 1)}{12}}$

$n_1 = $ size of the sample from which the rank sum $R$ is found

$n_2 = $ size of the other sample

$R = $ sum of ranks of the sample with size $n_1$

The expression for $\mu_R$ is a variation of a result of mathematical induction, which states that the sum of the first $n$ positive integers is given by $1 + 2 + 3 + \cdots + n = n(n + 1)/2$, and the expression for $\sigma_R$ is a variation of a result that states that the integers $1, 2, 3, \ldots, n$ have standard deviation $\sqrt{(n^2 - 1)/12}$.

For the salary data given in the table we have already found that the rank sum $R$ for the Massachusetts salaries is 210. Since there are 11 Massachusetts salaries, we have $n_1 = 11$. Also, $n_2 = 18$ since there are 18 Pennsylvania salaries. We can now determine the values of $\mu_R$, $\sigma_R$, and $z$.

$$\mu_R = \frac{n_1(n_1 + n_2 + 1)}{2} = \frac{11(11 + 18 + 1)}{2} = 165.00$$

$$\sigma_R = \sqrt{\frac{n_1 n_2(n_1 + n_2 + 1)}{12}} = \sqrt{\frac{(11)(18)(11 + 18 + 1)}{12}} = 22.25$$

$$z = \frac{R - \mu_R}{\sigma_R} = \frac{210 - 165.00}{22.25} = 2.02$$

A large positive value of $z$ would indicate that the higher ranks are disproportionately found in the Massachusetts salaries, while a large negative value of $z$ would indicate that Massachusetts has a disproportionate share of lower ranks. In either case, we would have strong evidence against the claim that the Massachusetts and Pennsylvania salaries are identical. The test is therefore two-tailed.

*continued*

**Solution** *continued*

The significance of the test statistic $z$ can now be treated in the same manner as in previous chapters. We are now testing (with $\alpha = 0.05$) the hypothesis that the two populations are the same, so we have a two-tailed test with critical $z$ values of 1.96 and $-1.96$. The test statistic of $z = 2.02$ falls within the critical region and we therefore reject the null hypothesis that the salaries are the same in both states. Massachusetts appears to have significantly higher teacher salaries than Pennsylvania.

We can verify that if we interchange the two sets of salaries, we will find that $R = 225$, $\mu_R = 270.0$, $\sigma_R = 22.25$, and $z = -2.02$, so that the same conclusion will be reached.

Like the Wilcoxon signed-ranks test, this test also considers the relative magnitudes of the sample data, whereas the sign test does not. In the sign test, a weight loss of 1 lb or 50 lb receives the same sign, so the actual magnitude of the loss is ignored. While rank-sum tests do not directly involve quantitative differences between data from two samples, changes in magnitude do cause changes in rank, and these in turn affect the value of the test statistic. For example, if we change the Pennsylvania salary of \$30,800 to \$40,800, then the value of the rank-sum $R$ will change, and the value of the $z$ test statistic will also change.

## 12–4 | Exercises A

In Exercise 1–16, use the Wilcoxon rank-sum test.

1. Random samples of teachers' salaries (in hundreds of dollars) from California and Maryland are as follows. (The data are based on a survey by the National Education Association.) At the 0.05 significance level, test the claim that salaries of teachers are the same in both states.

| California | 271 | 306 | 323 | 336 | 364 | 390 | 391 | 405 | 408 | 409 | 417 | 464 |
|---|---|---|---|---|---|---|---|---|---|---|---|---|
| Maryland | 287 | 312 | 323 | 326 | 334 | 339 | 341 | 344 | 387 | 396 | 403 | 405 |
| | 439 | 443 | 459 | 478 | | | | | | | | |

2. The values listed in the table at the top of the following page are selling prices (in thousands of dollars) of randomly selected homes recently sold in Dutchess County, New York. Test the claim that the selling prices are the same in homes with seven rooms as they are in homes with eight rooms. Use a 0.05 significance level.

| Seven rooms | 154 | 142 | 119 | 160 | 136 | 122 | 114 | 135 | 127 | 138 | 134 |
|---|---|---|---|---|---|---|---|---|---|---|---|
| Eight rooms | 215 | 165 | 127 | 170 | 153 | 172 | 164 | 205 | 190 | 135 | 145 |
|  | 197 | 124 | 212 | | | | | | | | |

3. The following scores represent the reaction times (in seconds) of randomly selected subjects from two age groups. Use a 0.05 level of significance to test the claim that both groups have the same reaction times.

| 18 years old | 1.96 | 0.94 | 0.96 | 1.51 | 1.36 | 1.41 | 1.03 | 1.12 |
|---|---|---|---|---|---|---|---|---|
|  | 2.12 | 0.86 | 0.79 | 1.17 | 1.13 | 1.00 | 1.01 | |
| 50 years old | 1.03 | 1.42 | 1.75 | 2.01 | 0.93 | 1.92 | 2.00 | 1.87 |
|  | 2.09 | 1.73 | 1.49 | 1.82 | | | | |

4. Given below are heights (in centimeters) of randomly selected six-year-old children (based on data from the U.S. Department of Health and Human Services). Use a 0.05 significance level to test the claim that the heights of six-year-old boys are the same as the heights of six-year-old girls.

| Girls | 112.9 | 107.3 | 126.2 | 122.2 | 122.2 | 117.0 | 123.1 | 108.9 |
|---|---|---|---|---|---|---|---|---|
|  | 107.3 | 120.7 | 131.5 | 115.9 | 117.3 | 115.1 | 118.0 | 112.4 |
| Boys | 116.1 | 111.1 | 112.9 | 119.1 | 122.2 | 120.6 | 114.7 | 123.9 |
|  | 126.1 | 124.9 | 111.9 | 110.4 | 118.9 | 111.3 | 126.2 | 120.7 |

5. An auto parts supplier must send many shipments from the central warehouse to the city in which the assembly takes place, and she wants to determine the faster of two railroad routes. A search of past records provides the following data. (The shipment times are in hours.) At the 0.05 level of significance, determine whether or not there is a significant difference between the routes.

| Route A | 98 | 102 | 83 | 117 | 128 | 92 | 112 | 108 | 108 | 100 | 93 | 72 | 95 | 91 |
|---|---|---|---|---|---|---|---|---|---|---|---|---|---|---|
| Route B | 96 | 132 | 121 | 87 | 106 | 102 | 116 | 95 | 99 | 76 | 97 | 104 | 115 | 114 |

6. The BAC (blood alcohol concentration) levels at arrest of randomly selected convicted jail inmates are given below, categorized by the type of drink consumed (based on data from the U.S. Department of Justice). At the 0.05 significance level, test the claim that beer drinkers and liquor drinkers have the same BAC levels.

| Beer | 0.129 | 0.146 | 0.148 | 0.152 | 0.154 | 0.155 | |
|---|---|---|---|---|---|---|---|
|  | 0.187 | 0.212 | 0.203 | 0.190 | 0.164 | 0.165 | |
| Liquor | 0.220 | 0.225 | 0.185 | 0.182 | 0.253 | 0.241 | 0.227 |
|  | 0.205 | 0.247 | 0.224 | 0.226 | 0.234 | 0.190 | 0.257 |

7. A study is conducted to determine whether a drug affects eye movements. A standardized scale is developed and the drug is administered to one group, while a control group is given a placebo that produces no effects. The eye movement ratings of subjects are as follows. At the 0.01 level of significance, test the claim that the drug has no effect on eye movements.

| Drugged group | 652 | 512 | 711 | 621 | 508 | 603 | 787 | 747 | 516 | 624 | 627 | 777 | 729 |
|---|---|---|---|---|---|---|---|---|---|---|---|---|---|
| Control group | 674 | 676 | 821 | 830 | 565 | 821 | 837 | 652 | 549 | 668 | 772 | 563 | 703 |
| | 789 | 800 | 711 | 598 | | | | | | | | | |

8. Two coffee-vending machines are studied to determine whether they distribute the same amounts. Samples are obtained and the contents (in liters) are as follows. At the 0.05 level of significance, test the claim that the machines distribute the same amount.

| Machine A | 0.210 | 0.213 | 0.206 | 0.195 | 0.180 | 0.250 | 0.212 | 0.217 |
|---|---|---|---|---|---|---|---|---|
| | 0.213 | 0.222 | 0.201 | 0.205 | 0.209 | | | |
| Machine B | 0.229 | 0.224 | 0.221 | 0.247 | 0.270 | 0.233 | 0.237 | 0.235 |
| | 0.238 | 0.200 | 0.198 | 0.216 | 0.241 | 0.273 | 0.205 | |

9. A large city police department offers a refresher course on arrest procedures. The effectiveness of this course is examined by testing 15 randomly selected officers who have recently completed the course. The same test is given to 15 randomly selected officers who have not had the refresher course. The results are as follows. At the 0.05 level of significance, test the claim that the course has no effect on the test grades.

| Group completing the course | 173 | 141 | 219 | 157 | 163 | 165 | 178 | 200 |
|---|---|---|---|---|---|---|---|---|
| | 154 | 189 | 192 | 201 | 157 | 168 | 181 | |
| Group without the course | 159 | 124 | 170 | 148 | 135 | 133 | 137 | 189 |
| | 181 | 111 | 144 | 127 | 138 | 151 | 162 | |

10. In a study of longevity, two groups of adult males are randomly selected; their longevity data (in years) are summarized below. Test the claim that there is no difference between the two groups.

| Group A | 65 | 66 | 73 | 78 | 54 | 39 | 47 | 59 | 67 | 67 |
|---|---|---|---|---|---|---|---|---|---|---|
| | 69 | 71 | 74 | 77 | 62 | 73 | 75 | 76 | 68 | 67 |
| Group B | 64 | 67 | 73 | 70 | 58 | 69 | 72 | 71 | 63 | 64 |
| | 63 | 63 | 55 | 43 | 35 | 50 | 74 | 61 | 62 | 69 |

11. In a study of crop yields, two different fertilizer treatments are tested on parcels with the same area and soil conditions. Listed below are the yields (in bushels of corn) for sample plots. Use a 0.05 significance level and determine if there is a difference between the treatments.

| Treatment A | 132 | 137 | 129 | 142 | 160 | 139 | 143 | 147 | 145 | 140 | 131 | 136 |
|---|---|---|---|---|---|---|---|---|---|---|---|---|
| Treatment B | 162 | 180 | 149 | 157 | 159 | 159 | 152 | 167 | 163 | 165 | 180 | 156 |
|  | 158 | 151 | | | | | | | | | | |

12. A consumer investigator obtains prices from mail order companies and computer stores. Listed below are the prices (in dollars) quoted for boxes of ten floppy disks from various manufacturers. Use a 0.05 level of significance to test the claim that there is no difference between mail order and store prices.

| Mail order | 23.00 | 26.00 | 27.99 | 31.50 | 32.75 | 27.00 |
|---|---|---|---|---|---|---|
|  | 27.98 | 24.50 | 24.75 | 28.15 | 29.99 | 29.99 |
| Computer store | 30.99 | 33.98 | 37.75 | 38.99 | 35.79 | 33.99 |
|  | 34.79 | 32.99 | 29.99 | 33.00 | 32.00 | |

13. The effectiveness of mental training was tested in a military training program. In an antiaircraft artillary examination, scores for an experimental group and a control group were recorded. Test the claim that both groups come from populations with the same scores. Use a 0.05 significance level. (See "Routinization of Mental Training in Organizations: Effects on Performance and Well-Being" by Larsson, *Journal of Applied Psychology*, Vol. 72, No. 1.)

| Experimental | | | | Control | | | |
|---|---|---|---|---|---|---|---|
| 60.83 | 117.80 | 44.71 | 75.38 | 122.80 | 70.02 | 119.89 | 138.27 |
| 73.46 | 34.26 | 82.25 | 59.77 | 118.43 | 54.22 | 118.58 | 74.61 |
| 69.95 | 21.37 | 59.78 | 92.72 | 121.70 | 70.70 | 99.08 | 120.76 |
| 72.14 | 57.29 | 64.05 | 44.09 | 104.06 | 94.23 | 111.26 | 121.67 |
| 80.03 | 76.59 | 74.27 | 66.87 | | | | |

14. Sample data were collected in a study of calcium supplements and the effects on blood pressure. A placebo group and a calcium group began the study with measures of blood pressures. At the 0.05 significance level, test the claim that the two sample groups come from populations with the same blood pressure levels. (The data are based on "Blood Pressure and Metabolic Effects of Calcium Supplementation in Normotensive White and Black Men" by Lyle and others, *Journal of the American Medical Association*, Vol. 257, No. 13.)

| Placebo | | | | Calcium | | | |
|---|---|---|---|---|---|---|---|
| 124.6 | 104.8 | 96.5 | 116.3 | 129.1 | 123.4 | 102.7 | 118.1 |
| 106.1 | 128.8 | 107.2 | 123.1 | 114.7 | 120.9 | 104.4 | 116.3 |
| 118.1 | 108.5 | 120.4 | 122.5 | 109.6 | 127.7 | 108.0 | 124.3 |
| 113.6 | | | | 106.6 | 121.4 | 113.2 | |

15. In a study involving motivation and test scores, data were obtained for males and females. Use the following data to test the claim that both samples come from populations with the same scores. Use a 0.05 significance level. (See "Relationships Between Achievement-Related Motives, Extrinsic Conditions, and Task Performance" by Schroth, *Journal of Social Psychology*, Vol. 127, No. 1.)

| | Male | | | | Female | | |
|---|---|---|---|---|---|---|---|
| 12.27 | 39.53 | 32.56 | 23.93 | 31.13 | 18.71 | 14.34 | 23.90 |
| 19.54 | 25.73 | 32.20 | 19.84 | 13.96 | 13.88 | 29.85 | 20.15 |
| 20.20 | 23.01 | 25.63 | 17.98 | 6.66 | 19.20 | 15.89 | |
| 22.99 | 22.12 | 12.63 | 18.06 | | | | |

16. The arrangement of test items was studied for its effect on anxiety. Sample results are given below. At the 0.05 level of significance, test the claim that the two samples come from populations with the same scores. (The data are based on "Item Arrangement, Cognitive Entry Characteristics, Sex and Test Anxiety as Predictors of Achievement Examination Performance" by Klimko, *Journal of Experimental Education*, Vol. 52, No. 4.)

| | Easy to difficult | | | | Difficult to easy | | |
|---|---|---|---|---|---|---|---|
| 24.64 | 39.29 | 16.32 | 32.83 | 33.62 | 34.02 | 26.63 | 30.26 |
| 28.02 | 33.31 | 20.60 | 21.13 | 35.91 | 26.68 | 29.49 | 35.32 |
| 26.69 | 28.90 | 26.43 | 24.23 | 27.24 | 32.34 | 29.34 | 33.53 |
| 7.10 | 32.86 | 21.06 | 28.89 | 27.62 | 42.91 | 30.20 | 32.54 |
| 28.71 | 31.73 | 30.02 | 21.96 | | | | |
| 25.49 | 38.81 | 27.85 | 30.29 | | | | |
| 30.72 | | | | | | | |

# 12–4 Exercises B

17. a. The *ranks* for group A are 1, 2, . . . , 15 and the *ranks* for group B are 16, 17, . . . , 30. At the 0.05 level of significance, use the Wilcoxon rank-sum test to test the claim that both groups come from the same population.
    b. The *ranks* for group A are 1, 3, 5, 7, . . . , 29 and the *ranks* for group B are 2, 4, 6, . . . , 30. At the 0.05 level of significance, use the Wilcoxon rank-sum test to test the claim that both groups come from the same population.
    c. Compare parts *a* and *b*.
    d. What changes occur when the rankings of the two groups in part *a* are interchanged?
    e. Use the two groups in part *a* and interchange the ranks of 1 and 30 and then note the changes that occur.

18. The Mann-Whitney U test is equivalent to the Wilcoxon rank-sum test for independent samples in the sense that they both apply to the same situations and they always lead to the same conclusions. In the Mann-Whitney U test we calculate

$$z = \frac{U - \frac{n_1 n_2}{2}}{\sqrt{\frac{n_1 n_2 (n_1 + n_2 + 1)}{12}}}$$

where

$$U = n_1 n_2 + \frac{n_1 (n_1 + 1)}{2} - R$$

Show that if the expression for $U$ is substituted into the preceding expression for $z$, we get the same test statistic (with opposite sign) used in the Wilcoxon rank-sum test for two inedependent samples.

19. Assume that we have two treatments (A and B) that produce measurable results, and we have only two observations for treatment A and two observations for treatment B. We cannot use Formula 12–2 because both sample sizes do not exceed 10.

   a. Complete the table below by listing the other five rows corresponding to the other five cases, and enter the corresponding rank sums for treatment A.

   | Rank | | | | (Rank sum for |
   |---|---|---|---|---|
   | 1 | 2 | 3 | 4 | treatment A) |
   | A | A | B | B | 3 |

   b. List the possible values of $R$ along with their corresponding probabilities. (Assume that the rows of the table from part *a* are equally likely.)

   c. Is it possible, at the 0.10 significance level, to reject the null hypothesis that there is no difference between treatments A and B? Explain.

20. Do Exercise 19 for the case involving a sample of size three for treatment A and a sample of size three for treatment B.

# 12–5 | Kruskal-Wallis Test

In Chapter 11 we used one-way analysis of variance to test hypotheses that differences among several samples are due to chance. That particular $F$ test requires that all the involved populations possess normal distributions with variances that are approximately equal. In this section we introduce the

**Kruskal-Wallis test** (also called the **H test**) as a nonparametric alternative that does not require normal distributions. The Kruskal-Wallis test is used under the conditions given below.

## Assumptions

1. We have at least three samples, all of which are random.

2. We want to test the null hypothesis that the samples come from the same or identical populations.

3. Each sample has at least five observations. For cases involving samples with fewer than five observations, refer to more advanced books for special tables of critical values.

4. Unlike the corresponding one-way analysis of variance method used in Chapter 11, the Kruskal-Wallis test does *not* require normally distributed populations. It does require equal variances, so this test shouldn't be used if the different samples have variances that are very far apart.

5. Unlike the corresponding one-way analysis of variance method in Chapter 11, the Kruskal-Wallis test can be used with data at the ordinal level of measurement, such as data consisting of ranks.

In applying the Kruskal-Wallis test, we compute the test statistic **H,** **which has a distribution that can be approximated by the chi-square distribution as long as each sample has at least five observations.** When we use the chi-square distribution in this context, the number of degrees of freedom is $k - 1$, where $k$ is the number of samples. (For a quick review of the key features of the chi-square distribution, see Section 6–4.)

$$\text{degrees of freedom} = k - 1$$

**Formula 12–3**     $$H = \frac{12}{N(N + 1)}\left(\frac{R_1^2}{n_1} + \frac{R_2^2}{n_2} + \cdots + \frac{R_k^2}{n_k}\right) - 3(N + 1)$$

where $N$ = total number of observations in all samples combined
$R_1$ = sum of ranks for the first sample
$R_2$ = sum of ranks for the second sample
$R_k$ = sum of ranks for the $k$th sample
$k$ = the number of samples

In using the Kruskal-Wallis test, we replace the original scores by their corresponding ranks. We then proceed to calculate the test statistic $H$, which is basically a measure of the variance of the rank sums $R_1, R_2, \ldots, R_k$. If the ranks are distributed evenly among the sample groups, then $H$ should be a relatively small number. If the samples are very different, then the ranks will be excessively low in some groups and high in others, with the net effect that $H$ will be large. Consequently, only large values of $H$ lead to rejection of the null hypothesis that the samples come from identical populations. **The Kruskal-Wallis test is therefore a right-tailed test.**

Begin by considering all observations together, and then assign a rank to each one. We rank from lowest to highest, and we also treat ties as we did in the previous sections of this chapter—the mean value of the ranks is assigned to each of the tied observations. Then take each individual sample and find the sum of the ranks and the corresponding sample size. We will illustrate the Kruskal-Wallis test in the following example. The Kruskal-Wallis test is especially appropriate here because there may be some doubt that home selling prices are normally distributed. We will see similarities between the Kruskal-Wallis test and Wilcoxon's rank-sum test, since both are based on rank sums.

## Example

A real estate investor randomly selects homes recently sold in three different zones of the same county. The selling prices (based on data from homes recently sold in Dutchess County, New York) are listed here. At the 0.05 significance level, test the claim that selling prices are the same in all three zones.

Selling Prices (dollars)

| Zone 1 | | Zone 4 | | Zone 7 | |
|---|---|---|---|---|---|
| $147,000 | (15.5) | $160,000 | (17.5) | $215,000 | (24) |
| 160,000 | (17.5) | 140,000 | (14) | 127,000 | (8) |
| 128,000 | (9) | 173,000 | (21) | 98,000 | (2) |
| 162,000 | (19) | 113,000 | (4) | 147,000 | (15.5) |
| 135,000 | (12) | 85,000 | (1) | 184,000 | (23) |
| 132,000 | (10) | 120,000 | (7) | 109,000 | (3) |
| 181,000 | (22) | 285,000 | (25) | 169,000 | (20) |
| 138,000 | (13) | 117,000 | (5) | | |
| | | 133,000 | (11) | | |
| | | 119,000 | (6) | | |

$$n_1 = 8 \qquad n_2 = 10 \qquad n_3 = 7$$
$$R_1 = 118 \qquad R_2 = 111.5 \qquad R_3 = 95.5$$

## Solution

$H_0$: The populations are identical.
$H_1$: The populations are not identical.

*Step 1.* Rank the combined samples from lowest to highest. Begin with the lowest observation of $85,000, which is assigned a rank of 1. Ranks are shown in parentheses with the original data in the preceding list.

*continued*

**Solution** *continued*

*Step 2.* For each individual sample, find the number of observations and the sum of the ranks. The first sample (Zone 1) has eight observations, so $n_1 = 8$. Also, $R_1 = 15.5 + 17.5 + \cdots + 13 = 118$. The values of $n_2$, $R_2$, $n_3$, and $R_3$ are shown above. Since the total number of observations is 25, we have $N = 25$.

*Step 3.* Compute the value of the test statistic $H$. Using Formula 12–3 for the given data, we get

$$H = \frac{12}{N(N+1)}\left(\frac{R_1^2}{n_1} + \frac{R_2^2}{n_2} + \frac{R_3^2}{n_3}\right) - 3(N+1)$$

$$= \frac{12}{25(26)}\left(\frac{118^2}{8} + \frac{111.5^2}{10} + \frac{95.5^2}{7}\right) - 3(26)$$

$$= \frac{12}{650}(1740.5 + 1243.225 + 1302.893) - 78$$

$$= 1.138$$

*Step 4.* Since each sample has at least five observations, the distribution of $H$ is approximately a chi-square distribution with $k - 1$ degrees of freedom. The number of samples is $k = 3$, so we get $3 - 1$, or 2, degrees of freedom. Refer to Table A–4 to find the critical value of 5.991, which corresponds to 2 degrees of freedom and a significance level of $\alpha = 0.05$. (This use of the chi-square distribution is always right-tailed, since only large values of $H$ reflect disparity in the distribution of ranks among the samples.) If any sample has fewer than five observations, use the special tables found in texts devoted exclusively to nonparametric statistics. Such cases are not included in this text.

*Step 5.* The test statistic $H = 1.138$ is less than the critical value of 5.991, so we fail to reject the null hypothesis of identical populations. We reject the null hypothesis of identical populations only when $H$ exceeds the critical value. The three zones appear to have selling prices that are not significantly different.

The test statistic $H$, as described in Formula 12–3, is the rank version of the test statistic $F$ used in the analysis of variance discussed in Chapter 11. When dealing with ranks $R$ instead of raw scores $x$, many components are predetermined. For example, the sum of all ranks can be expressed as $N(N + 1)/2$, where $N$ is the total number of scores in all samples combined. The expression

$$H = \frac{12}{N(N+1)} \Sigma n_i(\overline{R}_i - \overline{\overline{R}})^2$$

combines weighted variances of ranks in a test statistic that is algebraically equivalent to Formula 12–3, which is easier to work with. (In this expression, $\overline{R}_i = R_i/n_i$ and $\overline{R} = (\Sigma R_i)/(\Sigma n_i)$.)

In comparing the procedures of the parametric $F$ test for analysis of variance and the nonparametric Kruskal-Wallis test, we see that the Kruskal-Wallis test is much simpler to apply. We need not compute the sample variances and sample means. We do not require normal population distributions. Life becomes so much easier. However, the Kruskal-Wallis test is not as efficient as the $F$ test, and it may require more dramatic differences for the null hypothesis to be rejected.

# 12–5 | Exercises A

In Exercises 1–12, use the Kruskal-Wallis test.

1. An experiment involves raising samples of corn under identical conditions except for the type of fertilizer used. The yields are obtained for three different fertilizers, and those values are ranked with the results shown here. Find the value of the test statistic $H$, where $H$ is given in Formula 12–3.

2. Do Exercise 1 after replacing the given ranks with those listed below.

| Treatment | | |
|---|---|---|
| A | B | C |
| 1 | 2 | 3 |
| 6 | 4 | 5 |
| 7 | 8 | 9 |
| 12 | 11 | 10 |
| 14 | 15 | 13 |

| Treatment | | |
|---|---|---|
| A | B | C |
| 1 | 6 | 11 |
| 2 | 7 | 12 |
| 3 | 8 | 13 |
| 4 | 9 | 14 |
| 5 | 10 | 15 |

3. A sociologist randomly selects subjects from three different types of family structure: stable families, divorced families, and families in transition. The selected subjects are interviewed and rated for their social adjustment, and the sample results are given below. (The numbers in parentheses are the ranks.) At the 0.05 level of significance, test the claim that the samples come from identical populations.

| Stable | | Divorced | | Transition | |
|---|---|---|---|---|---|
| 108 | (8.5) | 113 | (10) | 92 | (1) |
| 104 | (4) | 106 | (7) | 123 | (13) |
| 103 | (3) | 108 | (8.5) | 126 | (14) |
| 97 | (2) | 127 | (15) | 105 | (5.5) |
| 118 | (12) | 114 | (11) | 105 | (5.5) |

4. Readability studies are conducted to determine the clarity of four different texts, and the sample scores follow. (The numbers in parentheses are the ranks.) At the 0.05 level of significance, test the claim that the four texts have the same readability level.

| Text A | Text B | Text C | Text D |
|---|---|---|---|
| 50  (3.5) | 59  (10.5) | 45  (1) | 62  (13) |
| 50  (3.5) | 60  (12) | 48  (2) | 64  (15) |
| 53  (6) | 63  (14) | 51  (5) | 68  (18) |
| 58  (9) | 65  (16) | 54  (7) | 70  (19) |
| 59  (10.5) | 67  (17) | 55  (8) | 72  (20) |

5. Three different transmission types are installed on different cars, all having 6 cylinders and an engine size of 3.0 liters. The fuel consumption (in mi/gal) is found for highway conditions with the results given here (based on data from the Environmental Protection Agency). The letters $A$, $M$, $L$ represent automatic, manual, and lockup torque converter, respectively. At the 0.05 significance level, test the claim that the fuel consumption values are the same for the three transmission types.

| A | M | L |
|---|---|---|
| 23.2 | 27.0 | 24.1 |
| 23.2 | 29.8 | 26.1 |
| 20.7 | 25.7 | 24.8 |
| 21.6 | 23.5 | 25.6 |
| 22.9 | 24.3 | 26.5 |

6. A store owner records the gross receipts for days randomly selected from periods during which she used only newspaper advertising, only radio advertising, or no advertising. The results are listed here. At the 0.05 level of significance, test the claim that the receipts are the same, regardless of advertising.

| Newspaper | Radio | None |
|---|---|---|
| 845 | 811 | 612 |
| 907 | 782 | 574 |
| 639 | 749 | 539 |
| 883 | 863 | 641 |
| 806 | 872 | 666 |

7. Three methods of instruction are used to train air traffic controllers. With method A, an experienced controller is assigned to a trainee for practical field experience. With method B, trainees are given extensive classroom instruction and are then placed without supervision. Method C requires a moderate amount of classroom training, and then several trainees are supervised on the job by an experienced instructor. A standardized test is used to measure levels of competency, and sample results are given below. At the 0.01 level of significance, test the claim that the three methods are equally effective.

| Method A | Method B | Method C |
|---|---|---|
| 195 | 187 | 193 |
| 198 | 210 | 212 |
| 223 | 222 | 215 |
| 240 | 238 | 231 |
| 251 | 256 | 252 |
|  |  | 260 |
|  |  | 267 |

8. Randomly selected teachers' salaries from four different states are given at the top of the following page. (The data are based on a survey by the National Education Association.) At the 0.05 significance level, test the claim that the salaries of teachers are the same in the four states.

| Georgia | Vermont | Texas | Oregon |
|---------|---------|-------|--------|
| $28,550 | $29,987 | $32,090 | $35,157 |
| 25,869 | 24,008 | 30,498 | 32,199 |
| 33,602 | 27,169 | 33,933 | 34,949 |
| 26,303 | 27,697 | 36,406 | 35,833 |
| 31,563 | 27,585 | 23,350 | 28,759 |
| 26,058 | 33,095 | 27,720 | 33,956 |
| 35,912 | | 28,755 | 30,918 |
| | | 37,063 | |

9. The accompanying data represent 25 different homes sold in Dutchess County, New York. The zones (1, 4, 7) correspond to different geographic regions of the county. LA values are living areas in hundreds of square feet, Acres values are lot sizes in acres, and Taxes values are the annual tax bills in thousands of dollars. At the 0.05 significance level, test the claim that living areas are the same in all three zones.

10. Use the data from Exercise 9. At the 0.05 level of significance, test the claim that lot sizes (as measured in acres) are the same in all three zones.

11. Use the data from Exercise 9. At the 0.05 level of significance, test the claim that taxes are the same in all three zones.

12. Refer to the data from Exercise 9 and change the zone 7 Tax values to the following: 3.0, 1.8, 1.1, 1.4, 2.3, 2.0, 2.4. Note that the mean Tax amount does not change, but these values vary more. Now repeat Exercise 11 with this modified data set and note the effect of increasing the spread of the values in one of the samples.

| Zone | LA | Acres | Taxes |
|------|-----|-------|-------|
| 1 | 20 | 0.50 | 1.9 |
| 1 | 18 | 1.00 | 2.4 |
| 1 | 27 | 1.05 | 1.5 |
| 1 | 17 | 0.42 | 1.6 |
| 1 | 18 | 0.84 | 1.6 |
| 1 | 13 | 0.33 | 1.5 |
| 1 | 24 | 0.90 | 1.7 |
| 1 | 15 | 0.83 | 2.2 |
| 4 | 18 | 0.55 | 2.8 |
| 4 | 20 | 0.46 | 1.8 |
| 4 | 19 | 0.94 | 3.2 |
| 4 | 12 | 0.29 | 2.1 |
| 4 | 9 | 0.26 | 1.4 |
| 4 | 18 | 0.33 | 2.1 |
| 4 | 28 | 1.70 | 4.2 |
| 4 | 10 | 0.50 | 1.7 |
| 4 | 15 | 0.43 | 1.8 |
| 4 | 12 | 0.25 | 1.6 |
| 7 | 21 | 3.04 | 2.7 |
| 7 | 16 | 1.09 | 1.9 |
| 7 | 14 | 0.23 | 1.3 |
| 7 | 23 | 1.00 | 1.7 |
| 7 | 17 | 6.20 | 2.2 |
| 7 | 17 | 0.46 | 2.0 |
| 7 | 20 | 3.20 | 2.2 |

## 12–5 | Exercises B

13. a. Simplify Formula 12-3 for the special case of eight samples, all consisting of exactly six observations each.
    b. In general, how is the value of the test statistic $H$ affected if a constant is added to (or subtracted from) each score?
    c. In general, how is the value of the test statistic $H$ affected if each score is multiplied (or divided) by a positive constant?

14. For three samples, each of size five, what are the largest and smallest possible values of $H$?

15. In using the Kruskal-Wallis test, there is a correction factor that should be applied whenever there are many ties: Divide $H$ by

$$1 - \frac{\Sigma T}{N^3 - N}$$

Here $T = t^3 - t$. For each group of tied scores, find the number of observations that are tied and represent this number by $t$. Then compute $t^3 - t$ to find the value of $T$. Repeat this procedure for all cases of ties and find the total of the $T$ values, which is $\Sigma T$. For the example presented in this section, use this procedure to find the corrected value of $H$.

16. Show that for the case of two samples, the Kruskal-Wallis test is equivalent to the Wilcoxon rank-sum test. This can be done by showing that for the case of two samples, the test statistic $H$ equals the square of the test statistic $z$ used in the Wilcoxon rank-sum test. Also note that with one degree of freedom, the critical values of $\chi^2$ correspond to the square of the critical $z$ score.

# 12–6   Rank Correlation

In Chapter 9 we considered the concept of correlation, and we introduced the *linear correlation coefficient* as a measure of the strength of the association between two variables. In this section we will study rank correlation, the nonparametric counterpart of that parametric measure. In Chapter 9 we computed values for the linear correlation coefficient $r$, but in this section we will be computing values for the **rank correlation coefficient.** The method presented in this section has some distinct advantages over the parametric methods discussed in Chapter 9.

## Advantages

1. With rank correlation, we can analyze some types of data that can be ranked, but not measured; yet such data could not be considered with the parametric linear correlation coefficient $r$ of Chapter 9.

2. Rank correlation can be used to detect some relationships that are not linear. An example illustrating this will be given later in this section.

3. The computations for rank correlation are much simpler than those for the linear correlation coefficient $r$. This can be readily seen by comparing Formula 12–4 (given later in this section) to Formula 9–1. With many calculators, you can get the value of $r$ easily, but if you do not have a calculator or computer, you would probably find that the rank correlation coefficient is easier to compute.

4. Rank correlation can be used when some of the more restrictive requirements of the linear correlation approach are not met. That is, the nonparametric approach can be used in a wider variety of circumstances than can the parametric method. For example, the parametric approach requires that the involved populations have normal distributions; the nonparametric approach does not require normality. We do assume that we have a random sample. If a sample is not random, it may be totally worthless.

As an example, consider the data in Table 12–6, which is based on results in *The Jobs Rated Almanac*. Randomly selected jobs are ranked according to salary and stress. The first entry (stockbroker) has the second highest salary and the second highest stress level.

| TABLE 12–6 | | |
| --- | --- | --- |
| Job | Salary Rank | Stress Rank |
| Stockbroker | 2 | 2 |
| Zoologist | 6 | 7 |
| Electrical engineer | 3 | 6 |
| School principal | 5 | 4 |
| Hotel manager | 7 | 5 |
| Bank officer | 10 | 8 |
| Occupational safety inspector | 9 | 9 |
| Home economist | 8 | 10 |
| Psychologist | 4 | 3 |
| Commercial airline pilot | 1 | 1 |

Is there a relationship between salary and stress? We can't use the linear correlation coefficient $r$ because the data consist of ranks and therefore do not satisfy the normal distribution requirement described in Section 9–2. Instead, we can use the following **rank correlation coefficient.**

**Formula 12–4**

$$r_s = 1 - \frac{6\Sigma d^2}{n(n^2 - 1)}$$

where $r_s$ = rank correlation coefficient

$n$ = number of *pairs* of data

$d$ = difference between ranks for the two observations within a pair

We use the notation $r_s$ for the *rank* correlation coefficient so that we don't confuse it with the *linear* correlation coefficient $r$. The subscript $s$ is commonly used in honor of Charles Spearman (1863–1945), who originated the rank correlation approach. In fact, $r_s$ is often called **Spearman's rank correlation coefficient.** The subscript $s$ has nothing to do with standard deviation, and for that we should be thankful. Just as $r$ is a sample statistic that can be considered an estimate of the population parameter $\rho$, we can also consider $r_s$ to be an estimate of $\rho_s$.

If the data lead to ties in ranks, we proceed as in previous sections: Calculate the mean of the ranks involved in the tie and then assign that mean rank to each of the tied items. Formula 12–4 leads to an exact value of $r_s$ only if no ties occur. With a relatively low number of ties, Formula 12–4

## TV RATINGS

■ Estimating the number of people that watch different television shows has become much more difficult, now that we have VCRs, satellite receiving dishes, cable networks, and local stations. With $25 billion in advertising revenue at stake, ratings companies such as Nielsen Media Research are becoming more sophisticated. Viewers once filled out diaries. Now about 4000 participants press buttons on hand-held "people meters." A new device consists of a very sophisticated camera that can track different viewers, sense their eye movements, and record exactly who is watching the different shows. The data can be sent by telephone lines to Nielsen's computers for analysis. ■

results in a good approximation of $r_s$. (With ties, we can get an exact value of $r_s$ by ranking the data and using the parametric Formula 9–1; after using Formula 9–1 to evaluate $r_s$, we should continue with the procedures in this section.)

We will now test the null hypothesis $H_0: \rho_s = 0$ against the alternative hypothesis $H_1: \rho_s \neq 0$; we will use a 0.05 level of significance. The calculation of the test statistic $r_s$ is shown in Table 12–7. Since there are ten pairs of data, $n = 10$. We obtain the differences $d$ for each pair by subtracting the lower rank from the higher rank.

| TABLE 12–7 | | | | |
|---|---|---|---|---|
| Job | Salary Rank | Stress Rank | Difference $d$ | $d^2$ |
| Stockbroker | 2 | 2 | 0 | 0 |
| Zoologist | 6 | 7 | 1 | 1 |
| Electrical engineer | 3 | 6 | 3 | 9 |
| School principal | 5 | 4 | 1 | 1 |
| Hotel manager | 7 | 5 | 2 | 4 |
| Bank officer | 10 | 8 | 2 | 4 |
| Occupational safety inspector | 9 | 9 | 0 | 0 |
| Home economist | 8 | 10 | 2 | 4 |
| Psychologist | 4 | 3 | 1 | 1 |
| Commercial airline pilot | 1 | 1 | 0 | 0 |
| | | | Total: $\Sigma d^2 =$ | 24 |

With $n = 10$ and $\Sigma d^2 = 24$, we get

$$r_s = 1 - \frac{6\Sigma d^2}{n(n^2 - 1)}$$
$$= 1 - \frac{6(24)}{10(10^2 - 1)}$$
$$= 1 - 0.145$$
$$= 0.855$$

If we now refer to Table A–9 for the critical value, we see that with $n = 10$ and $\alpha = 0.05$, the critical value of $r_s$ is 0.648. Since the test statistic $r_s = 0.855$ exceeds the critical value of 0.648, we reject the null hypothesis. There appears to be a positive correlation between salary and stress.

The hypothesis testing procedure we use here is very similar to the one used in Section 9–2, except for the calculation of the test statistic and the table used for the critical value. See Figure 12–4, which summarizes this procedure.

Rank Correlation
$H_0 : \rho = 0$
$H_1 : \rho \neq 0$

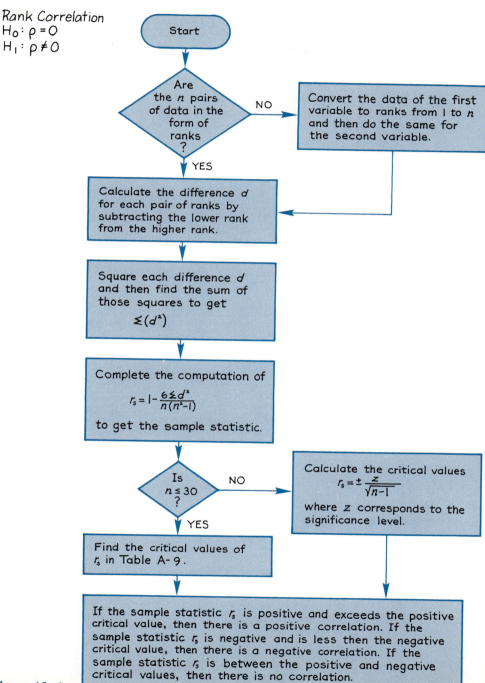

**Figure 12–4**

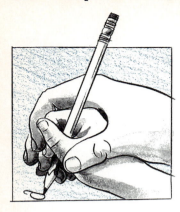

## STUDY CRITICIZED AS MISLEADING

■ A study sponsored by the National Association of Elementary School Principals and the Kettering Foundation suggested that there was a *correlation* between children with one parent and children who were low achievers. The media coverage of the final report implied that the *cause* of lower achievement was the absence of one of the parents. Critics charged that the media confused correlation with cause and effect. Also, some conclusions were based on samples too small to be statistically significant. ■

For practical reasons, we are omitting the theoretical derivation of Formula 12–4, but we can gain some insight by considering the following three cases. If we intuitively examine Formula 12–4, we can see that strong agreement between the two sets of ranks will lead to values of $d$ near 0, so that $r_s$ will be close to 1 (see Case I). Conversely, when the ranks of one set tend to be at opposite extremes when compared to the ranks of the second set, then the values of $d$ tend to be high, which will cause $r_s$ to be near $-1$ (see Case II). If there is no relationship between the two sets of ranks, then the values of $d$ will be neither high nor low and $r_s$ will tend to be near 0 (see Case III).

Cases I and II illustrate the most extreme cases, so that the following property applies.

$$-1 \leq r_s \leq 1$$

When the number of pairs of ranks, $n$, exceeds 30, the sampling distribution of $r_s$ is approximately a normal distribution with mean 0 and standard deviation $1/\sqrt{n-1}$. We therefore get

$$z = \frac{r_s - 0}{\dfrac{1}{\sqrt{n-1}}} = r_s\sqrt{n-1}$$

In a two-tailed case we would use the positive and negative $z$ values. Solving for $r_s$, we then get the critical values by evaluating

**Formula 12–5**
$$r_s = \frac{\pm z}{\sqrt{n-1}} \qquad \text{(when } n > 30\text{)}$$

The value of $z$ would correspond to the significance level.

---

### Example

Find the critical values of Spearman's rank correlation coefficient $r_s$ when the data consist of 40 pairs of ranks. Assume a two-tailed case with a 0.05 significance level.

### Solution

Since there are 40 pairs of data, $n = 40$. Because $n$ exceeds 30, we use Formula 12–5 instead of Table A–9. With $\alpha = 0.05$ in two tails, we let $z = 1.96$ to get

$$r_s = \frac{\pm 1.96}{\sqrt{40-1}} = \pm 0.314$$

## Case I: Perfect Positive Correlation

| Rank x | Rank y | Difference d | $d^2$ |
|--------|--------|--------------|-------|
| 1 | 1 | 0 | 0 |
| 3 | 3 | 0 | 0 |
| 5 | 5 | 0 | 0 |
| 4 | 4 | 0 | 0 |
| 2 | 2 | 0 | 0 |
| | | $\Sigma d^2 = 0$ | |

$$r_s = 1 - \frac{6(0)}{5(5^2 - 1)} = 1$$

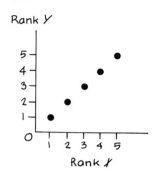

## Case II: Perfect Negative Correlation

| Rank x | Rank y | Difference d | $d^2$ |
|--------|--------|--------------|-------|
| 1 | 5 | 4 | 16 |
| 2 | 4 | 2 | 4 |
| 3 | 3 | 0 | 0 |
| 4 | 2 | 2 | 4 |
| 5 | 1 | 4 | 16 |
| | | $\Sigma d^2 = 40$ | |

$$r_s = 1 - \frac{6(40)}{5(5^2 - 1)} = -1$$

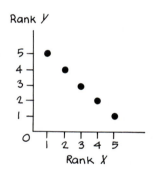

## Case III: No Correlation

| Rank x | Rank y | Difference d | $d^2$ |
|--------|--------|--------------|-------|
| 1 | 2 | 1 | 1 |
| 2 | 5 | 3 | 9 |
| 3 | 3 | 0 | 0 |
| 4 | 1 | 3 | 9 |
| 5 | 4 | 1 | 1 |
| | | $\Sigma d^2 = 20$ | |

$$r_s = 1 - \frac{6(20)}{5(5^2 - 1)} = 0$$

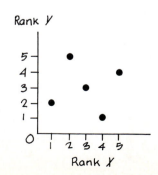

In the next example, the data are graphed in the scatter diagram shown in Figure 12–5. The pattern is not linear, but watch what happens.

## Example

Ten different students study for a test; the table below lists the number of hours studied ($x$) and the corresponding number of correct answers ($y$). At the 0.05 level of significance, use Spearman's rank correlation approach to determine if there is a relationship between hours studied and the number of correct answers.

| $x$ | 5 | 9 | 17 | 1 | 2 | 21 | 3 | 29 | 7 | 100 |
|---|---|---|---|---|---|---|---|---|---|---|
| $y$ | 6 | 16 | 18 | 1 | 3 | 21 | 7 | 20 | 15 | 22 |

## Solution

$H_0$: $\rho_s = 0$
$H_1$: $\rho_s \neq 0$

Refer to Figure 12–4, which we will follow in this solution. The given data are not ranks, so we convert them into ranks as shown in Table 12–8. (Section 12–1 describes the procedure for converting scores into ranks.)

**TABLE 12–8**

| $x$ | 4 | 6 | 7 | 1 | 2 | 8 | 3 | 9 | 5 | 10 |
|---|---|---|---|---|---|---|---|---|---|---|
| $y$ | 3 | 6 | 7 | 1 | 2 | 9 | 4 | 8 | 5 | 10 |
| $d$ | 1 | 0 | 0 | 0 | 0 | 1 | 1 | 1 | 0 | 0 |
| $d^2$ | 1 | 0 | 0 | 0 | 0 | 1 | 1 | 1 | 0 | 0 |

After expressing all data as ranks, we next calculate the differences, $d$, and then we square them. The sum of the $d^2$ values is 4. We now calculate

$$r_s = 1 - \frac{6\Sigma d^2}{n(n^2 - 1)}$$

$$= 1 - \frac{6(4)}{10(10^2 - 1)}$$

$$= 1 - 0.024$$

$$= 0.976$$

*continued*

**Solution** *continued*

Proceeding with Figure 12–4, $n = 10$, so we answer yes when asked if $n \le 30$. We use Table A–9 to get the critical values of $-0.648$ and $0.648$. Finally, the sample statistic of $0.976$ exceeds $0.648$, so we conclude that there is significant positive correlation. More hours of study appear to be associated with higher grades. (You didn't really think we would suggest otherwise, did you?)

If we compute the linear correlation coefficient $r$ (using Formula 9–1) for the original data in this last example, we get $r = 0.629$, which leads to the conclusion that there is no significant *linear* correlation at the 0.05 level of significance. If we examine the scatter diagram in Figure 12–5, we can see that there does seem to be a relationship, but it's not linear. This last example is intended to illustrate two advantages of the nonparametric approach over the parametric approach. We have already noted the advantage of detecting some relationships that are not linear. This last example also illustrates this additional advantage: *Spearman's rank correlation coefficient* $r_s$ *is less sensitive to a value that is very far out of line,* such as the 100 hours in the preceding data.

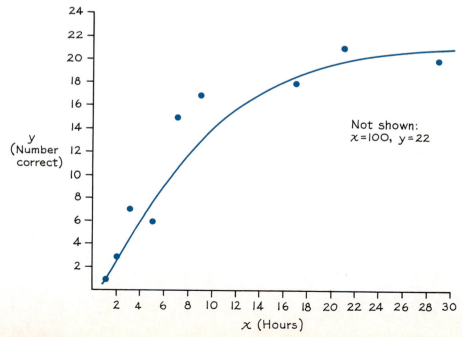

**Figure 12–5**

Like other nonparametric methods, the rank correlation approach does have the disadvantage of being less efficient. In Section 12–1 we noted that rank correlation has a 0.91 efficiency rating when compared to the parametric linear correlation method from Chapter 9. If the conditions required for the parametric approach are satisfied, the nonparametric rank correlation approach would require 100 sample observations to achieve the same result as 91 sample observations analyzed through the parametric linear correlation approach in Chapter 9.

## 12–6 | Exercises A

In Exercises 1 and 2, find the critical value for $r_s$ by using Table A–9 or Formula 12–5, as appropriate. Assume two-tailed cases where $\alpha$ represents the level of significance and $n$ represents the number of pairs of data.

1.  a.  $n = 20, \alpha = 0.05$
    b.  $n = 50, \alpha = 0.05$
    c.  $n = 40, \alpha = 0.02$
    d.  $n = 25, \alpha = 0.02$
    e.  $n = 37, \alpha = 0.04$

2.  a.  $n = 82, \alpha = 0.04$
    b.  $n = 15, \alpha = 0.01$
    c.  $n = 50, \alpha = 0.01$
    d.  $n = 37, \alpha = 0.01$
    e.  $n = 43, \alpha = 0.05$

In Exercises 3 and 4, rewrite the table of paired data so that each entry is replaced by its corresponding rank in accordance with the procedure described in this section.

3.
| $x$ | 2 | 5 | 8 | 6 | 10 |
|---|---|---|---|---|---|
| $y$ | 30 | 25 | 20 | 15 | 5 |

4.
| $x$ | 15 | 14 | 10 | 9 | 3 |
|---|---|---|---|---|---|
| $y$ | 3 | 7 | 10 | 11 | 20 |

In Exercises 5–8, compute the sample statistic $r_s$ by using Formula 12–4.

5.
| $x$ | 1 | 2 | 3 | 4 |
|---|---|---|---|---|
| $y$ | 4 | 2 | 3 | 1 |

6.
| $x$ | 10 | 20 | 30 | 40 |
|---|---|---|---|---|
| $y$ | 40 | 20 | 30 | 10 |

7.
| $x$ | 63 | 68 | 71 | 55 | 70 | 75 |
|---|---|---|---|---|---|---|
| $y$ | 43 | 44 | 39 | 30 | 28 | 20 |

8.
| $x$ | 28 | 28 | 35 | 37 | 40 |
|---|---|---|---|---|---|
| $y$ | 16 | 17 | 12 | 19 | 20 |

In Exercises 9–24:

    a.  Compute the rank correlation coefficient $r_s$ for the given sample data.

    b.  Assume that $\alpha = 0.05$ and find the critical value of $r_s$ from Table A–9 or Formula 12–5.

    c.  Based on the results of parts *a* and *b*, decide whether there is significant positive correlation, significant negative correlation, or no significant correlation. In each case, assume a significance level of 0.05.

9.  Table 12–6 in this section includes paired salary and stress level ranks for 10 randomly selected subjects. The physical demands of those jobs were also ranked; the salary and physical demand ranks are given below (based on data from *The Jobs Rated Almanac*).

| Salary | 2 | 6 | 3 | 5 | 7 | 10 | 9 | 8 | 4 | 1 |
|---|---|---|---|---|---|---|---|---|---|---|
| Physical demand | 5 | 2 | 3 | 8 | 10 | 9 | 1 | 7 | 6 | 4 |

10.  Ten jobs are randomly selected and ranked according to stress level and physical demand, with the results given below (based on data from *The Jobs Rated Almanac*).

| Stress level | 2 | 7 | 6 | 4 | 5 | 8 | 9 | 10 | 3 | 1 |
|---|---|---|---|---|---|---|---|---|---|---|
| Physical demand | 5 | 2 | 3 | 8 | 10 | 9 | 1 | 7 | 6 | 4 |

11.  The following table ranks 8 states according to teachers' salaries and SAT scores (based on data from the U.S. Department of Education).

|  | N.Y. | Calif. | Fla. | N.J. | Tex. | N.C. | Md. | Ore. |
|---|---|---|---|---|---|---|---|---|
| Teacher's salary | 1 | 2 | 7 | 4 | 6 | 8 | 3 | 5 |
| SAT score | 4 | 3 | 5 | 6 | 7 | 8 | 2 | 1 |

12.  The following table ranks 8 states according to SAT scores and cost per student (based on data from the U.S. Department of Education).

|  | N.Y. | Calif. | Fla. | N.J. | Tex. | N.C. | Md. | Ore. |
|---|---|---|---|---|---|---|---|---|
| SAT score | 4 | 3 | 5 | 6 | 7 | 8 | 2 | 1 |
| Cost per student | 1 | 5 | 6 | 2 | 7 | 8 | 3 | 4 |

13. In studying the effects of heredity and environment on intelligence, scientists have learned much by analyzing IQs of identical twins that were separated soon after birth. Identical twins share identical genes inherited from the same fertilized egg. By studying identical twins raised apart, we can eliminate the variable of heredity and we can better isolate the effects of environment. The following are the IQs of identical twins (older twins are $x$) raised apart (based on data from "IQ's of Identical Twins Reared Apart," by Arthur Jensen, *Behavioral Genetics*).

| $x$ | 107 | 96 | 103 | 90 | 96 | 113 | 86 | 99 | 109 | 105 | 96 | 89 |
|---|---|---|---|---|---|---|---|---|---|---|---|---|
| $y$ | 111 | 97 | 116 | 107 | 99 | 111 | 85 | 108 | 102 | 105 | 100 | 93 |

14. *Consumer Reports* tested VHS tapes used in VCRs. Given below are performance scores and prices (in dollars) or randomly selected tapes.

| Performance | 91 | 92 | 82 | 85 | 87 | 80 | 94 | 97 |
|---|---|---|---|---|---|---|---|---|
| Price | 4.56 | 6.48 | 5.99 | 7.92 | 5.36 | 3.32 | 7.32 | 5.27 |

15. A study was conducted to determine whether there is any relationship between age (in years) and the BAC (blood alcohol concentration) measured when convicted DWI jail inmates were first arrested. Sample data for randomly selected subjects are given below (based on data from the Dutchess County STOP-DWI Program).

| Age | 17.2 | 43.5 | 30.7 | 53.1 | 37.2 | 21.0 | 27.6 | 46.3 |
|---|---|---|---|---|---|---|---|---|
| BAC | 0.19 | 0.20 | 0.26 | 0.16 | 0.24 | 0.20 | 0.18 | 0.23 |

16. Randomly selected subjects ride a bicycle at 5.5 mi/h for one minute. Their weights (in pounds) are given with the numbers of calories used (based on data from *Diet Free* by Kuntzlemann).

| Weight | 167 | 191 | 112 | 129 | 140 | 173 | 119 |
|---|---|---|---|---|---|---|---|
| Calories | 4.23 | 4.69 | 3.21 | 3.47 | 3.72 | 4.45 | 3.36 |

17. At one point in a recent season of the National Basketball Association, *USA Today* reported the current statistics. Given below are the total minutes played and the total points scored for 9 randomly selected players.

| Minutes | 1364 | 53 | 457 | 717 | 384 | 1432 | 365 | 1626 | 840 |
|---|---|---|---|---|---|---|---|---|---|
| Points | 652 | 20 | 163 | 210 | 175 | 821 | 143 | 1098 | 459 |

18. For randomly selected states, the following table lists the per capita beer consumption (in gallons) and the per capita wine consumption (in gallons) (based on data from *Statistical Abstract of the United States*).

| Beer | 32.2 | 29.4 | 35.3 | 34.9 | 29.9 | 28.7 | 26.8 | 41.4 |
|---|---|---|---|---|---|---|---|---|
| Wine | 3.1 | 4.4 | 2.3 | 1.7 | 1.4 | 1.2 | 1.2 | 3.0 |

19. Loads were added to a hanging copper wire, causing the wire to stretch. The loads (in Newtons) and increases in length (in centimeters) are given here. (The table is based on data from *College Physics* by Sears, Zemansky, and Young.)

| Added load | 0 | 10 | 20 | 30 | 40 | 50 | 60 | 70 |
|---|---|---|---|---|---|---|---|---|
| Increase in length | 0 | 0.05 | 0.10 | 0.15 | 0.20 | 0.25 | 0.30 | 1.25 |

20. The following table lists the number of registered automatic weapons (in thousands) along with the murder rate (in murders per 100,000) for randomly selected states. (The data are from the FBI and the Bureau of Alcohol, Tobacco, and Firearms).

| Automatic weapons | 11.6 | 8.3 | 3.6 | 0.6 | 6.9 | 2.5 | 2.4 | 2.6 |
|---|---|---|---|---|---|---|---|---|
| Murder rate | 13.1 | 10.6 | 10.1 | 4.4 | 11.5 | 6.6 | 3.6 | 5.3 |

21. For randomly selected homes recently sold in Dutchess County, New York, the living areas (in hundreds of square feet) are listed along with the annual tax amounts (in thousands of dollars).

| Living area | 15 | 38 | 23 | 16 | 16 | 13 | 20 | 24 |
|---|---|---|---|---|---|---|---|---|
| Taxes | 1.9 | 3.0 | 1.4 | 1.4 | 1.5 | 1.8 | 2.4 | 4.0 |

22. For randomly selected homes recently sold in Dutchess County, New York, the annual tax amounts (in thousands of dollars) are listed along with the selling prices (in thousands of dollars)

| Taxes | 1.9 | 3.0 | 1.4 | 1.4 | 1.5 | 1.8 | 2.4 | 4.0 |
|---|---|---|---|---|---|---|---|---|
| Selling price | 145 | 228 | 150 | 130 | 160 | 114 | 142 | 265 |

23. There are many regions where the winter accumulation of snowfall is a primary source of water. Several investigations of snowpack characteristics have used satellite observations from the Landsat series along with measurements taken on the ground. The following table lists ground measurements of snow depth (in centimeters) along with the corresponding temperatures (in degrees Celsius) (data based on information in Kastner's *Space Mathematics*, published by NASA).

| Temperature (°C) | −62 | −41 | −36 | −26 | −33 | −56 | −50 | −66 |
|---|---|---|---|---|---|---|---|---|
| Snow depth (cm) | 21 | 13 | 12 | 3 | 6 | 22 | 14 | 19 |

24. In a study of employee stock ownership plans, data on satisfaction with the plan and the amount of organizational commitment were collected at 8 companies. Results are given below. (See "Employee Stock Ownership and Employee Attitudes: A Test of Three Models" by Klein, *Journal of Applied Psychology*, Vol. 72, No. 2.)

| Satisfaction | 5.05 | 4.12 | 5.39 | 4.17 | 4.00 | 4.49 | 5.40 | 4.86 |
|---|---|---|---|---|---|---|---|---|
| Commitment | 5.37 | 4.49 | 5.42 | 4.45 | 4.24 | 5.34 | 5.62 | 4.90 |

## 12–6 | Exercises B

25. Two judges each rank 3 contestants, and the ranks for the first judge are 1, 2, and 3, respectively.
    a. List all possible ways that the second judge can rank the same three contestants. (No ties allowed.)
    b. Compute $r_s$ for each of the cases found in part *a*.
    c. Assuming that all of the cases from part *a* are equally likely, find the probability that the sample statistic, $r_s$, is greater than 0.9.

26. One alternative to using Table A–9 involves an approximation of critical values for $r_s$ given as

$$r_s = \pm\sqrt{\frac{t^2}{t^2 + n - 2}}$$

   Here $t$ is the $t$ score from Table A–3 corresponding to the significance level and $n - 2$ degrees of freedom. Apply this approximation to find critical values of $r_s$ for the following cases.
    a. $n = 8$, $\alpha = 0.05$            b. $n = 15$, $\alpha = 0.05$
    c. $n = 30$, $\alpha = 0.05$           d. $n = 30$, $\alpha = 0.01$
    e. $n = 8$, $\alpha = 0.01$

27. a. Given the bivariate data depicted in the scatter diagram, which would be more likely to detect the relationship between $x$ and $y$: the linear correlation coefficient $r$ or the rank correlation coefficient $r_s$? Explain.
    b. How is $r_s$ affected if one variable is ranked from low to high while the other variable is ranked from high to low?
    c. One researcher ranks both variables from low to high, while another researcher ranks both variables from high to low. How will their values of $r_s$ compare?
    d. Using the job salary/stress level data given in Table 12–6, test the claim that there is a *positive* correlation. That is, test the claim that $\rho_s > 0$. Use a 0.05 significance level.

28. Assume that a set of paired data has been converted to ranks according to the procedure described in this section, and also assume that no ties occur. Show that Formula 12–4 (for $r_s$) and Formula 9–1 (for $r$) will provide the same result when the ranks are used as the $x$ and $y$ values.

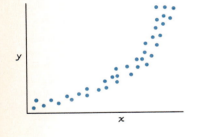

## 12–7 | Runs Test for Randomness

A classic example of the misuse of statistics involved a company president who was convinced that employees were stealing some of the pantyhose being produced. Further investigations showed that production figures were based on samples obtained with newly serviced machinery, which produced a finer mesh and more pantyhose. The production level dropped when the machinery became worn; there was no employee theft. The initial sampling was not random, and it led to misleading results and embarassment for the poor president who proclaimed that pantyhose were being pilfered.

In many of the examples and exercises in this book, we assumed that data were randomly selected. In this section we describe the **runs test,** which is a systematic and standard procedure for testing the randomness of data.

> ### Definition
>
> A **run** is a sequence of data that exhibit the same characteristic; the sequence is preceded and followed by different data or no data at all.

As an example, consider the political party of the sequence of ten voters interviewed by a pollster. We let $D$ denote a Democrat, while $R$ indicates a Republican. The following sequence contains exactly four runs.

### Example 1

| $D, D, D, D,$ | $R, R,$ | $D, D, D,$ | $R$ |
|:---:|:---:|:---:|:---:|
| 1st run | 2nd run | 3rd run | 4th run |

We would use the runs test in this situation to test for the randomness with which Democrats and Republicans occur. Let's use common sense to see how runs relate to randomness. Examine the sequence in Example 2 and then stop to consider how randomly Democrats and Republicans occur. Also count the number of runs.

### Example 2

$$D, D, D, D, D, D, D, D, D, D, R, R, R, R, R, R, R, R, R, R$$

In Example 2, it is reasonable to conclude that Democrats and Republicans occur in a sequence that is *not* random. Note that in the sequence of 20 data, there are only two runs. This example might suggest that if the number of runs is very low, randomness may be lacking. Now consider the sequence of 20 data given in Example 3. Try again to form your own conclusion about randomness before you continue reading.

> ### Example 3
>
> $$D, R, D, R, D, R, D, R, D, R, D, R, D, R, D, R, D, R, D, R$$

In Example 3, it should be apparent that the sequence of Democrats and Republicans is again *not* random, since there is a distinct, predictable pattern. In this case, the number of runs is 20; this example suggests that *randomness is lacking when the number of runs is too high.*

It is important to note that **this test for randomness is based on the order in which the data occur. This runs test is *not* based on the *frequency* of the data.** For example, a particular sequence containing 3 Democrats and 20 Republicans might lead to the conclusion that the sequence is random. The issue of whether or not 3 Democrats and 20 Republicans is a *biased* sample is another issue not addressed by the runs test.

The sequences in Examples 2 and 3 are obvious in their lack of randomness, but most sequences are not so obvious; we therefore need more sophisticated techniques for analysis. We begin by introducing some notation.

> ### Notation
>
> | | |
> |---|---|
> | $n_1$ | Number of elements in the sequence that have the same characteristic |
> | $n_2$ | Number of elements in the sequence that have the other characteristic |
> | $G$ | Number of runs |

We use $G$ to represent the number of runs because $n$ and $r$ have already been used for other statistics, and $G$ is a relatively innocuous letter that deserves more attention.

> ### Example
>
> In the sequence $D, D, D, D, R, R, D, D, D, R, R, R, R, D$, we obtain the following values for $n_1$, $n_2$, and G.
>
> | | |
> |---|---|
> | $n_1 = 8$ | since there are 8 Democrats |
> | $n_2 = 6$ | since there are 6 Republicans |
> | $G = 5$ | since there are 5 runs |

We can now revert to our standard procedure for hypothesis testing. **The null hypothesis $H_0$ will be the claim that the sequence is random;** the alternative hypothesis $H_1$ will be the claim that the sequence is *not* random. The flowchart in Figure 12–6 on the following page summarizes the mechanics of the procedure. That flowchart directs us to a table (Table A–10) of critical $G$ values when the following three conditions are all met:

1.  $\alpha = 0.05$, and
2.  $n_1 \leq 20$, and
3.  $n_2 \leq 20$.

If all these conditions are not satisfied, we use the fact that $G$ has a distribution that is approximately normal with mean and standard deviation as follows.

**Formula 12–6**
$$\mu_G = \frac{2n_1 n_2}{n_1 + n_2} + 1$$

**Formula 12–7**
$$\sigma_G = \sqrt{\frac{(2n_1 n_2)(2n_1 n_2 - n_1 - n_2)}{(n_1 + n_2)^2 (n_1 + n_2 - 1)}}$$

When this normal approximation is used, the test statistic is

**Formula 12–8**
$$z = \frac{G - \mu_G}{\sigma_G}$$

and the critical values are found by using the procedures introduced in Chapter 6. This normal approximation is quite good. If the entire table of critical values (Table A–10) had been computed using this normal approximation, no critical value would be off by more than one unit.

We now illustrate the use of the runs test for randomness by presenting examples of complete tests of hypotheses.

## SPORTS HOT STREAKS

■ It is a common belief that athletes often have "hot streaks," that is, brief periods of extraordinary success. Stanford University psychologist Amos Tversky and other researchers used statistics to analyze the thousands of shots taken by the Philadelphia 76ers for one full season and half of another. They found that the number of "hot streaks" was no different than you would expect from random trials with the outcome of each trial independent of any preceding results. That is, the probability of a hit doesn't depend on the preceding hit or miss. ■

> **Example**
>
> The president of an investment firm has observed that men and women have been hired in the following sequence: *M, M, M, W, M, M, M, M, W, W, W, M.* At the 0.05 level of significance, test the personnel officer's claim that the sequence of men and women is random. (Note that we are not testing for a *bias* in favor of one sex over the other. There are 8 men and 4 women, but we are testing only for the *randomness* in the way they appear in the given sequence.)
>
> *Solution follows on page 641*

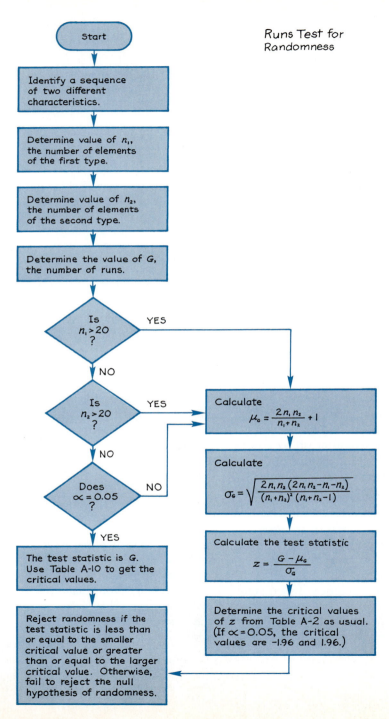

**Figure 12–6**

## Solution

The null hypothesis is the claim of randomness, so we get

$H_0$:   The 8 men and 4 women have been hired in a random sequence.

$H_1$:   The sequence is not random.

The significance level is $\alpha = 0.05$. Figure 12–6 summarizes the procedure for the runs test, so we refer to that flowchart. We now determine the values of $n_1$, $n_2$ and $G$ for the given sequence.

$$n_1 = \text{number of men} = 8$$

$$n_2 = \text{number of women} = 4$$

$$G = \text{number of runs} = 5$$

Continuing with Figure 12–6, we answer no when asked if $n_1 > 20$ (since $n_1 = 8$), no when asked if $n_2 > 20$ (since $n_2 = 4$), and yes when asked if $\alpha = 0.05$. The test statistic is $G = 5$, and we refer to Table A–10 to find the critical values of 3 and 10. Figure 12–7 shows that the test statistic $G = 5$ does not fall in the critical region. We therefore fail to reject the null hypothesis that the given sequence of men and women is random.

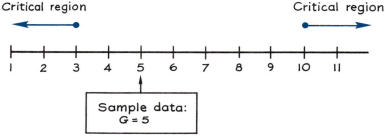

**Figure 12–7**

   The next example will illustrate the procedure to be followed when Table A–10 cannot be used and the normal approximation must be used instead.

## Example

A machine produces defective ($D$) items and acceptable ($A$) items in the following sequence.

   $D, A, A, A, A, D, A, A, A, A, D, A, A, A, A,$

   $D, A, A, A, A, D, A, A, A, A, D, A, A, A, A$

*continued*

**Example** *continued*

At the 0.05 level of significance, test the claim that the sequence is random.

$H_0$: The sequence is random. $H_1$: The sequence is not random. Here $\alpha = 0.05$, and the statistic relevant to this test is $G$. Referring to the flowchart in Figure 12–6, we determine the values of $n_1$, $n_2$, and $G$ as follows:

$$n_1 = \text{number of defective items} = 6$$

$$n_2 = \text{number of acceptable items} = 24$$

$$G = \text{number of runs} = 12$$

Continuing with the flowchart, we answer no when asked if $n_1 > 20$ (since $n_1 = 6$) and yes when asked if $n_2 > 20$ (since $n_2 = 24$). The flowchart now directs us to calculate $\mu_G$. Because of the sample sizes involved, we are on our way to using a normal approximation.

$$\mu_G = \frac{2n_1n_2}{n_1 + n_2} + 1$$

$$= \frac{2(6)(24)}{6 + 24} + 1$$

$$= \frac{288}{30} + 1 = 9.6 + 1 = 10.6$$

The next step requires the computation of $\sigma_G$.

$$\sigma_G = \sqrt{\frac{(2n_1n_2)(2n_1n_2 - n_1 - n_2)}{(n_1 + n_2)^2(n_1 + n_2 - 1)}}$$

$$= \sqrt{\frac{2(6)(24)[2(6)(24) - 6 - 24]}{(6 + 24)^2(6 + 24 - 1)}}$$

$$= \sqrt{\frac{288(258)}{900(29)}} = \sqrt{\frac{74{,}304}{26{,}100}}$$

$$= \sqrt{2.847} = 1.69$$

We now calculate the test statistic $z$.

$$z = \frac{G - \mu_G}{\sigma_G} = \frac{12 - 10.6}{1.69} = 0.83$$

At the 0.05 level of significance, the critical values of $z$ are $-1.96$ and $1.96$. The test statistic $z = 0.83$ does not fall in the critical region, so we fail to reject the null hypothesis that the sequence is random. This is a good example of why we do not say "accept the null hypothesis" because examination of the data reveals that there is a consistent pattern that is clearly not random.

# Randomness Above and Below the Mean or Median

In each of the preceding examples, the data clearly fit into two categories, but we can also test for randomness in the way numerical data fluctuate above or below a mean or median. That is, in addition to analyzing sequences of nominal data of the type already discussed, we can also analyze sequences of data at the interval or ratio levels of measurement, as in the following example.

## Example

An investor wants to analyze long-term trends in the stock market. The annual high points of the Dow Jones Industrial Average for a recent sequence of 10 years are given below.

> 1000, 1024, 1071, 1287, 1287, 1553, 1956, 2722, 2184, 2791

At the 0.05 significance level, test for randomness above and below the median.

## Solution

For the given data, the median is 1420. Let $B$ denote a value *below* the median of 1420 and let $A$ represent a value *above* 1420. We can rewrite the given sequence as follows. (If a value is equal to the median, we simply delete it from the sequence.)

| $B$ | $B$ | $B$ | $B$ | $B$ | $A$ | $A$ | $A$ | $A$ | $A$ |
|---|---|---|---|---|---|---|---|---|---|
| ↓ | ↓ | ↓ | ↓ | ↓ | ↓ | ↓ | ↓ | ↓ | ↓ |
| 1000 | 1024 | 1071 | 1287 | 1287 | 1553 | 1956 | 2722 | 2184 | 2791 |

It is helpful to write the $A$'s and $B$'s directly above the numbers they represent. This makes checking easier and also reduces the chance of having the wrong number of letters. After finding the sequence of letters, we can proceed to apply the runs test in the usual way. We get

$$n_1 = \text{number of } B\text{'s} = 5$$

$$n_2 = \text{number of } A\text{'s} = 5$$

$$G = \text{number of runs} = 2$$

From Table A–10, the critical values of $G$ are found to be 2 and 10. At the 0.05 level of significance, we reject a null hypothesis of randomness

*continued*

**Solution** *continued*

above and below the median if the number of runs is 2 or lower, or 10 or more. Since $G = 2$, we reject the null hypothesis of randomness above and below the median. It isn't necessary to be a financial wizard to recognize that the yearly high levels of the Dow Jones Industrial Average appear to be following an upward trend. This example illustrates one way to recognize such trends, especially those that are not so obvious.

We could also test for randomness above and below the *mean* by following the same procedure (after we have deleted from the sequence any values that are equal to the mean).

Economists use the runs test for randomness above and below the median in an attempt to identify trends or cycles. An upward economic trend would contain a predominance of $B$'s in the beginning and $A$'s at the end, so that the number of runs would be small, as in the preceding example. A downward trend would have $A$'s dominating the beginning and $B$'s at the end, with a low number of runs. A cyclical pattern would yield a sequence that systematically changes, so that the number of runs would tend to be large.

## 12–7 | Exercises A

In Exercises 1–8, use the given sequence to determine the values of $n_1$, $n_2$, the number of runs, $G$, and the appropriate critical values from Table A–10. (Assume a 0.05 significance level.)

1. *A, B, B, B, A, A, A, A,*
2. *A, A, A, B, A, B, A, A, B, B*
3. *A, A, A, A, B, B, B, B, B, B*
4. *A, A, B, B, A, B, A, B, A, B, A*
5. *A, A, A, A, A, A, A, B, B*
6. *A, A, A, B, A, A, B, B, A, B, B, B, B*
7. *A, B, B, B, A, B, B, B, B, B, A, B, A, A, A, A, A, B, A, A*
8. *A, A, A, B, B, B, B, A, A, A, A, A, A, B, B, B, B, B, A, A, A, B, B, B, B, B,* *B, A, A, A, A, A*

In Exercises 9–12, use the given sequence to answer the following.

    a.  Find the mean.
    b.  Let $B$ represent a value below the mean and let $A$ represent a value above the mean. Rewrite the given numerical sequence as a sequence of $A$'s and $B$'s.

   c. Find the values of $n_1$, $n_2$, and $G$.

   d. Assuming a 0.05 level of significance, use Table A–10 to find the appropriate critical values of $G$.

   e. What do you conclude about the randomness of the values above and below the mean?

9. 3, 8, 7, 7, 9, 12, 10, 16, 20, 18

10. 2, 2, 3, 2, 4, 4, 5, 8, 4, 6, 7, 9, 9, 12

11. 15, 12, 12, 10, 17, 11, 8, 7, 7, 5, 6, 5, 5, 9

12. 3, 3, 4, 4, 4, 8, 8, 8, 9, 7, 10, 4, 4, 3, 2, 4, 4, 10, 10, 12, 9, 10, 10, 11, 5, 4, 3, 3, 4, 5

In Exercises 13–16, use the given sequence to answer the following.

   a. Find the median.

   b. Let $B$ represent a value below the median and let $A$ represent a value above the median. Rewrite the given numerical sequence as a sequence of $A$'s and $B$'s.

   c. Find the values of $n_1$, $n_2$, and $G$.

   d. Assuming a 0.05 level of significance, use Table A–10 to find the appropriate critical values of $G$.

   e. What do you conclude about the randomness of the values above and below the median?

13. 3, 8, 7, 7, 9, 12, 10, 16, 20, 18

14. 2, 2, 3, 2, 4, 4, 5, 8, 4, 6, 7, 9, 9, 12

15. 3, 3, 4, 5, 8, 8, 8, 9, 9, 9, 10, 15, 14, 13, 16, 18, 20

16. 19, 17, 3, 2, 2, 15, 4, 12, 14, 13, 9, 7, 7, 7, 8, 8, 8

17. A professional football team's record in a recent season was eight wins and eight losses. These occurred in the order given below. Use a 0.05 significance level to test the claim that the wins and losses occurred in a random order.

W, L, L, L, W, W, L, W, W, L, L, W, W, W, L, L

18. At the 0.05 level of significance, test for randomness of odd $(O)$ and even $(E)$ numbers on a roulette wheel that yielded the results given in the sequence below.

O, E, O, O, O, O, E, E, O, O, E, O, O, O, O, O, O, E, E, E,<br>O, O, O, E, O, O, O

19. A machine produces defective $(D)$ items and acceptable $(A)$ items in the sequence given below. At the 0.05 level of significance, test the claim that the sequence is random.

A, A, A, D, A, A, A, A, D, A, A, A, A, A, D, A,<br>A, A, A, A, A, D, A, A, A, A, A, A, A, D, D, D,<br>A, D, D, D, D, A, A, A, A, D, D, D, D

20. The Apgar rating scale is used to rate the health of newborn babies. A sequence of births results in the following values.

9, 6, 9, 8, 7, 9, 5, 8, 9, 10, 8, 9, 7, 5, 9, 8, 8

At the 0.05 level of significance, test the claim that the ratings above and below the sample mean are random.

21. The number of housing starts for the more recent years in one particular county are listed below. At the 0.05 level of significance, test the claim of randomness of those values above and below the *median*.

973, 1067, 856, 971, 1456, 903, 899, 905,
812, 630, 720, 676, 731, 655, 598, 617

22. The gold reserves of the United States (in millions of fine troy ounces) for a recent 12-year sequence are given here (based on data from the International Monetary Fund). At the 0.05 significance level, test for randomness above (*A*) and below (*B*) the median. (The data are arranged in chronological order by rows.)

274.71   274.68   277.55   276.41   264.60   264.32
264.11   264.03   263.39   262.79   262.65   262.04

23. An elementary school nurse records the grade levels of students that develop chicken pox; the sequential list is given here. Consider children in grades 1, 2, and 3 to be younger and those in grades 4, 5, and 6 to be older. At the 0.05 level of significance, test the claim that the sequence is random with respect to the categories of younger and older.

3, 3, 3, 2, 3, 1, 4, 3, 6, 5, 3, 3, 2, 1, 4,
4, 4, 4, 4, 4, 4, 4, 4, 5, 5, 5, 5, 4, 3, 2,
1, 6, 6, 6, 6, 1, 3, 4, 2, 5, 6, 6, 6, 6, 6

24. A teacher develops a true-false test with these answers:

T, F, F, F, F, T, T, T, T, T, T, F, F, F,
F, T, T, T, T, T, T, T, F, F, F, F, F, F,
F, F, F, T, T, T, T, T, T, F, F, F

At the 0.05 level of significance, test the claim that the sequence of answers is random.

25. Test the claim that the sequence of World Series wins by American League and National League teams is random. Use a 0.05 level of significance. Given below are recent results with American and National league teams represented by *A* and *N*, respectively.

A N A N N N A A A A N A A A A N A N
N A A N N A A A A N A N N A A A A A
N A N A N A N A A A A A A N N A N
A N N A A N N N A N A N A N A A A N
N A A N N N N A A A N A N A N

26. A sequence of dates is selected through a process that has the outward appearance of being random. At the 0.05 level of significance, test the resulting sequence for randomness before and after the middle of the year.

> Nov. 27,  July 7,  Aug. 3,  Oct. 19,  Dec. 19,  Sept. 21,
> Apr.  1,  Mar. 5,  June 10,  May 21,  June 27,  Jan.  5

27. A *New York Times* article about the calculation of decimal places of $\pi$ noted that "mathematicians are pretty sure that the digits of $\pi$ are indistinguishable from any random sequence." Given below are the first 100 decimal places of $\pi$. At the 0.05 level of significance, test for randomness of odd ($O$) and even ($E$) digits.

> 1  4  1  5  9  2  6  5  3  5  8  9  7  9  3  2  3  8  4  6
> 2  6  4  3  3  8  3  2  7  9  5  0  2  8  8  4  1  9  7  1
> 6  9  3  9  9  3  7  5  1  0  5  8  2  0  9  7  4  9  4  4
> 5  9  2  3  0  7  8  1  6  4  0  6  2  8  6  2  0  8  9  9
> 8  6  2  8  0  3  4  8  2  5  3  4  2  1  1  7  0  6  7  9

28. Use the 100 decimal digits for $\pi$ given in Exercise 27. At the 0.05 level of significance, test for randomness above ($A$) and below ($B$) the value of 4.5.

## 12–7  Exercises B

29. Using $A$, $A$, $B$, $B$, what is the minimum number of possible runs that can be arranged? What is the maximum number of runs? Now refer to Table A–10 to find the critical $G$ values for $n_1 = n_2 = 2$. What do you conclude about this case?

30. Let $z = 1.96$ and $n_1 = n_2 = 20$, and then compute $\mu_G$ and $\sigma_G$. Use those values in Formula 12–8 to solve for $G$. What is the importance of this result? How does this result compare to the corresponding value found in Table A–10? How do you explain any discrepancy?

31. a. Using all of the elements $A$, $A$, $A$, $B$, $B$, $B$, $B$, $B$, $B$, list the 84 different possible sequences.
    b. Find the number of runs for each of the 84 sequences.
    c. Assuming that each sequence has a probability of 1/84, find $P(2 \text{ runs})$, $P(3 \text{ runs})$, $P(4 \text{ runs})$, and so on.
    d. Use the results of part c to establish your own critical $G$ values in a two-tailed test with 0.025 in each tail.
    e. Compare your results to those given in Table A–10.
    f. Assuming that the 84 sequences from part a are all equally likely, use the results from part b to find the mean number of runs. Compare your result to the result obtained by using Formula 12–6.

32. Using all of the elements $A$, $A$, $A$, $B$, $B$, $B$, $B$, $B$, $B$, $B$, $B$, $B$, it is possible to arrange 220 different sequences.
    a. List all of those sequences having exactly 3 runs.
    b. Using your result from part $a$, find $P(3 \text{ runs})$.
    c. Using your answer to part $b$, should $G = 3$ be in the critical region?
    d. Find the lower critical value from Table A–10.
    e. Find the lower critical value by using the normal approximation.
33. Refer to the example in this section that involved defective $(D)$ and acceptable $(A)$ items. Assume that the given sequence repeats itself four more times so that $n_1 = 30$ and $n_2 = 120$. Now test the claim that the sequence is random.

 *Vocabulary List*

*Define and give an example of each term.*

parametric methods                        Kruskal-Wallis test
nonparametric methods                     $H$ test
distribution-free tests                   rank correlation coefficient
efficiency                                Spearman's rank correlation
rank                                          coefficient
sign test                                 runs test
Wilcoxon signed-ranks test                run
Wilcoxon rank-sum test

 *Review*

In this chapter we examined six different **nonparametric** methods for analyzing statistics. Besides excluding involvement with population parameters like $\mu$ and $\sigma$, nonparametric methods are not encumbered by many of the restrictions placed on parametric methods, such as the requirement that data must come from normally distributed populations. While nonparametric methods generally lack the sensitivity of their parametric counterparts, they can be used in a wider variety of circumstances and can accommodate more types of data. Also, the computations required in nonparametric tests are generally much simpler than the computations required in the corresponding parametric tests.

In Table 12–9, we list the nonparametric tests presented in this chapter, along with their functions. The table also lists the corresponding parametric tests. Following Table 12–9 is a listing of the important formulas from this chapter.

## TABLE 12–9

| Nonparametric Test | Function | Parametric Test |
|---|---|---|
| Sign test (Section 12–2) | Test for claimed value of average with one sample | $t$ test or $z$ test (Sections 7–2, 7–3, 7–4) |
|  | Test for a difference between two dependent samples | $t$ test or $z$ test (Section 8–3) |
| Wilcoxon signed-ranks test (Section 12–3) | Test for difference between two dependent samples | $t$ test or $z$ test (Section 8–3) |
| Wilcoxon rank-sum test (Section 12–4) | Test for difference between two independent samples | $t$ test or $z$ test (Section 8–3) |
| Kruskal-Wallis test (Section 12–5) | Test for more than two independent samples coming from identical populations | Analysis of variance (Sections 11–2, 11–3) |
| Rank correlation (Section 12–6) | Test for a relationship between two variables | Linear correlation (Section 9–2) |
| Runs test (Section 12–7) | Test for randomness of sample data | (No parametric test) |

## Important Formulas

| Test | Test Statistic | Distribution |
|---|---|---|
| Sign test | $x =$ number of times the less frequent sign occurs. Test statistic when $n > 25$: $$z = \frac{(x + 0.5) - \left(\frac{n}{2}\right)}{\frac{\sqrt{n}}{2}}$$ | If $n \leq 25$, see Table A–7. If $n > 25$, use the normal distribution (Table A–2). |
| Wilcoxon signed-ranks test for two dependent samples | $T =$ smaller of the rank sums. Test statistic when $n > 30$: $$z = \frac{T - \frac{n(n + 1)}{4}}{\sqrt{\frac{n(n + 1)(2n + 1)}{24}}}$$ | If $n \leq 30$, see Table A–8. If $n > 30$, use the normal distribution (Table A–2). |

*continued*

| | Important Formulas | | |
|---|---|---|---|
| **Test** | **Test Statistic** | | **Distribution** |
| **Wilcoxon rank-sum test for two independent samples** | $z = \dfrac{R - \mu_R}{\sigma_R}$ <br> where <br> $R$ = sum of ranks of the sample with size $n_1$ <br> $\mu_R = \dfrac{n_1(n_1 + n_2 + 1)}{2}$ <br> $\sigma_R = \sqrt{\dfrac{n_1 n_2(n_1 + n_2 + 1)}{12}}$ | | Normal distribution (Table A–2) <br><br> (Requires that $n_1 > 10$ and $n_2 > 10$.) |
| **Kruskal-Wallis test** | $H = \dfrac{12}{N(N + 1)}\left(\dfrac{R_1^2}{n_1} + \dfrac{R_2^2}{n_2} + \cdots + \dfrac{R_k^2}{n_k}\right) - 3(N + 1)$ <br> where $N$ = total number of sample values <br> $R_k$ = sum of ranks for the $k$th sample <br> $k$ = the number of samples | | Chi-square with $k - 1$ degrees of freedom <br><br> (Requires that each sample has at least five values.) |
| **Rank correlation** | $r_s = 1 - \dfrac{6\Sigma d^2}{n(n^2 - 1)}$ | | If $n \leq 30$, use Table A–9. If $n > 30$, critical values are <br> $r_s = \dfrac{\pm z}{\sqrt{n - 1}}$ <br><br> where $z$ is from the normal distribution. |
| **Runs test for randomness** | Test statistic when $n > 20$: <br> $z = \dfrac{G - \mu_G}{\sigma_G}$ <br> where $\mu_G = \dfrac{2n_1 n_2}{n_1 + n_2} + 1$ <br> $\sigma_G = \sqrt{\dfrac{(2n_1 n_2)(2n_1 n_2 - n_1 - n_2)}{(n_1 + n_2)^2(n_1 + n_2 - 1)}}$ | | If $n_1 \leq 20$ and $n_2 \leq 20$, use Table A–10. If $n_1 > 20$ or $n_2 > 20$, use the normal distribution (Table A–2). |

# 2 *Review Exercises*

In Exercises 1–24, use the indicated test. If no particular test is specified, use the appropriate nonparametric test from this chapter.

1. A telephone solicitor writes a report listing $N$ for unanswered calls and $Y$ for answered calls. At the 0.05 level of significance, test the claim that the sample listed below is random.

   N, N, N, N, N, N, N, N, Y, Y, N, N, N, N, N, Y, Y,
   Y, Y, Y, Y, N, N, N, N, N, N, N, N, N, N, N, N

2. In a study of techniques used to measure lung volumes, physiological data were collected for 10 subjects. The values given in the table are in liters and they represent the measured forced vital capacities of the 10 subjects in a sitting position and in a supine (lying) position. (See "Validation of Esophageal Balloon Technique at Different Lung Volumes and Postures," by Baydur, Cha, and Sassoon, *Journal of Applied Physiology*, Vol. 62, No. 1.) At the 0.05 significance level, use the sign test to test the claim that there is no significant difference between the measurements from the two positions.

   | Sitting | 4.66 | 5.70 | 5.37 | 3.34 | 3.77 | 7.43 | 4.15 | 6.21 | 5.90 | 5.77 |
   |---|---|---|---|---|---|---|---|---|---|---|
   | Supine | 4.63 | 6.34 | 5.72 | 3.23 | 3.60 | 6.96 | 3.66 | 5.81 | 5.61 | 5.33 |

3. A ranking of randomly selected cities according to population density (people per square mile) and crime rate (crimes per 100,000 people) yielded the results given below (based on data from *USA Today*). At the 0.05 level of significance, use the rank correlation coefficient to determine if there is a positive correlation, a negative correlation, or no correlation.

   | City | Austin | Long Beach | San Diego | Detroit | Baltimore | Tampa | Boston |
   |---|---|---|---|---|---|---|---|
   | Population density | 1 | 4 | 3 | 5 | 6 | 2 | 7 |
   | Crime rate | 4 | 2 | 1 | 6 | 3 | 7 | 5 |

4. A study is made of the response time of police cars after dispatching occurs. Sample results for three precincts are given below in seconds. At the 0.05 level of significance, test the claim that precincts 1 and 2 have the same response times.

   | Precinct 1 | 160 | 172 | 176 | 176 | 178 | 191 | 183 | 177 | 173 | 179 | 180 | 185 |
   |---|---|---|---|---|---|---|---|---|---|---|---|---|
   | Precinct 2 | 165 | 174 | 180 | 181 | 184 | 186 | 190 | 200 | 176 | 192 | 195 | 201 |
   | Precinct 3 | 162 | 175 | 177 | 179 | 187 | 195 | 210 | 215 | 216 | 220 | 222 | |

5. Using the sample data given in Exercise 4, test the claim that the three precincts have the same response times.

6. An index of civic awareness is obtained for 10 randomly selected subjects before and after they take a short course. The results follow. Use the Wilcoxon signed-ranks test to test the claim that the course does not affect the index. Use a 0.05 level of significance.

| Before course | 63 | 65 | 67 | 70 | 71 | 77 | 78 | 80 | 85 | 91 |
|---|---|---|---|---|---|---|---|---|---|---|
| After course | 73 | 64 | 72 | 74 | 77 | 80 | 79 | 82 | 96 | 91 |

7. Test the claim that the median age of a teacher in Montana is 38 years. The ages of a sample group of teachers from Montana are 35, 31, 27, 42, 38, 39, 56, 64, 61, 33, 35, 24, 25, 28, 37, 36, 40, 43, 54, and 50. Use a 0.05 significance level.

8. A dose of the drug captopril, designed to lower systolic blood pressure, is administered to 10 randomly selected volunteers. The results follow. Use the Wilcoxon signed-ranks test at the 0.05 significance level to test the claim that systolic blood pressure is not affected by the pill.

| Before pill | 120 | 136 | 160 | 98 | 115 | 110 | 180 | 190 | 138 | 128 |
|---|---|---|---|---|---|---|---|---|---|---|
| After pill | 118 | 122 | 143 | 105 | 98 | 98 | 180 | 175 | 105 | 112 |

9. A police academy gives an entrance exam; sample results for applicants from two different counties follow. Use the Wilcoxon rank-sum test to test the claim that there is no difference between the scores from the two counties. Assume a significance level of 0.05.

| Orange County | 63 | 39 | 26 | 14 | 75 | 60 | 62 | 79 | 86 | 70 | 66 | |
|---|---|---|---|---|---|---|---|---|---|---|---|---|
| Westchester County | 54 | 35 | 39 | 27 | 40 | 78 | 17 | 7 | 5 | 10 | 48 | 50 | 49 |

10. Final exam grades are listed below according to the order in which exams were completed. At the 0.05 level of significance, test for randomness above and below the mean.

45, 50, 92, 87, 79, 89, 93, 75, 76, 74, 76,
73, 65, 68, 69, 70, 78, 60, 60, 60, 55, 100

11. The following table ranks 8 states according to SAT scores and monetary aid to families with dependent children (based on data from the U.S. Department of Education). At the 0.05 significance level, test for an association between these two variables.

| | N.Y. | Calif. | Fla. | N.J. | Tex. | N.C. | Md. | Ore. |
|---|---|---|---|---|---|---|---|---|
| SAT score | 4 | 3 | 5 | 6 | 7 | 8 | 2 | 1 |
| Aid per family with dependent children | 2 | 1 | 7 | 3 | 8 | 6 | 5 | 4 |

12. Three machines are programmed to produce keychains, and the daily outputs are listed below for randomly selected days. Test the claim that the machines produce the same amount. Use a 0.05 level of significance.

Machine  A:  660  690  690  672  683
Machine  B:  590  588  560  570  592
Machine  C:  520  572  578  553  564

13. Randomly selected cars are tested for fuel consumption and then re-tested after a tune-up. The measures of fuel consumption follow. At the 0.05 significance level, use the Wilcoxon signed-ranks test to test the claim that the tune-up has no effect on fuel consumption.

| Before tune-up | 16 | 23 | 12 | 13 | 7 | 31 | 27 | 18 | 19 | 19 | 19 | 11 | 9 | 15 |
|---|---|---|---|---|---|---|---|---|---|---|---|---|---|---|
| After tune-up | 18 | 23 | 16 | 17 | 8 | 29 | 31 | 21 | 19 | 20 | 24 | 13 | 14 | 18 |

14. Refer to the data given in Exercise 13. At the 0.05 level of significance, use rank correlation to test for a relationship between the before and after values.

15. Refer to the data given in Exercise 13. At the 0.05 level of significance, use the sign test to test the claim that the tune-up has no effect on fuel consumption.

16. Three groups of adult men were selected for an experiment designed to measure their blood alcohol levels after consuming five drinks. Members of group A were tested after one hour, members of group B were tested after two hours, and members of group C were tested after four hours. The results are given below. At the 0.05 level of significance, use the Kruskal-Wallis test to test the claim that the three groups come from identical populations.

| A | B | C |
|---|---|---|
| 0.11 | 0.08 | 0.04 |
| 0.10 | 0.09 | 0.04 |
| 0.09 | 0.07 | 0.05 |
| 0.09 | 0.07 | 0.05 |
| 0.10 | 0.06 | 0.06 |
|  |  | 0.04 |
|  |  | 0.05 |

17. Samples of equal amounts of two brands of a food substance were randomly selected and the amounts of carbohydrates were measured (in grams). The results are listed below. Use the Wilcoxon rank-sum test to test the claim that both brands are the same when compared on the basis of carbohydrate content. Use a 0.05 level of significance.

| Brand X | 20.3 | 21.2 | 19.3 | 19.2 | 19.1 | 19.0 | 22.6 | 23.6 | 22.9 | 20.7 | 20.7 |
|---|---|---|---|---|---|---|---|---|---|---|---|
| Brand Y | 18.9 | 18.8 | 19.1 | 21.0 | 20.0 | 18.6 | 20.4 | 23.3 | 20.1 | 17.9 | 17.7 |

18. A pollster is hired to collect data from 30 randomly selected adults. As the data are turned in, the sexes of the interviewed subjects are noted and the sequence below is obtained. At the 0.05 level of significance, test the claim that the sequence is random.

M, M, M, M, M, M, M, M, F, M, M, M, M, F, F, F,
F, F, F, F, F, F, M, M, M, M, M, M, M, M

19. A unit on business law is taught to 5 different classes of randomly selected students. A different teaching method is used for each group, and sample final test data are given below. The scores represent the final averages of the individual students. Test the claim that the 5 methods are equally effective.

| Traditional | Programmed | Audio | Audiovisual | Visual |
|---|---|---|---|---|
| 76.2 | 85.2 | 67.3 | 75.8 | 50.5 |
| 78.3 | 74.3 | 60.1 | 81.6 | 70.2 |
| 85.1 | 76.5 | 55.4 | 90.3 | 88.8 |
| 63.7 | 80.3 | 72.3 | 78.0 | 67.1 |
| 91.6 | 67.4 | 40.0 | 67.8 | 77.7 |
| 87.2 | 67.9 | | 57.6 | 73.9 |
| | 72.1 | | | |
| | 60.4 | | | |

20. The following scores were randomly selected from last year's college entrance examination scores. Use the Wilcoxon rank-sum test to test the claim that the performance of New Yorkers equals that of Californians. Assume a significance level of 0.05.

| New York | 520 | 490 | 571 | 398 | 602 | 475 | 557 | 621 | 737 | 403 | 511 | 598 |
|---|---|---|---|---|---|---|---|---|---|---|---|---|
| California | 508 | 563 | 385 | 617 | 704 | 401 | 409 | 527 | 393 | 478 | 521 | 536 |

21. Two judges grade entries in a science fair. The grades for eight randomly selected contestants are given below. Use the data to test the claim that there is no difference between their scoring. Use the Wilcoxon signed-ranks test at the 0.05 level of significance.

| Judge A | 6.3 | 7.2 | 6.6 | 8.5 | 9.7 | 7.0 | 7.3 | 8.8 |
|---|---|---|---|---|---|---|---|---|
| Judge B | 7.1 | 6.5 | 8.2 | 8.6 | 9.0 | 6.1 | 6.3 | 8.8 |

22. An annual art award was won by a woman in 30 out of 40 presentations. At the 0.05 significance level, use the sign test to test the claim that men and women are equal in their abilities to win this award.

23. A medical researcher selects blood samples and tests for the presence (P) or absence (A) of a certain virus. Test the claim that the following sequence of results is random.

A, A, A, A, A, A, P, P, A, A, A, A, A,
A, A, A, A, P, P, P, A, A, A, A, A

24. Several cities were ranked according to the number of hotel rooms and the amount of office space. The results are as follows. With a 0.05 significance level, use rank correlation to test for a relationship between office space and hotel rooms.

| City | NY | Ch | SF | Ph | LA | At | Mi | KC | NO | Da | Ba | Bo | Se | Ho | SL |
|------|-----|-----|-----|-----|-----|-----|-----|-----|-----|-----|-----|-----|-----|-----|-----|
| Office rank | 1 | 3 | 2 | 7 | 6 | 10 | 14 | 15 | 11 | 9 | 12 | 4 | 8 | 5 | 13 |
| Hotel rank | 1 | 2 | 3 | 8 | 6 | 7 | 14 | 15 | 4 | 10 | 13 | 5 | 9 | 11 | 12 |

# Computer Project

Most computer systems have a random number generator. Use such a system to generate 100 numbers and then use the runs test to test for randomness above and below the mean. Turn off the computer, restart it, then generate another 100 random numbers. Use the rank correlation coefficient to test for a relationship between the original 100 numbers and the second sequence of 100 numbers. Use the Wilcoxon rank-sum test to test the claim that the two samples come from the same population. Now generate a third set of 100 numbers and use the Kruskal-Wallis test to test the claim that the three samples come from the same population. (Because of the calculations involved in these tests, it would be helpful to use a statistics software package such as STATDISK or Minitab.) Based on these results, does the computer's number generator appear to be random?

# Applied Projects

1. Refer to Data Set I in Appendix B. Use the sign test to test the claim that the median home selling price is $150,000.
2. Refer to Data Set I in Appendix B. Use the Wilcoxon rank-sum test to test the claim that homes with fewer than two baths have the same selling price as those with two or more baths.
3. Refer to Data Set II in Appendix B. Use the Wilcoxon rank-sum test to test the claim that men and women have the same sitting heights.
4. Refer to Data Set II in Appendix B. Use the runs test to test for randomness of ages above and below the median.

5. Dr. Frank Drake of Cornell University developed a radio communication of pulses and gaps intended to be a message we might send to intelligent life from beyond our solar system. It was basically a sequence of 1271 zeros and ones, as listed below. If extraterrestrial beings analyzed this message with the runs test, would they conclude that the sequence of ones and zeros is random?

```
1 0 0 0 0 0 0 0 0 0 0 0 0 0 0 0 0 0 0 0 0 0 0 0 0 0 0 0 0 0 0 0 0 0 0 0 0 1 0 0 0 0 1
1 1 0 0 0 0 0 0 0 0 0 0 1 0 0 0 0 1 0 0 0 0 0 1 0 0 0 0 1 0 0 0 0 0 0 1 0 0 0 1 0 0
0 0 0 0 0 0 0 0 0 1 0 1 0 0 0 0 0 0 0 0 0 0 1 0 1 0 0 0 0 0 0 0 0 1 0 0 0 1 0 0 0 0 1 0 0
0 0 0 1 1 0 0 1 0 0 0 0 0 0 0 0 1 1 0 0 1 0 0 0 0 0 0 0 1 0 0 0 1 0 0 0 0 1 0 0 0 0 0 0
1 0 0 0 0 0 0 0 0 1 0 0 0 1 0 0 0 1 0 0 0 0 0 0 0 1 1 1 0 0 0 0 0 0 0 0 0 0 0 1 0 0 1 1 0 0
0 0 0 0 0 0 1 0 0 1 1 0 0 0 0 0 0 0 0 0 0 0 0 0 0 0 0 0 0 0 0 0 1 0 1 0 0 0 0 0 0 0 0
0 0 1 0 1 0 0 0 0 0 0 0 0 0 0 0 0 0 0 0 0 0 0 0 0 1 0 0 0 0 0 1 0 0 0 0 0 0 1 0 0 0 0
0 1 0 0 0 0 1 1 0 0 0 1 0 0 0 0 0 0 0 0 0 0 0 0 0 0 0 0 0 0 0 0 0 0 0 0 0 0 0 0 0 0 0
0 0 0 0 0 0 0 0 0 0 0 1 1 0 0 0 0 1 1 0 0 0 0 1 1 0 0 0 0 1 1 0 0 0 0 1 1 0 0 0 0 1 0 0 0 0
0 0 0 0 0 1 0 0 1 0 0 1 0 0 1 0 0 1 0 0 1 0 0 1 0 0 1 0 0 1 0 0 1 0 0 1 0 1 0 1 0 0 1 0 0 1
0 0 0 0 1 1 0 0 0 0 1 1 0 0 0 0 1 1 0 0 0 0 1 1 0 0 0 0 1 1 0 0 0 0 0 0 0 0 0 1 0 0 0 0 0 0
0 0 0 0 0 1 1 1 1 1 0 1 0 0 0 0 0 0 0 0 0 0 0 0 0 0 0 0 0 0 0 0 1 0 0 0 0 0 0 0 0 0 0 0 1
0 0 0 0 0 1 0 0 0 0 0 0 0 0 0 0 0 1 0 1 1 0 1 1 1 0 0 1 0 0 0 0 0 0 0 0 0 0 0 0 0 1 1 1 1 1
0 1 0 0 0 0 0 0 0 0 0 0 0 0 0 0 0 0 0 0 0 0 0 0 0 0 0 0 0 0 0 1 0 0 0 0 0 0 0 0 0 0 0 0 0 0
0 0 1 0 0 0 1 0 0 1 1 1 0 0 0 0 0 0 0 0 0 0 0 0 0 1 0 1 0 0 0 0 0 0 0 0 0 0 0 0 0 0 1 0 1 0
0 1 0 0 0 0 1 1 0 0 1 0 1 0 1 1 1 0 0 1 0 1 0 0 0 0 0 0 0 0 0 0 0 0 0 0 1 0 1 0 0 1 0 0 0
0 1 0 0 0 0 0 0 0 0 0 0 1 0 0 1 0 0 0 0 0 0 0 0 0 0 0 0 0 0 0 1 0 0 1 0 0 0 0 0 1 0 0 0
0 0 0 0 0 0 0 0 1 1 1 1 1 0 0 0 0 0 0 0 0 0 0 0 0 0 1 1 1 1 1 0 0 0 0 0 0 0 1 1 1 0 1 0 1 0 0
0 0 0 0 1 0 1 0 1 0 0 0 0 0 0 0 0 0 0 0 1 0 1 0 1 0 0 0 0 0 0 0 1 0 0 0 0 0 0 0 0 0 0 1 0
0 0 1 0 1 0 0 0 0 0 0 0 0 0 1 0 1 0 0 0 1 0 0 0 0 0 0 0 0 0 0 0 0 0 0 0 0 0 1 0 0 0 1 0 0
1 0 0 0 1 0 0 0 1 0 0 1 1 0 1 1 0 0 1 1 1 0 1 1 0 1 1 0 1 0 0 0 0 0 1 0 0 0 1 0 0 0 1 0 1 0
1 0 1 0 0 0 1 0 0 0 1 0 0 0 0 0 0 0 0 0 0 0 0 0 0 0 0 0 1 0 0 0 1 0 0 0 1 0 0 1 0 0 1 0 0
0 1 0 0 0 1 0 0 0 0 0 0 1 0 0 0 0 0 0 0 0 0 0 0 1 1 1 0 0 0 0 0 1 1 1 1 1 0 0 0 0 0 1 1 1
0 0 0 0 0 0 0 1 1 1 1 1 0 1 0 0 0 0 0 1 0 1 0 1 0 0 0 0 0 1 0 1 0 0 0 0 0 1 0 0 0 1 0 0 0 0
0 0 1 0 0 0 0 0 0 0 0 0 0 1 0 0 0 0 0 1 0 0 0 0 1 1 1 0 0 0 0 1 0 0 0 0 0 1 0 0 0 0 0 1 1 0
0 0 0 0 0 0 0 0 1 0 0 0 0 0 1 0 0 0 1 0 0 0 1 0 0 0 1 0 0 0 0 0 1 0 0 0 0 0 1 0 0 0 0 1 0 0 0 0 1 1 0
0 0 0 1 0 0 0 0 0 1 0 0 0 1 0 0 0 1 0 0 0 1 0 0 0 0 0 1 0 0 0 0 0 1 1 0 0 0 0 0 0 0 0 0 1 1 0
0 0 0 0 1 1 0 1 1 0 0 0 1 1 0 1 1 0 0 0 0 0 1 1 0 0 1 1 1
```

6. Listed on the following page are the 366 priority numbers selected in the 1970 Selective Service draft lottery.
   a. Use the runs test to test the sequence for randomness above and below the median of 183.5.
   b. Use the Kruskal-Wallis test to test the claim that the 12 months have priority numbers drawn from the same population.
   c. Calculate the 12 monthly means. Then plot those 12 means on a graph. (The horizontal scale lists the 12 months and the vertical scale ranges from 100 to 260.) Note any pattern suggesting that the original priority numbers are not randomly selected.

Jan: 305 159 251 215 101 224 306 199 194 325 329 221 318 238 017 121
235 140 058 280 186 337 118 059 052 092 355 077 349 164 211
Feb: 086 144 297 210 214 347 091 181 338 216 150 068 152 004 089 212
189 292 025 302 363 290 057 236 179 365 205 299 285
Mar: 108 029 267 275 293 139 122 213 317 323 136 300 259 354 169 166
033 332 200 239 334 265 256 258 343 170 268 223 362 217 030
Apr: 032 271 083 081 269 253 147 312 219 218 014 346 124 231 273 148
260 090 336 345 062 316 252 002 351 340 074 262 191 208
May: 330 298 040 276 364 155 035 321 197 065 037 133 295 178 130 055
112 278 075 183 250 326 319 031 361 357 296 308 226 103 313
Jun: 249 228 301 020 028 110 085 366 335 206 134 272 069 356 180 274
073 341 104 360 060 247 109 358 137 022 064 222 353 209
Jul: 093 350 115 279 188 327 050 013 277 284 248 015 042 331 322 120
098 190 227 187 027 153 172 023 067 303 289 088 270 287 193
Aug: 111 045 261 145 054 114 168 048 106 021 324 142 307 198 102 044
154 141 311 344 291 339 116 036 286 245 352 167 061 333 011
Sep: 225 161 049 232 082 006 008 184 263 071 158 242 175 001 113 207
255 246 177 063 204 160 119 195 149 018 233 257 151 315
Oct: 359 125 244 202 024 087 234 283 342 220 237 072 138 294 171 254
288 005 241 192 243 117 201 196 176 007 264 094 229 038 079
Nov: 019 034 348 266 310 076 051 097 080 282 046 066 126 127 131 107
143 146 203 185 156 009 182 230 132 309 047 281 099 174
Dec: 129 328 157 165 056 010 012 105 043 041 039 314 163 026 320 096
304 128 240 135 070 053 162 095 084 173 078 123 016 003 100

# Writing Projects

1. A friend of yours has just completed an introductory statistics course that did not include nonparametric methods. Write a description that summarizes the general nature of such methods.
2. In your own words, compare parametric and nonparametric statistical methods relative to their advantages and disadvantages.
3. Write a summary of one of the programs listed below.

# Videotapes

In the preceding eleven chapters, we recommended most of the videotape programs from the series *Against All Odds: Inside Statistics* because they related reasonably well to the chapter topics. Although the remaining programs do not relate directly to the topics in this chapter, we recommend them for their general interest. They are Programs 6, 7, 10, 14, and 26 from the series.

# Epilogue

## I   What Procedure Applies?

When attempting to implement a statistical analysis, one of the most difficult tasks is determining the specific procedure that is most appropriate. This text includes a wide variety of different procedures that apply to many different circumstances. We generally need to begin by clearly identifying the questions that need to be answered and by evaluating the quality of the sampling procedure. We must also answer questions such as these: What is the level of measurement (nominal, ordinal, interval, ratio) of the data? Is there a claim to be tested or a parameter to be estimated? Does the study involve one, two, or more populations? What is the relevant parameter (such as mean, standard deviation, proportion)? Is the sample large ($n > 30$) or small? Is there reason to believe that the population is normally distributed? The accompanying figure should help in determining the relevant text section.

## II   A Perspective

Upon reaching this point in the text, you have already studied at least several of the preceding chapters. It is quite natural for you to believe that you have not mastered the course to the extent necessary to make you a serious user of statistics. To keep a proper perspective, you should recognize that nobody expects a single introductory statistics course to transform you into an expert statistician. There are many important topics we have not even discussed because they are too advanced for this introductory level. It is important to know that professional help is available from those who have much more extensive training and experience with statistical methods. Your statistics course will help you open a dialogue with one of these statisticians.

Although you are not an expert statistician, you should know and understand some very important concepts that make you a better-educated person

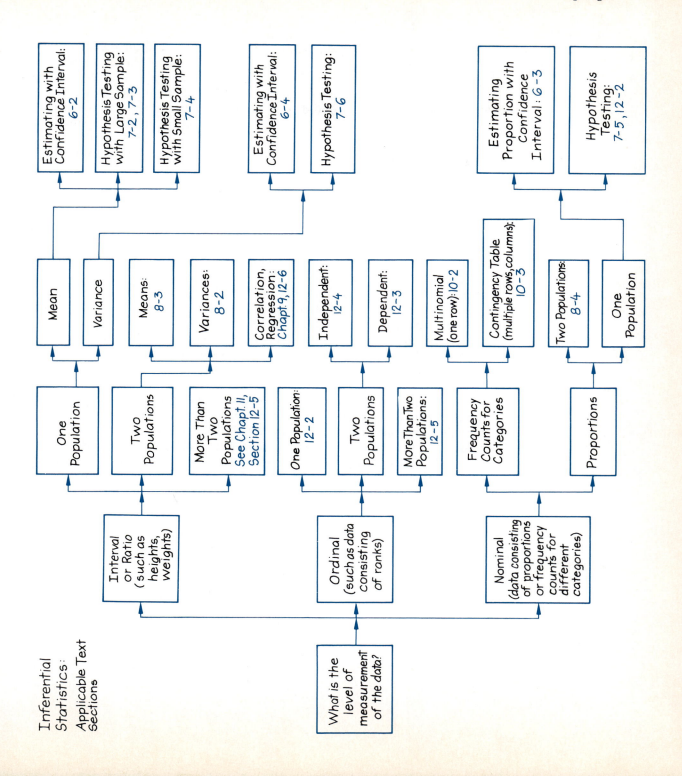

Inferential
Statistics:

Applicable Text
Sections

with improved job marketability. You should know and understand the basic concepts of probability and chance. You should know that in attempting to gain insight into a set of data, it is generally important to investigate measures of central tendency (such as mean and median), measures of dispersion (such as range and standard deviation), and the nature of the distribution (via a frequency table or graph). You should know and understand the importance of estimating population parameters (such as a mean, standard deviation, or proportion) as well as testing claims made about population parameters. You should realize that the nature and configuration of the data can have a dramatic effect on the particular statistical procedures that are used.

In several different places throughout this text, we've tried to emphasize the importance of good sampling. You should recognize that a *bad* sample may be beyond repair by even the most expert statisticians using the most sophisticated techniques. There are many mail, magazine, and telephone call-in surveys that allow respondents to be "self selected." The results of such surveys are generally worthless when judged according to the criteria of sound statistical methodology. We should keep this in mind when we are exposed to such self-selected surveys so that we don't let them affect our beliefs and decisions. In contrast, we should also recognize that many surveys and polls obtain very good results, even though the sample sizes might seem to be relatively small. Although many people refuse to believe it, a typical nationwide survey of only 1700 voters can provide good results if the sampling is carefully planned and executed.

At one time a person was considered educated if he or she could read, but our society has become highly complex and much more demanding. A modern education typically provides students with specific skills, such as the ability to read, write, understand the significance of the Renaissance, operate a computer, and do algebra. A larger picture involves several disciplines that use different approaches in a common goal—seeking the truth. The study of statistics helps us see the truth that is concealed by data that are disorganized or perhaps not yet collected. The study of statistics helps us see the truth that is sometimes distorted by others. The study of statistics is now essential for a very large and growing number of future employees, employers, and citizens. Congratulations for your success in the study of statistics.

# Appendix A: Tables

**TABLE A-1** (0+ represents a positive probability less than 0.0005)

### Binomial Probabilities

| | | | | | | | | $p$ | | | | | | | | |
|---|---|---|---|---|---|---|---|---|---|---|---|---|---|---|---|---|
| $n$ | $x$ | .01 | .05 | .10 | .20 | .30 | .40 | .50 | .60 | .70 | .80 | .90 | .95 | .99 | $x$ |
| 2 | 0 | 980 | 902 | 810 | 640 | 490 | 360 | 250 | 160 | 090 | 040 | 010 | 002 | 0+ | 0 |
| | 1 | 020 | 095 | 180 | 320 | 420 | 480 | 500 | 480 | 420 | 320 | 180 | 095 | 020 | 1 |
| | 2 | 0+ | 002 | 010 | 040 | 090 | 160 | 250 | 360 | 490 | 640 | 810 | 902 | 980 | 2 |
| 3 | 0 | 970 | 857 | 729 | 512 | 343 | 216 | 125 | 064 | 027 | 008 | 001 | 0+ | 0+ | 0 |
| | 1 | 029 | 135 | 243 | 384 | 441 | 432 | 375 | 288 | 189 | 096 | 027 | 007 | 0+ | 1 |
| | 2 | 0+ | 007 | 027 | 096 | 189 | 288 | 375 | 432 | 441 | 384 | 243 | 135 | 029 | 2 |
| | 3 | 0+ | 0+ | 001 | 008 | 027 | 064 | 125 | 216 | 343 | 512 | 729 | 857 | 970 | 3 |
| 4 | 0 | 961 | 815 | 656 | 410 | 240 | 130 | 062 | 026 | 008 | 002 | 0+ | 0+ | 0+ | 0 |
| | 1 | 039 | 171 | 292 | 410 | 412 | 346 | 250 | 154 | 076 | 026 | 004 | 0+ | 0+ | 1 |
| | 2 | 001 | 014 | 049 | 154 | 265 | 346 | 375 | 346 | 265 | 154 | 049 | 014 | 001 | 2 |
| | 3 | 0+ | 0+ | 004 | 026 | 076 | 154 | 250 | 346 | 412 | 410 | 292 | 171 | 039 | 3 |
| | 4 | 0+ | 0+ | 0+ | 002 | 008 | 026 | 062 | 130 | 240 | 410 | 656 | 815 | 961 | 4 |
| 5 | 0 | 951 | 774 | 590 | 328 | 168 | 078 | 031 | 010 | 002 | 0+ | 0+ | 0+ | 0+ | 0 |
| | 1 | 048 | 204 | 328 | 410 | 360 | 259 | 156 | 077 | 028 | 006 | 0+ | 0+ | 0+ | 1 |
| | 2 | 001 | 021 | 073 | 205 | 309 | 346 | 312 | 230 | 132 | 051 | 008 | 001 | 0+ | 2 |
| | 3 | 0+ | 001 | 008 | 051 | 132 | 230 | 312 | 346 | 309 | 205 | 073 | 021 | 001 | 3 |
| | 4 | 0+ | 0+ | 0+ | 006 | 028 | 077 | 156 | 259 | 360 | 410 | 328 | 204 | 048 | 4 |
| | 5 | 0+ | 0+ | 0+ | 0+ | 002 | 010 | 031 | 078 | 168 | 328 | 590 | 774 | 951 | 5 |
| 6 | 0 | 941 | 735 | 531 | 262 | 118 | 047 | 016 | 004 | 001 | 0+ | 0+ | 0+ | 0+ | 0 |
| | 1 | 057 | 232 | 354 | 393 | 303 | 187 | 094 | 037 | 010 | 002 | 0+ | 0+ | 0+ | 1 |
| | 2 | 001 | 031 | 098 | 246 | 324 | 311 | 234 | 138 | 060 | 015 | 001 | 0+ | 0+ | 2 |
| | 3 | 0+ | 002 | 015 | 082 | 185 | 276 | 312 | 276 | 185 | 082 | 015 | 002 | 0+ | 3 |
| | 4 | 0+ | 0+ | 001 | 015 | 060 | 138 | 234 | 311 | 324 | 246 | 098 | 031 | 001 | 4 |
| | 5 | 0+ | 0+ | 0+ | 002 | 010 | 037 | 094 | 187 | 303 | 393 | 354 | 232 | 057 | 5 |
| | 6 | 0+ | 0+ | 0+ | 0+ | 001 | 004 | 016 | 047 | 118 | 262 | 531 | 735 | 941 | 6 |
| 7 | 0 | 932 | 698 | 478 | 210 | 082 | 028 | 008 | 002 | 0+ | 0+ | 0+ | 0+ | 0+ | 0 |
| | 1 | 066 | 257 | 372 | 367 | 247 | 131 | 055 | 017 | 004 | 0+ | 0+ | 0+ | 0+ | 1 |
| | 2 | 002 | 041 | 124 | 275 | 318 | 261 | 164 | 077 | 025 | 004 | 0+ | 0+ | 0+ | 2 |
| | 3 | 0+ | 004 | 023 | 115 | 227 | 290 | 273 | 194 | 097 | 029 | 003 | 0+ | 0+ | 3 |
| | 4 | 0+ | 0+ | 003 | 029 | 097 | 194 | 273 | 290 | 227 | 115 | 023 | 004 | 0+ | 4 |
| | 5 | 0+ | 0+ | 0+ | 004 | 025 | 077 | 164 | 261 | 318 | 275 | 124 | 041 | 002 | 5 |
| | 6 | 0+ | 0+ | 0+ | 0+ | 004 | 017 | 055 | 131 | 247 | 367 | 372 | 257 | 066 | 6 |
| | 7 | 0+ | 0+ | 0+ | 0+ | 0+ | 002 | 008 | 028 | 082 | 210 | 478 | 698 | 932 | 7 |
| 8 | 0 | 923 | 663 | 430 | 168 | 058 | 017 | 004 | 001 | 0+ | 0+ | 0+ | 0+ | 0+ | 0 |
| | 1 | 075 | 279 | 383 | 336 | 198 | 090 | 031 | 008 | 001 | 0+ | 0+ | 0+ | 0+ | 1 |
| | 2 | 003 | 051 | 149 | 294 | 296 | 209 | 109 | 041 | 010 | 001 | 0+ | 0+ | 0+ | 2 |
| | 3 | 0+ | 005 | 033 | 147 | 254 | 279 | 219 | 124 | 047 | 009 | 0+ | 0+ | 0+ | 3 |
| | 4 | 0+ | 0+ | 005 | 046 | 136 | 232 | 273 | 232 | 136 | 046 | 005 | 0+ | 0+ | 4 |
| | 5 | 0+ | 0+ | 0+ | 009 | 047 | 124 | 219 | 279 | 254 | 147 | 033 | 005 | 0+ | 5 |
| | 6 | 0+ | 0+ | 0+ | 001 | 010 | 041 | 109 | 209 | 296 | 294 | 149 | 051 | 003 | 6 |
| | 7 | 0+ | 0+ | 0+ | 0+ | 001 | 008 | 031 | 090 | 198 | 336 | 383 | 279 | 075 | 7 |
| | 8 | 0+ | 0+ | 0+ | 0+ | 0+ | 001 | 004 | 017 | 058 | 168 | 430 | 663 | 923 | 8 |

*continued*

## TABLE A-1    (continued)

### Binomial Probabilities

| n | x | .01 | .05 | .10 | .20 | .30 | .40 | p .50 | .60 | .70 | .80 | .90 | .95 | .99 | x |
|---|---|-----|-----|-----|-----|-----|-----|-------|-----|-----|-----|-----|-----|-----|---|
| 9 | 0 | 914 | 630 | 387 | 134 | 040 | 010 | 002 | 0+ | 0+ | 0+ | 0+ | 0+ | 0+ | 0 |
|   | 1 | 083 | 299 | 387 | 302 | 156 | 060 | 018 | 004 | 0+ | 0+ | 0+ | 0+ | 0+ | 1 |
|   | 2 | 003 | 063 | 172 | 302 | 267 | 161 | 070 | 021 | 004 | 0+ | 0+ | 0+ | 0+ | 2 |
|   | 3 | 0+ | 008 | 045 | 176 | 267 | 251 | 164 | 074 | 021 | 003 | 0+ | 0+ | 0+ | 3 |
|   | 4 | 0+ | 001 | 007 | 066 | 172 | 251 | 246 | 167 | 074 | 017 | 001 | 0+ | 0+ | 4 |
|   | 5 | 0+ | 0+ | 001 | 017 | 074 | 167 | 246 | 251 | 172 | 066 | 007 | 001 | 0+ | 5 |
|   | 6 | 0+ | 0+ | 0+ | 003 | 021 | 074 | 164 | 251 | 267 | 176 | 045 | 008 | 0+ | 6 |
|   | 7 | 0+ | 0+ | 0+ | 0+ | 004 | 021 | 070 | 161 | 267 | 302 | 172 | 063 | 003 | 7 |
|   | 8 | 0+ | 0+ | 0+ | 0+ | 0+ | 004 | 018 | 060 | 156 | 302 | 387 | 299 | 083 | 8 |
|   | 9 | 0+ | 0+ | 0+ | 0+ | 0+ | 0+ | 002 | 010 | 040 | 134 | 387 | 630 | 914 | 9 |
| 10 | 0 | 904 | 599 | 349 | 107 | 028 | 006 | 001 | 0+ | 0+ | 0+ | 0+ | 0+ | 0+ | 0 |
|   | 1 | 091 | 315 | 387 | 268 | 121 | 040 | 010 | 002 | 0+ | 0+ | 0+ | 0+ | 0+ | 1 |
|   | 2 | 004 | 075 | 194 | 302 | 233 | 121 | 044 | 011 | 001 | 0+ | 0+ | 0+ | 0+ | 2 |
|   | 3 | 0+ | 010 | 057 | 201 | 267 | 215 | 117 | 042 | 009 | 001 | 0+ | 0+ | 0+ | 3 |
|   | 4 | 0+ | 001 | 011 | 088 | 200 | 251 | 205 | 111 | 037 | 006 | 0+ | 0+ | 0+ | 4 |
|   | 5 | 0+ | 0+ | 001 | 026 | 103 | 201 | 246 | 201 | 103 | 026 | 001 | 0+ | 0+ | 5 |
|   | 6 | 0+ | 0+ | 0+ | 006 | 037 | 111 | 205 | 251 | 200 | 088 | 011 | 001 | 0+ | 6 |
|   | 7 | 0+ | 0+ | 0+ | 001 | 009 | 042 | 117 | 215 | 267 | 201 | 057 | 010 | 0+ | 7 |
|   | 8 | 0+ | 0+ | 0+ | 0+ | 001 | 011 | 044 | 121 | 233 | 302 | 194 | 075 | 004 | 8 |
|   | 9 | 0+ | 0+ | 0+ | 0+ | 0+ | 002 | 010 | 040 | 121 | 268 | 387 | 315 | 091 | 9 |
|   | 10 | 0+ | 0+ | 0+ | 0+ | 0+ | 0+ | 001 | 006 | 028 | 107 | 349 | 599 | 904 | 10 |
| 11 | 0 | 895 | 569 | 314 | 086 | 020 | 004 | 0+ | 0+ | 0+ | 0+ | 0+ | 0+ | 0+ | 0 |
|   | 1 | 099 | 329 | 384 | 236 | 093 | 027 | 005 | 001 | 0+ | 0+ | 0+ | 0+ | 0+ | 1 |
|   | 2 | 005 | 087 | 213 | 295 | 200 | 089 | 027 | 005 | 001 | 0+ | 0+ | 0+ | 0+ | 2 |
|   | 3 | 0+ | 014 | 071 | 221 | 257 | 177 | 081 | 023 | 004 | 0+ | 0+ | 0+ | 0+ | 3 |
|   | 4 | 0+ | 001 | 016 | 111 | 220 | 236 | 161 | 070 | 017 | 002 | 0+ | 0+ | 0+ | 4 |
|   | 5 | 0+ | 0+ | 002 | 039 | 132 | 221 | 226 | 147 | 057 | 010 | 0+ | 0+ | 0+ | 5 |
|   | 6 | 0+ | 0+ | 0+ | 010 | 057 | 147 | 226 | 221 | 132 | 039 | 002 | 0+ | 0+ | 6 |
|   | 7 | 0+ | 0+ | 0+ | 002 | 017 | 070 | 161 | 236 | 220 | 111 | 016 | 001 | 0+ | 7 |
|   | 8 | 0+ | 0+ | 0+ | 0+ | 004 | 023 | 081 | 177 | 257 | 221 | 071 | 014 | 0+ | 8 |
|   | 9 | 0+ | 0+ | 0+ | 0+ | 001 | 005 | 027 | 089 | 200 | 295 | 213 | 087 | 005 | 9 |
|   | 10 | 0+ | 0+ | 0+ | 0+ | 0+ | 001 | 005 | 027 | 093 | 236 | 384 | 329 | 099 | 10 |
|   | 11 | 0+ | 0+ | 0+ | 0+ | 0+ | 0+ | 0+ | 004 | 020 | 086 | 314 | 569 | 895 | 11 |
| 12 | 0 | 886 | 540 | 282 | 069 | 014 | 002 | 0+ | 0+ | 0+ | 0+ | 0+ | 0+ | 0+ | 0 |
|   | 1 | 107 | 341 | 377 | 206 | 071 | 017 | 003 | 0+ | 0+ | 0+ | 0+ | 0+ | 0+ | 1 |
|   | 2 | 006 | 099 | 230 | 283 | 168 | 064 | 016 | 002 | 0+ | 0+ | 0+ | 0+ | 0+ | 2 |
|   | 3 | 0+ | 017 | 085 | 236 | 240 | 142 | 054 | 012 | 001 | 0+ | 0+ | 0+ | 0+ | 3 |
|   | 4 | 0+ | 002 | 021 | 133 | 231 | 213 | 121 | 042 | 008 | 001 | 0+ | 0+ | 0+ | 4 |
|   | 5 | 0+ | 0+ | 004 | 053 | 158 | 227 | 193 | 101 | 029 | 003 | 0+ | 0+ | 0+ | 5 |
|   | 6 | 0+ | 0+ | 0+ | 016 | 079 | 177 | 226 | 177 | 079 | 016 | 0+ | 0+ | 0+ | 6 |
|   | 7 | 0+ | 0+ | 0+ | 003 | 029 | 101 | 193 | 227 | 158 | 053 | 004 | 0+ | 0+ | 7 |
|   | 8 | 0+ | 0+ | 0+ | 001 | 008 | 042 | 121 | 213 | 231 | 133 | 021 | 002 | 0+ | 8 |
|   | 9 | 0+ | 0+ | 0+ | 0+ | 001 | 012 | 054 | 142 | 240 | 236 | 085 | 017 | 0+ | 9 |
|   | 10 | 0+ | 0+ | 0+ | 0+ | 0+ | 002 | 016 | 064 | 168 | 283 | 230 | 099 | 006 | 10 |
|   | 11 | 0+ | 0+ | 0+ | 0+ | 0+ | 0+ | 003 | 017 | 071 | 206 | 377 | 341 | 107 | 11 |
|   | 12 | 0+ | 0+ | 0+ | 0+ | 0+ | 0+ | 0+ | 002 | 014 | 069 | 282 | 540 | 886 | 12 |

NOTE: 0+ represents a positive probability less than 0.0005.

continued

## TABLE A-1    (continued)

### Binomial Probabilities

| n | x | .01 | .05 | .10 | .20 | .30 | .40 | .50 | .60 | .70 | .80 | .90 | .95 | .99 | x |
|---|---|-----|-----|-----|-----|-----|-----|-----|-----|-----|-----|-----|-----|-----|---|
| 13 | 0 | 878 | 513 | 254 | 055 | 010 | 001 | 0+ | 0+ | 0+ | 0+ | 0+ | 0+ | 0+ | 0 |
|    | 1 | 115 | 351 | 367 | 179 | 054 | 011 | 002 | 0+ | 0+ | 0+ | 0+ | 0+ | 0+ | 1 |
|    | 2 | 007 | 111 | 245 | 268 | 139 | 045 | 010 | 001 | 0+ | 0+ | 0+ | 0+ | 0+ | 2 |
|    | 3 | 0+ | 021 | 100 | 246 | 218 | 111 | 035 | 006 | 001 | 0+ | 0+ | 0+ | 0+ | 3 |
|    | 4 | 0+ | 003 | 028 | 154 | 234 | 184 | 087 | 024 | 003 | 0+ | 0+ | 0+ | 0+ | 4 |
|    | 5 | 0+ | 0+ | 006 | 069 | 180 | 221 | 157 | 066 | 014 | 001 | 0+ | 0+ | 0+ | 5 |
|    | 6 | 0+ | 0+ | 001 | 023 | 103 | 197 | 209 | 131 | 044 | 006 | 0+ | 0+ | 0+ | 6 |
|    | 7 | 0+ | 0+ | 0+ | 006 | 044 | 131 | 209 | 197 | 103 | 023 | 001 | 0+ | 0+ | 7 |
|    | 8 | 0+ | 0+ | 0+ | 001 | 014 | 066 | 157 | 221 | 180 | 069 | 006 | 0+ | 0+ | 8 |
|    | 9 | 0+ | 0+ | 0+ | 0+ | 003 | 024 | 087 | 184 | 234 | 154 | 028 | 003 | 0+ | 9 |
|    | 10 | 0+ | 0+ | 0+ | 0+ | 001 | 006 | 035 | 111 | 218 | 246 | 100 | 021 | 0+ | 10 |
|    | 11 | 0+ | 0+ | 0+ | 0+ | 0+ | 001 | 010 | 045 | 139 | 268 | 245 | 111 | 007 | 11 |
|    | 12 | 0+ | 0+ | 0+ | 0+ | 0+ | 0+ | 002 | 011 | 054 | 179 | 367 | 351 | 115 | 12 |
|    | 13 | 0+ | 0+ | 0+ | 0+ | 0+ | 0+ | 0+ | 001 | 010 | 055 | 254 | 513 | 878 | 13 |
| 14 | 0 | 869 | 488 | 229 | 044 | 007 | 001 | 0+ | 0+ | 0+ | 0+ | 0+ | 0+ | 0+ | 0 |
|    | 1 | 123 | 359 | 356 | 154 | 041 | 007 | 001 | 0+ | 0+ | 0+ | 0+ | 0+ | 0+ | 1 |
|    | 2 | 008 | 123 | 257 | 250 | 113 | 032 | 006 | 001 | 0+ | 0+ | 0+ | 0+ | 0+ | 2 |
|    | 3 | 0+ | 026 | 114 | 250 | 194 | 085 | 022 | 003 | 0+ | 0+ | 0+ | 0+ | 0+ | 3 |
|    | 4 | 0+ | 004 | 035 | 172 | 229 | 155 | 061 | 014 | 001 | 0+ | 0+ | 0+ | 0+ | 4 |
|    | 5 | 0+ | 0+ | 008 | 086 | 196 | 207 | 122 | 041 | 007 | 0+ | 0+ | 0+ | 0+ | 5 |
|    | 6 | 0+ | 0+ | 001 | 032 | 126 | 207 | 183 | 092 | 023 | 002 | 0+ | 0+ | 0+ | 6 |
|    | 7 | 0+ | 0+ | 0+ | 009 | 062 | 157 | 209 | 157 | 062 | 009 | 0+ | 0+ | 0+ | 7 |
|    | 8 | 0+ | 0+ | 0+ | 002 | 023 | 092 | 183 | 207 | 126 | 032 | 001 | 0+ | 0+ | 8 |
|    | 9 | 0+ | 0+ | 0+ | 0+ | 007 | 041 | 122 | 207 | 196 | 086 | 008 | 0+ | 0+ | 9 |
|    | 10 | 0+ | 0+ | 0+ | 0+ | 001 | 014 | 061 | 155 | 229 | 172 | 035 | 004 | 0+ | 10 |
|    | 11 | 0+ | 0+ | 0+ | 0+ | 0+ | 003 | 022 | 085 | 194 | 250 | 114 | 026 | 0+ | 11 |
|    | 12 | 0+ | 0+ | 0+ | 0+ | 0+ | 001 | 006 | 032 | 113 | 250 | 257 | 123 | 008 | 12 |
|    | 13 | 0+ | 0+ | 0+ | 0+ | 0+ | 0+ | 001 | 007 | 041 | 154 | 356 | 359 | 123 | 13 |
|    | 14 | 0+ | 0+ | 0+ | 0+ | 0+ | 0+ | 0+ | 001 | 007 | 044 | 229 | 488 | 869 | 14 |
| 15 | 0 | 860 | 463 | 206 | 035 | 005 | 0+ | 0+ | 0+ | 0+ | 0+ | 0+ | 0+ | 0+ | 0 |
|    | 1 | 130 | 366 | 343 | 132 | 031 | 005 | 0+ | 0+ | 0+ | 0+ | 0+ | 0+ | 0+ | 1 |
|    | 2 | 009 | 135 | 267 | 231 | 092 | 022 | 003 | 0+ | 0+ | 0+ | 0+ | 0+ | 0+ | 2 |
|    | 3 | 0+ | 031 | 129 | 250 | 170 | 063 | 014 | 002 | 0+ | 0+ | 0+ | 0+ | 0+ | 3 |
|    | 4 | 0+ | 005 | 043 | 188 | 219 | 127 | 042 | 007 | 001 | 0+ | 0+ | 0+ | 0+ | 4 |
|    | 5 | 0+ | 001 | 010 | 103 | 206 | 186 | 092 | 024 | 003 | 0+ | 0+ | 0+ | 0+ | 5 |
|    | 6 | 0+ | 0+ | 002 | 043 | 147 | 207 | 153 | 061 | 012 | 001 | 0+ | 0+ | 0+ | 6 |
|    | 7 | 0+ | 0+ | 0+ | 014 | 081 | 177 | 196 | 118 | 035 | 003 | 0+ | 0+ | 0+ | 7 |
|    | 8 | 0+ | 0+ | 0+ | 003 | 035 | 118 | 196 | 177 | 081 | 014 | 0+ | 0+ | 0+ | 8 |
|    | 9 | 0+ | 0+ | 0+ | 001 | 012 | 061 | 153 | 207 | 147 | 043 | 002 | 0+ | 0+ | 9 |
|    | 10 | 0+ | 0+ | 0+ | 0+ | 003 | 024 | 092 | 186 | 206 | 103 | 010 | 001 | 0+ | 10 |
|    | 11 | 0+ | 0+ | 0+ | 0+ | 001 | 007 | 042 | 127 | 219 | 188 | 043 | 005 | 0+ | 11 |
|    | 12 | 0+ | 0+ | 0+ | 0+ | 0+ | 002 | 014 | 063 | 170 | 250 | 129 | 031 | 0+ | 12 |
|    | 13 | 0+ | 0+ | 0+ | 0+ | 0+ | 0+ | 003 | 022 | 092 | 231 | 267 | 135 | 009 | 13 |
|    | 14 | 0+ | 0+ | 0+ | 0+ | 0+ | 0+ | 0+ | 005 | 031 | 132 | 343 | 366 | 130 | 14 |
|    | 15 | 0+ | 0+ | 0+ | 0+ | 0+ | 0+ | 0+ | 0+ | 005 | 035 | 206 | 463 | 860 | 15 |

NOTE: 0+ represents a positive probability less than 0.0005.

Frederick Mosteller, Robert E. K. Rourke, and George B. Thomas, Jr., *Probability with Statistical Applications*, 2nd ed. (Reading, Mass.: Addison-Wesley, 1961 and 1970). Reprinted with permission of the publisher.

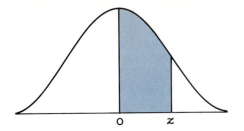

Notes:  1.  For values of z above 3.09, use
            0.4999 for the area.

        2.* Use these common values that
            result from interpolation:

| z score | area |
|---------|--------|
| 1. 645  | 0.4500 |
| 2.575   | 0.4950 |

## TABLE A-2    The Standard Normal (z) Distribution

| z | .00 | .01 | .02 | .03 | .04 | .05 | .06 | .07 | .08 | .09 |
|-----|------|------|------|------|------|------|------|------|------|------|
| 0.0 | .0000 | .0040 | .0080 | .0120 | .0160 | .0199 | .0239 | .0279 | .0319 | .0359 |
| 0.1 | .0398 | .0438 | .0478 | .0517 | .0557 | .0596 | .0636 | .0675 | .0714 | .0753 |
| 0.2 | .0793 | .0832 | .0871 | .0910 | .0948 | .0987 | .1026 | .1064 | .1103 | .1141 |
| 0.3 | .1179 | .1217 | .1255 | .1293 | .1331 | .1368 | .1406 | .1443 | .1480 | .1517 |
| 0.4 | .1554 | .1591 | .1628 | .1664 | .1700 | .1736 | .1772 | .1808 | .1844 | .1879 |
| 0.5 | .1915 | .1950 | .1985 | .2019 | .2054 | .2088 | .2123 | .2157 | .2190 | .2224 |
| 0.6 | .2257 | .2291 | .2324 | .2357 | .2389 | .2422 | .2454 | .2486 | .2517 | .2549 |
| 0.7 | .2580 | .2611 | .2642 | .2673 | .2704 | .2734 | .2764 | .2794 | .2823 | .2852 |
| 0.8 | .2881 | .2910 | .2939 | .2967 | .2995 | .3023 | .3051 | .3078 | .3106 | .3133 |
| 0.9 | .3159 | .3186 | .3212 | .3238 | .3264 | .3289 | .3315 | .3340 | .3365 | .3389 |
| 1.0 | .3413 | .3438 | .3461 | .3485 | .3508 | .3531 | .3554 | .3577 | .3599 | .3621 |
| 1.1 | .3643 | .3665 | .3686 | .3708 | .3729 | .3749 | .3770 | .3790 | .3810 | .3830 |
| 1.2 | .3849 | .3869 | .3888 | .3907 | .3925 | .3944 | .3962 | .3980 | .3997 | .4015 |
| 1.3 | .4032 | .4049 | .4066 | .4082 | .4099 | .4115 | .4131 | .4147 | .4162 | .4177 |
| 1.4 | .4192 | .4207 | .4222 | .4236 | .4251 | .4265 | .4279 | .4292 | .4306 | .4319 |
| 1.5 | .4332 | .4345 | .4357 | .4370 | .4382 | .4394 | .4406 | .4418 | .4429 | .4441 |
| 1.6 | .4452 | .4463 | .4474 | .4484 | .4495 * | .4505 | .4515 | .4525 | .4535 | .4545 |
| 1.7 | .4554 | .4564 | .4573 | .4582 | .4591 | .4599 | .4608 | .4616 | .4625 | .4633 |
| 1.8 | .4641 | .4649 | .4656 | .4664 | .4671 | .4678 | .4686 | .4693 | .4699 | .4706 |
| 1.9 | .4713 | .4719 | .4726 | .4732 | .4738 | .4744 | .4750 | .4756 | .4761 | .4767 |
| 2.0 | .4772 | .4778 | .4783 | .4788 | .4793 | .4798 | .4803 | .4808 | .4812 | .4817 |
| 2.1 | .4821 | .4826 | .4830 | .4834 | .4838 | .4842 | .4846 | .4850 | .4854 | .4857 |
| 2.2 | .4861 | .4864 | .4868 | .4871 | .4875 | .4878 | .4881 | .4884 | .4887 | .4890 |
| 2.3 | .4893 | .4896 | .4898 | .4901 | .4904 | .4906 | .4909 | .4911 | .4913 | .4916 |
| 2.4 | .4918 | .4920 | .4922 | .4925 | .4927 | .4929 | .4931 | .4932 | .4934 | .4936 |
| 2.5 | .4938 | .4940 | .4941 | .4943 | .4945 | .4946 | .4948 | .4949 * | .4951 | .4952 |
| 2.6 | .4953 | .4955 | .4956 | .4957 | .4959 | .4960 | .4961 | .4962 | .4963 | .4964 |
| 2.7 | .4965 | .4966 | .4967 | .4968 | .4969 | .4970 | .4971 | .4972 | .4973 | .4974 |
| 2.8 | .4974 | .4975 | .4976 | .4977 | .4977 | .4978 | .4979 | .4979 | .4980 | .4981 |
| 2.9 | .4981 | .4982 | .4982 | .4983 | .4984 | .4984 | .4985 | .4985 | .4986 | .4986 |
| 3.0 | .4987 | .4987 | .4987 | .4988 | .4988 | .4989 | .4989 | .4989 | .4990 | .4990 |

Frederick Mosteller and Robert E. K. Rourke, *Sturdy Statistics* Table A–1 (Reading, Mass.: Addison-Wesley, 1973). Reprinted
with permission.

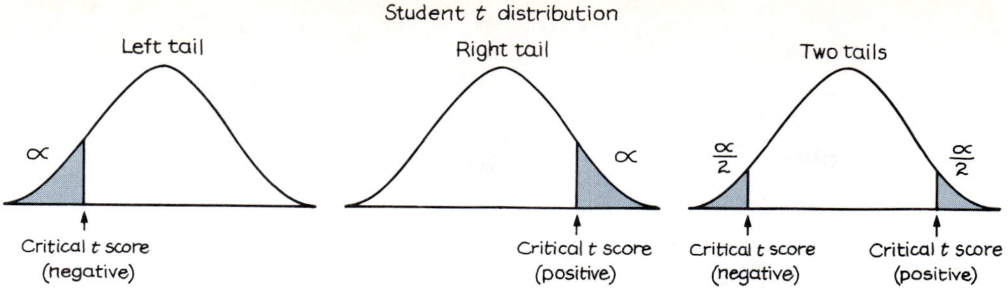
Student t distribution
Left tail
Right tail
Two tails
∝
∝
∝/2
∝/2
Critical t score (negative)
Critical t score (positive)
Critical t score (negative)
Critical t score (positive)

**TABLE A-3** *t* Distribution

| | $\alpha$ | | | | | |
|---|---|---|---|---|---|---|
| Degrees of freedom | .005 (one tail) .01 (two tails) | .01 (one tail) .02 (two tails) | .025 (one tail) .05 (two tails) | .05 (one tail) .10 (two tails) | .10 (one tail) .20 (two tails) | .25 (one tail) .50 (two tails) |
| 1 | 63.657 | 31.821 | 12.706 | 6.314 | 3.078 | 1.000 |
| 2 | 9.925 | 6.965 | 4.303 | 2.920 | 1.886 | .816 |
| 3 | 5.841 | 4.541 | 3.182 | 2.353 | 1.638 | .765 |
| 4 | 4.604 | 3.747 | 2.776 | 2.132 | 1.533 | .741 |
| 5 | 4.032 | 3.365 | 2.571 | 2.015 | 1.476 | .727 |
| 6 | 3.707 | 3.143 | 2.447 | 1.943 | 1.440 | .718 |
| 7 | 3.500 | 2.998 | 2.365 | 1.895 | 1.415 | .711 |
| 8 | 3.355 | 2.896 | 2.306 | 1.860 | 1.397 | .706 |
| 9 | 3.250 | 2.821 | 2.262 | 1.833 | 1.383 | .703 |
| 10 | 3.169 | 2.764 | 2.228 | 1.812 | 1.372 | .700 |
| 11 | 3.106 | 2.718 | 2.201 | 1.796 | 1.363 | .697 |
| 12 | 3.054 | 2.681 | 2.179 | 1.782 | 1.356 | .696 |
| 13 | 3.012 | 2.650 | 2.160 | 1.771 | 1.350 | .694 |
| 14 | 2.977 | 2.625 | 2.145 | 1.761 | 1.345 | .692 |
| 15 | 2.947 | 2.602 | 2.132 | 1.753 | 1.341 | .691 |
| 16 | 2.921 | 2.584 | 2.120 | 1.746 | 1.337 | .690 |
| 17 | 2.898 | 2.567 | 2.110 | 1.740 | 1.333 | .689 |
| 18 | 2.878 | 2.552 | 2.101 | 1.734 | 1.330 | .688 |
| 19 | 2.861 | 2.540 | 2.093 | 1.729 | 1.328 | .688 |
| 20 | 2.845 | 2.528 | 2.086 | 1.725 | 1.325 | .687 |
| 21 | 2.831 | 2.518 | 2.080 | 1.721 | 1.323 | .686 |
| 22 | 2.819 | 2.508 | 2.074 | 1.717 | 1.321 | .686 |
| 23 | 2.807 | 2.500 | 2.069 | 1.714 | 1.320 | .685 |
| 24 | 2.797 | 2.492 | 2.064 | 1.711 | 1.318 | .685 |
| 25 | 2.787 | 2.485 | 2.060 | 1.708 | 1.316 | .684 |
| 26 | 2.779 | 2.479 | 2.056 | 1.706 | 1.315 | .684 |
| 27 | 2.771 | 2.473 | 2.052 | 1.703 | 1.314 | .684 |
| 28 | 2.763 | 2.467 | 2.048 | 1.701 | 1.313 | .683 |
| 29 | 2.756 | 2.462 | 2.045 | 1.699 | 1.311 | .683 |
| Large (*z*) | 2.575 | 2.327 | 1.960 | 1.645 | 1.282 | .675 |

**TABLE A-4**    The Chi-Square $(\chi^2)$ Distribution

### Area to the Right of the Critical Value

| Degrees of freedom | 0.995 | 0.99 | 0.975 | 0.95 | 0.90 | 0.10 | 0.05 | 0.025 | 0.01 | 0.005 |
|---|---|---|---|---|---|---|---|---|---|---|
| 1 | — | — | 0.001 | 0.004 | 0.016 | 2.706 | 3.841 | 5.024 | 6.635 | 7.879 |
| 2 | 0.010 | 0.020 | 0.051 | 0.103 | 0.211 | 4.605 | 5.991 | 7.378 | 9.210 | 10.597 |
| 3 | 0.072 | 0.115 | 0.216 | 0.352 | 0.584 | 6.251 | 7.815 | 9.348 | 11.345 | 12.838 |
| 4 | 0.207 | 0.297 | 0.484 | 0.711 | 1.064 | 7.779 | 9.488 | 11.143 | 13.277 | 14.860 |
| 5 | 0.412 | 0.554 | 0.831 | 1.145 | 1.610 | 9.236 | 11.071 | 12.833 | 15.086 | 16.750 |
| 6 | 0.676 | 0.872 | 1.237 | 1.635 | 2.204 | 10.645 | 12.592 | 14.449 | 16.812 | 18.548 |
| 7 | 0.989 | 1.239 | 1.690 | 2.167 | 2.833 | 12.017 | 14.067 | 16.013 | 18.475 | 20.278 |
| 8 | 1.344 | 1.646 | 2.180 | 2.733 | 3.490 | 13.362 | 15.507 | 17.535 | 20.090 | 21.955 |
| 9 | 1.735 | 2.088 | 2.700 | 3.325 | 4.168 | 14.684 | 16.919 | 19.023 | 21.666 | 23.589 |
| 10 | 2.156 | 2.558 | 3.247 | 3.940 | 4.865 | 15.987 | 18.307 | 20.483 | 23.209 | 25.188 |
| 11 | 2.603 | 3.053 | 3.816 | 4.575 | 5.578 | 17.275 | 19.675 | 21.920 | 24.725 | 26.757 |
| 12 | 3.074 | 3.571 | 4.404 | 5.226 | 6.304 | 18.549 | 21.026 | 23.337 | 26.217 | 28.299 |
| 13 | 3.565 | 4.107 | 5.009 | 5.892 | 7.042 | 19.812 | 22.362 | 24.736 | 27.688 | 29.819 |
| 14 | 4.075 | 4.660 | 5.629 | 6.571 | 7.790 | 21.064 | 23.685 | 26.119 | 29.141 | 31.319 |
| 15 | 4.601 | 5.229 | 6.262 | 7.261 | 8.547 | 22.307 | 24.996 | 27.488 | 30.578 | 32.801 |
| 16 | 5.142 | 5.812 | 6.908 | 7.962 | 9.312 | 23.542 | 26.296 | 28.845 | 32.000 | 34.267 |
| 17 | 5.697 | 6.408 | 7.564 | 8.672 | 10.085 | 24.769 | 27.587 | 30.191 | 33.409 | 35.718 |
| 18 | 6.265 | 7.015 | 8.231 | 9.390 | 10.865 | 25.989 | 28.869 | 31.526 | 34.805 | 37.156 |
| 19 | 6.844 | 7.633 | 8.907 | 10.117 | 11.651 | 27.204 | 30.144 | 32.852 | 36.191 | 38.582 |
| 20 | 7.434 | 8.260 | 9.591 | 10.851 | 12.443 | 28.412 | 31.410 | 34.170 | 37.566 | 39.997 |
| 21 | 8.034 | 8.897 | 10.283 | 11.591 | 13.240 | 29.615 | 32.671 | 35.479 | 38.932 | 41.401 |
| 22 | 8.643 | 9.542 | 10.982 | 12.338 | 14.042 | 30.813 | 33.924 | 36.781 | 40.289 | 42.796 |
| 23 | 9.260 | 10.196 | 11.689 | 13.091 | 14.848 | 32.007 | 35.172 | 38.076 | 41.638 | 44.181 |
| 24 | 9.886 | 10.856 | 12.401 | 13.848 | 15.659 | 33.196 | 36.415 | 39.364 | 42.980 | 45.559 |
| 25 | 10.520 | 11.524 | 13.120 | 14.611 | 16.473 | 34.382 | 37.652 | 40.646 | 44.314 | 46.928 |
| 26 | 11.160 | 12.198 | 13.844 | 15.379 | 17.292 | 35.563 | 38.885 | 41.923 | 45.642 | 48.290 |
| 27 | 11.808 | 12.879 | 14.573 | 16.151 | 18.114 | 36.741 | 40.113 | 43.194 | 46.963 | 49.645 |
| 28 | 12.461 | 13.565 | 15.308 | 16.928 | 18.939 | 37.916 | 41.337 | 44.461 | 48.278 | 50.993 |
| 29 | 13.121 | 14.257 | 16.047 | 17.708 | 19.768 | 39.087 | 42.557 | 45.772 | 49.588 | 52.336 |
| 30 | 13.787 | 14.954 | 16.791 | 18.493 | 20.599 | 40.256 | 43.773 | 46.979 | 50.892 | 53.672 |
| 40 | 20.707 | 22.164 | 24.433 | 26.509 | 29.051 | 51.805 | 55.758 | 59.342 | 63.691 | 66.766 |
| 50 | 27.991 | 29.707 | 32.357 | 34.764 | 37.689 | 63.167 | 67.505 | 71.420 | 76.154 | 79.490 |
| 60 | 35.534 | 37.485 | 40.482 | 43.188 | 46.459 | 74.397 | 79.082 | 83.298 | 88.379 | 91.952 |
| 70 | 43.275 | 45.442 | 48.758 | 51.739 | 55.329 | 85.527 | 90.531 | 95.023 | 100.425 | 104.215 |
| 80 | 51.172 | 53.540 | 57.153 | 60.391 | 64.278 | 96.578 | 101.879 | 106.629 | 112.329 | 116.321 |
| 90 | 59.196 | 61.754 | 65.647 | 69.126 | 73.291 | 107.565 | 113.145 | 118.136 | 124.116 | 128.299 |
| 100 | 67.328 | 70.065 | 74.222 | 77.929 | 82.358 | 118.498 | 124.342 | 129.561 | 135.807 | 140.169 |

Donald B. Owen, *Handbook of Statistical Tables*, U.S. Department of Energy (Reading, Mass.: Addison-Wesley, 1962). Reprinted with permission of the publisher.

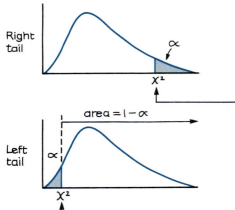

Right tail

To find this value, use the column with the area $\alpha$ given at the top of the table.

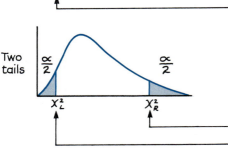

Left tail

To find this value, determine the area of the region to the right of this boundary (the unshaded area) and use the column with this value at the top. If the left tail has area $\alpha$, use the column with the value of $1-\alpha$ at the top of the table.

Two tails

To find this value, use the column with area $\alpha/2$ at the top of the table.

To find this value, use the column with area $1-\alpha/2$ at the top of the table.

**TABLE A-5**  F Distribution ($\alpha = 0.01$ in the right tail)

|  |  | Numerator degrees of freedom | | | | | | | |
|---|---|---|---|---|---|---|---|---|---|
| $df_2$ | 1 | 2 | 3 | 4 | 5 | 6 | 7 | 8 | 9 |
| 1 | 4052.2 | 4999.5 | 5403.4 | 5624.6 | 5763.6 | 5859.0 | 5928.4 | 5981.1 | 6022.5 |
| 2 | 98.503 | 99.000 | 99.166 | 99.249 | 99.299 | 99.333 | 99.356 | 99.374 | 99.388 |
| 3 | 34.116 | 30.817 | 29.457 | 28.710 | 28.237 | 27.911 | 27.672 | 27.489 | 27.345 |
| 4 | 21.198 | 18.000 | 16.694 | 15.977 | 15.522 | 15.207 | 14.976 | 14.799 | 14.659 |
| 5 | 16.258 | 13.274 | 12.060 | 11.392 | 10.967 | 10.672 | 10.456 | 10.289 | 10.158 |
| 6 | 13.745 | 10.925 | 9.7795 | 9.1483 | 8.7459 | 8.4661 | 8.2600 | 8.1017 | 7.9761 |
| 7 | 12.246 | 9.5466 | 8.4513 | 7.8466 | 7.4604 | 7.1914 | 6.9928 | 6.8400 | 6.7188 |
| 8 | 11.259 | 8.6491 | 7.5910 | 7.0061 | 6.6318 | 6.3707 | 6.1776 | 6.0289 | 5.9106 |
| 9 | 10.561 | 8.0215 | 6.9919 | 6.4221 | 6.0569 | 5.8018 | 5.6129 | 5.4671 | 5.3511 |
| 10 | 10.044 | 7.5594 | 6.5523 | 5.9943 | 5.6363 | 5.3858 | 5.2001 | 5.0567 | 4.9424 |
| 11 | 9.6460 | 7.2057 | 6.2167 | 5.6683 | 5.3160 | 5.0692 | 4.8861 | 4.7445 | 4.6315 |
| 12 | 9.3302 | 6.9266 | 5.9525 | 5.4120 | 5.0643 | 4.8206 | 4.6395 | 4.4994 | 4.3875 |
| 13 | 9.0738 | 6.7010 | 5.7394 | 5.2053 | 4.8616 | 4.6204 | 4.4410 | 4.3021 | 4.1911 |
| 14 | 8.8616 | 6.5149 | 5.5639 | 5.0354 | 4.6950 | 4.4558 | 4.2779 | 4.1399 | 4.0297 |
| 15 | 8.6831 | 6.3589 | 5.4170 | 4.8932 | 4.5556 | 4.3183 | 4.1415 | 4.0045 | 3.8948 |
| 16 | 8.5310 | 6.2262 | 5.2922 | 4.7726 | 4.4374 | 4.2016 | 4.0259 | 3.8896 | 3.7804 |
| 17 | 8.3997 | 6.1121 | 5.1850 | 4.6690 | 4.3359 | 4.1015 | 3.9267 | 3.7910 | 3.6822 |
| 18 | 8.2854 | 6.0129 | 5.0919 | 4.5790 | 4.2479 | 4.0146 | 3.8406 | 3.7054 | 3.5971 |
| 19 | 8.1849 | 5.9259 | 5.0103 | 4.5003 | 4.1708 | 3.9386 | 3.7653 | 3.6305 | 3.5225 |
| 20 | 8.0960 | 5.8489 | 4.9382 | 4.4307 | 4.1027 | 3.8714 | 3.6987 | 3.5644 | 3.4567 |
| 21 | 8.0166 | 5.7804 | 4.8740 | 4.3688 | 4.0421 | 3.8117 | 3.6396 | 3.5056 | 3.3981 |
| 22 | 7.9454 | 5.7190 | 4.8166 | 4.3134 | 3.9880 | 3.7583 | 3.5867 | 3.4530 | 3.3458 |
| 23 | 7.8811 | 5.6637 | 4.7649 | 4.2636 | 3.9392 | 3.7102 | 3.5390 | 3.4057 | 3.2986 |
| 24 | 7.8229 | 5.6136 | 4.7181 | 4.2184 | 3.8951 | 3.6667 | 3.4959 | 3.3629 | 3.2560 |
| 25 | 7.7698 | 5.5680 | 4.6755 | 4.1774 | 3.8550 | 3.6272 | 3.4568 | 3.3239 | 3.2172 |
| 26 | 7.7213 | 5.5263 | 4.6366 | 4.1400 | 3.8183 | 3.5911 | 3.4210 | 3.2884 | 3.1818 |
| 27 | 7.6767 | 5.4881 | 4.6009 | 4.1056 | 3.7848 | 3.5580 | 3.3882 | 3.2558 | 3.1494 |
| 28 | 7.6356 | 5.4529 | 4.5681 | 4.0740 | 3.7539 | 3.5276 | 3.3581 | 3.2259 | 3.1195 |
| 29 | 7.5977 | 5.4204 | 4.5378 | 4.0449 | 3.7254 | 3.4995 | 3.3303 | 3.1982 | 3.0920 |
| 30 | 7.5625 | 5.3903 | 4.5097 | 4.0179 | 3.6990 | 3.4735 | 3.3045 | 3.1726 | 3.0665 |
| 40 | 7.3141 | 5.1785 | 4.3126 | 3.8283 | 3.5138 | 3.2910 | 3.1238 | 2.9930 | 2.8876 |
| 60 | 7.0771 | 4.9774 | 4.1259 | 3.6490 | 3.3389 | 3.1187 | 2.9530 | 2.8233 | 2.7185 |
| 120 | 6.8509 | 4.7865 | 3.9491 | 3.4795 | 3.1735 | 2.9559 | 2.7918 | 2.6629 | 2.5586 |
| ∞ | 6.6349 | 4.6052 | 3.7816 | 3.3192 | 3.0173 | 2.8020 | 2.6393 | 2.5113 | 2.4073 |

Denominator degrees of freedom

*continued*

From Maxine Merrington and Catherine M. Thompson, "Tables of Percentage Points of the Inverted Beta ($F$) Distribution," *Biometrika* 33 (1943): 80–84. Reproduced by permission of Professor E. S. Pearson.

Numerator degrees of freedom

| $df_2$ \ $df_1$ | 10 | 12 | 15 | 20 | 24 | 30 | 40 | 60 | 120 | ∞ |
|---|---|---|---|---|---|---|---|---|---|---|
| 1 | 6055.8 | 6106.3 | 6157.3 | 6208.7 | 6234.6 | 6260.6 | 6286.8 | 6313.0 | 6339.4 | 6365.9 |
| 2 | 99.399 | 99.416 | 99.433 | 99.449 | 99.458 | 99.466 | 99.474 | 99.482 | 99.491 | 99.499 |
| 3 | 27.229 | 27.052 | 26.872 | 26.690 | 26.598 | 26.505 | 26.411 | 26.316 | 26.221 | 26.125 |
| 4 | 14.546 | 14.374 | 14.198 | 14.020 | 13.929 | 13.838 | 13.745 | 13.652 | 13.558 | 13.463 |
| 5 | 10.051 | 9.8883 | 9.7222 | 9.5526 | 9.4665 | 9.3793 | 9.2912 | 9.2020 | 9.1118 | 9.0204 |
| 6 | 7.8741 | 7.7183 | 7.5590 | 7.3958 | 7.3127 | 7.2285 | 7.1432 | 7.0567 | 6.9690 | 6.8800 |
| 7 | 6.6201 | 6.4691 | 6.3143 | 6.1554 | 6.0743 | 5.9920 | 5.9084 | 5.8236 | 5.7373 | 5.6495 |
| 8 | 5.8143 | 5.6667 | 5.5151 | 5.3591 | 5.2793 | 5.1981 | 5.1156 | 5.0316 | 4.9461 | 4.8588 |
| 9 | 5.2565 | 5.1114 | 4.9621 | 4.8080 | 4.7290 | 4.6486 | 4.5666 | 4.4831 | 4.3978 | 4.3105 |
| 10 | 4.8491 | 4.7059 | 4.5581 | 4.4054 | 4.3269 | 4.2469 | 4.1653 | 4.0819 | 3.9965 | 3.9090 |
| 11 | 4.5393 | 4.3974 | 4.2509 | 4.0990 | 4.0209 | 3.9411 | 3.8596 | 3.7761 | 3.6904 | 3.6024 |
| 12 | 4.2961 | 4.1553 | 4.0096 | 3.8584 | 3.7805 | 3.7008 | 3.6192 | 3.5355 | 3.4494 | 3.3608 |
| 13 | 4.1003 | 3.9603 | 3.8154 | 3.6646 | 3.5868 | 3.5070 | 3.4253 | 3.3413 | 3.2548 | 3.1654 |
| 14 | 3.9394 | 3.8001 | 3.6557 | 3.5052 | 3.4274 | 3.3476 | 3.2656 | 3.1813 | 3.0942 | 3.0040 |
| 15 | 3.8049 | 3.6662 | 3.5222 | 3.3719 | 3.2940 | 3.2141 | 3.1319 | 3.0471 | 2.9595 | 2.8684 |
| 16 | 3.6909 | 3.5527 | 3.4089 | 3.2587 | 3.1808 | 3.1007 | 3.0182 | 2.9330 | 2.8447 | 2.7528 |
| 17 | 3.5931 | 3.4552 | 3.3117 | 3.1615 | 3.0835 | 3.0032 | 2.9205 | 2.8348 | 2.7459 | 2.6530 |
| 18 | 3.5082 | 3.3706 | 3.2273 | 3.0771 | 2.9990 | 2.9185 | 2.8354 | 2.7493 | 2.6597 | 2.5660 |
| 19 | 3.4338 | 3.2965 | 3.1533 | 3.0031 | 2.9249 | 2.8442 | 2.7608 | 2.6742 | 2.5839 | 2.4893 |
| 20 | 3.3682 | 3.2311 | 3.0880 | 2.9377 | 2.8594 | 2.7785 | 2.6947 | 2.6077 | 2.5168 | 2.4212 |
| 21 | 3.3098 | 3.1730 | 3.0300 | 2.8796 | 2.8010 | 2.7200 | 2.6359 | 2.5484 | 2.4568 | 2.3603 |
| 22 | 3.2576 | 3.1209 | 2.9779 | 2.8274 | 2.7488 | 2.6675 | 2.5831 | 2.4951 | 2.4029 | 2.3055 |
| 23 | 3.2106 | 3.0740 | 2.9311 | 2.7805 | 2.7017 | 2.6202 | 2.5355 | 2.4471 | 2.3542 | 2.2558 |
| 24 | 3.1681 | 3.0316 | 2.8887 | 2.7380 | 2.6591 | 2.5773 | 2.4923 | 2.4035 | 2.3100 | 2.2107 |
| 25 | 3.1294 | 2.9931 | 2.8502 | 2.6993 | 2.6203 | 2.5383 | 2.4530 | 2.3637 | 2.2696 | 2.1694 |
| 26 | 3.0941 | 2.9578 | 2.8150 | 2.6640 | 2.5848 | 2.5026 | 2.4170 | 2.3273 | 2.2325 | 2.1315 |
| 27 | 3.0618 | 2.9256 | 2.7827 | 2.6316 | 2.5522 | 2.4699 | 2.3840 | 2.2938 | 2.1985 | 2.0965 |
| 28 | 3.0320 | 2.8959 | 2.7530 | 2.6017 | 2.5223 | 2.4397 | 2.3535 | 2.2629 | 2.1670 | 2.0642 |
| 29 | 3.0045 | 2.8685 | 2.7256 | 2.5742 | 2.4946 | 2.4118 | 2.3253 | 2.2344 | 2.1379 | 2.0342 |
| 30 | 2.9791 | 2.8431 | 2.7002 | 2.5487 | 2.4689 | 2.3860 | 2.2992 | 2.2079 | 2.1108 | 2.0062 |
| 40 | 2.8005 | 2.6648 | 2.5216 | 2.3689 | 2.2880 | 2.2034 | 2.1142 | 2.0194 | 1.9172 | 1.8047 |
| 60 | 2.6318 | 2.4961 | 2.3523 | 2.1978 | 2.1154 | 2.0285 | 1.9360 | 1.8363 | 1.7263 | 1.6006 |
| 120 | 2.4721 | 2.3363 | 2.1915 | 2.0346 | 1.9500 | 1.8600 | 1.7628 | 1.6557 | 1.5330 | 1.3805 |
| ∞ | 2.3209 | 2.1847 | 2.0385 | 1.8783 | 1.7908 | 1.6964 | 1.5923 | 1.4730 | 1.3246 | 1.0000 |

Denominator degrees of freedom

*continued*

**TABLE A-5**    *F* Distribution ($\alpha = 0.025$ in the right tail)

| $df_2$ \ $df_1$ | 1 | 2 | 3 | 4 | 5 | 6 | 7 | 8 | 9 |
|---|---|---|---|---|---|---|---|---|---|
| 1 | 647.79 | 799.50 | 864.16 | 899.58 | 921.85 | 937.11 | 948.22 | 956.66 | 963.28 |
| 2 | 38.506 | 39.000 | 39.165 | 39.248 | 39.298 | 39.331 | 39.335 | 39.373 | 39.387 |
| 3 | 17.443 | 16.044 | 15.439 | 15.101 | 14.885 | 14.735 | 14.624 | 14.540 | 14.473 |
| 4 | 12.218 | 10.649 | 9.9792 | 9.6045 | 9.3645 | 9.1973 | 9.0741 | 8.9796 | 8.9047 |
| 5 | 10.007 | 8.4336 | 7.7636 | 7.3879 | 7.1464 | 6.9777 | 6.8531 | 6.7572 | 6.6811 |
| 6 | 8.8131 | 7.2599 | 6.5988 | 6.2272 | 5.9876 | 5.8198 | 5.6955 | 5.5996 | 5.5234 |
| 7 | 8.0727 | 6.5415 | 5.8898 | 5.5226 | 5.2852 | 5.1186 | 4.9949 | 4.8993 | 4.8232 |
| 8 | 7.5709 | 6.0595 | 5.4160 | 5.0526 | 4.8173 | 4.6517 | 4.5286 | 4.4333 | 4.3572 |
| 9 | 7.2093 | 5.7147 | 5.0781 | 4.7181 | 4.4844 | 4.3197 | 4.1970 | 4.1020 | 4.0260 |
| 10 | 6.9367 | 5.4564 | 4.8256 | 4.4683 | 4.2361 | 4.0721 | 3.9498 | 3.8549 | 3.7790 |
| 11 | 6.7241 | 5.2559 | 4.6300 | 4.2751 | 4.0440 | 3.8807 | 3.7586 | 3.6638 | 3.5879 |
| 12 | 6.5538 | 5.0959 | 4.4742 | 4.1212 | 3.8911 | 3.7283 | 3.6065 | 3.5118 | 3.4358 |
| 13 | 6.4143 | 4.9653 | 4.3472 | 3.9959 | 3.7667 | 3.6043 | 3.4827 | 3.3880 | 3.3120 |
| 14 | 6.2979 | 4.8567 | 4.2417 | 3.8919 | 3.6634 | 3.5014 | 3.3799 | 3.2853 | 3.2093 |
| 15 | 6.1995 | 4.7650 | 4.1528 | 3.8043 | 3.5764 | 3.4147 | 3.2934 | 3.1987 | 3.1227 |
| 16 | 6.1151 | 4.6867 | 4.0768 | 3.7294 | 3.5021 | 3.3406 | 3.2194 | 3.1248 | 3.0488 |
| 17 | 6.0420 | 4.6189 | 4.0112 | 3.6648 | 3.4379 | 3.2767 | 3.1556 | 3.0610 | 2.9849 |
| 18 | 5.9781 | 4.5597 | 3.9539 | 3.6083 | 3.3820 | 3.2209 | 3.0999 | 3.0053 | 2.9291 |
| 19 | 5.9216 | 4.5075 | 3.9034 | 3.5587 | 3.3327 | 3.1718 | 3.0509 | 2.9563 | 2.8801 |
| 20 | 5.8715 | 4.4613 | 3.8587 | 3.5147 | 3.2891 | 3.1283 | 3.0074 | 2.9128 | 2.8365 |
| 21 | 5.8266 | 4.4199 | 3.8188 | 3.4754 | 3.2501 | 3.0895 | 2.9686 | 2.8740 | 2.7977 |
| 22 | 5.7863 | 4.3828 | 3.7829 | 3.4401 | 3.2151 | 3.0546 | 2.9338 | 2.8392 | 2.7628 |
| 23 | 5.7498 | 4.3492 | 3.7505 | 3.4083 | 3.1835 | 3.0232 | 2.9023 | 2.8077 | 2.7313 |
| 24 | 5.7166 | 4.3187 | 3.7211 | 3.3794 | 3.1548 | 2.9946 | 2.8738 | 2.7791 | 2.7027 |
| 25 | 5.6864 | 4.2909 | 3.6943 | 3.3530 | 3.1287 | 2.9685 | 2.8478 | 2.7531 | 2.6766 |
| 26 | 5.6586 | 4.2655 | 3.6697 | 3.3289 | 3.1048 | 2.9447 | 2.8240 | 2.7293 | 2.6528 |
| 27 | 5.6331 | 4.2421 | 3.6472 | 3.3067 | 3.0828 | 2.9228 | 2.8021 | 2.7074 | 2.6309 |
| 28 | 5.6096 | 4.2205 | 3.6264 | 3.2863 | 3.0626 | 2.9027 | 2.7820 | 2.6872 | 2.6106 |
| 29 | 5.5878 | 4.2006 | 3.6072 | 3.2674 | 3.0438 | 2.8840 | 2.7633 | 2.6686 | 2.5919 |
| 30 | 5.5675 | 4.1821 | 3.5894 | 3.2499 | 3.0265 | 2.8667 | 2.7460 | 2.6513 | 2.5746 |
| 40 | 5.4239 | 4.0510 | 3.4633 | 3.1261 | 2.9037 | 2.7444 | 2.6238 | 2.5289 | 2.4519 |
| 60 | 5.2856 | 3.9253 | 3.3425 | 3.0077 | 2.7863 | 2.6274 | 2.5068 | 2.4117 | 2.3344 |
| 120 | 5.1523 | 3.8046 | 3.2269 | 2.8943 | 2.6740 | 2.5154 | 2.3948 | 2.2994 | 2.2217 |
| ∞ | 5.0239 | 3.6889 | 3.1161 | 2.7858 | 2.5665 | 2.4082 | 2.2875 | 2.1918 | 2.1136 |

Numerator degrees of freedom

Denominator degrees of freedom

continued

Numerator degrees of freedom

| $df_2$ \ $df_1$ | 10 | 12 | 15 | 20 | 24 | 30 | 40 | 60 | 120 | ∞ |
|---|---|---|---|---|---|---|---|---|---|---|
| 1 | 968.63 | 976.71 | 984.87 | 993.10 | 997.25 | 1001.4 | 1005.6 | 1009.8 | 1014.0 | 1018.3 |
| 2 | 39.398 | 39.415 | 39.431 | 39.448 | 39.456 | 39.465 | 39.473 | 39.481 | 39.490 | 39.498 |
| 3 | 14.419 | 14.337 | 14.253 | 14.167 | 14.124 | 14.081 | 14.037 | 13.992 | 13.947 | 13.902 |
| 4 | 8.8439 | 8.7512 | 8.6565 | 8.5599 | 8.5109 | 8.4613 | 8.4111 | 8.3604 | 8.3092 | 8.2573 |
| 5 | 6.6192 | 6.5245 | 6.4277 | 6.3286 | 6.2780 | 6.2269 | 6.1750 | 6.1225 | 6.0693 | 6.0153 |
| 6 | 5.4613 | 5.3662 | 5.2687 | 5.1684 | 5.1172 | 5.0652 | 5.0125 | 4.9589 | 4.9044 | 4.8491 |
| 7 | 4.7611 | 4.6658 | 4.5678 | 4.4667 | 4.4150 | 4.3624 | 4.3089 | 4.2544 | 4.1989 | 4.1423 |
| 8 | 4.2951 | 4.1997 | 4.1012 | 3.9995 | 3.9472 | 3.8940 | 3.8398 | 3.7844 | 3.7279 | 3.6702 |
| 9 | 3.9639 | 3.8682 | 3.7694 | 3.6669 | 3.6142 | 3.5604 | 3.5055 | 3.4493 | 3.3918 | 3.3329 |
| 10 | 3.7168 | 3.6209 | 3.5217 | 3.4185 | 3.3654 | 3.3110 | 3.2554 | 3.1984 | 3.1399 | 3.0798 |
| 11 | 3.5257 | 3.4296 | 3.3299 | 3.2261 | 3.1725 | 3.1176 | 3.0613 | 3.0035 | 2.9441 | 2.8828 |
| 12 | 3.3736 | 3.2773 | 3.1772 | 3.0728 | 3.0187 | 2.9633 | 2.9063 | 2.8478 | 2.7874 | 2.7249 |
| 13 | 3.2497 | 3.1532 | 3.0527 | 2.9477 | 2.8932 | 2.8372 | 2.7797 | 2.7204 | 2.6590 | 2.5955 |
| 14 | 3.1469 | 3.0502 | 2.9493 | 2.8437 | 2.7888 | 2.7324 | 2.6742 | 2.6142 | 2.5519 | 2.4872 |
| 15 | 3.0602 | 2.9633 | 2.8621 | 2.7559 | 2.7006 | 2.6437 | 2.5850 | 2.5242 | 2.4611 | 2.3953 |
| 16 | 2.9862 | 2.8890 | 2.7875 | 2.6808 | 2.6252 | 2.5678 | 2.5085 | 2.4471 | 2.3831 | 2.3163 |
| 17 | 2.9222 | 2.8249 | 2.7230 | 2.6158 | 2.5598 | 2.5020 | 2.4422 | 2.3801 | 2.3153 | 2.2474 |
| 18 | 2.8664 | 2.7689 | 2.6667 | 2.5590 | 2.5027 | 2.4445 | 2.3842 | 2.3214 | 2.2558 | 2.1869 |
| 19 | 2.8172 | 2.7196 | 2.6171 | 2.5089 | 2.4523 | 2.3937 | 2.3329 | 2.2696 | 2.2032 | 2.1333 |
| 20 | 2.7737 | 2.6758 | 2.5731 | 2.4645 | 2.4076 | 2.3486 | 2.2873 | 2.2234 | 2.1562 | 2.0853 |
| 21 | 2.7348 | 2.6368 | 2.5338 | 2.4247 | 2.3675 | 2.3082 | 2.2465 | 2.1819 | 2.1141 | 2.0422 |
| 22 | 2.6998 | 2.6017 | 2.4984 | 2.3890 | 2.3315 | 2.2718 | 2.2097 | 2.1446 | 2.0760 | 2.0032 |
| 23 | 2.6682 | 2.5699 | 2.4665 | 2.3567 | 2.2989 | 2.2389 | 2.1763 | 2.1107 | 2.0415 | 1.9677 |
| 24 | 2.6396 | 2.5411 | 2.4374 | 2.3273 | 2.2693 | 2.2090 | 2.1460 | 2.0799 | 2.0099 | 1.9353 |
| 25 | 2.6135 | 2.5149 | 2.4110 | 2.3005 | 2.2422 | 2.1816 | 2.1183 | 2.0516 | 1.9811 | 1.9055 |
| 26 | 2.5896 | 2.4908 | 2.3867 | 2.2759 | 2.2174 | 2.1565 | 2.0928 | 2.0257 | 1.9545 | 1.8781 |
| 27 | 2.5676 | 2.4688 | 2.3644 | 2.2533 | 2.1946 | 2.1334 | 2.0693 | 2.0018 | 1.9299 | 1.8527 |
| 28 | 2.5473 | 2.4484 | 2.3438 | 2.2324 | 2.1735 | 2.1121 | 2.0477 | 1.9797 | 1.9072 | 1.8291 |
| 29 | 2.5286 | 2.4295 | 2.3248 | 2.2131 | 2.1540 | 2.0923 | 2.0276 | 1.9591 | 1.8861 | 1.8072 |
| 30 | 2.5112 | 2.4120 | 2.3072 | 2.1952 | 2.1359 | 2.0739 | 2.0089 | 1.9400 | 1.8664 | 1.7867 |
| 40 | 2.3882 | 2.2882 | 2.1819 | 2.0677 | 2.0069 | 1.9429 | 1.8752 | 1.8028 | 1.7242 | 1.6371 |
| 60 | 2.2702 | 2.1692 | 2.0613 | 1.9445 | 1.8817 | 1.8152 | 1.7440 | 1.6668 | 1.5810 | 1.4821 |
| 120 | 2.1570 | 2.0548 | 1.9450 | 1.8249 | 1.7597 | 1.6899 | 1.6141 | 1.5299 | 1.4327 | 1.3104 |
| ∞ | 2.0483 | 1.9447 | 1.8326 | 1.7085 | 1.6402 | 1.5660 | 1.4835 | 1.3883 | 1.2684 | 1.0000 |

Denominator degrees of freedom

continued

**TABLE A-5**   $F$ Distribution $(\alpha = 0.05$ in the right tail)

Numerator degrees of freedom

| $df_2$ \ $df_1$ | 1 | 2 | 3 | 4 | 5 | 6 | 7 | 8 | 9 |
|---|---|---|---|---|---|---|---|---|---|
| 1 | 161.45 | 199.50 | 215.71 | 224.58 | 230.16 | 233.99 | 236.77 | 238.88 | 240.54 |
| 2 | 18.513 | 19.000 | 19.164 | 19.247 | 19.296 | 19.330 | 19.353 | 19.371 | 19.385 |
| 3 | 10.128 | 9.5521 | 9.2766 | 9.1172 | 9.0135 | 8.9406 | 8.8867 | 8.8452 | 8.8123 |
| 4 | 7.7086 | 6.9443 | 6.5914 | 6.3882 | 6.2561 | 6.1631 | 6.0942 | 6.0410 | 5.9988 |
| 5 | 6.6079 | 5.7861 | 5.4095 | 5.1922 | 5.0503 | 4.9503 | 4.8759 | 4.8183 | 4.7725 |
| 6 | 5.9874 | 5.1433 | 4.7571 | 4.5337 | 4.3874 | 4.2839 | 4.2067 | 4.1468 | 4.0990 |
| 7 | 5.5914 | 4.7374 | 4.3468 | 4.1203 | 3.9715 | 3.8660 | 3.7870 | 3.7257 | 3.6767 |
| 8 | 5.3177 | 4.4590 | 4.0662 | 3.8379 | 3.6875 | 3.5806 | 3.5005 | 3.4381 | 3.3881 |
| 9 | 5.1174 | 4.2565 | 3.8625 | 3.6331 | 3.4817 | 3.3738 | 3.2927 | 3.2296 | 3.1789 |
| 10 | 4.9646 | 4.1028 | 3.7083 | 3.4780 | 3.3258 | 3.2172 | 3.1355 | 3.0717 | 3.0204 |
| 11 | 4.8443 | 3.9823 | 3.5874 | 3.3567 | 3.2039 | 3.0946 | 3.0123 | 2.9480 | 2.8962 |
| 12 | 4.7472 | 3.8853 | 3.4903 | 3.2592 | 3.1059 | 2.9961 | 2.9134 | 2.8486 | 2.7964 |
| 13 | 4.6672 | 3.8056 | 3.4105 | 3.1791 | 3.0254 | 2.9153 | 2.8321 | 2.7669 | 2.7144 |
| 14 | 4.6001 | 3.7389 | 3.3439 | 3.1122 | 2.9582 | 2.8477 | 2.7642 | 2.6987 | 2.6458 |
| 15 | 4.5431 | 3.6823 | 3.2874 | 3.0556 | 2.9013 | 2.7905 | 2.7066 | 2.6408 | 2.5876 |
| 16 | 4.4940 | 3.6337 | 3.2389 | 3.0069 | 2.8524 | 2.7413 | 2.6572 | 2.5911 | 2.5377 |
| 17 | 4.4513 | 3.5915 | 3.1968 | 2.9647 | 2.8100 | 2.6987 | 2.6143 | 2.5480 | 2.4943 |
| 18 | 4.4139 | 3.5546 | 3.1599 | 2.9277 | 2.7729 | 2.6613 | 2.5767 | 2.5102 | 2.4563 |
| 19 | 4.3807 | 3.5219 | 3.1274 | 2.8951 | 2.7401 | 2.6283 | 2.5435 | 2.4768 | 2.4227 |
| 20 | 4.3512 | 3.4928 | 3.0984 | 2.8661 | 2.7109 | 2.5990 | 2.5140 | 2.4471 | 2.3928 |
| 21 | 4.3248 | 3.4668 | 3.0725 | 2.8401 | 2.6848 | 2.5727 | 2.4876 | 2.4205 | 2.3660 |
| 22 | 4.3009 | 3.4434 | 3.0491 | 2.8167 | 2.6613 | 2.5491 | 2.4638 | 2.3965 | 2.3419 |
| 23 | 4.2793 | 3.4221 | 3.0280 | 2.7955 | 2.6400 | 2.5277 | 2.4422 | 2.3748 | 2.3201 |
| 24 | 4.2597 | 3.4028 | 3.0088 | 2.7763 | 2.6207 | 2.5082 | 2.4226 | 2.3551 | 2.3002 |
| 25 | 4.2417 | 3.3852 | 2.9912 | 2.7587 | 2.6030 | 2.4904 | 2.4047 | 2.3371 | 2.2821 |
| 26 | 4.2252 | 3.3690 | 2.9752 | 2.7426 | 2.5868 | 2.4741 | 2.3883 | 2.3205 | 2.2655 |
| 27 | 4.2100 | 3.3541 | 2.9604 | 2.7278 | 2.5719 | 2.4591 | 2.3732 | 2.3053 | 2.2501 |
| 28 | 4.1960 | 3.3404 | 2.9467 | 2.7141 | 2.5581 | 2.4453 | 2.3593 | 2.2913 | 2.2360 |
| 29 | 4.1830 | 3.3277 | 2.9340 | 2.7014 | 2.5454 | 2.4324 | 2.3463 | 2.2783 | 2.2229 |
| 30 | 4.1709 | 3.3158 | 2.9223 | 2.6896 | 2.5336 | 2.4205 | 2.3343 | 2.2662 | 2.2107 |
| 40 | 4.0847 | 3.2317 | 2.8387 | 2.6060 | 2.4495 | 2.3359 | 2.2490 | 2.1802 | 2.1240 |
| 60 | 4.0012 | 3.1504 | 2.7581 | 2.5252 | 2.3683 | 2.2541 | 2.1665 | 2.0970 | 2.0401 |
| 120 | 3.9201 | 3.0718 | 2.6802 | 2.4472 | 2.2899 | 2.1750 | 2.0868 | 2.0164 | 1.9588 |
| $\infty$ | 3.8415 | 2.9957 | 2.6049 | 2.3719 | 2.2141 | 2.0986 | 2.0096 | 1.9384 | 1.8799 |

Denominator degrees of freedom

continued

Numerator degrees of freedom

| $df_2$ \ $df_1$ | 10 | 12 | 15 | 20 | 24 | 30 | 40 | 60 | 120 | ∞ |
|---|---|---|---|---|---|---|---|---|---|---|
| 1 | 241.88 | 243.91 | 245.95 | 248.01 | 249.05 | 250.10 | 251.14 | 252.20 | 253.25 | 254.31 |
| 2 | 19.396 | 19.413 | 19.429 | 19.446 | 19.454 | 19.462 | 19.471 | 19.479 | 19.487 | 19.496 |
| 3 | 8.7855 | 8.7446 | 8.7029 | 8.6602 | 8.6385 | 8.6166 | 8.5944 | 8.5720 | 8.5494 | 8.5264 |
| 4 | 5.9644 | 5.9117 | 5.8578 | 5.8025 | 5.7744 | 5.7459 | 5.7170 | 5.6877 | 5.6581 | 5.6281 |
| 5 | 4.7351 | 4.6777 | 4.6188 | 4.5581 | 4.5272 | 4.4957 | 4.4638 | 4.4314 | 4.3985 | 4.3650 |
| 6 | 4.0600 | 3.9999 | 3.9381 | 3.8742 | 3.8415 | 3.8082 | 3.7743 | 3.7398 | 3.7047 | 3.6689 |
| 7 | 3.6365 | 3.5747 | 3.5107 | 3.4445 | 3.4105 | 3.3758 | 3.3404 | 3.3043 | 3.2674 | 3.2298 |
| 8 | 3.3472 | 3.2839 | 3.2184 | 3.1503 | 3.1152 | 3.0794 | 3.0428 | 3.0053 | 2.9669 | 2.9276 |
| 9 | 3.1373 | 3.0729 | 3.0061 | 2.9365 | 2.9005 | 2.8637 | 2.8259 | 2.7872 | 2.7475 | 2.7067 |
| 10 | 2.9782 | 2.9130 | 2.8450 | 2.7740 | 2.7372 | 2.6996 | 2.6609 | 2.6211 | 2.5801 | 2.5379 |
| 11 | 2.8536 | 2.7876 | 2.7186 | 2.6464 | 2.6090 | 2.5705 | 2.5309 | 2.4901 | 2.4480 | 2.4045 |
| 12 | 2.7534 | 2.6866 | 2.6169 | 2.5436 | 2.5055 | 2.4663 | 2.4259 | 2.3842 | 2.3410 | 2.2962 |
| 13 | 2.6710 | 2.6037 | 2.5331 | 2.4589 | 2.4202 | 2.3803 | 2.3392 | 2.2966 | 2.2524 | 2.2064 |
| 14 | 2.6022 | 2.5342 | 2.4630 | 2.3879 | 2.3487 | 2.3082 | 2.2664 | 2.2229 | 2.1778 | 2.1307 |
| 15 | 2.5437 | 2.4753 | 2.4034 | 2.3275 | 2.2878 | 2.2468 | 2.2043 | 2.1601 | 2.1141 | 2.0658 |
| 16 | 2.4935 | 2.4247 | 2.3522 | 2.2756 | 2.2354 | 2.1938 | 2.1507 | 2.1058 | 2.0589 | 2.0096 |
| 17 | 2.4499 | 2.3807 | 2.3077 | 2.2304 | 2.1898 | 2.1477 | 2.1040 | 2.0584 | 2.0107 | 1.9604 |
| 18 | 2.4117 | 2.3421 | 2.2686 | 2.1906 | 2.1497 | 2.1071 | 2.0629 | 2.0166 | 1.9681 | 1.9168 |
| 19 | 2.3779 | 2.3080 | 2.2341 | 2.1555 | 2.1141 | 2.0712 | 2.0264 | 1.9795 | 1.9302 | 1.8780 |
| 20 | 2.3479 | 2.2776 | 2.2033 | 2.1242 | 2.0825 | 2.0391 | 1.9938 | 1.9464 | 1.8963 | 1.8432 |
| 21 | 2.3210 | 2.2504 | 2.1757 | 2.0960 | 2.0540 | 2.0102 | 1.9645 | 1.9165 | 1.8657 | 1.8117 |
| 22 | 2.2967 | 2.2258 | 2.1508 | 2.0707 | 2.0283 | 1.9842 | 1.9380 | 1.8894 | 1.8380 | 1.7831 |
| 23 | 2.2747 | 2.2036 | 2.1282 | 2.0476 | 2.0050 | 1.9605 | 1.9139 | 1.8648 | 1.8128 | 1.7570 |
| 24 | 2.2547 | 2.1834 | 2.1077 | 2.0267 | 1.9838 | 1.9390 | 1.8920 | 1.8424 | 1.7896 | 1.7330 |
| 25 | 2.2365 | 2.1649 | 2.0889 | 2.0075 | 1.9643 | 1.9192 | 1.8718 | 1.8217 | 1.7684 | 1.7110 |
| 26 | 2.2197 | 2.1479 | 2.0716 | 1.9898 | 1.9464 | 1.9010 | 1.8533 | 1.8027 | 1.7488 | 1.6906 |
| 27 | 2.2043 | 2.1323 | 2.0558 | 1.9736 | 1.9299 | 1.8842 | 1.8361 | 1.7851 | 1.7306 | 1.6717 |
| 28 | 2.1900 | 2.1179 | 2.0411 | 1.9586 | 1.9147 | 1.8687 | 1.8203 | 1.7689 | 1.7138 | 1.6541 |
| 29 | 2.1768 | 2.1045 | 2.0275 | 1.9446 | 1.9005 | 1.8543 | 1.8055 | 1.7537 | 1.6981 | 1.6376 |
| 30 | 2.1646 | 2.0921 | 2.0148 | 1.9317 | 1.8874 | 1.8409 | 1.7918 | 1.7396 | 1.6835 | 1.6223 |
| 40 | 2.0772 | 2.0035 | 1.9245 | 1.8389 | 1.7929 | 1.7444 | 1.6928 | 1.6373 | 1.5766 | 1.5089 |
| 60 | 1.9926 | 1.9174 | 1.8364 | 1.7480 | 1.7001 | 1.6491 | 1.5943 | 1.5343 | 1.4673 | 1.3893 |
| 120 | 1.9105 | 1.8337 | 1.7505 | 1.6587 | 1.6084 | 1.5543 | 1.4952 | 1.4290 | 1.3519 | 1.2539 |
| ∞ | 1.8307 | 1.7522 | 1.6664 | 1.5705 | 1.5173 | 1.4591 | 1.3940 | 1.3180 | 1.2214 | 1.0000 |

Denominator degrees of freedom

## TABLE A-6

Critical Values of the Pearson Correlation Coefficient $r$

| $n$ | $\alpha = .05$ | $\alpha = .01$ |
|-----|------|------|
| 4 | .950 | .999 |
| 5 | .878 | .959 |
| 6 | .811 | .917 |
| 7 | .754 | .875 |
| 8 | .707 | .834 |
| 9 | .666 | .798 |
| 10 | .632 | .765 |
| 11 | .602 | .735 |
| 12 | .576 | .708 |
| 13 | .553 | .684 |
| 14 | .532 | .661 |
| 15 | .514 | .641 |
| 16 | .497 | .623 |
| 17 | .482 | .606 |
| 18 | .468 | .590 |
| 19 | .456 | .575 |
| 20 | .444 | .561 |
| 25 | .396 | .505 |
| 30 | .361 | .463 |
| 35 | .335 | .430 |
| 40 | .312 | .402 |
| 45 | .294 | .378 |
| 50 | .279 | .361 |
| 60 | .254 | .330 |
| 70 | .236 | .305 |
| 80 | .220 | .286 |
| 90 | .207 | .269 |
| 100 | .196 | .256 |

To test $H_0$: $\rho = 0$ against $H_1$: $\rho \neq 0$, reject $H_0$ if the absolute value of $r$ is greater than the critical value in the table.

| | α | | | |
|---|---|---|---|---|
| **TABLE A-7** Critical Values for the Sign Test | | | | |
| $n$ | .005 (one tail) .01 (two tails) | .01 (one tail) .02 (two tails) | .025 (one tail) .05 (two tails) | .05 (one tail) .10 (two tails) |
| 1 | * | * | * | * |
| 2 | * | * | * | * |
| 3 | * | * | * | * |
| 4 | * | * | * | * |
| 5 | * | * | * | 0 |
| 6 | * | * | 0 | 0 |
| 7 | * | 0 | 0 | 0 |
| 8 | 0 | 0 | 0 | 1 |
| 9 | 0 | 0 | 1 | 1 |
| 10 | 0 | 0 | 1 | 1 |
| 11 | 0 | 1 | 1 | 2 |
| 12 | 1 | 1 | 2 | 2 |
| 13 | 1 | 1 | 2 | 3 |
| 14 | 1 | 2 | 2 | 3 |
| 15 | 2 | 2 | 3 | 3 |
| 16 | 2 | 2 | 3 | 4 |
| 17 | 2 | 3 | 4 | 4 |
| 18 | 3 | 3 | 4 | 5 |
| 19 | 3 | 4 | 4 | 5 |
| 20 | 3 | 4 | 5 | 5 |
| 21 | 4 | 4 | 5 | 6 |
| 22 | 4 | 5 | 5 | 6 |
| 23 | 4 | 5 | 6 | 7 |
| 24 | 5 | 5 | 6 | 7 |
| 25 | 5 | 6 | 7 | 7 |

*NOTES:*
1. * indicates that it is not possible to get a value in the critical region.
2. The null hypothesis is rejected if the number of the less frequent sign $(x)$ is less than or equal to the value in the table.
3. For values of $n$ greater than 25, a normal approximation is used with

$$z = \frac{(x + 0.5) - \left(\frac{n}{2}\right)}{\frac{\sqrt{n}}{2}}$$

## TABLE A-8

### Critical Values of $T$ for the Wilcoxon Signed-Rank Test

| | $\alpha$ | | | |
|---|---|---|---|---|
| $n$ | .005 (one tail) .01 (two tails) | .01 (one tail) .02 (two tails) | .025 (one tail) .05 (two tails) | .05 (one tail) .10 (two tails) |
| 5 | | | | 1 |
| 6 | | | 1 | 2 |
| 7 | | 0 | 2 | 4 |
| 8 | 0 | 2 | 4 | 6 |
| 9 | 2 | 3 | 6 | 8 |
| 10 | 3 | 5 | 8 | 11 |
| 11 | 5 | 7 | 11 | 14 |
| 12 | 7 | 10 | 14 | 17 |
| 13 | 10 | 13 | 17 | 21 |
| 14 | 13 | 16 | 21 | 26 |
| 15 | 16 | 20 | 25 | 30 |
| 16 | 19 | 24 | 30 | 36 |
| 17 | 23 | 28 | 35 | 41 |
| 18 | 28 | 33 | 40 | 47 |
| 19 | 32 | 38 | 46 | 54 |
| 20 | 37 | 43 | 52 | 60 |
| 21 | 43 | 49 | 59 | 68 |
| 22 | 49 | 56 | 66 | 75 |
| 23 | 55 | 62 | 73 | 83 |
| 24 | 61 | 69 | 81 | 92 |
| 25 | 68 | 77 | 90 | 101 |
| 26 | 76 | 85 | 98 | 110 |
| 27 | 84 | 93 | 107 | 120 |
| 28 | 92 | 102 | 117 | 130 |
| 29 | 100 | 111 | 127 | 141 |
| 30 | 109 | 120 | 137 | 152 |

Reject the null hypothesis if the test statistic $T$ is less than or equal to the critical value found in this table. Fail to reject the null hypothesis if the test statistic $T$ is greater than the critical value found in this table.

## TABLE A-9

### Critical Values of Spearman's Rank Correlation Coefficient $r_s$

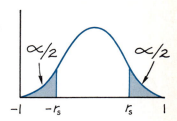

| $n$ | $\alpha = 0.10$ | $\alpha = 0.05$ | $\alpha = 0.02$ | $\alpha = 0.01$ |
|---|---|---|---|---|
| 5 | .900 | — | — | — |
| 6 | .829 | .886 | .943 | — |
| 7 | .714 | .786 | .893 | — |
| 8 | .643 | .738 | .833 | .881 |
| 9 | .600 | .683 | .783 | .833 |
| 10 | .564 | .648 | .745 | .794 |
| 11 | .523 | .623 | .736 | .818 |
| 12 | .497 | .591 | .703 | .780 |
| 13 | .475 | .566 | .673 | .745 |
| 14 | .457 | .545 | .646 | .716 |
| 15 | .441 | .525 | .623 | .689 |
| 16 | .425 | .507 | .601 | .666 |
| 17 | .412 | .490 | .582 | .645 |
| 18 | .399 | .476 | .564 | .625 |
| 19 | .388 | .462 | .549 | .608 |
| 20 | .377 | .450 | .534 | .591 |
| 21 | .368 | .438 | .521 | .576 |
| 22 | .359 | .428 | .508 | .562 |
| 23 | .351 | .418 | .496 | .549 |
| 24 | .343 | .409 | .485 | .537 |
| 25 | .336 | .400 | .475 | .526 |
| 26 | .329 | .392 | .465 | .515 |
| 27 | .323 | .385 | .456 | .505 |
| 28 | .317 | .377 | .448 | .496 |
| 29 | .311 | .370 | .440 | .487 |
| 30 | .305 | .364 | .432 | .478 |

For $n > 30$ use $r_s = \pm z/\sqrt{n-1}$, where $z$ corresponds to the level of significance. For example, if $\alpha = 0.05$, then $z = 1.96$.

> To test $H_0: \rho_s = 0$
> against $H_1: \rho_s \neq 0$

E. G. Olds, "Distribution of sums of squares of rank differences to small numbers of individuals," *Annals of Statistics* 9 (1938): 133–148, and E. G. Olds, with amendment in *Annals of Statistics* 20 (1949): 117–118. Reprinted with permission.

## TABLE A-10    Critical Values for Number of Runs $G$

| | | Value of $n_2$ | | | | | | | | | | | | | | | | | |
|---|---|---|---|---|---|---|---|---|---|---|---|---|---|---|---|---|---|---|---|
| | | 2 | 3 | 4 | 5 | 6 | 7 | 8 | 9 | 10 | 11 | 12 | 13 | 14 | 15 | 16 | 17 | 18 | 19 | 20 |
| Value of $n_1$ | 2 | 1 | 1 | 1 | 1 | 1 | 1 | 1 | 1 | 1 | 1 | 2 | 2 | 2 | 2 | 2 | 2 | 2 | 2 | 2 |
| | | 6 | 6 | 6 | 6 | 6 | 6 | 6 | 6 | 6 | 6 | 6 | 6 | 6 | 6 | 6 | 6 | 6 | 6 | 6 |
| | 3 | 1 | 1 | 1 | 1 | 2 | 2 | 2 | 2 | 2 | 2 | 2 | 2 | 2 | 3 | 3 | 3 | 3 | 3 | 3 |
| | | 6 | 8 | 8 | 8 | 8 | 8 | 8 | 8 | 8 | 8 | 8 | 8 | 8 | 8 | 8 | 8 | 8 | 8 | 8 |
| | 4 | 1 | 1 | 1 | 2 | 2 | 2 | 3 | 3 | 3 | 3 | 3 | 3 | 3 | 3 | 4 | 4 | 4 | 4 | 4 |
| | | 6 | 8 | 9 | 9 | 9 | 10 | 10 | 10 | 10 | 10 | 10 | 10 | 10 | 10 | 10 | 10 | 10 | 10 | 10 |
| | 5 | 1 | 1 | 2 | 2 | 3 | 3 | 3 | 3 | 3 | 4 | 4 | 4 | 4 | 4 | 4 | 4 | 5 | 5 | 5 |
| | | 6 | 8 | 9 | 10 | 10 | 11 | 11 | 12 | 12 | 12 | 12 | 12 | 12 | 12 | 12 | 12 | 12 | 12 | 12 |
| | 6 | 1 | 2 | 2 | 3 | 3 | 3 | 3 | 4 | 4 | 4 | 4 | 5 | 5 | 5 | 5 | 5 | 5 | 6 | 6 |
| | | 6 | 8 | 9 | 10 | 11 | 12 | 12 | 13 | 13 | 13 | 13 | 14 | 14 | 14 | 14 | 14 | 14 | 14 | 14 |
| | 7 | 1 | 2 | 2 | 3 | 3 | 3 | 4 | 4 | 5 | 5 | 5 | 5 | 5 | 6 | 6 | 6 | 6 | 6 | 6 |
| | | 6 | 8 | 10 | 11 | 12 | 13 | 13 | 14 | 14 | 14 | 14 | 15 | 15 | 15 | 16 | 16 | 16 | 16 | 16 |
| | 8 | 1 | 2 | 3 | 3 | 3 | 4 | 4 | 5 | 5 | 5 | 6 | 6 | 6 | 6 | 6 | 7 | 7 | 7 | 7 |
| | | 6 | 8 | 10 | 11 | 12 | 13 | 14 | 14 | 15 | 15 | 16 | 16 | 16 | 16 | 17 | 17 | 17 | 17 | 17 |
| | 9 | 1 | 2 | 3 | 3 | 4 | 4 | 5 | 5 | 5 | 6 | 6 | 6 | 7 | 7 | 7 | 7 | 8 | 8 | 8 |
| | | 6 | 8 | 10 | 12 | 13 | 14 | 14 | 15 | 16 | 16 | 16 | 17 | 17 | 18 | 18 | 18 | 18 | 18 | 18 |
| | 10 | 1 | 2 | 3 | 3 | 4 | 5 | 5 | 5 | 6 | 6 | 7 | 7 | 7 | 7 | 8 | 8 | 8 | 8 | 9 |
| | | 6 | 8 | 10 | 12 | 13 | 14 | 15 | 16 | 16 | 17 | 17 | 18 | 18 | 18 | 19 | 19 | 19 | 20 | 20 |
| | 11 | 1 | 2 | 3 | 4 | 4 | 5 | 5 | 6 | 6 | 7 | 7 | 7 | 8 | 8 | 8 | 9 | 9 | 9 | 9 |
| | | 6 | 8 | 10 | 12 | 13 | 14 | 15 | 16 | 17 | 17 | 18 | 19 | 19 | 19 | 20 | 20 | 20 | 21 | 21 |
| | 12 | 2 | 2 | 3 | 4 | 4 | 5 | 6 | 6 | 7 | 7 | 7 | 8 | 8 | 8 | 9 | 9 | 9 | 10 | 10 |
| | | 6 | 8 | 10 | 12 | 13 | 14 | 16 | 16 | 17 | 18 | 19 | 19 | 20 | 20 | 21 | 21 | 21 | 22 | 22 |
| | 13 | 2 | 2 | 3 | 4 | 5 | 5 | 6 | 6 | 7 | 7 | 8 | 8 | 9 | 9 | 9 | 10 | 10 | 10 | 10 |
| | | 6 | 8 | 10 | 12 | 14 | 15 | 16 | 17 | 18 | 19 | 19 | 20 | 20 | 21 | 21 | 22 | 22 | 23 | 23 |
| | 14 | 2 | 2 | 3 | 4 | 5 | 5 | 6 | 7 | 7 | 8 | 8 | 9 | 9 | 9 | 10 | 10 | 10 | 11 | 11 |
| | | 6 | 8 | 10 | 12 | 14 | 15 | 16 | 17 | 18 | 19 | 20 | 20 | 21 | 22 | 22 | 23 | 23 | 23 | 24 |
| | 15 | 2 | 3 | 3 | 4 | 5 | 6 | 6 | 7 | 7 | 8 | 8 | 9 | 9 | 10 | 10 | 11 | 11 | 11 | 12 |
| | | 6 | 8 | 10 | 12 | 14 | 15 | 16 | 18 | 18 | 19 | 20 | 21 | 22 | 22 | 23 | 23 | 24 | 24 | 25 |
| | 16 | 2 | 3 | 4 | 4 | 5 | 6 | 6 | 7 | 8 | 8 | 9 | 9 | 10 | 10 | 11 | 11 | 11 | 12 | 12 |
| | | 6 | 8 | 10 | 12 | 14 | 16 | 17 | 18 | 19 | 20 | 21 | 21 | 22 | 23 | 23 | 24 | 25 | 25 | 25 |
| | 17 | 2 | 3 | 4 | 4 | 5 | 6 | 7 | 7 | 8 | 9 | 9 | 10 | 10 | 11 | 11 | 11 | 12 | 12 | 13 |
| | | 6 | 8 | 10 | 12 | 14 | 16 | 17 | 18 | 19 | 20 | 21 | 22 | 23 | 23 | 24 | 25 | 25 | 26 | 26 |
| | 18 | 2 | 3 | 4 | 5 | 5 | 6 | 7 | 8 | 8 | 9 | 9 | 10 | 10 | 11 | 11 | 12 | 12 | 13 | 13 |
| | | 6 | 8 | 10 | 12 | 14 | 16 | 17 | 18 | 19 | 20 | 21 | 22 | 23 | 24 | 25 | 25 | 26 | 26 | 27 |
| | 19 | 2 | 3 | 4 | 5 | 6 | 6 | 7 | 8 | 8 | 9 | 10 | 10 | 11 | 11 | 12 | 12 | 13 | 13 | 13 |
| | | 6 | 8 | 10 | 12 | 14 | 16 | 17 | 18 | 20 | 21 | 22 | 23 | 23 | 24 | 25 | 26 | 26 | 27 | 27 |
| | 20 | 2 | 3 | 4 | 5 | 6 | 6 | 7 | 8 | 9 | 9 | 10 | 10 | 11 | 12 | 12 | 13 | 13 | 13 | 14 |
| | | 6 | 8 | 10 | 12 | 14 | 16 | 17 | 18 | 20 | 21 | 22 | 23 | 24 | 25 | 25 | 26 | 27 | 27 | 28 |

The entries in this table are the critical $G$ values assuming a two-tailed test with a significance level of $\alpha = 0.05$. The null hypothesis of randomness is rejected if the total number of runs $G$ is less than or equal to the smaller entry or greater than or equal to the larger entry.

Adapted from C. Eisenhardt and F. Swed, "Tables for testing randomness of grouping in a sequence of alternatives," *Annals of Statistics* 14 (1943): 83–86. Reprinted with permission.

# Appendix B

## DATA SET I: REAL ESTATE DATA
## (150 Randomly Selected Homes Recently
## Sold in Dutchess County, N.Y.)

Data sets in this appendix are available on disk in both Minitab and STAT-DISK formats.

| Selling Price (dollars) | Living Area (sq. ft) | Lot Size (acres) | Number of Rooms | Number of Baths |
|---|---|---|---|---|
| 179,000 | 3,060 | 0.75 | 8 | 2 |
| 126,500 | 1,600 | 0.26 | 8 | 1.5 |
| 134,500 | 2,000 | 0.7 | 8 | 1 |
| 125,000 | 1,300 | 0.65 | 5 | 1 |
| 142,000 | 2,000 | 0.75 | 9 | 1.5 |
| 164,000 | 1,956 | 0.5 | 8 | 2.5 |
| 146,000 | 2,400 | 0.4 | 7 | 2.5 |
| 129,000 | 1,200 | 0.33 | 6 | 1 |
| 141,900 | 1,632 | 3 | 6 | 3 |
| 135,000 | 1,800 | 0.5 | 7 | 2 |
| 118,500 | 1,248 | 0.25 | 7 | 1 |
| 160,000 | 2,025 | 1.1 | 7 | 2 |
| 89,900 | 1,660 | 0.21 | 7 | 1 |
| 169,900 | 2,858 | 0.79 | 9 | 3 |
| 127,500 | 1,296 | 0.5 | 9 | 1 |
| 162,500 | 1,848 | 0.5 | 7 | 2.5 |
| 152,000 | 1,800 | 0.68 | 7 | 1.5 |
| 122,500 | 1,100 | 0.37 | 7 | 1 |
| 220,000 | 3,000 | 1.15 | 10 | 3.5 |
| 141,000 | 2,000 | 0.65 | 7 | 1 |
| 80,500 | 922 | 0.3 | 5 | 1 |
| 152,000 | 1,450 | 0.3 | 6 | 1.5 |
| 231,750 | 2,981 | 1.3 | 10 | 3.5 |
| 180,000 | 1,800 | 1.52 | 8 | 2.5 |

*continued*

| Selling Price (dollars) | Living Area (sq. ft) | Lot Size (acres) | Number of Rooms | Number of Baths |
|---|---|---|---|---|
| 185,000 | 2,600 | 0.75 | 8 | 2 |
| 265,000 | 2,400 | 2 | 7 | 2 |
| 135,000 | 1,625 | 0.36 | 7 | 1.5 |
| 203,000 | 2,653 | 1.8 | 9 | 3 |
| 141,000 | 3,500 | 1 | 10 | 2.5 |
| 159,000 | 1,728 | 0.5 | 8 | 1.5 |
| 182,000 | 2,400 | 0.5 | 8 | 2.5 |
| 208,000 | 2,288 | 1.2 | 8 | 2.5 |
| 96,000 | 864 | 0.32 | 4 | 1 |
| 156,000 | 2,300 | 0.65 | 7 | 3 |
| 185,500 | 2,800 | 1.68 | 9 | 1.5 |
| 275,000 | 2,820 | 1 | 9 | 2.5 |
| 144,900 | 1,900 | 0.44 | 6 | 2 |
| 155,000 | 2,100 | 0.58 | 8 | 1.5 |
| 110,000 | 1,450 | 0.3 | 6 | 2 |
| 154,000 | 1,800 | 0.679 | 7 | 2 |
| 151,500 | 1,900 | 0.75 | 7 | 2 |
| 141,000 | 1,575 | 0.25 | 7 | 1.5 |
| 119,000 | 1,200 | 0.25 | 7 | 1 |
| 108,500 | 1,540 | 0.18 | 7 | 2 |
| 126,500 | 1,700 | 0.3037 | 8 | 2 |
| 302,000 | 2,130 | 11.91 | 8 | 1.5 |
| 130,000 | 1,800 | 0.3 | 7 | 1.5 |
| 140,000 | 1,650 | 0.5 | 8 | 2.5 |
| 123,500 | 1,362 | 0.4 | 7 | 2 |
| 153,500 | 3,700 | 1.1 | 10 | 3 |
| 194,900 | 2,080 | 1 | 8 | 2.5 |
| 165,000 | 2,320 | 0.4 | 8 | 2.5 |
| 179,900 | 2,790 | 0.75 | 13 | 2.5 |
| 194,500 | 2,544 | 0.28 | 9 | 2.5 |
| 127,500 | 1,850 | 0.26 | 9 | 2 |
| 170,000 | 2,277 | 0.8 | 8 | 3 |
| 160,000 | 1,900 | 1 | 8 | 2.5 |
| 135,000 | 1,400 | 0.35 | 6 | 2 |
| 117,000 | 1,248 | 0.3 | 6 | 1 |
| 235,000 | 3,150 | 0.3 | 11 | 4 |
| 223,000 | 1,680 | 14.37 | 8 | 2 |
| 163,500 | 2,276 | 1 | 8 | 2.5 |
| 78,000 | 821 | 2.3 | 4 | 1 |
| 187,000 | 2,080 | 1.23 | 8 | 2.5 |
| 133,000 | 1,100 | 0.33 | 6 | 1 |
| 125,000 | 1,200 | 0.33 | 5 | 1 |
| 116,000 | 1,100 | 1.1 | 6 | 1 |
| 135,000 | 1,800 | 1 | 8 | 2.5 |
| 194,500 | 2,300 | 0.91 | 8 | 2.5 |
| 99,500 | 1,000 | 0.49 | 4 | 1 |
| 152,500 | 1,786 | 0.3 | 8 | 2 |
| 141,900 | 1,950 | 0.75 | 8 | 2.5 |
| 139,900 | 1,839 | 2.6 | 7 | 1.5 |
| 117,500 | 1,300 | 0.29 | 6 | 1 |
| 150,000 | 1,564 | 0.3328 | 6 | 2 |

*continued*

| Selling Price (dollars) | Living Area (sq. ft) | Lot Size (acres) | Number of Rooms | Number of Baths |
|---|---|---|---|---|
| 177,000 | 2,010 | 0.68 | 8 | 1.5 |
| 136,000 | 1,300 | 0.3 | 7 | 1 |
| 158,000 | 1,500 | 0.54 | 5 | 2.5 |
| 211,900 | 2,310 | 0.46 | 8 | 2.5 |
| 165,000 | 1,725 | 1.528 | 8 | 2.5 |
| 183,000 | 2,016 | 0.78 | 8 | 2.5 |
| 85,000 | 875 | 0.26 | 5 | 1 |
| 126,500 | 1,092 | 0.259 | 6 | 1 |
| 162,000 | 2,496 | 0.75 | 9 | 2.5 |
| 169,000 | 1,930 | 3 | 9 | 3 |
| 175,000 | 2,400 | 0.7 | 8 | 3 |
| 267,000 | 1,950 | 18.7 | 7 | 2.5 |
| 150,000 | 1,122 | 3.09 | 5 | 2 |
| 115,000 | 1,080 | 0.31 | 5 | 1 |
| 126,500 | 1,500 | 0.5 | 7 | 1.5 |
| 215,000 | 2,100 | 0.5 | 8 | 2.5 |
| 190,000 | 2,300 | 5.63 | 7 | 2.5 |
| 190,000 | 2,473 | 1.25 | 9 | 2.5 |
| 113,500 | 1,624 | 1.8 | 7 | 1.5 |
| 116,300 | 1,050 | 0.43 | 5 | 1.5 |
| 190,000 | 2,100 | 1.3 | 8 | 1.5 |
| 145,000 | 1,800 | 0.658 | 8 | 2.5 |
| 269,900 | 2,500 | 0.92 | 8 | 3 |
| 135,500 | 1,526 | 0.3 | 7 | 1.5 |
| 190,000 | 1,745 | 0.58 | 7 | 2.5 |
| 98,000 | 1,165 | 0.12 | 6 | 1 |
| 137,900 | 1,856 | 0.33 | 7 | 1.5 |
| 108,000 | 1,036 | 0.948 | 6 | 1 |
| 120,500 | 1,600 | 0.4 | 6 | 2 |
| 128,500 | 1,344 | 0.936 | 6 | 2 |
| 142,500 | 1,552 | 0.46 | 6 | 1.5 |
| 72,000 | 600 | 0.5 | 3 | 1 |
| 124,900 | 1,248 | 0.22 | 7 | 1 |
| 134,000 | 1,502 | 0.35 | 7 | 1.5 |
| 205,406 | 2,465 | 1.55 | 8 | 2.5 |
| 217,000 | 3,100 | 0.54 | 10 | 3.5 |
| 94,000 | 850 | 0.11 | 4 | 1 |
| 189,900 | 2,464 | 0.43 | 8 | 2.5 |
| 168,500 | 1,900 | 1.0636 | 7 | 2.5 |
| 133,000 | 2,000 | 0.5 | 8 | 2 |
| 180,000 | 2,272 | 0.41 | 9 | 2.5 |
| 139,500 | 1,610 | 0.45 | 8 | 1.5 |
| 210,000 | 2,100 | 0.5 | 8 | 2.5 |
| 126,500 | 1,050 | 1 | 5 | 1 |
| 285,000 | 2,516 | 8.1 | 7 | 2.5 |
| 195,000 | 2,265 | 0.85 | 8 | 2.5 |
| 97,000 | 1,300 | 0.37 | 5 | 1 |
| 117,000 | 1,008 | 0.5 | 6 | 1 |
| 150,000 | 1,600 | 1.84 | 7 | 2 |
| 180,500 | 2,000 | 0.6 | 9 | 2.5 |

*continued*

| Selling Price (dollars) | Living Area (sq. ft) | Lot Size (acres) | Number of Rooms | Number of Baths |
|---|---|---|---|---|
| 160,000 | 1,760 | 0.05 | 7 | 2 |
| 181,500 | 2,250 | 0.33 | 9 | 2.5 |
| 124,000 | 1,783 | 0.22 | 8 | 1.5 |
| 125,900 | 1,118 | 0.56 | 7 | 1.5 |
| 165,000 | 2,680 | 0.5 | 9 | 3 |
| 122,000 | 1,950 | 0.5 | 7 | 1.5 |
| 132,000 | 2,000 | 0.108 | 8 | 2 |
| 145,900 | 1,680 | 0.5 | 6 | 1.5 |
| 156,000 | 3,000 | 0.5 | 11 | 2.5 |
| 136,000 | 1,750 | 0.5 | 7 | 2 |
| 142,000 | 1,500 | 0.41 | 7 | 1 |
| 140,000 | 1,403 | 0.5 | 6 | 2 |
| 144,900 | 1,450 | 0.3 | 7 | 1 |
| 133,000 | 1,908 | 0.46 | 7 | 2 |
| 196,800 | 1,960 | 1.33 | 8 | 2.5 |
| 121,900 | 1,300 | 0.78 | 6 | 1 |
| 126,000 | 1,232 | 0.314 | 6 | 2 |
| 164,900 | 1,980 | 0.7 | 8 | 2.5 |
| 172,000 | 2,100 | 1 | 8 | 2.5 |
| 100,000 | 1,338 | 0.12 | 6 | 1 |
| 129,900 | 1,070 | 1.69 | 5 | 1 |
| 110,000 | 1,289 | 0.25 | 6 | 1 |
| 131,000 | 1,066 | 0.33 | 5 | 1 |
| 107,000 | 1,100 | 0.17 | 5 | 1 |
| 165,900 | 1,840 | 1.162 | 8 | 2 |

# Data Set II: Health Survey Data

(Based on data from the Second National Health and Examination Survey, U.S. National Center for Health Statistics.)

Notes on data:

Sex:   1 = male; 2 = female
Age:   Age in years; data restricted to subjects 18 and older
Emp:   0 = blank; 1 = employed; 2 = unemployed; 8 = blank
Cars:   Number of motor vehicles owned or regularly used by family
Wt (lb):   Weight in pounds
Ht (in):   Height in inches
Sit:   Sitting height in millimeters
Hand:   1 = right handed; 2 = left handed; 3 = uses both hands; 8 = blank
Bitro:   Bitrochanteric breadth (body breadth at hips) in millimeters
Elbow:   Breadth of right elbow joint in millimeters
UAG:   Upper arm girth (right arm) in millimeters

| Sex | Age | Emp | Cars | Wt (lb) | Ht (in) | Sit | Hand | Bitro | Elbow | UAG |
|-----|-----|-----|------|---------|---------|-----|------|-------|-------|-----|
| 2 | 48 | 1 | 2 | 119 | 61.6 | 844 | 1 | 306 | 61 | 264 |
| 2 | 52 | 1 | 4 | 136 | 61.9 | 847 | 1 | 327 | 63 | 295 |
| 2 | 62 | 0 | 1 | 115 | 65.0 | 842 | 1 | 288 | 60 | 238 |
| 1 | 67 | 1 | 2 | 151 | 69.8 | 864 | 1 | 278 | 71 | 300 |
| 1 | 72 | 0 | 3 | 182 | 63.5 | 886 | 1 | 344 | 70 | 346 |
| 1 | 64 | 1 | 2 | 127 | 66.7 | 857 | 1 | 314 | 72 | 278 |
| 2 | 35 | 1 | 3 | 134 | 64.6 | 896 | 1 | 287 | 62 | 320 |
| 2 | 50 | 0 | 1 | 156 | 63.1 | 816 | 1 | 341 | 69 | 321 |
| 2 | 42 | 0 | 2 | 140 | 63.8 | 903 | 2 | 348 | 57 | 270 |
| 2 | 38 | 0 | 2 | 147 | 65.5 | 882 | 1 | 326 | 63 | 313 |
| 1 | 64 | 0 | 1 | 180 | 71.0 | 909 | 1 | 321 | 82 | 330 |
| 1 | 28 | 1 | 1 | 182 | 68.7 | 891 | 1 | 317 | 74 | 348 |
| 1 | 69 | 0 | 1 | 185 | 67.0 | 939 | 1 | 345 | 73 | 360 |
| 1 | 21 | 1 | 3 | 156 | 68.0 | 880 | 1 | 292 | 70 | 300 |
| 2 | 20 | 1 | 2 | 153 | 67.7 | 932 | 1 | 309 | 61 | 306 |
| 1 | 19 | 0 | 3 | 203 | 75.4 | 988 | 2 | 354 | 75 | 296 |
| 2 | 67 | 0 | 1 | 107 | 60.8 | 831 | 1 | 303 | 61 | 252 |
| 2 | 65 | 0 | 0 | 174 | 63.0 | 877 | 1 | 308 | 59 | 345 |
| 2 | 47 | 1 | 3 | 159 | 63.0 | 853 | 1 | 306 | 62 | 310 |
| 2 | 41 | 0 | 2 | 141 | 65.0 | 903 | 1 | 332 | 62 | 307 |
| 2 | 67 | 0 | 1 | 150 | 58.0 | 783 | 1 | 303 | 59 | 335 |
| 2 | 68 | 0 | 1 | 162 | 59.9 | 823 | 1 | 328 | 64 | 361 |
| 1 | 74 | 0 | 1 | 133 | 65.4 | 841 | 1 | 304 | 73 | 267 |
| 1 | 69 | 1 | 2 | 164 | 72.2 | 956 | 1 | 350 | 81 | 310 |
| 1 | 71 | 0 | 1 | 133 | 70.6 | 927 | 1 | 325 | 70 | 247 |
| 2 | 36 | 0 | 1 | 138 | 66.5 | 883 | 1 | 316 | 63 | 292 |
| 1 | 31 | 0 | 2 | 246 | 70.8 | 922 | 1 | 354 | 76 | 366 |
| 1 | 33 | 1 | 2 | 179 | 71.9 | 985 | 1 | 332 | 70 | 328 |
| 1 | 39 | 1 | 2 | 187 | 69.2 | 898 | 1 | 320 | 75 | 345 |

*continued*

| Sex | Age | Emp | Cars | Wt (lb) | Ht (in) | Sit | Hand | Bitro | Elbow | UAG |
|-----|-----|-----|------|---------|---------|-----|------|-------|-------|-----|
| 2 | 34 | 0 | 3 | 137 | 64.6 | 864 | 1 | 313 | 66 | 294 |
| 1 | 59 | 1 | 3 | 208 | 74.7 | 984 | 1 | 364 | 76 | 322 |
| 1 | 34 | 1 | 1 | 130 | 64.7 | 865 | 1 | 292 | 66 | 305 |
| 2 | 27 | 1 | 3 | 141 | 64.4 | 841 | 1 | 303 | 55 | 260 |
| 1 | 29 | 1 | 2 | 179 | 72.6 | 962 | 1 | 334 | 75 | 327 |
| 2 | 31 | 0 | 2 | 131 | 66.7 | 908 | 1 | 325 | 59 | 284 |
| 1 | 63 | 1 | 2 | 186 | 72.1 | 957 | 1 | 337 | 76 | 320 |
| 1 | 61 | 0 | 2 | 194 | 73.6 | 959 | 2 | 346 | 78 | 335 |
| 2 | 35 | 1 | 2 | 139 | 62.6 | 812 | 1 | 304 | 63 | 295 |
| 2 | 28 | 1 | 2 | 103 | 62.7 | 870 | 1 | 311 | 56 | 257 |
| 1 | 55 | 1 | 4 | 153 | 71.1 | 946 | 1 | 327 | 75 | 292 |
| 2 | 67 | 0 | 0 | 195 | 58.5 | 763 | 1 | 341 | 64 | 402 |
| 1 | 57 | 1 | 2 | 227 | 68.4 | 933 | 1 | 359 | 73 | 375 |
| 1 | 60 | 0 | 2 | 165 | 68.3 | 886 | 1 | 316 | 71 | 301 |
| 2 | 66 | 0 | 1 | 147 | 62.5 | 850 | 1 | 341 | 71 | 316 |
| 2 | 63 | 1 | 2 | 166 | 59.9 | 808 | 1 | 338 | 65 | 353 |
| 2 | 65 | 1 | 1 | 129 | 62.3 | 830 | 1 | 336 | 65 | 294 |
| 1 | 62 | 0 | 3 | 188 | 70.9 | 928 | 1 | 325 | 77 | 327 |
| 2 | 48 | 1 | 4 | 185 | 66.6 | 886 | 1 | 346 | 71 | 331 |
| 1 | 64 | 1 | 2 | 192 | 68.7 | 915 | 1 | 345 | 76 | 353 |
| 1 | 34 | 1 | 2 | 189 | 69.3 | 963 | 1 | 322 | 68 | 330 |
| 1 | 37 | 2 | 1 | 125 | 66.8 | 888 | 8 | 311 | 68 | 268 |
| 1 | 42 | 1 | 1 | 179 | 72.0 | 964 | 1 | 325 | 79 | 324 |
| 2 | 70 | 0 | 1 | 171 | 61.2 | 828 | 1 | 340 | 69 | 355 |
| 1 | 18 | 1 | 3 | 123 | 70.0 | 919 | 1 | 308 | 72 | 258 |
| 1 | 71 | 1 | 2 | 106 | 61.4 | 836 | 1 | 271 | 66 | 255 |
| 1 | 46 | 1 | 3 | 217 | 71.6 | 959 | 1 | 328 | 74 | 356 |
| 2 | 30 | 1 | 2 | 203 | 61.6 | 853 | 1 | 368 | 62 | 368 |
| 2 | 73 | 0 | 1 | 125 | 63.6 | 829 | 1 | 322 | 64 | 260 |
| 2 | 49 | 1 | 1 | 135 | 63.7 | 851 | 1 | 301 | 59 | 290 |
| 1 | 58 | 1 | 2 | 172 | 67.9 | 888 | 2 | 315 | 74 | 320 |
| 1 | 64 | 1 | 2 | 184 | 72.7 | 917 | 1 | 342 | 75 | 319 |
| 1 | 46 | 1 | 1 | 232 | 67.9 | 911 | 1 | 309 | 75 | 381 |
| 2 | 22 | 0 | 0 | 120 | 60.6 | 812 | 1 | 285 | 56 | 276 |
| 2 | 55 | 0 | 0 | 123 | 56.7 | 789 | 1 | 277 | 63 | 338 |
| 2 | 31 | 0 | 0 | 99 | 61.1 | 830 | 1 | 296 | 58 | 252 |
| 1 | 19 | 1 | 2 | 157 | 70.9 | 936 | 1 | 313 | 71 | 314 |
| 1 | 21 | 2 | 2 | 138 | 67.9 | 931 | 1 | 301 | 65 | 278 |
| 2 | 31 | 1 | 1 | 147 | 59.5 | 803 | 1 | 295 | 61 | 339 |
| 1 | 59 | 1 | 3 | 182 | 71.4 | 939 | 1 | 330 | 76 | 309 |
| 1 | 23 | 1 | 2 | 155 | 72.3 | 953 | 1 | 320 | 75 | 319 |
| 1 | 30 | 1 | 2 | 167 | 66.9 | 927 | 1 | 312 | 72 | 356 |
| 2 | 69 | 0 | 1 | 120 | 60.9 | 857 | 1 | 301 | 60 | 281 |
| 2 | 21 | 1 | 3 | 113 | 60.2 | 834 | 1 | 289 | 58 | 278 |
| 1 | 71 | 1 | 1 | 180 | 68.5 | 905 | 1 | 329 | 78 | 332 |
| 2 | 20 | 0 | 2 | 140 | 65.6 | 903 | 1 | 338 | 62 | 283 |
| 2 | 24 | 1 | 2 | 118 | 64.4 | 854 | 1 | 281 | 56 | 281 |
| 1 | 46 | 1 | 3 | 195 | 65.9 | 880 | 2 | 320 | 79 | 370 |
| 1 | 50 | 1 | 2 | 178 | 68.0 | 896 | 1 | 304 | 76 | 353 |
| 2 | 41 | 1 | 2 | 109 | 65.9 | 861 | 1 | 318 | 62 | 233 |
| 2 | 61 | 1 | 2 | 130 | 67.4 | 884 | 1 | 343 | 64 | 277 |
| 2 | 65 | 0 | 4 | 131 | 62.1 | 864 | 1 | 316 | 61 | 303 |
| 2 | 41 | 0 | 1 | 136 | 64.4 | 910 | 1 | 320 | 61 | 300 |
| 1 | 64 | 1 | 2 | 175 | 68.9 | 935 | 1 | 332 | 70 | 333 |
| 1 | 39 | 2 | 3 | 216 | 69.7 | 933 | 1 | 340 | 73 | 382 |

continued

| Sex | Age | Emp | Cars | Wt (lb) | Ht (in) | Sit | Hand | Bitro | Elbow | UAC |
|-----|-----|-----|------|---------|---------|-----|------|-------|-------|-----|
| 1 | 61 | 1 | 2 | 149 | 65.2 | 903 | 1 | 303 | 73 | 306 |
| 2 | 19 | 1 | 1 | 107 | 67.8 | 875 | 1 | 285 | 61 | 231 |
| 1 | 18 | 2 | 0 | 134 | 65.9 | 883 | 2 | 308 | 68 | 296 |
| 1 | 61 | 1 | 2 | 235 | 70.2 | 945 | 2 | 338 | 75 | 344 |
| 1 | 45 | 1 | 3 | 168 | 68.1 | 900 | 1 | 331 | 74 | 304 |
| 2 | 19 | 1 | 4 | 246 | 63.5 | 865 | 1 | 337 | 68 | 430 |
| 1 | 22 | 2 | 1 | 156 | 71.0 | 933 | 1 | 313 | 71 | 316 |
| 2 | 36 | 2 | 1 | 195 | 59.6 | 844 | 1 | 330 | 65 | 384 |
| 1 | 63 | 1 | 0 | 169 | 65.4 | 850 | 1 | 324 | 72 | 320 |
| 2 | 42 | 1 | 0 | 201 | 65.9 | 894 | 1 | 341 | 64 | 355 |
| 2 | 66 | 0 | 1 | 147 | 65.5 | 850 | 1 | 306 | 66 | 288 |
| 1 | 70 | 1 | 1 | 174 | 66.5 | 909 | 1 | 323 | 80 | 349 |
| 2 | 47 | 1 | 2 | 130 | 65.4 | 873 | 1 | 320 | 61 | 252 |
| 1 | 53 | 8 | 9 | 186 | 66.7 | 918 | 1 | 323 | 72 | 359 |
| 1 | 31 | 1 | 2 | 206 | 75.1 | 989 | 1 | 319 | 77 | 355 |
| 2 | 58 | 0 | 2 | 161 | 63.5 | 856 | 1 | 307 | 65 | 330 |
| 2 | 57 | 0 | 3 | 179 | 64.4 | 896 | 1 | 337 | 66 | 356 |
| 1 | 42 | 1 | 4 | 173 | 71.9 | 967 | 1 | 323 | 72 | 310 |
| 1 | 57 | 1 | 2 | 176 | 69.6 | 922 | 1 | 332 | 74 | 337 |
| 2 | 23 | 2 | 1 | 126 | 67.6 | 888 | 2 | 306 | 62 | 253 |
| 1 | 32 | 2 | 1 | 210 | 67.3 | 930 | 1 | 347 | 73 | 370 |
| 2 | 46 | 1 | 2 | 145 | 66.0 | 883 | 1 | 312 | 66 | 338 |
| 1 | 66 | 0 | 2 | 138 | 65.4 | 873 | 1 | 300 | 69 | 288 |
| 1 | 59 | 1 | 3 | 146 | 66.1 | 918 | 1 | 334 | 70 | 296 |
| 1 | 74 | 0 | 1 | 176 | 66.7 | 882 | 2 | 342 | 75 | 307 |
| 2 | 68 | 0 | 3 | 155 | 61.2 | 830 | 1 | 316 | 61 | 338 |
| 1 | 60 | 0 | 2 | 160 | 71.0 | 940 | 1 | 338 | 78 | 332 |
| 1 | 62 | 1 | 2 | 166 | 72.2 | 941 | 1 | 330 | 76 | 295 |
| 2 | 44 | 1 | 3 | 199 | 63.8 | 876 | 1 | 345 | 74 | 392 |
| 2 | 30 | 0 | 4 | 144 | 63.1 | 880 | 1 | 295 | 60 | 309 |
| 1 | 49 | 1 | 3 | 176 | 68.6 | 924 | 1 | 324 | 72 | 344 |
| 1 | 32 | 1 | 2 | 267 | 75.4 | 970 | 1 | 379 | 85 | 396 |
| 1 | 19 | 1 | 4 | 153 | 68.0 | 874 | 1 | 291 | 72 | 317 |
| 2 | 62 | 1 | 1 | 172 | 63.7 | 840 | 1 | 335 | 66 | 322 |
| 1 | 29 | 2 | 1 | 180 | 69.1 | 939 | 1 | 333 | 74 | 364 |
| 1 | 72 | 1 | 2 | 144 | 64.9 | 895 | 1 | 307 | 70 | 302 |
| 1 | 18 | 1 | 4 | 118 | 66.3 | 860 | 1 | 293 | 66 | 266 |
| 2 | 57 | 1 | 2 | 138 | 60.6 | 828 | 1 | 335 | 64 | 308 |
| 1 | 45 | 1 | 3 | 153 | 64.4 | 899 | 1 | 321 | 74 | 328 |
| 1 | 62 | 0 | 1 | 155 | 64.4 | 907 | 1 | 328 | 66 | 336 |
| 2 | 37 | 0 | 2 | 154 | 61.3 | 814 | 1 | 322 | 65 | 320 |
| 1 | 25 | 1 | 1 | 148 | 67.9 | 914 | 1 | 310 | 67 | 294 |
| 2 | 32 | 0 | 2 | 192 | 68.6 | 956 | 1 | 394 | 64 | 292 |
| 2 | 60 | 2 | 2 | 116 | 62.8 | 832 | 1 | 307 | 61 | 268 |
| 2 | 22 | 0 | 2 | 157 | 64.0 | 875 | 1 | 325 | 60 | 307 |
| 2 | 18 | 0 | 5 | 93 | 64.8 | 880 | 1 | 291 | 58 | 204 |
| 2 | 18 | 0 | 0 | 126 | 69.4 | 918 | 1 | 314 | 64 | 234 |
| 2 | 21 | 0 | 1 | 120 | 61.6 | 808 | 1 | 305 | 56 | 288 |
| 2 | 25 | 1 | 1 | 104 | 61.7 | 838 | 2 | 300 | 57 | 253 |
| 1 | 26 | 1 | 2 | 135 | 69.9 | 945 | 1 | 325 | 77 | 277 |
| 1 | 18 | 1 | 2 | 121 | 70.5 | 903 | 1 | 302 | 72 | 253 |
| 2 | 19 | 2 | 1 | 170 | 63.1 | 870 | 1 | 329 | 62 | 321 |
| 1 | 53 | 0 | 1 | 152 | 68.7 | 889 | 1 | 310 | 72 | 282 |
| 1 | 60 | 0 | 1 | 174 | 67.0 | 909 | 1 | 336 | 76 | 332 |
| 2 | 54 | 0 | 1 | 114 | 67.4 | 883 | 1 | 329 | 59 | 257 |

*continued*

| Sex | Age | Emp | Cars | Wt (lb) | Ht (in) | Sit | Hand | Bitro | Elbow | UAG |
|-----|-----|-----|------|---------|---------|-----|------|-------|-------|-----|
| 1 | 71 | 0 | 0 | 151 | 67.2 | 813 | 1 | 328 | 69 | 285 |
| 2 | 59 | 1 | 1 | 108 | 63.1 | 900 | 1 | 307 | 58 | 263 |
| 2 | 26 | 1 | 2 | 137 | 65.2 | 916 | 1 | 314 | 65 | 293 |
| 2 | 37 | 1 | 3 | 108 | 64.4 | 877 | 2 | 311 | 59 | 247 |
| 1 | 18 | 1 | 3 | 136 | 68.4 | 932 | 1 | 310 | 75 | 287 |
| 1 | 20 | 1 | 4 | 159 | 66.1 | 887 | 1 | 315 | 73 | 321 |
| 1 | 52 | 1 | 4 | 196 | 72.1 | 940 | 1 | 332 | 77 | 341 |
| 2 | 19 | 0 | 0 | 135 | 66.9 | 839 | 1 | 296 | 61 | 243 |
| 2 | 49 | 1 | 0 | 124 | 60.2 | 825 | 1 | 302 | 60 | 303 |
| 1 | 68 | 0 | 2 | 179 | 70.6 | 935 | 1 | 327 | 79 | 310 |
| 2 | 27 | 2 | 1 | 121 | 63.1 | 876 | 1 | 316 | 62 | 267 |
| 1 | 60 | 0 | 1 | 172 | 70.8 | 914 | 3 | 338 | 69 | 300 |
| 1 | 55 | 1 | 2 | 208 | 71.6 | 949 | 1 | 348 | 80 | 367 |
| 2 | 18 | 2 | 3 | 172 | 62.8 | 867 | 1 | 351 | 70 | 424 |
| 2 | 64 | 2 | 2 | 172 | 66.9 | 849 | 1 | 332 | 70 | 327 |
| 1 | 21 | 1 | 2 | 175 | 68.5 | 873 | 1 | 310 | 72 | 340 |
| 2 | 41 | 1 | 2 | 128 | 64.9 | 857 | 1 | 335 | 62 | 260 |
| 2 | 70 | 0 | 2 | 149 | 63.0 | 857 | 1 | 318 | 64 | 301 |
| 2 | 29 | 1 | 1 | 117 | 60.7 | 837 | 1 | 300 | 58 | 271 |
| 1 | 71 | 0 | 1 | 184 | 68.3 | 908 | 1 | 334 | 74 | 339 |
| 2 | 70 | 0 | 1 | 147 | 61.1 | 820 | 1 | 311 | 63 | 290 |
| 2 | 29 | 0 | 1 | 166 | 63.0 | 866 | 1 | 349 | 64 | 338 |
| 2 | 40 | 1 | 3 | 125 | 65.5 | 874 | 1 | 318 | 65 | 255 |
| 2 | 64 | 0 | 0 | 115 | 62.5 | 832 | 1 | 308 | 62 | 256 |
| 2 | 45 | 1 | 4 | 138 | 67.2 | 918 | 1 | 332 | 66 | 321 |
| 2 | 57 | 0 | 2 | 114 | 61.3 | 853 | 1 | 315 | 62 | 278 |
| 1 | 33 | 1 | 2 | 188 | 72.4 | 950 | 1 | 327 | 74 | 336 |
| 2 | 29 | 0 | 2 | 144 | 60.2 | 812 | 1 | 306 | 59 | 313 |
| 2 | 66 | 0 | 2 | 145 | 62.8 | 838 | 1 | 311 | 65 | 327 |
| 2 | 18 | 1 | 1 | 134 | 59.3 | 835 | 2 | 320 | 60 | 292 |
| 2 | 31 | 1 | 2 | 143 | 64.4 | 875 | 1 | 307 | 60 | 304 |
| 1 | 27 | 1 | 2 | 181 | 73.3 | 982 | 1 | 326 | 80 | 319 |
| 1 | 42 | 1 | 2 | 187 | 67.4 | 930 | 2 | 319 | 72 | 360 |
| 1 | 66 | 0 | 1 | 217 | 69.8 | 952 | 1 | 341 | 75 | 364 |
| 2 | 73 | 0 | 1 | 121 | 58.4 | 754 | 1 | 344 | 62 | 288 |
| 2 | 68 | 1 | 2 | 204 | 63.9 | 852 | 1 | 353 | 72 | 383 |
| 2 | 64 | 1 | 1 | 101 | 62.4 | 849 | 1 | 302 | 65 | 254 |
| 2 | 71 | 0 | 0 | 146 | 63.0 | 832 | 1 | 336 | 62 | 337 |
| 2 | 66 | 0 | 2 | 134 | 65.0 | 845 | 1 | 318 | 61 | 288 |
| 1 | 51 | 1 | 4 | 205 | 67.7 | 916 | 1 | 324 | 76 | 373 |
| 2 | 71 | 0 | 0 | 250 | 62.4 | 869 | 1 | 339 | 72 | 465 |
| 1 | 69 | 1 | 1 | 127 | 68.6 | 903 | 1 | 345 | 73 | 266 |
| 2 | 37 | 0 | 2 | 145 | 64.5 | 894 | 1 | 327 | 65 | 323 |
| 1 | 60 | 1 | 2 | 200 | 68.0 | 900 | 1 | 332 | 74 | 358 |
| 2 | 43 | 1 | 1 | 129 | 63.5 | 854 | 1 | 314 | 62 | 296 |
| 2 | 42 | 2 | 4 | 148 | 61.9 | 871 | 1 | 320 | 66 | 336 |
| 1 | 44 | 1 | 1 | 159 | 64.4 | 866 | 1 | 325 | 73 | 328 |
| 1 | 69 | 1 | 1 | 196 | 67.6 | 905 | 2 | 347 | 71 | 332 |
| 2 | 27 | 0 | 2 | 126 | 66.7 | 886 | 1 | 318 | 60 | 263 |
| 1 | 63 | 1 | 2 | 181 | 67.3 | 936 | 1 | 343 | 77 | 354 |
| 1 | 68 | 0 | 1 | 158 | 66.2 | 912 | 1 | 316 | 70 | 332 |
| 2 | 62 | 0 | 2 | 140 | 63.5 | 842 | 1 | 321 | 63 | 307 |
| 1 | 71 | 0 | 1 | 150 | 65.3 | 846 | 1 | 311 | 72 | 290 |

# Appendix C: Minitab

There are two different levels of software recommended for use with this text. STATDISK, developed specifically for this text, is provided free to colleges that adopt this book. It is an easy-to-use software package designed for students with little or no prior computer experience. It is available for the IBM PC, Apple IIe, and compatible models. A separate *STATDISK Manual/ Workbook* is available. We highly recommend STATDISK for those who can spare little or no class time for discussion of computer use. It is menu-driven, and students can easily use it on their own.

For those who incorporate computer usage as a major component of the course, we recommend Minitab, although other popular packages (such as BMDP, SAS, SPSS, or Microstat) can also be used with this text. An inexpensive student version of Minitab is available from Addison-Wesley Publishing Company, 1 Jacob Way, Reading, MA 01867.

Minitab uses commands to accomplish various tasks. This means that there are special words (commands) that are used to accomplish various tasks. We first list the important *general* commands that are used throughout the course, then we list *specific* commands that apply to each particular chapter. In the commands given below, we use CAPITAL letters to highlight the necessary part of the expression and we use lowercase letters to represent parts that may be omitted. When running Minitab, either capital letters or lowercase letters may be used. For example, the command "RANK C1 and put ranks in C2" may be entered in an abbreviated form as "RANK C1 C2" or "rank c1 c2."

# General Commands

| Minitab Command | Use |
|---|---|
| SET C1 | Enter data in a column. For example,<br><br>    SET C1<br>    66 65 64 64 63<br><br>causes the five scores to be stored in the column identified as C1. |
| READ C1 C2 | Enter data in two different columns. Use READ C1 C2 C3 for three sets of scores, and so on. (Note: Both SET and READ allow you to enter data, but SET is generally better with *one* set of data, while READ tends to be better with two or more sets of data. With SET, you enter all of the data in one operation; with READ you enter data one row at a time.) |
| ENDOFDATA | Signals the end of the entry of data. Can be abbreviated as simply END. |
| NAME C1 'HEIGHT' | Gives the column of data C1 the more meaningful name of 'HEIGHT'. When such a name is used, it must always be enclosed within single quotes. |
| PRINT C1 | Displays the data stored in the column C1. This is especially useful for verifying that data have been entered correctly. Also, PRINT 'HEIGHT' will display the scores in the column named 'HEIGHT'. |
| SAVE 'GRADES' | Saves the data in a computer file. To save on a disk in drive B, type: SAVE 'B:GRADES'. File names and column names should be different. |
| RETRIEVE 'GRADES' | Retrieves data previously saved in a computer file. To retrieve data from a disk in drive B, type: RETRIEVE 'B: GRADES'. |
| ERASE C1 | Erases the data stored in column C1. |
| STOP | Signals the end of the use of Minitab. |

In the sample run shown below and on the next page, five scores are entered in column C1 and printed. Then five pairs of data are entered in columns C1 and C2. They are named and printed. Everything immediately to the right of the symbol > is entered by the user; everything else is displayed by Minitab.

```
MTB ) SET C1
DATA) 66 65 64 64 63
DATA) ENDOFDATA
MTB ) PRINT C1

C1
    66   65   64   64   63
MTB ) READ C1 C2
DATA) 66 115
```

```
DATA⟩ 65 107
DATA⟩ 64 110
DATA⟩ 64 128
DATA⟩ 63 130
DATA⟩ ENDOFDATA
        5 ROWS READ
MTB ⟩ NAME C1 'HEIGHT'
MTB ⟩ NAME C2 'WEIGHT'
MTB ⟩ PRINT C1 C2
```

| ROW | HEIGHT | WEIGHT |
|-----|--------|--------|
| 1 | 66 | 115 |
| 2 | 65 | 107 |
| 3 | 64 | 110 |
| 4 | 64 | 128 |
| 5 | 63 | 130 |

Miscellaneous notes:

1. Don't use commas in numbers. For example, enter 734527.80 instead of 734,527.80.

2. To correct an error of entering a wrong number:
   a. If you haven't yet hit the RETURN key, backspace and type over.
   b. If you have already hit the RETURN key, you can reenter the correct data set. If you prefer to replace, delete, or insert a score, use the formats suggested by the following examples.

   LET C3(7) = 9     *Replaces* the 7th entry of column C3 with the number 9

   DELETE 3 C5     *Deletes* the entry in the 3rd row of column C5

   INSERT 5 6 C1
   9
   ENDOFDATA     *Inserts* a 9 in column C1 between rows 5 and 6

3. Use LET to do arithmetic with data. For example,

   LET C3 = C1 + C2     Column C3 is created by adding the column C1 values to those in column C2.

   LET C2 = C2/2     Each entry of column C2 is divided by 2.

   LET C5 = C2 − C1     Column C5 is created by subtracting the column C1 values from column C2.

4. Specific numbers have been used in many of the commands that follow, but other values may also be used. For example, BINOMIAL n = 5 p = .3 can be replaced by BINOMIAL n = 32 p = .75. The particular numbers used have been included only to illustrate typical cases.

# Specific Minitab Commands

**Chapter 2** (See page 90.)

| | |
|---|---|
| MEAN C1 | Gives the mean of the values in column C1 |
| MEDIAN C1 | Gives the median of the values in column C1 |
| STDEV C1 | Gives the standard deviation of the values in column C1 |
| SUM C1 | Gives the sum of the values in column C1 |
| SSQ C1 | Gives the sum of the squares of the values in column C1 |
| MAXIMUM C1 | Gives the maximum value found in column C1 |
| MINIMUM C1 | Gives the minimum value found in column C1 |
| HISTOGRAM C1 | Gives a histogram of the values found in column C1 |
| STEM-AND-LEAF C1 | Gives a stem-and-leaf plot of the values in column C1 |
| BOXPLOT C1 | Gives a boxplot of the data in column C1 |
| DESCRIBE C1 | Gives the mean, median, standard deviation, minimum, maximum, first and third quartiles, and the number of scores |

**Chapter 3** (See example below.)

RANDOM 50 C1;
INTEGERS randomly selected from 1 to 6.

Randomly generates 50 integers, where each is between 1 and 6 inclusive. (INTEGERS provides a distribution in which the numbers are equally likely. Some other distributions are also available.)

RANDOM 6 C1;
UNIFORM 1 5.

Randomly generates 6 numbers between 1 and 5. Numbers come from a population with a uniform distribution.

```
MTB  ) RANDOM 50 C1;
SUBC ) INTEGERS 1 6.
MTB  ) PRINT C1

C1
   4  4  2  5  2  3  1  3  4  5  2  2  2  4  5
   2  5  5  2  3  2  4  4  4  5  6  5  1  1  4
   6  1  3  3  1  2  2  3  1  2  6  4  6  1  2
   4  2  2  2  5

MTB  ) RANDOM 6 C1;
SUBC ) UNIFORM 1 5.
MTB  ) PRINT C1

C1
   3.33638  1.04711  2.88852  1.06512  1.14015  2.51822
```

## Chapter 4 (See page 201.)

PDF;
BINOMIAL n = 5 p = .3.

Gives probabilities for a binomial experiment in which $n = 5$ and $p = 0.3$. (PDF represents a "probability density function.")

## Chapter 5 (See example below.)

RANDOM 36 C1;
NORMAL mu = 100 sigma = 15.

Randomly generates 36 numbers from a normally distributed population with a mean of 100 and a standard deviation of 15

MTB   ⟩ RANDOM 36 C1;
SUBC ⟩ NORMAL mu = 100 sigma = 15.
MTB   ⟩ PRINT C1

C1

| | | | | | | |
|---|---|---|---|---|---|---|
| 102.789 | 94.374 | 115.105 | 115.908 | 93.704 | 102.303 | 107.026 |
| 89.323 | 108.569 | 98.781 | 84.783 | 74.635 | 81.170 | 90.567 |
| 97.233 | 111.450 | 138.779 | 86.234 | 85.722 | 82.985 | 119.258 |
| 97.943 | 84.047 | 100.943 | 73.494 | 103.590 | 98.168 | 96.171 |
| 98.488 | 102.445 | 96.174 | 104.018 | 97.192 | 95.563 | 87.498 |
| 114.200 | | | | | | |

## Chapter 6 (See page 294.)

ZINTERVAL 95 percent and sigma = 15 for the data in C1

Constructs the 95% confidence interval estimate of the mean using the data in column C1. Assumes that the population standard deviation is known to be 15.

TINTERVAL with 95 percent for data in C1

Constructs the 95% confidence interval estimate of the mean using the data from column C1.

## Chapter 7 (See example below.)

ZTEST mean = 100 sigma = 15 with data in C1

*Two-tailed* test of the null hypothesis that the mean equals 100 where the population standard deviation is known to be 15. Sample data are in C1.

ZTEST mean = 100 sigma = 15;
ALTERNATIVE = +1.

*Right-tailed* test

ZTEST mean = 100 sigma = 15;
ALTERNATIVE = −1.

*Left-tailed* test

TTEST mean = 100 with data in C1

*Two-tailed* t test of the null hypothesis that the mean equals 100. The sample data are in C1.

TTEST mean = 100 with data in  *Right-tailed* test
C1;
ALTERNATIVE = + 1.

TTEST mean = 100 with data in  *Left-tailed* test
C1;
ALTERNATIVE = − 1.

MTB ⟩ SET C1
DATA⟩ 106  103  102  101  99  103
DATA⟩ 102  105  100  104  97  107
DATA⟩ ENDOFDATA
MTB ⟩ TTEST mean = 100 with data in C1

TEST OF mu = 100.000 VS mu N.E. 100.000

|      | N  | MEAN    | STDEV | SE MEAN | T    | P VALUE |
|------|----|---------|-------|---------|------|---------|
| C1   | 12 | 102.417 | 2.906 | 0.839   | 2.88 | 0.015   |

### Chapter 8 (See page 407.)

TWOSAMPLE 95 percent C1 C2 — Construct the confidence interval and test the claim that two populations have the same mean. The first set of sample data is in C1 and the second set is in C2.

TWOSAMPLE 95 percent C1 C2;
POOLED. — If the two population variances are believed to be approximately equal, we use this format to get the confidence interval and to test the claim that the two means are equal.

TWOSAMPLE 95 percent C1 C2;
POOLED;
ALTERNATIVE = + 1. — Same as above, except that the test is right-tailed. Use − 1 for a left-tailed test.

### Chapter 9 (See pages 478 and 495.)

PLOT C1 VS C2 — Gives a scatter diagram.

CORRELATION C1 and C2 — Gives the value of the linear correlation coefficient $r$.

REGRESSION C2 1 C1 — Gives the equation of the regression line. The second variable is expressed in terms of the first.

REGRESSION C3 2 C2 C1 — Gives the multiple regression equation where the third variable is expressed in terms of the first two variables.

### Chapter 10 (See page 533.)

CHISQUARE analysis for C1 C2 C3 — Gives the chi-square test statistic for a contingency table with columns C1, C2, C3.

### Chapter 11

AOVONEWAY C1 C2 C3

Uses analysis of variance (one-way) to test the claim that the three sets of data (in columns C1, C2, C3) have the same mean. Can also be used with more than three sets of data.

TWOWAY C1 C2 C3

Uses analysis of variance (two-way) to test for effects from the row factor, the column factor, and interaction between factors. The data are in column C1, while columns C2 and C3 identify their locations according to row and column. See the examples of Minitab displays in Section 11–4.

### Chapter 12

LET C3 = C2 − C1
STEST difference in median = 0 for data in C3.

Sign test for paired data in columns C1 and C2.

LET C3 = C2 − C1
WTEST for data in C3

Wilcoxon signed-ranks test for paired data in columns C1 and C2.

MANN-WHITNEY C1 C2

Wilcoxon rank-sum test (equivalent to the Mann-Whitney test) for the data in columns C1 and C2.

KRUSKAL-WALLIS C1 C2

Kruskal-Wallis test for the data in columns C1 and C2. Column C1 contains all sample data while column C2 contains identifying numbers.

RANK C1 and put ranks in C3
RANK C2 and put ranks in C4
CORRELATION C3 and C4

Gives the rank correlation coefficient.

RUNS test above and below 1179 for data in C1

Does runs test for randomness above and below the median of 1179 for the data in C1. (The median can be found by entering MEDIAN C1.)

# Appendix D: Glossary

**Addition rule** Rule for determining the probability that, on a single trial, either event A occurs, or event B occurs, or they both occur.

**Alternative hypothesis** Denoted by $H_1$, the statement that is equivalent to the negation of the null hypothesis.

**Analysis of variance** A method analyzing population variance in order to make inferences about the population.

**ANOVA** See analysis of variance.

**Arithmetic mean** The sum of a set of scores divided by the number of scores.

**Attribute data** Data consisting of qualities, such as political party, religion, or sex.

**Average** Any one of several measures designed to reveal the central tendency of a collection of data.

**Bimodal** Having two modes.

**Binomial experiment** An experiment with a fixed number of independent trials. Each outcome falls into exactly one of two categories.

**Binomial probability formula** See Formula 4–6 in Section 4–4.

**Bivariate data** Data arranged in pairs.

**Bivariate normal distribution** With paired data, for any fixed value of one variable, the values of the other variable have a normal distribution.

**Boxplot** Graphic method of showing the spread of a set of data.

**Central limit theorem** Theorem stating that sample means tend to be normally distributed.

**Centroid** The point $(\overline{x}, \overline{y})$ determined from a collection of bivariate data.

**Chebyshev's theorem** Uses standard deviation to provide information about the distribution of data. See Section 2–5.

**Chi-square distribution** Continuous probability distribution with selected values (Table A–4).

**Class boundaries** Values obtained from a frequency table by increasing the upper class limits and decreasing the lower class limits by the same amount so that there are no gaps between consecutive classes.

**Classical approach to probability** Determining the probability of an event by dividing the number of ways the event can occur by the total number of possible outcomes.

**Class marks** The midpoints of the classes in a frequency table.

**Class width** The difference between two consecutive lower class limits in a frequency table.

**Cluster sampling** Population is divided into sections and a sampling of sections is randomly selected.

**Coefficient of determination** The amount of the variation in $y$ that is explained by the regression line.

**Combinations rule** Rule for determining the number of different combinations of selected items.

**Complement of an event** All outcomes in which the original event does not occur.

**Completely randomized design** In analysis of variance, each element is given the same chance of belonging to the different categories or treatments.

**Compound event** A combination of simple events.

**Conditional probability** The conditional probability of event $B$, given that event $A$ has already occurred, is $P(A \text{ and } B)/P(A)$.

**Confidence interval** A range of values used to estimate some population parameter with a specific level of confidence.

**Confidence interval limits** The two numbers that are used as the high and low boundaries of a confidence interval.

**Contingency table** A table of observed frequencies where the rows correspond to one variable of classification and the columns correspond to another variable of classification.

**Continuity correction** An adjustment made when a discrete random variable is being approximated by a continuous random variable. See Section 5–5.

**Continuous data** Data resulting from infinitely many possible values that can be associated with points on a continuous scale in such a way that there are no gaps or interruptions.

**Continuous random variable** A random variable with infinite values that can be associated with points on a continuous line interval.

**Control chart** Any one of several types of charts (such as Figure 5–14) used to depict some characteristic of a process in order to detect a trend over time.

**Convenience sampling** Sample data are selected because they are readily available.

**Correlation** Statistical association between two variables.

**Correlation coefficient** A measurement of the strength of the relationship between two variables.

**Countable set** A set with either a finite number of values or values that can be made to correspond to the positive integers.

**Critical region** The area under a curve containing the values that lead to rejection of the null hypothesis.

**Critical value** Value separating the critical region from the values of the test statistic that would not lead to rejection of the null hypothesis.

**Cumulative frequency table** Frequency table in which each class and frequency represents cumulative data up to and including that class.

**Data** The numbers or information collected in an experiment.

**Decile** The 9 deciles divide the ranked data into 10 groups with 10% of the scores in each group.

**Degree of confidence** Probability that a population parameter is contained within a particular confidence interval.

**Degrees of freedom** The number of values that are free to vary after certain restrictions have been imposed on all values.

**Denominator degrees of freedom** The degrees of freedom corresponding to the denominator of the $F$ test statistic.

**Dependent events** Events that are not independent. See independent events.

**Dependent samples** The values in one sample are related to the values in another sample.

**Descriptive statistics** The methods used to summarize the key characteristics of known population data.

**Discrete data** Data resulting from either a finite number of possible values or a countable number of possible values.

**Discrete random variable** A random variable with either a finite number of values or a countable number of values.

**Distribution-free tests** Tests not requiring a particular distribution, such as the normal distribution. See nonparametric methods.

**Efficiency** Measure of the sensitivity of a nonparametric test in comparison to a corresponding parametric test.

**Empirical approximation of probability** Estimated value of probability based on actual observations.

**Empirical rule** Uses standard deviation to provide information about data with a bell-shaped distribution. See Section 2–5.

**Event** A result or outcome of an experiment.

**Expected frequency** Theoretical frequency for a cell of a contingency table or multinomial table.

**Expected value** For a discrete random variable, the sum of the products obtained by multiplying each value of the random variable by the corresponding probability.

**Experiment** Process that allows observations to be made.

**Explained deviation** For one pair of values in a collection of bivariate data, the difference between the predicted $y$ value and the mean of the $y$ values.

**Explained variation** The sum of the squares of the explained deviations for all pairs of bivariate data in a sample.

**Exploratory data analysis (EDA)** Branch of statistics emphasizing the investigation of data.

**Factor** A property that is the basis for categorizing different groups of data.

**Factorial rule** $n$ different items can be arranged $n!$ different ways.

**$F$ distribution** A continuous probability distribution with selected values given in Table A–5.

**Finite population correction factor** Factor for correcting the standard error of the mean when a sample size exceeds 5% of the size of a finite population.

**5-number summary** The minimum score, maximum score, median, and the two hinges of a set of data.

**Frequency polygon** Graphical method for representing the distribution of data using connected straight-line segments.

**Frequency table** A list of categories of scores along with their corresponding frequencies.

**Fundamental counting rule** For a sequence of two events in which the first event can occur $m$ ways and the second can occur $n$ ways, the events together can occur a total of $m \cdot n$ ways.

**Goodness-of-fit test**   Test for how well some observed frequency distribution fits some theoretical distribution.

**Hinge**   The median value of the bottom (or top) half of a set of ranked data.

**Histogram**   A graph of vertical bars representing the frequency distribution of a set of data.

**H test**   See the Kruskal-Wallis test.

**Hypothesis**   A statement or claim that some population characteristic is true.

**Hypothesis test**   A method for testing claims made about populations. Also called test of significance.

**Independent events**   The case when the occurrence of any one of the events does not affect the probabilities of the occurrences of the other events.

**Independent samples**   The values in one sample are not related to the values in another sample.

**Inferential statistics**   The methods of using sample data to make generalizations or inferences about a population.

**Interaction**   In two-way analysis of variance, the effect due to the combination of the two factors.

**Interquartile range**   The difference between the first and third quartiles.

**Interval**   Level of measurement of data: data can be arranged in order, and differences between data values are meaningful.

**Interval estimate**   See confidence interval.

**Kruskal-Wallis test**   A nonparametric hypothesis test used to compare three or more independent samples.

**Least-squares property**   For a regression line, the sum of the squares of the vertical deviations of the sample points from the regression line is the smallest sum possible.

**Left-tailed test**   Hypothesis test in which the critical region is located in the extreme left area of the probability distribution.

**Linear correlation coefficient**   Measure of strength of relationship between two variables.

**Lower class limits**   The smallest numbers that can actually belong to the different classes in a frequency table.

**Margin of error**   See maximum error of estimate.

**Maximum error of estimate**   The largest difference between a point estimate and the true value of a population parameter.

**Mean**   The sum of a set of scores divided by the number of scores.

**Mean deviation**   The measure of dispersion equal to the sum of the deviations of each score from the mean, divided by the number of scores.

**Measure of central tendency**   Value intended to indicate the center of the scores in a collection of data.

**Measure of dispersion**   Any of several measures designed to reflect the amount of variability among a set of scores.

**Median**   The middle value of a set of scores arranged in order of magnitude.

**Midquartile**   One-half of the sum of the first and third quartiles.

**Midrange**   One-half the sum of the highest and lowest scores.

**Mode**   The score that occurs most frequently.

**Multimodal**   Having more than two modes.

**Multinomial experiment**   An experiment with a fixed number of independent trials and each outcome falls into exactly one of several categories.

**Multiple coefficient of determination**   Measure of how well a multiple regression equation fits the sample data.

**Multiple comparison procedures**   Procedures for identifying which particular means are different, after concluding that three or more means are not all equal.

**Multiple regression**   Study of linear relationships among three or more variables.

**Multiplication rule**   Rule for determining the probability that event A will occur on one trial while event B occurs on a second trial.

**Mutually exclusive events**   Events that cannot occur simultaneously.

**Nominal**   Level of measurement of data: data consist of names, labels, or categories only.

**Nonparametric methods**   Statistical procedures for testing hypotheses or estimating parameters; they are not based on population parameters and do not require many of the restrictions of parametric tests.

**Nonsampling errors**   Errors from external factors not related to sampling.

**Normal distribution**   A bell-shaped probability distribution described algebraically by Equation 5–1 in Section 5–1.

**Null hypothesis**   Denoted by $H_0$, it is the claim made about some population characteristic. It usually involves the case of no difference.

**Numerator degrees of freedom**   The degrees of freedom corresponding to the numerator of the $F$ test statistic.

**Numerical data**   Data consisting of numbers representing counts or measurements.

**Observed frequency** The actual frequency count recorded in one cell of a contingency table or multinomial table.

**Odds against** The odds against event $A$ are obtained by finding $P(\overline{A})/P(A)$, usually expressed in the form of $a:b$ where $a$ and $b$ are integers having no common factors.

**Odds in favor** The odds in favor of event $A$ are obtained by finding $P(A)/P(\overline{A})$, usually expressed as the ratio of two integers with no common factors.

**Ogive** Graphical method of representing a cumulative frequency table.

**One-way analysis of variance** Analysis of variance involving data classified into groups according to a single criterion only.

**Ordinal** Level of measurement of data: data may be arranged in order, but differences between data values either cannot be determined or they are meaningless.

**Parameter** A measured characteristic of a population.

**Parametric methods** Statistical procedures for testing hypotheses or estimating parameters; based on population parameters.

**Pearson's product moment** See linear correlation coefficient.

**Percentile** The 99 percentiles divide the ranked data into 100 groups with 1% of the scores in each group.

**Permutations rule** Rule for determining the number of different arrangements of selected items.

**Pie chart** Graphical method for representing data in the form of a circle containing wedges.

**Point estimate** A single value that serves as an estimate of a population parameter.

**Pooled estimate of $p_1$ and $p_2$** The probability obtained by combining the data from two sample proportions and dividing the total number of successes by the total number of observations.

**Population** The complete and entire collection of elements to be studied.

**Predicted value** Using a regression equation, the value of one variable given a value for the other variable.

**Probability** A measure of the likelihood that a given event will occur. Mathematical probabilities are expressed as numbers between 0 and 1.

**Probability distribution** Collection of values of a random variable along with their corresponding probabilities.

**Probability histogram** Histogram with outcomes listed along the horizontal axis and probabilities listed along the vertical axis.

**P-value** The probability that a test statistic in a hypothesis test is at least as extreme as the one actually obtained.

**Quartile** The three quartiles divide the ranked data into four groups with 25% of the scores in each group.

**Randomized block experiment** An experimental design in which data are collected for each of several treatments. See Section 11–4.

**Random sample** A sample selected in a way that allows every member of the population to have the same chance of being chosen.

**Random selection** Sample elements are selected in such a way that all elements available for selection have the same chance of being selected.

**Random variable** The values that correspond to the numbers associated with events in a sample space.

**Range** The measure of dispersion that is the difference between the highest and lowest scores.

**Rank** The numerical position of an item in a sample set arranged in order.

**Rank correlation coefficient** Measure of the strength of the relationship between two variables; based on the ranks of the values.

**Ratio** Level of measurement of data: data can be arranged in order, differences between data values are meaningful, and there is an inherent zero starting point.

**Regression line** A straight line that summarizes the relationship between two variables.

**Relative frequency histogram** Variation of the basic histogram in which frequencies are replaced by relative frequencies. See relative frequency table.

**Relative frequency table** Variation of the basic frequency table in which the frequency for each class is divided by the total of all frequencies.

**Right-tailed test** Hypothesis test in which the critical region is located in the extreme right area of the probability distribution.

**Rigorously controlled design** In analysis of variance, all factors are forced to be constant so that effects of extraneous factors are eliminated.

**Run** Used in the runs test for randomness, a sequence of data exhibiting the same characteristic.

**Runs test** Nonparametric method used to test for randomness.

**Sample** A subset of a population.

**Sample space** In an experiment, the set of all possible outcomes or events that cannot be further broken down.

**Sampling errors** Errors resulting from the sampling process itself.

**Scatter diagram** Graphical method for displaying bivariate data.

**Semi-interquartile range**   One-half of the difference between the first and third quartiles.

**Significance level**   The probability that serves as a cutoff between results attributed to chance and results attributed to significant differences.

**Sign test**   A nonparametric hypothesis test used to compare samples from two populations.

**Simple event**   An experimental outcome that cannot be further broken down.

**Single factor analysis of variance**   See one-way analysis of variance.

**Slope**   Measure of steepness of a straight line.

**Spearman's rank correlation coefficient**   See rank correlation coefficient.

**Standard deviation**   The measure of dispersion equal to the square root of the variance.

**Standard error of estimate**   Measure of spread of sample points about the regression line.

**Standard error of the mean**   The standard deviation of all possible sample means $\bar{x}$.

**Standard normal distribution**   A normal distribution with a mean of 0 and a standard deviation equal to 1.

**Standard score**   Also called $z$ score, it is the number of standard deviations that a given value is above or below the mean.

**Statistic**   A measured characteristic of a sample.

**Statistical Process Control (SPC)**   The use of statistical techniques such as control charts to analyze a process or its outputs so as to take appropriate actions to achieve and maintain a state of statistical control and to improve the process capability.

**Statistics**   The collection, organization, description, and analysis of data.

**Stem-and-leaf plot**   Method of sorting and arranging data to reveal the distribution.

**Stepwise regression**   In multiple regression, the process of using different combinations of variables until the best model is obtained.

**Stratified sampling**   Samples are drawn from each stratum (class).

**Student $t$ distribution**   See $t$ distribution.

**Systematic sampling**   Every $k$th element is selected for a sample.

**$t$ distribution**   A bell-shaped distribution usually associated with small sample experiments. Also called the Student $t$ distribution.

**10–90 percentile range**   The difference between the 10th and 90th percentiles.

**Test of significance**   See hypothesis test.

**Test statistic**   Used in hypothesis testing, it is the sample statistic based on the sample data.

**Total deviation**   The sum of the explained deviation and unexplained deviation for a given pair of values in a collection of bivariate data.

**Total variation**   The sum of the squares of the total deviation for all pairs of bivariate data in a sample.

**Tree diagram**   Graphical depiction of the different possible outcomes in a compound event.

**Two-tailed test**   Hypothesis test: the critical region is divided between the left and right extreme areas of the probability distribution.

**Two-way analysis of variance**   Analysis of variance involving data classified according to two different factors.

**Two-way table**   See contingency table.

**Type I error**   The mistake of rejecting the null hypothesis when it is true.

**Type II error**   The mistake of failing to reject the null hypothesis when it is false.

**Unexplained deviation**   For one pair of values in a collection of bivariate data: difference between $y$ coordinate and predicted value.

**Unexplained variation**   The sum of the squares of the unexplained deviations for all pairs of bivariate data in a sample.

**Uniform distribution**   A distribution of values evenly distributed over the range of possibilities.

**Upper class limits**   Largest numbers that can belong to the different classes in a frequency table.

**Variance**   The measure of dispersion found by using Formula 2–5 in Section 2–5.

**Variance between samples**   In analysis of variance, the variation among the different samples.

**Variation due to error**   In analysis of variance, the variation within samples that is due to chance.

**Variation due to treatment**   See variance between samples.

**Variation within samples**   In analysis of variance, the variation that is due to chance.

**Weighted mean**   Mean of a collection of scores that have been assigned different degrees of importance.

**Wilcoxon rank-sum test**   A nonparametric hypothesis test used to compare two independent samples.

**Wilcoxon signed-ranks test**   A nonparametric hypothesis test used to compare two dependent samples.

**$y$-intercept**   Point at which a straight line crosses the $y$-axis.

**$z$ score**   See standard score.

# Appendix E: Bibliography

**\*Denotes recommended reading. Other books are recommended for reference.**

Beyer, W. 1991. *CRC Standard Probability and Statistics Tables and Formulae.* Boca Raton, Fla: CRC Press.

\*Brook, R., and others, eds. 1986. *The Fascination of Statistics.* New York: Dekker.

Byrkit, D. 1987. *Statistics Today: A Comprehensive Introduction.* Menlo Park, Calif.: Benjamin/Cummings.

\*Campbell, S. 1974. *Flaws and Fallacies in Statistical Thinking.* Englewood Cliffs, N.J.: Prentice-Hall.

Cochran, W. 1982. *Contributions to Statistics.* New York: Wiley.

Conover, W. 1980. *Practical Nonparametric Statistics.* 2nd ed. New York: Wiley.

Devore, J., and R. Peck. 1986. *Statistics: The Exploration and Analysis of Data.* St. Paul, Minn.: West Publishing.

\*Fairley, W., and F. Mosteller. 1977. *Statistics and Public Policy.* Reading, Mass.: Addison-Wesley.

Fisher, R. 1966. *The Design of Experiments.* 8th ed. New York: Hafner.

\*Freedman, D., R. Pisani, and R. Purves. 1978. *Statistics.* New York: Norton.

Freund, J. 1988. *Modern Elementary Statistics.* 7th ed. Englewood Cliffs, N.J.: Prentice-Hall.

Hauser, P. 1975. *Social Statistics in Use.* New York: Russell Sage Foundation.

Heerman, E., and L. Braskam. 1970. *Readings in Statistics for the Behavioral Sciences.* Englewood Cliffs, N.J.: Prentice-Hall.

Hoaglin, D., F. Mosteller, and J. Tukey, eds. 1983. *Understanding Robust and Exploratory Data Analysis.* New York: Wiley.

Hollander, M., and D. Wolfe. 1973. *Nonparametric Statistical Methods.* New York: Wiley.

\*Hollander, M., and F. Proschan. 1984. *The Statistical Exorcist: Dispelling Statistics Anxiety.* New York: Marcel Dekker.

\*Holmes, C. 1990. *The Honest Truth About Lying with Statistics.* Springfield, IL: Charles C. Thomas.

\*Hooke, R. 1983. *How to Tell the Liars from the Statisticians.* New York: Dekker.

\*Huff, D. 1954. *How to Lie with Statistics.* New York: Norton.

\*Jaffe, A., and H. Spirer. 1987. *Misused Statistics.* New York: Marcel Dekker.

\*Kimble, G. 1978. *How to Use (and Misuse) Statistics.* Englewood Cliffs, N.J.: Prentice-Hall.

King, R., and B. Julstrom. 1982. *Applied Statistics Using the Computer.* Sherman Oaks, Calif.: Alfred.

Kirk, R., ed. 1972. *Statistical Issues: A Reader for the Behavioral Sciences.* Belmont, Calif.: Brooks/Cole.

Kotz, S., and D. Stroup. 1983. *Educated Guessing—How to Cope in an Uncertain World.* New York: Dekker.

Mason, D. 1992. *Student Solutions Manual to Accompany Elementary Statistics.* 5th ed. Reading, MA: Addison-Wesley.

\*Moore, D. 1991. *Statistics: Concepts and Controversies.* 3rd ed. San Francisco: Freeman.

Mosteller, F., and R. Rourke. 1973. *Sturdy Statistics, Nonparametrics and Order Statistics.* Reading, Mass.: Addison-Wesley.

Mosteller, F., R. Rourke, and G. Thomas, Jr. 1970. *Probability with Statistical Applications.* 2nd ed. Reading, Mass.: Addison-Wesley.

Neter, J., W. Wasserman, and M. Kutner. 1985. *Applied Linear Statistical Models.* Homewood, Ill.: Irwin.

Noether, G. 1976. *Elements of Nonparametric Statistics.* 2nd ed. New York: Wiley.

Ott, L., and W. Mendenhall. 1985. *Understanding Statistics.* 4th ed. Boston: Duxbury Press.

Owen, D. 1962. *Handbook of Statistical Tables.* Reading, Mass.: Addison-Wesley.

*Paulos, J. 1988. *Innumeracy: Mathematical Illiteracy and its Consequences.* New York: Hill and Wang.

*Reichard, R. 1974. *The Figure Finaglers.* New York: McGraw-Hill.

*Reichmann, W. 1962. *Use and Abuse of Statistics.* New York: Oxford University Press.

*Runyon, R. 1977. *Winning with Statistics.* Reading, Mass.: Addison-Wesley.

Ryan, T., B. Joiner, and B. Ryan. 1985. *Minitab Student Handbook.* 2nd ed. Boston: Duxbury.

Schmid, C. 1983. *Statistical Graphics.* New York: Wiley.

Siegal, S. 1956. *Nonparametric Statistics for the Behavioral Sciences.* New York: McGraw-Hill.

*Stigler, S. 1986. *The History of Statistics.* Cambridge, Mass.: Harvard University Press.

*Tanur, J., ed. 1978. *Statistics: A Guide to the Unknown.* 2nd ed. San Francisco: Holden-Day.

Triola, M. 1992. *Minitab Student Laboratory Manual and Workbook.* 3rd ed. Reading, MA: Addison-Wesley.

Triola, M. 1992. *Statdisk Student Laboratory Manual and Workbook.* 3rd ed. Reading, MA: Addison-Wesley.

Tukey, J. 1977. *Exploratory Data Analysis.* Reading, Mass.: Addison-Wesley.

Velleman, P., and D. Hoaglin. 1981. *Applications, Basics, and Computing of Exploratory Data Analysis.* Boston: Duxbury Press.

Wayne, D. 1978. *Applied Nonparametric Statistics.* Boston: Houghton Mifflin.

Weisberg, S. 1985. *Applied Linear Regression.* 2nd ed. New York: Wiley.

Wonnacott, R. and T. Wonnacott. 1985. *Introductory Statistics.* 4th ed. New York: Wiley.

Zeisel, H. 1968. *Say It with Figures.* 5th ed. New York: Harper & Row.

# Appendix F: Answers to Odd-Numbered Exercises

## 1–2 Answers

1. Because the graph does not start at the zero point, the differences are exaggerated.

3. The maker of shoe polish has an obvious interest in the importance of the product and there are many ways this could affect the survey results.

5. 62% of 8% of 1875 is 93.

7. One answer: In recent years a large proportion of women with no prior experience has entered the job market for the first time.

9. Unlisted numbers would be excluded and your sample could be biased and not representative.

11. Alumni with lower salaries would be less inclined to respond, so the reported salaries will tend to be disproportionately higher. Also, those who cheat on their income tax forms might not want to reveal their true incomes. Others might want to retain privacy.

13. 50%. Not necessarily. No matter how good the operating levels are, about 50% will be below average.

15. a, c, d

17. The 6-year figure is equivalent to almost 2 hours each day. Take your daily eating time and project it to a lifetime of about 75 years.

19. The bars are not drawn in their proper proportions.

21. Since there are groups of 20 subjects each, all percentages of success should be multiples of 5. The given percentages cannot be correct.

23. According to the *New York Times*, "It would have to remove all the plaque, remove it again and then remove it for a third time plus some more still."

## 1–3 Answers

| | | | |
|---|---|---|---|
| 1. | Discrete | 3. | Continuous |
| 5. | Continuous | 7. | Discrete |
| 9. | Continuous | 11. | Ordinal |
| 13. | Nominal | 15. | Nominal |
| 17. | Ratio | 19. | Interval |

21. a. Ratio

    b. Interval (or possibly ordinal)

23. Fahrenheit temperatures are at the interval level of measurement so that ratios are not meaningful. Three times 300° F is not the same as 900° F.

## 1–4 Answers

| | | | |
|---|---|---|---|
| 1. | Systematic | 3. | Cluster |
| 5. | Convenience | 7. | Stratified |
| 9. | Cluster | 11. | Random |

13. People often don't know the correct values or they round off. Also, they often give responses that are wrong but create a more favorable impression.

15. Poorer people will be less inclined to spend their money and the sample group will tend to have a disproportionate number of wealthier people. People who feel very strongly about the topic will be more inclined to call and they might not be representative of the population.

17. An advantage is that open questions provide the subject and the interviewer with a much wider variety of responses. A disadvantage is that open questions can be very difficult to analyze. An advantage of closed questions is that they reduce the

chance of misinterpretation of the topic; a disadvantage is that closed questions prevent the inclusion of valid responses that the pollster might not have considered. It is easier to analyze closed questions with formal statistical procedures.

## Chapter 1 Review Exercises

1. a. Continuous    b. Ratio    c. Stratified
3. a. Nominal    b. Ordinal    c. Interval
   d. Ratio    e. Ordinal
5. The figure is very precise, but it is probably not very accurate. The use of such a precise number may incorrectly suggest that it is also accurate.
7. Respondents often tend to round off to a nice even number like 50.

## 2–2 Answers

1. Class width: 8. Class marks: 83.5, 91.5, 99.5, 107.5, 115.5. Class boundaries: 79.5 87.5, 95.5, 103.5, 111.5, 119.5.
3. Class width: 5.0. Class marks: 18.65, 23.65, 28.65, 33.65, 38.65. Class boundaries: 16.15, 21.15, 26.15, 31.15, 36.15, 41.15.

5.
| IQ | Cumulative Frequency |
|---|---|
| Less than  88 | 16 |
| Less than  96 | 53 |
| Less than 104 | 103 |
| Less than 112 | 132 |
| Less than 120 | 146 |

7.
| Weight | Cumulative Frequency |
|---|---|
| Less than 21.2 | 16 |
| Less than 26.2 | 31 |
| Less than 31.2 | 43 |
| Less than 36.2 | 51 |
| Less than 41.2 | 54 |

9.
| IQ | Relative Frequency |
|---|---|
| 80– 87 | 0.110 |
| 88– 95 | 0.253 |
| 96–103 | 0.342 |
| 104–111 | 0.199 |
| 112–119 | 0.096 |

11.
| Weight | Relative Frequency |
|---|---|
| 16.2–21.1 | 0.296 |
| 21.2–26.1 | 0.278 |
| 26.2–31.1 | 0.222 |
| 31.2–36.1 | 0.148 |
| 36.2–41.1 | 0.056 |

13. 80–84, 85–89, 90–94, 95–99, 100–104, 105–109, 110–114, 115–119

15. 16.0–19.9, 20.0–23.9, 24.0–27.9, 28.0–31.9, 32.0–35.9, 36.0–39.9, 40.0–43.9

17. $0–$1999

19. 17.3–19.4

21.
| Age (years) | Frequency |
|---|---|
| 0.0– 1.9 | 3 |
| 2.0– 3.9 | 4 |
| 4.0– 5.9 | 3 |
| 6.0– 7.9 | 4 |
| 8.0– 9.9 | 1 |
| 10.0–11.9 | 3 |
| 12.0–13.9 | 3 |
| 14.0–15.9 | 2 |
| 16.0–17.9 | 4 |
| 18.0–19.9 | 1 |
| 20.0–21.9 | 2 |
| 22.0–23.9 | 6 |
| 24.0–25.9 | 2 |
| 26.0–27.9 | 2 |

23.
| Energy (kWh) | Frequency |
|---|---|
| 700–734 | 4 |
| 735–769 | 4 |
| 770–804 | 6 |
| 805–839 | 10 |
| 840–874 | 4 |
| 875–909 | 3 |

25. The relative frequencies for men are: 0.019, 0.071, 0.118, 0.171, 0.087, 0.273, 0.142, 0.118. The relative frequencies for women are: 0.010, 0.072, 0.173, 0.265, 0.042, 0.279, 0.060, 0.100.

27. The third guideline is clearly violated since the class width varies. While the class limits don't appear to be convenient numbers (fifth guideline), they do correspond to special classes, including preschool (under 5), elementary school (5–13), and high school (14–17).

**2–3 Answers**

1.
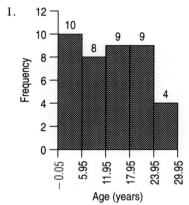

3.
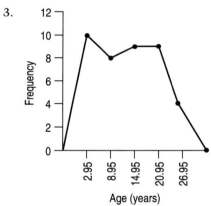

5.
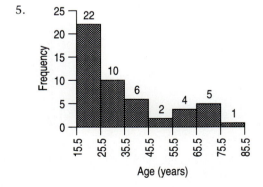

7.
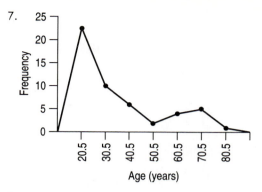

9.
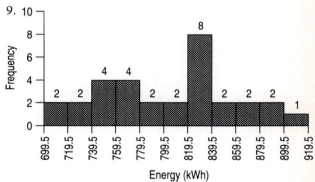

11.
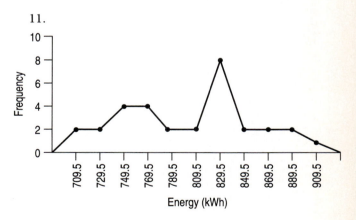

13.

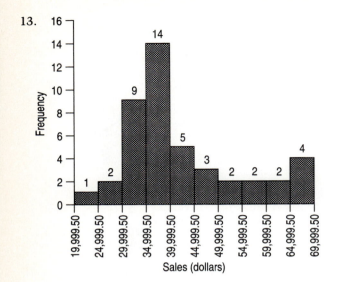

15.

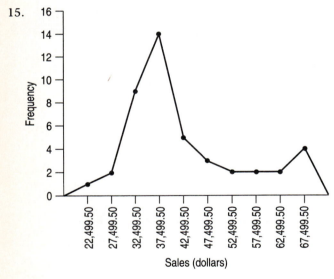

17.    a.

| Score | Frequency |
|-------|-----------|
| 15 | 1 |
| 16 | 3 |
| 17 | 2 |
| 18 | 7 |
| 19 | 6 |
| 20 | 9 |
| 21 | 8 |
| 22 | 3 |
| 23 | 5 |
| 24 | 2 |
| 25 | 1 |

b.

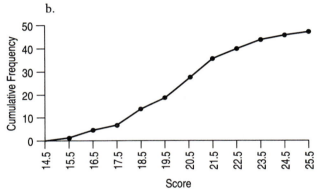

c.  28
d.  8.5%

19.    a.   The histogram for $\pi$ is more uniform.

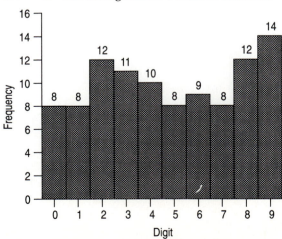

b. $\pi$ is irrational whereas 22/7 is rational. The decimal form of $\pi$ is non-repeating, whereas the decimal form of 22/7 has the digits 142857 repeating.

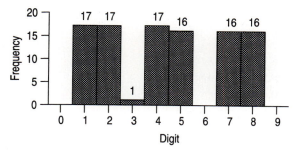

21. a.

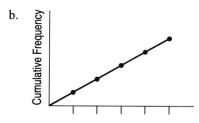

b.

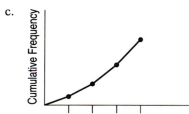

c.

**2–4 Answers**

| | Mean | Median | Mode | Midrange |
|---|---|---|---|---|
| 1. | 3.1 | 3.0 | 3 | 3.5 |
| 3. | 71.5 | 72.0 | 77 | 71.0 |
| 5. | 99.7 | 54.0 | None | 171.0 |

7.    0.187     0.170     0.16, 0.17     0.205

9.    8.33     8.40     8.4     8.40

11.    1.3339     1.2780     None     1.4185

13.    13.41     13.05     23.1     13.70

15.    806.9     818.0     752,774     811.0

17.    a. 19.5, 59.5, 99.5, 139.5, 179.5      b. 95.7

19.    a. 2, 3, 4, 5, 6, 7, 8      b. 3.2

21.    a. 18, 25, 34.5, 44.5, 54.5, 64.5      b. 40.6

23.    a.   1249.5, 4999.5, 9999.5,      b. 25,032.8
          13,749.5, 17,499.5, 22,499.5,
          29,999.5, 42,499.5

25.    a. 95.990, 96.050, no mode, 95.985
      b. 0.990, 1.050, no mode, 0.985
      c. When a constant $k$ is added to or subtracted from every score, the averages change by that same amount $k$.
      d. 1,070,900,000;     876,000,000;    no    mode; 1,270,000,000
      e. 1070.9, 876.0, no mode, 1270.0
      f. The mean, median, mode, and midrange are divided or multiplied by the same constant $k$.

27.    The averages are the same for both sets and do not reveal any differences, but the lists are different in the degree of variation among the scores.

29.    No. For the scores 1, 2, 3, 4, 5, $\log \bar{x} = \log 3 = 0.477$, but the mean of the logarithms of these scores is 0.416.

31.    1.092

33.    Skewed right: 1 and 2. Skewed left: 3 and 4.

35.    $147,606

**2–5 Answers**

| | Range | Variance | Standard Deviation |
|---|---|---|---|
| 1. | 3.0 | 0.8 | 0.9 |
| 3. | 12.0 | 22.7 | 4.8 |
| 5. | 330.0 | 11,871.1 | 109.0 |
| 7. | 0.170 | 0.003 | 0.051 |
| 9. | 6.40 | 3.83 | 1.96 |
| 11. | 0.8110 | 0.0814 | 0.2854 |
| 13. | 27.20 | 68.64 | 8.28 |
| 15. | 194.0 | 2763.1 | 52.6 |

| | Range | Variance | Standard Deviation |
|---|---|---|---|
| 17. | 12.0 | 22.7 | 4.8 (Same results) |
| 19. | 120.0 | 2272.2 | 47.7 |

(Range and standard deviation are multiplied by 10, but the variance is multiplied by 100.)

| | | | |
|---|---|---|---|
| 21. | | 1451.0 | 38.1 |
| 23. | | 1.6 | 1.3 |
| 25. | | 219.6 | 14.8 |
| 27. | | 162,750,559.3 | 12,757.4 |

29.    The statistics students are a more homogeneous group and should therefore have a smaller standard deviation.

31.    The standard deviation will be zero whenever all scores are the same, but it can never be a negative number.

33.    Group C: Range is 19.0; standard deviation is 6.0. Group D: Range is 16.0; standard deviation is 6.5.

35.    a. 68%    b. 95%    c. 44.0, 68.0

37.    $\sigma = 4.52$, $R = 12$ and $4.52 \leq 6$.

39.    Mean is 100.0 and standard deviation is 47.6.

41.    $\bar{x} = 4.56$, $s = 1.67$, $s^2 = 2.78$

43.    a. 95%    b. 1.05

## 2–6 Answers

1.    a. 2.00    b. −2.00    c. 0.50

3.    0.55                    5.    −1.62

7.    −1.50

9.    Test b since $z = 0.60$ is greater than $z = 0.30$.

11.    Test b since $z = 3.00$ is greater than 2.00 or 2.67.

13.    9                        15.    73

17.    $117,500                19.    $184,000

21.    $126,500                23.    $179,000

25.    8                        27.    72

29.    6400                    31.    13,450

33.    7470                    35.    7930

37.    a. $52,500    b. $152,750    c. $96,703
       d. Yes; yes    e. No

39.    Answer varies.

## 2–7 Answers

1.    20, 20, 23, 25, 28

3.    406, 406, 407, 408, 410, 419, 419, 419, 419, 421, 423, 424, 426, 426, 430, 438, 438

5.    6 | 48
      7 | 5679
      8 | 000112334444569
      9 | 0012233447

7.    0.5 | 000008
      0.6 | 55588
      0.7 | 055589
      0.8 | 0
      0.9 | 12

9.    17.2    18.7  19.3  20.5                    26.3

11.    0.02          0.08  0.095  0.11            0.19

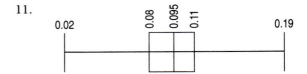

13.    130          159     171.5   181          198

15.    2      14.5          27      34.5          45

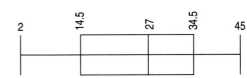

17.    a.

b.
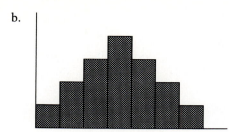

c.
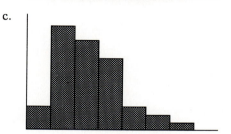

3.

| Number | Frequency |
|--------|-----------|
| 0–9 | 2 |
| 10–19 | 4 |
| 20–29 | 10 |
| 30–39 | 5 |
| 40–49 | 7 |
| 50–59 | 2 |
| 60–69 | 6 |
| 70–79 | 2 |
| 80–89 | 7 |
| 90–99 | 5 |

5.
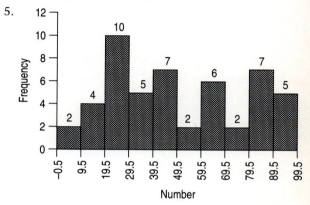

19. a.

| Actors' Ages | Stem | Actresses' Ages |
|---|---|---|
| | 2 | 146667 |
| 998753221 | 3 | 00113344455778 |
| 8876543322100 | 4 | 11129 |
| 6651 | 5 | |
| 210 | 6 | 011 |
| 6 | 7 | 4 |
| | 8 | 0 |

b.
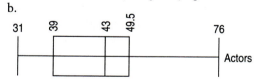

c. Females who win Oscars tend to be younger than males.

7. Mean is 50.1 and standard deviation is 27.8.

9.
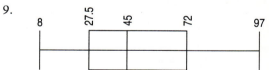

11. 1 | 200 248 300 600 632 800 956
2 | 000 000 025 400
3 | 060

13.
1200   1600 1878 2000   3060

**Chapter 2 Review Exercises**

1. a. 16.0   b. 17.0   c. 8   d. 16.5
   e. 17.0   f. 39.1   g. 6.3

15.

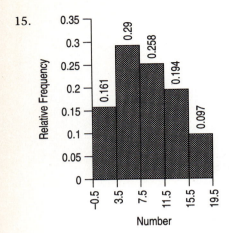

17.

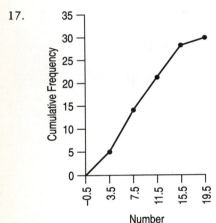

19.  a. 179.3    b. 181.0    c. 181, 183    d. 179.5
     e. 23.0     f. 46.0     g. 6.8

21.  16 | 8
     17 | 045
     18 | 112334
     19 | 1

23.  a. 0.766        b. 0.575        c. 0.50, 0.75
     d. 1.625        e. 2.750        f. 0.555
     g. 0.745

**3–2 Answers**

1.   1.2, 77/75, − 1/2, 5, 1.001, $\sqrt{2}$

3.   0.750                    5.   0.382
7.   1/365                    9.   0.979
11.  0.230                    13.  0.0900
15.  237/600                  17.  0.340
19.  0.0896                   21.  1/20
23.  0.0460
25.  b. 1/4        c. 1/2
27.  b. 1/8        c. 1/8        d. 1/2
29.  a. 4/1461     b. 400/146,097
31.  1

**3–3 Answers**

1.   Mutually exclusive: d, e
3.   4/5            5.   0.520
7.   30/43          9.   0.217
11.  0.824          13.  0.438
15.  0.710          17.  69/758
19.  453/758        21.  0.500
23.  0.490          25.  0.550
27.  0.530          29.  17/60
31.  1/8
33.  a. $P(A \text{ or } B) = 0.9$   b. $P(A \text{ or } B) < 0.9$
     c. $0.4 \leq P(B) \leq 0.8$ and they may or may not be
        mutually exclusive.
35.  $P(A \text{ or } B) = P(A) + P(B) - 2P(A \text{ and } B)$

**3–4 Answers**

1.   Independent: a, b, e
3.   0.000216; no              5.   0.00766
7.   1/4                       9.   0.384
11.  0.0000000206             13.   55/96
15.  0.000000250
17.  a. 0.995   b. 0.995   19.  0.787
21.  0.00000256             23.   0.0000225
25.  0.0896                 27.   0.477
29.  a. 5/8      b. 3/8      c. 5/18     d. 5/8
31.  a. 4/5      b. 1/5      c. 2/5      d. 3/5
     e. 8/25     f. 16           g. Yes; 24, 4, 6

## 3–5 Answers

1. The system works successfully.
3. All answers are "true."
5. 1/10; 9/10      7.   3/5; 2/5
9. 0.0728      11.   0.999957
13. 0.779      15.   0.999999179
17. a. $P(\overline{A} \text{ or } \overline{B}) = 1 - P(A) - P(B) + P(A \text{ and } B)$
    b. $P(\overline{A} \text{ or } \overline{B}) = 1 - P(A \text{ and } B)$
    c. Different
19. Produce the 10¢ resistors, since the true cost of a good resistor is only 20¢, which is less than the 55.6¢ cost of good resistors by the other method.

## 3–6 Answers

1. 5040      3.   4830
5. 720      7.   30
9. 120      11.   1326
13. $n!$      15.   1
17. 6      19.   5040
21. 646,646      23.   $10! = 3,628,800$
25. a. 5040   b. 1/5040
27. 1,000,000,000      29.   43,680
31. 125,000 permutations      33.   18,720
35. a. 2002   b. 0.675
37. 1/13,983,816 You have a much better chance of being struck by lightning.
39. $6.20 \times 10^{23}$      41.   120
43. 2,095,681,645,538 (about 2 trillion)
45. a. 2   b. $(n - 1)!$
47. Calculator: $3.0414093 \times 10^{64}$; approximation: $3.0363452 \times 10^{64}$.

## Chapter 3 Review Exercises

1. 1/4      3.   7/15      5.   0.923
7. 0.694      9.   10,000      11.   0.507
13. 1/18, 595, 558, 800
15. 0.999     17.   1/32      19.   1/120
21. a. 1/13   b. 57/65
23. a. 40,320   b. 20,160   c. 45   d. 3160
25. 1/8     27.   0.0268

## 4–2 Answers

1. Probability distribution
3. No, $\Sigma P(x) \neq 1$
5. Probability distribution
7. Probability distribution
9. Probability distribution
11. Probability distribution
13. a. 0, 1, 2, 3   b. 0.125, 0.375, 0.375, 0.125
    c. 

| x | P(x) |
|---|------|
| 0 | 0.125 |
| 1 | 0.375 |
| 2 | 0.375 |
| 3 | 0.125 |

d.

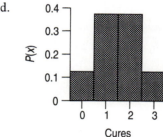

e. 0.125, 0.375, 0.375, 0.125
15. a. 0, 1, 2   b. 1/3, 1/3, 1/3
    c.

| x | P(x) |
|---|------|
| 0 | 1/3 |
| 1 | 1/3 |
| 2 | 1/3 |

d.

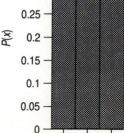

e. 1/3, 1/3, 1/3

17.  a. 0, 1, 2, 3
     b. 1/8, 3/8, 3/8, 1/8

     c.

| x | P(x) |
|---|------|
| 0 | 1/8  |
| 1 | 3/8  |
| 2 | 3/8  |
| 3 | 1/8  |

     d.

Girls

     e. 1/8, 3/8, 3/8, 1/8

19.  a. 0, 1, 2   b. 0.0225, 0.255, 0.7225
     c.

| x | P(x)   |
|---|--------|
| 0 | 0.0225 |
| 1 | 0.2550 |
| 2 | 0.7225 |

     d.

Number of
Shots Made

     e. 0.0225, 0.255, 0.7225

21. Yes
23. 0.179, 0.111, 0.015

**4–3 Answers**

| | Mean | Variance | Standard Deviation |
|---|------|----------|--------------------|
| 1.  | 1.1  | 0.5   | 0.7  |
| 3.  | 2.3  | 0.6   | 0.8  |
| 5.  | 21.3 | 304.7 | 17.5 |
| 7.  | 6.0  | 5.9   | 2.4  |
| 9.  | 0.7  | 0.8   | 0.9  |
| 11. | 2.8  | 6.4   | 2.5  |
| 13. | 0.8  | 0.5   | 0.7  |
| 15. | 2.0  |       | 1.1  |
| 17. | 0.4  | 0.3   | 0.6  |
| 19. | 5.9  | 1.1   | 1.1  |

21. $275     23. − $70
25. $122,500     27. − $100,000
29.  a.  bbbbb bbbbg bbbgb bbbgg bbgbb bbgbg bbggb
         bbggg
         bgbbb bgbbg bgbgb bgbgg bggbb bggbg bgggb
         bgggg
         gbbbb gbbbg gbbgb gbbgg gbgbb gbgbg gbggb
         gbggg
         ggbbb ggbbg ggbgb ggbgg gggbb gggbg ggggb
         ggggg
     b.  1/32, 5/32, 10/32, 10/32, 5/32, 1/32
     c.  2.5          d.  1.25          e.  1.12

31.

$$\mu = \Sigma x \cdot P(x) = \left(1 \cdot \frac{1}{n}\right) + \left(2 \cdot \frac{1}{n}\right) + \ldots + \left(n \cdot \frac{1}{n}\right)$$

$$= \frac{1}{n}(1 + 2 + \ldots + n)$$

$$= \frac{1}{n} \cdot \frac{n(n+1)}{2} = \frac{n+1}{2}.$$

$$\sigma^2 = \Sigma x^2 \cdot P(x) - \mu^2$$

$$= \left(1^2 \cdot \frac{1}{n}\right) + \left(2^2 \cdot \frac{1}{n}\right) + \ldots + \left(n^2 \cdot \frac{1}{n}\right) - \left(\frac{n+1}{2}\right)^2$$

$$= \frac{1}{n}(1^2 + 2^2 + \ldots + n^2) - \left(\frac{n+1}{2}\right)^2$$

$$= \frac{1}{n} \cdot \frac{n(n+1)(2n+1)}{6} - \frac{n^2+2n+1}{4}$$

$$= \frac{2n^2+3n+1}{6} - \frac{n^2+2n+1}{4}$$

$$= \frac{4n^2+6n+2-(3n^2+6n+3)}{12} = \frac{n^2-1}{12}$$

## 4–4 Answers

1. a, c, d, e
3. a. 0.117    b. 0.215    c. 0.478    d. 0+    e. 0+
5. a. 10    b. 1    c. 8    d. 56    e. 1
7. a. 45/512    b. 8/27    c. 10/243
9. 0.205                    11.    0.346
13. 0.989                   15.    0.033
17. 0.200                   19.    0.0743
21. a. 0.105    b. 0.243    c. 0.274
23. 0.000297
25.

| x | P(x) |
|---|---|
| 0 | 0.296 |
| 1 | 0.444 |
| 2 | 0.222 |
| 3 | 0.037 |

27. 0.00274
29. a. 0.225    b. 0.115
31. a. 0.736    b. 0.318
33. 0.0524    35.    0.417

## 4–5 Answers

| | Mean | Variance | Standard Deviation |
|---|---|---|---|
| 1. | 32.0 | 16.0 | 4.0 |
| 3. | 4.8 | 1.9 | 1.4 |
| 5. | 9.0 | 6.8 | 2.6 |
| 7. | 92.4 | 76.4 | 8.7 |
| 9. | 3.2 | 2.6 | 1.6 |
| 11. | 168.7 | 56.2 | 7.5 |
| 13. | 25.0 | 12.5 | 3.5 |
| 15. | 2.0 | 1.0 | 1.0 |
| 17. | 13.5 | | 2.5 |
| 19. | 4.5 | | 2.0 |
| 21. | 8.8 | 4.9 | 2.2 |
| 23. | 94.0 | 49.8 | 7.1 |
| 25. | 4.4 | | 1.7 |
| 27. | 338.0 | | 17.1 |

| | Mean | Variance | Standard Deviation |
|---|---|---|---|
| 29. | 730.0 | | 14.0 |
| 31. | 6.9 | | 2.4 |

33. a. Yes    b. No
35. 516.8, 583.2; no

## Chapter 4 Review Exercises

1. a. Mean is 7.9 and standard deviation is 1.6
   b. 0.238
3. a. 0.60    b. 1.0    c. 1.0
5. No, $\Sigma P(x) \neq 1$
7. a.

| x | P(x) |
|---|---|
| 1 | 0.2 |
| 2 | 0.2 |
| 3 | 0.2 |
| 4 | 0.2 |
| 5 | 0.2 |

   b. 3.0    c. 2.0
9. a. 0.010    b. 0.230    c. 2.0
   d. 1.2    e. 1.1
11. No, $\Sigma P(x) \neq 1$.
13. a. 0.682    b. 1.1    c. 1.0
15. a. 0.196    b. 8.0    c. 4.0

## 5–2 Answers

| | | | | |
|---|---|---|---|---|
| 1. | 0.0987 | | 3. | 0.4332 |
| 5. | 0.3413 | | 7. | 0.4599 |
| 9. | 0.1587 | | 11. | 0.0336 |
| 13. | 0.1587 | | 15. | 0.1814 |
| 17. | 0.8413 | | 19. | 0.9989 |
| 21. | 0.8185 | | 23. | 0.9104 |
| 25. | 0.1359 | | 27. | 0.0954 |
| 29. | 0.1017 | | 31. | 0.3128 |
| 33. | 0.5 | | 35. | 0.3753 |
| 37. | 0.0049 | | 39. | 0.8997 |
| 41. | 0.2843 | | 43. | 0.0843 |

45.    0.3621                    47.    0.9929
49.    a. 68.26%    b. 95.00%    c. 99.74%
       d. 81.85%    e. 4.56%
51.    Trapezoid area = 0.3217; table value = 0.3413

### 5–3 Answers

1.    0.4987                    3.    0.0038
5.    0.1596                    7.    0.5
9.    0.0038; She either has a problem or a very under-
       standing husband.
11.    0.6406                   13.    0.0918
15.    0.1915                   17.    41.68%
19.    0.0329                   21.    0.2063
23.    3.59%                    25.    94.84%
27.    0.2186                   29.    34.04
31.    24.25%; 97.0
33.    a. 200.46    b. 0.29          c. 7.5%
       d. 2.94%     e. Yes
35.    Production is out of control.

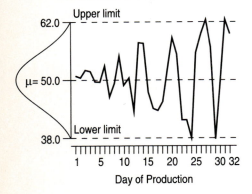

### 5–4 Answers

1.    1.645                     3.    − 1.645
5.    − 1.645, 1.645            7.    − 1.645, 1.28
9.    0.25                      11.    0.67
13.    153.4 cm                 15.    135
17.    $15,595                  19.    12,576 mi
21.    87.4, 112.6             23.    2.432

25.    24.87 mm, 25.40 mm       27.    3.470, 5.756
29.    62.8, 55.2, 44.8, 37.2   31.    29.6 min

### 5–5 Answers

1.    Yes; $\mu$ = 6.25; $\sigma$ = 2.17
3.    No; $\mu$ = 79.8; $\sigma$ = 2.00
5.    a. 0.121    b. 0.1173
7.    a. 0.194    b. 0.1922
9.    0.0287                    11.    0.0019
13.    0.0043                   15.    0.5871
17.    0.0023                   19.    0.1034
21.    0.0099                   23.    0.0869
25.    0.0222                   27.    0.2206
29.    0.6119                   31.    0.0198
33.    a. 0.782    b. 0.782722294    35.    262
       c. 0.7852

### 5–6 Answers

1.    a. 0.1179    b. 0.4641
3.    a. 0.6844    b. 0.9987
5.    a. 0.1293    b. 0.4772
7.    0.4699          9.    0.7734
11.    0.0668         13.    0.1075
15.    0.2766         17.    It is halved.
19.    a. 1.06    b. 0.454    c. 0.628
       d. 0.217   e. 2.12
21.    0.9898          23. 0.8805
25.    a. $\mu$ = 8.0, $\sigma$ = 5.4
       b. 2, 3   2, 6   2, 8   2, 11   2, 18   3, 6   3, 8
           3, 11   3, 18   6, 8   6, 11   6, 18   8, 11
           8, 18   11, 18
       c. 2.5, 4.0, 5.0, 6.5, 10.0, 4.5, 5.5, 7.0, 10.5, 7.0,
           8.5, 12.0, 9.5, 13.0, 14.5
       d. $\mu_{\bar{x}}$ = 8.0, $\sigma_{\bar{x}}$ = 3.4
27.    $\mu$ = 28.44; $\sigma$ = 9.44; 0.1210

### Chapter 5 Review Exercises

1.    0.7292          3.    a. 0.4090    b. 0.0764
5.    a. 0.6293       b. 0.9706         c. 0.9798

7.    0.3557      9.     5.5

11.    a. 0.6568      b. 0.3707        c. 0.0336
      d. 124.7       e. 84.4

13.    a. 0.2704      b. 0.0392        c. 0.8599

15.    0.1398

## 6–2 Answers

1.    a. 1.96   b. 2.33   c. 2.05   d. 2.093   e. 2.977

3.    2.94        5.     20.6        7.     3.8625

9.    $68.8 < \mu < 72.0$     11.    $98.1 < \mu < 99.1$

13.    $40.0 < \mu < 43.6$     15.    $\$25.17 < \mu < \$37.47$

17.    $1.79 < \mu < 3.01$     19.    $20.8 < \mu < 27.6$

21.    110               23.    5968

25.    $20.12 < \mu < 33.56$     27.    $13.9 < \mu < 18.5$

29.    $3.84 < \mu < 4.04$     31.    $5.08 < \mu < 5.22$

33.    208               35.    61

37.    $7.29 < \mu < 9.37$     39.    $3.0 < \mu < 3.4$

41.    a. $424 < \mu < 476$   b. 450   c. 102   d. 92%

43.    b. 52

## 6–3 Answers

1.    a. 0.2       b. 0.8      c. 0.2      d. 0.0351

3.    a. 0.304    b. 0.696    c. 0.304    d. 0.0276

5.    $0.208 < p < 0.292$

7.    $0.553 < p < 0.654$

9.    625

11.    $0.0527 < p < 0.0812$

13.    $0.243 < p < 0.311$

15.    $70.3\% < p < 73.7\%$

17.    553

19.    4145

21.    $0.158 < p < 0.442$

23.    $0.663 < p < 0.677$

25.    $0.828 < p < 0.852$; yes

27.    $15.1\% < p < 21.5\%$

29.    b. 305

31.    89%

33.    $p > 0.818$; 81.8%

## 6–4 Answers

1.    a. 12          b. 97      c. 9.886, 45.559

3.    a. 3.247, 20.483     b. 1335    c. 1.44

5.    $4.0 < \sigma^2 < 11.3$        7.     $58.5 < \sigma^2 < 208.5$

9.    $14.01 < \sigma < 23.88$    11.     $7.3 < \sigma < 12.3$

13.    $6.6 < \sigma < 11.9$       15.     $3.13 < \sigma < 4.54$

17.    $0.064 < \sigma < 0.406$ $(s^2 = 0.0136)$

19.    $1.73 < \sigma < 3.33$ $(s = 2.2814)$

21.    98%              23.     $2.7 < \sigma < 2.9$

## Chapter 6 Review Exercises

1.    a. 83.2      b. $82.2 < \mu < 84.2$

3.    423

5.    404

7.    $1.4 < \sigma < 11.9$

9.    $0.0696 < p < 0.110$

11.    a. 1.645    b. 2.700, 19.023    c. 2.262    d. 0.25

13.    a. 16.81    b. $3.0 < \sigma < 6.3$

15.    $0.215 < p < 0.265$    17.     271

19.    $31.4 < \mu < 37.6$      21.     8687

23.    $71.29 < \mu < 74.04$

## 7–2 Answers

1.    a. $H_0: \mu \leq 30$, $H_1: \mu > 30$.
      b. $H_0: \mu \leq 100$; $H_1: \mu > 100$.
      c. $H_0: \mu \geq 100$; $H_1: \mu < 100$.
      d. $H_0: \mu \geq 12{,}300$; $H_1: \mu < 12{,}300$.
      e. $H_0: \mu = 3271$, $H_1: \mu \neq 3271$.

3.    a. Type I error: Reject the claim that the mean age of professors is 30 years or less when their mean is actually 30 years or less.
Type II error: Fail to reject the claim that the mean age of professors is 30 years or less when that mean is actually greater than 30 years.
      b. Type I error: Reject the claim that the mean IQ of criminals is 100 or less when their mean IQ is actually 100 or less.
Type II error: Fail to reject the claim that the mean IQ of criminals is 100 or less when that mean is actually greater than 100.
      c. Type I error: Reject the claim that the mean IQ of college students is at least 100 when it really is at least 100.

Type II error: Fail to reject the claim that the mean IQ of college students is at least 100 when it really is less than 100.

d. Type I error: Reject the claim that the mean annual household income is at least $12,300 when it really is at least that much.
Type II error: Fail to reject the claim that the mean annual household income is at least $12,300 when it is actually less than that amount.

e. Type I error: Reject the claim that the mean monthly cost equals $3271 when it does equal that amount.
Type II error: Fail to reject the claim that the mean monthly cost equals $3271 when it does not equal that amount.

5. Right-tailed: a, b. Left-tailed: c, d. Two-tailed: e.

7. a. 1645     b. 2.33     c. $\pm 1.96$
d. $\pm 2.575$     e. $-1.645$

9. Test statistic: $z = 1.44$. Critical value: $z = 2.33$. Fail to reject $H_0$: $\mu \le 100$.
There is not sufficient evidence to warrant rejection of the claim that the mean is less than or equal to 100.

11. Test statistic: $z = -4.33$. Critical value: $z = -1.645$. Reject $H_0$: $\mu \ge 20$.
There is sufficient evidence to warrant rejection of the claim that the mean is at least 20.

13. Test statistic: $z = 1.73$. Critical value: $z = \pm 1.645$. Reject $H_0$: $\mu = 500$.
There is sufficient evidence to warrant rejection of the claim that the mean is equal to 500.

15. Test statistic: $z = 1.77$. Critical value: $z = 1.645$. Reject $H_0$: $\mu \le 40$.
There is sufficient evidence to support the claim that the mean is greater than 40.

17. Test statistic: $z = -0.45$. Critical values: $z = \pm 1.96$. Fail to reject $H_0$: $\mu = 14.00$.
There is not sufficient evidence to warrant rejection of the claim that the mean age is equal to 14 years.

19. Test statistic: $z = 2.47$. Critical value: $z = 1.645$. Reject $H_0$: $\mu \le 90,000$.
There is sufficient evidence to support the claim that the mean is more than 90,000 mi.

21. Test statistic: $z = -1.31$. Critical value: $z = -1.645$ (assuming a 0.05 significance level). Fail to reject $H_0$: $\mu \ge 7.5$.
There is not sufficient evidence to support the claim that the mean is less than 7.5 years.

23. Test statistic: $z = 1.60$. Critical values: $z = \pm 1.96$. Fail to reject $H_0$: $\mu = 35.2$.
There is not sufficient evidence to warrant rejection of the claim that baking does not affect hardness. Based on the available sample evidence, it appears that longer baking does not affect hardness of the paint.

25. Test statistic: $z = -8.63$. Critical value: $z = 1.645$. Fail to reject $H_0$: $\mu \le 420$.
There is not sufficient evidence to support the claim of improved reliability. In fact, it appears that reliability actually deteriorated.

27. Test statistic: $z = 0.26$. The critical value depends on the significance level chosen, but the test statistic will not be in the critical region for any reasonable choice. Fail to reject $H_0$: $\mu = 180$.
There is not sufficient evidence to warrant rejection of the claim that the population mean is 180.

29. $\bar{x} = 31.946$, $s = 0.533$.
Test statistic: $z = -0.61$. Critical values: $z = \pm 1.96$. Fail to reject $H_0$: $\mu = 32$.
There is not sufficient evidence to warrant rejection of the brewery claim that the mean is equal to 32 oz.

31. (Note that $\sigma \approx (115 - 99)/(2(0.67)) = 11.9$, where 0.67 corresponds to an area of 0.25 in Table A–2.)
Test statistic: $z = 1.62$. Critical values: $z = \pm 2.17$. Fail to reject $H_0$: $\mu = 105$.
There is not sufficient evidence to warrant rejection of the claim that the population has a mean IQ score equal to 105.

33. 13.25

## 7–3 Answers

1. Reject the null hypothesis.

3. Fail to reject the null hypothesis.

5. P-value: 0.0808. Fail to reject the null hypothesis.

7. P-value: 0.0262. Reject the null hypothesis.

9. P-value: 0.0970. Fail to reject the null hypothesis.

11. P-value: 0.0524. Fail to reject the null hypothesis.

13. Test statistic: $z = -2.24$. P-value: 0.0125.
Reject $H_0$: $\mu \ge 100$.
There is sufficient evidence to warrant rejection of the claim that the mean is greater than or equal to 100.

15. Test statistic: $z = 2.40$. P-value: 0.0164.
Fail to reject $H_0$: $\mu = 75.6$.

There is not sufficient evidence to warrant rejection of the claim that the mean is 75.6.

17. Test statistic: $z = -0.45$. $P$-value: 0.6528. Fail to reject $H_0$: $\mu = 14.00$. There is not sufficient evidence to warrant rejection of the claim that the mean is equal to 14 years.

19. Test statistic: $z = 2.47$. $P$-value: 0.0068. Reject $H_0$: $\mu \leq 90,000$.
There is sufficient evidence to support the claim that the mean is more than 90,000 mi.

21. 11.0

23. 0.1052

## 7–4 Answers

1. a. $\pm 2.056$   b. 1.337   c. $-3.365$
   d. $\pm 4.032$   e. $-2.467$

3. Test statistic: $t = 1.500$. Critical value: $t = 1.860$. Fail to reject $H_0$: $\mu \leq 10$.
There is not sufficient evidence to warrant rejection of the claim that the mean is 10 or less.

5. Test statistic: $t = -2.121$. Critical value: $t = -2.110$. Reject $H_0$: $\mu \geq 98.6$.
There is sufficient evidence to warrant rejection of the claim that the mean is 98.6 or more.

7. Test statistic: $t = 2.014$. Critical values: $t = \pm 2.145$. Fail to reject $H_0$: $\mu = 75$.
There is not sufficient evidence to warrant rejection of the claim that the mean equals 75.

9. Test statistic: $t = -6.241$. Critical values: $t = \pm 2.500$. Reject $H_0$: $\mu = 157.6$.
There is sufficient evidence to warrant rejection of the claim that the mean is equal to 157.6 cm.

11. Test statistic: $t = -4.845$. Critical value: $t = -2.861$. Reject $H_0$: $\mu \geq 154$. The new tires will be purchased.

13. Test statistic: $t = 1.826$. Critical values: $t = \pm 2.462$. Fail to reject $H_0$: $\mu = 20$. The pills appear to be acceptable.

15. Test statistic: $t = 1.518$. Critical value: $t = 2.064$. Fail to reject $H_0$: $\mu \leq 1500$.
There is not sufficient evidence to warrant rejection of the claim that the mean amount charged is greater than $1500.

17. Test statistic: $t = 0.714$. Critical value: $t = 1.711$. Fail to reject $H_0$: $\mu \leq 10.7$.
There is not sufficient evidence to warrant rejection of the claim that the mean is greater than 10.7.

19. Test statistic: $t = 0.472$. Critical value: $t = 1.314$. Fail to reject $H_0$: $\mu \leq 5$.
There is not sufficient evidence to support the claim that the mean is greater than 5 years.

21. Test statistic: $t = -4.021$. Critical values: $t = \pm 2.977$. Reject $H_0$: $\mu = 41.9$.
There is sufficient evidence to conclude that the population mean for third-graders is different than 41.9.

23. Test statistic: $t = -0.100$. Critical value: $t = -2.015$ (assuming a 0.05 significance level).
Fail to reject $H_0$: $\mu \geq 4.00$.
There is not sufficient evidence to warrant rejection of the claim that the mean is less than 4.00 L for 1 minute.

25. $\bar{x} = 78.0500$ and $s = 5.5958$.
Test statistic: $t = 2.438$. Critical value: $t = 1.729$. Reject $H_0$: $\mu \leq 75$.
There is sufficient evidence to support the claim that the class is above average.

27. Test statistic: $t = -1.027$. Critical values: $t = \pm 2.447$ (assuming a 0.05 significance level).
Fail to reject $H_0$: $\mu = 11,000$.
There is not sufficient evidence to warrant rejection of the claim that the mean equals 11,000 kWh.

29. The critical $z$ score from Table A–2 will be less than the corresponding $t$ score from Table A–3 so that the critical region will be larger than it should be, and you are more likely to reject the null hypothesis.

31. Using $z = 1.645$, the table and the approximation both result in $t = 1.833$.

## 7–5 Answers

1. Test statistic: $z = 3.27$. Critical values: $z = \pm 1.96$. Reject $H_0$: $p = 0.3$.
There is sufficient evidence to warrant rejection of the claim that the proportion of defects is equal to 0.3. In fact, it seems to be significantly higher.

3. Test statistic: $z = 3.59$. Critical value: $z = 2.33$. Reject $H_0$: $p \leq 0.7$.
There is sufficient evidence to support the claim that more than 70% of the population supports the ban.

5. Test statistic: $z = 1.48$. Critical values: $z = \pm 1.645$. Fail to reject $H_0$: $p = 0.71$.

There is not sufficient evidence to warrant rejection of the claim that the actual percentage is 71%.

7. Test statistic: $z = 6.91$. Critical value: $z = 2.05$.
Reject $H_0$: $p \leq 1/4$.
There is sufficient evidence to support the claim that more than 1/4 of all white-collar criminals have attended college.

9. Test statistic: $z = -4.37$. Critical value: $z = -2.33$. Reject $H_0$: $p \geq 0.10$.
There is sufficient evidence to support the claim that less than 10% prefer pediatrics.

11. Test statistic: $z = 3.73$. Critical value: $z = 2.33$.
Reject $H_0$: $p \leq 0.784$.
There is sufficient evidence to support the claim that the on-time arrival rate is higher than 78.4%.

13. Test statistic: $z = -1.75$. Critical value: $z = -1.645$ (assuming a 0.05 significance level).
Reject $H_0$: $p \geq 0.07$.
There is sufficient evidence to support the claim that the no-show rate is lower than 7%.

15. Test statistic: $z = -4.40$. Critical value: $z = -2.33$. Reject $H_0$: $p \geq 1/2$.
There is sufficient evidence to support the claim that fewer than 1/2 of San Francisco residential telephones have unlisted numbers.

17. Test statistic: $z = 2.69$. Critical values: $z = \pm 2.33$. Reject $H_0$: $p = 0.2$.
There is sufficient evidence to warrant rejection of the claim that computers are in 20% of all households. In fact, they appear to be in more than 20% of all households.

19. Test statistic: $z = -0.25$. Critical values: $z = \pm 1.645$. Fail to reject $H_0$: $p = 0.98$.
There is not sufficient evidence to warrant rejection of the claim that the recognition rate is equal to 98%.

21. 1.977%

23. 93.55%

## 7–6 Answers

1. a. 8.907, 32.852    b. 10.117    c. 8.643, 42.796
   d. 40.289    e. 4.075

3. Test statistic: $\chi^2 = 50.440$. Critical value: $\chi^2 = 38.885$. Reject $H_0$: $\sigma^2 \leq 100$.
There is sufficient evidence to support the claim that the variance is greater than 100.

5. Test statistic: $\chi^2 = 35.743$. Critical values: $\chi^2 = 5.697, 35.718$. Reject $H_0$: $\sigma = 10.0$.
There is sufficient evidence to warrant rejection of the claim that the standard deviation is equal to 10.0.

7. Test statistic: $\chi^2 = 108.889$. Critical value: $\chi^2 = 101.879$. Reject $H_0$: $\sigma^2 \leq 9.00$.
There is sufficient evidence to warrant rejection of the claim that the variance is equal to or less than 9.00.

9. Test statistic: $\chi^2 = 114.586$. Critical values: $\chi^2 = 57.153, 106.629$. Reject $H_0$: $\sigma = 43.7$.
There is sufficient evidence to warrant rejection of the claim that the standard deviation is equal to 43.7 ft.

11. Test statistic: $\chi^2 = 44.800$. Critical value: $\chi^2 = 51.739$. Reject $H_0$: $\sigma^2 \geq 0.0225$.
There is sufficient evidence to support the claim that the new machine produces less variance.

13. Test statistic: $\chi^2 = 69.135$. Critical value: $\chi^2 = 67.505$. Reject $H_0$: $\sigma \leq 19.7$.
There is sufficient evidence to support the claim that women have a larger standard deviation.

15. Test statistic: $\chi^2 = 35.490$. Critical value: $\chi^2 = 35.479$. Reject $H_0$: $\sigma \leq 50$.
There is sufficient evidence to support the claim that the standard deviation of the hardness indices is greater than 50.0

17. $n = 12$, $\overline{x} = 33.05$, $s = 1.129$.
Test statistic: $\chi^2 = 3.505$. Critical value: $\chi^2 = 3.816$. Reject $H_0$: $\sigma \geq 2.0$.
There is sufficient evidence to support the claim that the standard deviation is less than 2.0 mg.

19. The sample standard deviation is $s = 0.657$. Test statistic: $\chi^2 = 29.311$. Critical values: $\chi^2 = 6.262, 27.488$ (assuming a 0.05 significance level).
Reject $H_0$: $\sigma = 0.470$ kg.
There is sufficient evidence to warrant rejection of the claim that the standard deviation is equal to 0.470 kg.

21. a. Estimated values: 73.772, 129.070
       Table A–4 values: 74.222, 129.561
    b. 116.643, 184.199

23. a. $0.01 < P\text{-value} < 0.025$
    b. $0.005 < P\text{-value} < 0.01$
    c. $P\text{-value} < 0.01$

**Chapter 7 Review Exercises**

1. a. $z = -1.645$    b. $z = 2.33$    c. $t = \pm 3.106$
   d. $\chi^2 = 10.856$    e. $\chi^2 = 16.047, 45.722$

3. a. $H_0: \mu \geq 20.0$
   b. Left-tailed
   c. Rejecting the claim that the mean is at least 20.0 min when it really is at least 20.0 min.
   d. Failing to reject the claim that the mean is at least 20.0 min when it really is less than 20.0 min.
   e. 0.01

5. Test statistic: $z = -1.03$. Critical values: $z = \pm 1.96$. Fail to reject $H_0: p = 0.15$.
   There is not sufficient evidence to warrant rejection of the claim that 15% of U.S. families have incomes below the poverty level.

7. Test statistic: $t = 2.432$. Critical value: $t = 2.492$. Fail to reject $H_0: \mu \leq 25.5$.
   There is not sufficient evidence to warrant rejection of the claim that the mean is less than or equal to 25.5 ft.

9. Test statistic: $z = -2.73$. Critical value: $z = -2.33$. Reject $H_0: \mu \geq 5.00$.
   There is sufficient evidence to support the claim that the mean radiation dosage is below 5.00 milliroentgens.

11. Test statistic: $z = 13.50$. Critical value: $z = 1.96$. Reject $H_0: \mu \leq 55.0$.
    There is sufficient evidence to support the claim that the mean speed is above 55.0 mi/h.

13. Test statistic: $\chi^2 = 77.906$. Critical value: $\chi^2 = 74.397$. (Interpolated critical value is $\chi^2 = 73.274$.) Reject $H_0: \sigma^2 \leq 6410$.
    There is sufficient evidence to support the counselor's claim that the current group has more varied aptitudes.

15. Test statistic: $z = -5.42$. Critical values: $z = \pm 2.575$. Reject $H_0: p = 0.95$.
    There is sufficient evidence to warrant rejection of the claim that 95% recognize Columbus. In fact, the rate appears to be lower than 95%.

17. $\bar{x} = 98.000$, $s = 8.409$, $n = 15$.
    Test statistic: $t = -0.922$. Critical values: $t = \pm 2.977$. Fail to reject $H_0: \mu = 100$.
    There is not sufficient evidence to warrant rejection of the claim that the mean is equal to 100.

19. Test statistic: $z = -1.36$. Critical value: $z = -1.645$. Fail to reject $H_0: \mu \geq 0.700$.
    There is not sufficient evidence to support the claim that the mean reaction time is less than 0.700 s.

**8–2 Answers**

1. Test statistic: $F = 2.0000$. Critical value: $F = 4.0260$. Fail to reject $H_0: \sigma_1^2 = \sigma_2^2$.
   There is not sufficient evidence to warrant rejection of the claim that the variances are equal.

3. Test statistic: $F = 169.0000$. Critical value: $F = 3.8919$. Reject $H_0: \sigma_1^2 = \sigma_2^2$.
   There is sufficient evidence to warrant rejection of the claim that the variances are equal.

5. Test statistic: $F = 4.0000$. Critical value: $F = 3.1789$. Reject $H_0: \sigma_1^2 \leq \sigma_2^2$.
   There is sufficient evidence to warrant rejection of the claim that the variance of Population A exceeds that of Population B.

7. Test statistic: $F = 1.4603$. Critical value: $F = 1.6664$. Fail to reject $H_0: \sigma_1^2 \leq \sigma_2^2$.
   There is not sufficient evidence to support the claim that the variance of Population A exceeds that of Population B.

9. Test statistic: $F = 1.4246$. Critical value: $F = 2.1540$. Fail to reject $H_0: \sigma_1^2 = \sigma_2^2$.
   There is not sufficient evidence to warrant rejection of the claim that the two production methods yield batteries with the same variance.

11. Test statistic: $F = 2.2500$. Critical value: $F = 4.9621$. Fail to reject $H_0: \sigma_1^2 = \sigma_2^2$.
    There is not sufficient evidence to warrant rejection of the claim that the variances are equal.

13. Test statistic: $F = 1.7923$. Critical value: $F = 1.6373$. Reject $H_0: \sigma_1 = \sigma_2$.
    There is sufficient evidence to warrant rejection of the claim that the two sample groups come from populations with the same standard deviation.

15. Test statistic: $F = 1.3786$. Critical value: $F = 1.8363$. Fail to reject $H_0: \sigma_1 = \sigma_2$.
    There is not sufficient evidence to warrant rejection of the claim that the two groups come from populations with the same standard deviation.

17. Test statistic: $F = 1.3478$. Critical value: $F = 2.6171$. Fail to reject $H_0: \sigma_1^2 = \sigma_2^2$.
    There is not sufficient evidence to warrant rejection of the claim that both groups come from populations with the same variance.

19. Test statistic: $F = 1.2478$. Critical value: $F = 3.0502$. Fail to reject $H_0$: $\sigma_1 = \sigma_2$.
There is not sufficient evidence to warrant rejection of the claim that both groups come from populations with the same standard deviation.

21. a. $F_L = 0.2484$, $F_R = 4.0260$
b. $F_L = 0.2315$, $F_R = 5.5234$
c. $F_L = 0.1810$, $F_R = 4.3197$
d. $F_L = 0.3071$, $F_R = 4.7290$
e. $F_L = 0.2115$, $F_R = 3.2560$

23. a. No     b. No     c. No

## 8–3 Answers

1. Test statistic: $z = -1.67$. Critical values: $z = \pm 1.96$. Fail to reject $H_0$: $\mu_1 = \mu_2$.
There is not sufficient evidence to warrant rejection of the claim that the two population means are equal.

3. $F$-test results: Test statistic is $F = 4.1927$. Critical value: $F = 2.8621$. Reject $H_0$: $\sigma_1^2 = \sigma_2^2$.
Test of means: Test statistic is $t = 0.334$. Critical values: $t = \pm 2.132$. Fail to reject $H_0$: $\mu_1 = \mu_2$.
There is not sufficient evidence to warrant rejection of the claim that the two population means are equal.

5. $-10.0 < \mu_1 - \mu_2 < 0.8$

7. $-4.3 < \mu_1 - \mu_2 < 5.9$

9. Test statistic: $t = -9.214$. Critical values: $t = \pm 2.365$. Reject $H_0$: $\mu_1 = \mu_2$.
There is sufficient evidence to warrant rejection of the claim that the before and after responses have the same mean.

11. Test statistic: $z = -1.45$. Critical values: $z = \pm 1.645$. Fail to reject $H_0$: $\mu_1 = \mu_2$.
There is not sufficient evidence to warrant rejection of the claim that the two populations have the same mean.

13. $27 < \mu_1 - \mu_2 < 121$

15. $F$-test results: Test statistic is $F = 3.0785$. Critical value of $F$ is between 2.8442 and 2.7608.
Reject $H_0$: $\sigma_1^2 = \sigma_2^2$. Test of means: Test statistic is $t = 1.900$. Critical values: $t = \pm 2.540$. Fail to reject $H_0$: $\mu_1 = \mu_2$.
There is not sufficient evidence to warrant rejection of the claim that both divisions have the same mean.

17. $F$-test results: Test statistic is $F = 1.3010$. Critical value of $F$ is between 3.5257 and 3.4296.
Fail to reject $H_0$: $\sigma_1^2 = \sigma_2^2$.
Test of means: Test statistic is $t = -4.337$. Critical values: $t = \pm 2.074$. Reject $H_0$: $\mu_1 = \mu_2$.
There is sufficient evidence to warrant rejection of the claim that the mean is the same for each brand.

19. $3.9 < \mu_d < 19.9$

21. $-2.5 < \mu_1 - \mu_2 < -1.1$

23. $F$-test results: Test statistic is $F = 6.9125$. Critical value of $F$ is between 3.0502 and 2.9493.
Reject $H_0$: $\sigma_1^2 = \sigma_2^2$.
Test of means: Test statistic is $t = -1.908$. Critical value: $t = -1.761$. Reject $H_0$: $\mu_1 \geq \mu_2$.
There is sufficient evidence to support the claim that System 2 has a larger mean.

25. $-31.9 < \mu_d < -3.3$

27. $8.4 < \mu_1 - \mu_2 < 15.6$

29. Test statistic: $t = 2.301$. Critical values: $t = \pm 2.262$. Reject $H_0$: $\mu_1 = \mu_2$.
There is sufficient evidence to warrant rejection of the claim that pretraining and posttraining weights have the same mean.

31. $F$-test results: Test statistic is $F = 1.3477$. Critical value of $F$ is 2.6171. Fail to reject $H_0$: $\sigma_1^2 = \sigma_2^2$.
Test of means: Test statistic is $t = -4.951$. Critical values: $t = \pm 1.96$. Reject $H_0$: $\mu_1 = \mu_2$.
There is sufficient evidence to warrant rejection of the claim that the experimental and control groups have the same mean.

33. $df = 22$; $-4.2 < \mu_1 - \mu_2 < 5.8$

35. a. 50/3     b. 2/3     c. 52/3

## 8–4 Answers

1. a. $-2.05$     b. $\pm 1.96$     c. Reject $H_0$: $p_1 = p_2$.
d. 0.0404     e. $-0.244 < p_1 - p_2 < -0.006$

3. Test statistic: $z = -1.92$. Critical values: $z = \pm 1.96$. Fail to reject $H_0$: $p_1 = p_2$.
There is not sufficient evidence to warrant rejection of the claim that there is no difference between the proportions of Democrats and Republicans who feel that the Government should regulate airline prices.

5. $-0.271 < p_1 - p_2 < -0.089$

7. Test statistic: $z = 12.86$. Critical values: $z = \pm 2.575$. Reject $H_0$: $p_1 = p_2$.
There is sufficient evidence to warrant rejection of the claim that vinyl and latex gloves have the same virus leak rates.

9. Test statistic: $z = 4.46$. Critical values: $z = \pm 2.575$. Reject $H_0$: $p_1 = p_2$.
There is sufficient evidence to warrant rejection of the claim that the central city refusal rate is the same as the refusal rate in other areas.

11. $-0.0144 < p_1 - p_2 < 0.0086$

13. $-0.0045 < p_1 - p_2 < 0.0445$

15. Test statistic: $z = -1.33$. Critical values: $z = \pm 1.96$. Fail to reject $H_0$: $p_1 = p_2$.
There is not sufficient evidence to warrant rejection of the claim that Pittsburgh and Chicago have the same proportion of taxis with seat belts at least partially visible.

17. Test statistic: $z = 1.61$. Critical values: $z = \pm 1.96$. Fail to reject $H_0$: $p_1 = p_2$.
There is not sufficient evidence to warrant rejection of the claim that 4-year public and private colleges have the same freshman dropout rate.

19. Test statistic: $z = -6.74$. Critical value: $z = -2.33$. Reject $H_0$: $p_1 \geq p_2$.
There is sufficient evidence to support the claim that the Salk vaccine is effective.

21. $-0.205 < p_1 - p_2 < -0.135$

23. $0.158 < p_1 - p_2 < 0.322$

25. Test statistic: $z = -2.40$. Critical values: $z = \pm 1.96$. Reject $H_0$: $p_1 = p_2$.
There is sufficient evidence to warrant rejection of the claim that the California percentage exceeds the New York percentage by an amount equal to 25%.

27. 2135

## Chapter 8 Review Exercises

1. a. Test statistic: $z = -1.50$. Critical values: $z = \pm 1.645$. Fail to reject $H_0$: $p_1 = p_2$.
There is not sufficient evidence to warrant rejection of the claim that the proportion of speeding convictions is the same for both counties.
b. $-0.0118 < p_1 - p_2 < 0.0005$

3. Test statistic: $t = 1.185$. Critical values: $t = \pm 2.262$. Fail to reject $H_0$: $\mu_1 = \mu_2$.

There is not sufficient evidence to warrant rejection of the claim that both positions have the same mean.

5. Test statistic: $F = 1.0271$. Critical value: $F = 3.8049$. Fail to reject $H_0$: $\sigma_1 = \sigma_2$.
There is not sufficient evidence to warrant rejection of the claim that both groups come from populations with the same standard deviation.

7. a. Test statistic: $F = 27.5625$. Critical value of $F$ is close to 2.1952. Reject $H_0$: $\sigma_1 = \sigma_2$.
There is sufficient evidence to warrant rejection of the claim that the standard deviations are equal.
b. $F$-test results: Test statistic is $F = 27.5625$. Critical value of $F$ is close to 2.1952. Reject $H_0$: $\sigma_1^2 = \sigma_2^2$.
Test of means: Test statistic is $t = -4.843$. Critical values: $t = \pm 2.093$. Reject $H_0$: $\mu_1 = \mu_2$.
There is sufficient evidence to warrant rejection of the claim that the means are equal.
c. $-3.3 < \mu_1 - \mu_2 < -1.3$

9. a. Test statistic: $t = -19.19$. Critical value: $t = -1.645$. Reject $H_0$: $\mu_1 \geq \mu_2$.
There is sufficient evidence to support the claim that theaters are warmer than stores.
b. $-4.3 < \mu_1 - \mu_2 < -3.5$

11. $F$-test results: Test statistic is $F = 1.8906$. Critical value: $F = 5.2852$. Fail to reject $H_0$: $\sigma_1^2 = \sigma_2^2$.
Test of means: Test statistic is $t = -7.511$. Critical values: $t = \pm 2.179$. Reject $H_0$: $\mu_1 = \mu_2$.
There is sufficient evidence to warrant rejection of the claim that there is no difference in the mean noise levels.

13. Test statistic: $z = 3.67$. Critical value: $z = 1.645$. Reject $H_0$: $p_1 \leq p_2$.
There is sufficient evidence to support the claim that the proportion of correct answers by good students is greater.

15. a. Test statistic: $z = -8.06$. Critical values: $z = \pm 2.24$. Reject $H_0$: $\mu_1 = \mu_2$.
There is sufficient evidence to warrant rejection of the claim that there is no difference between the 2 population means. There does appear to be a difference.
b. $-10.7 < \mu_1 - \mu_2 < -6.1$

## 9–2 Answers

1. a. Negative correlation   b. Positive correlation
   c. Positive correlation   d. Negative correlation
   e. No correlation

3.  a. Significant positive linear correlation
    b. Significant negative linear correlation
    c. No significant linear correlation
    d. No significant linear correlation
    e. No significant linear correlation

5.  b. 4    c. 7    d. 15    e. 49    f. 20    g. $-0.191$

7.  b. 5    c. 9    d. 31    e. 81    f. 47    g. 0.917

9.  b. $-0.069$    c. 0.707
    d. No significant linear correlation

11. b. 0.967    c. 0.811
    d. Significant positive linear correlation

13. b. 0.885    c. 0.707
    d. Significant positive linear correlation

15. b. 0.845    c. 0.707
    d. Significant positive linear correlation

17. b. 0.883    c. 0.707
    d. Significant positive linear correlation

19. b. 0.617    c. 0.632
    d. No significant linear correlation

21. b. $-0.926$    c. 0.707
    d. Significant negative linear correlation

23. b. 0.506    c. 0.576
    d. No significant linear correlation

25. The conclusion implies that there is a significant negative correlation, while the correlation is actually close to zero.

27. Although there is no *linear* correlation, the two variables may be related in some other nonlinear way.

29. a. 0.972    b. 0.905    c. 0.999    d. 0.992
    e. $-0.984$    (Part c results in the largest value of r.)

31. In attempting to calculate r, we get a denominator of zero so a real value of r does not exist. However, it should be obvious that the value of x is not at all related to the value of y.

33. r changes from $-0.926$ to $-0.631$. The effect of an extreme value can be minimal or severe, depending on the other data.

35. With $r = 0.963$ and $n = 11$, it is reasonable to conclude that there is a significant positive linear correlation. This section of the parabola can be approximated by a straight line.

## 9–3 Answers

1.  $y' = -0.36x + 3.64$
3.  $y' = 0.62x + 3.08$
5.  $y' = 2.09x + 2.22$
7.  $y' = 0.407x + 1.51$
9.  $y' = -0.000182x + 0.214$
11. $y' = 1.49x - 14.1$
13. $y' = 0.85x + 4.05$
15. $y' = 30.1x - 8.22$
17. $y' = 49.9x + 58.3$
19. $y' = 0.0164x + 0.601$
21. $y' = -0.438x - 6.50$
23. $y' = 3.52x + 783$
25. a. 110.0    b. 25.0    c. 110.0
    d. 25.0    e. 110.0
27. a. 6    b. 17    c. $-13$
    d. 6    e. 6
29. $y' = 0.000351x - 182$; $y' = 0.351x - 182$
    The slope is multiplied by 1000 and the y-intercept doesn't change. If each y entry is divided by 1000; the slope and y-intercept are both divided by 1000.
31. The sum of the squares using the regression line $(y' = -2.00x + 7.25)$ is 0.75. The sum of the squares using the line $y' = -x + 6$ is 5.0.
33. Note that $s_x$ and $s_y$ are never negative.
35. With $M = 0$, the regression line is horizontal and different values of x result in the same y value, so there is no correlation between x and y.

## 9–4 Answers

1.  0.111; 11.1%    3.    0.640; 64.0%
5.  a. 154.8    b. 0    c. 154.8    d. 1    e. 0
7.  a. 140.4973    b. 0.7027    c. 141.2000
    d. 0.9950    e. 0.4840
9.  a. 14    b. Since the standard error of estimate is zero, $E = 0$ and there is no "interval" estimate.
11. a. 18.9    b. $17.1 < y < 20.8$
13. $\$83,000 < $ selling price $ < \$291,000$
15. $\$123,000 < $ selling price $ < \$298,000$

17. $1.735 < M < 2.451; 0.416 < B < 2.700$

19. a. $(n - 2)s_e^2$    b. $\dfrac{r^2 \cdot (\text{unexplained variation})}{1 - r^2}$

    c. $-0.949$

## 9–5 Answers

1. 85      3.    75

5. $y' = -128 + 1.03x_1 + 12.4x_2 + 2.30x_3 - 1.06x_4$

7. 255

9. $y' = 4.73 + 1.07x_1 + 0.869x_2$ and $R^2 = 0.230$; 14.7

11. $y' = 3.72 + 1.27x_2 + 0.297x_3$ and $R^2 = 0.164$; 15.4

13. a. Taxes $= -0.399 + 0.0172(\text{SP}) - 0.0143(\text{LA})$
    b. 0.787

15. a. Taxes $= 2.78 + 0.320(\text{LA}) - 0.939(\text{Acreage}) - 0.756(\text{Rooms})$
    b. 0.619

17. $y' = 2.17 + 2.44x + 0.464x^2$
Since $R^2 = 1$, the parabola fits perfectly.

19. With no variable in $x_2$ the equations become
$$\Sigma_y = bn + m\Sigma x_1$$
$$\Sigma x_1 y = b\Sigma x_1 + m\Sigma x_1^2$$

## Chapter 9 Review Exercises

1. a. Correlation    b. Regression    c. Regression
    d. Correlation    e. Correlation

3. a. 0.444      b. 0.576
    c. No significant linear correlation
    d. $y' = 0.330x + 44.6$

5. a. 0.565      b. 0.514
    c. Significant positive linear correlation
    d. $y' = 0.30x + 21.6$

7. a. 0.926      b. 0.811
    c. Significant positive linear correlation
    d. $y' = 1.14x + 9.52$

9. a. 0.964      b. 0.666
    c. Significant positive linear correlation
    d. $y' = 0.000187x + 10.1$

11. a. 0.951      b. 0.666
    c. Significant positive linear correlation
    d. $y' = 0.0149x + 20.2$

13. a. $y' = 22.6 + 0.0154x_1 - 0.00066x_2$
    b. 0.906    c. Very well

15. a. $y' = -467 + 0.0144x_1 - 0.0139x_2$
    b. 0.986    c. Very well

17. a. 4.0720    b. 10.6780    c. 14.7500
    d. 0.2761    e. 2.3106

19. a. 14.75    b. 0    c. 14.75    d. 1    e. 0

## 10–2 Answers

1. The expected frequencies are 20, 20, 20, 20, 20.
    a. 5.900      b. 13.277
    c. Fail to reject the claim that absences occur on the 5 days with equal frequency.

3. Test statistic: $\chi^2 = 4.200$. Critical value: $\chi^2 = 16.919$.
Fail to reject the claim that the digits are uniformly distributed.

5. Test statistic: $\chi^2 = 156.500$. Critical value: $\chi^2 = 21.666$.
Reject the claim that the last digits occur with the same frequency.

7. Test statistic: $\chi^2 = 4.323$. Critical value: $\chi^2 = 7.815$.
Fail to reject the claim that the customer preference rates are as stated.

9. Test statistic: $\chi^2 = 0.264$. Critical value: $\chi^2 = 5.991$.
Fail to reject the claim that the correct identifications are equally distributed among the 3 types of balls.

11. Test statistic: $\chi^2 = 1.954$. Critical value: $\chi^2 = 12.592$.
Fail to reject the claim that the given percentages are correct.

13. Test statistic: $\chi^2 = 22.600$. Critical value: $\chi^2 = 19.675$.
Reject the claim that months were selected with equal frequencies.

15. Test statistic: $\chi^2 = 9.013$. Critical value: $\chi^2 = 9.488$.
Fail to reject the claim that the given percentages are correct.

17. Test statistic: $\chi^2 = 12.807$. Critical value: $\chi^2 = 14.067$ (assuming a 0.05 significance level).
Fail to reject the claim that fatal accidents occur at the different times of day with the same rate.

19. 180; reject the claim of equal frequencies.

21.    It's greater than 0.10 and less than 0.90.

23.    a.  Critical value: $\chi^2 = 3.841$.

$$\chi^2 = \frac{\left(f_1 - \frac{f_1 + f_2}{2}\right)^2}{\frac{f_1 + f_2}{2}} + \frac{\left(f_2 - \frac{f_1 + f_2}{2}\right)^2}{\frac{f_1 + f_2}{2}}$$

$$= \frac{(f_1 - f_2)^2}{f_1 + f_2}$$

b.  Critical value: $z^2 = 1.96^2 = 3.842$.

$$z^2 = \frac{\left(\frac{f_1}{f_1 + f_2} - 0.5\right)^2}{\frac{1/4}{f_1 + f_2}} = \frac{(f_1 - f_2)^2}{f_1 + f_2}$$

25.    Combining time slots 1 and 2, 5 and 6, 7 and 8, and 9 and 10, we get a test statistic of $\chi^2 = 4.118$ and a critical value of $\chi^2 = 11.071$ (assuming that $\alpha = 0.05$). Fail to reject the claim that observed and expected frequencies are compatible.

## 10–3 Answers

1.    Test statistic: $\chi^2 = 1.174$. Critical value: $\chi^2 = 3.841$.
      Fail to reject the claim that the treatment is independent of the reaction.

3.    Test statistic: $\chi^2 = 12.321$. Critical value: $\chi^2 = 3.841$.
      Reject the claim that success and group are independent.

5.    Test statistic: $\chi^2 = 0.202$. Critical value: $\chi^2 = 5.991$.
      Fail to reject the claim that payer and group are independent.

7.    Test statistic: $\chi^2 = 1.505$. Critical value: $\chi^2 = 3.841$.
      Fail to reject the claim that the accident rate is independent of the use of cellular phones.

9.    Test statistic: $\chi^2 = 1.358$. Critical value: $\chi^2 = 7.815$.
      Fail to reject the claim that the amount of smoking is independent of seat belt use.

11.   Test statistic: $\chi^2 = 3.271$. Critical value: $\chi^2 = 5.991$.
      Fail to reject the claim that the cause of death is independent of the determining source.

13.   Test statistic: $\chi^2 = 0.615$. Critical value: $\chi^2 = 7.815$.
      Fail to reject the claim that smoking is independent of age group.

15.   Test statistic: $\chi^2 = 13.143$. Critical value: $\chi^2 = 12.017$.
      Reject the claim that the time of day of fatal accidents is the same in New York City as in all other New York state locations.

17.   It's greater than 0.10 and less than 0.90.

19.   The test statistic is multiplied by the same constant, but the critical value doesn't change.

## Chapter 10 Review Exercises

1.    Test statistic: $\chi^2 = 31.368$. Critical value: $\chi^2 = 5.991$.
      Reject the claim that those who experience drowsiness are equally distributed among the 3 categories.

3.    Test statistic: $\chi^2 = 28.309$. Critical value: $\chi^2 = 6.635$.
      Reject the claim that the sentence is independent of the plea.

5.    Test statistic: $\chi^2 = 6.780$. Critical value: $\chi^2 = 12.592$.
      Fail to reject the claim that gunfire death rates are the same for the different days of the week.

7.    Test statistic: $\chi^2 = 7.848$. Critical value: $\chi^2 = 5.991$.
      Reject the claim that the proportion is the same in all 3 groups.

## 11–2 Answers

1.    Test statistic: $F = 1.6373$. Critical value: $F = 3.2317$.
      Fail to reject the claim of equal means.

3.    Test statistic: $F = 1.5430$. Critical value: $F = 2.6060$.
      Fail to reject the claim of equal means.

5.    Test statistic: $F = 1.6500$. Critical value: $F = 3.0556$.
      Fail to reject the claim of equal means.

7.    Test statistic: $F = 8.2086$. Critical value: $F = 3.8853$.
      Reject the claim of equal means.

9. Test statistic: $F = 0.1587$. Critical value: $F = 3.3541$.
Fail to reject the claim of equal means.

11. Test statistic: $F = 5.0793$. Critical value: $F = 3.3541$.
Reject the claim of equal means.

13. a. With test statistic $z = -8.41$ and critical values $z = \pm 1.96$, reject the claim that $\mu_1 = \mu_2$.
b. With test statistic $z = 8.58$ and critical values $z = \pm 1.96$, reject the claim that $\mu_2 = \mu_3$.
c. With test statistic $z = 1.21$ and critical values $z = \pm 1.96$, fail to reject the claim that $\mu_1 = \mu_3$.
d. With test statistic $F = 48.2442$ and critical value $F = 3.0000$ (approx.), reject the claim of equal means.

15. a. The test statistic does not change.
b. The test statistic is multiplied by the square of the constant.

## 11–3 Answers

1. Test statistic: $F = 15.8140$. Critical value: $F = 3.6823$.
Reject the claim of equal means.

3. Test statistic: $F = 9.8683$. Critical value: $F = 3.6823$, assuming that the significance level is $\alpha = 0.05$.
Reject the claim of equal means.

5. Test statistic: $F = 3.7239$. Critical value: $F = 6.2262$.
Fail to reject the claim of equal means.

7. Test statistic: $F = 11.6744$. Critical value: $F = 4.6755$.
Reject the claim of equal means.

9. $SS(\text{treatments}) = 2.17$, $df(\text{treatments}) = 2$, $df(\text{error}) = 16$, $df(\text{total}) = 18$, $MS(\text{treatments}) = 1.085$, $MS(\text{error}) = 7.036$, $F = 0.1542$.

11. The $SS$ and $MS$ values all increase dramatically. The test statistic changes from 11.6744 to 1.3502. In general, the changes caused by the single outlier are dramatic.

13. a. Since all samples have the same size $n$, the factor $n_i$ can be replaced by the constant $n$ that can be put before the summation symbol $\Sigma$. The factor $n$ is then multiplied by an expression that represents $s_{\bar{x}}^2$.
b. With all $n_i = n$, the numerator becomes $\Sigma(n - 1)s_i^2 = (n - 1)\Sigma s_i^2$.

With all $n_i = n$, the denominator becomes $\Sigma(n - 1) = k(n - 1)$.
The numerator and denominator both have factors of $(n - 1)$ that can be divided out to produce the desired result.

## 11–4 Answers

1. a. 0.17    b. 1.44    c. 12.50    d. 126.17

3. Test statistic: $F = 87.6181$. Critical value: $F = 3.8853$, assuming that the significance level is $\alpha = 0.05$.
It appears that engine size does have an effect on mileage.

5. Test statistic: $F = 1.0000$. Critical value: $F = 18.513$.
Do not reject the hypothesis that transmission type does not have an effect on mileage.

7. Test statistic: $F = 17.5851$. Critical value: $F = 4.7571$. It appears that the typist does have an effect on the time.

9. Transposing the table does not change the results or conclusions.

11. The $SS$ and $MS$ values are multiplied by 100, but the values of the test statistics do not change.

## Chapter 11 Review Exercises

1. Test statistic: $F = 10.8266$. Critical value: $F = 3.0718$.
Reject the claim of equal means.

3. Test statistic: $F = 4.3124$. Critical value: $F = 4.8740$.
Fail to reject the claim of equal means.

5. Test statistic: $F = 0.5920$. Critical value: $F = 2.9277$.
Fail to reject the hypothesis that interaction between employee and machine does not affect the number of items produced.

7. Test statistic: $F = 37.7701$. Critical value: $F = 3.5546$.
The choice of employee does appear to affect the number of items produced.

## 12–2 Answers

1. The test statistic $x = 0$ is less than or equal to the critical value of 0. Reject the claim that the before and after responses are the same.

3. The test statistic $x = 4$ is not less than or equal to the critical value of 2. Fail to reject the claim that the $x$ and $y$ samples come from the same population.

5. The test statistic $x = 1$ is less than or equal to the critical value of 1. Reject the null hypothesis of no effect. The pill appears to lower blood pressure.

7. The test statistic $x = 0$ is less than or equal to the critical value of 1. Reject the claim of no difference.

9. The test statistic $x = 4$ is not less than or equal to the critical value of 0. Fail to reject the null hypothesis of no difference. There appears to be no difference between the ages of husbands and wives.

11. The test statistic $x = 7$ is not less than or equal to the critical value of 5. Fail to reject the null hypothesis that both parties are preferred equally. There is not a significant difference.

13. The statistic $x = 32$ is converted to $z = -1.15$. The critical value is $z = -1.645$. Since the test statistic is not less than or equal to the critical value, fail to reject the null hypothesis that the median life is at least 40 hours.

15. The statistic $x = 18$ is converted to the test statistic $z = -1.84$. The critical value is $z = -1.645$. Since the test statistic is less than or equal to the critical value, reject the null hypothesis that production was unchanged or lowered by the new fertilizer. It appears that production was increased.

17. The statistic $x = 6$ is converted to the test statistic $z = -2.83$. The critical values are $z = \pm 2.575$. Reject the null hypothesis that the drink had no effect.

19. The statistic $x = 22$ is converted to the test statistic $z = -0.71$. The critical value is $z = -1.28$. Fail to reject the null hypothesis that at most half of the voters favor the bill. There is not sufficient evidence to support the claim that the majority favor the bill.

21. With $k = 3$ we get $6 < M < 17$.

23. 78

## 12–3 Answers

1. The given entries 5, 8, 10, 12, 15 correspond to ranks 1, 2, 3, 4, 5.

3. The given entries 50, 100, 150, 200, 400, 600 correspond to ranks 1, 2, 3, 4, 5, 6.

5. The given entries 6, 8, 8, 9, 12, 20 correspond to ranks 1, 2.5, 2.5, 4, 5, 6.

7. The given entries 12, 13, 13, 13, 14, 15, 16, 16, 18 correspond to ranks 1, 3, 3, 3, 5, 6, 7.5, 7.5, 9.

9. a. 3, $-7$, $-2$, $-4$, $-1$, 0 (discard), $-10$, $-15$
   b. 3, 5, 2, 4, 1, 6, 7
   c. $+3$, $-5$, $-2$, $-4$, $-1$, $-6$, $-7$
   d. $T = 3$ (the smaller of 3 and 25)

11. a. 1, $-1$, 2, $-3$, $-6$, 5, $-10$, $-20$, 0 (discard)
    b. 1.5, 1.5, 3, 4, 6, 5, 7, 8
    c. $+1.5$, $-1.5$, $+3$, $-4$, $-6$, $+5$, $-7$, $-8$
    d. $T = 9.5$ (the smaller of 9.5 and 26.5)

13. a. 25; reject $H_0$    b. 25; reject $H_0$

15. a. 35; reject $H_0$    b. 8; reject $H_0$

17. $T = 6.5$, $n = 9$. The test statistic $T = 6.5$ is greater than the critical value of 6 so fail to reject the null hypothesis of equal results.

19. $T = 15$, $n = 10$. The test statistic $T = 15$ is greater than the critical value of 8 so fail to reject the null hypothesis that both positions have the same distribution.

21. $T = 4.5$, $n = 8$. The test statistic $T = 4.5$ is greater than the critical value of 4 so fail to reject the null hypothesis of equal perceptions of depth.

23. $T = 11.5$, $n = 14$. The test statistic $T = 11.5$ is less than or equal to the critical value of 21 so reject the null hypothesis of equal anxiety levels.

25. $T = 3$, $n = 10$. The test statistic $T = 3$ is less than or equal to the critical value of 8 so reject the null hypothesis of equal test results.

27. $T = 0$, $n = 8$. The test statistic $T = 0$ is less than or equal to the critical value of 4 so reject the null hypothesis that the before and after responses have the same distribution.

29. a. $T = 51.5$. $n = 36$. Test statistic $z = -4.42$ is less than or equal to the critical value of $z = -2.575$, so reject the null hypothesis that both systems require the same times. (It appears that the scanner system is faster.)
    b. $T = 51.5$, $n = 36$. Test statistic $z = -4.42$ and the critical value is $z = -2.33$, so reject the null hypothesis that the distribution of scanner times is the same as (or to the right of) the distribution of times for the manual system. There is sufficient evidence to support the claim that the distribution of scanner times is shifted to the left of the distribution of times

for the manual system. It appears that the scanner times are lower.

31. 1954

## 12–4 Answers

1. $\mu_R = 174.0$, $\sigma_R = 21.54$, $R = 175$, $z = 0.05$.
   Test statistic: $z = 0.05$. Critical values: $z = \pm 1.96$.
   Fail to reject the null hypothesis that the states have the same salaries.

3. $\mu_R = 210$, $\sigma_R = 20.49$, $R = 160.5$, $z = -2.42$.
   Test statistic: $z = -2.42$. Critial values: $z = \pm 1.96$.
   Reject the null hypothesis of no difference. The groups appear to have different reaction times.

5. $\mu_R = 203$, $\sigma_R = 21.76$, $R = 184$, $z = -0.87$.
   Test statistic: $z = -0.87$. Critical values: $z = \pm 1.96$.
   Fail to reject the null hypothesis of equal times.

7. $\mu_R = 201.5$, $\sigma_R = 23.89$, $R = 163$, $z = -1.61$.
   Test statistic: $z = -1.61$. Critical values: $z = \pm 2.575$.
   Fail to reject the null hypothesis of equality.

9. $\mu_R = 232.5$, $\sigma_R = 24.11$, $R = 307$, $z = 3.09$.
   Test statistic: $z = 3.09$. Critical values: $z = \pm 1.96$.
   Reject the null hypothesis of no difference.

11. $\mu_R = 162$, $\sigma_R = 19.44$, $R = 86$, $z = -3.91$.
    Test statistic: $z = -3.91$. Critical values: $z = \pm 1.96$.
    Reject the null hypothesis of no difference between the 2 treatments.

13. $\mu_R = 370$, $\sigma_R = 31.411$, $R = 254$, $z = -3.69$.
    Test statistic: $z = -3.69$. Critical values: $z = \pm 1.96$.
    Reject the null hypothesis of equal scores.

15. $\mu_R = 224$, $\sigma_R = 20.265$, $R = 254$, $z = 1.48$.
    Test statistic: $z = 1.48$. Critical values: $z = \pm 1.96$.
    Fail to reject the null hypothesis that both populations are the same.

17. a. $\mu_R = 232.5$, $\sigma_R = 24.11$, $R = 120$, $z = -4.67$.
    Test statistic: $z = -4.67$. Critical values: $z = \pm 1.96$.
    Reject the null hypothesis that both groups come from the same population.
    b. $\mu_R = 232.5$, $\sigma_R = 24.11$, $R = 225$, $z = -0.31$.
    Test statistic: $z = -0.31$. Critical values: $z = \pm 1.96$.

Fail to reject the null hypothesis that both groups come from the same population.

19. a.
| | |
|---|---|
| ABAB | 4 |
| ABBA | 5 |
| BBAA | 7 |
| BAAB | 5 |
| BABA | 6 |

b.
| $R$ | $p$ |
|---|---|
| 3 | 1/6 |
| 4 | 1/6 |
| 5 | 2/6 |
| 6 | 1/6 |
| 7 | 1/6 |

c. No, the most extreme rank distribution has a probability of at least 1/6 and we can never get into a critical region with a probability of 0.10 or less.

## 12–5 Answers

1. 0

3. Test statistic: $H = 2.435$. Critical value: $\chi^2 = 5.991$.
   Fail to reject the null hypothesis that the samples come from identical populations.

5. Test statistic: $H = 9.420$. Critical value: $\chi^2 = 5.991$.
   Reject the null hypothesis that the fuel consumption values are the same for the 3 transmission types.

7. Test statistic: $H = 0.775$. Critical value: $\chi^2 = 9.210$.
   Fail to reject the null hypothesis of equally effective methods.

9. Test statistic: $H = 1.589$. Critical value: $\chi^2 = 5.991$.
   Fail to reject the null hypothesis that the living areas are the same in all 3 zones.

11. Test statistic: $H = 1.732$. Critical value: $\chi^2 = 5.991$.
    Fail to reject the null hypothesis that the taxes are the same in all 3 zones.

13. a. $H = \dfrac{1}{1176}(R_1^2 + R_2^2 + \ldots + R_8^2) - 147$
    b. No change
    c. No change

15. Dividing the unrounded value of $H$ by the correction factor results in $1.1375604 \div 0.99923077$, which is rounded off to 1.138, the same value obtained without the correction factor.

## 12–6 Answers

1. a. $\pm 0.450$   b. $\pm 0.280$   c. $\pm 0.373$
   d. $\pm 0.475$   e. $\pm 0.342$

3. $\dfrac{x \mid 1\ 2\ 4\ 3\ 5}{y \mid 5\ 4\ 3\ 2\ 1}$

5. $d = 3, 0, 0, 3;\ \Sigma d^2 = 18;\ n = 4;\ r_s = -0.8$

7. $d = 3, 3, 1, 2, 2, 5;\ \Sigma d^2 = 52;\ n = 6;\ r_s = -0.486$

9. $r_s = 0.261$. Critical values: $r_s = \pm 0.648$.
   No significant correlation.

11. $r_s = 0.571$. Critical values: $r_s = \pm 0.738$.
    No significant correlation.

13. $r_s = 0.715$. Critical values: $r_s = \pm 0.591$.
    Significant positive correlation.

15. $r_s = 0.006$. Critical values: $r_s = \pm 0.738$.
    No significant correlation.

17. $r_s = 0.983$. Critical values: $r_s = \pm 0.683$.
    Significant positive correlation.

19. $r_s = 1.000$. Critical values: $r_s = \pm 0.738$.
    Significant positive correlation.

21. $r_s = 0.458$. Critical values: $r_s = \pm 0.738$.
    No significant correlation.

23. $r_s = -0.905$. Critical values: $r_s = \pm 0.738$.
    Significant negative correlation.

25. a. 123, 132, 213, 231, 312, 321
    b. 1, 0.5, 0.5, $-0.5$, $-0.5$, $-1$
    c. 1/6 or 0.167

27. a. The rank correlation coefficient, because the trend is not linear.
    b. Sign changes.
    c. Both will be the same.
    d. Test statistic: $r_s = 0.855$. Critical value: $r_s = 0.564$.
       Reject $H_0$: $\rho_s \le 0$. There is sufficient evidence to support the claim that there is a positive correlation.

## 12–7 Answers

1. $n_1 = 5, n_2 = 3, G = 3$, critical values: 1, 8.

3. $n_1 = 4, n_2 = 6, G = 2$, critical values: 2, 9.

5. $n_1 = 7, n_2 = 2, G = 2$, critical values: 1, 6.

7. $n_1 = 10, n_2 = 10, G = 9$, critical values: 6, 16.

9. $\bar{x} = 11.0$; BBBBBABAAA; $n_1 = 6, n_2 = 4, G = 4$. The critical values are 2, 9. Fail to reject randomness.

11. $\bar{x} = 9.2$; $n_1 = 6, n_2 = 8, G = 2$. The critical values are 3, 12. Reject randomness.

13. Median is 9.5, $n_1 = 5, n_2 = 5, G = 2$. Critical values are 2, 10. Reject randomness.

15. Median is 9 so delete the three 9's to get BBBBBBBAAAAAAA. $n_1 = 7, n_2 = 7, G = 2$. Critical values are 3, 13. Reject randomness.

17. $n_1 = 8, n_2 = 8, G = 8$. Critical values are 4, 14. Fail to reject randomness.

19. $n_1 = 30, n_2 = 15, G = 14, \mu_G = 21, \sigma_G = 2.94$. Test statistic: $z = -2.38$. Critical values: $z = \pm 1.96$. Reject randomness.

21. Median is 834. $n_1 = 8, n_2 = 8, G = 2$. Critical values are 4, 14. Reject randomness. (There appears to be a downward trend.)

23. $n_1 = 17, n_2 = 28, G = 12, \mu_G = 22.16, \sigma_G = 3.11$. Test statistic: $z = -3.27$. Critical values: $z = \pm 1.96$. Reject randomness.

25. $n_1 = 50, n_2 = 37, G = 48, \mu_G = 43.53, \sigma_G = 4.532$. Test statitistic: $z = 0.99$. Critical values: $z = \pm 1.96$. Fail to reject randomness.

27. $n_1 = 49, n_2 = 51, G = 43, \mu_G = 50.98, \sigma_G = 4.9727$. Test statistic: $z = -1.60$. Critical values: $z = \pm 1.96$. Fail to reject randomness.

29. Minimum is 2, maximum is 4. Critical values of 1 and 6 can never be realized so that the null hypothesis of randomness can never be rejected.

31. The 84 sequences yield two runs of 2, seven runs of 3, twenty runs of 4, twenty-five runs of 5, twenty runs of 6, and ten runs of 7 so that $P(2 \text{ runs}) = 2/84$, $P(3 \text{ runs}) = 7/84$, $P(4 \text{ runs}) = 20/84$, $P(5 \text{ runs}) = 25/84$, $P(6 \text{ runs}) = 20/84$, and $P(7 \text{ runs}) = 10/84$. From this we conclude that the $G$ values of 3, 4, 5, 6, 7 can easily occur by chance, while $G = 2$ is unlikely since $P(2 \text{ runs})$ is less than 0.025. The lower critical $G$ value is therefore 2 and this agrees with Table A–10. The table lists 8 as the upper critical value, but it is impossible to get 8 runs using the given elements. In part $f$ you get the same result of 5.

33. $n_1 = 30, n_2 = 120, G = 60, \mu_G = 49.0, \sigma_G = 3.89$. Test statistic: $z = 2.83$. Critical values: $z = \pm 1.96$. Reject randomness.

## Chapter 12 Review Exercises

1. $n_1 = 25$, $n_2 = 8$, $G = 5$, $\mu_G = 13.12$, $\sigma_G = 2.05$. Test statistic: $z = -3.96$. Critical values: $z = \pm 1.96$.
   Reject randomness.

3. $r_s = 0$. Critical value: $r_s = \pm 0.786$.
   No significant correlation.

5. Test statistic: $H = 6.181$. Critical value: $\chi^2 = 5.991$ (assuming a 0.05 significance level). Reject the null hypothesis that the 3 precincts have the same response times.

7. Using the sign test, discard the 0 to get 10 negative signs and 9 positive signs so that $x = 9$, which is not less than or equal to the critical value of 4 found from Table A–7. Fail to reject the null hypothesis that the median is 38.

9. $\mu_R = 137.50$, $\sigma_R = 17.26$, $R = 177.5$, $z = 2.32$. Test statistic: $z = 2.32$. Critical values: $z = \pm 1.96$.
   Reject the null hypothesis of no difference.

11. $r_S = 0.476$. Critical values: $r_s = \pm 0.738$.
    No significant correlation.

13. $T = 4$, $n = 12$. The test statistic of $T = 4$ is less than or equal to the critical value of 14, so reject the null hypothesis that the tune-up has no effect on fuel consumption.

15. Discard the two zeros to get 11 negative signs and 1 positive sign so that $x = 1$, which is less than the critical value of 2 found from Table A–7. Reject the null hypothesis of no effect.

17. $\mu_R = 126.5$, $\sigma_R = 15.23$, $R = 155.5$, $z = 1.90$. Test statistic: $z = 1.90$. Critical values: $z = \pm 1.96$.
    Fail to reject the null hypothesis that both brands are the same.

19. Test statistic: $H = 8.756$. Critical value: $\chi^2 = 9.488$ (assuming a 0.05 level of significance). Fail to reject the null hypothesis that the methods are equally effective.

21. $T = 12$, $n = 7$. The test statistic of $T = 12$ is greater than the critical value of 2, so fail to reject the null hypothesis that there is no difference between their scoring.

23. $n_1 = 20$, $n_2 = 5$, $G = 5$. Critical values are 5, 12. Reject randomness.

# Index